Student Solutions Manual

Calculus: An Applied Approach (+Brief)

NINTH EDITION

Ron Larson
The Pennsylvania State University,
The Behrend College

Australia • Brazil • Japan • Korea • Mexico • Singapore • Spain • United Kingdom • United States

For product information and technology assistance, contact us at **Cengage Learning Customer & Sales Support, 1-800-354-9706**

For permission to use material from this text or product, submit all requests online at **www.cengage.com/permissions**
Further permissions questions can be emailed to **permissionrequest@cengage.com**

ISBN-13: 978-1-133-11279-2
ISBN-10: 1-133-11279-X

Brooks/Cole
20 Channel Center Street
Boston, MA 02210
USA

Cengage Learning is a leading provider of customized learning solutions with office locations around the globe, including Singapore, the United Kingdom, Australia, Mexico, Brazil, and Japan. Locate your local office at: **www.cengage.com/global**

Cengage Learning products are represented in Canada by Nelson Education, Ltd.

To learn more about Brooks/Cole, visit **www.cengage.com/brookscole**

Purchase any of our products at your local college store or at our preferred online store **www.cengagebrain.com**

Printed in the United States of America
1 2 3 4 5 6 7 15 14 13 12

CONTENTS

Preface

This *Student Solutions Manual* is designed as a supplement to *Calculus: An Applied Approach*, Ninth Edition, and *Brief Calculus: An Applied Approach*, Ninth Edition, by Ron Larson. All references to chapters, theorems, and exercises relate to the main text. Solutions to every odd-numbered exercise in the text are given with all essential algebraic steps included. Although this supplement is not a substitute for good study habits, it can be valuable when incorporated into a well planned course of study.

We have made every effort to see that the solutions are correct. However, we would appreciate hearing about any errors or other suggestions for improvement. Good luck with your study of calculus.

Ron Larson
Larson Texts, Inc.

CHAPTER 1
Functions, Graphs, and Limits

CHAPTER 1

Functions, Graphs, and Limits

Section 1.1 The Cartesian Plane and the Distance Formula

Skills Warm Up

1. $\sqrt{(3-6)^2 + [1-(-5)]^2} = \sqrt{(-3)^2 + 6^2}$

$= \sqrt{9+36}$

$= \sqrt{45}$

$= 3\sqrt{5}$

2. $\sqrt{(-2-0)^2 + [-7-(-3)]^2} = \sqrt{(-2)^2 + (-4)^2}$

$= \sqrt{4+16}$

$= \sqrt{20}$

$= 2\sqrt{5}$

3. $\dfrac{5+(-4)}{2} = \dfrac{1}{2}$

4. $\dfrac{-3+(-1)}{2} = \dfrac{-4}{2} = -2$

5. $\sqrt{27} + \sqrt{12} = 3\sqrt{3} + 2\sqrt{3} = 5\sqrt{3}$

6. $\sqrt{8} - \sqrt{18} = 2\sqrt{2} - 3\sqrt{2} = -\sqrt{2}$

7.
$$\begin{aligned}
\sqrt{(3-x)^2 + (7-4)^2} &= \sqrt{45} \\
\left(\sqrt{(3-x)^2 + (7-4)^2}\right)^2 &= \left(\sqrt{45}\right)^2 \\
(3-x)^2 + (7-4)^2 &= 45 \\
(3-x)^2 + 3^2 &= 45 \\
(3-x)^2 + 9 &= 45 \\
(3-x)^2 &= 36 \\
3-x &= \pm 6 \\
-x &= -3 \pm 6 \\
x &= 3 \mp 6 \\
x &= -3, 9
\end{aligned}$$

8.
$$\begin{aligned}
\sqrt{(6-2)^2 + (-2-y)^2} &= \sqrt{52} \\
\left(\sqrt{(6-2)^2 + (-2-y)^2}\right)^2 &= \left(\sqrt{52}\right)^2 \\
(6-2)^2 + (-2-y)^2 &= 52 \\
4^2 + (-2-y)^2 &= 52 \\
16 + (-2-y)^2 &= 52 \\
(-2-y)^2 &= 36 \\
-2-y &= \pm 6 \\
-y &= \pm 6 + 2 \\
y &= \mp 6 - 2 \\
y &= -8, 4
\end{aligned}$$

9.
$$\begin{aligned}
\frac{x+(-5)}{2} &= 7 \\
x + (-5) &= 14 \\
x &= 19
\end{aligned}$$

10.
$$\begin{aligned}
\frac{-7+y}{2} &= -3 \\
-7 + y &= -6 \\
y &= 1
\end{aligned}$$

1.

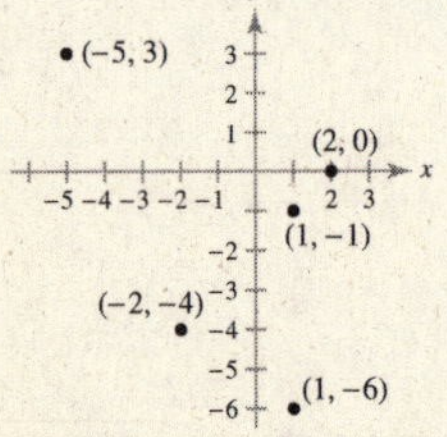

(−5, 3)
(2, 0)
(1, −1)
(−2, −4)
(1, −6)

3. (a)

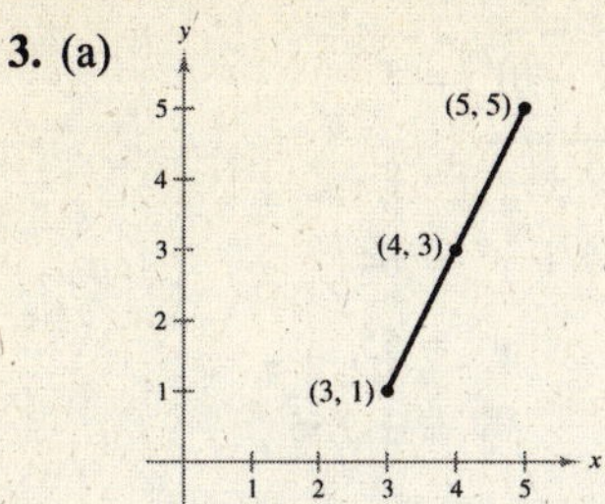

(b) $d = \sqrt{(5-3)^2 + (5-1)^2} = \sqrt{4+16} = 2\sqrt{5}$

(c) Midpoint $= \left(\dfrac{3+5}{2}, \dfrac{1+5}{2}\right) = (4, 3)$

5. (a)

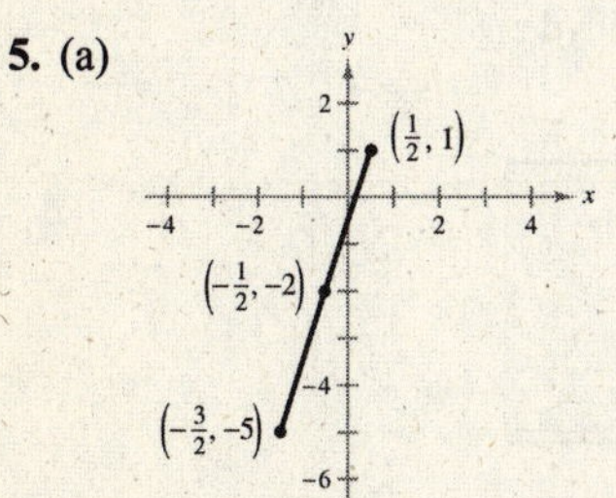

(b) $d = \sqrt{\left[(3/2) - (1/2)\right]^2 + (5-1)^2}$

$= \sqrt{4 + 36}$

$= 2\sqrt{10}$

(c) Midpoint $= \left(\dfrac{(1/2) + (-3/2)}{2}, \dfrac{1 + (-5)}{2}\right) = \left(-\dfrac{1}{2}, -2\right)$

7. (a)

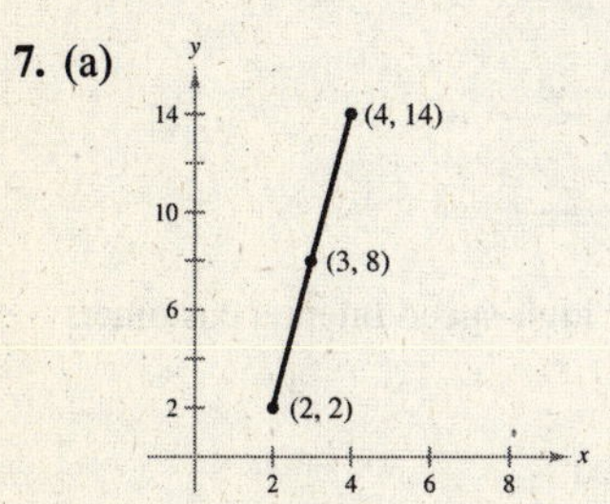

(b) $d = \sqrt{(4-2)^2 + (14-2)^2}$

$= \sqrt{4 + 144}$

$= 2\sqrt{37}$

(c) Midpoint $= \left(\dfrac{2+4}{2}, \dfrac{2+14}{2}\right) = (3, 8)$

9. (a)

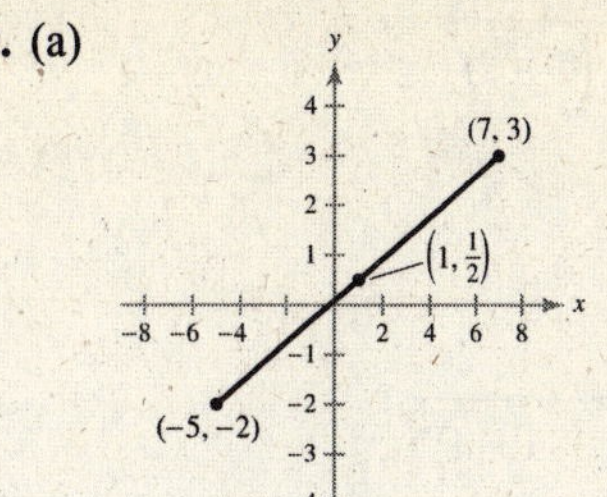

(b) $d = \sqrt{\left(7 - (-5)\right)^2 + \left(3 - (-2)^2\right)}$

$= \sqrt{12^2 + 5^2}$

$= \sqrt{144 + 25}$

$= \sqrt{169} = 13$

(c) Midpoint $= \left(\dfrac{7 + (-5)}{2}, \dfrac{3 + (-2)}{2}\right)$

$= \left(\dfrac{2}{2}, \dfrac{1}{2}\right)$

$= \left(1, \dfrac{1}{2}\right)$

11. (a)

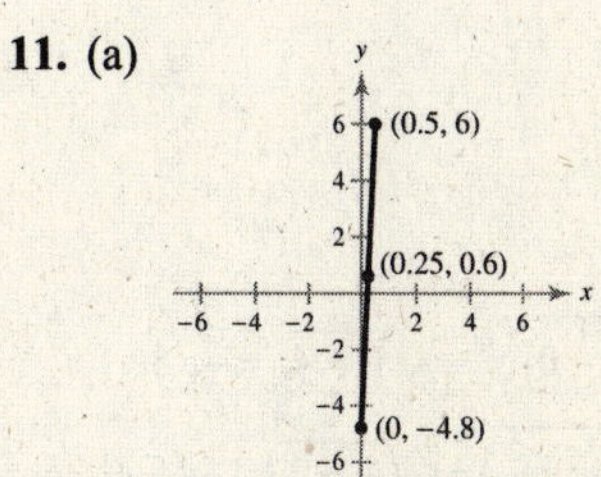

(b) $d = \sqrt{(0.5 - 0)^2 + \left(6 - (-4.8)\right)^2}$

$= \sqrt{0.25 + 116.64}$

$= \sqrt{116.89}$

(c) Midpoint $= \left(\dfrac{0 + 0.5}{2}, \dfrac{-4.8 + 6}{2}\right) = (0.25, 0.6)$

13. (a) $a = 4$

$b = 3$

$c = \sqrt{(4-0)^2 + (3-0)^2} = \sqrt{16+9} = 5$

(b) $a^2 + b^2 = 16 + 9 = 25 = c^2$

15. (a) $a = 10$

$b = 3$

$c = \sqrt{(7+3)^2 + (4-1)^2} = \sqrt{100+9} = \sqrt{109}$

(b) $a^2 + b^2 = 100 + 9 = 109 = c^2$

17. $d_1 = \sqrt{(3-0)^2 + (7-1)^2}$
$= \sqrt{9 + 36}$
$= \sqrt{45}$
$= 3\sqrt{5}$

$d_2 = \sqrt{(4-0)^2 + (-1-1)^2}$
$= \sqrt{16 + 4}$
$= \sqrt{20}$
$= 2\sqrt{5}$

$d_3 = \sqrt{(3-4)^2 + [7-(-1)]^2}$
$= \sqrt{1 + 64}$
$= \sqrt{65}$

Because $d_1^2 + d_2^2 = d_3^2$, the figure is a right triangle.

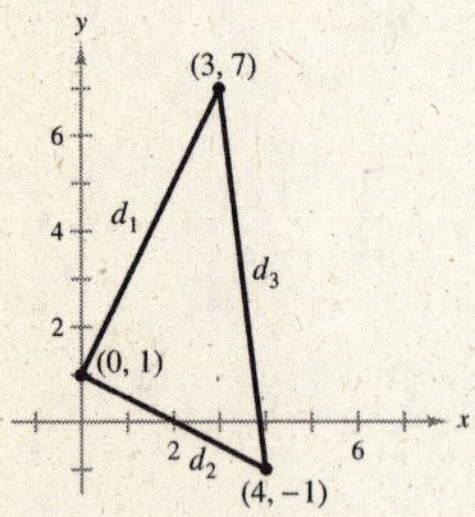

19. $d_1 = \sqrt{(1-0)^2 + (2-0)^2} = \sqrt{1+4} = \sqrt{5}$
$d_2 = \sqrt{(3-1)^2 + (3-2)^2} = \sqrt{4+1} = \sqrt{5}$
$d_3 = \sqrt{(2-3)^2 + (1-3)^2} = \sqrt{1+4} = \sqrt{5}$
$d_4 = \sqrt{(0-2)^2 + (0-1)^2} = \sqrt{4+1} = \sqrt{5}$

Because $d_1 = d_2 = d_3 = d_4$, the figure is a parallelogram.

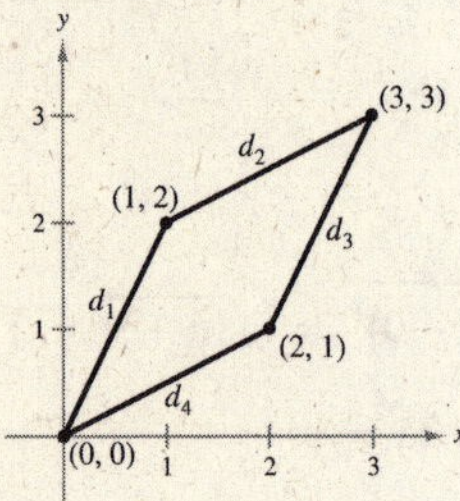

21. $d = \sqrt{(x-1)^2 + (-4-0)^2} = 5$
$\sqrt{x^2 - 2x + 17} = 5$
$x^2 - 2x + 17 = 25$
$x^2 - 2x - 8 = 0$
$(x-4)(x+2) = 0$
$x = 4, -2$

23. $d = \sqrt{(3-0)^2 + (y-0)^2} = 8$
$\sqrt{9 + y^2} = 8$
$9 + y^2 = 64$
$y^2 = 55$
$y = \pm\sqrt{55}$

25. $d = \sqrt{(50-12)^2 + (42-18)^2}$
$= \sqrt{38^2 + 24^2}$
$= \sqrt{2020}$
$= 2\sqrt{505} \approx 44.9$ yd

27.

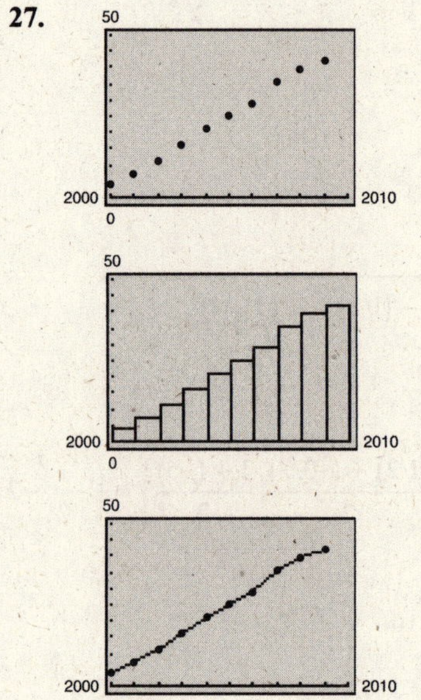

The number of cable high-speed Internet customers increases each year.

29. (a) March 2009: 7600
July 2009: 9200
July 2010: 10,500

(b) April 2010: 11,000
May 2010: 10,100

Decrease: $|10{,}100 - 11{,}000| = 900$

Percent decrease: $\dfrac{900}{11{,}000} \approx 0.082 = 8.2\%$

31. (a) Revenue $= \left(\frac{2007 + 2009}{2}, \frac{329.7 + 538.9}{2}\right) = (2008, 434.3)$

Revenue estimate for 2008: \$434.3 million

Profit $= \left(\frac{2007 + 2009}{2}, \frac{19.7 + 30.7}{2}\right) = (2008, 25.2)$

Profit estimate for 2008: \$25.2 million

(b) Actual 2008 revenue: \$422.4 million

Actual 2008 profit: \$24.4 million

(c) Yes, the revenue and profit increased in a linear pattern from 2007 to 2009.

(d) 2007 expense: $329.7 - 19.7 = \$310$ million

2008 expense: $422.4 - 24.4 = \$398$ million

2009 expense: $538.9 - 30.7 = \$508.2$ million

(e) Answers will vary.

33. (a)

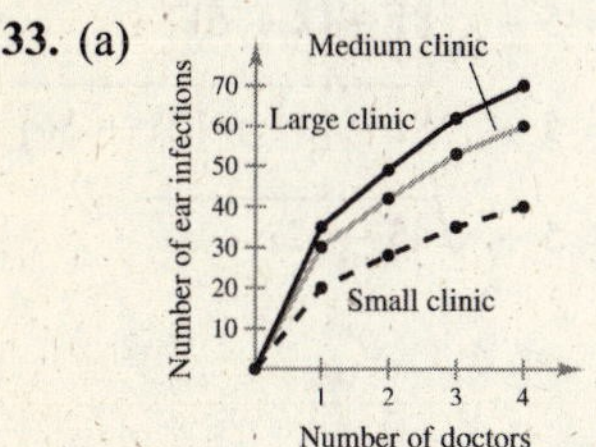

(b) The larger the clinic, the more patients a doctor can treat.

35. The vertex $(-3, -1)$ is translated to $(-6, -6)$.

The vertex $(0, 0)$ is translated to $(-3, -5)$.

The vertex $(-1, -2)$ is translated to $(-4, -7)$.

37. Midpoint $= \left(\frac{x_1 + x_2}{2}, \frac{y_1 + y_2}{2}\right)$

The point one-fourth of the way between (x_1, y_1) and (x_2, y_2) is the midpoint of the line segment from (x_1, y_1) to $\left(\frac{x_1 + x_2}{2}, \frac{y_1 + y_2}{2}\right)$, which is

$$\left(\frac{x_1 + \frac{x_1 + x_2}{2}}{2}, \frac{y_1 + \frac{y_1 + y_2}{2}}{2}\right) = \left(\frac{3x_1 + x_2}{4}, \frac{3y_1 + y_2}{4}\right).$$

The point three-fourths of the way between (x_1, y_1) and (x_2, y_2) is the midpoint of the line segment from $\left(\frac{x_1 + x_2}{2}, \frac{y_1 + y_2}{2}\right)$ to (x_2, y_2), which is

$$\left(\frac{\frac{x_1 + x_2}{2} + x_2}{2}, \frac{\frac{y_1 + y_2}{2} + y_2}{2}\right) = \left(\frac{x_1 + 3x_2}{4}, \frac{y_1 + 3y_2}{4}\right).$$

Thus,

$\left(\frac{3x_1 + x_2}{4}, \frac{3y_1 + y_2}{4}\right), \left(\frac{x_1 + x_2}{2}, \frac{y_1 + y_2}{2}\right)$, and $\left(\frac{x_1 + 3x_2}{4}, \frac{y_1 + 3y_2}{4}\right)$

are the three points that divide the line segment joining (x_1, y_1) and (x_2, y_2) into four equal parts.

39. To show $\left(\frac{2x_1 + x_2}{3}, \frac{2y_1 + y_2}{3}\right)$ is a point of trisection of the line segment joining (x_1, y_1) and (x_2, y_2), we must show that $d_1 = \frac{1}{2}d_2$ and $d_1 + d_2 = d_3$.

$$d_1 = \sqrt{\left(\frac{2x_1 + x_2}{3} - x_1\right)^2 + \left(\frac{2y_1 + y_2}{3} - y_1\right)^2}$$
$$= \sqrt{\left(\frac{x_2 - x_1}{3}\right)^2 + \left(\frac{y_2 - y_1}{3}\right)^2}$$
$$= \frac{1}{3}\sqrt{(x_2 - x_1)^2 + (y_2 - y_1)^2}$$

$$d_2 = \sqrt{\left(x_2 - \frac{2x_1 + x_2}{3}\right)^2 + \left(y_2 - \frac{2y_1 + y_2}{3}\right)^2}$$
$$= \sqrt{\left(\frac{2x_2 - 2x_1}{3}\right)^2 + \left(\frac{2y_2 - 2y_1}{3}\right)^2}$$
$$= \frac{2}{3}\sqrt{(x_2 - x_1)^2 + (y_2 - y_1)^2}$$

$$d_3 = \sqrt{(x_2 - x_1)^2 + (y_2 - y_1)^2}$$

Therefore, $d_1 = \frac{1}{2}d_2$ and $d_1 + d_2 = d_3$. The midpoint of the line segment joining $\left(\frac{2x_1 + x_2}{3}, \frac{2y_1 + y_2}{3}\right)$ and (x_2, y_2) is

$$\text{Midpoint} = \left(\frac{\frac{2x_1 + x_2}{3} + x_2}{2}, \frac{\frac{2y_1 + y_2}{3} + y_2}{2}\right)$$
$$= \left(\frac{x_1 + 2x_2}{3}, \frac{y_1 + 2y_2}{3}\right).$$

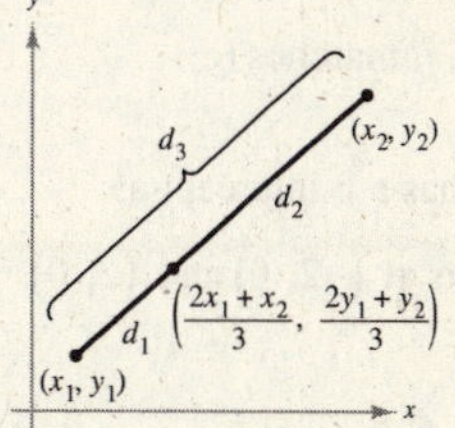

Section 1.2 Graphs of Equations

Skills Warm Up

1. $5y - 12 = x$
   $5y = x + 12$
   $y = \frac{x + 12}{5}$

2. $-y = 15 - x$
   $y = x - 15$

3. $x^3y + 2y = 1$
   $y(x^3 + 2) = 1$
   $y = \frac{1}{x^3 + 2}$

4. $x^2 + x - y^2 - 6 = 0$
   $-y^2 = 6 - x^2 - x$
   $y^2 = x^2 + x - 6$
   $y = \sqrt{x^2 + x - 6}$

5. $(x - 2)^2 + (y + 1)^2 = 9$
   $(y + 1)^2 = 9 - (x - 2)^2$
   $y + 1 = \sqrt{9 - (x - 2)^2}$
   $y = \left(\sqrt{9 - (x - 2)^2}\right) - 1$
   $= \sqrt{9 - (x^2 - 4x + 4)} - 1$
   $= \sqrt{5 + 4x - x^2} - 1$

6. $(x + 6)^2 + (y - 5)^2 = 81$
   $(y - 5)^2 = 81 - (x + 6)^2$
   $y - 5 = \sqrt{81 - (x + 6)^2}$
   $y = 5 + \sqrt{81 - (x + 6)^2}$
   $= 5 + \sqrt{81 - (x^2 + 12x + 36)}$
   $= 5 + \sqrt{45 - 12x - x^2}$

7. $y = 5(-2) = -10$

8. $y = 3(3) - 4 = 5$

9. $y = 2(2)^2 + 1 = 9$

10. $y = (-4)^2 + 2(-4) - 7 = 1$

11. $x^2 - 3x + 2$
    $(x - 1)(x - 2)$

12. $x^2 + 5x + 6$
    $(x + 2)(x + 3)$

13. $y^2 - 3y + \frac{9}{4}$
    $\left(y - \frac{3}{2}\right)^2$

14. $y^2 - 7y + \frac{49}{4}$
    $\left(y - \frac{7}{2}\right)^2$

1. The graph of $y = x - 2$ is a straight line with y-intercept at $(0, -2)$. So, it matches (e).

3. The graph of $y = x^2 + 2x$ is a parabola opening up with vertex at $(-1, -1)$. So, it matches (c).

5. The graph of $y = |x| - 2$ has a y-intercept at $(0, -2)$ and has x-intercepts at $(-2, 0)$ and $(2, 0)$. So, it matches (a).

7. $y = 2x + 3$

x	-2	$-\frac{3}{2}$	-1	0	1	2
y	-1	0	1	3	5	7

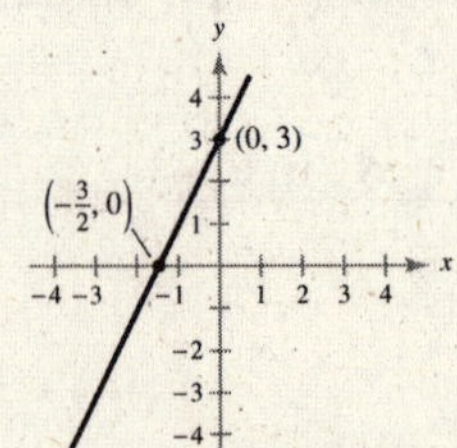

9. $y = x^2 - 3$

x	-2	-1	0	1	2	3
y	1	-2	-3	-2	1	6

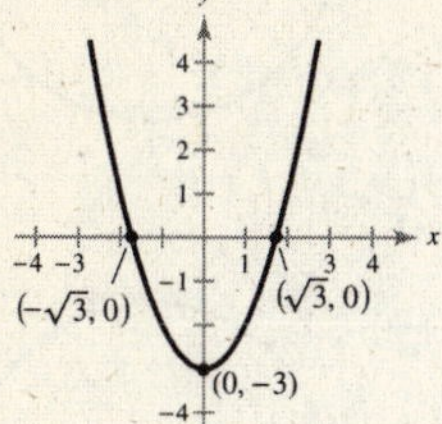

11. $y = (x - 1)^2$

x	-2	-1	0	1	2
y	9	4	1	0	1

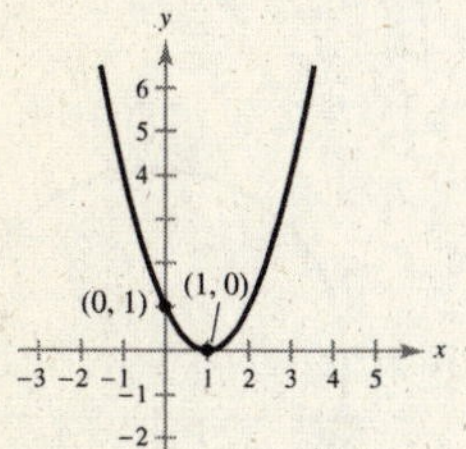

13. $y = x^3 + 2$

x	-2	-1	0	1	2
y	-6	1	2	3	10

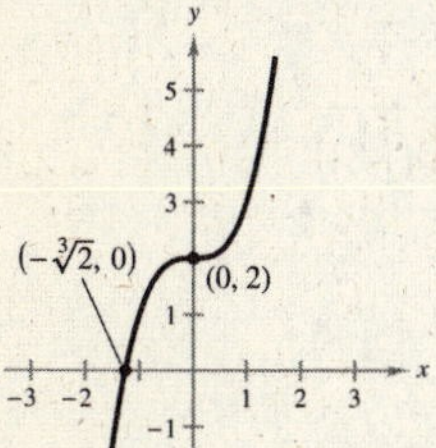

15. $y = -\sqrt{x - 1}$

x	1	2	3	4	5
y	0	-1	-1.41	-1.73	-2

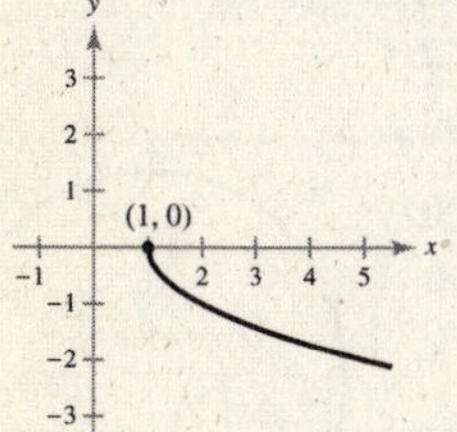

17. $y = |x + 1|$

x	-3	-2	-1	0	1
y	2	1	0	1	2

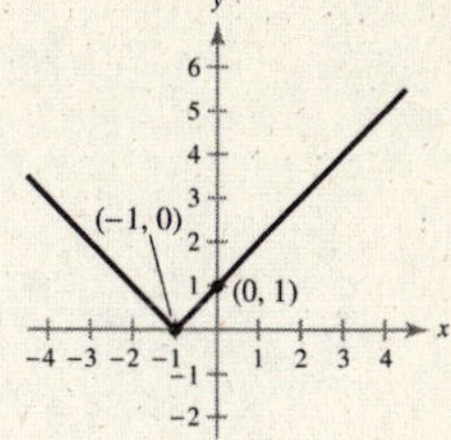

19. $y = \dfrac{1}{x - 3}$

x	-1	0	1	2	2.5	3.5	4	5	6
y	$-\frac{1}{4}$	$-\frac{1}{3}$	$-\frac{1}{2}$	-1	-2	2	1	$\frac{1}{2}$	$\frac{1}{3}$

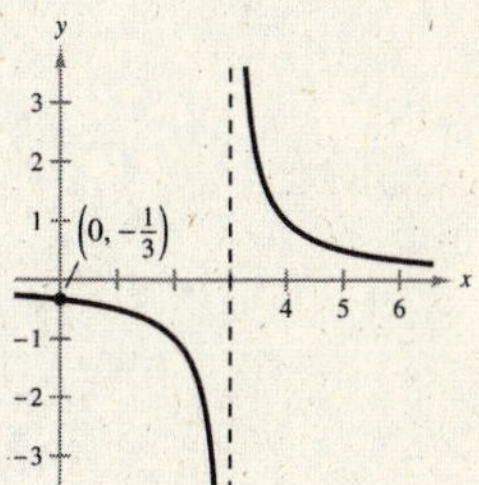

21. $x = y^2 - 4$

x	5	0	-3	-4
y	± 3	± 2	± 1	0

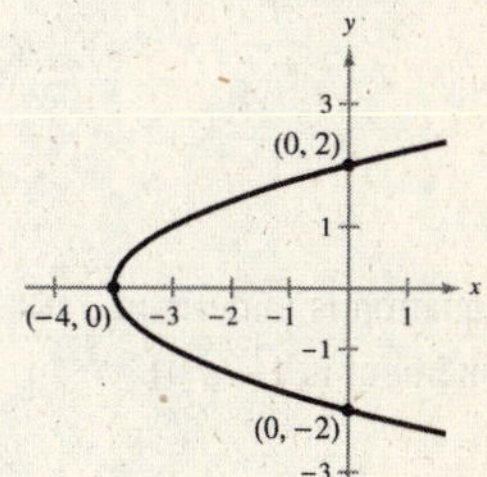

23. Let $y = 0$. Then,

$$2x - (0) - 3 = 0$$
$$x = \tfrac{3}{2}.$$

Let $x = 0$. Then,

$$2(0) - y - 3 = 0$$
$$y = -3.$$

x-intercept: $\left(\frac{3}{2}, 0\right)$

y-intercept: $(0, -3)$

25. Let $y = 0$. Then,

$$0 = x^2 + x - 2$$
$$0 = (x + 2)(x - 1)$$
$$x = -2, 1.$$

Let $x = 0$. Then,

$$y = (0)^2 + (0) - 2$$
$$y = -2$$

x-intercepts: $(-2, 0), (1, 0)$

y-intercept: $(0, -2)$

27. Let $y = 0$. Then,

$$0 = \sqrt{4 - x^2}$$
$$x^2 = 4$$
$$x = \pm 2.$$

Let $x = 0$. Then,

$$y = \sqrt{4 - (0)^2}$$
$$y = 2.$$

x-intercepts: $(-2, 0), (2, 0)$

y-intercept: $(0, 2)$

29. Let $y = 0$. Then,

$$0 = \frac{x^2 - 4}{x - 2}$$
$$0 = (x - 2)(x + 2)$$
$$x = \pm 2.$$

Let $x = 0$. Then,

$$y = \frac{(0)^2 - 4}{(0) - 2}$$
$$y = 2.$$

x-intercept: Because the equation is undefined when $x = 2$, the only x-intercept is $(-2, 0)$.

y-intercept: $(0, 2)$

31. Let $y = 0$. Then,

$$x^2(0) - x^2 + 4(0) = 0$$
$$x^2 = 0$$
$$x = 0.$$

Let $x = 0$. Then,

$$(0)^2 y - (0)^2 + 4y = 0$$
$$y = 0.$$

x-intercept: $(0, 0)$

y-intercept: $(0, 0)$

33. $(x - 0)^2 + (y - 0)^2 = 4^2$

$$x^2 + y^2 = 16$$

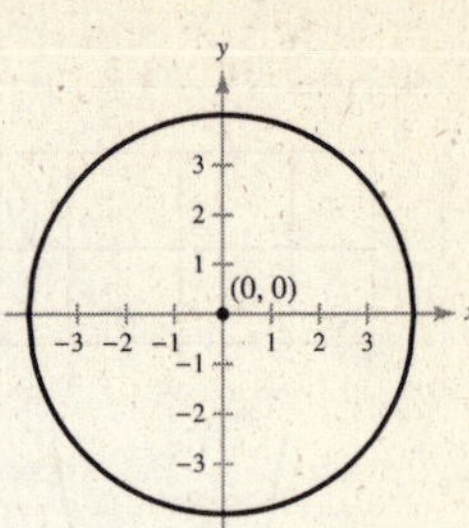

35. $(x - 2)^2 + (y - (-1))^2 = 3^2$

$$(x - 2)^2 + (y + 1)^2 = 9$$

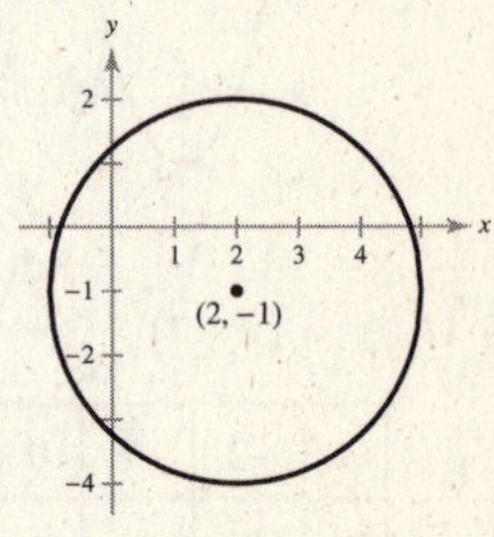

37. The radius is the distance between $(-1, 5)$ and $(-1, 1)$.

$$r = \sqrt{(-1 - (-1))^2 + (5 - 1)^2}$$
$$= \sqrt{0^2 + 4^2}$$
$$= \sqrt{16} = 4$$

Using the center $(-1, 1)$ and the radius $r = 4$:

$$(x - (-1))^2 + (y - 1)^2 = 4^2$$
$$(x + 1)^2 + (y - 1)^2 = 16$$

39. The diameter is the distance between $(-6, -8)$ and $(6, 8)$.

$$d = \sqrt{(6 - (-6))^2 + (8 - (-8))^2}$$
$$= \sqrt{12^2 + 16^2}$$
$$= \sqrt{144 + 256}$$
$$= \sqrt{400}$$
$$= 20$$

The radius is one-half the diameter: $r = \frac{20}{2} = 10$.

The center is the midpoint of the diameter: $\left(\frac{-6 + 6}{2}, \frac{-8 + 8}{2}\right) = (0, 0)$.

$$(x - 0)^2 + (y - 0)^2 = 10^2$$
$$x^2 + y^2 = 100$$

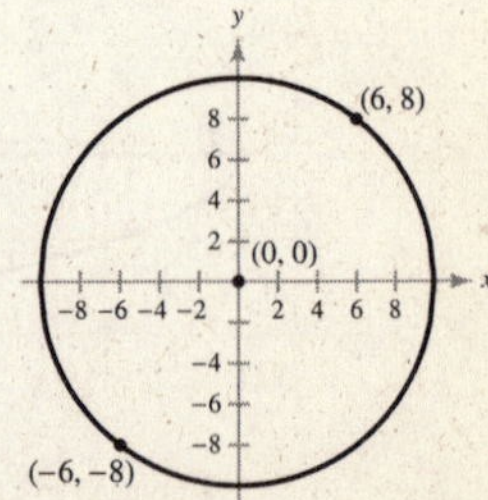

41. Set the two equations equal to each other.

$$\begin{aligned} -x + 2 &= 2x - 1 \\ -3x &= -3 \\ x &= 1 \end{aligned}$$

Substitute $x = 1$ into one of the equations.

$$y = (-1) + 2 = 1$$

The point of intersection is $(1, 1)$.

43. Set the two equations equal to each other.

$$\begin{aligned} -x^2 + 15 &= 3x + 11 \\ -x^2 - 3x + 4 &= 0 \\ x^2 + 3x - 4 &= 0 \\ (x + 4)(x - 1) &= 0 \end{aligned}$$

$$x + 4 = 0 \qquad x - 1 = 0$$
$$x = -4 \qquad x = -1$$

Substitute $x = -4$:

$$\begin{aligned} y &= -(-4)^2 + 15 \\ y &= -16 + 15 \\ y &= -1 \end{aligned}$$

Substitute $x = 1$:

$$\begin{aligned} y &= -(1)^2 + 15 \\ y &= -1 + 15 \\ y &= 14 \end{aligned}$$

The points of intersection are $(-4, -1)$ and $(1, 14)$.

45. By equating the y-values for the two equations, we have

$$\begin{aligned} x^3 &= 2x \\ x^3 - 2x &= 0 \\ x(x^2 - 2) &= 0 \\ x &= 0, \pm\sqrt{2}. \end{aligned}$$

The corresponding y-values are $y = 0$, $y = -2\sqrt{2}$, and $y = 2\sqrt{2}$, so the points of intersection are $(0, 0)$, $(-\sqrt{2}, -2\sqrt{2})$, and $(\sqrt{2}, 2\sqrt{2})$.

47. By equating the y-values for the two equations, we have

$$\begin{aligned} x^4 - 2x^2 + 1 &= 1 - x^2 \\ x^4 - x^2 &= 0 \\ x^2(x + 1)(x - 1) &= 0 \\ x &= 0, \pm 1. \end{aligned}$$

The corresponding y-values are $y = 1$, 0, and 0, so the points of intersection are $(-1, 0)$, $(0, 1)$, and $(1, 0)$.

49. To find the break-even point, set $R = C$.

$$\begin{aligned} 1.55x &= 0.85x + 35{,}000 \\ 0.7x &= 35{,}000 \\ x &= \frac{35{,}000}{0.7} = 50{,}000 \text{ units} \end{aligned}$$

51. To find the break-even point, set $R = C$.

$$\begin{aligned} 9950x &= 8650x + 250{,}000 \\ 1300x &= 250{,}000 \\ x &= \frac{250{,}000}{1300} \approx 193 \text{ units} \end{aligned}$$

53. To find the break-even point, set $R = C$.

$$\begin{aligned} 10x &= 6x + 5000 \\ 4x &= 5000 \\ x &= \frac{5000}{4} \approx 1250 \text{ units} \end{aligned}$$

55. (a) $C = 11.8x + 15{,}000$

$R = 19.30x$

(b)
$$\begin{aligned} C &= R \\ 11.8x + 15{,}000 &= 19.30x \\ 15{,}000 &= 7.5x \\ x &= 2000 \text{ units} \end{aligned}$$

(c)
$$\begin{aligned} P &= R - C \\ 1000 &= 19.3x - (11.8x + 15{,}000) \\ 16{,}000 &= 7.5x \\ x &\approx 2133.3 \end{aligned}$$

So, 2134 units would yield a profit of \$1000.

57.
$$\begin{aligned} 240 - 4x &= 135 + 3x \\ 105 &= 7x \\ 15 &= x \end{aligned}$$

Equilibrium point $(x, p) = (15, 180)$

59. (a)

Year	2004	2005	2006	2007	2008	2009
Amount	30	44	54	67	113	313
Model	26	55	48	56	126	308

The model fits the data well. Answers will vary.

(b) Let $t = 14$ (2014).

$$y = 8.148(14)^3 - 139.71(14)^2 + 789.0(14) - 1416 \approx \$4605 \text{ million}$$

61. (a)

Year	2004	2005	2006	2007	2008	2012
Degrees	667.4	692	713.6	732.2	747.8	780.2

(b) Answers will vary.

(c) Let $t = 16$ (2016).

$$y = -1.50(16)^2 + 38.1(16) + 539 = 764.6 \text{ degrees}$$

The prediction is not valid because the number of associate's degrees should keep increasing over time, and not decrease.

63.

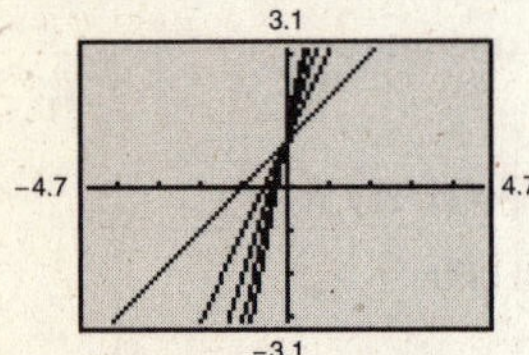

The greater the value of c, the steeper the line.

65.

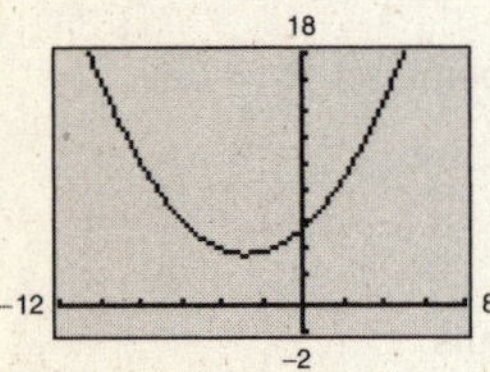

Intercept: $(0, 5.36)$

67.

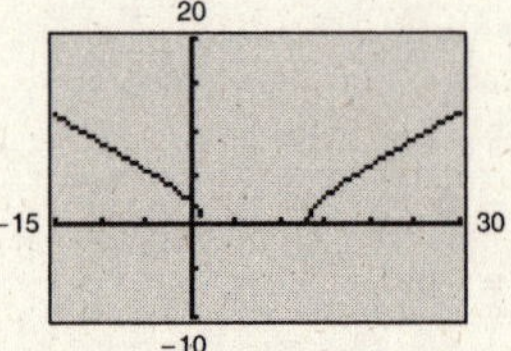

Intercepts: $(1.4780, 0)$, $(12.8553, 0)$, $(0, 2.3875)$

69.

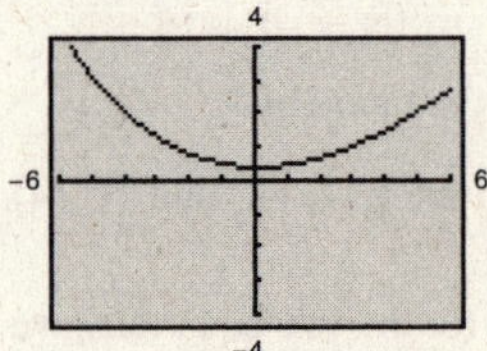

Intercept: $\left(0, \frac{5}{12}\right) \approx (0, 0.4167)$

71. Answers will vary.

Section 1.3 Lines in the Plane and Slope

Skills Warm Up

1. $\dfrac{5 - (-2)}{-3 - 4} = \dfrac{7}{-7} = -1$

2. $\dfrac{-7 - (-10)}{4 - 1} = 1$

3. $-\dfrac{1}{m}$, $m = -3$

$$-\frac{1}{-3} = \frac{1}{3}$$

4. $-\dfrac{1}{m}$, $m = \dfrac{6}{7}$

$$-\frac{1}{\frac{6}{7}} = -\frac{7}{6}$$

5. $-4x + y = 7$

$$y = 4x + 7$$

6. $3x - y = 7$

$$-y = 7 - 3x$$
$$y = 3x - 7$$

7. $y - 2 = 3(x - 4)$

$$y = 3(x - 4) + 2$$
$$y = 3x - 12 + 2$$
$$y = 3x - 10$$

Skills Warm Up —*continued*—

8. $y - (-5) = -1[x - (-2)]$

$y + 5 = -x - 2$

$y = -x - 7$

9. $y - (-3) = \dfrac{4 - (-3)}{2 - 1}(x - 2)$

$y + 3 = \dfrac{7}{1}(x - 2)$

$y + 3 = 7x - 14$

$y = 7x - 17$

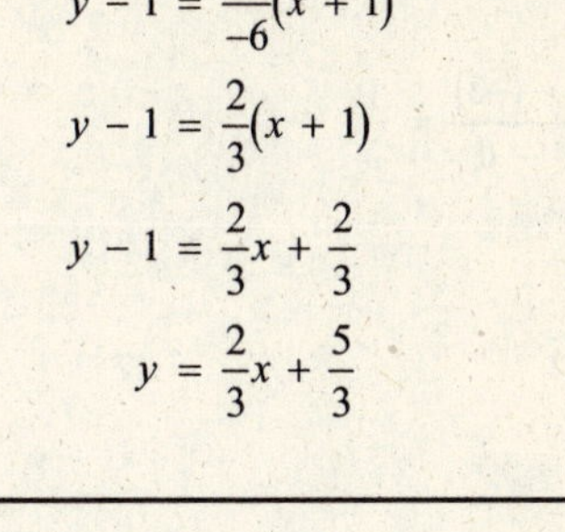

10. $y - 1 = \dfrac{-3 - 1}{-7 - (-1)}[x - (-1)]$

$y - 1 = \dfrac{-4}{-6}(x + 1)$

$y - 1 = \dfrac{2}{3}(x + 1)$

$y - 1 = \dfrac{2}{3}x + \dfrac{2}{3}$

$y = \dfrac{2}{3}x + \dfrac{5}{3}$

1. The slope is $m = 1$ because the line rises one unit vertically for each unit the line moves to the right.

3. The slope is $m = 0$ because the line is horizontal.

5. $y = x + 7$

So, the slope is $m = 1$, and the y-intercept is $(0, 7)$.

7. $5x + y = 20$

$y = -5x + 20$

So, the slope is $m = -5$, and the y-intercept is $(0, 20)$.

9. $7x + 6y = 30$

$y = -\frac{7}{6}x + 5$

So, the slope is $m = -\frac{7}{6}$, and the y-intercept is $(0, 5)$.

11. $3x - y = 15$

$y = 3x - 15$

So, the slope is $m = 3$, and the y-intercept is $(0, -15)$.

13. $x = 4$

Because the line is vertical, the slope is undefined. There is no y-intercept.

15. $y - 4 = 0$

$y = 4$

So, the slope is $m = 0$, and the y-intercept is $(0, 4)$.

17. $y = -2$

x	−2	−1	0	1
y	−2	−2	−2	−2

19. $y = -2x + 1$

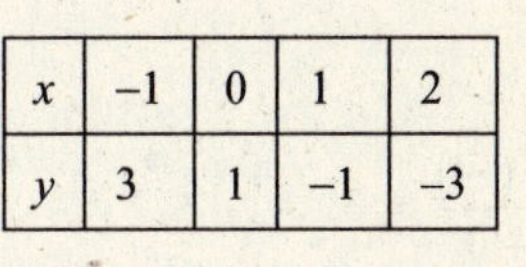

x	−1	0	1	2
y	3	1	−1	−3

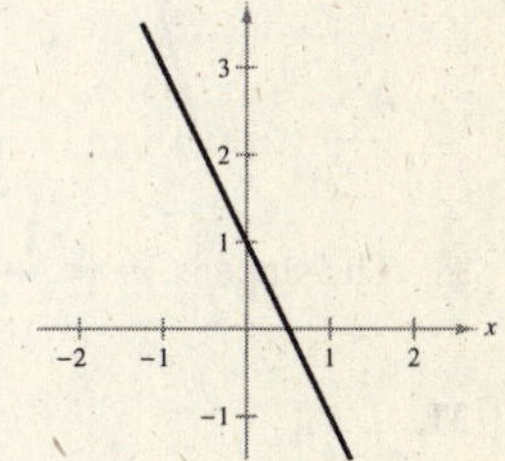

21. $6x + 3y = 18$

$3y = -6x + 18$

$y = -2x + 6$

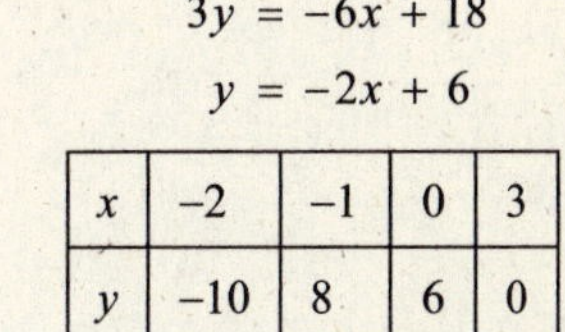

x	−2	−1	0	3
y	−10	8	6	0

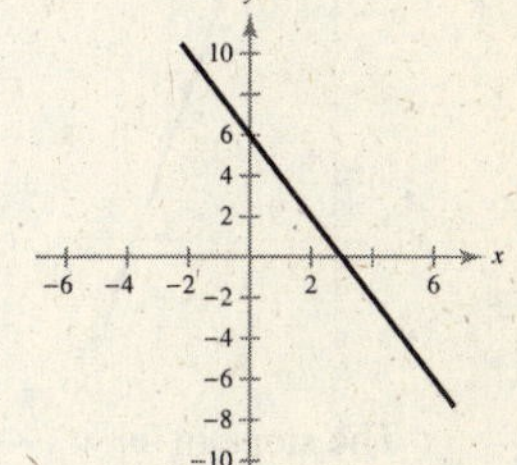

23. $y = 2x - 3$

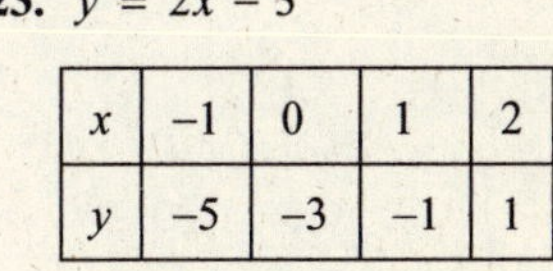

x	−1	0	1	2
y	−5	−3	−1	1

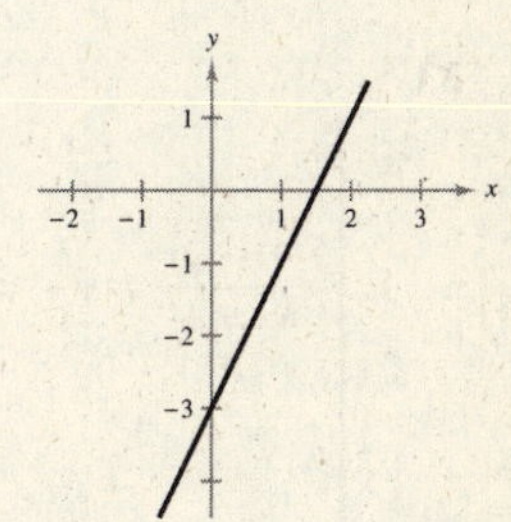

25. $y = -\frac{3}{5}x - 3$

x	0	2	4	5
y	−3	$-\frac{21}{5}$	$-\frac{27}{5}$	−6

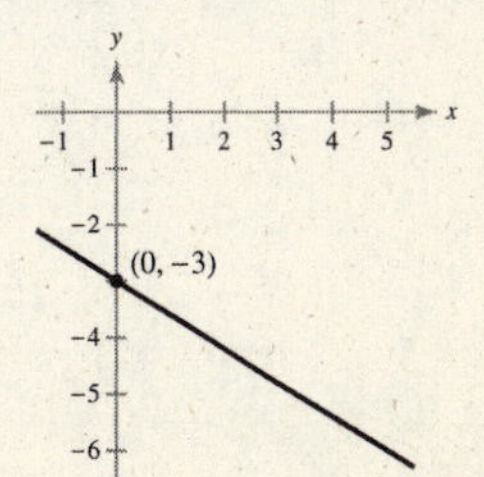

27.

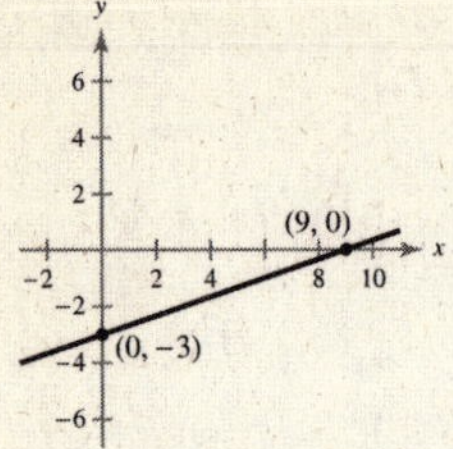

The slope is $m = \dfrac{0 - (-3)}{9 - 0} = \dfrac{1}{3}$.

29.

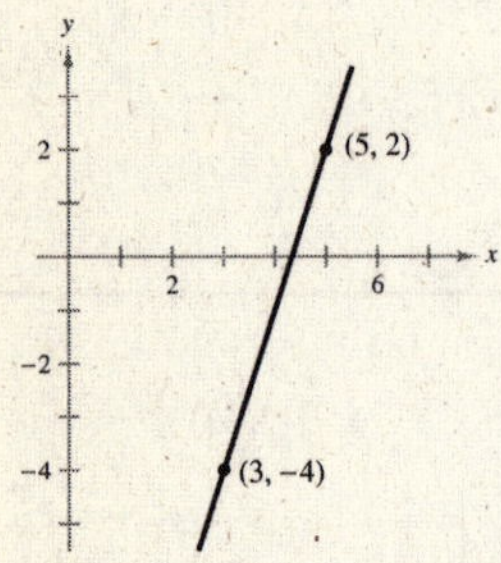

The slope is $m = \dfrac{2 - (-4)}{5 - 3} = 3$.

31.

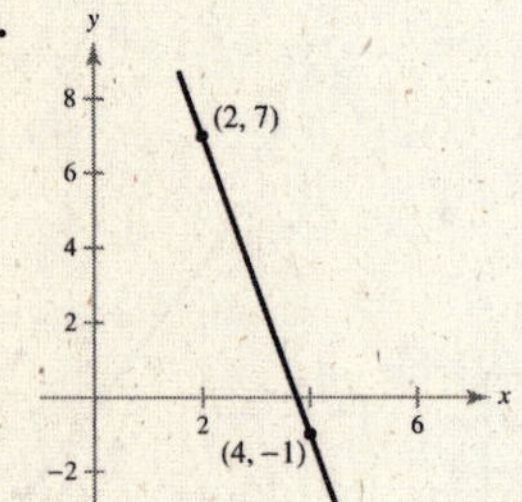

The slope of $m = \dfrac{7 - (-1)}{2 - 4} = \dfrac{8}{-2} = -4$.

33.

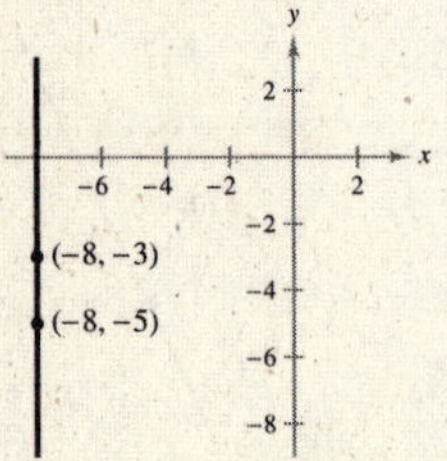

The slope is undefined because $m = \dfrac{-5 - (-3)}{-8 - (-8)}$ and division by zero is undefined. So, the line is vertical.

35.

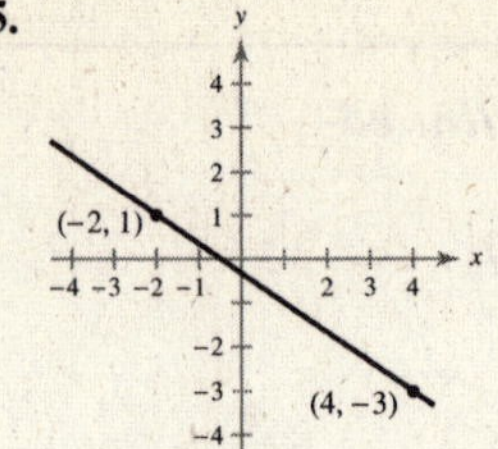

The slope is $m = \dfrac{-3 - 1}{4 - (-2)} = \dfrac{-4}{6} = -\dfrac{2}{3}$.

37.

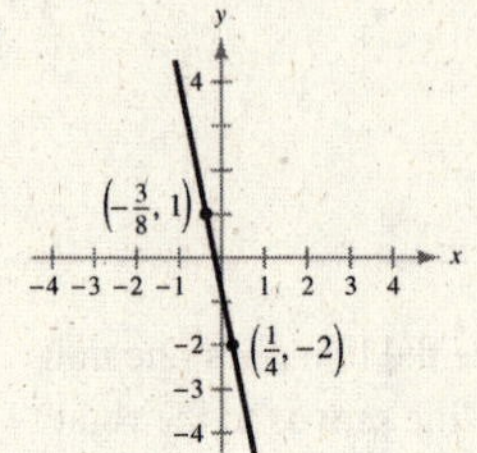

The slope is $m = \dfrac{1 - (-2)}{-\frac{3}{8} - \frac{1}{4}} = \dfrac{3}{-\frac{5}{8}} = -\dfrac{24}{5}$.

39.

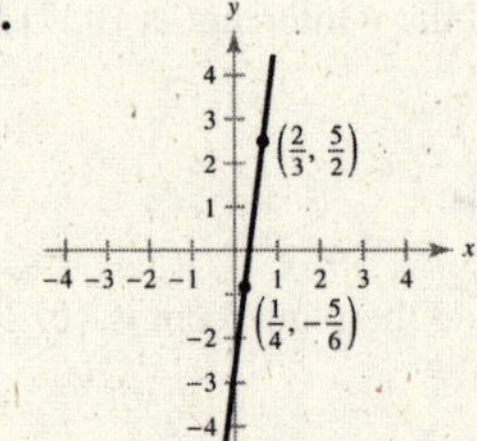

The slope is $m = \dfrac{\frac{5}{2} - \left(-\frac{5}{6}\right)}{\frac{2}{3} - \frac{1}{4}} = \dfrac{\frac{10}{3}}{\frac{5}{12}} = \dfrac{10}{3} \cdot \dfrac{12}{5} = 8$.

41. The equation of this horizontal line is $y = 1$. So, three additional points are $(0, 1)$, $(1, 1)$, and $(3, 1)$.

43. The equation of the line is

$$y - 7 = -3(x - 1)$$
$$y = -3x + 10.$$

So, three additional points are $(0, 10)$, $(2, 4)$, and $(3, 1)$.

45. The equation of this line is

$$y + 4 = \frac{2}{3}(x - 6)$$
$$y = \frac{2}{3}x - 8.$$

So, three additional points are $(3, -6)$, $(9, -2)$, and $(12, 0)$.

47. The equation of this vertical line is $x = -8$. So, three additional points are $(-8, 0)$, $(-8, 2)$, and $(-8, 3)$.

49. The slope of the line joining $(-2, 1)$ and $(-1, 0)$ is

$$\frac{1-0}{-2-(-1)} = \frac{1}{-1} = -1.$$

The slope of the line joining $(-1, 0)$ and $(2, -2)$ is

$$\frac{0-(-2)}{-1-2} = \frac{2}{-3} = -\frac{2}{3}.$$

Because the slopes are different, the points are not collinear.

51. The slope of the line joining $(2, 7)$ and $(-2, -1)$ is

$$\frac{-1-7}{-2-2} = 2.$$

The slope of the line joining $(0, 3)$ and $(-2, -1)$ is

$$\frac{-1-3}{-2-0} = 2.$$

Because the slopes are equal and both lines pass through $(-2, -1)$, the three points are collinear.

53. Using the slope-intercept form, we have $y = \frac{3}{4}x + 3$.

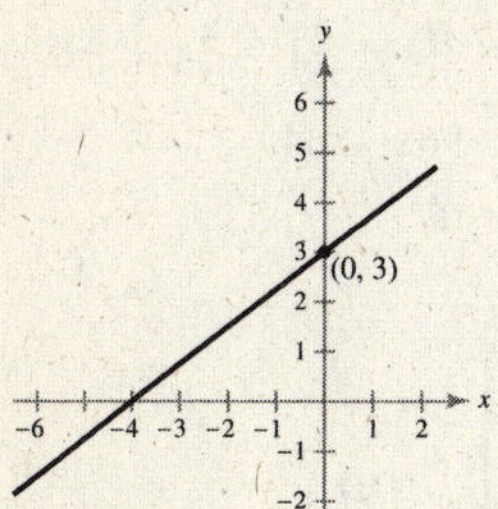

55. Because the slope is undefined, the line is vertical and its equation is $x = -1$.

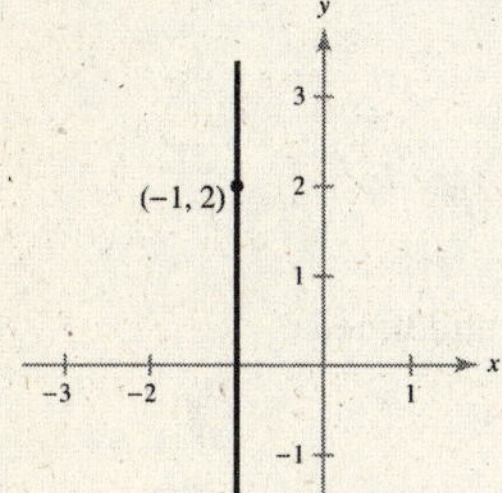

57. Because the slope is 0, the line is horizontal and its equation is $y = 7$.

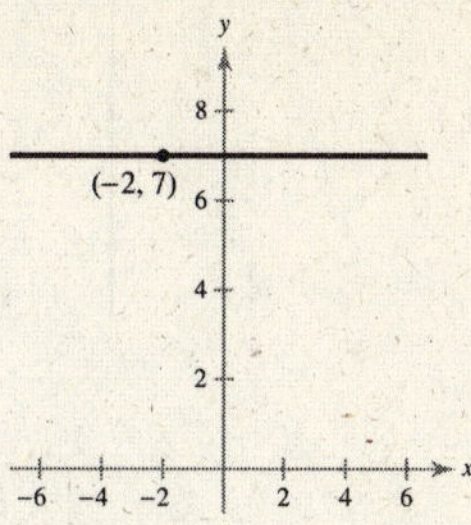

59. Using the point-slope form, we have

$$y + 2 = -4(x - 0)$$
$$y = -4x - 2$$
$$4x + y + 2 = 0.$$

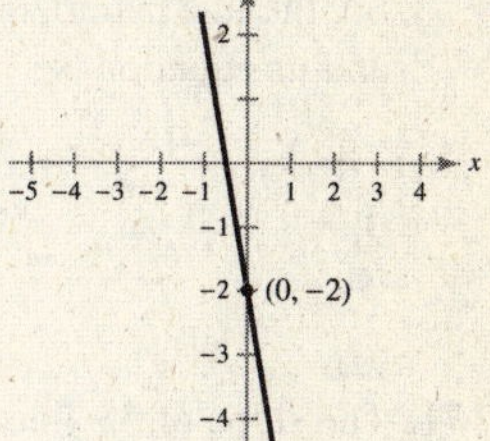

61. Using the point-slope form, we have

$$y - \frac{2}{3} = \frac{3}{4}(x - 0)$$
$$y = \frac{3}{4}x + \frac{2}{3}$$
$$0 = 9x - 12y + 8.$$

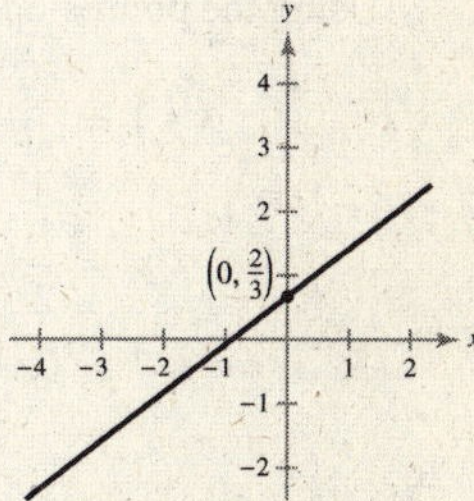

63. The slope of the line is

$$m = \frac{3-(-5)}{4-0} = 2.$$

Using the point-slope form, we have

$$y + 5 = 2(x - 0)$$
$$y = 2x - 5$$
$$0 = 2x - y - 5.$$

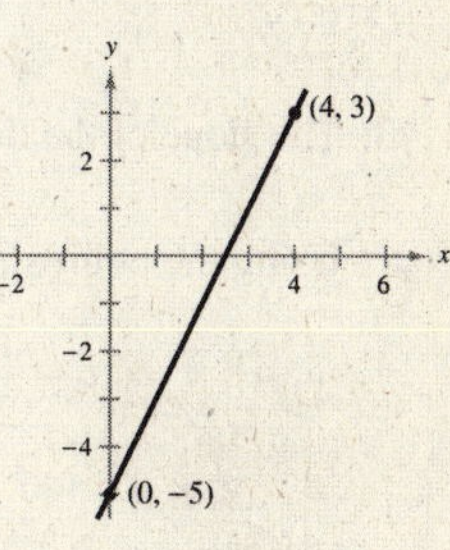

65. The slope of the line is $m = \dfrac{3-0}{-1-0} = -3$.

Using the point-slope form, we have

$$y = -3x$$
$$3x + y = 0.$$

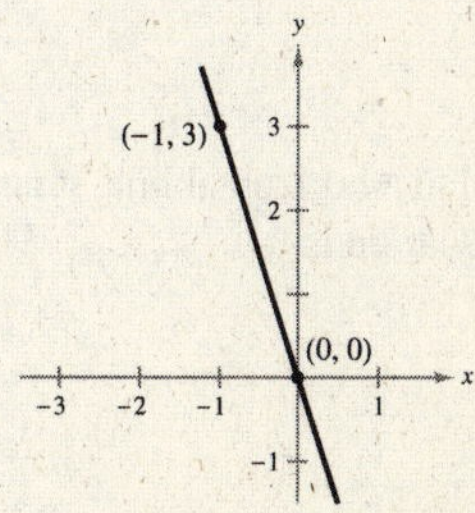

67. The slope of the line is

$$m = \frac{-2 - 3}{2 - 2} = \text{undefined.}$$

So, the line is vertical, and its equation is

$$x = 2$$
$$x - 2 = 0.$$

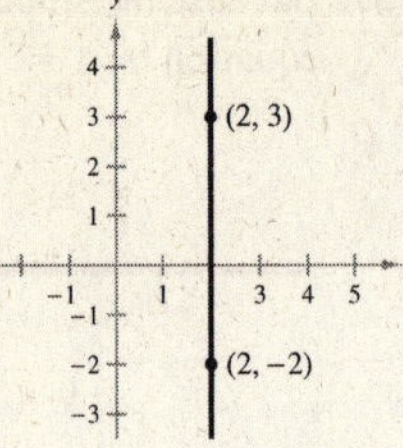

69. The slope of the line is

$$m = \frac{-1 - (-1)}{-2 - 3} = 0.$$

So, the line is horizontal, and its equation is

$$y = -1$$
$$y + 1 = 0.$$

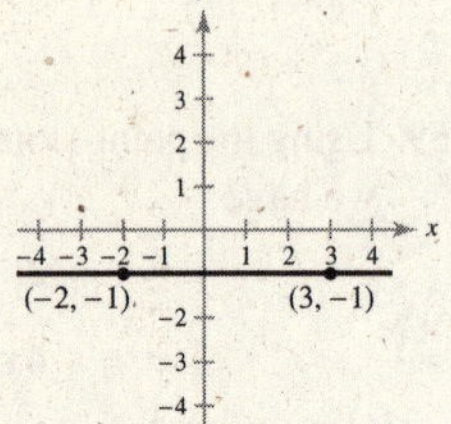

71. The slope of the line is $m = \dfrac{1 - 5/6}{(-1/3) + 2/3} = \dfrac{1}{2}.$

Using the point-slope form, we have

$$y - 1 = \frac{1}{2}\left(x + \frac{1}{3}\right)$$
$$y = \frac{1}{2}x + \frac{7}{6}$$
$$3x - 6y + 7 = 0.$$

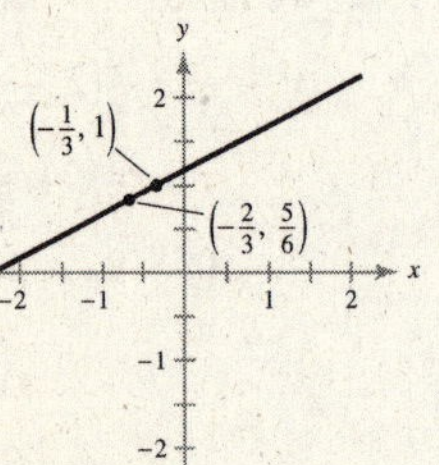

73. The slope of the line is $m = \dfrac{8 - 4}{1/2 + 1/2} = 4.$

Using the point-slope form, we have

$$y - 8 = 4\left(x - \frac{1}{2}\right)$$
$$y = 4x + 6$$
$$0 = 4x - y + 6.$$

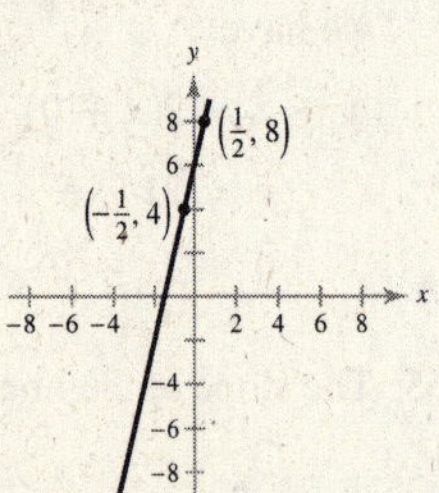

75. Because the line is vertical, it has an undefined slope, and its equation is

$$x = 3$$
$$x - 3 = 0.$$

77. Because the line is parallel to a horizontal line, it has a slope of $m = 0$, and its equation is

$$y = -10.$$

79. Given line: $y = -x + 7$, $m = -1$

(a) Parallel: $m_1 = -1$

$$y - 2 = -1(x + 3)$$
$$x + y + 1 = 0$$

(b) Perpendicular: $m_2 = 1$

$$y - 2 = 1(x + 3)$$
$$x - y + 5 = 0$$

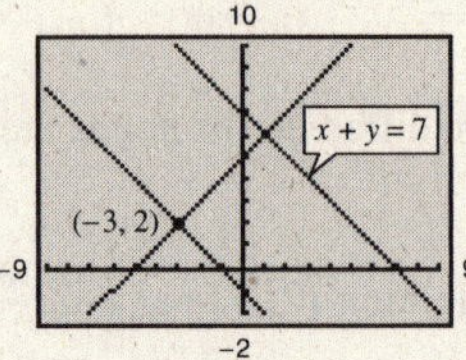

81. Given line: $y = -\frac{3}{4}x + \frac{7}{4}$, $m = -\frac{3}{4}$

(a) Parallel: $m_1 = -\frac{3}{4}$

$$y - \tfrac{7}{8} = -\tfrac{3}{4}\left(x + \tfrac{2}{3}\right) = -\tfrac{3}{4}x - \tfrac{1}{2}$$
$$8y - 7 = -6x - 4$$
$$6x + 8y - 3 = 0$$

(b) Perpendicular: $m_2 = \frac{4}{3}$

$$y - \tfrac{7}{8} = \tfrac{4}{3}\left(x + \tfrac{2}{3}\right) = \tfrac{4}{3}x + \tfrac{8}{9}$$
$$72y - 63 = 96x + 64$$
$$96x - 72y + 127 = 0$$

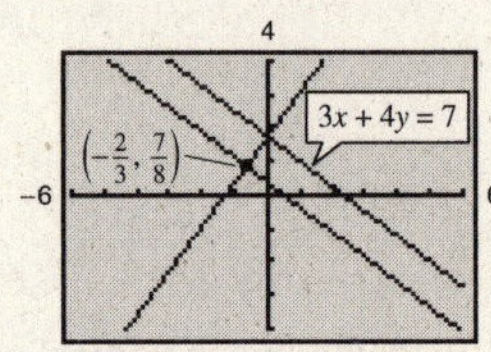

83. Given line: $y = -3$ is horizontal, $m = 0$

(a) Parallel: $m_1 = 0$

$$y - 0 = 0(x + 1)$$
$$y = 0$$

(b) Perpendicular: m_2 is undefined

$$x = -1$$

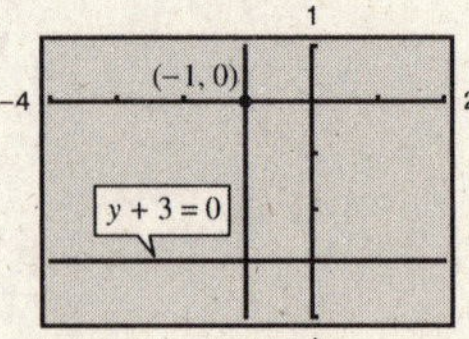

85. Given line: $x - 2 = 0$ is vertical, m is undefined

(a) Parallel: m_1 is undefined, $x = 1$

(b) Perpendicular: $m_2 = 0,\ y - 1 = 0(x - 1),\ y = 1$

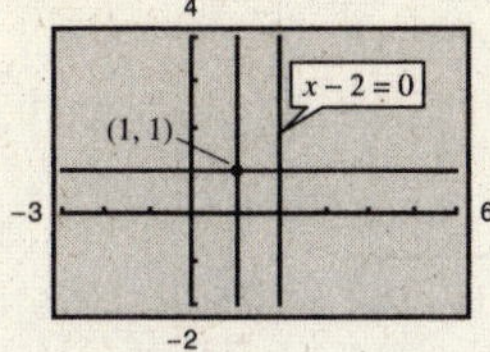

87. (a) The average salary increased the most from 2006 to 2008 and increased the least from 2002 to 2004.

(b) 1996: $(6, 69{,}277)$ and 2008: $(18, 97{,}486)$

$$m = \frac{97{,}486 - 69{,}277}{18 - 6} = \frac{28{,}209}{12} = \$2350.75/\text{yr}$$

(c) The average salary increased \$2350.75 per year over the 12 years between 1996 and 2008.

89. $(0, 32)$, $(100, 212)$

$$F - 32 = \frac{212 - 32}{100 - 0}(C - 0)$$

$$F = 1.8C + 32 = \frac{9}{5}C + 32$$

or

$$C = \frac{5}{9}(F - 32)$$

91. (a) 2004: $(4, 5511)$ and 2009: $(9, 5655)$

$$m = \frac{5655 - 5511}{9 - 4} = \frac{144}{5} = 28.8$$

$$y - y_1 = m(t - t_1)$$
$$y - 5511 = 28.8(t - 4)$$
$$y - 5511 = 28.8t - 115.2$$
$$y = 28.8t + 5395.8$$

The slope is 28.8 and indicates the population increases 28.8 thousand per year from 2004 to 2009.

(b) Let $t = 6$.

$$y = 28.8(6) + 5395.8$$
$$y = 5568.6$$

The population was 5568.6 thousand or 5,568,600 in 2006.

(c) The actual population was 5,572,000.

The model's estimate was very close to the actual population.

(d) The model could possibly be used to predict the population in 2015 if the population continues to grow at the same linear rate.

93. (a) 2004: $(4, 9937)$ and 2009: $(9, 12{,}175)$

$$m = \frac{12{,}175 - 9937}{9 - 4} = \frac{2238}{5} = 447.6$$

$$y - y_1 = m(t - t_1)$$
$$y - 9937 = 447.6(t - 4)$$
$$y - 9937 = 447.6t - 1790.4$$
$$y = 447.6t + 8146.6$$

(b) Let $t = 6$.

$$y = 447.6(6) + 8146.6 = \$10{,}832.2 \text{ billion}$$

(c) Let $t = 11$.

$$y = 447.6(11) + 8146.6 = \$13{,}070.2 \text{ billion}$$

(d) Answers will vary.

95. (a) $C = 50x + 350{,}000$

$R = 120x$

(b) $P = R - C = 120x - (50x + 350{,}000) = 70x - 350{,}000$

(c) If $x = 13{,}000$, then the profit is $P = 70(13{,}000) - 350{,}000 = \$560{,}000$.

Chapter 1 Quiz Yourself

1. (a)

(−3, 1) $\left(0, -\frac{1}{2}\right)$ (3, −2)

(b) $d = \sqrt{(-3 - 3)^2 + (1 - (-2))^2}$

$= \sqrt{36 + 9}$

$= 3\sqrt{5}$

(c) Midpoint $= \left(\frac{-3 + 3}{2}, \frac{1 - 2}{2}\right) = \left(0, -\frac{1}{2}\right)$

2. (a)

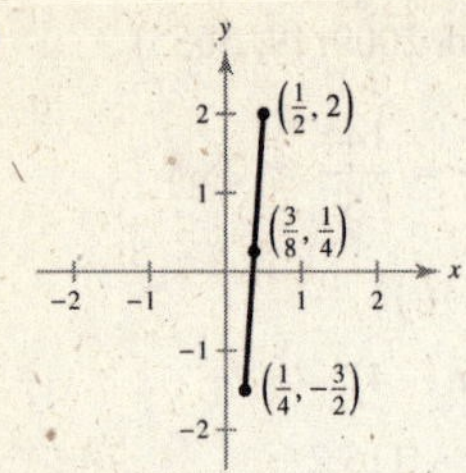

(b) $d = \sqrt{\left(\frac{1}{2} - \frac{1}{4}\right)^2 + \left(2 - \left(-\frac{3}{2}\right)\right)^2}$

$= \sqrt{\frac{1}{16} + \frac{49}{4}}$

$= \sqrt{\frac{197}{16}}$

$= \frac{1}{4}\sqrt{197}$

(c) Midpoint $= \left(\frac{\frac{1}{2} + \frac{1}{4}}{2}, \frac{2 - \frac{3}{2}}{2}\right) = \left(\frac{3}{8}, \frac{1}{4}\right)$

3. (a)

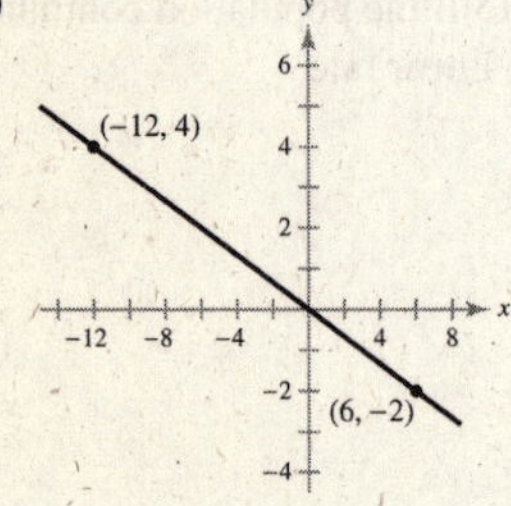

(b) $d = \sqrt{(6 - (-12))^2 + (-2 - 4)^2}$

$= \sqrt{18^2 + (-6)^2}$

$= \sqrt{324 + 36}$

$= \sqrt{360}$

$= 6\sqrt{10}$

≈ 18.97

(c) Midpoint $= \left(\frac{-12 + 6}{2}, \frac{4 + (-2)}{2}\right) = (-3, 1)$

4.

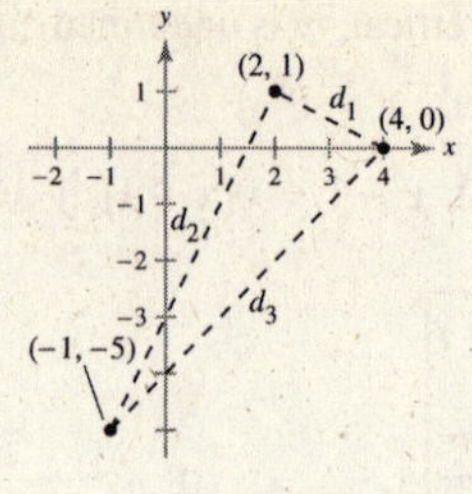

$a = \sqrt{(2 - 4)^2 + (1 - 0)^2} = \sqrt{5}$

$b = \sqrt{(2 - (-1))^2 + (1 - (-5))^2} = 3\sqrt{5}$

$c = \sqrt{(-1 - 4)^2 + (-5 - 0)^2} = 5\sqrt{2}$

$a^2 + b^2 = (\sqrt{5})^2 + (3\sqrt{5})^2 = (5\sqrt{2})^2 = c^2$

5. $(2007, 9534)$ and $(2009, 9829)$

Midpoint $= \left(\frac{2007 + 2009}{2}, \frac{9534 + 9829}{2}\right)$

$= (2008, 9681.5)$

The population in 2008 was approximately 9681.5 thousand or 9,681,500.

6. $y = 5x + 2$

x	$-\frac{2}{5}$	0	$\frac{1}{5}$	1
y	0	2	3	7

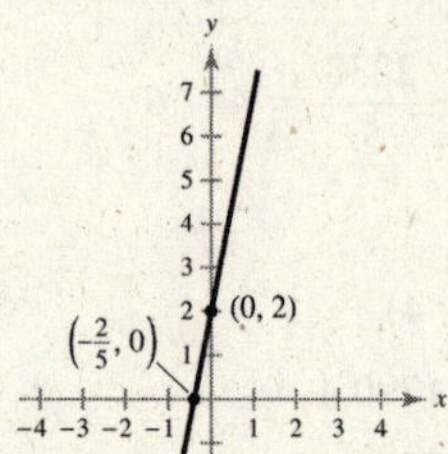

7. $y = x^2 + x - 6$

x	−3	−2	−1	−0.5	0	1	2
y	0	−4	−6	−6.25	−6	−4	0

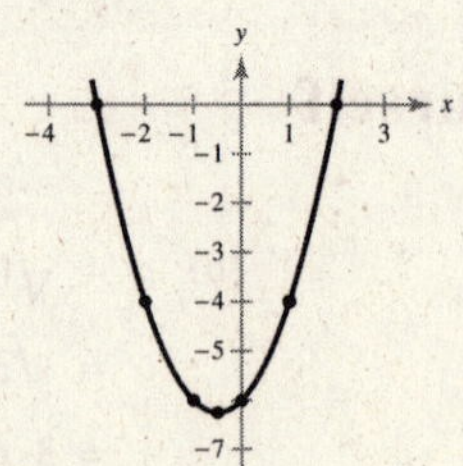

8. $y = |x - 3|$

x	0	1	2	3	4	5	6
y	3	2	1	0	1	2	3

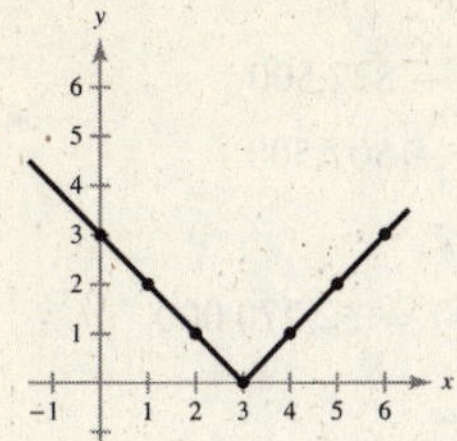

9. $(x - 0)^2 + (y - 0)^2 = 9^2$

$$x^2 + y^2 = 81$$

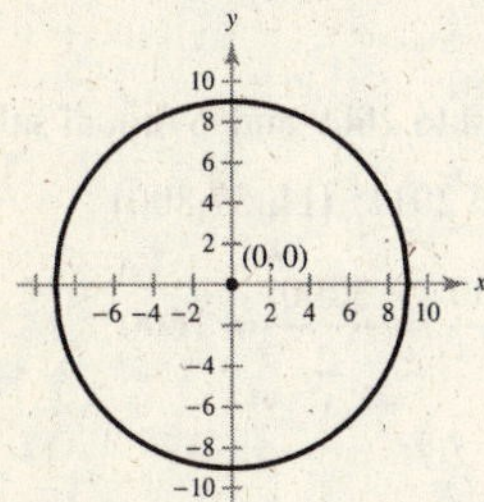

10. $(x - (-1))^2 + (y - 0)^2 = 6^2$

$$(x + 1)^2 + y^2 = 36$$

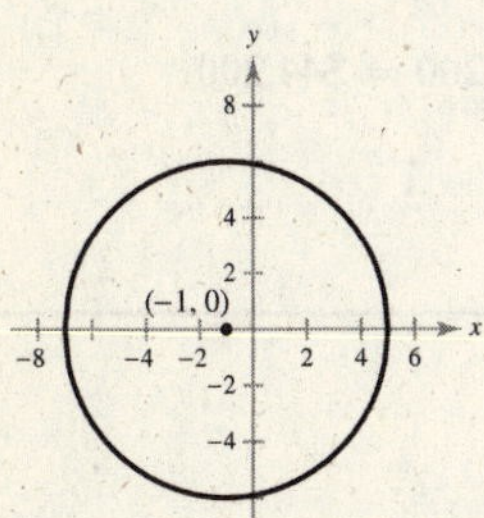

11. The radius is the distance between $(2, -2)$ and $(-1, 2)$.

$$r = \sqrt{(-1 - 2)^2 + (2 - (-2)^2)}$$

$$r = \sqrt{(-3)^2 + 4^2} = \sqrt{9 + 16} = \sqrt{25} = 5$$

Using the center $(2, -2)$ and radius $r = 5$:

$$(x - 2)^2 + (y - (-2))^2 = 5^2$$

$$(x - 2)^2 + (y + 2)^2 = 25$$

12. $C = 4.55x + 12{,}500$

$$R = 7.19x$$

$$R = C$$

$$7.19x = 4.55x + 12{,}500$$

$$2.64x = 12{,}500$$

$$x \approx 4734.8$$

The company must sell 4735 units to break even.

13. $y = mx + b$

$$y = \tfrac{1}{2}x - 3$$

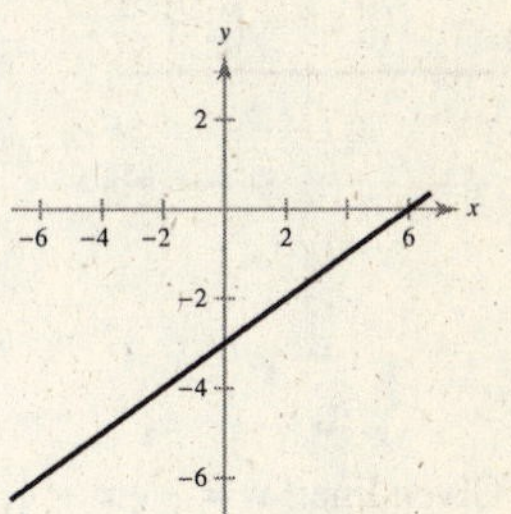

14. $y - y_1 = m(x - x_1)$

$$y - 1 = 2(x - 1)$$

$$y - 1 = 2x - 2$$

$$y = 2x - 1$$

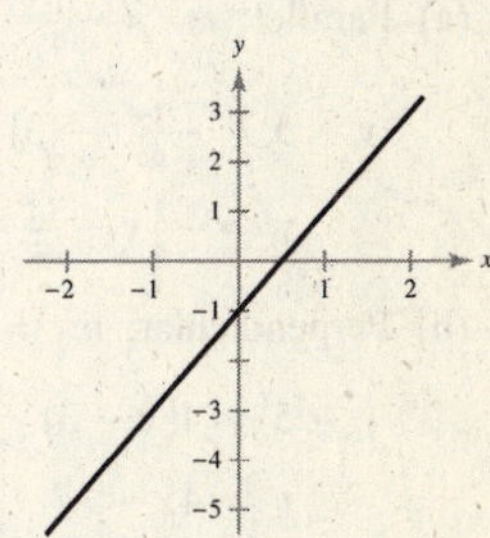

15. $y - y_1 = m(x - x_1)$

$$y - 5 = -\tfrac{1}{3}(x - 6)$$

$$y - 5 = -\tfrac{1}{3}x + 2$$

$$y = -\tfrac{1}{3}x + 7$$

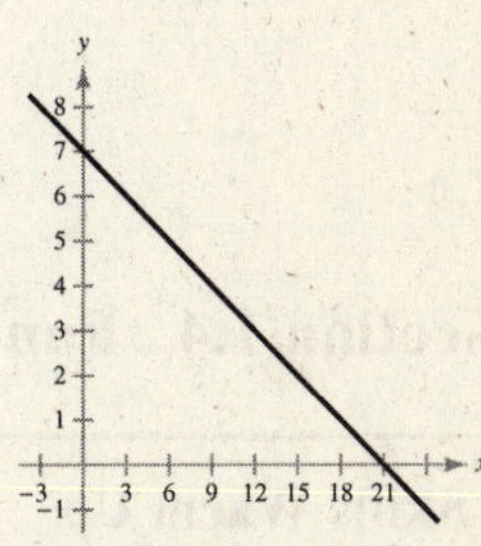

16. $(1, -1), (-4, 5)$

$$m = \frac{5 + 1}{-4 - 1} = -\frac{6}{5}$$

$$y + 1 = -\frac{6}{5}(x - 1)$$

$$y = -\frac{6}{5}x + \frac{1}{5}$$

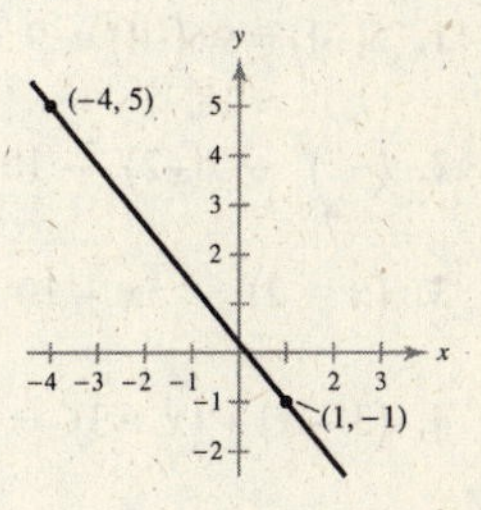

17. $(-2, 3), (-2, 2)$

$$m = \frac{2 - 3}{-2 + 2} = \text{undefined}$$

Because the slope is undefined, the line is vertical and its equation is $x = -2$.

18. $\left(\frac{5}{2}, 2\right), (0, 2)$

$$m = \frac{2-2}{0-\frac{5}{2}} = 0$$

Because the slope is 0, the line is horizontal and its equation is $y = 2$.

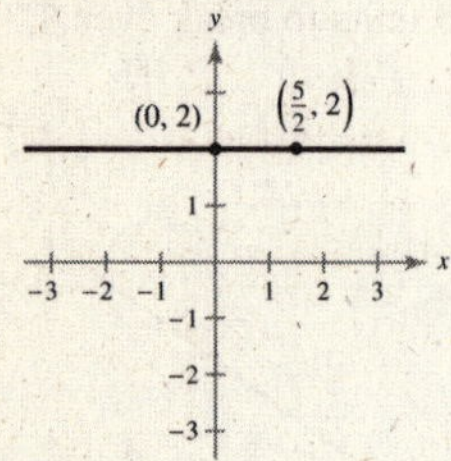

19. Given line: $y = -\frac{1}{4}x - \frac{1}{2}$, $m = -\frac{1}{4}$

(a) Parallel: $m_1 = -\frac{1}{4}$

$$y + 5 = -\tfrac{1}{4}(x - 3)$$
$$y = -\tfrac{1}{4}x - \tfrac{17}{4}$$

(b) Perpendicular: $m_2 = 4$

$$y + 5 = 4(x - 3)$$
$$y = 4x - 17$$

20. Let $t = 7$ correspond to 2007.

$(7, 1{,}330{,}000), (11, 1{,}800{,}000)$

$$m = \frac{1{,}800{,}000 - 1{,}330{,}000}{11 - 7} = \frac{470{,}000}{4} = 117{,}500$$

$$y - 1{,}330{,}000 = 117{,}500(x - 7)$$
$$y - 1{,}330{,}000 = 117{,}500x - 822{,}500$$
$$y = 117{,}500x + 507{,}500$$

For 2015, let $t = 15$.

$$y = 117{,}500(15) + 507{,}500 = \$2{,}270{,}000$$

For 2018, let $t = 18$.

$$y = 117{,}500(18) + 507{,}500 = \$2{,}622{,}500$$

21. The daily cost equals the cost for lodging and meals plus the cost per mile driven, x.

$$C = 175 + 0.55x$$

22. (a) Let $t = 9$ correspond to 2009 and S equal salary.

2009: $(9, 34{,}600)$ and 2011: $(11, 37{,}800)$

$$m = \frac{37{,}800 - 34{,}600}{11 - 9} = \frac{3200}{2} = 1600$$

$$S - S_1 = m(t - t_1)$$
$$S - 34{,}600 = 1600(t - 9)$$
$$S - 34{,}600 = 1600t - 14{,}400$$
$$S = 1600t + 20{,}200$$

(b) For 2015, let $t = 15$.

$$S = 1600(15) + 20{,}200 = \$44{,}200$$

Section 1.4 Functions

Skills Warm Up

1. $5(-1)^2 - 6(-1) + 9 = 5(1) + 6 + 9 = 20$

2. $(-2)^3 + 7(-2)^2 - 10 = -8 + 7(4) - 10 = -18 + 28 = 10$

3. $(x - 2)^2 + 5x - 10 = x^2 - 4x + 4 + 5x - 10 = x^2 + x - 6$

4. $(3 - x) + (x + 3)^3 = (3 - x) + (x + 3)(x^2 + 6x + 9) = (3 - x) + x^3 + 3x^2 + 6x^2 + 18x + 9x + 27 = x^3 + 9x^2 + 26x + 30$

5. $\dfrac{1}{1 - (1 - x)} = \dfrac{1}{1 - 1 + x} = \dfrac{1}{x}$

6. $1 + \dfrac{x - 1}{x} = \dfrac{x}{x} + \dfrac{x - 1}{x} = \dfrac{x + x - 1}{x} = \dfrac{2x - 1}{x}$

7. $2x + y - 6 = 11$

$$y = -2x + 17$$

8. $5y - 6x^2 - 1 = 0$

$$5y = 6x^2 + 1$$
$$y = \frac{6x^2 + 1}{5} = \frac{6}{5}x^2 + \frac{1}{5}$$

Skills Warm Up —*continued*—

9. $(y-3)^2 = 5 + (x+1)^2$

$y - 3 = \sqrt{5 + (x+1)^2}$

$y - 3 = \sqrt{5 + x^2 + 2x + 1}$

$y = \sqrt{x^2 + 2x + 6} + 3$

10. $y^2 - 4x^2 = 2$

$y^2 = 2 + 4x^2$

$y = \sqrt{2 + 4x^2}$

11. $x = \dfrac{2y-1}{4}$

$4x = 2y - 1$

$4x + 1 = 2y$

$\dfrac{4x+1}{2} = y$

$2x + \dfrac{1}{2} = y$

12. $x = \sqrt[3]{2y - 1}$

$x^3 = 2y - 1$

$-2y = -x^3 - 1$

$y = \frac{1}{2}x^3 + \frac{1}{2}$

1. $x^2 + y^2 = 16$

$y^2 = 16 - x^2$

$y = \pm\sqrt{16 - x^2}$

y is *not* a function of x since there are two values of y for some x.

3. $\frac{1}{2}x - 6y = -3$

$y = \frac{1}{12}x + \frac{1}{2}$

y *is* a function of x since there is only one value of y for each x.

5. $y = 4 - x^2$

y *is* a function of x since there is only one value of y for each x.

7. $y = |x + 2|$

y *is* a function of x since there is only one value of y for each x.

9. y is *not* a function of x.

11. y *is* a function of x.

13. Domain: $(-\infty, \infty)$

Range: $(-\infty, \infty)$

15. Domain: $(-\infty, \infty)$

Range: $(-\infty, 4]$

17.

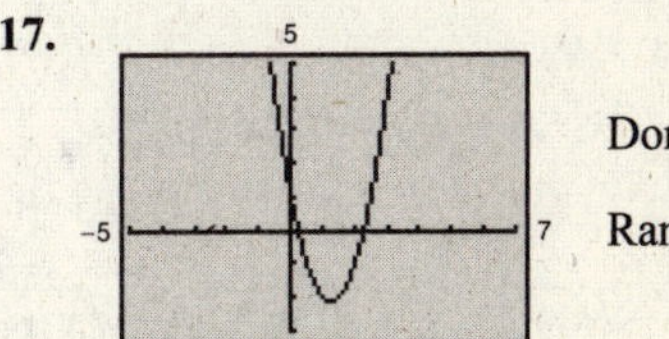

Domain: $(-\infty, \infty)$

Range: $[-2.125, \infty)$

19.

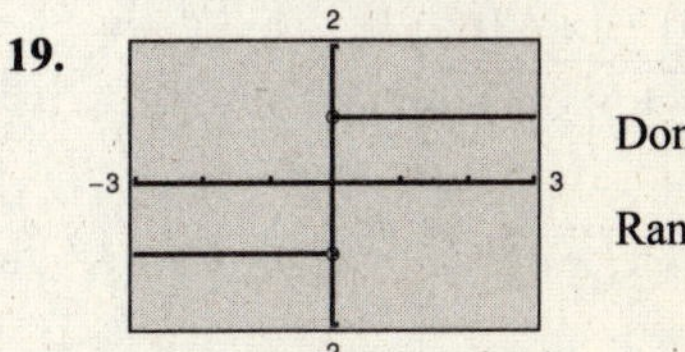

Domain: $(-\infty, 0) \cup (0, \infty)$

Range: $\{-1, 1\}$

21.

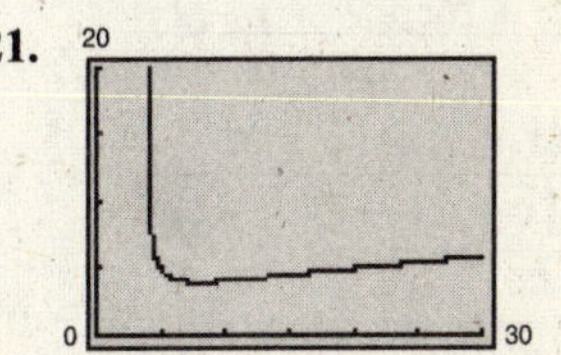

Domain: $(4, \infty)$

Range: $[4, \infty)$

23.

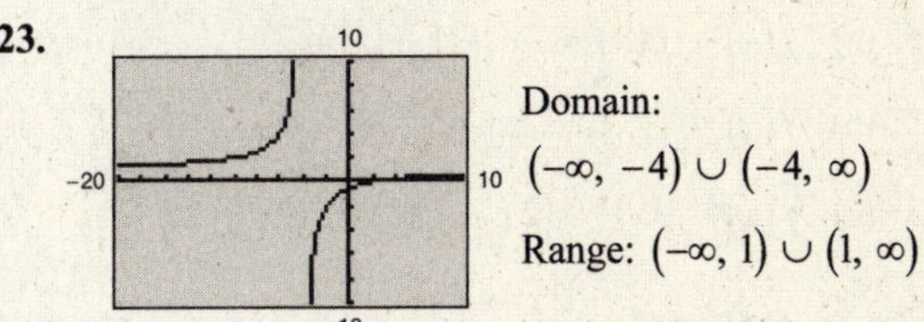

Domain: $(-\infty, -4) \cup (-4, \infty)$

Range: $(-\infty, 1) \cup (1, \infty)$

25. $f(x) = 3x - 2$

(a) $f(0) = 3(0) - 2 = -2$

(b) $f(5) = 3(5) - 2 = 13$

(c) $f(x-1) = 3(x-1) - 2 = 3x - 3 - 2 = 3x - 5$

27. $g(x) = \frac{1}{x}$

(a) $g\left(\frac{1}{4}\right) = \frac{1}{\frac{1}{4}} = 4$

(b) $g(-4) = \frac{1}{-4} = -\frac{1}{4}$

(c) $g(x + 4) = \frac{1}{x + 4}$

29. $\frac{f(x + \Delta x) - f(x)}{\Delta x}$

$= \frac{(x + \Delta x)^2 - 5(x + \Delta x) + 2 - (x^2 - 5x + 2)}{\Delta x}$

$= \frac{[x^2 + 2x\Delta x + (\Delta x)^2 - 5x + 5\Delta x + 2] - [x^2 - 5x + 2]}{\Delta x}$

$= \frac{2x\Delta x + (\Delta x)^2 + 5\Delta x}{\Delta x}$

$= 2x + \Delta x + 5, \ \Delta x \neq 0$

31. $\frac{g(x + \Delta x) - g(x)}{\Delta x}$

$= \frac{\sqrt{x + \Delta x + 1} - \sqrt{x + 1}}{\Delta x} \cdot \frac{\sqrt{x + \Delta x + 1} + \sqrt{x + 1}}{\sqrt{x + \Delta x + 1} + \sqrt{x + 1}}$

$= \frac{(x + \Delta x + 1) - (x + 1)}{\Delta x\left[\sqrt{x + \Delta x + 1} + \sqrt{x + 1}\right]}$

$= \frac{1}{\sqrt{x + \Delta x + 1} + \sqrt{x + 1}}, \ \Delta x \neq 0$

33. $\frac{f(x + \Delta x) - f(x)}{\Delta x} = \frac{\frac{1}{x + \Delta x - 2} - \frac{1}{x - 2}}{\Delta x}$

$= \frac{(x - 2) - (x + \Delta x - 2)}{(x + \Delta x - 2)(x - 2)\Delta x}$

$= \frac{-1}{(x + \Delta x - 2)(x - 2)}, \ \Delta x \neq 0$

35. (a) $f(x) + g(x) = 2x - 5 + 5 = 2x$

(b) $f(x) - g(x) = 2x - 5 - 5 = 2x - 10$

(c) $f(x) \cdot g(x) = (2x - 5)(5) = 10x - 25$

(d) $f(x)/g(x) = \frac{2x - 5}{5}$

(e) $f(g(x)) = f(5) = 2(5) - 5 = 5$

(f) $g(f(x)) = g(2x - 5) = 5$

37. (a) $f(x) + g(x) = x^2 + 1 + x - 1 = x^2 + x$

(b) $f(x) - g(x) = x^2 + 1 - x + 1 = x^2 - x + 2$

(c) $f(x) \cdot g(x) = (x^2 + 1)(x - 1) = x^3 - x^2 + x - 1$

(d) $f(x)/g(x) = \frac{x^2 + 1}{x - 1}, x \neq 1$

(e) $f(g(x)) = (x - 1)^2 + 1 = x^2 - 2x + 2$

(f) $g(f(x)) = x^2 + 1 - 1 = x^2$

39. (a) $f(g(1)) = f(1^2 - 1) = f(0) = 0$

(b) $g(f(1)) = g(\sqrt{1}) = g(1) = 0$

(c) $g(f(0)) = g(0) = -1$

(d) $f(g(-4)) = f(15) = \sqrt{15}$

(e) $f(g(x)) = f(x^2 - 1) = \sqrt{x^2 - 1}$

(f) $g(f(x)) = g(\sqrt{x}) = x - 1, \ x \geq 0$

41. $f(x) = 4x$

$f^{-1}(x) = \frac{1}{4}x$

$f(f^{-1}(x)) = 4\left(\frac{1}{4}x\right) = x$

$f^{-1}(f(x)) = \frac{1}{4}(4x) = x$

43. $f(x) = x + 12$

$f^{-1}(x) = x - 12$

$f(f^{-1}(x)) = (x - 12) + 12 = x$

$f^{-1}(f(x)) = (x + 12) - 12 = x$

45. $f(x) = 2x - 3 = y$

$2y - 3 = x$

$2y = x + 3$

$y = \frac{x + 3}{2}$

$f'(x) = \frac{1}{2}x + \frac{3}{2}$

47. $f(x) = \frac{3}{2}x + 1 = y$

$\frac{3}{2}y + 1 = x$

$3y + 2 = 2x$

$3y = 2x - 2$

$y = \frac{2}{3}x - \frac{2}{3}$

$f^{-1}(x) = \frac{2}{3}x - \frac{2}{3}$

49. $f(x) = x^5 = y$

$y^5 = x$

$y = \sqrt[5]{x}$

$f^{-1}(x) = \sqrt[5]{x}$

51. $f(x) = \dfrac{1}{x} = y$

$\dfrac{1}{y} = x$

$y = \dfrac{1}{x}$

$f^{-1}(x) = \dfrac{1}{x}$

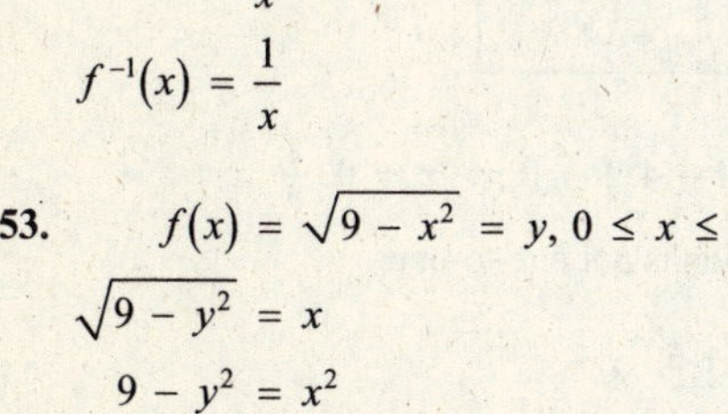

53. $f(x) = \sqrt{9 - x^2} = y, 0 \le x \le 3$

$\sqrt{9 - y^2} = x$

$9 - y^2 = x^2$

$y^2 = 9 - x^2$

$y = \sqrt{9 - x^2}$

$f^{-1}(x) = \sqrt{9 - x^2},\ \ 0 \le x \le 3$

55. $f(x) = x^{2/3} = y,\ x \ge 0$

$y^{2/3} = x$

$y = x^{3/2}$

$f^{-1}(x) = x^{3/2}$

57.

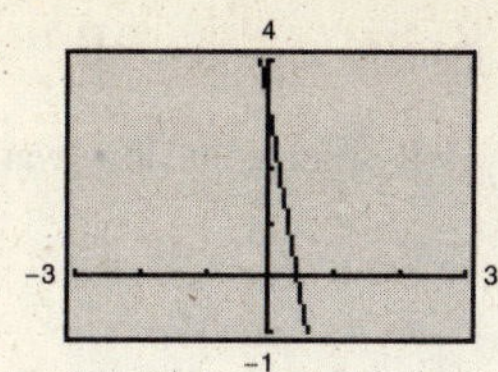

$f(x) = 3 - 7x$ is one-to-one.

$y = 3 - 7x$

$x = 3 - 7y$

$y = \dfrac{3 - x}{7}$

59.

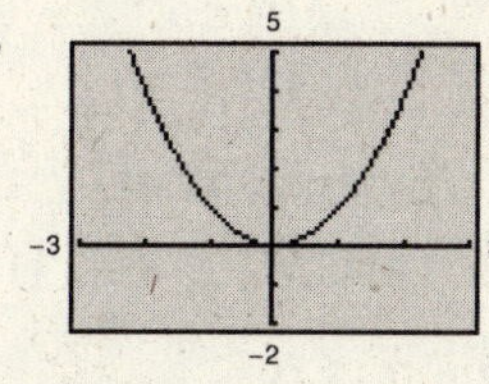

$f(x) = x^2$

f is *not* one-to-one because $f(1) = 1 = f(-1)$.

61. $f(x) = |x + 3|$

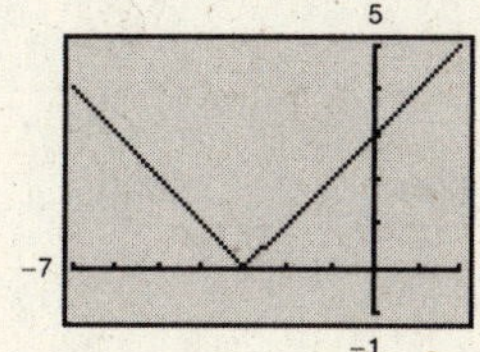

f is *not* one-to-one because $f(-5) = 2 = f(-1)$.

63. (a) $y = \sqrt{x} + 2$

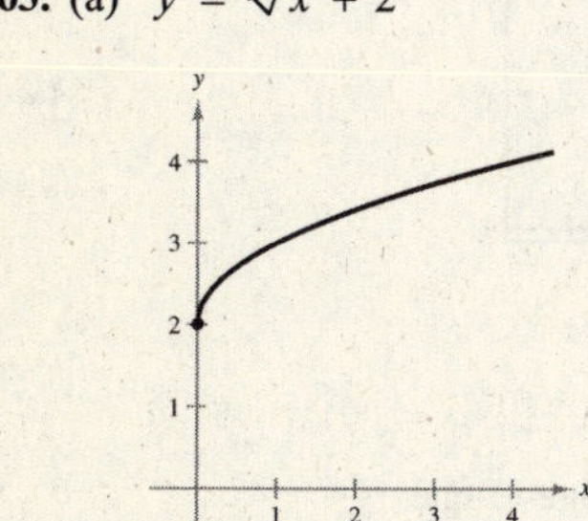

(b) $y = -\sqrt{x}$

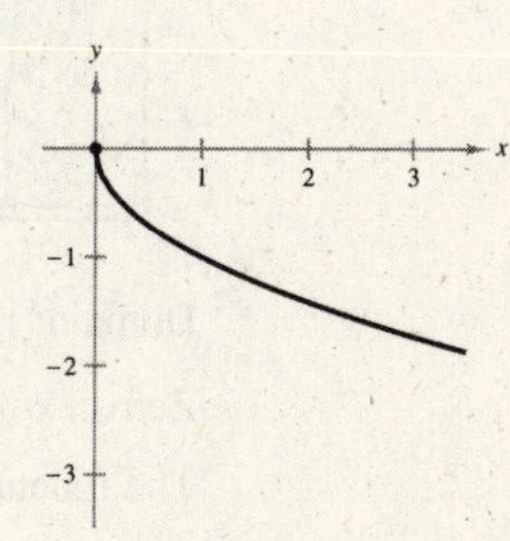

(c) $y = \sqrt{x - 2}$

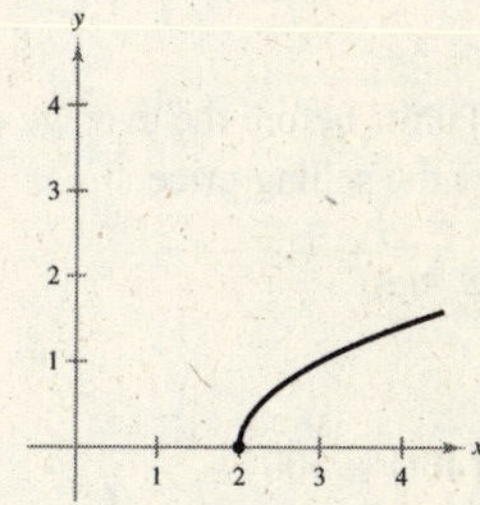

(d) $y = \sqrt{x + 3}$

(e) $y = \sqrt{x - 4} - 1$

(f) $y = 2\sqrt{x}$

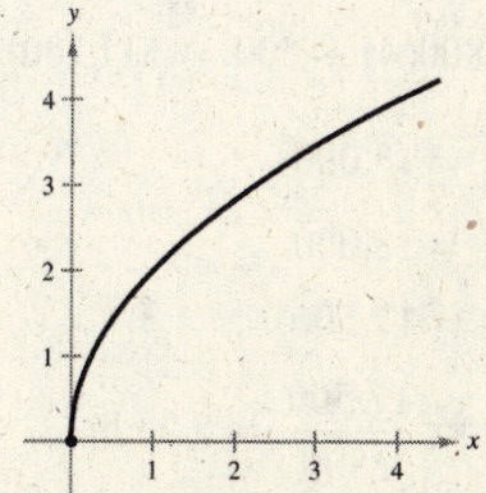

65. (a) Shifted three units to the left: $y = (x + 3)^2$

(b) Shifted six units to the left, three units downward, and reflected: $y = -(x + 6)^2 - 3$

67. (a) 2000: $120 billion
2003: 170 billion
2007: $225 billion

(b) 2000: $121.3 billion
2003: $173.8 billion
2007: $225.6 billion

The model fits the data well.

69. $R_{\text{TOTAL}} = R_1 + R_2$
$= 690 - 8t - 0.8t^2 + 458 + 0.78t$
$= -0.8t^2 - 7.22t + 1148, t = 5, 6, \ldots, 11$

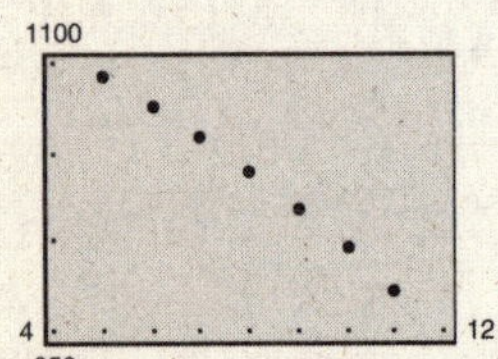

71. (a) $C = 1.95x + 6000$

(b) $\overline{C} = \dfrac{C}{x} = \dfrac{1.95x + 6000}{x} = 1.95 + \dfrac{6000}{x}$

(c) $1.95 + \dfrac{6000}{x} < 4.95$

$\dfrac{6000}{x} < 3$

$\dfrac{6000}{3} < x$ because $x > 0$.

$2000 < x$

Must sell 2000 units before the average cost per unit falls below the selling price.

73. (a) $C(x) = 70x + 500$
$x(t) = 40t$
$C(x(t)) = 70(40t) + 500$
$= 2800t + 500$

C is the weekly cost per t hours of production.

(b) $C(x(4)) = 2800(4) + 500 = \$11{,}700$

(c) $C(x(t)) = 18{,}000$
$2800t + 500 = 18{,}000$
$2800t = 17{,}500$
$t = \dfrac{17{,}500}{2800} = 6.25$ hr

75. (a) Cost $= C = 98{,}000 + 12.30x$

(b) Revenue $= R = 17.98x$

(c) Profit $= R - C$
$= 17.98x - (12.30x + 98{,}000)$
$= 5.68x - 98{,}000$

77. $f(x) = 9x - 4x^2$

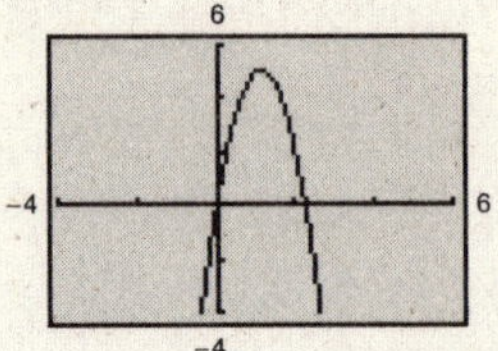

Zeros: $x(9 - 4x) = 0 \Rightarrow x = 0, \frac{9}{4}$

The function is not one-to-one.

79. $g(t) = \dfrac{t + 3}{1 - t}$

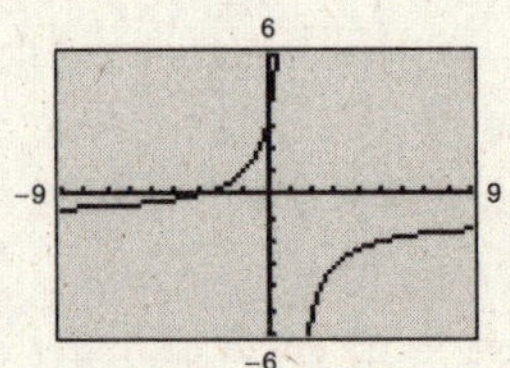

Zero: $t = -3$

The function is one-to-one

81. $g(x) = x^2\sqrt{x^2 - 4}$

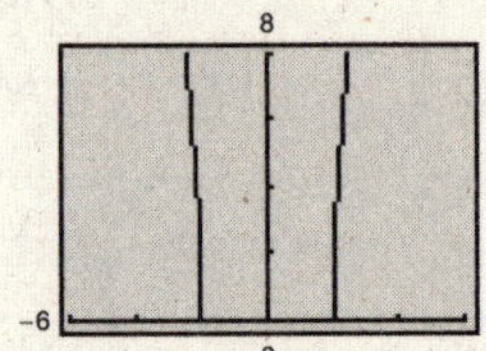

Domain: $|x| \geq 2$

Zeros: $x = \pm 2$

The function is not one-to-one.

83. Answers will vary.

Section 1.5 Limits

Skills Warm Up

1. $\dfrac{2x^3 + x^2}{6x} = \dfrac{x^2(2x + 1)}{6x} = \dfrac{x(2x + 1)}{6} = \dfrac{1}{6}x(2x + 1)$

2. $\dfrac{x^5 + 9x^4}{x^2} = \dfrac{x^4(x + 9)}{x^2} = x^3(x + 9)$

3. $\dfrac{x^2 - 3x - 28}{x - 7} = \dfrac{(x - 7)(x + 4)}{x - 7} = x + 4$

4. $\dfrac{x^2 + 11x + 30}{x + 5} = \dfrac{(x + 6)(x + 5)}{x + 5} = x + 6$

5. $f(x) = x^2 - 3x + 3$

   (a) $f(-1) = (-1)^2 - 3(-1) + 3 = 1 + 3 + 3 = 7$

   (b) $f(c) = c^2 - 3c + 3$

   (c) $f(x + h) = (x + h)^2 - 3(x + h) + 3$
   $= x^2 + 2xh + h^2 - 3x - 3h + 3$

6. $f(x) = \begin{cases} 2x - 2, & x < 1 \\ 3x + 1, & x \ge 1 \end{cases}$

   (a) $f(-1) = 2(-1) - 2 = -2 - 2 = -4$

   (b) $f(3) = 3(3) + 1 = 9 + 1 = 10$

   (c) $f(t^2 + 1) = 3(t^2 + 1) + 1$
   $= 3t^2 + 3 + 1$
   $= 3t^2 + 4$

7. $f(x) = x^2 - 2x + 2$

   $\dfrac{f(1 + h) - f(1)}{h}$
   $= \dfrac{(1 + h)^2 - 2(1 + h) + 2 - (1^2 - 2(1) + 2)}{h}$
   $= \dfrac{1 + 2h + h^2 - 2 - 2h + 2 - 1 + 2 - 2}{h}$
   $= \dfrac{h^2}{h}$
   $= h$

8. $f(x) = 4x$

   $\dfrac{f(2 + h) - f(2)}{h} = \dfrac{4(2 + h) - 4(2)}{h}$
   $= \dfrac{8 + 4h - 8}{h}$
   $= \dfrac{4h}{h}$
   $= 4$

9. $h(x) = -\dfrac{5}{x}$

   Domain: $(-\infty, 0) \cup (0, \infty)$

   Range: $(-\infty, 0) \cup (0, \infty)$

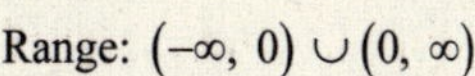

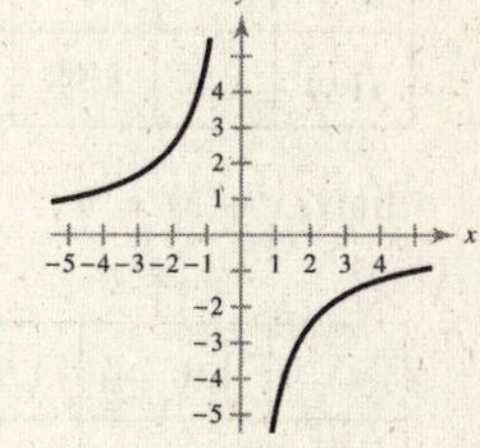

10. $g(x) = \sqrt{25 - x^2}$

    Domain: $[-5, 5]$

    Range: $[0, 5]$

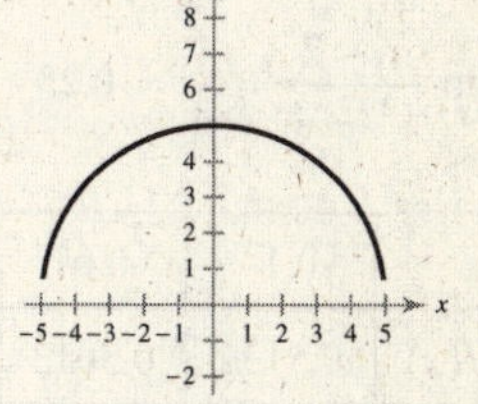

11. $f(x) = |x - 3|$

    Domain: $(-\infty, \infty)$

    Range: $[0, \infty)$

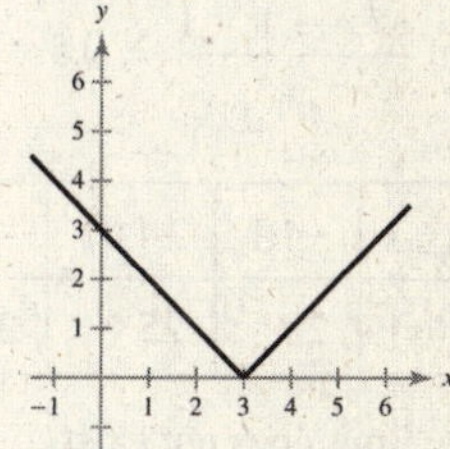

12. $f(x) = \dfrac{|x|}{x}$

    Domain: $(-\infty, 0) \cup (0, \infty)$

    Range: $y = -1,\ y = 1$

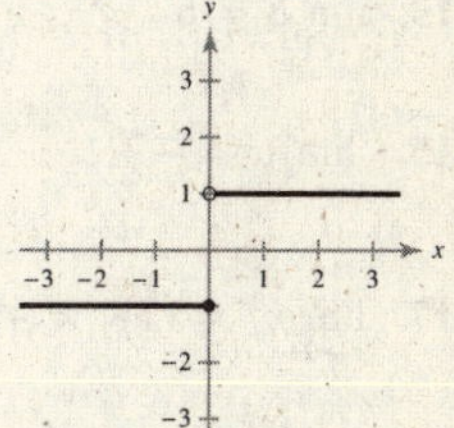

13. $9x^2 + 4y^2 = 49$
    $4y^2 = 49 - 9x^2$
    $y^2 = \dfrac{49 - 9x^2}{4}$
    $y = \dfrac{\pm\sqrt{49 - 9x^2}}{2}$

    Not a function of x (fails the vertical line test).

14. $2x^2y + 8x = 7y$
    $2x^2y - 7y = -8x$
    $y(2x^2 - 7) = -8x$
    $y = -\dfrac{8x}{2x^2 - 7}$

    Yes, y is a function of x.

1. (a) $\lim_{x \to 0} f(x) = 1$

(b) $\lim_{x \to -1} f(x) = 3$

3. (a) $\lim_{x \to 0} g(x) = 1$

(b) $\lim_{x \to -1} g(x) = 3$

5.

x	1.9	1.99	1.999	2	2.001	2.01	2.1
$f(x)$	8.8	8.98	8.998	?	9.002	9.02	9.2

$\lim_{x \to 2}(2x + 5) = 9$

7.

x	1.9	1.99	1.999	2	2.001	2.01	2.1
$f(x)$	0.2564	0.2506	0.2501	?	0.2499	0.2494	0.2439

$\lim_{x \to 2} \frac{x - 2}{x^2 - 4} = \frac{1}{4} = 0.25$

9.

x	−0.1	−0.01	−0.001	0	0.001	0.01	0.1
$f(x)$	0.5132	0.5013	0.5001	?	0.4999	0.4988	0.4881

$\lim_{x \to 0} \frac{\sqrt{x + 1} - 1}{x} = 0.5$

11.

x	−4.1	−4.01	−4.001	−4	−3.999	−3.99	−3.9
$f(x)$	2.5	25	250	?	−250	−25	−2.5

The limit does not exist.

13. $\lim_{x \to 3} 6 = 6$

15. $\lim_{x \to -2} x = -2$

17. $\lim_{x \to 7} x^2 = (7)^2 = 49$

19. $\lim_{x \to 16} \sqrt{x} = \sqrt{16} = 4$

21. (a) $\lim_{x \to c}\left[f(x) + g(x)\right] = \lim_{x \to c} f(x) + \lim_{x \to c} g(x)$

$= 3 + 9$

$= 12$

(b) $\lim_{x \to c}\left[f(x)g(x)\right] = \left[\lim_{x \to c} f(x)\right]\left[\lim_{x \to c} g(x)\right]$

$= 3 \cdot 9$

$= 27$

(c) $\lim_{x \to c} \frac{f(x)}{g(x)} = \frac{\lim_{x \to c} f(x)}{\lim_{x \to c} g(x)} = \frac{3}{9} = \frac{1}{3}$

23. (a) $\lim_{x \to c} \sqrt{f(x)} = \sqrt{16} = 4$

(b) $\lim_{x \to c}\left[3f(x)\right] = 3(16) = 48$

(c) $\lim_{x \to c}\left[f(x)\right]^2 = 16^2 = 256$

25. $\lim_{x \to -3}(2x + 5) = \lim_{x \to -3} 2x + \lim_{x \to -3} 5 = 2(-3) + 5 = -1$

27. $\lim_{x \to 1}(1 - x^2) = \lim_{x \to 1} 1 - \lim_{x \to 1} x^2 = 1 - 1^2 = 0$

29. $\lim_{x \to 3} \sqrt{x + 6} = \sqrt{3 + 6} = 3$

31. $\lim_{x \to -3} \frac{2}{x + 2} = \frac{2}{-3 + 2} = -2$

33. $\lim_{x \to -2} \frac{x^2 - 1}{2x} = \frac{(-2)^2 - 1}{2(-2)} = \frac{3}{-4} = -\frac{3}{4}$

35. $\lim_{x \to 5} \frac{\sqrt{x + 11} + 6}{x} = \frac{\sqrt{5 + 11} + 6}{5}$

$= \frac{\sqrt{16} + 6}{5}$

$= \frac{4 + 6}{5} = \frac{10}{5} = 2$

37. $$\lim_{x\to-3}\frac{x^2-9}{x+3}=\lim_{x\to-3}\frac{(x+3)(x-3)}{x+3}$$
$$=\lim_{x\to-3}(x-3)=-6$$

39. $$\lim_{x\to2}\frac{2-x}{x^2-4}=\lim_{x\to2}\frac{-(x-2)}{(x+2)(x-2)}$$
$$=\lim_{x\to2}\frac{-1}{x+2}=-\frac{1}{4}$$

41. $$\lim_{x\to-2}\frac{x^3+8}{x+2}=\lim_{x\to-2}\frac{(x+2)(x^2-2x+4)}{x+2}$$
$$=\lim_{x\to-2}(x^2-2x+4)=12$$

43. $$\lim_{\Delta x\to0}\frac{2(x+\Delta x)-2x}{\Delta x}=\lim_{\Delta x\to0}\frac{2x+2\Delta x-2x}{\Delta x}$$
$$=\lim_{\Delta x\to0}2=2$$

45. $$\lim_{\Delta t\to0}\frac{(t+\Delta t)^2-5(t+\Delta t)-(t^2-5t)}{\Delta t}=\lim_{\Delta t\to0}\frac{t^2+2t(\Delta t)+(\Delta t)^2-5t-5(\Delta t)-t^2+5t}{\Delta t}$$
$$=\lim_{\Delta t\to0}\frac{2t(\Delta t)+(\Delta t)^2-5(\Delta t)}{\Delta t}$$
$$=\lim_{\Delta t\to0}2t+(\Delta t)-5$$
$$=2t-5$$

47. $$\lim_{x\to4}\frac{\sqrt{x+5}-3}{x-4}=\lim_{x\to4}\frac{\sqrt{x+5}-3}{x-4}\cdot\frac{\sqrt{x+5}+3}{\sqrt{x+5}+3}$$
$$=\lim_{x\to4}\frac{(x+5)-9}{(x-4)\sqrt{x+5}+3}$$
$$=\lim_{x\to4}\frac{x-4}{(x-4)\left(\sqrt{x+5}+3\right)}$$
$$=\lim_{x\to4}\frac{1}{\sqrt{x+5}+3}=\frac{1}{6}$$

49. $$\lim_{x\to0}x=\frac{\sqrt{x+5}-\sqrt{5}}{x}=\lim_{x\to0}\frac{\sqrt{x+5}-\sqrt{5}}{x}\cdot\frac{\sqrt{x+5}+\sqrt{5}}{\sqrt{x+5}+\sqrt{5}}$$
$$=\lim_{x\to0}\frac{(x+5)-5}{x\left(\sqrt{x+5}+\sqrt{5}\right)}$$
$$=\lim_{x\to0}\frac{x}{x\left(\sqrt{x+5}+\sqrt{5}\right)}$$
$$=\lim_{x\to0}\frac{1}{\sqrt{x+5}+\sqrt{5}}=\frac{1}{2\sqrt{5}}$$

51. $$\lim_{x\to2^-}(4-x)=2$$
$$\lim_{x\to2^+}(4-x)=2$$
So, $\lim_{x\to2}f(x)=2$

53. $$\lim_{x\to3^-}f(x)=\lim_{x\to3^-}\left(\frac{1}{3}x-2\right)=-1$$
$$\lim_{x\to3^+}f(x)=\lim_{x\to3^+}(-2x+5)=-1$$
So, $\lim_{x\to3}f(x)=-1$.

55. $$\lim_{x\to-4}\frac{2}{x+4}=\frac{2}{0}$$
The limit does not exist.

57. $$\lim_{x\to2}\frac{x-2}{x^2-4x+4}=\lim_{x\to2}\frac{x-2}{(x-2)(x-2)}$$
$$=\lim_{x\to2}\frac{1}{x-2}$$
The limit does not exist.

59.

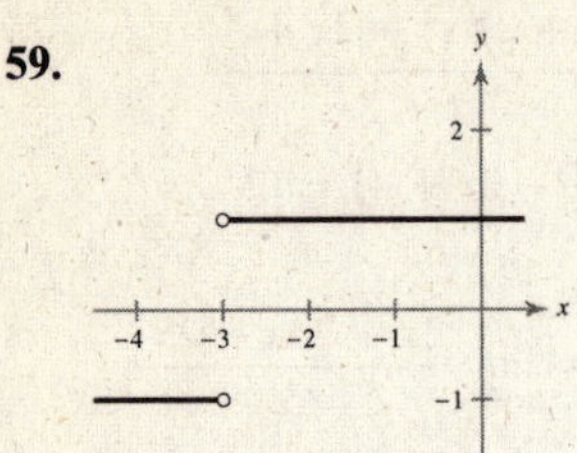

$$\lim_{x \to -3^-} \frac{|x + 3|}{x + 3} = -1 \text{ and}$$

$$\lim_{x \to -3^+} \frac{|x + 3|}{x + 3} = 1$$

61.

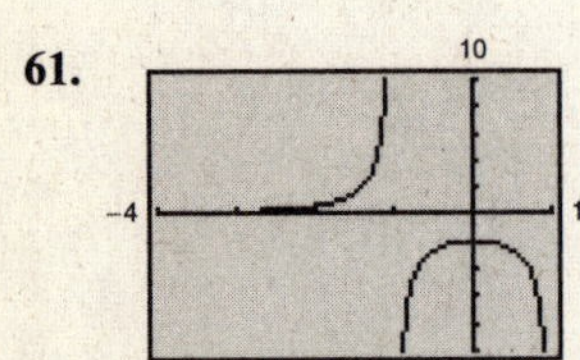

$$\lim_{x \to 1^-} \frac{2}{x^2 - 1} = -\infty$$

x	0	0.5	0.9	0.99	0.999	0.9999	1
$f(x)$	−2	−2.67	−10.53	−100.5	−1000.5	−10,000.5	undefined

Because $f(x) = \dfrac{2}{x^2 - 1}$ decreases without bound as x tends to 1 from the left, the limit is $-\infty$.

63.

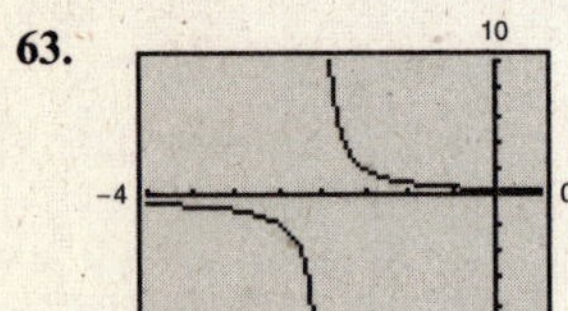

$$\lim_{x \to -2^-} \frac{1}{x + 2} = -\infty$$

x	−3	−2.5	−2.1	−2.01	−2.001	−2.0001	−2
$f(x)$	−1	−2	−10	−100	−1000	−10,000	undefined

Because $f(x) = \dfrac{1}{x + 2}$ decreases without bound as x tends to −2 from the left, the limit is $-\infty$.

65. (a) $\lim_{x \to 3^+} f(x) = 1$

(b) $\lim_{x \to 3^-} f(x) = 1$

(c) $\lim_{x \to 3} f(x) = 1$

67. (a) $\lim_{x \to 3^+} f(x) = 0$

(b) $\lim_{x \to 3^-} f(x) = 0$

(c) $\lim_{x \to 3} f(x) = 0$

69. (a) $\lim_{x \to 3^+} f(x) = 3$

(b) $\lim_{x \to 3^-} f(x) = -3$

(c) $\lim_{x \to 3} f(x)$ does not exist.

71.

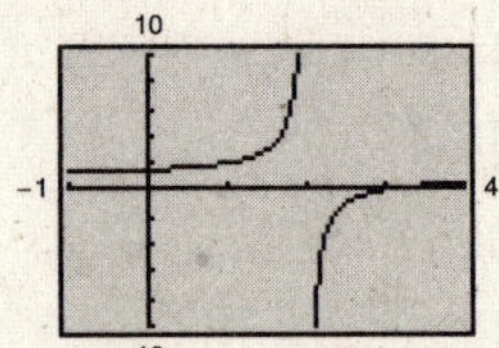

$$\lim_{x \to 2} \frac{x^2 - 5x + 6}{x^2 - 4x + 4} \text{ does not exist.}$$

73.

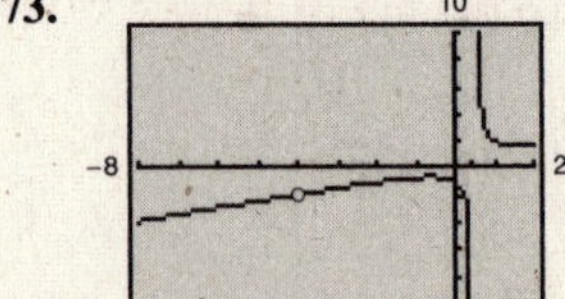

$$\lim_{x \to -4} \frac{x^3 + 4x^2 + x + 4}{2x^2 + 7x - 4} \approx -1.889$$

75. $C = \dfrac{25p}{100 - p},\ 0 \le p < 100$

(a) $C(50) = \dfrac{25(50)}{100 - 50} = \$25$ thousand

(b) Find p for $C = 100$.

$$100 = \frac{25p}{100 - p}$$
$$100(100 - p) = 25p$$
$$10{,}000 - 100p = 25p$$
$$10{,}000 = 125p$$
$$80 = p, \text{ or } 80\%$$

(c) $\displaystyle\lim_{p\to 100^-} = \lim_{p\to 100^-} \frac{25p}{100 - p} = \infty$

The cost function increases without bound as x approaches 100 from the left. Therefore, according to the model, it is not possible to remove 100% of the pollutants.

77. (a)

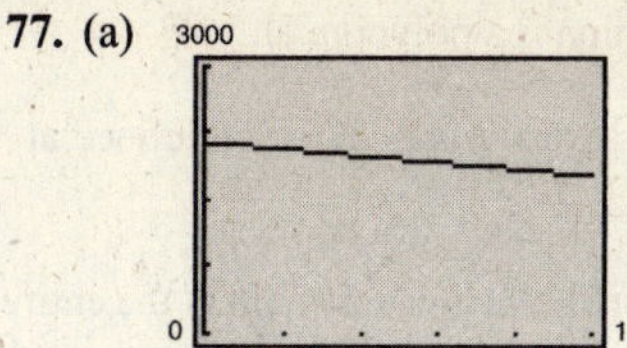

(b) When $x = 0.25$: $A = 2685.06$

When $x = \frac{1}{365}$: $A = 2717.91$

(c) Using the *zoom* and *trace* features, $\displaystyle\lim_{x\to 0^+} A \approx \$2718.28$. Because x, the length of the compounding period, is approaching 0, this limit represents the balance with continuous compounding.

Section 1.6 Continuity

Skills Warm Up

1. $\dfrac{x^2 + 6x + 8}{x^2 - 6x - 16} = \dfrac{(x + 4)(x + 2)}{(x - 8)(x + 2)} = \dfrac{x + 4}{x - 8}$

2. $\dfrac{x^2 - 5x - 6}{x^2 - 9x + 18} = \dfrac{(x - 6)(x + 1)}{(x - 6)(x - 3)} = \dfrac{x + 1}{x - 3}$

3.
$$\frac{2x^2 - 2x - 12}{4x^2 - 24x + 36} = \frac{2(x^2 - x - 6)}{4(x^2 - 6x + 9)} = \frac{2(x - 3)(x + 2)}{4(x - 3)(x - 3)} = \frac{x + 2}{2(x - 3)}$$

4.
$$\frac{x^3 - 16x}{x^3 + 2x^2 - 8x} = \frac{x(x^2 - 16)}{x(x^2 + 2x - 8)} = \frac{x(x^2 - 16)}{x(x + 4)(x - 2)} = \frac{x(x + 4)(x - 4)}{x(x + 4)(x - 2)} = \frac{x - 4}{x - 2}$$

5.
$$x^2 + 7x = 0$$
$$x(x + 7) = 0$$
$$x = 0$$
$$x + 7 = 0 \Rightarrow x = -7$$

6.
$$x^2 + 4x - 5 = 0$$
$$(x + 5)(x - 1) = 0$$
$$x + 5 = 0 \Rightarrow x = -5$$
$$x - 1 = 0 \Rightarrow x = 1$$

7.
$$3x^2 + 8x + 4 = 0$$
$$(3x + 2)(x + 2) = 0$$
$$3x + 2 = 0 \Rightarrow x = -\tfrac{2}{3}$$
$$x + 2 = 0 \Rightarrow x = -2$$

8.
$$x^3 + 5x^2 - 24x = 0$$
$$x(x^2 + 5x - 24) = 0$$
$$x(x - 3)(x + 8) = 0$$
$$x = 0$$
$$x - 3 = 0 \Rightarrow x = 3$$
$$x + 8 = 0 \Rightarrow x = -8$$

9.
$$\lim_{x\to 3}(2x^2 - 3x + 4) = 2(3^2) - 3(3) + 4 = 2(9) - 9 + 4 = 13$$

10.
$$\lim_{x\to -2}(3x^3 - 8x + 7) = 3(-2)^3 - 8(-2) + 7 = 3(-8) + 16 + 7 = -24 + 23 = -1$$

1. Continuous; The function is a polynomial.

3. Not continuous; The rational function is not defined at $x = \pm 4$.

5. Continuous; The rational function's domain is the entire real line.

7. Not continuous; The rational function is not defined at $x = 3$ or $x = 5$.

9. Not continuous; The rational function is not defined at $x = \pm 2$.

11. $f(x) = \dfrac{x^2 - 1}{x}$ is continuous on $(-\infty, 0)$ and $(0, \infty)$ because the domain of f consists of all real numbers except $x = 0$. There is a discontinuity at $x = 0$ because $f(0)$ is not defined and $\lim_{x\to 0} f(x)$ does not exist.

13. $f(x) = \dfrac{x^2 - 1}{x + 1}$ is continuous on $(-\infty, -1)$ and $(-1, \infty)$ because the domain of f consists of all real numbers except $x = -1$. There is a discontinuity at $x = -1$ because $f(-1)$ is not defined and $\lim_{x\to -1} f(x) \neq f(-1)$.

15. $f(x) = x^2 - 2x + 1$ is continuous on $(-\infty, \infty)$ because the domain of f consists of all real numbers.

17. $f(x) = \dfrac{x}{x^2 - 1} = \dfrac{x}{(x + 1)(x - 1)}$ is continuous on $(-\infty, -1)$, $(-1, 1)$, and $(1, \infty)$ because the domain of f consists of all real numbers except $x = \pm 1$. There are discontinuities at $x = \pm 1$ because $f(1)$ and $f(-1)$ are not defined and $\lim_{x\to 1} f(x)$ and $\lim_{x\to -1} f(x)$ do not exist.

19. $f(x) = \dfrac{x}{x^2 + 1}$ is continuous on $(-\infty, \infty)$ because the domain of f consists of all real numbers.

21. $f(x) = \dfrac{x - 5}{x^2 - 9x + 20} = \dfrac{x - 5}{(x - 5)(x - 4)}$ is continuous on $(-\infty, 4)$, $(4, 5)$, and $(5, \infty)$ because the domain of f consists of all real numbers except $x = 4$ and $x = 5$. There is a discontinuity at $x = 4$ and $x = 5$ because $f(4)$ and $f(5)$ are not defined and $\lim_{x\to 4} f(x)$ does not exist and $\lim_{x\to 5} f(x) \neq f(5)$.

23. $f(x) = \sqrt{4 - x}$ is continuous on $(-\infty, 4]$ because the domain of f consists of all real $x \le 4$.

25. $f(x) = \sqrt{x} + 2$ is continuous on $[0, \infty)$ because the domain of f consists of all real $x \ge 0$.

27. $f(x) = \begin{cases} -2x + 3, & -1 \le x \le 1 \\ x^2, & 1 < x \le 3 \end{cases}$ is continuous on $[-1, 3]$.

29. $f(x) = \begin{cases} 3 + x, & x \le 2 \\ x^2 + 1, & x > 2 \end{cases}$

is continuous on $(-\infty, \infty)$ because the domain of f consists of all real numbers, $f(2)$ is defined, $\lim_{x\to 2} f(x)$ exists, and $\lim_{x\to 2} f(x) = f(2)$.

31. $f(x) = \dfrac{|x + 1|}{x + 1}$ is continuous on $(-\infty, -1)$ and $(-1, \infty)$ because the domain of f consists of all real numbers except $x = -1$. There is a discontinuity at $x = -1$ because $f(-1)$ is not defined, and $\lim_{x\to -1} f(x)$ does not exist.

33. $f(x) = x\sqrt{x + 3}$ is continuous on $[-3, \infty)$.

35. $f(x) = [\![2x]\!] + 1$ is continuous on all intervals of the form $\left(\frac{1}{2}c, \frac{1}{2}c + \frac{1}{2}\right)$, where c is an integer. That is, f is continuous on …, $\left(-\frac{1}{2}, 0\right)$, $\left(0, \frac{1}{2}\right)$, $\left(\frac{1}{2}, 1\right)$, …. f is not continuous at all points $\frac{1}{2}c$, where c is an integer. There are discontinuities at $x = \dfrac{c}{2}$, where c is an integer, because $\lim_{x\to c/2} f(x)$ does not exist.

37. $f(x) = [\![x - 1]\!]$ is continuous on all intervals $(c, c + 1)$. There are discontinuities at $x = c$, where c is an integer, because $\lim_{x\to c} f(x)$ does not exist.

39. $h(x) = f(g(x)) = f(x - 1) = \dfrac{1}{\sqrt{x - 1}}, \; x > 1$

h is continuous on its entire domain $(1, \infty)$.

41. $f(x) = \dfrac{x^2 - 16}{x - 4} = \dfrac{(x + 4)(x - 4)}{x - 4} = x + 4, \; x \neq 4$

f has a removable discontinuity at $x = 4$; Continuous on $(-\infty, 4)$ and $(4, \infty)$.

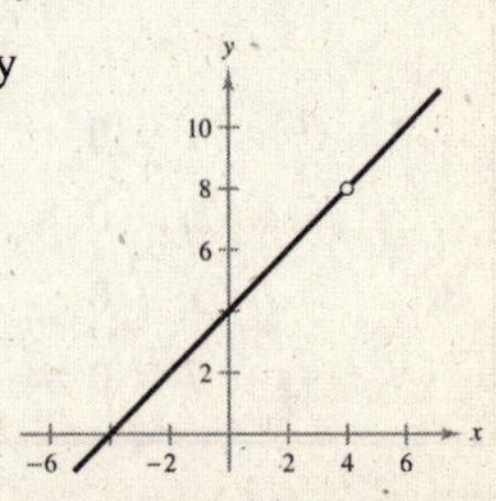

43. $f(x) = \dfrac{x + 4}{3x^2 - 12}$

Continuous on $(-\infty, 2)$, $(-2, 2)$, and $(2, \infty)$.

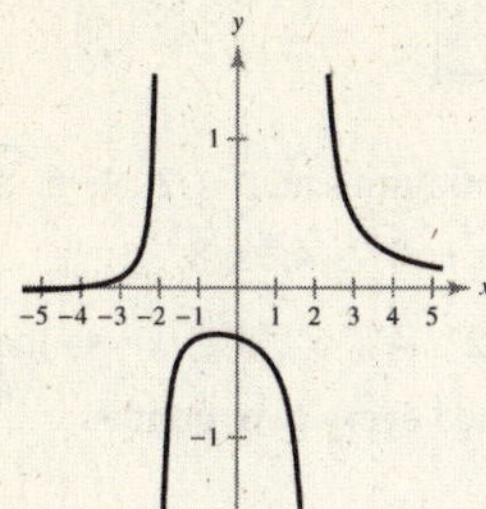

45. $f(x) = \begin{cases} x^2 + 1, & x < 0 \\ x - 1, & x \ge 0 \end{cases}$

f has a nonremovable discontinuity at $x = 0$; Continuous on $(-\infty, 0)$ and $(0, \infty)$.

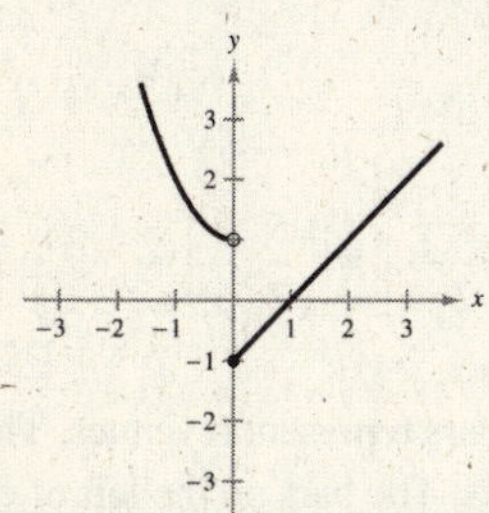

47. Continuous on $[-1, 5]$ because $f(x) = x^2 - 4x - 5$ is a polynomial.

49. Continuous on $[1, 2)$ and $(2, 4]$ because $f(x) = \dfrac{1}{x - 2}$ has a nonremovable discontinuity at $x = 2$.

51.

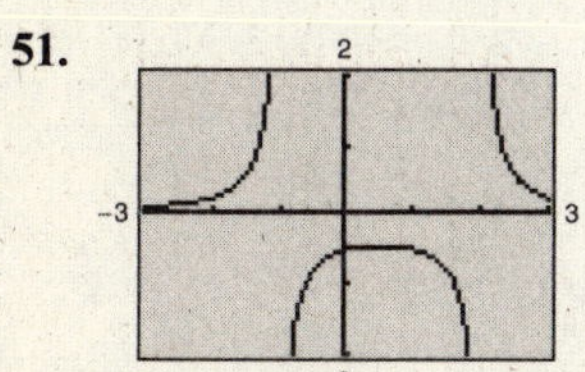

From the graph, you can see that $h(2)$ and $h(-1)$ are not defined, so h is not continuous at $x = 2$ and $x = -1$.

53.

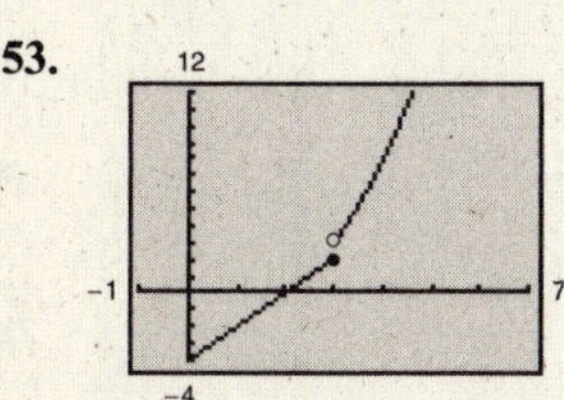

From the graph, you can see that $\lim_{x\to 3} f(x)$ does not exist, so f is not continuous at $x = 3$.

55.

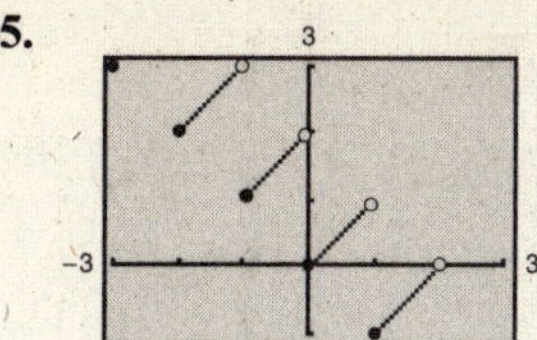

From the graph, you can see that $\lim_{x\to c} (x - 2[\![x]\!])$, where c is an integer, does not exist. So f is not continuous at all integers c.

57. $\lim_{x\to 2^-} f(x) = \lim_{x\to 2^-} x^3 = 8$

$\lim_{x\to 2^+} f(x) = \lim_{x\to 2^+} ax^2 = 4a$

So, $8 = 4a$ and $a = 2$.

59.

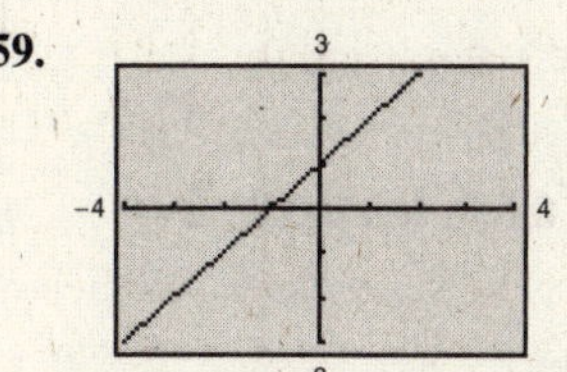

$f(x) = \dfrac{x^2 + x}{x} = \dfrac{x(x + 1)}{x}$ appears to be continuous on $[-4, 4]$. But it is not continuous at $x = 0$ (removable discontinuity). Examining a function analytically can reveal removable discontinuities that are difficult to find just from analyzing its graph.

61. (a) $[0, 100)$; Negative x-values and values greater than 100 do not make sense in this context. Also, $C(100)$ is undefined.

(b) C is continuous on its domain because all rational functions are continuous on their domains.

(c) For $x = 75$,

$$C = \frac{2(75)}{100 - 75} = \frac{150}{25} = 6 \text{ million dollars.}$$

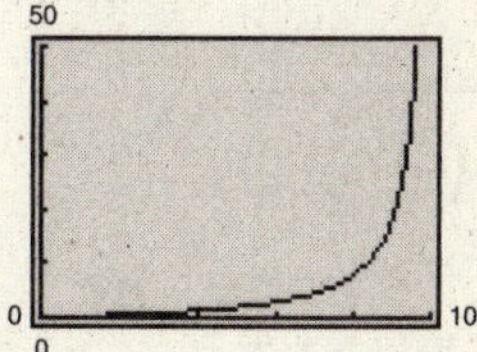

63.

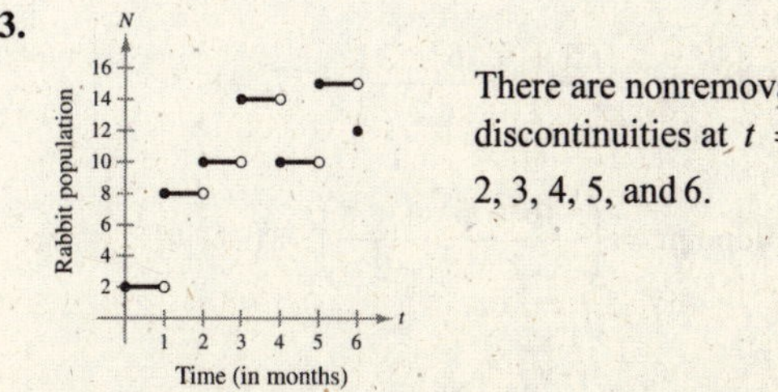

There are nonremovable discontinuities at $t = 1, 2, 3, 4, 5$, and 6.

65. (a)

Discontinuities at $x = 1, x = 2, x = 3$
Explanations will vary.

(b) $C(2.5) = \$0.84$

67. (a) The graph has nonremovable discontinuities at $t = \frac{1}{4}, \frac{1}{2}, \frac{3}{4}, 1, \frac{5}{4}, \ldots$

(b) Let $t = 2$.

$A = 7500(1.015)^{[\![4 \cdot 2]\!]} \approx \$8448.69$

(c) Let $t = 7$.

$A = 7500(1.015)^{[\![4 \cdot 7]\!]} \approx \$11{,}379.17$

69. (a)

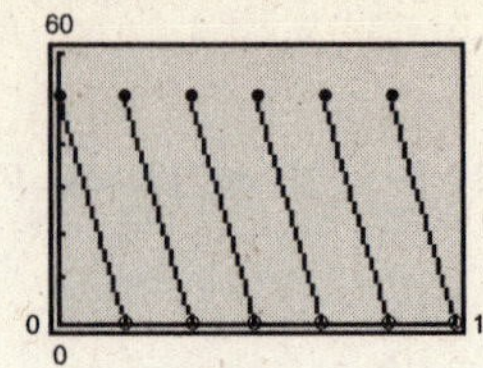

Nonremovable discontinuities at $t = 2, 4, 6, 8, \ldots$;
N is not continuous at $t = 2, 4, 6, 8, \ldots$.

(b) $N \to 0$ when $t \to 2^-, 4^-, 6^-, 8^-, \ldots$, so the inventory is replenished every two months.

Review Exercises for Chapter 1

1.

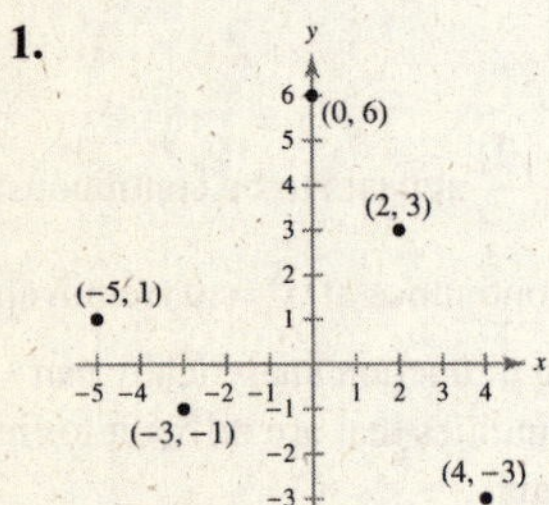

3. Distance $= \sqrt{(0-5)^2 + (0-2)^2}$

$= \sqrt{25 + 4}$

$= \sqrt{29}$

5. Distance $= \sqrt{[-1-(-4)]^2 + (3-6)^2}$

$= \sqrt{9 + 9}$

$= 3\sqrt{2}$

7. $d = \sqrt{\left(\frac{3}{4} - \frac{1}{4}\right)^2 + (-6-(-8))^2}$

$= \sqrt{\left(\frac{1}{2}\right)^2 + (2)^2} = \sqrt{\frac{1}{4} + 4} = \sqrt{\frac{17}{4}} = \frac{\sqrt{17}}{2}$

9. Midpoint $= \left(\frac{5+9}{2}, \frac{6+2}{2}\right) = (7, 4)$

11. Midpoint $= \left(\frac{-10-6}{2}, \frac{4+8}{2}\right) = (-8, 6)$

13. Midpoint $= \left(\frac{-1+6}{2}, \frac{\frac{1}{5}+\frac{3}{5}}{2}\right) = \left(\frac{5}{2}, \frac{2}{5}\right)$

15. $P = R - C$. The tallest bars represent revenues. The middle bars represent costs. The bars on the left of each group represent profits because $P = R - C$.

17. $(1, 3)$ translates to $(-2, 7)$.

$(2, 4)$ translates to $(-1, 8)$.

$(4, 1)$ translates to $(1, 5)$.

19. $y = 4x - 12$

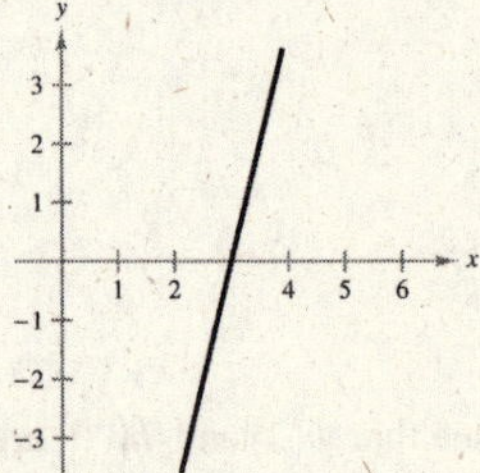

21. $y = x^2 + 5$

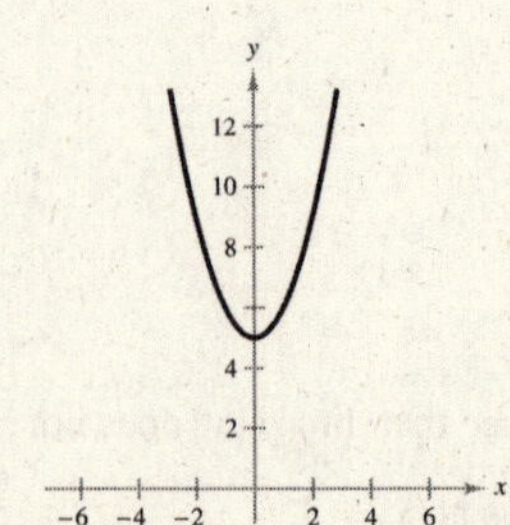

23. 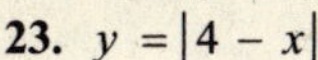 $y = |4 - x|$

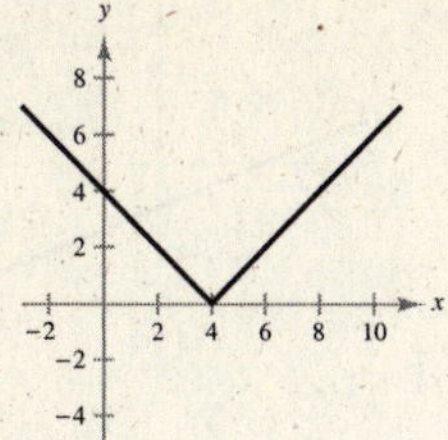

25. $y = x^3 + 4$

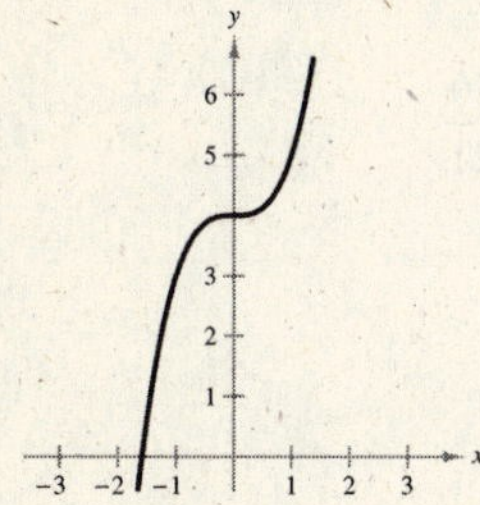

27. $y = \sqrt{4x + 1}$

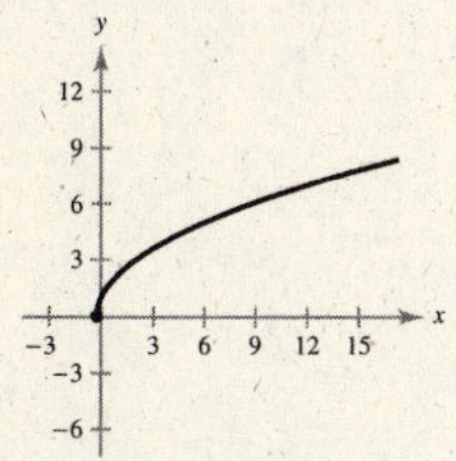

29. Let $y = 0$. Then,

$$4x + 0 + 3 = 0$$
$$x = -\tfrac{3}{4}.$$

Let $x = 0$. Then,

$$4(0) + y + 3 = 0$$
$$y = -3.$$

x-intercept: $\left(-\frac{3}{4}, 0\right)$

y-intercept: $(0, -3)$

31. Let $y = 0$. Then,

$$0 = x^2 + 2x - 8$$
$$0 = (x + 4)(x - 2)$$
$$x + 4 = 0 \quad \text{or} \quad x - 2 = 0$$
$$x = -4 \qquad\qquad x = 2.$$

Let $x = 0$. Then,

$$y = (0)^2 + 2(0) - 8$$
$$y = -8.$$

x-intercepts: $(-4, 0)$, $(2, 0)$

y-intercept: $(0, -8)$

33. $(x - 0)^2 + (y - 0)^2 = 8^2$

$$x^2 + y^2 = 64$$

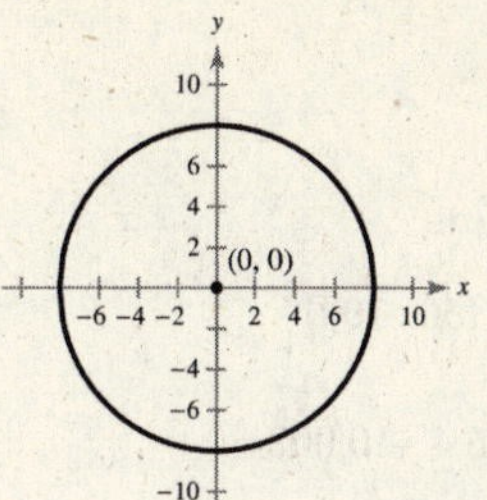

35. $(x - 0)^2 + (y - 0)^2 = r^2$

$$x^2 + y^2 = r^2$$
$$2^2 + \left(\sqrt{5}\right)^2 = r^2$$
$$9 = r^2$$
$$x^2 + y^2 = 9$$

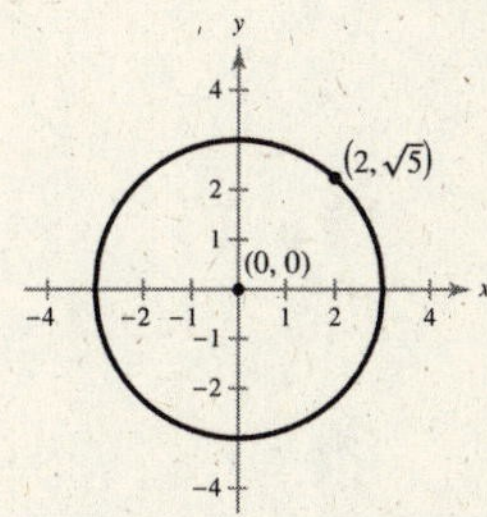

37. $y = 2x + 13$ and $y = -5x - 1$

Set the two equations equal to each other.

$$2x + 13 = -5x - 1$$
$$7x = -14$$
$$x = -2$$

Substitute $x = -2$ into one of the equations.

$$y = 2(-2) + 13 = 9$$

39. By equating the y-values for the two equations, we have

$$x^3 = x$$
$$x(x^2 - 1) = 0$$
$$x = -1, 0, 1$$

The corresponding y-values are $y = -1$, $y = 0$, and $y = 1$, so the points of intersection are $(-1, -1)$, $(0, 0)$, and $(1, 1)$.

41. (a) $C = 200 + 2x + 8x = 200 + 10x$

$R = 14x$

(b) $C = R$

$200 + 10x = 14x$

$200 = 4x$

$x = 50$ shirts

$(x, R) = (x, C) = (50, 700)$

43. $p = 91.4 - 0.009x = 6.4 + 0.008x$

$85 = 0.017x$

$x = 5000$ units

Equilibrium point $(x, p) = (5000, 46.40)$

45. $y = -x + 12$

Slope: $m = -1$

y-intercept: $(0, 12)$

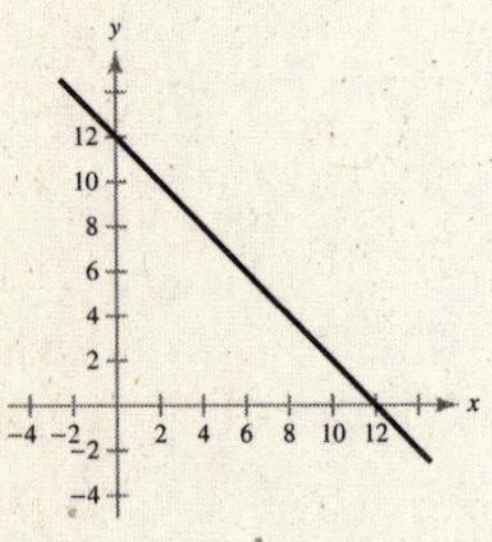

47. $3x + y = -2$

$y = -3x - 2$

Slope: $m = -3$

y-intercept: $(0, -2)$

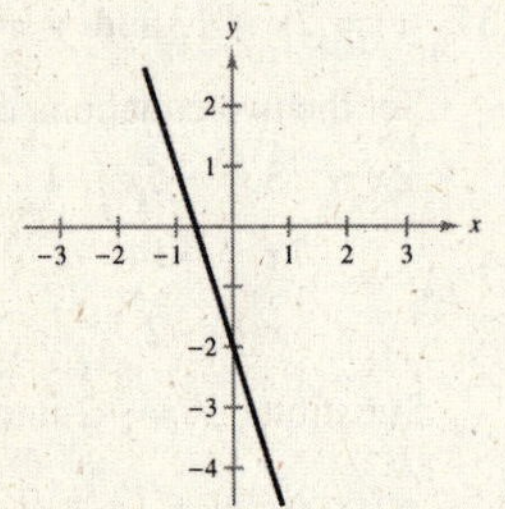

49. $y = -\frac{5}{3}$

Slope: $m = 0$ (horizontal line)

y-intercept: $\left(0, -\frac{5}{3}\right)$

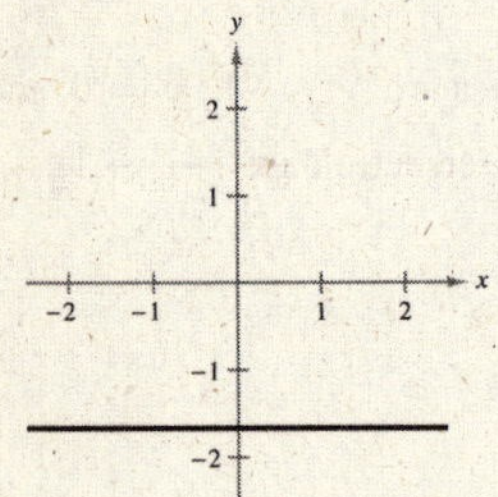

51. $-2x - 5y - 5 = 0$

$5y = -2x - 5$

$y = -\frac{2}{5}x - 1$

Slope: $m = -\frac{2}{5}$

y-intercept: $(0, -1)$

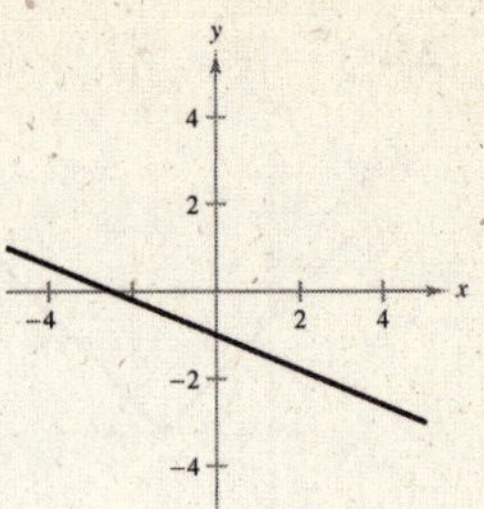

53. Slope $= \dfrac{6 - 0}{7 - 0} = \dfrac{6}{7}$

55. Slope $= \dfrac{17 - (-3)}{10 - (-11)} = \dfrac{20}{21}$

57. $y - (-1) = -2(x - 3)$

$y + 1 = -2x + 6$

$y = -2x + 5$

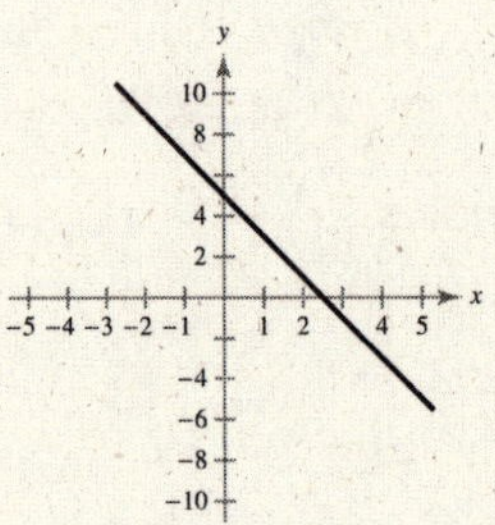

59. $m = 0$: horizontal line through $(1.5, -4)$

$y = -4$

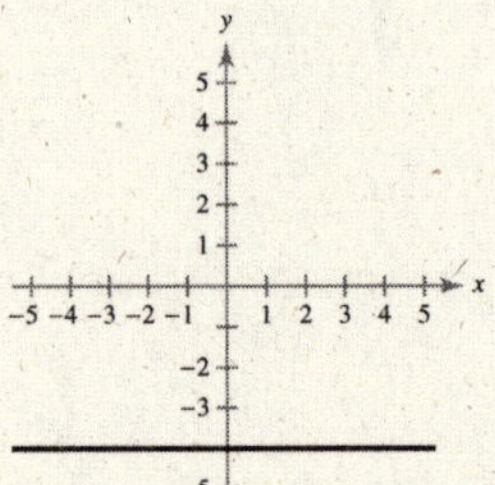

61. $m = \dfrac{5 - (-7)}{7 - 1} = \dfrac{12}{6} = 2$

$y - (-7) = 2(x - 1)$

$y + 7 = 2x - 2$

$y = 2x - 9$

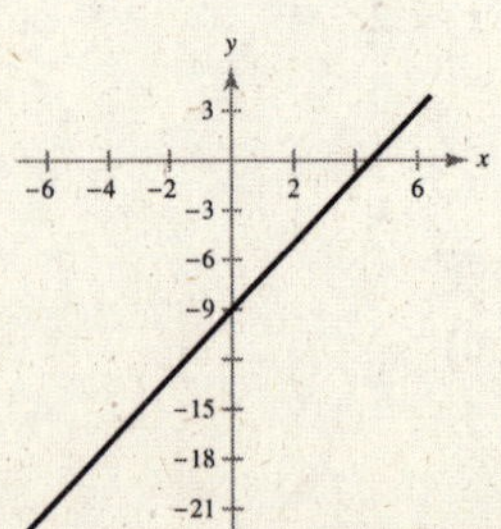

63. $m = \frac{14 - 7}{5 - 5} = \frac{7}{0} \Rightarrow m$ is undefined.

Vertical line through $(5, 7)$

$x = 5$

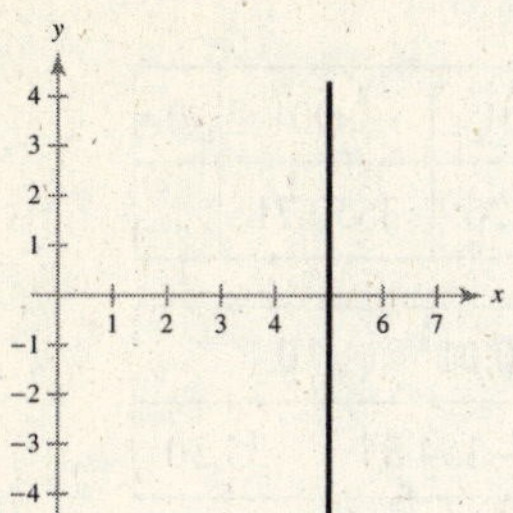

65. (a) $y - 6 = \frac{7}{8}[x - (-3)]$

$y = \frac{7}{8}x + \frac{69}{8}$

$7x - 8y + 69 = 0$

(b) $4x + 2y = 7 \Rightarrow y = -2x + \frac{7}{2}$; slope $= -2$

$y - 6 = -2[x - (-3)]$

$y = -2x$

$2x + y = 0$

(c) The line through $(0, 0)$ and $(-3, 6)$ has slope

$\frac{6}{-3} = -2$

$y = -2x$

$2x + y = 0$

(d) $3x - 2y = 2 \Rightarrow y = \frac{3}{2}x - 1$

Slope of perpendicular is $-\frac{2}{3}$.

$y - 6 = -\frac{2}{3}[x - (-3)]$

$y = -\frac{2}{3}x + 4$

$2x + 3y - 12 = 0$

67. $(32, 750), (37, 700)$

$m = \frac{750 - 700}{32 - 37} = \frac{50}{-5} = -10$

(a) $x - 750 = -10(p - 32)$

$x = -10p + 1070$

(b) If $p = 34.50$, $x = -10(34.50) + 1070 = 725$ units

(c) If $p = 42.00$, $x = -10(42.00) + 1070 = 650$ units

69. Yes, $y = -x^2 + 2$ is a function of x.

71. No, $y^2 - \frac{1}{4}x^2 = 4$ is not a function of x.

73.

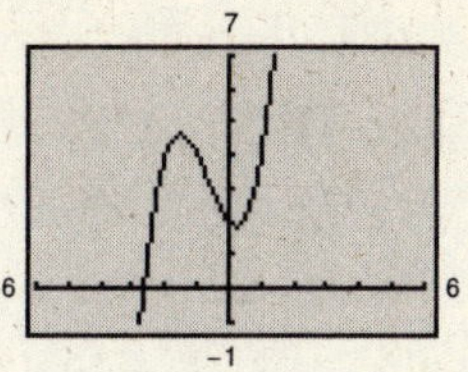

Domain: $(-\infty, \infty)$

Range: $(-\infty, \infty)$

75.

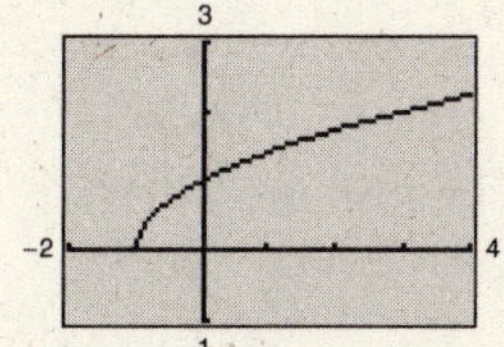

Domain: $[-1, \infty)$

Range: $[0, \infty)$

77.

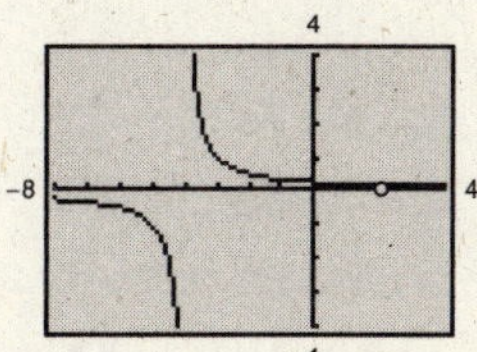

$$f(x) = \frac{x - 3}{x^2 + x - 12} = \frac{(x - 3)}{(x - 3)(x + 4)} = \frac{1}{x + 4}, \quad x \neq 3$$

Domain: $(-\infty, -4) \cup (-4, 3) \cup (3, \infty)$

Range: $(-\infty, 0) \cup \left(0, \frac{1}{7}\right) \cup \left(\frac{1}{7}, \infty\right)$

79. $f(x) = 3x + 4$

(a) $f(1) = 3(1) + 4 = 7$

(b) $f(-5) = 3(-5) + 4 = -11$

(c) $f(x + 1) = 3(x + 1) + 4 = 3x + 7$

81. (a) $f(x) + g(x) = (1 + x^2) + (2x - 1)$
$= x^2 + 2x$

(b) $f(x) - g(x) = (1 + x^2) - (2x - 1)$
$= x^2 - 2x + 2$

(c) $f(x)g(x) = (1 + x^2)(2x - 1)$
$= 2x^3 - x^2 + 2x - 1$

(d) $\dfrac{f(x)}{g(x)} = \dfrac{1 + x^2}{2x - 1}$

(e) $f(g(x)) = f(2x - 1)$
$= 1 + (2x - 1)^2$
$= 4x^2 - 4x + 2$

(f) $g(f(x)) = g(1 + x^2)$
$= 2(1 + x^2) - 1$
$= 2x^2 + 1$

83.

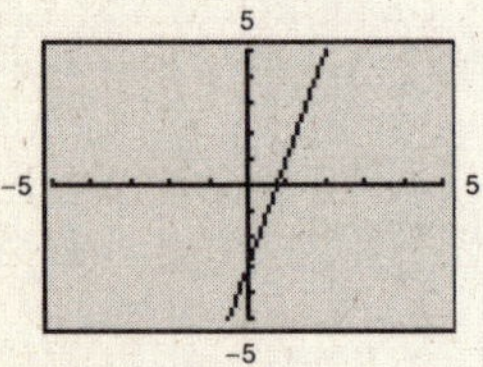

$f(x)$ is one-to-one.

$$f(x) = 4x - 3 = y$$
$$4y - 3 = x$$
$$4y = x + 3$$
$$y = \tfrac{1}{4}x + \tfrac{3}{4}$$
$$f^{-1}(x) = \tfrac{1}{4}x + \tfrac{3}{4}$$

85.

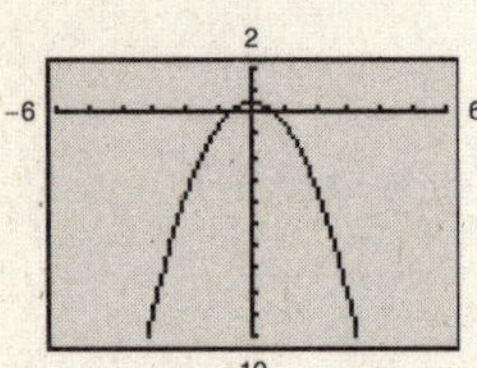

$f(x)$ does not have an inverse function.

87.

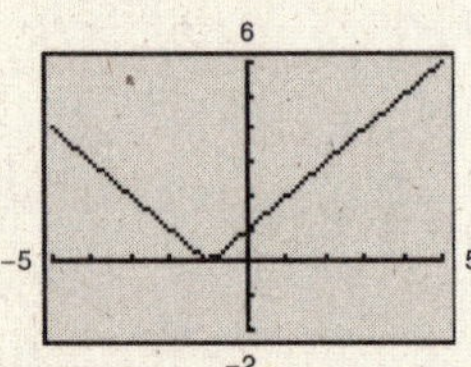

$f(x)$ does not have an inverse function.

89.

x	0.9	0.99	0.999	1	1.001	1.01	1.1
$f(x)$	0.6	0.96	0.996	?	1.004	1.04	1.4

$\lim_{x\to1}(4x - 3) = 1$

91.

x	−0.1	−0.01	−0.001	0
$f(x)$	35.71	355.26	3550.71	?

x	0.001	0.01	0.1
$f(x)$	−3550.31	−354.85	−35.30

$\lim_{x\to0}\dfrac{\sqrt{x + 6} - 6}{x}$ does not exist.

93. $\lim_{x\to3} 8 = 8$

95. $\lim_{x\to2}(5x - 3) = 5(2) - 3 = 7$

97. $\lim_{x\to-1}\dfrac{x + 3}{6x + 1} = \dfrac{-1 + 3}{6(-1) + 1} = -\dfrac{2}{5}$

99. $\lim_{t\to0^-}\dfrac{t^2 + 1}{t} = -\infty$

$\lim_{t\to0^+}\dfrac{t^2 + 1}{t} = \infty$

$\lim_{t\to0}\dfrac{t^2 + 1}{t}$ does not exist.

101. $\lim_{x\to-2}\dfrac{x + 2}{x^2 - 4} = \lim_{x\to-2}\dfrac{x + 2}{(x + 2)(x - 2)}$
$= \lim_{x\to-2}\dfrac{1}{x - 2}$
$= -\dfrac{1}{4}$

103. $\lim_{x\to0^+}\left(x - \dfrac{1}{x}\right) = \lim_{x\to0^+}\dfrac{x^2 - 1}{x} = -\infty$

105. $\lim_{x\to0}\dfrac{[1/(x - 2)] - 1}{x} = \lim_{x\to0}\dfrac{1 - (x - 2)}{x(x - 2)}$
$= \lim_{x\to0}\dfrac{3 - x}{x(x - 2)}$ does not exist.

107. $\lim_{x\to0} f(x)$, where $f(x) = \begin{cases} x + 5, & x \neq 0 \\ 3, & x = 0 \end{cases}$

$\lim_{x\to0} f(x) = \lim_{x\to0}(x + 5) = 0 + 5 = 5$

109. $$\lim_{\Delta x\to 0}\frac{(x+\Delta x)^3-(x+\Delta x)-(x^3-x)}{\Delta x}=\lim_{\Delta x\to 0}\frac{x^3+3x^2\Delta x+3x(\Delta x)^2+(\Delta x)^3-x-\Delta x-x^3+x}{\Delta x}$$
$$=\lim_{\Delta x\to 0}\frac{3x^2\Delta x+3x(\Delta x)^2+(\Delta x)^3-\Delta x}{\Delta x}$$
$$=\lim_{\Delta x\to 0}\left[3x^2+3x\Delta x+(\Delta x)^2-1\right]$$
$$=3x^2-1$$

111. $f(x)=x+b$ is continuous on $(-\infty,\infty)$ because the domain of f consists of all real x.

113. $f(x)=\dfrac{1}{(x+4)^2}$ is continuous on the intervals $(-\infty,-4)$ and $(-4,\infty)$ because the domain of f consists of all real numbers except $x=-4$. There is a discontinuity at $x=-4$ because $f(4)$ is not defined.

115. $f(x)=\dfrac{3}{x+1}$ is continuous on the intervals $(-\infty,-1)$ and $(-1,\infty)$ because the domain of f consists of all real numbers except $x=-1$. There is a discontinuity at $x=-1$ because $f(-1)$ is not defined.

117. $f(x)=[\![x+3]\!]$ is continuous on all intervals of the form $(c, c+1)$, where c is an integer. There are discontinuities at all integer values c because $\lim_{x\to c} f(x)$ does not exist.

119. $f(x)=\begin{cases}x, & x\le 0\\ x+1, & x>0\end{cases}$ is continuous on the intervals $(-\infty,0)$ and $(0,\infty)$. There is a discontinuity at $x=0$ because $\lim_{x\to 0} f(x)$ does not exist.

121. $\lim_{x\to 3^-} f(x)=\lim_{x\to 3^-}(-x+1)=-2$

$\lim_{x\to 3^+} f(x)=\lim_{x\to 3^+}(ax-8)=3a-8$

So, $-2=3a-8$ and $a=2$.

123. (a)

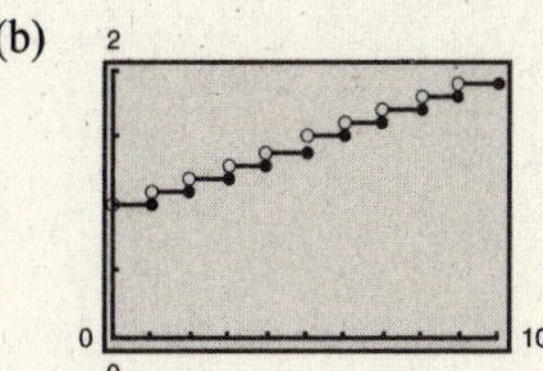

Explanations will vary. The function is defined for all values of x greater than zero. The function is not continuous at $x=5$, $x=10$, and $x=15$.

(b) $C(10)=4.99(10)=\$49.90$

125. (a) $C(t)=\begin{cases}1+0.1[\![t]\!], & t>0,\ t\text{ not an integer}\\ 1+0.1[\![t-1]\!], & t>0,\ t\text{ an integer}\end{cases}$

(b)

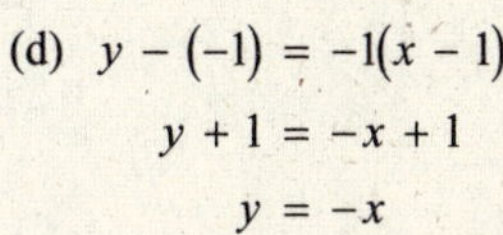

C is not continuous at $t=1, 2, 3, \ldots$.

Chapter 1 Test Yourself

1. (a) $d=\sqrt{(-4-1)^2+(4-(-1))^2}$
$=\sqrt{(-5)^2+(5)^2}=\sqrt{50}=5\sqrt{2}$

(b) Midpoint $=\left(\dfrac{1+(-4)}{2},\dfrac{-1+4}{2}\right)=\left(-\dfrac{3}{2},\dfrac{3}{2}\right)$

(c) $m=\dfrac{4-(-1)}{-4-1}=\dfrac{5}{-5}=-1$

(d) $y-(-1)=-1(x-1)$
$y+1=-x+1$
$y=-x$

(e)

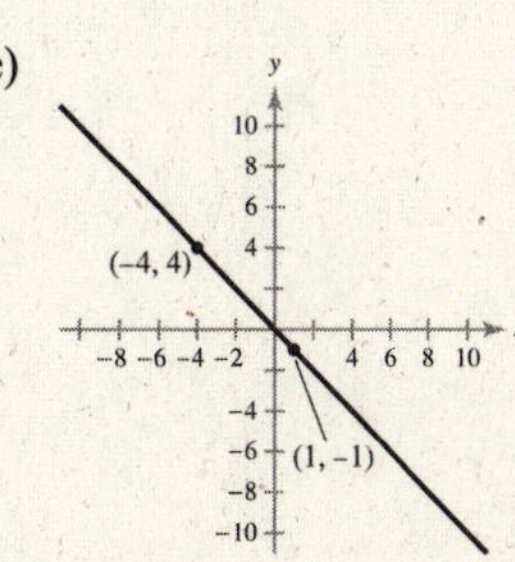

2. (a) $d = \sqrt{\left(0 - \frac{5}{2}\right)^2 + (2 - 2)^2}$

$= \sqrt{\left(-\frac{5}{2}\right)^2 + 0^2} = \frac{5}{2}$

(b) Midpoint $= \left(\frac{\frac{5}{2} + 0}{2}, \frac{2 + 2}{2}\right) = \left(\frac{5}{4}, 2\right)$

(c) $m = \frac{2 - 2}{0 - \frac{5}{2}} = 0$

(d) Horizontal line: $y = 2$

(e)

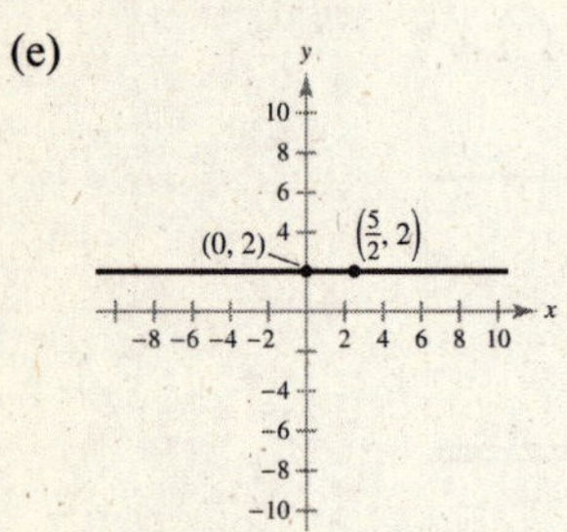

3. (a) $d = \sqrt{(-4 - 2)^2 + (1 - 3)^2}$

$= \sqrt{(-6)^2 + (-2)^2} = \sqrt{40}$

$= 2\sqrt{10}$

(b) Midpoint $= \left(\frac{2 + (-4)}{2}, \frac{3 + 1}{2}\right)$

$= \left(-\frac{2}{2}, \frac{4}{2}\right)$

$= (-1, 2)$

(c) $m = \frac{1 - 3}{-4 - 2} = \frac{-2}{-6} = \frac{1}{3}$

(d) $y - 3 = \frac{1}{3}(x - 2)$

$y - 3 = \frac{1}{3}x - \frac{2}{3}$

$y = \frac{1}{3}x + \frac{7}{3}$

(e)

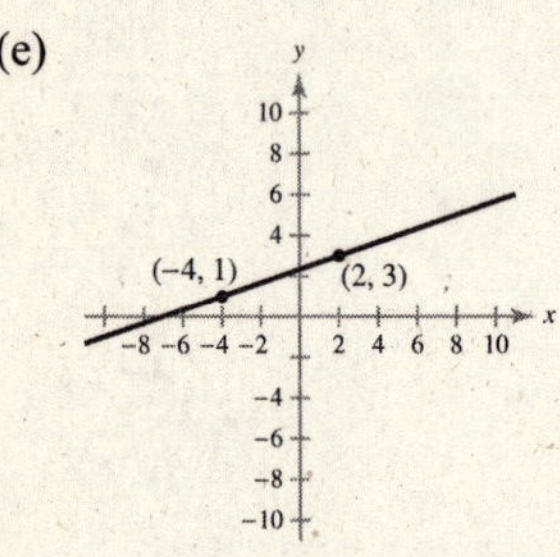

4. $65 - 2.1x = 43 + 1.9x$

$-4x = -22$

$x = 5.5$

Equilibrium point $(x, p) = (5.5, 5500)$

The equilibrium point occurs when the demand and supply are each 5500 units.

5. $m = \frac{1}{5}$

When $x = 0$: $y = \frac{1}{5}(0) - 2 = -2$

y-intercept: $(0, -2)$

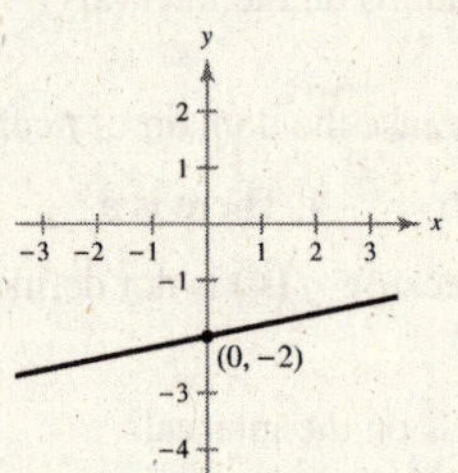

6. The line is vertical, so its slope is undefined, and it has no y-intercept.

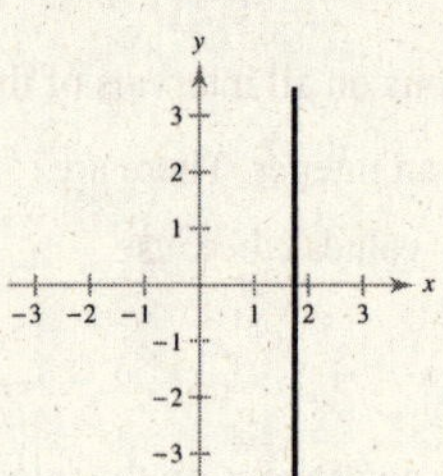

7. $y = -2.5x + 6.25$

$m = -2.5$

When $x = 0$: $y = -2.5(0) + 6.25 = 6.25$

y-intercept: $(0, 6.25)$

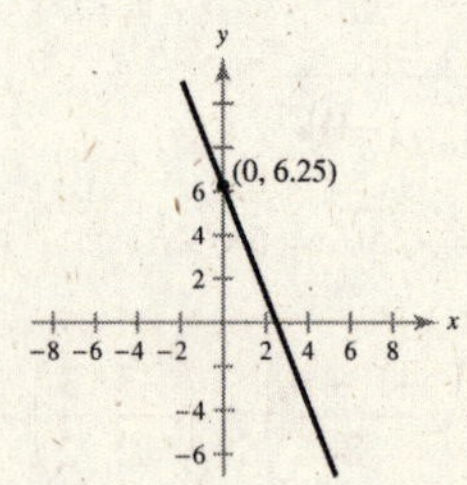

8. The slope of the given line $-4x + y = 8 \Rightarrow y = 4x + 8$ is $m = 4$. The slope of the line perpendicular is $m = -\frac{1}{4}$.

Using the point $(-1, -7)$ and $m = -\frac{1}{4}$, the equation is:

$$\begin{aligned} y - (-7) &= -\tfrac{1}{4}(x - (-1)) \\ y + 7 &= -\tfrac{1}{4}(x + 1) \\ y + 7 &= -\tfrac{1}{4}x - \tfrac{1}{4} \\ y &= -\tfrac{1}{4}x - \tfrac{29}{4} \end{aligned}$$

9. The slope of the given line $5x - 2y = 8 \Rightarrow -2y = -5x + 8 \Rightarrow y = \frac{5}{2}x - 4$ is $m = \frac{5}{2}$. The slope of the line parallel is $m = \frac{5}{2}$.

Using the point $(2, 1)$ and $m = \frac{5}{2}$, the equation is:

$$\begin{aligned} y - 1 &= \tfrac{5}{2}(x - 2) \\ y - 1 &= \tfrac{5}{2}x - 5 \\ y &= \tfrac{5}{2}x - 4 \end{aligned}$$

10. (a)

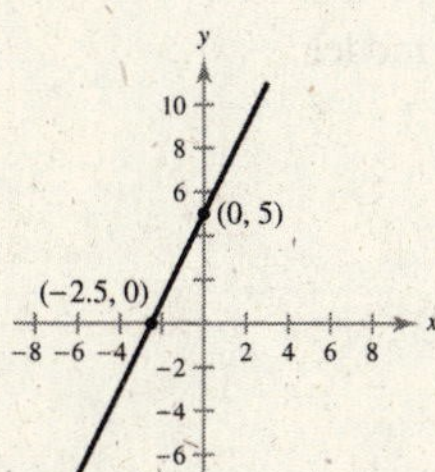

(b) Domain: $(-\infty, \infty)$

Range: $(-\infty, \infty)$

(c)

x	-3	-2	3
$f(x)$	-1	1	11

(d) The function is one-to-one.

11. (a)

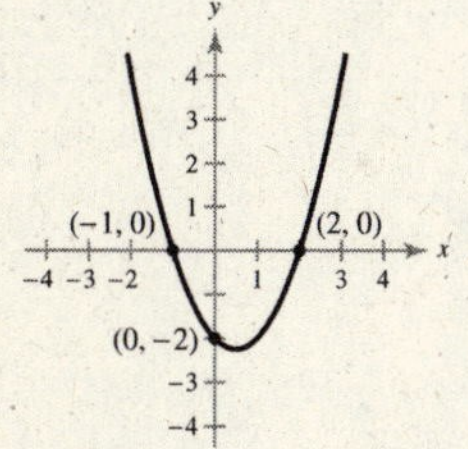

(b) Domain: $(-\infty, \infty)$

Range: $\left[-\frac{9}{4}, \infty\right)$

(c)

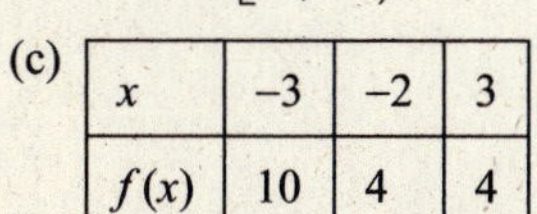

x	-3	-2	3
$f(x)$	10	4	4

(d) The function is not one-to-one.

12. (a)

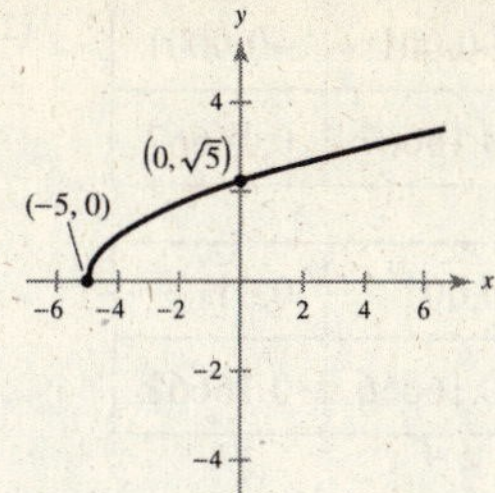

(b) Domain: $[-5, \infty)$

Range: $[0, \infty)$

(c)

x	-3	-2	3
$f(x)$	$\sqrt{2}$	$\sqrt{3}$	$2\sqrt{2}$

(d) The function is one-to-one.

13.
$$\begin{aligned} f(x) = 4x + 6 &= y \\ 4y + 6 &= x \\ 4y &= x - 6 \\ y &= \tfrac{1}{4}x - \tfrac{3}{2} \\ f^{-1}(x) &= \tfrac{1}{4}x - \tfrac{3}{2} \end{aligned}$$

14.
$$\begin{aligned} f(x) = \sqrt[3]{8 - 3x} &= y \\ \sqrt[3]{8 - 3y} &= x \\ 8 - 3y &= x^3 \\ -3y &= x^3 - 8 \\ y &= -\tfrac{1}{3}x^3 + \tfrac{8}{3} \\ f^{-1}(x) &= -\tfrac{1}{3}x^3 + \tfrac{8}{3} \end{aligned}$$

15. $\displaystyle \lim_{x \to 0} \frac{x - 2}{x + 2} = \frac{0 - 2}{0 + 2} = -1$

16.

x	4.9	4.99	4.999	5.001	5.01	5.1
$f(x)$	−99	−999	−9999	10,001	1001	101

$$\lim_{x \to 5^-} \frac{x + 5}{x - 5} = -\infty$$

$$\lim_{x \to 5^+} \frac{x + 5}{x - 5} = \infty$$

$\displaystyle \lim_{x \to 5} f(x)$ does not exist.

17.
$$\begin{aligned} \lim_{x \to -3} \frac{x^2 + 2x - 3}{x^2 + 4x + 3} &= \lim_{x \to -3} \frac{(x - 1)(x + 3)}{(x + 1)(x + 3)} \\ &= \lim_{x \to -3} \frac{x - 1}{x + 1} \\ &= 2 \end{aligned}$$

18.

x	-0.01	-0.001	-0.0001
$f(x)$	0.16671	0.16667	0.16667

x	0.0001	0.001	0.01
$f(x)$	0.16667	0.16666	0.16662

$$\lim_{x\to 0}\frac{\sqrt{x+9}-3}{x} \approx 0.16667$$

19. $f(x) = \dfrac{x^2 - 16}{x - 4}$ is continuous on the intervals $(-\infty, 4)$ and $(4, \infty)$ because the domain of f consists of all real numbers except $x = 4$. There is a discontinuity at $x = 4$ because $f(4)$ is not defined.

20. $f(x) = \sqrt{5 - x}$ is continuous on the interval $(5, \infty)$ because the domain of f consists of all $x > 5$.

21. $\lim\limits_{x\to 1^-} f(x) = \lim\limits_{x\to 1^-}(1 - x) = 0$

$\lim\limits_{x\to 1^+} f(x) = \lim\limits_{x\to 1^+}\left(x - x^2\right) = 0$

So, $\lim\limits_{x\to 1} f(x) = 0.$

Because $f(1)$ is defined, $\lim\limits_{x\to 1} f(x)$ exists, and $\lim\limits_{x\to 1} f(x) = f(1)$, the function is continuous on the interval $(-\infty, \infty)$.

22. (a)

t	4	5	6	7	8	9
y (actual)	8149	7591	7001	7078	8924	14,265
y (model)	8113.4	7696.8	6936.4	6995.8	9038.3	14,227.2

The model fits the data well because the actual values are close to the values given by the model.

(b) Let $t = 14$.

$y \approx 128{,}087.392$ or 128,087,392 workers

CHAPTER 2
Differentiation

CHAPTER 2
Differentiation

Section 2.1 The Derivative and the Slope of a Graph

Skills Warm Up

1. $P(2, 1), Q(2, 4)$

   $x = 2$

2. $P(2, 2), Q = (-5, 2)$

   $m = \dfrac{2 - 2}{-5 - 2} = 0$

   $y - 2 = 0(x - 2)$

   $y = 2$

3. $P(2, 0), Q = (3, -1)$

   $m = \dfrac{-1 - 0}{3 - 2} = -1$

   $y - 0 = -1(x - 2)$

   $y = 2 - x$

4. $P(3, 5), Q(-1, -7)$

   $m = \dfrac{-7 - 5}{-1 - 3} = \dfrac{-12}{-4} = 3$

   $y - 5 = 3(x - 3)$

   $y = 3x - 4$

5. $\lim\limits_{\Delta x \to 0} \dfrac{2x\Delta x + (\Delta x)^2}{\Delta x} = \lim\limits_{\Delta x \to 0} \dfrac{\Delta x(2x + \Delta x)}{\Delta x}$

   $= \lim\limits_{\Delta x \to 0} 2x + \Delta x$

   $= 2x$

6. $\lim\limits_{\Delta x \to 0} \dfrac{3x^2\Delta x + 3x(\Delta x)^2 + (\Delta x)^3}{\Delta x}$

   $= \lim\limits_{\Delta x \to 0} \dfrac{\Delta x\left[3x^2 + 3x\Delta x + (\Delta x)^2\right]}{\Delta x}$

   $= \lim\limits_{\Delta x \to 0} 3x^2 + 3x\Delta x + (\Delta x)^2$

   $= 3x^2$

7. $\lim\limits_{\Delta x \to 0} \dfrac{1}{x(x + \Delta x)} = \dfrac{1}{x^2}$

8. $\lim\limits_{\Delta x \to 0} \dfrac{(x + \Delta x)^2 - x^2}{\Delta x}$

   $= \lim\limits_{\Delta x \to 0} \dfrac{x^2 + 2x\Delta x + (\Delta x)^2 - x^2}{\Delta x}$

   $= \lim\limits_{\Delta x \to 0} \dfrac{2x\Delta x + (\Delta x)^2}{\Delta x}$

   $= \lim\limits_{\Delta x \to 0} \dfrac{\Delta x(2x + \Delta x)}{\Delta x}$

   $= 2x$

9. $f(x) = 3x$

   Domain: $(-\infty, \infty)$

10. $f(x) = \dfrac{1}{x - 1}$

    Domain: $(-\infty, 1) \cup (1, \infty)$

11. $f(x) = \frac{1}{5}x^3 - 2x^2 + \frac{1}{3}x - 1$

    Domain: $(-\infty, \infty)$

12. $f(x) = \dfrac{6x}{x^3 + x}$

    Domain: $(-\infty, 0) \cup (0, \infty)$

1.

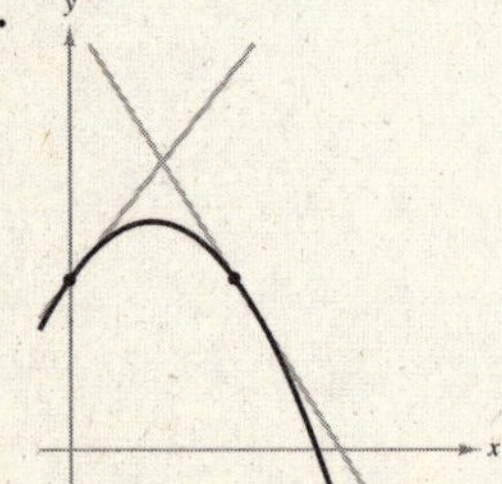

3.

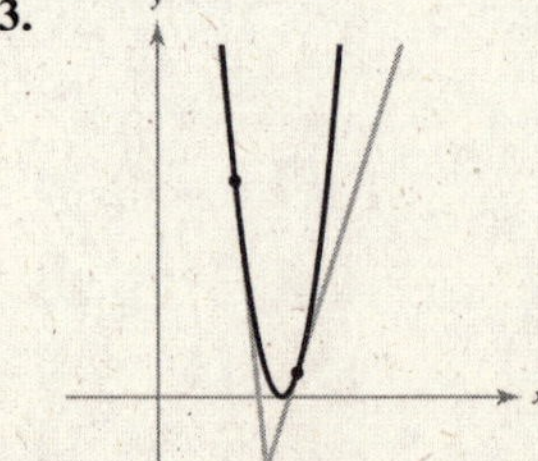

5.

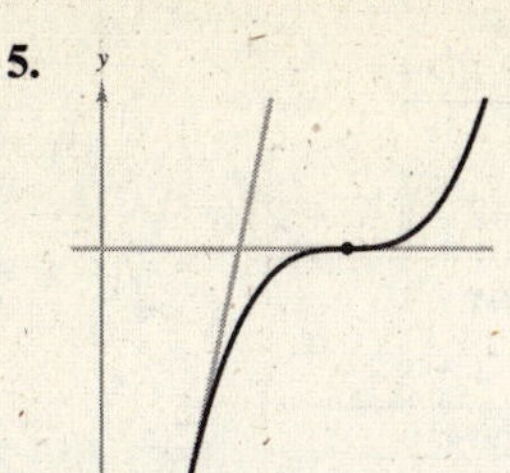

7. The slope is $m = 1$.

9. The slope is $m = 0$.

11. The slope is $m = -\frac{1}{3}$.

13. 2005: $m \approx 119$

2007: $m \approx 161$

The slope is the rate of change of revenue at the given point in time.

15. $t = 3: m \approx 9$

$t = 7: m \approx 0$

$t = 10: m \approx -10$

The slope is the rate of change of the average temperature at the given point in time.

17.
$$\begin{aligned} m_{\text{sec}} &= \frac{f(0 + \Delta x) - f(0)}{\Delta x} \\ &= \frac{-1 - (-1)}{\Delta x} \\ &= \frac{0}{\Delta x} \\ &= 0 \end{aligned}$$

$$m = \lim_{\Delta x \to 0} m_{\text{sec}} = \lim_{\Delta x \to 0} 0 = 0$$

19.
$$\begin{aligned} m_{\text{sec}} &= \frac{f(2 + \Delta x) - f(2)}{\Delta x} \\ &= \frac{\left[8 - 3(2 + \Delta x)\right] - 2}{\Delta x} \\ &= \frac{8 - 6 - 3\Delta x - 2}{\Delta x} \\ &= \frac{-3\Delta x}{\Delta x} \\ &= -3 \end{aligned}$$

$$m = \lim_{\Delta x \to 0} m_{\text{sec}} = \lim_{\Delta x \to 0} -3 = -3$$

21.
$$\begin{aligned} m_{\text{sec}} &= \frac{f(2 + \Delta x) - f(2)}{\Delta x} \\ &= \frac{\left[2(2 + \Delta x)^2 - 3\right] - 5}{\Delta x} \\ &= \frac{\left[2\left(4 + 4\Delta x + (\Delta x)^2\right) - 3\right] - 5}{\Delta x} \\ &= \frac{\left[8 + 8\Delta x + 2(\Delta x)^2 - 3\right] - 5}{\Delta x} \\ &= \frac{2\Delta x(4 + \Delta x)}{\Delta x} \\ &= 2(4 + \Delta x) \end{aligned}$$

$$m = \lim_{\Delta x \to 0} m_{\text{sec}} = \lim_{\Delta x \to 0} \left(2(4 + \Delta x)\right) = 8$$

23.
$$\begin{aligned} m_{\text{sec}} &= \frac{f(2 + \Delta x) - f(2)}{\Delta x} \\ &= \frac{(2 + \Delta x)^3 - (2 + \Delta x) - \left[(2)^3 - (2)\right]}{\Delta x} \\ &= \frac{8 + 12\Delta x + 6(\Delta x)^2 + (\Delta x)^3 - 2 - \Delta x - 6}{\Delta x} \\ &= \frac{11\Delta x + 6(\Delta x)^2 + (\Delta x)^3}{\Delta x} \\ &= \frac{\Delta x\left(11 + 6\Delta x + (\Delta x)^2\right)}{\Delta x} \\ &= 11 + 6\Delta x + (\Delta x)^2 \end{aligned}$$

$$m = \lim_{\Delta x \to 0} m_{\text{sec}} = \lim_{\Delta x \to 0} \left(11 + 6\Delta x + (\Delta x)^2\right) = 11$$

25.
$$\begin{aligned} m_{\text{sec}} &= \frac{f(4 + \Delta x) - f(4)}{\Delta x} \\ &= \frac{2\sqrt{4 + \Delta x} - 2\sqrt{4}}{\Delta x} \\ &= \frac{2\sqrt{4 + \Delta x} - 4}{\Delta x} \cdot \frac{2\sqrt{4 + \Delta x} + 4}{2\sqrt{4 + \Delta x} + 4} \\ &= \frac{4(4 + \Delta x) - 16}{\Delta x\left(2\sqrt{4 + \Delta x} + 4\right)} \\ &= \frac{16 + 4\Delta x - 16}{\Delta x\left(2\sqrt{4 + \Delta x} + 4\right)} \\ &= \frac{4\Delta x}{\Delta x\left(2\sqrt{4 + \Delta x} + 4\right)} \\ &= \frac{4}{2\sqrt{4 + \Delta x} + 4} \end{aligned}$$

$$\begin{aligned} m = \lim_{\Delta x \to 0} m_{\text{sec}} &= \lim_{\Delta x \to 0} \frac{4}{2\sqrt{4 + \Delta x} + 4} \\ &= \frac{4}{2\sqrt{4} + 4} = \frac{1}{2} \end{aligned}$$

27. $$\begin{aligned} f'(x) &= \lim_{\Delta x\to 0} \frac{f(x+\Delta x) - f(x)}{\Delta x} \\ &= \lim_{\Delta x\to 0} \frac{3-3}{\Delta x} \\ &= \lim_{\Delta x\to 0} \frac{0}{\Delta x} \\ &= \lim_{\Delta x\to 0} 0 \\ &= 0 \end{aligned}$$

29. $$\begin{aligned} f'(x) &= \lim_{\Delta x\to 0} \frac{f(x+\Delta x) - f(x)}{\Delta x} \\ &= \lim_{\Delta x\to 0} \frac{-5(x+\Delta x) - (-5x)}{\Delta x} \\ &= \lim_{\Delta x\to 0} \frac{-5x - 5\Delta x + 5x}{\Delta x} \\ &= \lim_{\Delta x\to 0} \frac{-5\Delta x}{\Delta x} \\ &= \lim_{\Delta x\to 0} -5 \\ &= -5 \end{aligned}$$

31. $$\begin{aligned} g'(s) &= \lim_{\Delta s\to 0} \frac{g(s+\Delta s) - g(s)}{\Delta s} \\ &= \lim_{\Delta s\to 0} \frac{\frac{1}{3}(s+\Delta s) + 2 - \left(\frac{1}{3}s + 2\right)}{\Delta s} \\ &= \lim_{\Delta s\to 0} \frac{\frac{1}{3}s + \frac{1}{3}\Delta s + 2 - \frac{1}{3}s - 2}{\Delta s} \\ &= \lim_{\Delta s\to 0} \frac{\frac{1}{3}\Delta s}{\Delta s} \\ &= \lim_{\Delta s\to 0} \frac{1}{3} \\ &= \frac{1}{3} \end{aligned}$$

33. $$\begin{aligned} f'(x) &= \lim_{\Delta x\to 0} \frac{f(x+\Delta x) - f(x)}{\Delta x} \\ &= \lim_{\Delta x\to 0} \frac{\left[4(x+\Delta x)^2 - 5(x+\Delta x)\right] - \left(4x^2 - 5x\right)}{\Delta x} \\ &= \lim_{\Delta x\to 0} \frac{\left[4\left(x^2 + 2x\Delta x + (\Delta x)^2\right) - 5x - 5\Delta x\right] - 4x^2 + 5x}{\Delta x} \\ &= \lim_{\Delta x\to 0} \frac{4x^2 + 8x\Delta x + 4(\Delta x)^2 - 5x - 5\Delta x - 4x^2 + 5x}{\Delta x} \\ &= \lim_{\Delta x\to 0} \frac{4\Delta x\left(2x + \Delta x - \frac{5}{4}\right)}{\Delta x} \\ &= \lim_{\Delta x\to 0} 4\left(2x + \Delta x - \frac{5}{4}\right) = 8x - 5 \end{aligned}$$

35. $$\begin{aligned} h'(t) &= \lim_{\Delta t\to 0} \frac{h(t+\Delta t) - h(t)}{t} \\ &= \lim_{\Delta t\to 0} \frac{\sqrt{t+\Delta t-1} - \sqrt{t-1}}{\Delta t} \\ &= \lim_{\Delta t\to 0} \frac{\sqrt{t+\Delta t-1} - \sqrt{t-1}}{\Delta t} \cdot \frac{\sqrt{t+\Delta t-1} + \sqrt{t-1}}{\sqrt{t+\Delta t-1} + \sqrt{t-1}} \\ &= \lim_{\Delta t\to 0} \frac{t + \Delta t - 1 - (t-1)}{\Delta t\left(\sqrt{t+\Delta t-1} + \sqrt{t-1}\right)} \\ &= \lim_{\Delta t\to 0} \frac{\Delta t}{\Delta t\left(\sqrt{t+\Delta t-1} + \sqrt{t-1}\right)} \\ &= \lim_{\Delta t\to 0} \frac{1}{\sqrt{t+\Delta t-1} + \sqrt{t-1}} \\ &= \frac{1}{2\sqrt{t-1}} \end{aligned}$$

37. $$f'(t) = \lim_{\Delta t \to 0} \frac{f(t + \Delta t) - f(t)}{\Delta t}$$
$$= \lim_{\Delta t \to 0} \frac{(t + \Delta t)^3 - 12(t + \Delta t) - (t^3 - 12t)}{\Delta t}$$
$$= \lim_{\Delta t \to 0} \frac{t^3 + 3t^2\Delta t + 3t(\Delta t)^2 + (\Delta t)^3 - 12t - 12\Delta t - t^3 + 12t}{\Delta t}$$
$$= \lim_{\Delta t \to 0} \frac{3t^2\Delta t + 3t(\Delta t)^2 + (\Delta t)^3 - 12\Delta t}{\Delta t}$$
$$= \lim_{\Delta t \to 0} \frac{\Delta t\left(3t^2 + 3t\Delta t + (\Delta t)^2 - 12\right)}{\Delta t}$$
$$= \lim_{\Delta t \to 0} \left(3t^2 + 3t\Delta t + (\Delta t)^2 - 12\right)$$
$$= 3t^2 - 12$$

39. $$f'(x) = \lim_{\Delta x \to 0} \frac{f(x + \Delta x) - f(x)}{\Delta x}$$
$$= \lim_{\Delta x \to 0} \frac{\dfrac{1}{x + \Delta x + 2} - \dfrac{1}{x + 2}}{\Delta x}$$
$$= \lim_{\Delta x \to 0} \frac{\dfrac{1}{x + \Delta x + 2} \cdot \dfrac{x + 2}{x + 2} - \dfrac{1}{x + 2} \cdot \dfrac{x + \Delta x + 2}{x + \Delta x + 2}}{\Delta x}$$
$$= \lim_{\Delta x \to 0} \frac{\dfrac{x + 2 - (x + \Delta x + 2)}{(x + \Delta x + 2)(x + 2)}}{\Delta x}$$
$$= \lim_{\Delta x \to 0} \frac{-\Delta x}{\Delta x(x + \Delta x + 2)(x + 2)}$$
$$= \lim_{\Delta x \to 0} \frac{-1}{(x + \Delta x + 2)(x + 2)}$$
$$= -\frac{1}{(x + 2)^2}$$

41. $$f'(x) = \lim_{\Delta x \to 0} \frac{f(x + \Delta x) - f(x)}{\Delta x}$$
$$= \lim_{\Delta x \to 0} \frac{\frac{1}{2}(x + \Delta x)^2 - \frac{1}{2}x^2}{\Delta x}$$
$$= \lim_{\Delta x \to 0} \frac{\frac{1}{2}\left(x^2 + 2x\Delta x + (\Delta x)^2\right) - \frac{1}{2}x^2}{\Delta x}$$
$$= \lim_{\Delta x \to 0} \frac{x\Delta x + (\Delta x)^2}{\Delta x}$$
$$= \lim_{\Delta x \to 0} \frac{\Delta x(x + \Delta x)}{\Delta x}$$
$$= \lim_{\Delta x \to 0} (x + \Delta x)$$
$$= x$$

$$m = f'(2) = 2$$
$$y - 2 = 2(x - 2)$$
$$y = 2x - 2$$

43. $f'(x) = \lim_{\Delta x \to 0} \frac{f(x + \Delta x) - f(x)}{\Delta x}$

$= \lim_{\Delta x \to 0} \frac{(x + \Delta x - 1)^2 - (x - 1)^2}{\Delta x}$

$= \lim_{\Delta x \to 0} \frac{x^2 + 2x\Delta x - 2x + (\Delta x)^2 - 2\Delta x + 1 - x^2 + 2x - 1}{\Delta x}$

$= \lim_{\Delta x \to 0} \frac{2x\Delta x + (\Delta x)^2 - 2\Delta x}{\Delta x}$

$= \lim_{\Delta x \to 0} \frac{\Delta x(2x + \Delta x - 2)}{\Delta x}$

$= \lim_{\Delta x \to 0} (2x + \Delta x - 2)$

$= 2x - 2$

$m = f'(-2) = 2(-2) - 2 = -6$

$y - 9 = -6[x - (-2)]$

$y = -6x - 3$

45. $f'(x) = \lim_{\Delta x \to 0} \frac{f(x + \Delta x) - f(x)}{\Delta x}$

$= \lim_{\Delta x \to 0} \frac{\sqrt{x + \Delta x} + 1 - (\sqrt{x} + 1)}{\Delta x}$

$= \lim_{\Delta x \to 0} \frac{\sqrt{x + \Delta x} - \sqrt{x}}{\Delta x} \cdot \frac{\sqrt{x + \Delta x} + \sqrt{x}}{\sqrt{x + \Delta x} + \sqrt{x}}$

$= \lim_{\Delta x \to 0} \frac{x + \Delta x - x}{\Delta x(\sqrt{x + \Delta x} + \sqrt{x})}$

$= \lim_{\Delta x \to 0} \frac{\Delta x}{\Delta x(\sqrt{x + \Delta x} + \sqrt{x})}$

$= \lim_{\Delta x \to 0} \frac{1}{\sqrt{x + \Delta x} + \sqrt{x}}$

$= \frac{1}{2\sqrt{x}}$

$m = f'(4) = \frac{1}{2\sqrt{4}} = \frac{1}{4}$

$y - 3 = \frac{1}{4}(x - 4)$

$y = \frac{1}{4}x + 2$

47. $f'(x) = \lim_{\Delta x \to 0} \frac{f(x + \Delta x) - f(x)}{\Delta x}$

$= \lim_{\Delta x \to 0} \frac{\frac{1}{x + \Delta x} - \frac{1}{x}}{\Delta x}$

$= \lim_{\Delta x \to 0} \frac{\frac{1}{x + \Delta x} \cdot \frac{x}{x} - \frac{1}{x} \cdot \frac{x + \Delta x}{x + \Delta x}}{\Delta x}$

$= \lim_{\Delta x \to 0} \frac{\frac{x - (x + \Delta x)}{x(x + \Delta x)}}{\Delta x}$

$= \lim_{\Delta x \to 0} \frac{-\Delta x}{(\Delta x)(x)(x + \Delta x)}$

$= \lim_{\Delta x \to 0} \frac{-1}{x(x + \Delta x)}$

$= -\frac{1}{x^2}$

$m = f'(1) = -\frac{1}{(1)^2} = -1$

$y - 1 = -1(x - 1)$

$y = -x + 2$

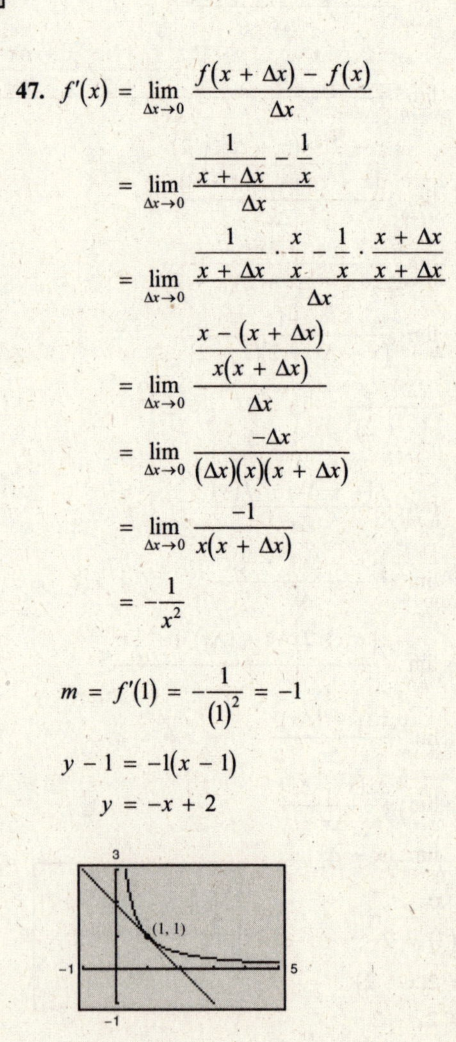

49. $$\begin{aligned} f'(x) &= \lim_{\Delta x \to 0} \frac{f(x + \Delta x) - f(x)}{\Delta x} \\ &= \lim_{\Delta x \to 0} \frac{-\frac{1}{4}(x + \Delta x)^2 - \left(-\frac{1}{4}x^2\right)}{\Delta x} \\ &= \lim_{\Delta x \to 0} \frac{-\frac{1}{4}x^2 - \frac{1}{2}x\Delta x - \frac{1}{4}(\Delta x)^2 + \frac{1}{4}x^2}{\Delta x} \\ &= \lim_{\Delta x \to 0} \frac{-\frac{1}{2}x\Delta x - \frac{1}{4}(\Delta x)^2}{\Delta x} \\ &= \lim_{\Delta x \to 0} \frac{\Delta x\left(-\frac{1}{2}x - \frac{1}{4}\Delta x\right)}{\Delta x} \\ &= \lim_{\Delta x \to 0}\left(-\frac{1}{2}x - \frac{1}{4}\Delta x\right) \\ &= -\frac{1}{2}x \end{aligned}$$

Since the slope of the given line is -1,

$$\begin{aligned} -\tfrac{1}{2}x &= -1 \\ x &= 2 \text{ and } f(2) = -1. \end{aligned}$$

At the point $(2, -1)$, the tangent line parallel to $x + y = 0$ is

$$\begin{aligned} y - (-1) &= -1(x - 2) \\ y &= -x + 1. \end{aligned}$$

51. $$\begin{aligned} f'(x) &= \lim_{\Delta x \to 0} \frac{f(x + \Delta x) - f(x)}{\Delta x} \\ &= \lim_{\Delta x \to 0} \frac{-\frac{1}{3}(x + \Delta x)^3 - \left(-\frac{1}{3}x^3\right)}{\Delta x} \\ &= \lim_{\Delta x \to 0} \frac{-\frac{1}{3}\left(x^3 + 3x^2\Delta x + 3x(\Delta x)^2 + (\Delta x)^3\right) + \frac{1}{3}x^3}{\Delta x} \\ &= \lim_{\Delta x \to 0} \frac{-\frac{1}{3}x^3 - x^2\Delta x - x(\Delta x)^2 - \frac{1}{3}(\Delta x)^3 + \frac{1}{3}x^3}{\Delta x} \\ &= \lim_{\Delta x \to 0} \frac{\Delta x\left(-x^2 - x\Delta x - \frac{1}{3}(\Delta x)^2\right)}{\Delta x} \\ &= \lim_{\Delta x \to 0}\left(-x^2 - x\Delta x - \frac{1}{3}(\Delta x)^2\right) = -x^2 \end{aligned}$$

Since the slope of the given line is -9,

$$\begin{aligned} -x^2 &= -9 \\ x^2 &= 9 \\ x &= \pm 3 \text{ and } f(3) = -9 \text{ and } f(-3) = 9. \end{aligned}$$

At the point $(3, -9)$, the tangent line parallel to $9x + y - 6 = 0$ is

$$\begin{aligned} y - (-9) &= -9(x - 3) \\ y &= -9x + 18. \end{aligned}$$

At the point $(-3, 9)$, the tangent line parallel to $9x + y - 6 = 0$ is

$$\begin{aligned} y - 9 &= -9(x - (-3)) \\ y &= -9x - 18. \end{aligned}$$

53. y is differentiable for all $x \neq -3$.

At $(-3, 0)$ the graph has a node.

55. y is differentiable for all $x \neq 3$.

At (3, 0) the graph has a cusp.

57. y is differentiable for all $x \neq \pm 2$.

The function is not defined at $x = \pm 2$.

59. Since $f'(x) = -3$ for all x, f is a line of the form

$f(x) = -3x + b$.

$f(0) = 2$, so $2 = (-3)(0) + b$, or $b = 2$.

Thus, $f(x) = -3x + 2$.

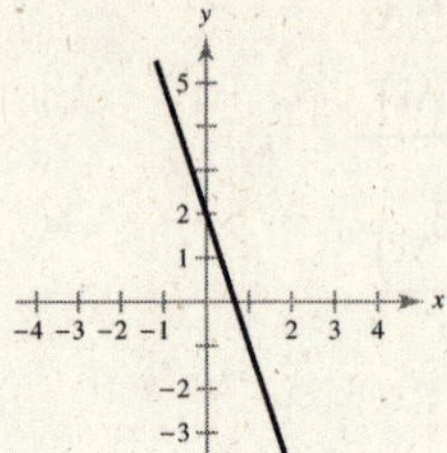

61.

x	-2	$-\frac{3}{2}$	-1	$-\frac{1}{2}$	0	$\frac{1}{2}$	1	$\frac{3}{2}$	2
$f(x)$	-2	-0.8438	-0.25	-0.0313	0	0.0313	0.25	0.8438	2
$f'(x)$	3	1.6875	0.75	0.1875	0	0.1875	0.75	1.6875	3

Analytically, the slope of $f(x) = \frac{1}{4}x^3$ is

$$\begin{aligned} m &= \lim_{\Delta x \to 0} \frac{f(x + \Delta x) - f(x)}{\Delta x} \\ &= \lim_{\Delta x \to 0} \frac{\frac{1}{4}(x + \Delta x)^3 - \frac{1}{4}x^3}{\Delta x} \\ &= \lim_{\Delta x \to 0} \frac{\frac{1}{4}\left[3x^2\Delta x + 3x(\Delta x)^2 + (\Delta x)^3\right]}{\Delta x} \\ &= \lim_{\Delta x \to 0} \frac{1}{4}\left[3x^2 + 3x\Delta x + (\Delta x)^2\right] \\ &= \frac{3}{4}x^2. \end{aligned}$$

63.

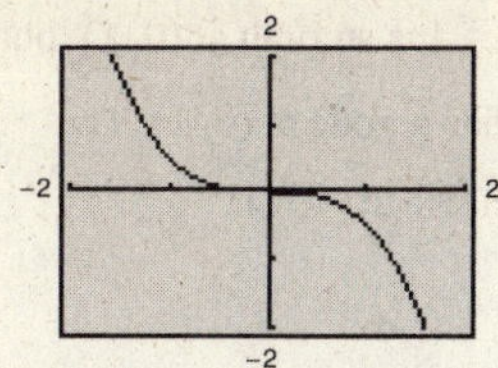

x	-2	$-\frac{3}{2}$	-1	$-\frac{1}{2}$	0	$\frac{1}{2}$	1	$\frac{3}{2}$	2
$f(x)$	4	1.6875	0.5	0.0625	0	−0.0625	−0.5	−1.6875	−4
$f'(x)$	−6	−3.375	−1.5	−0.375	0	−0.375	−1.5	−3.375	−6

Analytically, the slope of $f(x) = -\frac{1}{2}x^3$ is

$$\begin{aligned} m &= \lim_{\Delta x \to 0} \frac{f(x + \Delta x) - f(x)}{\Delta x} \\ &= \lim_{\Delta x \to 0} \frac{-\frac{1}{2}(x + \Delta x)^3 + \frac{1}{2}x^3}{\Delta x} \\ &= \lim_{\Delta x \to 0} \frac{-\frac{1}{2}\left[x^3 + 3x^2\Delta x + 3x(\Delta x)^2 + (\Delta x)^3\right] + \frac{1}{2}x^3}{\Delta x} \\ &= \lim_{\Delta x \to 0} \frac{-\frac{1}{2}\left[3x^2\Delta x + 3x(\Delta x)^2 + (\Delta x)^3\right]}{\Delta x} \\ &= \lim_{\Delta x \to 0} -\frac{1}{2}\left[3x^2 + 3x(\Delta x) + (\Delta x)^2\right] \\ &= -\frac{3}{2}x^2. \end{aligned}$$

65. $$\begin{aligned} f'(x) &= \lim_{\Delta x \to 0} \frac{f(x + \Delta x) - f(x)}{\Delta x} \\ &= \lim_{\Delta x \to 0} \frac{(x + \Delta x)^2 - 4(x + \Delta x) - (x^2 - 4x)}{\Delta x} \\ &= \lim_{\Delta x \to 0} \frac{2x\Delta x + (\Delta x)^2 - 4\Delta x}{\Delta x} \\ &= \lim_{\Delta x \to 0} (2x + \Delta x - 4) \\ &= 2x - 4 \end{aligned}$$

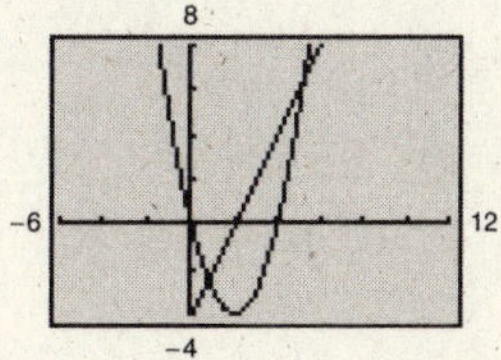

The x-intercept of the derivative indicates a point of horizontal tangency for f.

67. $$\begin{aligned} f'(x) &= \lim_{\Delta x \to 0} \frac{f(x + \Delta x) - f(x)}{\Delta x} \\ &= \lim_{\Delta x \to 0} \frac{(x + \Delta x)^3 - 3(x + \Delta x) - (x^3 - 3x)}{\Delta x} \\ &= \lim_{\Delta x \to 0} \frac{x^3 + 3x^2\Delta x + 3x(\Delta x)^2 + (\Delta x)^3 - 3x - 3\Delta x - x^3 + 3x}{\Delta x} \\ &= \lim_{\Delta x \to 0} \frac{3x^2\Delta x + 3x(\Delta x)^2 + (\Delta x)^3 - 3\Delta x}{\Delta x} \\ &= \lim_{\Delta x \to 0} \left(3x^2 + 3x\Delta x + (\Delta x)^2 - 3\right) \\ &= 3x^2 - 3 \end{aligned}$$

The x-intercepts of the derivative indicate points of horizontal tangency for f.

69. True. The slope of the graph is given by $f'(x) = 2x$, which is different for each different x value.

71. True. See page 122.

73. The graph of $f(x) = x^2 + 1$ is smooth at (0, 1), but the graph of $g(x) = |x| + 1$ has a node at (0, 1). The function g is not differentiable at (0, 1).

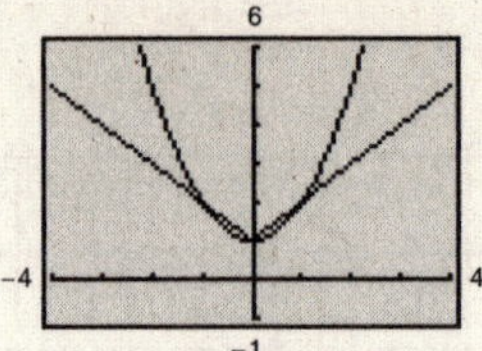

Section 2.2 Some Rules for Differentiation

Skills Warm Up

1. (a) $2x^2,\ x = 2$

$2(2^2) = 2(4) = 8$

(b) $(2x)^2,\ x = 2$

$[2(2)]^2 = 4^2 = 16$

(c) $2x^{-2},\ x = 2$

$2(2)^{-2} = 2\left(\frac{1}{4}\right) = \frac{1}{2}$

2. (a) $\dfrac{1}{(3x)^2},\ x = 2$

$\dfrac{1}{[3(2)]^2} = \dfrac{1}{6^2} = \dfrac{1}{36}$

(b) $\dfrac{1}{4x^3},\ x = 2$

$\dfrac{1}{4(2^3)} = \dfrac{1}{4(8)} = \dfrac{1}{32}$

(c) $\dfrac{(2x)^{-3}}{4x^{-2}},\ x = 2$

$\dfrac{[2(2)]^{-3}}{4(2)^{-2}} = \dfrac{4^{-3}}{4(2)^{-2}} = \dfrac{2^2}{4(4^3)} = \dfrac{1}{64}$

3. $4(3)x^3 + 2(2)x = 12x^3 + 4x = 4x(3x^2 + 1)$

4. $\frac{1}{2}(3)x^2 - \frac{3}{2}x^{1/2} = \frac{3}{2}x^2 - \frac{3}{2}\sqrt{x} = \frac{3}{2}x^{1/2}(x^{3/2} - 1)$

5. $\left(\dfrac{1}{4}\right)x^{-3/4} = \dfrac{1}{4x^{3/4}}$

6. $\dfrac{1}{3}(3)x^2 - 2\left(\dfrac{1}{2}\right)x^{-1/2} + \dfrac{1}{3}x^{-2/3} = x^2 - x^{-1/2} + \dfrac{1}{3}x^{-2/3}$

$= x^2 - \dfrac{1}{\sqrt{x}} + \dfrac{1}{3\sqrt[3]{x^2}}$

7. $3x^2 + 2x = 0$

$x(3x + 2) = 0$

$x = 0$

$3x + 2 = 0 \rightarrow x = -\frac{2}{3}$

8. $x^3 - x = 0$

$x(x^2 - 1) = 0$

$x(x + 1)(x - 1) = 0$

$x = 0$

$x + 1 = 0 \rightarrow x = -1$

$x - 1 = 0 \rightarrow x = 1$

9. $x^2 + 8x - 20 = 0$

$(x + 10)(x - 2) = 0$

$x + 10 = 0 \rightarrow x = -10$

$x - 2 = 0 \rightarrow x = 2$

10. $x^2 - 10x - 24 = 0$

$(x - 12)(x + 2) = 0$

$x - 12 = 0 \rightarrow x = 12$

$x + 2 = 0 \rightarrow x = -2$

1. $y = 3$

$y' = 0$

3. $y = x^5$

$y' = 5x^4$

5. $h(x) = 3x^3$

$h'(x) = 9x^2$

7. $y = \dfrac{2x^3}{3}$

$y' = 2x^2$

9. $f(x) = 4x$

$f'(x) = 4$

11. $y = 8 - x^3$

$y = -3x^2$

13. $f(x) = 4x^2 - 3x$

$f'(x) = 8x - 3$

15. $f(t) = -3t^2 + 2t - 4$

$f'(t) = -6t + 2$

17. $s(t) = t^3 - 2t + 4$

$s'(t) = 3t^2 - 2$

19. $g(x) = x^{2/3}$

$g'(x) = \dfrac{2}{3}x^{-1/3} = \dfrac{2}{3x^{1/3}}$

21. $y = 4t^{4/3}$

$y' = 4\left(\frac{4}{3}\right)t^{1/3} = \frac{16}{3}t^{1/3}$

23. $y = 4x^{-2} + 2x^2$

$y' = -8x^{-3} + 4x^1 = -\dfrac{8}{x^3} + 4x$

	Function	*Rewrite*	*Differentiate*	*Simplify*
25.	$y = \dfrac{2}{7x^4}$	$y = \dfrac{2}{7}x^{-4}$	$y' = \dfrac{-8}{7}x^{-5}$	$y' = -\dfrac{8}{7x^5}$
27.	$y = \dfrac{1}{(4x)^3}$	$y = \dfrac{1}{64}x^{-3}$	$y' = -\dfrac{3}{64}x^{-4}$	$y' = -\dfrac{3}{64x^4}$
29.	$y = \dfrac{4}{(2x)^{-5}}$	$y = 128x^5$	$y' = 128(5)x^4$	$y' = 640x^4$
31.	$y = 6\sqrt{x}$	$y = 6x^{1/2}$	$y' = 6\left(\frac{1}{2}\right)x^{-1/2}$	$y' = \dfrac{3}{\sqrt{x}}$
33.	$y = \dfrac{1}{5\sqrt[5]{x}}$	$y = \dfrac{1}{5}x^{-1/5}$	$y' = \dfrac{1}{5}\left(-\dfrac{1}{5}\right)x^{-6/5}$	$y' = -\dfrac{1}{25\sqrt[5]{x^6}}$
35.	$y = \sqrt{3x}$	$y = (3x)^{1/2}$	$y' = \dfrac{1}{2}(3x)^{-1/2}(3)$	$y' = \dfrac{\sqrt{3x}}{2x}$

37. $y = x^{3/2}$

$y' = \frac{3}{2}x^{1/2}$

At the point $(1, 1)$, $y' = \frac{3}{2}(1)^{1/2} = \frac{3}{2} = m$.

39. $f(t) = t^2$

$f'(t) = 2t$

At the point $(4, 16)$, $f'(4) = 2(4) = 8 = m$.

41. $f(x) = 2x^3 + 8x^2 - x - 4$

$f'(x) = 6x^2 + 16x - 1$

At the point $(-1, 3)$, $f'(-1) = 6(-1)^2 + 16(-1) - 1 = -11 = m$.

43. $f(x) = -\frac{1}{2}x(1 + x^2) = -\frac{1}{2}x - \frac{1}{2}x^3$

$f'(x) = -\frac{1}{2} - \frac{3}{2}x^2$

$f'(1) = -\frac{1}{2} - \frac{3}{2} = -2$

45. (a) $y = -2x^4 + 5x^2 - 3$

$y' = -8x^3 + 10x$

$m = y'(1) = -8 + 10 = 2$

The equation of the tangent line is

$y - 0 = 2(x - 1)$

$y = 2x - 2.$

(b) and (c)

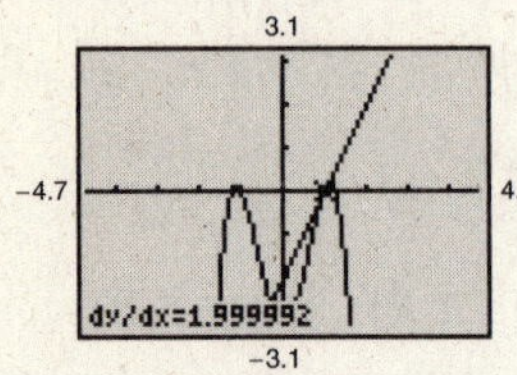

47. (a) $f(x) = \sqrt[3]{x} + \sqrt[5]{x} = x^{1/3} + x^{1/5}$

$f'(x) = \frac{1}{3}x^{-2/3} + \frac{1}{5}x^{-4/5} = \frac{1}{3x^{2/3}} + \frac{1}{5x^{4/5}}$

$m = f'(1) = \frac{1}{3} + \frac{1}{5} = \frac{8}{15}$

The equation of the tangent line is

$y - 2 = \frac{8}{15}(x - 1)$

$y = \frac{8}{15}x + \frac{22}{15}.$

(b) and (c)

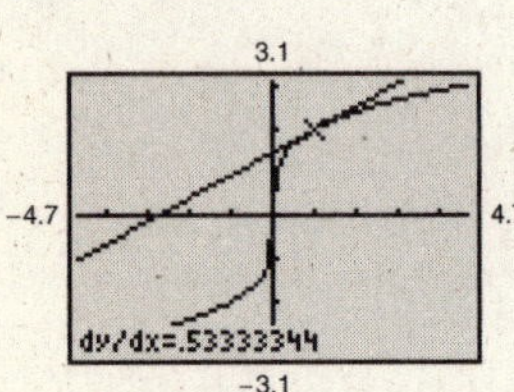

49. (a) $y = 3x\left(x^2 - \frac{2}{x}\right)$

$y = 3x^3 - 6$

$y' = 9x^2$

$m = y' = 9(2)^2 = 36$

The equation of the tangent line is

$y - 18 = 36(x - 2)$

$y = 36x - 54.$

(b) and (c)

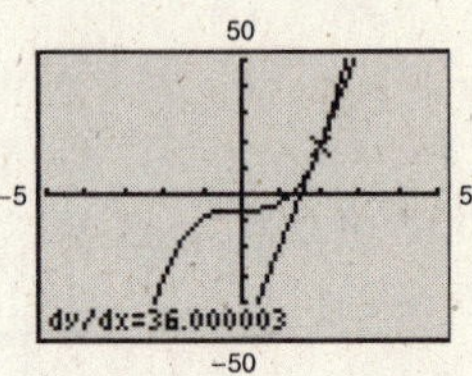

51. $f(x) = x^2 - 4x^{-1} - 3x^{-2}$

$f'(x) = 2x + 4x^{-2} + 6x^{-3} = 2x + \frac{4}{x^2} + \frac{6}{x^3}$

53. $f(x) = x^2 - 2x - \frac{2}{x^4} = x^2 - 2x - 2x^{-4}$

$f'(x) = 2x - 2 + 8x^{-5} = 2x - 2 + \frac{8}{x^5}$

55. $f(x) = x^{4/5} + x$

$f'(x) = \frac{4}{5}x^{-1/5} + 1 = \frac{4}{5x^{1/5}} + 1$

57. $f(x) = x(x^2 + 1) = x^3 + x$

$f'(x) = 3x^2 + 1$

59. $f(x) = \frac{2x^3 - 4x^2 + 3}{x^2} = 2x - 4 + 3x^{-2}$

$f'(x) = 2 - 6x^{-3} = 2 - \frac{6}{x^3} = \frac{2x^3 - 6}{x^3} = \frac{2(x^3 - 3)}{x^3}$

61. $f(x) = \frac{4x^3 - 3x^2 + 2x + 5}{x^2} = 4x - 3 + 2x^{-1} + 5x^{-2}$

$f'(x) = 4 - 2x^{-2} - 10x^{-3} = 4 - \frac{2}{x^2} - \frac{10}{x^3} = \frac{4x^3 - 2x - 10}{x^3}$

63. $y' = -4x^3 + 6x = 2x(3 - 2x^2) = 0$ when $x = 0, \pm\frac{\sqrt{6}}{2}$

If $x = \pm\frac{\sqrt{6}}{2}$, then $y = -\left(\pm\frac{\sqrt{6}}{2}\right)^4 + 3\left(\pm\frac{\sqrt{6}}{2}\right)^2 - 1 = -\frac{9}{4} + 3\left(\frac{3}{2}\right) - 1 = \frac{5}{4}.$

The function has horizontal tangent lines at the points

$(0, -1)$, $\left(-\frac{\sqrt{6}}{2}, \frac{5}{4}\right)$, and $\left(\frac{\sqrt{6}}{2}, \frac{5}{4}\right)$.

65. $y' = x + 5 = 0$ when $x = -5$.

The function has a horizontal tangent line at the point $\left(-5, -\frac{25}{2}\right)$.

67. (a)

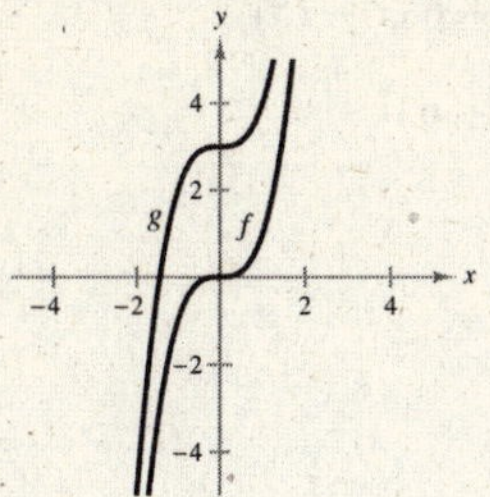

(b) $f'(x) = g'(x) = 3x^2$

$f'(1) = g'(1) = 3$

(c) Tangent line to f at $x = 1$:

$f(1) = 1$

$y - 1 = 3(x - 1)$

$y = 3x - 2$

Tangent line to g at $x = 1$:

$g(1) = 4$

$y - 4 = 3(x - 1)$

$y = 3x + 1$

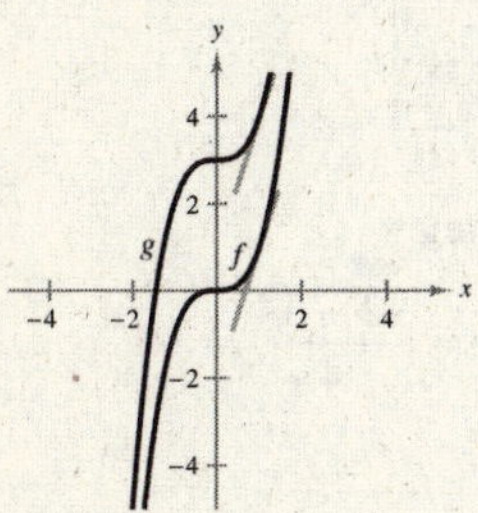

(d) f' and g' are the same.

69. If $g(x) = f(x) + 6$, then $g'(x) = f'(x)$ because the derivative of a constant is 0, $g'(x) = f'(x)$.

71. If $g(x) = -5f(x)$, then $g'(x) = -5f'(x)$ because of the Constant Multiple Rule.

73. (a) 2005: $m \approx 119.2$; 2007: $m \approx 161$

(b) These results are close to the estimates in Exercise 13 in Section 2.1.

(c) The slope of the graph at time t is the rate at which sales are increasing in millions of dollars per year.

75. (a) More men and women seem to suffer from migraines between 30 and 40 years old. More females than males suffer from migraines. Fewer people whose income is greater than or equal to \$30,000 suffer from migraines than people whose income is less than \$10,000.

(b) The derivatives are positive up to approximately 37 years old and negative after about 37 years of age. The percent of adults suffering from migraines increases up to about 37 years old, then decreases. The units of the derivative are percent of adults suffering from migraines per year.

77. $C = 7.75x + 500$

$C' = 7.75$, which equals the variable cost.

79. $f(x) = 4.1x^3 - 12x^2 + 2.5x$

$f'(x) = 12.3x^2 - 24x + 2.5$

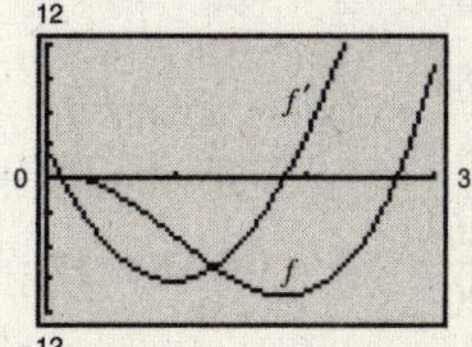

f has horizontal tangents at $(0.110, 0.135)$ and $(1.841, -10.486)$.

81. False. Let $f(x) = x$ and $g(x) = x + 1$.

Then $f'(x) = g'(x) = 1$, but $f(x) \neq g(x)$.

Section 2.3 Rates of Change: Velocity and Marginals

Skills Warm Up

1. $\dfrac{-63 - (-105)}{21 - 7} = \dfrac{42}{14} = 3$

2. $\dfrac{-37 - 54}{16 - 3} = \dfrac{-91}{13} = -7$

3. $\dfrac{24 - 33}{9 - 6} = \dfrac{-9}{3} = -3$

4. $\dfrac{40 - 16}{18 - 8} = \dfrac{24}{10} = \dfrac{12}{5}$

Skills Warm Up *—continued—*

5. $y = 4x^2 - 2x + 7$

$y' = 8x - 2$

6. $y = -3t^3 + 2t^2 - 8$

$y' = -9t^2 + 4t$

7. $s = -16t^2 + 24t + 30$

$s' = -32t + 24$

8. $y = -16x^2 + 54x + 70$

$y' = -32x + 54$

9. $A = \frac{1}{10}\left(-2r^3 + 3r^2 + 5r\right)$

$A' = \frac{1}{10}\left(-6r^2 + 6r + 5\right)$

$A' = -\frac{3}{5}r^2 + \frac{3}{5}r + \frac{1}{2}$

10. $y = \frac{1}{9}\left(6x^3 - 18x^2 + 63x - 15\right)$

$y' = \frac{1}{9}\left(18x^2 - 36x + 63\right)$

$y' = 2x^2 - 4x + 7$

11. $y = 12x - \dfrac{x^2}{5000}$

$y' = 12 - \dfrac{2x}{5000}$

$y' = 12 - \dfrac{x}{2500}$

12. $y = 138 + 74x - \dfrac{x^3}{10,000}$

$y' = 74 - \dfrac{3x^2}{10,000}$

1. (a) 1980–1985: $\dfrac{115 - 63}{5 - 0} = \$10.4$ billion/yr

(b) 1985–1990: $\dfrac{152 - 115}{10 - 5} = \$7.4$ billion/yr

(c) 1990–1995: $\dfrac{184 - 152}{15 - 10} = \$6.4$ billion/yr

(d) 1995–2000: $\dfrac{267 - 184}{20 - 15} = \$16.6$ billion/yr

(e) 2000–2005: $\dfrac{322 - 267}{25 - 20} = \$11.0$ billion/yr

(f) 1980–2008: $\dfrac{398 - 63}{28 - 0} = \$11.96$ billion/yr

(g) 1990–2008: $\dfrac{398 - 152}{28 - 10} = \$13.67$ billion/yr

(h) 2000–2008: $\dfrac{398 - 267}{28 - 20} = \$16.38$ billion/yr

3.

12
(2, 11)
(1, 8)
−10
11
−2

Average rate of change:

$$\frac{\Delta y}{\Delta t} = \frac{f(2) - f(1)}{2 - 1} = \frac{11 - 8}{1} = 3$$

$f'(t) = 3$

Instantaneous rates of change: $f'(1) = 3$, $f'(2) = 3$

5. 

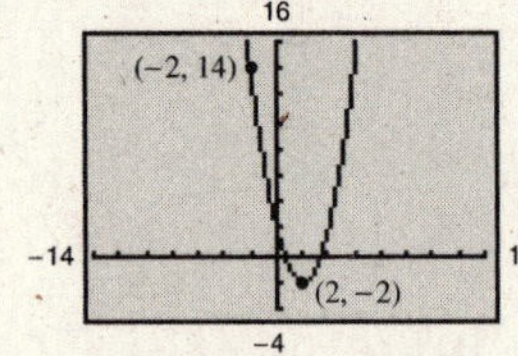

Average rate of change:

$$\frac{\Delta y}{\Delta x} = \frac{h(2) - h(-2)}{2 - (-2)} = \frac{-2 - 14}{4} = -4$$

$h'(x) = 2x - 4$

Instantaneous rates of change: $h'(-2) = -8$, $h'(2) = 0$

7.

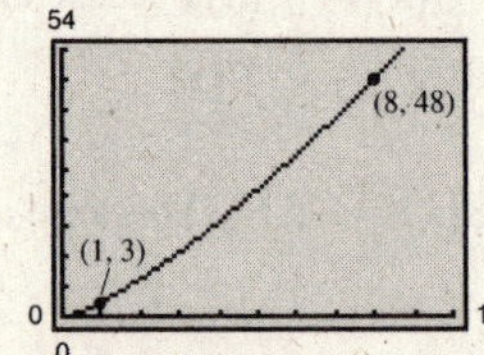

Average rate of change:

$$\frac{\Delta y}{\Delta x} = \frac{f(8) - f(1)}{8 - 1} = \frac{48 - 3}{7} = \frac{45}{7}$$

$f'(x) = 4x^{1/3}$

Instantaneous rates of change: $f'(1) = 4$, $f'(8) = 8$

9.

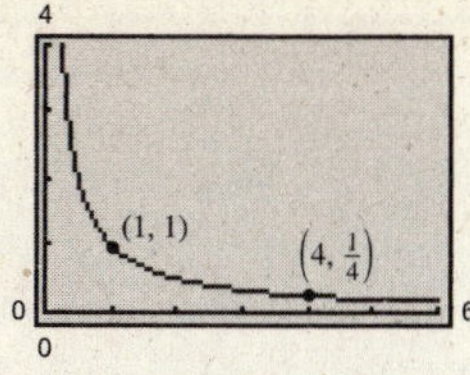

Average rate of change:

$$\frac{\Delta y}{\Delta x} = \frac{f(4) - f(1)}{4 - 1} = \frac{(1/4) - 1}{3} = -\frac{1}{4}$$

$$f'(x) = -\frac{1}{x^2}$$

Instantaneous rates of change:

$$f'(1) = -1,\ f'(4) = -\frac{1}{16}$$

11.

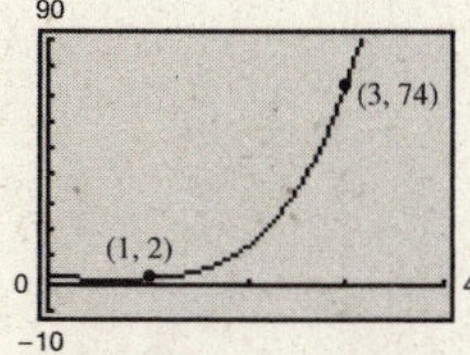

Average rate of change:

$$\frac{\Delta y}{\Delta x} = \frac{g(3) - g(1)}{3 - 1} = \frac{74 - 2}{2} = 36$$

$$g'(x) = 4x^3 - 2x$$

Instantaneous rates of change:

$$g'(1) = 2,\ g'(3) = 102$$

13. (a) $\approx \dfrac{0 - 1400}{3} \approx -467$

The number of visitors to the park is decreasing at an average rate of 467 people per month from September to December.

(b) Answers will vary. Sample answer: $[4, 11]$

Both the instantaneous rate of change at $t = 8$ and the average rate of change on $[4, 11]$ are about zero.

15. $s = -16t^2 + 30t + 250$

Instantaneous: $v(t) = s'(t) = -32t + 30$

(a) Average:

$$[0, 1]: \frac{s(1) - s(0)}{1 - 0} = \frac{264 - 250}{1} = 14 \text{ ft/sec}$$

$v(0) = s'(0) = 30$ ft/sec

$v(1) = s'(1) = -2$ ft/sec

(b) Average:

$$[1, 2]: \frac{s(2) - s(1)}{2 - 1} = \frac{246 - 264}{1} = -18 \text{ ft/sec}$$

$v(1) = s'(1) = -2$ ft/sec

$v(2) = s'(2) = -34$ ft/sec

(c) Average:

$$[2, 3]: \frac{s(3) - s(2)}{3 - 2} = \frac{196 - 246}{1} = -50 \text{ ft/sec}$$

$v(2) = s'(2) = -34$ ft/sec

$v(3) = s'(3) = -66$ ft/sec

(d) Average:

$$[3, 4]: \frac{s(4) - s(3)}{4 - 3} = \frac{114 - 196}{1} = -82 \text{ ft/sec}$$

$v(3) = s'(3) = -66$ ft/sec

$v(4) = s'(4) = -98$ ft/sec

17. $s = -16t^2 + 555$

(a) Average velocity $= \dfrac{s(3) - s(2)}{3 - 2}$

$$= \frac{411 - 491}{1}$$

$= -80$ ft/sec

(b) $v = s'(t) = -32t,\ v(2) = -64$ ft/sec,

$v(3) = -96$ ft/sec

(c) $s = -16t^2 + 555 = 0$

$16t^2 = 555$

$$t^2 = \frac{555}{16}$$

$t \approx 5.89$ seconds

(d) $v(5.89) \approx -188.5$ ft/sec

19. $C = 205{,}000 + 9800x$

$$\frac{dC}{dx} = 9800$$

21. $C = 55{,}000 + 470x - 0.25x^2,\ 0 \le x \le 940$

$$\frac{dC}{dx} = 470 - 0.5x$$

23. $R = 50x - 0.5x^2$

$\dfrac{dR}{dx} = 50 - x$

25. $R = -6x^3 + 8x^2 + 200x$

$\dfrac{dR}{dx} = -18x^2 + 16x + 200$

27. $P = -2x^2 + 72x - 145$

$\dfrac{dP}{dx} = -4x + 72$

29. $P = 0.0013x^3 + 12x$

$\dfrac{dP}{dx} = 0.0039x^2 + 12$

31. $C = 3.6\sqrt{x} + 500$

(a) $C(10) - C(9) \approx \$0.584$

(b) $C'(x) = 1.8/\sqrt{x}$

$C'(9) = \$0.60$ per unit.

(c) The answers are close.

33. $P = -0.04x^2 + 25x - 1500$

(a) $\dfrac{\Delta P}{\Delta x} = \dfrac{P(151) - P(150)}{151 - 150} = \dfrac{1362.96 - 1350}{1} = \$12.96$

(b) $\dfrac{dP}{dx} = -0.08x + 25 = P'(x)$

$P'(150) = \$13$

(c) The results are close.

35. $P = -15.56t^2 + 802.1t + 117{,}001$

(a) $P(0) = 117{,}001.0$ thousand people

$P(5) = 120{,}622.5$ thousand people

$P(10) = 123{,}466.0$ thousand people

$P(15) = 125{,}531.5$ thousand people

$P(20) = 126{,}819.0$ thousand people

$P(25) = 127{,}328.5$ thousand people

$P(30) = 127{,}060.0$ thousand people

The population is growing from 1980 to 2005. It then begins to decline.

(b) $\dfrac{dP}{dt} = -31.12t + 802.1 = P'(t)$

(c) $P'(0) = 802.1$ thousand people per year

$P'(5) = 646.5$ thousand people per year

$P'(10) = 490.9$ thousand people per year

$P'(15) = 335.3$ thousand people per year

$P'(20) = 179.7$ thousand people per year

$P'(25) = 24.1$ thousand people per year

$P'(30) = -131.5$ thousand people per year

The rate of growth is decreasing.

37. (a) $TR = -10Q^2 + 160Q$

(b) $(TR)' = MR = -20Q + 160$

(c)

Q	0	2	4	6	8	10
Model	160	120	80	40	0	−40
Table	–	130	90	50	10	−30

39. (a) $(400, 1.75), (500, 1.50)$

$\text{Slope} = \dfrac{1.50 - 1.75}{500 - 400} = -0.0025$

$p - 1.75 = -0.0025(x - 400)$

$p = -0.0025x + 2.75$

$P = R - C = xp - c$

$= x(-0.0025x + 2.75) - (0.1x + 25)$

$= -0.0025x^2 + 2.65x - 25$

(b)

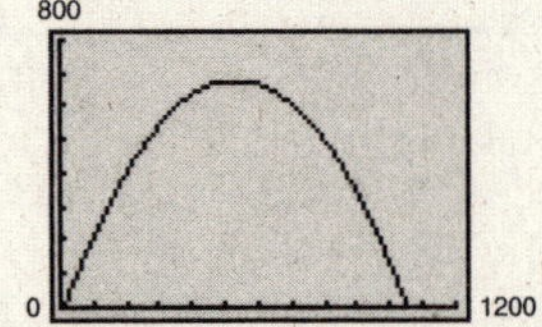

At $x = 300$, P has a positive slope.

At $x = 700$, P has a negative slope.

(c) $P'(x) = -0.005x + 2.65$

$P'(300) = \$1.15$ per unit

$P'(700) = -\$.85$ per unit

41. (a) $C(x) = \left(\frac{15{,}000 \text{ mi}}{\text{yr}}\right)\left(\frac{1 \text{ gal}}{x \text{ mi}}\right)\left(\frac{2.95 \text{ dollars}}{1 \text{ gal}}\right)$

$C(x) = \frac{44{,}250}{x} \frac{\text{dollars}}{\text{yr}}$

(b) $\frac{dC}{dx} = -\frac{44{,}250}{x^2} \frac{\frac{\text{dollars}}{\text{yr}}}{\text{mpg}}$

The marginal cost is the change of savings for a 1-mile per gallon increase in fuel efficiency.

(c)

x	10	15	20	25	30	35	40
C	4425.00	2950.00	2212.50	1770.00	1475.00	1264.29	1106.25
dC/dx	−442.5	−196.67	−110.63	−70.80	−49.17	−36.12	−27.66

(d) The driver who gets 15 miles per gallon would benefit more than the driver who gets 35 miles per gallon. The value of dC/dx is a greater savings for $x = 15$ than for $x = 35$.

43. (a) Average rate of change from 1995 to 2009: $\frac{\Delta p}{\Delta t} = \frac{10{,}428.05 - 5117.12}{19 - 5} \approx 379.35$ dollars per year

(b) Average rate of change from 1996 to 2000: $\frac{\Delta p}{\Delta t} = \frac{10{,}786.85 - 6448.26}{10 - 6} \approx \$1084.65$

So, the instantaneous rate of change for 1998 is $p'(8) \approx \$1084.65/\text{yr}$.

(c) Average rate of change from 1997 to1999: $\frac{\Delta p}{\Delta t} = \frac{11{,}497.12 - 7908.24}{9 - 7} \approx \$1794.44$

So, the instantaneous rate of change for 1998 is $p'(8) \approx \$1794.44/\text{yr}$.

(d) The average rate of change from 1997 to 1999 is a better estimate because the data is closer to the year in question.

Section 2.4 The Product and Quotient Rules

Skills Warm Up

1. $(x^2 + 1)(2) + (2x + 7)(2x) = 2x^2 + 2 + 4x^2 + 14x = 6x^2 + 14x + 2 = 2(3x^2 + 7x + 1)$

2. $(2x - x^3)(8x) + (4x^2)(2 - 3x^2) = 16x^2 - 8x^4 + 8x^2 - 12x^4 = 24x^2 - 20x^4 = 4x^2(6 - 5x^2)$

3. $x(4)(x^2 + 2)^3(2x) + (x^2 + 4)(1) = 8x^2(x^2 + 2)^3(x^2 + 4)$

4. $x^2(2)(2x + 1)(2) + (2x + 1)^4(2x) = 4x^2(2x + 1) + 2x(2x + 1)^4 = 2x(2x + 1)\left[2x + (2x + 1)^3\right]$

5. $\frac{(2x + 7)(5) - (5x + 6)(2)}{(2x + 7)^2} = \frac{10x + 35 - 10x - 12}{(2x + 7)^2} = \frac{23}{(2x + 7)^2}$

6. $\frac{(x^2 - 4)(2x + 1) - (x^2 + x)(2x)}{(x^2 - 4)^2} = \frac{2x^3 + x^2 - 8x - 4 - 2x^3 - 2x^2}{(x^2 - 4)^2} = \frac{-x^2 - 8x - 4}{(x^2 - 4)^2}$

7. $\frac{(x^2 + 1)(2) - (2x + 1)(2x)}{(x^2 + 1)^2} = \frac{2x^2 + 2 - 4x^2 - 2x}{(x^2 + 1)^2} = \frac{-2x^2 - 2x + 2}{(x^2 + 1)^2} = \frac{-2(x^2 + x - 1)}{(x^2 + 1)^2}$

Skills Warm Up —*continued*—

8. $\dfrac{(1-x^4)(4)-(4x-1)(-4x^3)}{(1-x^4)^2} = \dfrac{4-4x^4+16x^4-4x^3}{(1-x^4)^2} = \dfrac{12x^4-4x^3+4}{(1-x^4)^2} = \dfrac{4(3x^4-x^3+1)}{(1-x^4)^2}$

9. $(x^{-1}+x)(2)+(2x-3)(-x^{-2}+1) = 2x^{-1}+2x+(-2x^{-1}+2x+3x^{-2}-3)$

$= 4x+3x^{-2}-3$

$= 4x+\dfrac{3}{x^2}-3$

$= \dfrac{4x^3-3x^2+3}{x^2}$

10. $\dfrac{(1-x^{-1})(1)-(x-4)(x^{-2})}{(1-x^{-1})^2} = \left(\dfrac{1-x^{-1}-x^{-1}+4x^{-2}}{1-2x^{-1}+x^{-2}}\right)\left(\dfrac{x^2}{x^2}\right) = \dfrac{x^2-2x+4}{x^2-2x+1} = \dfrac{x^2-2x+4}{(x-1)^2}$

11. $f(x) = 3x^2-x+4$

$f'(x) = 6x-1$

$f'(2) = 6(2)-1 = 12-1 = 11$

12. $f(x) = -x^3+x^2+8x$

$f'(x) = -3x^2+2x+8$

$f'(2) = -3(2^2)+2(2)+8 = -3(4)+4+8 = 0$

13. $f(x) = \dfrac{1}{x}$

$f'(x) = -\dfrac{1}{x^2}$

$f'(2) = -\dfrac{1}{2^2} = -\dfrac{1}{4}$

14. $f(x) = x^2-\dfrac{1}{x^2}$

$f'(x) = 2x+\dfrac{2}{x^3}$

$f'(2) = 2(2)+\dfrac{2}{2^3} = 4+\dfrac{2}{8} = 4+\dfrac{1}{4} = \dfrac{17}{4}$

1. $f(x) = (2x-3)(1-5x)$

$f'(x) = (2x-3)(-5)+(1-5x)(2)$

$= -10x+15+2-10x$

$= -20x+17$

3. $f(x) = (6x-x^2)(4+3x)$

$f'(x) = (6x-x^2)(3)+(4+3x)(6-2x)$

$= 18x-3x^2+24-8x+18x-6x^2$

$= -9x^2+28x+24$

5. $f(x) = x(x^2+3)$

$f'(x) = x(2x)+(x^2+3)(1) = 2x^2+x^2+3 = 3x^2+3$

7. $h(x) = \left(\dfrac{2}{x}-3\right)(x^2+7) = (2x^{-1}-3)(x^2+7)$

$h'(x) = (2x^{-1}-3)(2x)+(x^2+7)(-2x^{-2})$

$= 4-6x-2-14x^{-2}$

$= -6x+2-\dfrac{14}{x^2}$

9. $g(x) = (x^2-4x+3)(x-2)$

$g'(x) = (x^2-4x+3)(1)+(x-2)(2x-4)$

$= x^2-4x+3+2x^2-4x-4x+8$

$= 3x^2-12x+11$

11. $h(x) = \dfrac{x}{x-5}$

$h'(x) = \dfrac{(x-5)(1)-x(1)}{(x-5)^2} = \dfrac{x-5-x}{(x-5)^2} = -\dfrac{5}{(x-5)^2}$

13. $f(t) = \dfrac{2t^2-3}{3t+1}$

$f'(t) = \dfrac{(3t+1)(4t)-(2t^2-3)(3)}{(3t+1)^2}$

$= \dfrac{12t^2+4t-6t^2+9}{(3t+1)^2}$

$= \dfrac{6t^2+4t+9}{(3t+1)^2}$

15. $f(t) = \dfrac{t^2 - 1}{t + 4}$

$$f'(t) = \frac{(t + 4)(2t) - (t^2 - 1)(1)}{(t + 4)^2}$$
$$= \frac{2t^2 + 8t - t^2 + 1}{(t + 4)^2}$$
$$= \frac{t^2 + 8t + 1}{(t + 4)^2}$$

17. $f(x) = \dfrac{x^2 + 6x + 5}{2x - 1}$

$$f'(x) = \frac{(2x - 1)(2x + 6) - (x^2 + 6x + 5)(2)}{(2x - 1)^2}$$
$$= \frac{4x^2 + 12x - 2x - 6 - 2x^2 - 12x - 10}{(2x - 1)^2}$$
$$= \frac{2x^2 - 2x - 16}{(2x - 1)^2}$$

19. $f(x) = \dfrac{6 + 2x^{-1}}{3x - 1}$

$$f'(x) = \frac{(3x - 1)(-2x^{-2}) - (6 + 2x^{-1})(3)}{(3x - 1)^2}$$
$$= \frac{-6x^{-1} + 2x^{-2} - 18 - 6x^{-1}}{(3x - 1)^2}$$
$$= \frac{2x^{-2} - 12x^{-1} - 18}{(3x - 1)^2}$$
$$= \frac{\frac{2}{x^2} - \frac{12}{x} - 18}{(3x - 1)^2}$$
$$= \frac{2 - 12x - 18x^2}{x^2(3x - 1)^2}$$

	Function	*Rewrite*	*Differentiate*	*Simplify*
21.	$f(x) = \dfrac{x^3 + 6x}{3}$	$f(x) = \dfrac{1}{3}x^3 + 2x$	$f'(x) = x^2 + 2$	$f'(x) = x^2 + 2$
23.	$y = \dfrac{x^2 + 2x}{3}$	$y = \dfrac{1}{3}(x^2 + 2x)$	$y' = \dfrac{1}{3}(2x + 2)$	$y' = \dfrac{2}{3}(x + 1)$
25.	$y = \dfrac{7}{3x^3}$	$y = \dfrac{7}{3}x^{-3}$	$y' = -7x^{-4}$	$y' = -\dfrac{7}{x^4}$
27.	$y = \dfrac{4x^2 - 3x}{8\sqrt{x}}$	$y = \dfrac{1}{2}x^{3/2} - \dfrac{3}{8}x^{1/2},\ x \neq 0$	$y' = \dfrac{3}{4}x^{1/2} - \dfrac{3}{16}x^{-1/2}$	$y' = \dfrac{3}{4}\sqrt{x} - \dfrac{3}{16\sqrt{x}}$
29.	$y = \dfrac{x^2 - 4x + 3}{2(x - 1)}$	$y = \dfrac{1}{2}(x - 3),\ x \neq 1$	$y' = \dfrac{1}{2}(1),\ x \neq 1$	$y' = \dfrac{1}{2},\ x \neq 1$

31. $f'(x) = (x^3 - 3x)(4x + 3) + (3x^2 - 3)(2x^2 + 3x + 5)$

$$= 4x^4 + 3x^3 - 12x^2 - 9x + 6x^4 + 9x^3 + 9x^2 - 9x - 15$$
$$= 10x^4 + 12x^3 - 3x^2 - 18x - 15$$

Product Rule

33. $f'(x) = \dfrac{(x^2 - 1)(3x^2 + 3) - (x^3 + 3x + 2)(2x)}{(x^2 - 1)^2}$

$$= \frac{3x^4 - 3 - 2x^4 - 6x^2 - 4x}{(x^2 - 1)^2}$$
$$= \frac{x^4 - 6x^2 - 4x - 3}{(x^2 - 1)^2}$$

Quotient Rule

35. $f(x) = \dfrac{x^2 - x - 20}{x + 4} = \dfrac{(x - 5)(x + 4)}{x + 4} = x - 5,\ x \neq -4$

$f'(x) = 1,\ x \neq -4$

Simple Power Rule

37. $g(t) = (2t^3 - 1)^2 = (2t^3 - 1)(2t^3 - 1)$

$g'(t) = (2t^3 - 1)(6t^2) + (2t^3 - 1)(6t^2) = 12t^2(2t^3 - 1)$

Product Rule

39. $g(s) = \dfrac{s^2 - 2s + 5}{\sqrt{s}} = \dfrac{s^2 - 2s + 5}{s^{1/2}}$

$$g'(s) = \frac{s^{1/2}(2s - 2) - (s^2 - 2s + 5)\left(\frac{1}{2}s^{-1/2}\right)}{s}$$

$$= \frac{2s^{3/2} - 2s^{1/2} - \frac{1}{2}s^{3/2} + s^{1/2} - \frac{5}{2}s^{-1/2}}{s}$$

$$= \frac{3}{2}s^{1/2} - s^{-1/2} - \frac{5}{2}s^{-3/2} = \frac{3s^2 - 2s - 5}{2s^{3/2}}$$

Quotient Rule

41. $f(x) = \dfrac{(x - 2)(3x + 1)}{4x + 2} = \dfrac{3x^2 - 5x - 2}{4x + 2}$

$$f'(x) = \frac{(4x + 2)(6x - 5) - (3x^2 - 5x - 2)(4)}{(4x + 2)^2}$$

$$= \frac{24x^2 - 8x - 10 - 12x^2 + 20x + 8}{(4x + 2)^2}$$

$$= \frac{12x^2 + 12x - 2}{4(2x + 1)^2}$$

$$= \frac{2(6x^2 + 6x - 1)}{2(2x + 1)^2}$$

$$= \frac{6x^2 + 6x - 1}{2(2x + 1)^2}$$

Quotient Rule

43. $g(x) = \dfrac{x - 3}{x + 4}(x^2 + 2x + 1) = \dfrac{x^3 - x^2 - 5x - 3}{x + 4}$

$$g'(x) = \frac{(x + 4)(3x^2 - 2x - 5) - (x^3 - x^2 - 5x - 3)(1)}{(x + 4)^2}$$

$$= \frac{3x^3 + 10x^2 - 13x - 20 - x^3 + x^2 + 5x + 3}{(x + 4)^2}$$

$$= \frac{2x^3 + 11x^2 - 8x - 17}{(x + 4)^2}$$

Quotient Rule

45. $f(x) = (5x + 2)(x^2 + x)$

$$f'(x) = (5x + 2)(2x + 1) + (x^2 + x)(5)$$

$$= 10x^2 + 9x + 2 + 5x^2 + 5x$$

$$= 15x^2 + 14x + 2$$

$$m = f'(-1) = 3$$

$$y - 0 = 3(x - (-1))$$

$$y = 3x + 3$$

47. $f(x) = x^3(x^2 - 4)$

$$f'(x) = x^2(2x) + (x^2 - 4)(3x^2)$$

$$= 2x^4 + 3x^4 - 12x^2$$

$$= 5x^4 - 12x^2$$

$$m = f'(1) = -7$$

$$y - (-3) = -7(x - 1)$$

$$y = -7x + 4$$

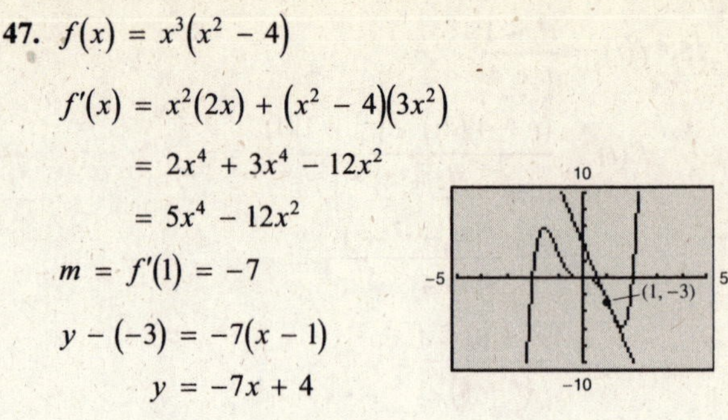

49. $f'(x) = \dfrac{(x + 1) - (x - 2)}{(x + 1)^2} = \dfrac{3}{(x + 1)^2}$

$$f'(1) = \frac{3}{4}$$

$$y + \frac{1}{2} = \frac{3}{4}(x - 1)$$

$$y = \frac{3}{4}x - \frac{5}{4}$$

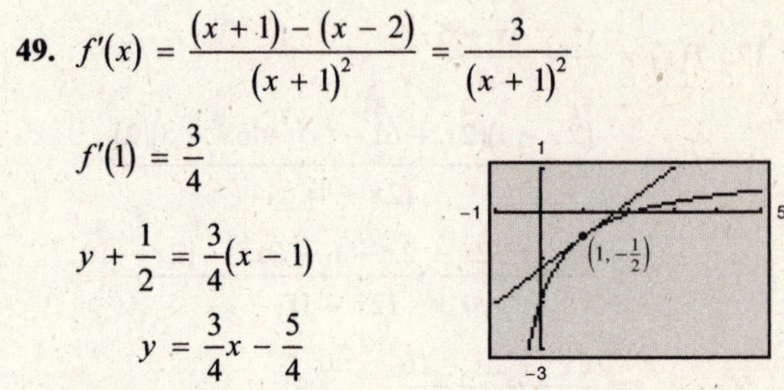

51. $f(x) = \dfrac{(3x - 2)(6x + 5)}{2x - 3} = \dfrac{18x^2 + 3x - 10}{2x - 3}$

$$f'(x) = \frac{(2x - 3)(36x + 3) - (18x^2 + 3x - 10)(2)}{(2x - 3)^2}$$

$$= \frac{72x^2 - 102x - 9 - 36x^2 - 6x - 20}{(2x - 3)^2}$$

$$= \frac{36x^2 - 108x + 11}{(2x - 3)^2}$$

$$m = f'(-1) = \frac{36(-1)^2 - 108(-1) + 11}{(2(-1) - 3)^2} = \frac{31}{5}$$

$$y - (-1) = \frac{31}{5}(x - (-1))$$

$$y = \frac{31}{5}x + \frac{26}{5}$$

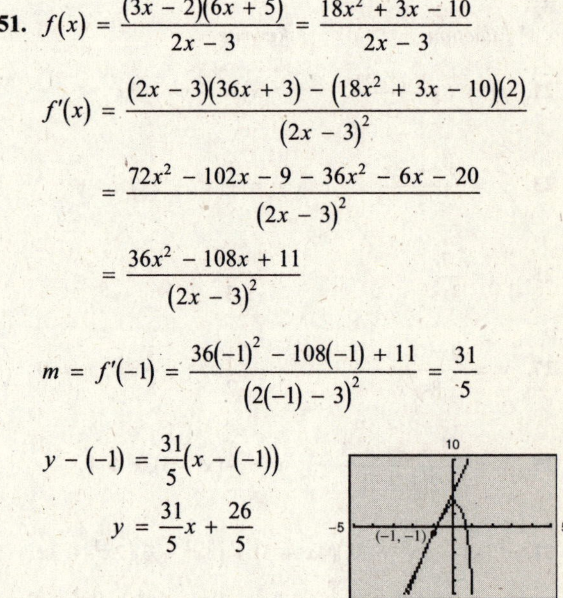

53. $f'(x) = \dfrac{(x - 1)(2x) - x^2(1)}{(x - 1)^2} = \dfrac{x^2 - 2x}{(x - 1)^2}$

$f'(x) = 0$ when $x^2 - 2x = x(x - 2) = 0$, which implies that $x = 0$ or $x = 2$. Thus, the horizontal tangent lines occur at $(0, 0)$ and $(2, 4)$.

55. $f'(x) = \dfrac{(x^3+1)(4x^3) - x^4(3x^2)}{(x^3+1)^2} = \dfrac{x^6+4x^3}{(x^3+1)^2}$

$f'(x) = 0$ when $x^6 + 4x^3 = x^3(x^3+4) = 0$, which implies that $x = 0$ or $x = \sqrt[3]{-4}$. Thus, the horizontal tangent lines occur at $(0, 0)$ and $(\sqrt[3]{-4}, -2.117)$.

57. $f(x) = x(x+1) = x^2 + x$

$f'(x) = 2x + 1$

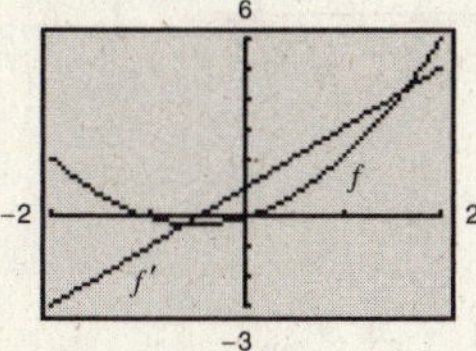

59. $f(x) = x(x+1)(x-1)$

$= x^3 - x$

$f'(x) = 3x^2 - 1$

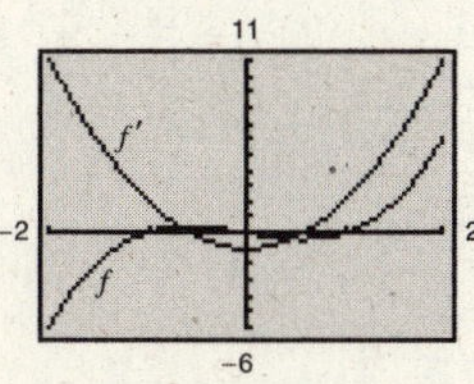

61. $x = 275\left(1 - \dfrac{3p}{5p+1}\right)$

$\dfrac{dx}{dp} = -275\left[\dfrac{(5p+1)3 - (3p)(5)}{(5p+1)^2}\right] = -275\left[\dfrac{3}{(5p+1)^2}\right]$

When $p = 4$, $\dfrac{dx}{dp} = -275\left[\dfrac{3}{(21)^2}\right] \approx -\$1.87$ per unit.

63. $P' = 500\left[\dfrac{(50+t^2)(4) - (4t)(2t)}{(50+t^2)^2}\right] = 500\left[\dfrac{200 - 4t^2}{(50+t^2)^2}\right]$

When $t = 2$, $P' = 500\left[\dfrac{184}{(54)^2}\right] \approx 31.55$ bacteria/hour.

65. $f(t) = \dfrac{t^2 - t + 1}{t^2 + 1}$

$f'(t) = \dfrac{(t^2+1)(2t-1) - (t^2-t+1)(2t)}{(t^2+1)^2} = \dfrac{t^2-1}{(t^2+1)^2}$

(a) $f'(0.5) = -0.480$/week

(b) $f'(2) = 0.120$/week

(c) $f'(8) = 0.015$/week

Each rate in parts (a), (b), and (c) is the rate at which the level of oxygen in the pond is changing at that particular time.

67. $C = x^3 - 15x^2 + 87x - 73,\ 4 \le x \le 9$

Marginal cost: $\dfrac{dC}{dx} = 3x^2 - 30x + 87$

Average cost: $\dfrac{C}{x} = x^2 - 15x + 87 - \dfrac{73}{x}$

(a)

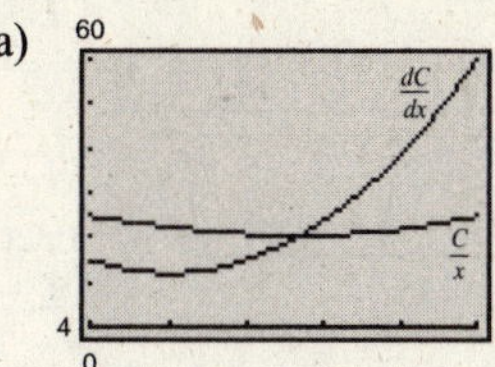

(b) Point of intersection:

$3x^2 - 30x + 87 = x^2 - 15x + 87 - \dfrac{73}{x}$

$2x^2 - 15x + \dfrac{73}{x} = 0$

$2x^3 - 15x^2 + 73 = 0$

$x \approx 6.683$

When $x = 6.683$, $\dfrac{C}{x} = \dfrac{dC}{dx} \approx 20.50$.

Thus, the point of intersection is (6.683, 20.50). At this point average cost is at a minimum.

69. $C = 100\left(\dfrac{200}{x^2} + \dfrac{x}{x+30}\right),\ x \ge 1$

$C' = 100\left[-2(200x^{-3}) + \dfrac{(x+30) - x}{(x+30)^2}\right]$

$= 100\left[-\dfrac{400}{x^3} + \dfrac{30}{(x+30)^2}\right]$

(a) $C'(10) = 100\left(-\dfrac{400}{10^3} + \dfrac{30}{40^2}\right) = -38.125$

(b) $C'(15) \approx -10.37$

(c) $C'(20) \approx -3.8$

Increasing the order size reduces the cost per item. An order size of 2000 should be chosen since the cost per item is the smallest at $x = 20$.

71. $f(x) = 2g(x) + h(x)$

$f'(x) = 2g'(x) + h'(x)$

$f'(2) = 2(-2) + 4 = 0$

73. $f(x) = g(x)h(x)$

$f'(x) = g(x)h'(x) + h(x)g'(x)$

$f'(2) = g(2)h'(2) + h(2)g'(2)$

$= (3)(4) + (-1)(-2)$

$= 14$

75. Answers will vary.

Chapter 2 Quiz Yourself

1. $f(x) = 5x + 3$

$$f'(x) = \lim_{\Delta x \to 0} \frac{f(x + \Delta x) - f(x)}{\Delta x}$$
$$= \lim_{\Delta x \to 0} \frac{[5(x + \Delta x) + 3] - (5x + 3)}{\Delta x}$$
$$= \lim_{\Delta x \to 0} \frac{5x + 5\Delta x + 3 - 5x - 3}{\Delta x}$$
$$= \lim_{\Delta x \to 0} \frac{5\Delta x}{\Delta x}$$
$$= \lim_{\Delta x \to 0} 5 = 5$$

At $(-2, -7)$: $m = 5$

2. $f(x) = \sqrt{x + 3}$

$$f'(x) = \lim_{\Delta x \to 0} \frac{f(x + \Delta x) - f(x)}{\Delta x}$$
$$= \lim_{\Delta x \to 0} \frac{\sqrt{x + \Delta x + 3} - \sqrt{x + 3}}{\Delta x}$$
$$= \lim_{\Delta x \to 0} \frac{x + \Delta x + 3 - (x + 3)}{\Delta x\left(\sqrt{x + \Delta x + 3} + \sqrt{x + 3}\right)}$$
$$= \lim_{\Delta x \to 0} \frac{1}{\sqrt{x + \Delta x + 3} + \sqrt{x + 3}}$$
$$= \frac{1}{2\sqrt{x + 3}}$$

At $(1, 2)$: $m = \dfrac{1}{2\sqrt{1 + 3}} = \dfrac{1}{4}$

3. $f(x) = x^2 - 2x$

$$f'(x) = \lim_{\Delta x \to 0} \frac{f(x + \Delta x) - f(x)}{\Delta x}$$
$$= \lim_{\Delta x \to 0} \frac{\left[(x + \Delta x)^2 - 2(x + \Delta x)\right] - (x^2 - 2x)}{\Delta x}$$
$$= \lim_{\Delta x \to 0} \frac{x^2 + 2x(\Delta x) + (\Delta x)^2 - 2x - 2\Delta x - x^2 + 2x}{\Delta x}$$
$$= \lim_{\Delta x \to 0} \frac{\Delta x(2x + \Delta x - 2)}{\Delta x}$$
$$= \lim_{\Delta x \to 0} (2x + \Delta x - 2) = 2x - 2$$

At $(3, 3)$: $m = 2(3) = -2 = 4$

4. $f(x) = 12$

$f'(x) = 0$

5. $f(x) = 19x + 9$

$f'(x) = 19$

6. $f(x) = 5 - 3x^2$

$f'(x) = -6x$

7. $f(x) = 12x^{1/4}$

$f'(x) = 3x^{-3/4} = \dfrac{3}{x^{3/4}}$

8. $f(x) = 4x^{-2}$

$f'(x) = -8x^{-3} = -\dfrac{8}{x^3}$

9. $f(x) = 2\sqrt{x} = 2x^{1/2}$

$f'(x) = x^{-1/2} = \dfrac{1}{\sqrt{x}}$

10. $f(x) = \dfrac{2x + 3}{3x + 2}$

$f'(x) = \dfrac{(3x + 2)(2) - (2x + 3)(3)}{(3x + 2)^2} = -\dfrac{5}{(3x + 2)^2}$

11. $f(x) = (x^2 + 1)(-2x + 4)$

$f'(x) = (x^2 + 1)(-2) + (-2x + 4)(2x)$

$f'(x) = -6x^2 + 8x - 2$

12. $f(x) = (x^2 + 3x + 4)(5x - 2)$

$$f'(x) = (x^2 + 3x + 4)(5) + (5x - 2)(2x + 3)$$
$$= 5x^2 + 15x + 20 + 10x^2 + 11x - 6$$
$$= 15x^2 + 26x + 14$$

13. $f(x) = \dfrac{4x}{x^2 + 3}$

$$f'(x) = \frac{(x^2 + 3)(4) - 4x(2x)}{(x^2 + 3)^2}$$
$$= \frac{4x^2 + 12 - 8x^2}{(x^2 + 3)^2}$$
$$= \frac{-4x^2 + 12}{(x^2 + 3)^2}$$
$$= \frac{-4(x^2 - 3)}{(x^2 + 3)^2}$$

14.

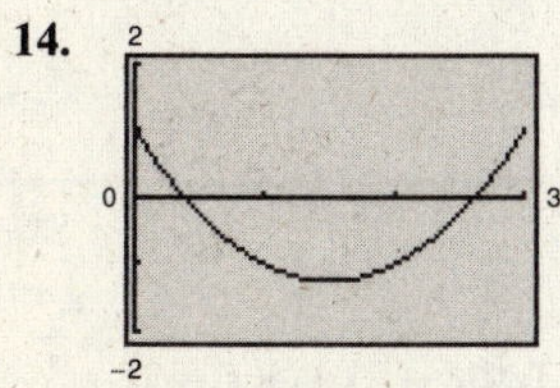

Average rate of change:

$$\frac{\Delta y}{\Delta x} = \frac{f(3) - f(0)}{3 - 0} = \frac{1 - 1}{3} = 0$$

$f'(x) = 2x - 3$

Instantaneous rates of change: $f'(0) = -3,\ f'(3) = 3$

15.

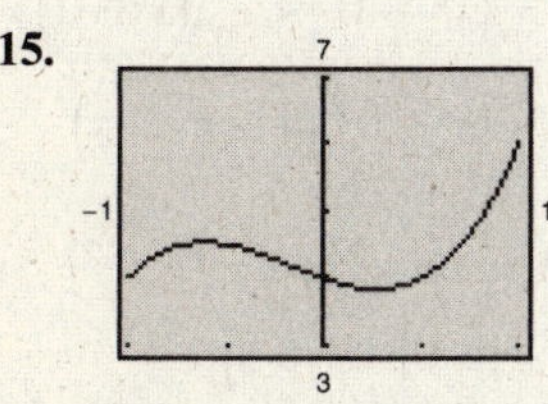

Average rate of change:

$$\frac{\Delta y}{\Delta x} = \frac{f(1) - f(-1)}{1 - (-1)} = \frac{6 - 4}{2} = 1$$

$f'(x) = 6x^2 + 2x - 1$

Instantaneous rates of change: $f'(-1) = 3,\ f'(1) = 7$

16.

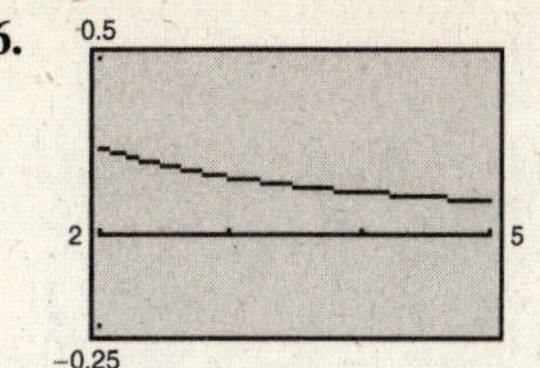

Average rate of change:

$$\frac{\Delta y}{\Delta x} = \frac{f(5) - f(2)}{5 - 2} = \frac{\frac{1}{10} - \frac{1}{4}}{3} = -\frac{1}{20}$$

$f(x) = \dfrac{1}{2x} = \dfrac{1}{2}x^{-1}$

$f'(x) = -\dfrac{1}{2x^2}$

Instantaneous rates of change:

$f'(2) = -\dfrac{1}{8},\ f'(5) = -\dfrac{1}{50}$

17.

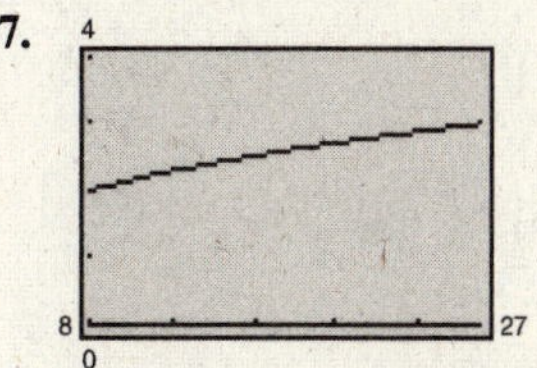

Average rate of change:

$$\frac{\Delta y}{\Delta x} = \frac{f(27) - f(8)}{27 - 8} = \frac{3 - 2}{19} = \frac{1}{19}$$

$f(x) = \sqrt[3]{x} = x^{1/3}$

$f'(x) = \dfrac{1}{3}x^{-2/3} = \dfrac{1}{3x^{2/3}}$

Instantaneous rates of change:

$f'(8) = \dfrac{1}{12},\ f'(27) = \dfrac{1}{27}$

18. $P = -0.0125x^2 + 16x - 600$

(a) $P(176) - P(175) = 1828.8 - 1817.1875$
$= \$11.6125$

(b) $\dfrac{dP}{dx} = -0.025x + 16$

When $x = 175$, $\dfrac{dP}{dx} = 11.625$

(c) The results are very similar.

19. $f(x) = 5x^2 + 6x - 1$

$f'(x) = 10x + 6$

At $(-1, -2)$, $m = -4$.

$y + 2 = -4(x + 1)$

$y = -4x - 6$

20. $f(x) = (x^2 + 1)(4x - 3)$

$f'(x) = (x^2 + 1)(4) + (4x - 3)(2x)$

$= 4x^2 + 4 + 8x^2 - 6x$

$= 12x^2 - 6x + 4$

$m = f'(1) = 12(1)^2 - 6(1) + 4 = 10$

$y - 2 = 10(x - 1)$

$y = 10x - 8$

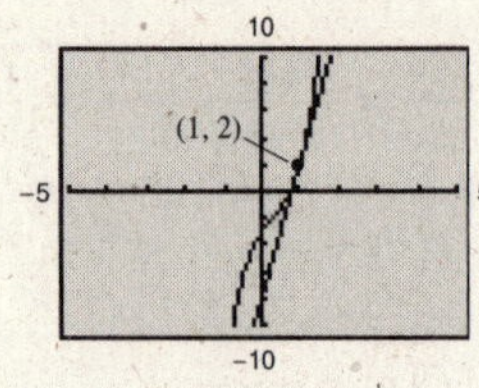

21. $S = -0.13556t^3 + 1.8682t^2 - 4.351t + 23.52$

(a) $\dfrac{dS}{dt} = S'(t) = -0.40668t^2 + 3.7364t - 4.351$

(b) 2004: $S'(4) \approx \$4.088/\text{yr}$

2007: $S'(7) \approx \$1.876/\text{yr}$

2008: $S'(8) \approx -\$0.487/\text{yr}$

Section 2.5 The Chain Rule

Skills Warm Up

1. $\sqrt[5]{(1 - 5x)^2} = (1 - 5x)^{2/5}$

2. $\sqrt[4]{(2x - 1)^3} = (2x - 1)^{3/4}$

3. $\dfrac{1}{\sqrt{4x^2 + 1}} = (4x^2 + 1)^{-1/2}$

4. $\dfrac{1}{\sqrt[3]{x - 6}} = (x - 6)^{-1/3}$

5. $\dfrac{\sqrt{x}}{\sqrt[3]{1 - 2x}} = x^{1/2}(1 - 2x)^{-1/3}$

6. $\dfrac{\sqrt{(3 - 7x)^3}}{2x} = \dfrac{(3 - 7x)^{3/2}}{2x} = (2x)^{-1}(3 - 7x)^{3/2}$

7. $3x^3 - 6x^2 + 5x - 10 = 3x^2(x - 2) + 5(x - 2)$

$= (3x^2 + 5)(x - 2)$

8. $5x\sqrt{x} - x - 5\sqrt{x} + 1 = x(5\sqrt{x} - 1) - 1(5\sqrt{x} - 1)$

$= (x - 1)(5\sqrt{x} - 1)$

9. $4(x^2 + 1)^2 - x(x^2 + 1)^3 = (x^2 + 1)^2\left[4 - x(x^2 + 1)\right]$

$= (x^2 + 1)^2(4 - x^3 - x)$

10. $-x^5 + 3x^3 + x^2 - 3 = x^3(-x^2 + 3) - 1(-x^2 + 3)$

$= (x^3 - 1)(-x^2 + 3)$

$= (x - 1)(x^2 + x + 1)(3 - x^2)$

	$y = f(g(x))$	$u = g(x)$	$y = f(u)$
1.	$y = (6x - 5)^4$	$u = 6x - 5$	$y = u^4$
3.	$y = \sqrt{5x - 2}$	$u = 5x - 2$	$y = \sqrt{u}$
5.	$y = (3x + 1)^{-1}$	$u = 3x + 1$	$y = u^{-1}$

7. $\dfrac{dy}{du} = 2u$

$\dfrac{du}{dx} = 4$

$\dfrac{dy}{dx} = (2u)(4) = 8(4x + 7) = 32x + 56$

9. $\dfrac{dy}{du} = \dfrac{1}{2}u^{-1/2}$

$\dfrac{du}{dx} = -2x$

$\dfrac{dy}{dx} = \left(\dfrac{1}{2}u^{-1/2}\right)(-2x)$

$= -(3 - x^2)^{-1/2}(x)$

$= -\dfrac{x}{\sqrt{3 - x^2}}$

11. $\dfrac{dy}{du} = \dfrac{2}{3}u^{-1/3}$

$\dfrac{du}{dx} = 20x^3 - 2$

$\dfrac{dy}{dx} = \dfrac{2}{3}u^{-1/3}\left(20x^3 - 2\right)$

$= \dfrac{2}{3}\left(5x^4 - 2x\right)^{-1/3}\left(20x^3 - 2\right)$

$= \dfrac{4\left(10x^3 - 1\right)}{3\sqrt[3]{5x^4 - 2x}}$

13. $f(x) = \dfrac{2}{1 - x^3} = 2\left(1 - x^3\right)^{-1}$; (c) General Power Rule

15. $f(x) = \sqrt[3]{8^2}$; (b) Constant Rule.

17. $f(x) = \dfrac{x^2 + 2}{x} = x + 2x^{-1}$; (a) Simple Power Rule

19. $f(x) = \dfrac{2}{x - 2} = 2(x - 2)^{-1}$;

(a) General Power Rule

21. $y' = 3(2x - 7)^2(2) = 6(2x - 7)^2$

23. $f(x) = \left(5x - x^2\right)^{3/2}$

$f'(x) = \frac{3}{2}\left(5x - x^2\right)^{1/2}(5 - 2x) = \frac{3}{2}(5 - 2x)\sqrt{5x - x^2}$

25. $h'(x) = 2\left(6x - x^3\right)\left(6 - 3x^2\right) = 6x\left(6 - x^2\right)\left(2 - x^2\right)$

27. $f(t) = \sqrt{t + 1} = (t + 1)^{1/2}$

$f'(t) = \dfrac{1}{2}(t + 1)^{-1/2}(1) = \dfrac{1}{2\sqrt{t + 1}}$

29. $s(t) = \sqrt{2t^2 + 5t + 2} = \left(2t^2 + 5t + 2\right)^{1/2}$

$s'(t) = \dfrac{1}{2}\left(2t^2 + 5t + 2\right)^{-1/2}(4t + 5) = \dfrac{4t + 5}{2\sqrt{2t^2 + 5t + 2}}$

31. $y = \sqrt[3]{9x^2 + 4} = \left(9x^2 + 4\right)^{1/3}$

$y' = \dfrac{1}{3}\left(9x^2 + 4\right)^{-2/3}(18x) = \dfrac{6x}{\left(9x^2 + 4\right)^{2/3}}$

33. $f(x) = 2(2 - 9x)^{-3}$

$f'(x) = 2(-3)(2 - 9x)^{-4}(-9) = \dfrac{54}{(2 - 9x)^4}$

35. $f(x) = \left(x^2 + 25\right)^{-1/2}$

$f'(x) = \left(-\dfrac{1}{2}\right)\left(x^2 + 25\right)^{-3/2}(2x) = -\dfrac{x}{\left(x^2 + 25\right)^{3/2}}$

37. $f'(x) = 2(3)\left(x^2 - 1\right)^2(2x) = 12x\left(x^2 - 1\right)^2$

$f'(2) = 24\left(3^2\right) = 216$

$f(2) = 54$

$y - 54 = 216(x - 2)$

$y = 216x - 378$

39. $f(x) = \sqrt{4x^2 - 7} = \left(4x^2 - 7\right)^{1/2}$

$f'(x) = \dfrac{1}{2}\left(4x^2 - 7\right)^{-1/2}(8x) = \dfrac{4x}{\sqrt{4x^2 - 7}}$

$f'(2) = \dfrac{8}{3}$

$f(2) = 3$

$y - 3 = \dfrac{8}{3}(x - 2)$

$y = \dfrac{8}{3}x - \dfrac{7}{3}$

41. $f(x) = \sqrt{x^2 - 2x + 1} = \left(x^2 - 2x + 1\right)^{1/2}$

$f'(x) = \dfrac{1}{2}\left(x^2 - 2x + 1\right)^{-1/2}(2x - 2)$

$= \dfrac{x - 1}{\sqrt{x^2 - 2x + 1}}$

$= \dfrac{x - 1}{|x - 1|}$

$f'(2) = 1$

$f(2) = 1$

$y - 1 = 1(x - 2)$

$y = x - 1$

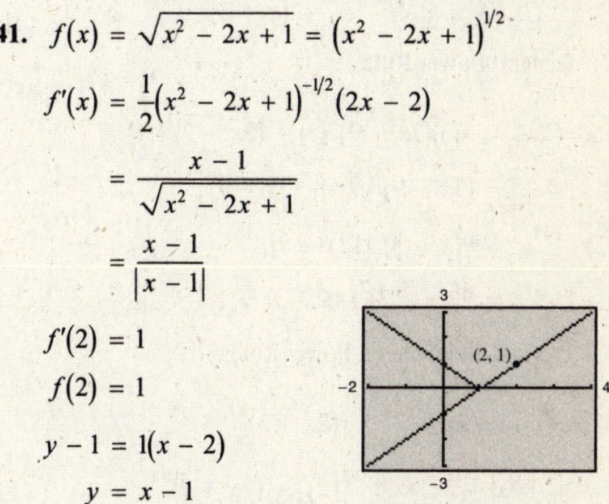

43. $f'(x) = \dfrac{1 - 3x^2 - 4x^{3/2}}{2\sqrt{x}\left(x^2 + 1\right)^2}$

f has a horizontal tangent when $f' = 0$.

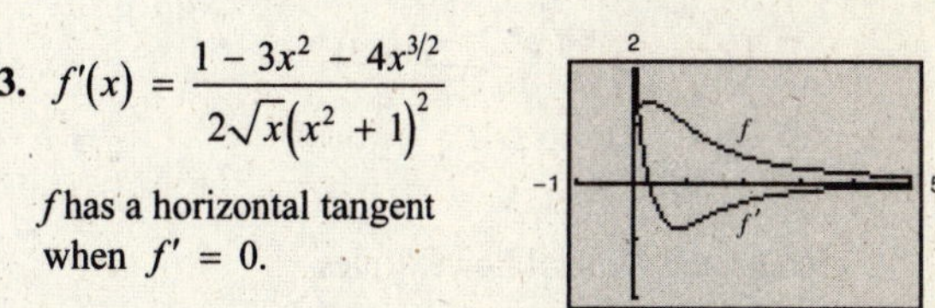

45. $f'(x) = -\dfrac{1}{2x^{3/2}\sqrt{x + 1}}$

f' is never 0.

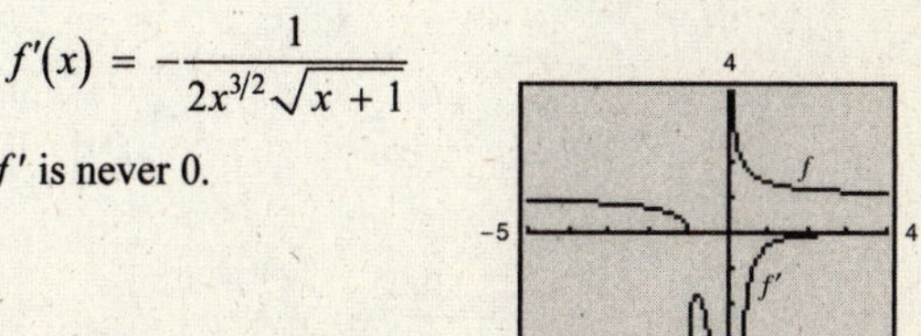

47. $y = (4 - x^2)^{-1}$

$y' = (-1)(4 - x^2)^{-2}(-2x)$

$= \dfrac{2x}{(4 - x^2)^2}$

General Power Rule

49. $y = -\dfrac{4}{(t + 2)^2} = -4(t + 2)^{-2}$

$y' = 8(t + 2)^{-3} = \dfrac{8}{(t + 2)^3}$

General Power Rule

51. $f(x) = (2x - 1)(9 - 3x^2)$

$f'(x) = (2x - 1)(-6x) + (9 - 3x^2)(2)$

$= -12x^2 + 6x + 18 - 6x^2$

$= 18 + 6x - 18x^2$

$= -6(3x^2 - 2x - 3)$

Product Rule

53. $y = \dfrac{1}{\sqrt{x + 2}} = (x + 2)^{-1/2}$

$y' = -\dfrac{1}{2}(x + 2)^{-3/2} = -\dfrac{1}{2(x + 2)^{3/2}}$

General Power Rule

55. $f'(x) = x(3)(3x - 9)^2(3) + (3x - 9)^3(1)$

$= (3x - 9)^2[9x + (3x - 9)]$

$= 9(x - 3)^2(12x - 9)$

$= 27(x - 3)^2(4x - 3)$

Product and General Power Rules

57. $y = x\sqrt{2x + 3} = x(2x + 3)^{1/2}$

$y' = x\left[\dfrac{1}{2}(2x + 3)^{-1/2}(2)\right] + (2x + 3)^{1/2}$

$= (2x + 3)^{-1/2}[x + (2x + 3)]$

$= \dfrac{3(x + 1)}{\sqrt{2x + 3}}$

Product and General Power Rules

59. $y = t^2\sqrt{t - 2} = t^2(t - 2)^{1/2}$

$y' = t^2\left[\dfrac{1}{2}(t - 2)^{-1/2}(1)\right] + 2t(t - 2)^{1/2}$

$= \dfrac{1}{2}(t - 2)^{-1/2}[t^2 + 4t(t - 2)]$

$= \dfrac{t^2 + 4t(t - 2)}{2\sqrt{t - 2}}$

$= \dfrac{t(5t - 8)}{2\sqrt{t - 2}}$

Product and General Power Rules

61. $y = \left(\dfrac{6 - 5x}{x^2 - 1}\right)^2$

$y' = 2\left(\dfrac{6 - 5x}{x^2 - 1}\right)\left[\dfrac{(x^2 - 1)(-5) - (6 - 5x)(2x)}{(x^2 - 1)^2}\right]$

$= \dfrac{2(6 - 5x)(5x^2 - 12x + 5)}{(x^2 - 1)^3}$

Quotient and General Power Rules

63. $f(t) = \dfrac{36}{(3 - t)^2} = 36(3 - t)^{-2}$

$f'(t) = -72(3 - t)^{-3}(-1) = \dfrac{72}{(3 - t)^3}$

$f'(0) = \dfrac{72}{27} = \dfrac{8}{3}$

$y - 4 = \dfrac{8}{3}(t - 0)$

$y = \dfrac{8}{3}t + 4$

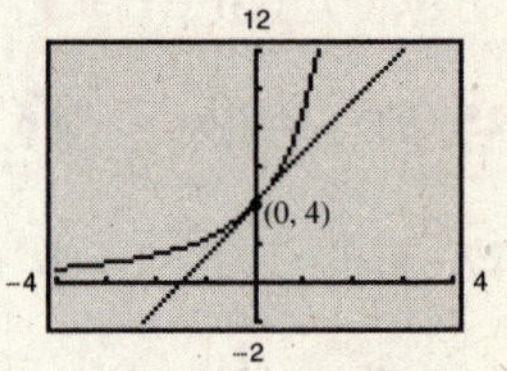

65. $f(x) = (3x^3 + 4x)^{1/5}$

$f'(x) = \dfrac{1}{5}(3x^3 + 4x)^{-4/5}(9x^2 + 4)$

$= \dfrac{9x^2 + 4}{5(3x^3 + 4x)^{4/5}}$

$m = f'(2) = \dfrac{1}{2}$

$y - 2 = \dfrac{1}{2}(x - 2)$

$y = \dfrac{1}{2}x + 1$

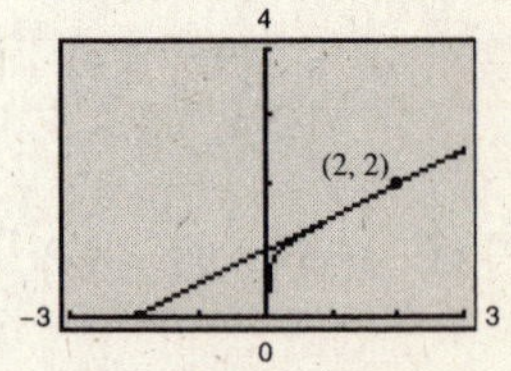

67. $f(t) = (t^2 - 9)\sqrt{t+2} = (t^2 - 9)(t+2)^{1/2}$

$$f'(t) = (t^2 - 9)\left[\frac{1}{2}(t+2)^{-1/2}\right] + (t+2)^{1/2}(2t)$$

$$= \frac{1}{2}(t^2 - 9)(t+2)^{-1/2} + 2t(t+2)^{1/2}$$

$$= (t+2)^{-1/2}\left[\frac{1}{2}(t^2 - 9) + 2t(t+2)\right]$$

$$= (t+2)^{-1/2}\left[\frac{1}{2}t^2 - \frac{9}{2} + 2t^2 + 4t\right]$$

$$= (t+2)^{-1/2}\left(\frac{5}{2}t^2 + 4t - \frac{9}{2}\right)$$

$$= \frac{\frac{5}{2}t^2 + 4t - \frac{9}{2}}{\sqrt{t+2}}$$

$f'(-1) = -6$

$y - (-8) = -6[t - (-1)]$

$y = -6t - 14$

69. $f(x) = \dfrac{x+1}{\sqrt{2x-3}} = \dfrac{x+1}{(2x-3)^{1/2}}$

$$f'(x) = \frac{(2x-3)^{1/2}(1) - (x+1)\left(\frac{1}{2}\right)(2x-3)^{-1/2}(2)}{(2x-3)}$$

$$= \frac{(2x-3) - (x+1)}{(2x-3)^{3/2}}$$

$$= \frac{x-4}{(2x-3)^{3/2}}$$

$f'(2) = \dfrac{1-3}{1} = -2$

$y - 3 = -2(x - 2)$

$y = -2x + 7$

71. $A' = 1000(60)\left(1 + \dfrac{r}{12}\right)^{59}\left(\dfrac{1}{12}\right) = 5000\left(1 + \dfrac{r}{12}\right)^{59}$

(a) $A'(0.08) = 50\left(1 + \dfrac{0.08}{12}\right)^{59}$

$\approx$ \$74.00 per percentage point

(b) $A'(0.10) = 50\left(1 + \dfrac{0.10}{12}\right)^{59}$

$\approx$ \$81.59 per percentage point

(c) $A'(0.12) = 50\left(1 + \dfrac{0.12}{12}\right)^{59}$

$\approx$ \$89.94 per percentage point

73. (a) $V = \dfrac{k}{\sqrt{t+1}}$

When $t = 0$, $V = 10{,}000$.

$10{,}000 = \dfrac{k}{\sqrt{0+1}} \Rightarrow k = 10{,}000$

$V = \dfrac{10{,}000}{\sqrt{t+1}}$

(b) $V = 10{,}000(t+1)^{-1/2}$

$\dfrac{dV}{dt} = -5000(t+1)^{-3/2}(1) = -\dfrac{5000}{(t+1)^{3/2}}$

When $t = 1$, $\dfrac{dV}{dt} = -\dfrac{5000}{(2)^{3/2}}$

$= -\dfrac{2500}{\sqrt{2}}$

$\approx -\$1767.77$ per year.

(c) When $t = 3$, $\dfrac{dV}{dt} = -\dfrac{5000}{(4)^{3/2}}$

$= -\$625.00$ per year.

75. (a) $r = \left(2.8557t^4 - 72.792t^3 + 676.14t^2 - 2706t + 4096\right)^{1/2}$

$$\frac{dr}{dt} = r'(t) = \frac{1}{2}\left(2.8557t^4 - 72.792t^3 + 676.14t^2 - 2706t + 4096\right)^{-1/2} \cdot \left(11.4228t^3 - 218.376t^2 + 1352.28t - 2706\right)$$

$$= \frac{11.4228t^3 - 218.376t^2 + 1352.28t - 2706}{2\sqrt{2.8557t^4 - 72.792t^3 + 676.14t^2 - 2706t + 4096}}$$

Chain Rule

(b)

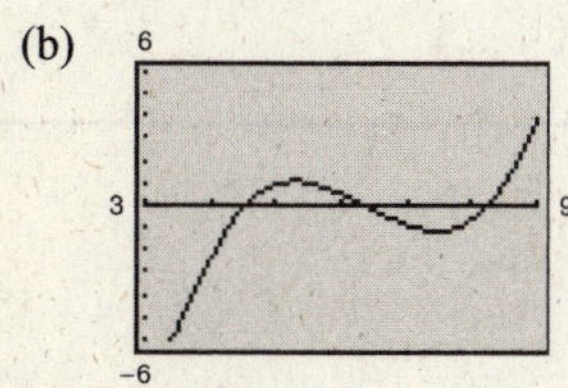

(c) $t = 3$ yr

(d) $t \approx 4.52$ yr, $t \approx 6.36$ yr, $t \approx 8.24$ yr

Section 2.6 Higher-Order Derivatives

Skills Warm Up

1. $-16t^2 + 24t = 0$

$t(-16t + 24) = 0$

$t = 0$

$-16t + 24 = 0 \rightarrow t = 1.5$

2. $-16t^2 + 80t + 224 = 0$

$-16(t^2 - 5t - 14) = 0$

$-16(t - 7)(t + 2) = 0$

$t - 7 = 0 \rightarrow t = 7$

$t + 2 = 0 \rightarrow t = -2$

3. $-16t^2 + 128t + 320 = 0$

$-16(t^2 - 8t - 20) = 0$

$-16(t - 10)(t + 2) = 0$

$t - 10 = 0 \rightarrow t = 10$

$t + 2 = 0 \rightarrow t = -2$

4. $-16t^2 + 9t + 1440 = 0$

$$t = \frac{-9 \pm \sqrt{9^2 - 4(-16)(1440)}}{2(-16)}$$

$$= \frac{-9 \pm \sqrt{92241}}{-32}$$

$$= \frac{9 \pm 3\sqrt{10249}}{32}$$

$t \approx -9.21$ and $t \approx 9.77$

5. $y = x^2(2x + 7)$

$$\frac{dy}{dx} = x^2(2) + 2x(2x + 7)$$

$$= 2x^2 + 4x^2 + 14x$$

$$= 6x^2 + 14x$$

6. $y = (x^2 + 3x)(2x^2 - 5)$

$$\frac{dy}{dx} = (x^2 + 3x)(4x) + (2x + 3)(2x^2 - 5)$$

$$= 4x^3 + 12x^2 + 4x^3 - 10x + 6x^2 - 15$$

$$= 8x^3 + 18x^2 - 10x - 15$$

7. $y = \dfrac{x^2}{2x + 7}$

$$\frac{dy}{dx} = \frac{(2x + 7)(2x) - (x^2)(2)}{(2x + 7)^2}$$

$$= \frac{4x^2 + 14x - 2x^2}{(2x + 7)^2}$$

$$= \frac{2x^2 + 14x}{(2x + 7)^2}$$

$$= \frac{2x(x + 7)}{(2x + 7)^2}$$

8. $y = \dfrac{x^2 + 3x}{2x^2 - 5}$

$$\frac{dy}{dx} = \frac{(2x^2 - 5)(2x + 3) - (x^2 + 3x)(4x)}{(2x^2 - 5)^2}$$

$$= \frac{4x^3 + 6x^2 - 10x - 15 - 4x^3 - 12x^2}{(2x^2 - 5)^2}$$

$$= \frac{-6x^2 - 10x - 15}{(2x^2 - 5)^2}$$

9. $f(x) = x^2 - 4$

Domain: $(-\infty, \infty)$

Range: $[-4, \infty)$

10. $f(x) = \sqrt{x - 7}$

Domain: $[7, \infty)$

Range: $[0, \infty)$

---

1. $f'(x) = -2$

$f''(x) = 0$

3. $f'(x) = 2x + 7$

$f''(x) = 2$

5. $g'(t) = t^2 - 8t + 2$

$g''(t) = 2t - 8$

7. $f(t) = \dfrac{3}{4t^2} = \dfrac{3}{4}t^{-2}$

$f'(t) = -\dfrac{3}{2}t^{-3}$

$f''(t) = \dfrac{9}{2}t^{-4} = \dfrac{9}{2t^4}$

9. $f'(x) = 9(2 - x^2)^2(-2x) = -18x(2 - x^2)^2$

$f''(x) = (-18x)2(2 - x^2)(-2x) + (2 - x^2)^2(-18)$

$= 18(2 - x^2)\left[4x^2 - (2 - x^2)\right]$

$= 18(2 - x^2)(5x^2 - 2)$

11. $f'(x) = \dfrac{(x - 1)(1) - (x + 1)(1)}{(x - 1)^2}$

$= -\dfrac{2}{(x - 1)^2} = -2(x - 1)^{-2}$

$f''(x) = 4(x - 1)^{-3}(1) = \dfrac{4}{(x - 1)^3}$

13. $f'(x) = 5x^4 - 12x^3$

$f''(x) = 20x^3 - 36x^2$

$f'''(x) = 60x^2 - 72x$

15. $f(x) = 5x(x + 4)^3$

$= 5x(x^3 + 12x^2 + 48x + 64)$

$= 5x^4 + 60x^3 + 240x^2 + 320x$

$f'(x) = 20x^3 + 180x^2 + 480x + 320$

$f''(x) = 60x^2 + 360x + 480$

$f'''(x) = 120x + 360$

17. $f(x) = \dfrac{3}{16x^2} = \dfrac{3}{16}x^{-2}$

$f'(x) = -\dfrac{3}{8}x^{-3}$

$f''(x) = \dfrac{9}{8}x^{-4}$

$f'''(x) = -\dfrac{9}{2}x^{-5} = -\dfrac{9}{2x^5}$

19. $g'(t) = 20t^3 + 20t$

$g''(t) = 60t^2 + 20$

$g''(2) = 60(4) + 20 = 260$

21. $f(x) = \sqrt{4 - x} = (4 - x)^{1/2}$

$f'(x) = -\dfrac{1}{2}(4 - x)^{-1/2}$

$f''(x) = -\dfrac{1}{4}(4 - x)^{-3/2}$

$f'''(x) = -\dfrac{3}{8}(4 - x)^{-5/2} = \dfrac{-3}{8(4 - x)^{5/2}}$

$f'''(-5) = \dfrac{-3}{8(9)^{5/2}} = -\dfrac{1}{648}$

23. $f(x) = (x^3 - 2x)^3 = x^9 - 6x^7 + 12x^5 - 8x^3$

$f'(x) = 9x^8 - 42x^6 + 60x^4 - 24x^2$

$f''(x) = 72x^7 - 252x^5 + 240x^3 - 48x$

$f''(1) = 12$

25. $f''(x) = 4x$

27. $f'''(x) = 2\sqrt{x - 1} = 2(x - 1)^{1/2}$

$f^{(4)}(x) = 2\left(\dfrac{1}{2}\right)(x - 1)^{-1/2} = \dfrac{1}{\sqrt{x - 1}}$

29. $f^{(5)}(x) = 2(x^2 + 1)(2x)$

$= 4x^3 + 4x$

$f^{(6)}(x) = 12x^2 + 4$

31. $f'(x) = 3x^2 - 18x + 27$

$f''(x) = 6x - 18$

$f''(x) = 0 \Rightarrow 6x = 18$

$x = 3$

33. $f(x) = x\sqrt{x^2 - 1} = x(x^2 - 1)^{1/2}$

$$f'(x) = x\frac{1}{2}(x^2 - 1)^{-1/2}(2x) + (x^2 - 1)^{1/2} = \frac{x^2}{(x^2 - 1)^{1/2}} + (x^2 - 1)^{1/2}$$

$$f''(x) = \frac{(x^2 - 1)^{1/2}(2x) - x^2\left(\frac{1}{2}\right)(x^2 - 1)^{-1/2}(2x)}{x^2 - 1} + \frac{1}{2}(x^2 - 1)^{-1/2}(2x)$$

$$= \frac{(x^2 - 1)(2x) - x^3}{(x^2 - 1)^{3/2}} + \frac{x}{(x^2 - 1)^{1/2}} \cdot \frac{x^2 - 1}{x^2 - 1}$$

$$= \frac{2x^3 - 3x}{(x^2 - 1)^{3/2}}$$

$$f''(x) = 0 \Rightarrow 2x^3 - 3x = x(2x^2 - 3) = 0$$

$$x = \pm\sqrt{\frac{3}{2}} = \pm\frac{\sqrt{6}}{2}$$

$x = 0$ is not in the domain of f.

35. (a) $s(t) = -16t^2 + 144t$

$v(t) = s'(t) = -32t + 144$

$a(t) = v'(t) = s''(t) = -32$

(b) $s(3) = 288$ ft

$v(3) = 48$ ft/sec

$a(3) = -32$ ft/sec^2

(c) $v(t) = 0$

$-32t + 144 = 0$

$-32t = -144$

$t = 4.5$ sec

$s(4.5) = 324$ ft

(d) $s(t) = 0$

$-16t^2 + 144t = 0$

$-16t(t - 9) = 0$

$t = 0$ sec $\quad t = 9$ sec

$v(9) = -32(9) + 144 = -144$ ft/sec

This is the same speed as the initial velocity.

37. $$\frac{d^2s}{dt^2} = \frac{(t + 10)(90) - (90t)(1)}{(t + 10)^2} = \frac{900}{(t + 10)^2}$$

t	0	10	20	30	40	50	60
$\frac{ds}{dt}$	0	45	60	67.5	72	75	77.14
$\frac{d^2s}{dt^2}$	9	2.25	1	0.56	0.36	0.25	0.18

As time increases, the acceleration decreases. After 1 minute, the automobile is traveling at about 77.14 feet per second.

39. $f(x) = x^2 - 6x + 6$

$f'(x) = 2x - 6$

$f''(x) = 2$

(a)

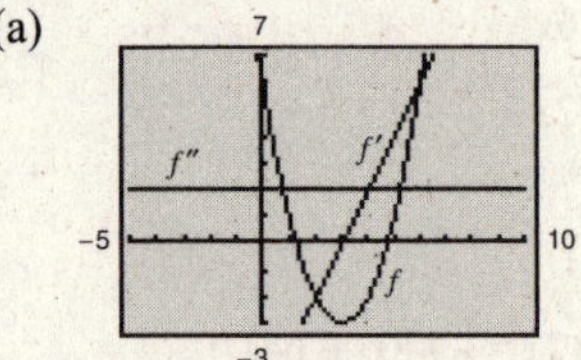

(b) The degree decreased by 1 for each successive derivative.

(c) $f(x) = 3x^2 - 9x$

$f'(x) = 6x - 9$

$f''(x) = 6$

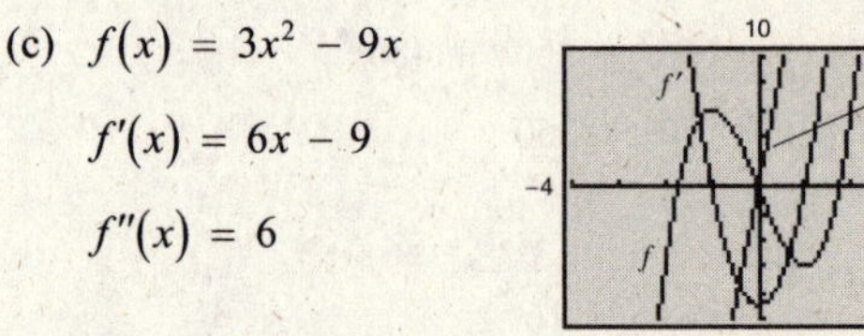

(d) The degree decreases by 1 for each successive derivative.

41. (a) $y = -68.991t^3 + 1208.34t^2 - 5445.4t + 10{,}145$

(b) $y' = -206.973t^2 + 2416.68t - 5445.4$

$y'' = -413.946t + 2416.68$

(c) Over the interval $5 \le t \le 8$, $y' > 0$, therefore y is increasing over $5 \le t \le 8$ or from 2005 to 2008.

(d) $y'' = 0$

$-413.946t + 2416.68 = 0$

$-413.946t = -2416.68$

$t \approx 5.84$ or 2005

43. False. The Product Rule is $\left[f(x)g(x)\right]' = f'(x)g(x) + f(x)g'(x)$.

45. Answers will vary.

Section 2.7 Implicit Differentiation

Skills Warm Up

1. $x - \dfrac{y}{x} = 2$

$$x^2 - y = 2x$$
$$-y = 2x - x^2$$
$$y = x^2 - 2x$$

2. $\dfrac{4}{x-3} = \dfrac{1}{y}$

$$4y = x - 3$$
$$y = \frac{x-3}{4}$$

3. $xy - x + 6y = 6$

$$xy + 6y = 6 + x$$
$$y(x+6) = 6 + x$$
$$y = \frac{6+x}{x+6}$$
$$y = 1,\ x \neq -6$$

4. $12 + 3y = 4x^2 + x^2y$

$$3y - x^2y = 4x^2 - 12$$
$$y(3 - x^2) = 4x^2 - 12$$
$$y = \frac{4x^2 - 12}{3 - x^2} = -4,\ x \neq \pm\sqrt{3}$$

5. $x^2 + y^2 = 5$

$$y^2 = 5 - x^2$$
$$y = \pm\sqrt{5 - x^2}$$

6. $x = \pm\sqrt{6 - y^2}$

$$x^2 = 6 - y^2$$
$$x^2 - 6 = -y^2$$
$$6 - x^2 = y^2$$
$$\pm\sqrt{6 - x^2} = y$$

7. $\dfrac{3x^2 - 4}{3y^2}$, $(2, 1)$

$$\frac{3(2^2) - 4}{3(1^2)} = \frac{3(4) - 4}{3} = \frac{8}{3}$$

8. $\dfrac{x^2 - 2}{1 - y}$, $(0, -3)$

$$\frac{0^2 - 2}{1 - (-3)} = \frac{-2}{4} = -\frac{1}{2}$$

9. $\dfrac{5x}{3y^2 - 12y + 5}$, $(-1, 2)$

$$\frac{5(-1)}{3(2^2) - 12(2) + 5} = \frac{-5}{3(4) - 24 + 5} = \frac{-5}{-7} = \frac{5}{7}$$

1. $xy = 4$

$$x\frac{dy}{dx} + y = 0$$
$$x\frac{dy}{dx} = -y$$
$$\frac{dy}{dx} = -\frac{y}{x}$$

3. $y^2 = 1 - x^2$

$$2y\frac{dy}{dx} = -2x$$
$$\frac{dy}{dx} = -\frac{x}{y}$$

5. $x^2y^2 - 2x = 3$

$$2x^2\frac{dy}{dx}y + 2xy^2 - 2 = 0$$
$$2x^2y\frac{dy}{dx} = 2 - 2xy^2$$
$$\frac{dy}{dx} = \frac{2 - 2xy^2}{2x^2y}$$
$$= \frac{1 - xy^2}{x^2y}$$

7.
$$4y^2 - xy = 2$$
$$8y\frac{dy}{dx} - x\frac{dy}{dx} - y = 0$$
$$(8y - x)\frac{dy}{dx} = y$$
$$\frac{dy}{dx} = \frac{y}{8y - x}$$

9.
$$\frac{xy - y^2}{y - x} = 1$$
$$xy - y^2 = y - x$$
$$y(x - y) = -(x - y)$$
$$y = -1$$
$$\frac{dy}{dx} = 0$$

11.
$$\frac{2y - x}{y^2 - 3} = 5$$
$$2y - x = 5y^2 - 15$$
$$2\frac{dy}{dx} - 1 = 10y\frac{dy}{dx}$$
$$-1 = (10y - 2)\frac{dy}{dx}$$
$$\frac{dy}{dx} = -\frac{1}{2(5y - 1)}$$

13.
$$x^2 + y^2 = 16$$
$$2x + 2y\frac{dy}{dx} = 0$$
$$2y\frac{dy}{dx} = -2x$$
$$\frac{dy}{dx} = -\frac{x}{y}$$

At $(0, 4)$, $\frac{dy}{dx} = -\frac{0}{4} = 0.$

15.
$$y + xy = 4$$
$$\frac{dy}{dx} + x\frac{dy}{dx} + y = 0$$
$$\frac{dy}{dx}(1 + x) = -y$$
$$\frac{dy}{dx} = -\frac{y}{x + 1}$$

At $(-5, -1)$, $\frac{dy}{dx} = -\frac{1}{4}.$

17.
$$x^3 - xy + y^2 = 4$$
$$3x^2 - x\frac{dy}{dx} - y + 2y\frac{dy}{dx} = 0$$
$$\frac{dy}{dx}(2y - x) = y - 3x^2$$
$$\frac{dy}{dx} = \frac{y - 3x^2}{2y - x}$$

At $(0, -2)$, $\frac{dy}{dx} = \frac{1}{2}.$

19.
$$x^3y^3 - y = x$$
$$3x^3y^2\frac{dy}{dx} + 3x^2y^3 - \frac{dy}{dx} = 1$$
$$\frac{dy}{dx}(3x^3y^2 - 1) = 1 - 3x^2y^3$$
$$\frac{dy}{dx} = \frac{1 - 3x^2y^3}{3x^3y^2 - 1}$$

At $(0, 0)$, $\frac{dy}{dx} = -1.$

21.
$$x^{1/2} + y^{1/2} = 9$$
$$\frac{1}{2}x^{-1/2} + \frac{1}{2}y^{-1/2}\frac{dy}{dx} = 0$$
$$x^{-1/2} + y^{-1/2}\frac{dy}{dx} = 0$$
$$\frac{dy}{dx} = \frac{-x^{-1/2}}{y^{-1/2}} = -\sqrt{\frac{y}{x}}$$

At $(16, 25)$, $\frac{dy}{dx} = -\frac{5}{4}.$

23.
$$\sqrt{xy} = x - 2y$$
$$\sqrt{x}\sqrt{y} = x - 2y$$
$$\sqrt{x}\left(\frac{1}{2}y^{-1/2}\frac{dy}{dx}\right) + \sqrt{y}\left(\frac{1}{2}x^{-1/2}\right) = 1 - 2\frac{dy}{dx}$$
$$\frac{\sqrt{x}}{2\sqrt{y}}\frac{dy}{dx} + 2\frac{dy}{dx} = 1 - \frac{\sqrt{y}}{2\sqrt{x}}$$
$$\frac{dy}{dx} = \frac{1 - \frac{\sqrt{y}}{2\sqrt{x}}}{\frac{\sqrt{x}}{2\sqrt{y}} + 2} \cdot \frac{2\sqrt{x}\sqrt{y}}{2\sqrt{x}\sqrt{y}}$$
$$= \frac{2\sqrt{xy} - y}{x + 4\sqrt{xy}}$$
$$= \frac{2(x - 2y) - y}{x + 4(x - 2y)}$$
$$= \frac{2x - 5y}{5x - 8y}$$

At $(4, 1)$, $\frac{dy}{dx} = \frac{1}{4}.$

25.
$$y^2(x^2 + y^2) = 2x^2$$
$$y^2\left(2x + 2y\frac{dy}{dx}\right) + (x^2 + y^2)\left(2y\frac{dy}{dx}\right) = 4x$$
$$2xy^2 + 2y^3\frac{dy}{dx} + 2x^2y\frac{dy}{dx} + 2y^3\frac{dy}{dx} = 4x$$
$$\frac{dy}{dx}(4y^3 + 2x^2y) = 4x - 2xy^2$$
$$\frac{dy}{dx} = \frac{2x(2 - y^2)}{2y(2y^2 + x^2)}$$
$$\frac{dy}{dx} = \frac{x(2 - y^2)}{y(2y^2 + x^2)}$$

At $(1, 1)$, $\frac{dy}{dx} = \frac{1}{3}$.

27. $3x^2 - 2y + 5 = 0$
$$6x - 2\frac{dy}{dx} = 0$$
$$\frac{dy}{dx} = 3x$$

At $(1, 4)$, $\frac{dy}{dx} = 3$.

29.
$$x^2 + y^2 = 4$$
$$2x + 2y\frac{dy}{dx} = 0$$
$$\frac{dy}{dx} = -\frac{x}{y}$$

At $(0, 2)$, $\frac{dy}{dx} = 0$.

31. $(4 - x)y^2 = x^3$
$$y^2 = \frac{x^3}{4 - x}$$
$$2y\frac{dy}{dx} = \frac{(4 - x)(3x^2) - x^3(-1)}{(4 - x)^2}$$
$$2y\frac{dy}{dx} = \frac{12x^2 - 3x^3 + x^3}{(4 - x)^2}$$
$$2y\frac{dy}{dx} = \frac{12x^2 - 2x^3}{(4 - x)^2}$$
$$\frac{dy}{dx} = -\frac{2x^2(x - 6)}{2y(4 - x)^2}$$
$$\frac{dy}{dx} = -\frac{x^2(x - 6)}{y(4 - x)^2}$$

At $(2, 7)$, $\frac{dy}{dx} = 2$.

33. Implicitly: $1 - 2y\frac{dy}{dx} = 0$
$$\frac{dy}{dx} = \frac{1}{2y}$$

Explicitly:
$$y = \pm\sqrt{x - 1}$$
$$= \pm(x - 1)^{1/2}$$
$$\frac{dy}{dx} = \pm\frac{1}{2}(x - 1)^{-1/2}(1)$$
$$= \pm\frac{1}{2\sqrt{x - 1}}$$
$$= \frac{1}{2\left(\pm\sqrt{x - 1}\right)}$$
$$= \frac{1}{2y}$$

At $(2, -1)$, $\frac{dy}{dx} = -\frac{1}{2}$.

35.
$$x^2 + y^2 = 100$$
$$2x + 2y\frac{dy}{dx} = 0$$
$$\frac{dy}{dx} = -\frac{x}{y}$$

At $(8, 6)$:
$$m = -\frac{4}{3}$$
$$y - 6 = -\frac{4}{3}(x - 8)$$
$$y = -\frac{4}{3}x + \frac{50}{3}$$

At $(-6, 8)$:
$$m = \frac{3}{4}$$
$$y - 8 = \frac{3}{4}(x + 6)$$
$$y = \frac{3}{4}x + \frac{25}{2}$$

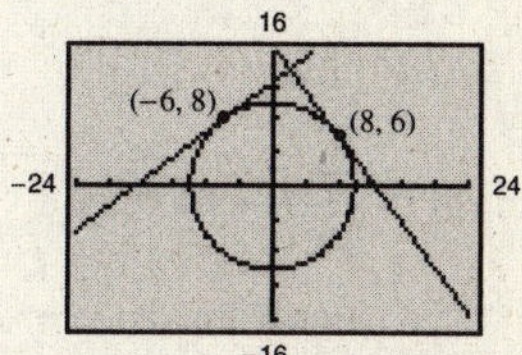

37. $y^2 = 5x^3$

$$2y\frac{dy}{dx} = 15x^2$$

$$\frac{dy}{dx} = \frac{15x^2}{2y}$$

At $(1, \sqrt{5})$:

$$m = \frac{15}{2\sqrt{5}} = \frac{3\sqrt{5}}{2}$$

$$y - \sqrt{5} = \frac{3\sqrt{5}}{2}(x - 1)$$

$$y = \frac{3\sqrt{5}}{2}x - \frac{\sqrt{5}}{2}$$

At $(1, -\sqrt{5})$:

$$m = \frac{-15}{2\sqrt{5}} = -\frac{3\sqrt{5}}{2}$$

$$y + \sqrt{5} = -\frac{3\sqrt{5}}{2}(x - 1)$$

$$y = -\frac{3\sqrt{5}}{2}x + \frac{\sqrt{5}}{2}$$

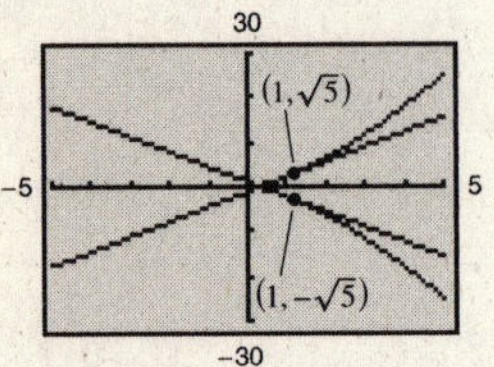

39. $x^3 + y^3 = 8$

$$3x^2 + 3y^2\frac{dy}{dx} = 0$$

$$3y^2\frac{dy}{dx} = -3x^2$$

$$\frac{dy}{dx} = -\frac{x^2}{y^2}$$

At $(0, 2)$:

$$m = \frac{dy}{dx} = 0$$

$$y - 2 = 0(x - 0)$$

$$y = 2$$

At $(2, 0)$:

$m = \frac{dy}{dx}$ is undefined.

The tangent line is $x = 2$.

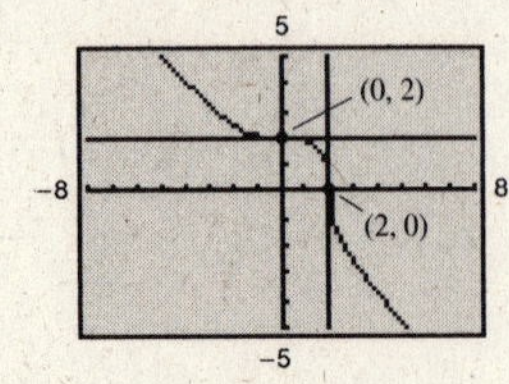

41. $x^2y - 8 = -4y$

$$x^2y + 4y = 8$$

$$y(x^2 + 4) = 8$$

$$y = \frac{8}{x^2 + 4} = 8(x^2 + 4)^{-1}$$

$$\frac{dy}{dx} = 8(-1)(x^2 + 4)^{-2}(2x)$$

$$\frac{dy}{dx} = -\frac{16x}{(x^2 + 4)^2}$$

At $(-2, 1)$:

$$m = \frac{dy}{dx} = -\frac{16(-2)}{((-2)^2 + 4)^2} = \frac{32}{64} = \frac{1}{2}$$

$$y - 1 = \frac{1}{2}(x - (-2))$$

$$y = \frac{1}{2}x + 2$$

At $\left(6, \frac{1}{5}\right)$:

$$m = \frac{dy}{dx} = -\frac{16(6)}{[(6)^2 + 4]^2} = -\frac{96}{1600} = -\frac{3}{50}$$

$$y - \frac{1}{5} = -\frac{3}{50}(x - 6)$$

$$y - \frac{1}{5} = -\frac{3}{50}x + \frac{9}{25}$$

$$y = -\frac{3}{50}x + \frac{14}{25}$$

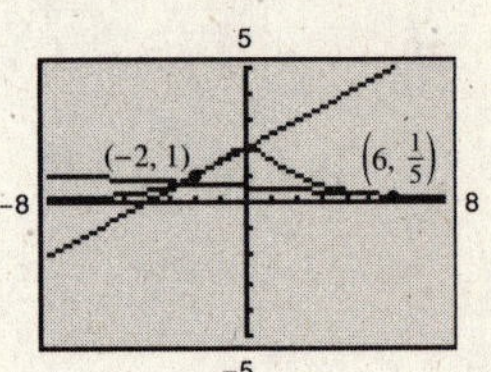

43. $p = \dfrac{2}{0.00001x^3 + 0.1x}, x \ge 0$

$$0.00001x^3 + 0.1x = \frac{2}{p}$$

$$0.00003x^2\frac{dx}{dp} + 0.1\frac{dx}{dp} = -\frac{2}{p^2}$$

$$(0.00003x^2 + 0.1)\frac{dx}{dp} = -\frac{2}{p^2}$$

$$\frac{dx}{dp} = -\frac{2}{p^2(0.00003x^2 + 0.1)}$$

45. $p = \sqrt{\dfrac{200 - x}{2x}}, 0 < x \le 200$

$$2xp^2 = 200 - x$$

$$2x(2p) + p^2\left(2\frac{dx}{dp}\right) = -\frac{dx}{dp}$$

$$(2p^2 + 1)\frac{dx}{dp} = -4xp$$

$$\frac{dx}{dp} = -\frac{4xp}{2p^2 + 1}$$

47. (a)

$$100x^{0.75}y^{0.25} = 135{,}540$$

$$100x^{0.75}\left(0.25y^{-0.75}\frac{dy}{dx}\right) + y^{0.25}\left(75x^{-0.25}\right) = 0$$

$$\frac{25x^{0.75}}{y^{0.75}} \cdot \frac{dy}{dx} = -\frac{75y^{0.25}}{x^{0.25}}$$

$$\frac{dy}{dx} = -\frac{3y}{x}$$

When $x = 1500$ and $y = 1000$, $\frac{dy}{dx} = -2$.

(b)

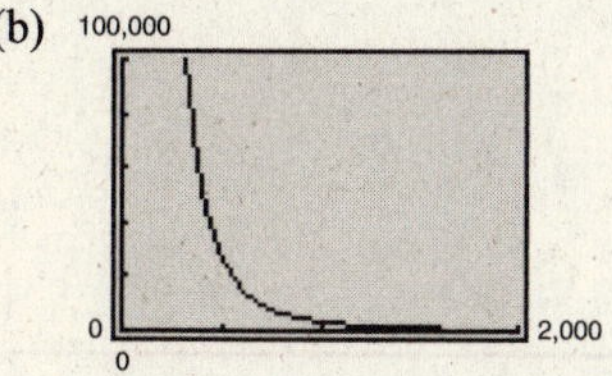

If more labor is used, then less capital is available.

If more capital is used, then less labor is available.

49.

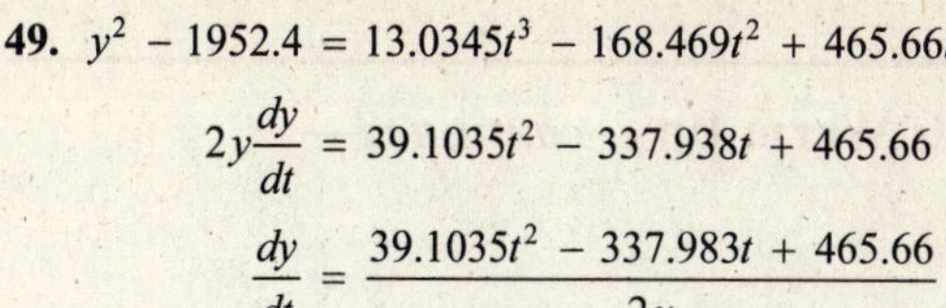

$$y^2 - 1952.4 = 13.0345t^3 - 168.469t^2 + 465.66t$$

$$2y\frac{dy}{dt} = 39.1035t^2 - 337.938t + 465.66$$

$$\frac{dy}{dt} = \frac{39.1035t^2 - 337.983t + 465.66}{2y}$$

(a)

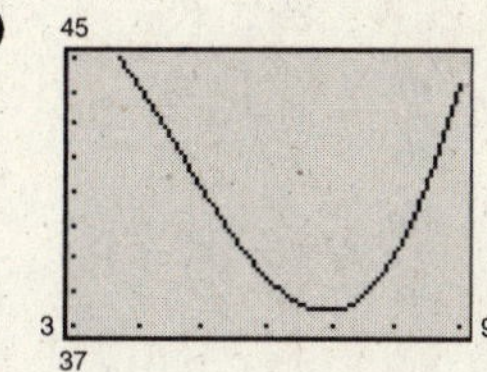

The number of cases of HIV/AIDS decreases from 2004 through 2007, then begins to increase.

(b) Using the graph, the number of reported cases was decreasing at the greatest rate during 2005.

(c)

t	4	5	6	7	8
y	44.11	41.06	38.46	37.50	39.21
y'	−2.95	−3.00	−2.01	0.22	3.38

The result is the same.

Section 2.8 Related Rates

Skills Warm Up

1. $A = \pi r^2$

2. $V = \frac{4}{3}\pi r^3$

3. $SA = 6s^2$

4. $V = s^3$

5. $V = \frac{1}{3}\pi r^2 h$

6. $A = \frac{1}{2}bh$

7.

$$x^2 + y^2 = 9$$

$$\frac{d}{dx}\left[x^2 + y^2\right] = \frac{d}{dx}[9]$$

$$2x + 2y\frac{dy}{dx} = 0$$

$$2y\frac{dy}{dx} = -2x$$

$$\frac{dy}{dx} = \frac{-2x}{2y}$$

$$= \frac{-x}{y}$$

8.

$$3xy - x^2 = 6$$

$$\frac{d}{dx}\left[3xy - x^2\right] = \frac{d}{dx}[6]$$

$$3y + 3x\frac{dy}{dx} - 2x = 0$$

$$3x\frac{dy}{dx} = 2x - 3y$$

$$\frac{dy}{dx} = \frac{2x - 3y}{3x}$$

9.

$$x^2 + 2y + xy = 12$$

$$\frac{d}{dx}\left[x^2 + 2y + xy\right] = \frac{d}{dx}(12)$$

$$2x + 2\frac{dy}{dx} + y + x\frac{dy}{dx} = 0$$

$$2\frac{dy}{dx} + x\frac{dy}{dx} = -y - 2x$$

$$\frac{dy}{dx}(2 + x) = -y - 2x$$

$$\frac{dy}{dx} = \frac{-y - 2x}{2 + x}$$

Skills Warm Up —continued—

10.
$$x + xy^2 - y^2 = xy$$
$$\frac{d}{dx}\left[x + xy^2 - y^2\right] = \frac{d}{dx}[xy]$$
$$1 + y^2 + 2xy\frac{dy}{dx} - 2y\frac{dy}{dx} = y + x\frac{dy}{dx}$$
$$2xy\frac{dy}{dx} - 2y\frac{dy}{dx} - x\frac{dy}{dx} = y - y^2 - 1$$
$$\frac{dy}{dx}(2xy - 2y - x) = y - y^2 - 1$$
$$\frac{dy}{dx} = \frac{y - y^2 - 1}{2xy - 2y - x}$$

1. $y = \sqrt{x}, \dfrac{dy}{dt} = \dfrac{1}{2}x^{-1/2}\dfrac{dx}{dt} = \dfrac{1}{2\sqrt{x}}\dfrac{dx}{dt}, \dfrac{dx}{dt} = 2\sqrt{x}\dfrac{dy}{dt}$

(a) When $x = 4$ and $\dfrac{dx}{dt} = 3$, $\dfrac{dy}{dt} = \left(\dfrac{1}{2\sqrt{4}}\right)(3) = \dfrac{3}{4}$.

(b) When $x = 25$ and $\dfrac{dy}{dt} = 2$, $\dfrac{dx}{dt} = 2\sqrt{25}(2) = 20$.

3. $xy = 4, x\dfrac{dy}{dt} + y\dfrac{dx}{dt} = 0, \dfrac{dy}{dt} = \left(-\dfrac{y}{x}\right)\dfrac{dx}{dt}, \dfrac{dx}{dt} = \left(-\dfrac{x}{y}\right)\dfrac{dy}{dt}$

(a) When $x = 8$, $y = \dfrac{1}{2}$, and $\dfrac{dx}{dt} = 10$, $\dfrac{dy}{dt} = -\dfrac{1/2}{8}(10) = -\dfrac{5}{8}$.

(b) When $x = 1$, $y = 4$, and $\dfrac{dy}{dt} = -6$, $\dfrac{dx}{dt} = -\dfrac{1}{4}(-6) = \dfrac{3}{2}$.

5. $A = \pi r^2, \dfrac{dr}{dt} = 3, \dfrac{dA}{dt} = 2\pi r\dfrac{dr}{dt} = 6\pi r$

(a) When $r = 6$, $\dfrac{dA}{dt} = 2\pi(6)(3) = 36\pi$ in.^{2}/min.

(b) When $r = 24$, $\dfrac{dA}{dt} = 2\pi(24)(3) = 144\pi$ in.^{2}/min.

7. $A = \pi r^2, \dfrac{dA}{dt} = 2\pi r\dfrac{dr}{dt}$

If $\dfrac{dr}{dt}$ is constant, then $\dfrac{dA}{dt}$ is not constant; $\dfrac{dA}{dt}$ is proportional to r.

9. $V = \dfrac{4}{3}\pi r^3, \dfrac{dV}{dt} = 10, \dfrac{dV}{dt} = 4\pi r^2\dfrac{dr}{dt}, \dfrac{dr}{dt} = \left(\dfrac{1}{4\pi r^2}\right)\dfrac{dV}{dt}$

(a) When $r = 1$, $\dfrac{dr}{dt} = \dfrac{1}{4\pi(1)^2}(10) = \dfrac{5}{2\pi}$ ft/min.

(b) When $r = 2$, $\dfrac{dr}{dt} = \dfrac{1}{4\pi(2)^2}(10) = \dfrac{5}{8\pi}$ ft/min.

11. (a) $\dfrac{dC}{dt} = 0.75\dfrac{dx}{dt} = 0.75(150)$

$= 112.5$ dollars per week

(b) $\dfrac{dR}{dt} = 250\dfrac{dx}{dt} - \dfrac{1}{5}x\dfrac{dx}{dt}$

$= 250(150) - \dfrac{1}{5}(1000)(150)$

$= 7500$ dollars per week

(c) $P = R - C$

$\dfrac{dP}{dt} = \dfrac{dR}{dt} - \dfrac{dC}{dt} = 7500 - 112.5$

$= 7387.5$ dollars per week

13. $V = x^3,\ \dfrac{dx}{dt} = 3,\ \dfrac{dV}{dt} = 3x^2\dfrac{dx}{dt}$

(a) When $x = 1$, $\dfrac{dV}{dt} = 3(1)^2(3) = 9\ \text{cm}^3/\text{sec}$.

(b) When $x = 10$, $\dfrac{dV}{dt} = 3(10)^2(3) = 900\ \text{cm}^3/\text{sec}$.

15. $y = x^2$

$\dfrac{dy}{dt} = 2x\dfrac{dx}{dt}$

(a) When $x = -3$ and $\dfrac{dx}{dt} = 3$ in./sec:

$\dfrac{dy}{dt} = 2(-3)(3) = -18$ in./sec

(b) When $x = 0$ and $\dfrac{dx}{dt} = 3$ in./sec:

$\dfrac{dy}{dt} = 2(0)(3) = 0$ in./sec

(c) When $x = 1$ and $\dfrac{dx}{dt} = 3$ in./sec:

$\dfrac{dy}{dt} = 2(1)(3) = 6$ in./sec

(d) When $x = 3$ and $\dfrac{dx}{dt} = 3$ in./sec:

$\dfrac{dy}{dt} = 2(3)(3) = 18$ in./sec

17. Let x be the distance from the boat to the dock and y be the length of the rope.

$12^2 + x^2 = y^2$

$\dfrac{dy}{dt} = -4$

$2x\dfrac{dx}{dt} = 2y\dfrac{dy}{dt}$

$\dfrac{dx}{dt} = \dfrac{y}{x}\dfrac{dy}{dt}$

When $y = 13$, $x = 5$ and

$\dfrac{dx}{dt} = \dfrac{13}{5}(-4) = -10.4$ ft/sec.

As $x \to 0$, $\dfrac{dx}{dt}$ increases.

19. $x^2 + 6^2 = s^2$

$2x\dfrac{dx}{dt} = 2s\dfrac{ds}{dt}$

$\dfrac{dx}{dt} = \dfrac{s}{x}\dfrac{ds}{dt}$

When $s = 10$, $x = 8$ and $\dfrac{ds}{dt} = 240$:

$\dfrac{dx}{dt} = \dfrac{10}{8}(-240) = 300$ mi/hr.

21. (a) $L^2 = x^2 + y^2$, $\dfrac{dx}{dt} = -450$, $\dfrac{dy}{dt} = -600$, and

$\dfrac{dL}{dt} = \dfrac{x(dx/dt) + y(dy/dt)}{L}$

When $x = 150$ and $y = 200$, $L = 250$ and

$\dfrac{dL}{dt} = \dfrac{150(-450) + 200(-600)}{250} = -750$ mph.

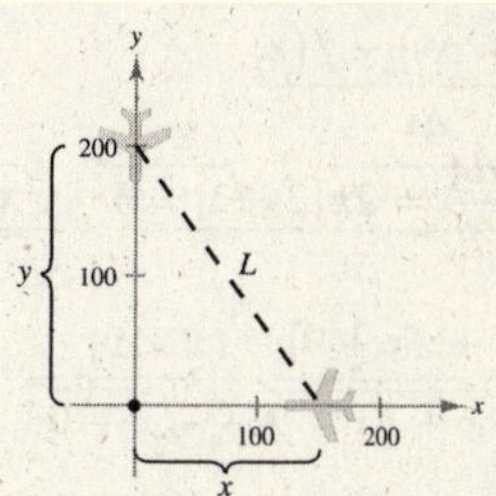

(b) $t = \dfrac{250}{750} = \dfrac{1}{3}$ hr $= 20$ min

23. $V = \pi r^2 h,\ h = 0.08,\ V = 0.08\pi r^2,\ \dfrac{dV}{dt} = 0.16\pi r\dfrac{dr}{dt}$

When $r = 150$ and $\dfrac{dr}{dt} = \dfrac{1}{2}$,

$\dfrac{dV}{dt} = 0.16\pi(150)\left(\dfrac{1}{2}\right) = 12\pi = 37.70\ \text{ft}^3/\text{min}.$

25. $P = R - C = xp - C = x(6000 - 25x) - (2400x + 5200)$

$$= -25x^2 + 3600x - 5200$$

$$\frac{dP}{dt} = -50x\frac{dx}{dt} + 3600\frac{dx}{dt}$$

$$\frac{dx}{dt} = \frac{1}{3600 - 50x}\frac{dP}{dt}$$

When $x = 44$ and $\frac{dP}{dt} = 5600$, $\frac{dx}{dt} = \frac{1}{3600 - 50(44)}(5600) = 4$ units per week.

Review Exercises for Chapter 2

1. Slope $\approx \frac{-4}{2} = -2$

3. Slope ≈ 0

5. Answers will vary. Sample answer:

$t = 4$; slope $\approx$ \$290 million/yr; Sales were increasing by about \$290 million/yr in 2004.

$t = 7$; slope $\approx$ \$320 million/yr; Sales were increasing by about \$320 million/yr in 2007.

7. Answers will vary. Sample answer:

$t = 1$: $m \approx 65$ hundred thousand visitors/month; The number of visitors to the national park is increasing at about 65,000,000/month in January.

$t = 8$: $m \approx 0$ visitors/month; The number of visitors to the national park is neither increasing nor decreasing in August.

$t = 12$: $m \approx -1000$ hundred thousand/month; The number of visitors to the national park is decreasing at about 1,000,000,000 visitors per month in December.

9. $f'(x) = \lim_{\Delta x \to 0} \frac{f(x + \Delta x) - f(x)}{\Delta x}$

$$= \lim_{\Delta x \to 0} \frac{-3(x + \Delta x) - 5 - (-3x - 5)}{\Delta x}$$

$$= \lim_{\Delta x \to 0} \frac{-3\Delta x}{\Delta x} = -3$$

$f'(-2) = -3$

11. $f'(x) = \lim_{\Delta x \to 0} \frac{f(x + \Delta x) - f(x)}{\Delta x}$

$$= \lim_{\Delta x \to 0} \frac{(x + \Delta x)^2 - 4(x + \Delta x) - (x^2 - 4x)}{\Delta x}$$

$$= \lim_{\Delta x \to 0} \frac{x^2 + 2x\Delta x + (\Delta x)^2 - 4\Delta x - x^2}{\Delta x}$$

$$= \lim_{\Delta x \to 0} \frac{2x\Delta x + (\Delta x)^2 - 4\Delta x}{\Delta x}$$

$$= \lim_{\Delta x \to 0} (2x + \Delta x - 4) = 2x - 4$$

$f'(1) = 2(1) - 4 = -2$

13. $f'(x) = \lim_{\Delta x \to 0} \frac{f(x + \Delta x) - f(x)}{\Delta x}$

$$= \lim_{\Delta x \to 0} \frac{\sqrt{x + \Delta x + 9} - \sqrt{x + 9}}{\Delta x} \cdot \frac{\sqrt{x + \Delta x + 9} + \sqrt{x + 9}}{\sqrt{x + \Delta x + 9} + \sqrt{x + 9}}$$

$$= \lim_{\Delta x \to 0} \frac{(x + \Delta x + 9) - (x + 9)}{\Delta x\left[\sqrt{x + \Delta x + 9} + \sqrt{x + 9}\right]}$$

$$= \lim_{\Delta x \to 0} \frac{1}{\sqrt{x + \Delta x + 9} + \sqrt{x + 9}} = \frac{1}{2\sqrt{x + 9}}$$

$f'(-5) = \frac{1}{4}$

15. $f'(x) = \lim_{\Delta x \to 0} \frac{f(x + \Delta x) - f(x)}{\Delta x}$

$= \lim_{\Delta x \to 0} \frac{\frac{1}{x + \Delta x - 5} - \frac{1}{x - 5}}{\Delta x}$

$= \lim_{\Delta x \to 0} \frac{(x - 5) - (x + \Delta x - 5)}{\Delta x(x + \Delta x - 5)(x - 5)}$

$= \lim_{\Delta x \to 0} \frac{-1}{(x + \Delta x - 5)(x - 5)} = -\frac{1}{(x - 5)^2}$

$f'(6) = -1$

17. $f(x) = 9x + 1$

$f'(x) = \lim_{\Delta x \to 0} \frac{f(x + \Delta x) - f(x)}{\Delta x}$

$= \lim_{\Delta x \to 0} \frac{[9(x + \Delta x) + 1] - (9x + 1)}{\Delta x}$

$= \lim_{\Delta x \to 0} \frac{9x + 9\Delta x + 1 - 9x - 1}{\Delta x}$

$= \lim_{\Delta x \to 0} \frac{9\Delta x}{\Delta x} = \lim_{\Delta x \to 0} 9 = 9$

19. $f(x) = -\frac{1}{2}x^2 + 2x$

$f'(x) = \lim_{\Delta x \to 0} \frac{f(x + \Delta x) - f(x)}{\Delta x}$

$= \lim_{\Delta x \to 0} \frac{\left[-\frac{1}{2}(x + \Delta x)^2 + 2(x + \Delta x)\right] - \left(-\frac{1}{2}x^2 + 2x\right)}{\Delta x}$

$= \lim_{\Delta x \to 0} \frac{-\frac{1}{2}x^2 - x(\Delta x) - \frac{1}{2}(\Delta x)^2 + 2x + 2\Delta x + \frac{1}{2}x^2 - 2x}{\Delta x}$

$= \lim_{\Delta x \to 0} \frac{\Delta x\left(-x - \frac{1}{2}\Delta x + 2\right)}{\Delta x}$

$= \lim_{\Delta x \to 0} \left(-x - \frac{1}{2}\Delta x + 2\right) = -x + 2$

21. $f(x) = \sqrt{x - 5}$

$f'(x) = \lim_{\Delta x \to 0} \frac{f(x + \Delta x) - f(x)}{\Delta x}$

$= \lim_{\Delta x \to 0} \frac{\sqrt{x + \Delta x - 5} - \sqrt{x - 5}}{\Delta x} \cdot \frac{\sqrt{x + \Delta x - 5} + \sqrt{x - 5}}{\sqrt{x + \Delta x - 5} + \sqrt{x - 5}}$

$= \lim_{\Delta x \to 0} \frac{(x + \Delta x - 5) - (x - 5)}{\Delta x\left(\sqrt{x + \Delta x - 5} + \sqrt{x - 5}\right)}$

$= \lim_{\Delta x \to 0} \frac{1}{\sqrt{x + \Delta x - 5} + \sqrt{x - 5}} = \frac{1}{2\sqrt{x - 5}}$

23. $f(x) = \frac{5}{x}$

$f'(x) = \lim_{\Delta x \to 0} \frac{f(x + \Delta x) - f(x)}{\Delta x}$

$= \lim_{\Delta x \to 0} \frac{\frac{5}{x + \Delta x} - \frac{5}{x}}{\Delta x}$

$= \lim_{\Delta x \to 0} \frac{\frac{5x - 5(x + \Delta x)}{x(x + \Delta x)}}{\Delta x}$

$= \lim_{\Delta x \to 0} \frac{-5\Delta x}{\Delta x[x(x + \Delta x)]}$

$= \lim_{\Delta x \to 0} -\frac{5}{x(x + \Delta x)} = -\frac{5}{x^2}$

25. y is not differentiable at $x = 1$, a discontinuity.

27. y is not differentiable at $x = 0$, a discontinuity.

29. $y = -6$

$y' = 0$

31. $f(x) = x^3$

$f'(x) = 3x^2$

33. $f(x) = 4x^2$

$f'(x) = 8x$

35. $f(x) = \dfrac{2x^4}{5}$

$f'(x) = \dfrac{8x^3}{5}$

37. $g(x) = 2x^4 + 3x^2$

$g'(x) = 8x^3 + 6x$

39. $y = x^2 + 6x - 7$

$y' = 2x + 6$

41. $f(x) = 2x^{-1/2}; (4, 1)$

$f'(x) = -x^{-3/2}$

$f'(4) = -(4)^{-3/2} = -0.125$

43. $g(x) = x^3 - 4x^2 - 6x + 8; (-1, 9)$

$g'(x) = 3x^2 - 8x - 6$

$g'(-1) = 3(-1)^2 - 8(-1) - 6 = 5$

45. $f'(x) = 4x - 3$

$f'(2) = 5$

$y - 3 = 5(x - 2)$

$y = 5x - 7$

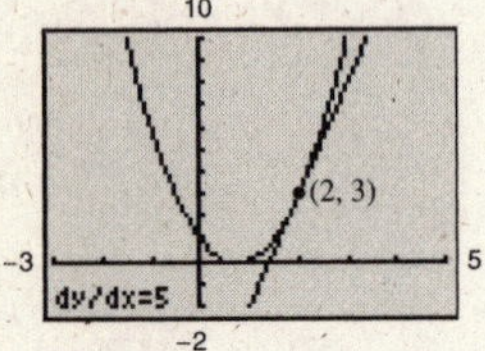

47. $f(x) = \sqrt{x} - \dfrac{1}{\sqrt{x}} = x^{1/2} - x^{-1/2}$

$f'(x) = \dfrac{1}{2}x^{-1/2} + \dfrac{1}{2}x^{-3/2}$

$= \dfrac{1}{2\sqrt{x}} + \dfrac{1}{2x^{3/2}}$

$f'(1) = 1$

$y - 0 = 1(x - 1)$

$y = x - 1$

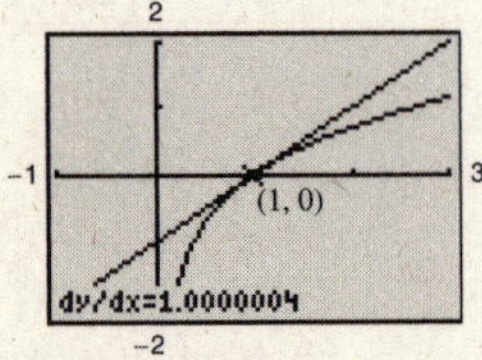

49. $S = -0.7500t^4 + 13.278t^3 - 74.50t^2 + 440.2t + 523$

$\dfrac{ds}{dt} = s'(t) = -3t^3 + 39.834t^2 - 149t + 440.2$

(a) 2004: $s'(4) = 289.544$

2007: $s'(7) = 320.066$

(b) Results should be similar.

(c) The slope shows the rate at which sales were increasing or decreasing in that particular year or value of t.

51.

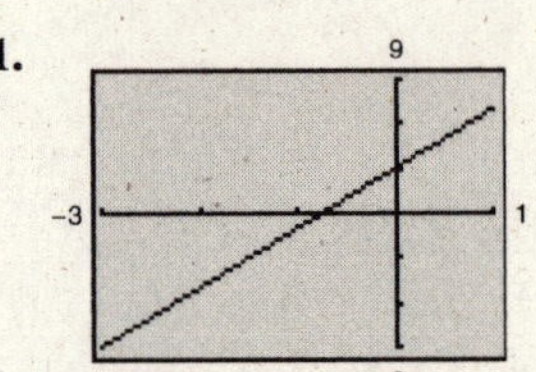

Average rate of change:

$\dfrac{f(1) - f(-3)}{1 - (-3)} = \dfrac{7 - (-9)}{4} = 4$

Instantaneous rate of change:

$f'(t) = 4$

$f'(1) = 4$

$f'(-3) = 4$

53.

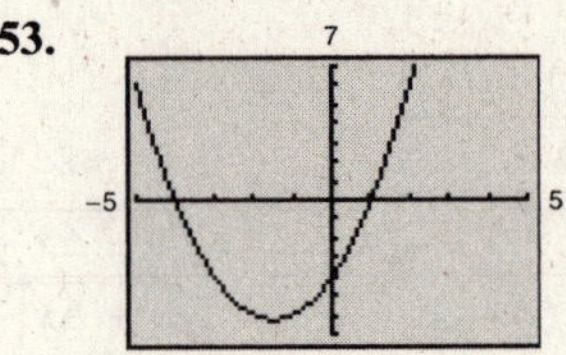

Average rate of change:

$\dfrac{f(1) - f(0)}{1 - 0} = \dfrac{0 - (-4)}{1} = 4$

Instantaneous rate of change:

$f'(x) = 2x + 3$

$f'(1) = 5$

$f'(0) = 3$

55. $s(t) = -16t^2 + 24t + 300$

(a) Average velocity $= \dfrac{s(2) - s(1)}{2 - 1} = \dfrac{284 - 308}{1} = -24$ ft/sec

(b) $v(t) = s'(t) = -32t + 24$

$v(1) = -8$ ft/sec

$v(3) = -72$ ft/sec

(c) $s(t) = 0$

$-16t^2 + 24t + 300 = 0$

$16t^2 - 24t - 300 = 0$

$$t = \frac{-(-24) \pm \sqrt{(-24)^2 - 4(16)(-300)}}{2(16)} = \frac{24 \pm \sqrt{19{,}776}}{32}$$

$t \approx 5.14$ sec

(d) $v(5.14) = -140.48$ ft/sec

57. $C = 2500 + 320x$

$\dfrac{dC}{dx} = 320$

59. $C = 370 + 2.55\sqrt{x} = 370 + 2.25x^{1/2}$

$\dfrac{dC}{dx} = \dfrac{1}{2}(2.55)(x^{-1/2}) = \dfrac{1.275}{\sqrt{x}}$

61. $R = 150x - 0.6x^2$

$\dfrac{dR}{dx} = 150 - 1.2x$

63. $R = -4x^3 + 2x^2 + 100x$

$\dfrac{dR}{dx} = -12x^2 + 4x + 100$

65. $P = -0.0002x^3 + 6x^2 - x - 2000$

$\dfrac{dP}{dx} = -0.0006x^2 + 12x - 1$

67. $P = -0.05x^2 + 20x - 1000$

(a) Find $\dfrac{\Delta P}{\Delta x}$ for $100 \le x \le 101$.

$\dfrac{P(101) - P(100)}{101 - 100} = 509.95 - 500 = \$9.95$

(b) Find $\dfrac{dP}{dx}$ when $x = 100$.

$\dfrac{dP}{dx} = -0.1x + 20 = P'(x)$

When $x = 100$, $\dfrac{dP}{dx} = P'(100) = \$10$.

(c) Parts (a) and (b) differ by only $0.05.

69. $f(x) = x^3(5 - 3x^2) = 5x^3 - 3x^5$

$f'(x) = 15x^2 - 15x^4 = 15x^2(1 - x^2)$

Simple Power Rule

71. $y = (4x - 3)(x^3 - 2x^2)$

$y' = (4x - 3)(3x^2 - 4x) + 4(x^3 - 2x^2)$

$= 12x^3 - 25x^2 + 12x + 4x^3 - 8x^2$

$= 16x^3 - 33x^2 + 12x$

Product Rule

73. $g(x) = \dfrac{x}{x + 3}$

$g'(x) = \dfrac{(x + 3)(1) - x(1)}{(x + 3)^2}$

$g'(x) = \dfrac{3}{(x + 3)^2}$

Quotient Rule

75. $f(x) = \dfrac{6x - 5}{x^2 + 1}$

$f'(x) = \dfrac{(x^2 + 1)(6) - (6x - 5)(2x)}{(x^2 + 1)^2}$

$= \dfrac{6 + 10x - 6x^2}{(x^2 + 1)^2}$

$= \dfrac{2(3 + 5x - 3x^2)}{(x^2 + 1)^2}$

Quotient Rule

77. $f(x) = (5x^2 + 2)^3$

$$f'(x) = 3(5x^2 + 2)^2(10x)$$
$$= 30x(5x^2 + 2)^2$$

General Power Rule

79. $h(x) = \dfrac{2}{\sqrt{x+1}} = 2(x + 1)^{-1/2}$

$$h'(x) = 2\left(-\frac{1}{2}\right)(x + 1)^{-3/2}$$
$$= -\frac{1}{(x + 1)^{3/2}}$$

General Power Rule

81. $g(x) = x\sqrt{x^2 + 1} = x(x^2 + 1)^{1/2}$

$$g'(x) = x\left[\frac{1}{2}(x^2 + 1)^{-1/2}(2x)\right] + (1)(x^2 + 1)^{1/2}$$
$$= (x^2 + 1)^{-1/2}\left[x^2 + (x^2 + 1)\right]$$
$$= \frac{2x^2 + 1}{\sqrt{x^2 + 1}}$$

Product and General Power Rules

83. $f(x) = x(1 - 4x^2)^2$

$$f'(x) = x(2)(1 - 4x^2)(-8x) + (1 - 4x^2)^2$$
$$= -16x^2(1 - 4x^2) + (1 - 4x^2)^2$$
$$= (1 - 4x^2)\left[-16x^2 + (1 - 4x^2)\right]$$
$$= (1 - 4x^2)(1 - 20x^2)$$

Product and General Power Rules

85. $h(x) = \left[x^2(2x + 3)\right]^3 = x^6(2x + 3)^3$

$$h'(x) = x^6\left[3(2x + 3)^2(2)\right] + 6x^5(2x + 3)^3$$
$$= 6x^5(2x + 3)^2\left[x + (2x + 3)\right]$$
$$= 18x^5(2x + 3)^2(x + 1)$$

Product and General Power Rules

87. $f(x) = x^2(x - 1)^5$

$$f'(x) = 5x^2(x - 1)^4 + 2x(x - 1)^5$$
$$= x(x - 1)^4\left[5x + 2(x - 1)\right]$$
$$= x(x - 1)^4(7x - 2)$$

Product and General Power Rules

89. $h(t) = \dfrac{\sqrt{3t + 1}}{(1 - 3t)^2} = \dfrac{(3t + 1)^{1/2}}{(1 - 3t)^2}$

$$h'(t) = \frac{(1 - 3t)^2(1/2)(3t + 1)^{-1/2}(3) - (3t + 1)^{1/2}(2)(1 - 3t)(-3)}{(1 - 3t)^4}$$
$$= \frac{(3t + 1)^{-1/2}\left[(1 - 3t)(3/2) + (3t + 1)6\right]}{(1 - 3t)^3}$$
$$= \frac{3(9t + 5)}{2\sqrt{3t + 1}(1 - 3t)^3}$$

Quotient and General Power Rules

91. $T = \dfrac{1300}{t^2 + 2t + 25} = 1300(t^2 + 2t + 25)^{-1}$

$$T'(t) = -1300(t^2 + 2t + 25)^{-2}(2t + 2) = -\frac{2600(t + 1)}{(t^2 + 2t + 25)^2}$$

(a) $T'(1) = -\dfrac{325}{49} \approx -6.63°\text{F/hr}$

$T'(3) = -\dfrac{13}{2} \approx -6.5°\text{F/hr}$

$T'(5) = -\dfrac{13}{3} \approx -4.33°\text{F/hr}$

$T'(10) = -\dfrac{1144}{841} \approx -1.36°\text{F/hr}$

(b)

60
0
0
24

The rate of decrease is approaching zero.

93. $f(x) = 3x^2 + 7x + 1$

$f'(x) = 6x + 7$

$f''(x) = 6$

95. $f'''(x) = -6x^{-4}$

$f^{(4)}(x) = 24x^{-5}$

$f^{(5)}(x) = -120x^{-6} = -\dfrac{120}{x^6}$

97. $f'(x) = 7x^{5/2}$

$f''(x) = \dfrac{35}{2}x^{3/2}$

99. $f''(x) = 6\sqrt[3]{x} = 6x^{1/3}$

$f'''(x) = 2x^{-2/3} = \dfrac{2}{x^{2/3}}$

101. (a) $s(t) = -16t^2 + 5t + 30$

(b) $s(t) = 0 = -16t^2 + 5t + 30$

Using the Quadratic Formula, $t \approx 1.534$ seconds.

(c) $v(t) = s'(t) = -32t + 5$

$v(1.534) \approx -44.09$ ft/sec

(d) $a(t) = v'(t) = -32$ ft/sec^2

103.

$$x^2 + 3xy + y^3 = 10$$

$$2x + 3x\frac{dy}{dx} + 3y + 3y^2\frac{dy}{dx} = 0$$

$$\frac{dy}{dx}\left(3x + 3y^2\right) = -2x - 3y$$

$$\frac{dy}{dx} = \frac{-2x - 3y}{3x + 3y^2} = -\frac{2x + 3y}{3\left(x + y^2\right)}$$

105. $y^2 - x^2 + 8x - 9y - 1 = 0$

$$2y\frac{dy}{dx} - 2x + 8 - 9\frac{dy}{dx} = 0$$

$$(2y - 9)\frac{dy}{dx} = 2x - 8$$

$$\frac{dy}{dx} = \frac{2x - 8}{2y - 9}$$

109. $y^2 - 2x = xy$

$$2y\frac{dy}{dx} - 2 = x\frac{dy}{dx} + y$$

$$(2y - x)\frac{dy}{dx} = y + 2$$

$$\frac{dy}{dx} = \frac{y + 2}{2y - x}$$

At $(1, 2)$, $\dfrac{dy}{dx} = \dfrac{4}{3}$.

$$y - 2 = \frac{4}{3}(x - 1)$$

$$y = \frac{4}{3}x + \frac{2}{3}$$

111. $A = \pi r^2$

$$\frac{dA}{dt} = 2\pi r\frac{dr}{dt}$$

(a) Find $\dfrac{dA}{dt}$ when $r = 3$ in. and $\dfrac{dr}{dt} = 2$ in./min.

$$\frac{dA}{dt} = 2\pi(3)(2) = 12\pi \text{ in.}^2/\text{min}$$

$$\approx 37.7 \text{ in.}^2/\text{min}$$

(b) Find $\dfrac{dA}{dt}$ when $r = 10$ in. and $\dfrac{dr}{dt} = 2$ in./min.

$$\frac{dA}{dt} = 2\pi(10)(2) = 40\pi \text{ in.}^2/\text{min}$$

$$\approx 125.7 \text{ in.}^2/\text{min}$$

113. Let b be the horizontal distance of the water and h be the depth of the water at the deep end.

Then $b = 8h$ for $0 \le h \le 5$.

$$V = \frac{1}{2}bh(20) = 10bh = 10(8h)h = 80h^2$$

$$\frac{dV}{dt} = 160h\frac{dh}{dt}$$

$$\frac{dh}{dt} = \frac{1}{160h}\frac{dV}{dt} = \frac{1}{160h}(10) = \frac{1}{16h}$$

When $h = 4$, $\dfrac{dh}{dt} = \dfrac{1}{16(4)} = \dfrac{1}{64}$ ft/min.

Chapter 2 Test Yourself

1. $f'(x) = \lim_{\Delta x \to 0} \frac{f(x + \Delta x) - f(x)}{\Delta x}$

$= \lim_{\Delta x \to 0} \frac{(x + \Delta x)^2 + 1 - (x^2 + 1)}{\Delta x}$

$= \lim_{\Delta x \to 0} \frac{2x\Delta x + (\Delta x)^2}{\Delta x}$

$= \lim_{\Delta x \to 0} 2x + \Delta x$

$= 2x$

At $(2, 5)$: $m = 2(2) = 4$

2. $f'(x) = \lim_{\Delta x \to 0} \frac{f(x + \Delta x) - f(x)}{\Delta x}$

$= \lim_{\Delta x \to 0} \frac{\sqrt{x + \Delta x} - 2 - (\sqrt{x} - 2)}{\Delta x}$

$= \lim_{\Delta x \to 0} \frac{\sqrt{x + \Delta x} - \sqrt{x}}{\Delta x}$

$= \lim_{\Delta x \to 0} \frac{1}{\sqrt{x + \Delta x} + \sqrt{x}}$

$= \frac{1}{2\sqrt{x}}$

At $(4, 0)$: $m = \frac{1}{2\sqrt{4}} = \frac{1}{4}$

3. $f(t) = t^3 + 2t$

$f'(t) = 3t^2 + 2$

4. $f(x) = 4x^2 - 8x + 1$

$f'(x) = 8x - 8$

5. $f(x) = x^{3/2}$

$f'(x) = \frac{3}{2}x^{1/2} = \frac{3}{2\sqrt{x}}$

6. $f(x) = (x + 3)(x^2 + 2x)$

$f(x) = x^3 + 5x^2 + 6x$

$f'(x) = 3x^2 + 10x + 6$

(Or use the Product Rule.)

7. $f(x) = -3x^{-3}$

$f'(x) = 9x^{-4} = \frac{9}{x^4}$

8. $f(x) = \sqrt{x}(5 + x) = 5x^{1/2} + x^{3/2}$

$f'(x) = \frac{5}{2}x^{-1/2} + \frac{3}{2}x^{1/2} = \frac{5}{2\sqrt{x}} + \frac{3\sqrt{x}}{2}$

9. $f(x) = (3x^2 + 4)^2$

$f'(x) = 2(3x^2 + 4)(6x)$

$= 36x^3 + 48x$

10. $f(x) = \sqrt{1 - 2x} = (1 - 2x)^{1/2}$

$f'(x) = \frac{1}{2}(1 - 2x)^{-1/2}(-2)$

$= -\frac{1}{\sqrt{1 - 2x}}$

11. $f(x) = \frac{(5x - 1)^3}{x}$

$f'(x) = \frac{x(3)(5x - 1)^2(5) - (5x - 1)^3}{x^2}$

$= \frac{(5x - 1)^2[15x - (5x - 1)]}{x^2}$

$= \frac{(5x - 1)^2(10x + 1)}{x^2}$

12. $f(x) = x - \frac{1}{x}$

$f'(x) = 1 + \frac{1}{x^2}$

$f'(1) = 1 + \frac{1}{1^2} = 2$

$y - 0 = 2(x - 1)$

$y = 2x - 2$

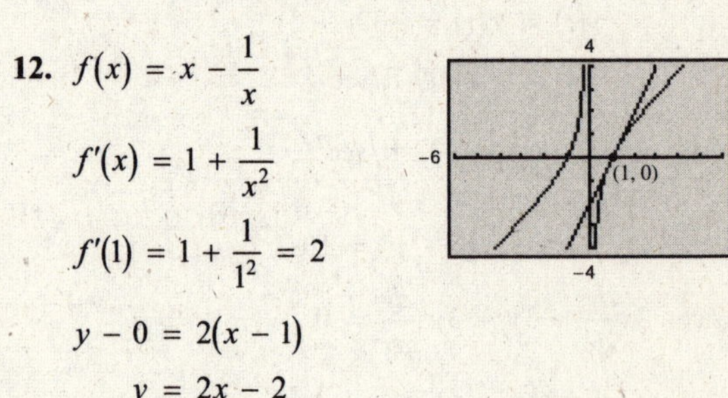

13. $S = -1.3241t^3 + 26.562t^3 - 155.81t + 314.3$

(a) $\frac{\Delta S}{\Delta t}$ for $5 \le t \le 8$

$\frac{S(8) - S(5)}{8 - 5} = \frac{89.8488 - 33.7875}{3}$

$= \$18.6871$ billion/yr

(b) $S'(t) = -3.9723t^2 + 53.124t - 155.81$

$S'(5) = \$10.5025$ billion/yr

$S'(8) = \$14.9548$ billion/yr

(c) The annual sales of CVS Caremark from 2005 to 2008 increased on average by about \$18.69 billion/year, and the instantaneous rates of change for 2005 and 2008 are \$10.50 billion/year and \$14.95 billion/year, respectively.

14. $P = 1700 - 0.016x,\ C = 715{,}000 + 240x$

Profit = Revenue − Cost

(a) Revenue: $R = xp$

$$R = x(1700 - 0.016x)$$

$$R = 1700x - 0.016x^2$$

$$P = R - C$$

$$P = (1700x - 0.016x^2) - (715{,}000 + 240x)$$

$$P = -0.016x^2 + 1460x - 715{,}000$$

(b) $\dfrac{dP}{dx} = -0.032x + 1460 = P'(x)$

$P'(700) = \$1437.60$

15. $f(x) = 2x^2 + 3x + 1$

$$f'(x) = 4x + 3$$

$$f''(x) = 4$$

$$f'''(x) = 0$$

16. $f(x) = \sqrt{3 - x} = (3 - x)^{1/2}$

$$f'(x) = \frac{1}{2}(3 - x)^{-1/2}(-1) = -\frac{1}{2}(3 - x)^{-1/2}$$

$$f''(x) = -\frac{1}{2}\left(-\frac{1}{2}\right)(3 - x)^{-3/2}(-1) = -\frac{1}{4}(3 - x)^{-3/2}$$

$$f'''(x) = -\frac{1}{4}\left(-\frac{3}{2}\right)(3 - x)^{-5/2}(-1)$$

$$= -\frac{3}{8}(3 - x)^{-5/2}$$

$$= -\frac{3}{8(3 - x)^{5/2}}$$

17. $f(x) = \dfrac{2x + 1}{2x - 1}$

$$f'(x) = \frac{(2x - 1)(2) - (2x + 1)(2)}{(2x - 1)^2}$$

$$= \frac{4}{(2x - 1)^2}$$

$$= -4(2x - 1)^{-2}$$

$$f''(x) = 8(2x - 1)^{-3}(2) = 16(2x - 1)^{-3}$$

$$f'''(x) = -48(2x - 1)^{-4}(2) = -\frac{96}{(2x - 1)^4}$$

18. $s(t) = -16t^2 + 30t + 75$

$$v(t) = s'(t) = -32t + 30$$

$$a(t) = v'(t) = s''(t) = -32$$

At $t = 2$: $s(2) = 71$ ft

$v(2) = -34$ ft/sec

$a(2) = -32$ ft/sec^2

19. $x + xy = 6$

$$1 + x\frac{dy}{dx} + y = 0$$

$$x\frac{dy}{dx} = -y - 1$$

$$\frac{dy}{dx} = -\frac{y + 1}{x}$$

20. $y^2 + 2x - 2y + 1 = 0$

$$2y\frac{dy}{dx} + 2 - 2\frac{dy}{dx} = 0$$

$$\frac{dy}{dx}(2y - 2) = -2$$

$$\frac{dy}{dx} = -\frac{1}{y - 1}$$

21. $x^2 - 2y^2 = 4$

$$2x - 4y\frac{dy}{dx} = 0$$

$$\frac{dy}{dx} = \frac{x}{2y}$$

22. $V = \pi r^2 h = 20\pi r^3$

$$\frac{dV}{dt} = 60\pi r^2\frac{dr}{dt}$$

(a) $\dfrac{dV}{dt} = 60\pi(0.5)^2(0.25) = 3.75\pi$ cm^3/min

(b) $\dfrac{dV}{dt} = 60\pi(1)^2(0.25) = 15\pi$ cm^3/min

CHAPTER 3
Applications of the Derivative

CHAPTER 3
Applications of the Derivative

Section 3.1 Increasing and Decreasing Functions

Skills Warm Up

1.
$$\begin{aligned} x^2 &= 8x \\ x^2 - 8x &= 0 \\ x(x-8) &= 0 \\ x &= 0 \\ x - 8 &= 0 \Rightarrow x = 8 \end{aligned}$$

2.
$$\begin{aligned} 15x &= \tfrac{5}{8}x^2 \\ 15x - \tfrac{5}{8}x^2 &= 0 \\ x\left(15 - \tfrac{5}{8}x\right) &= 0 \\ x &= 0 \\ 15 - \tfrac{5}{8}x &= 0 \Rightarrow x = 24 \end{aligned}$$

3.
$$\begin{aligned} \frac{x^2 - 25}{x^3} &= 0 \\ \frac{1}{x} - \frac{25}{x^3} &= 0 \\ \frac{1}{x} &= \frac{25}{x^3} \\ x^3 &= 25x \\ x^3 - 25x &= 0 \\ x\left(x^2 - 25\right) &= 0 \\ x(x+5)(x-5) &= 0 \\ x &= 0 \text{ Extraneous} \\ x + 5 &= 0 \Rightarrow x = -5 \\ x - 5 &= 0 \Rightarrow x = 5 \end{aligned}$$

4.
$$\begin{aligned} \frac{2x}{\sqrt{1 - x^2}} &= 0 \\ 2x &= 0 \\ x &= 0 \end{aligned}$$

5. The domain of $\dfrac{x+3}{x-3}$ is $(-\infty, 3) \cup (3, \infty)$.

6. The domain of $\dfrac{2}{\sqrt{1-x}}$ is $(-\infty, 1)$.

7. The domain of $\dfrac{2x+1}{x^2 - 3x - 10}$ is

$(-\infty, -2) \cup (-2, 5) \cup (5, \infty)$.

8. The domain of $\dfrac{3x}{\sqrt{9 - 3x^2}}$ is $\left(-\sqrt{3}, \sqrt{3}\right)$.

9. When $x = -2$: $-2(-2+1)(-2-1) = -6$

When $x = 0$: $-2(0+1)(0-1) = 2$

When $x = 2$: $-2(2+1)(2-1) = -6$

10. When $x = -2$: $4\left[2(-2)+1\right]\left[2(-2)-1\right] = 60$

When $x = 0$: $4(2 \cdot 0 + 1)(2 \cdot 0 - 1) = -4$

When $x = 2$: $4(2 \cdot 2 + 1)(2 \cdot 2 - 1) = 60$

11. When $x = -2$: $\dfrac{2(-2)+1}{(-2-1)^2} = -\dfrac{1}{3}$

When $x = 0$: $\dfrac{2 \cdot 0 + 1}{(0-1)^2} = 1$

When $x = 2$: $\dfrac{2 \cdot 2 + 1}{(2-1)^2} = 5$

12. When $x = -2$: $\dfrac{-2(-2+1)}{(-2-4)^2} = \dfrac{1}{18}$

When $x = 0$: $\dfrac{-2(0+1)}{(0-4)^2} = -\dfrac{1}{8}$

When $x = 2$: $\dfrac{-2(2+1)}{(2-4)^2} = -\dfrac{3}{2}$

1. f has a critical number at $x = -1$.

f is increasing on $(-\infty, -1)$.

f is decreasing on $(-1, \infty)$.

3. f has critical numbers at $x = \pm 1, 0$.

f is increasing on $(-1, 0) \cup (1, \infty)$.

f is decreasing on $(-\infty, -1) \cup (0, 1)$.

5. $f(x) = 4x^2 - 6x$

$f'(x) = 8x - 6 = 2(4x - 3)$

Set $f'(x) = 0$.

$2(4x - 3) = 0$

$4x - 3 = 0$

$x = \frac{3}{4}$

$f'(x)$ is never undefined.

Critical number: $x = \frac{3}{4}$

7. $y = x^4 + 4x^3 + 8$

$y' = 4x^3 + 12x^2 = 4x^2(x + 3)$

Set $y' = 0$.

$4x^2(x + 3) = 0$

$4x^2 = 0 \qquad x + 3 = 0$

$x = 0 \qquad x = -3$

y' is never undefined.

Critical numbers: $x = 0, -3$

9. $f(x) = \sqrt{x^2 - 4} = (x^2 - 4)^{1/2}$

$f'(x) = \frac{1}{2}(x^2 - 4)^{-1/2}(2x)$

$= \frac{x}{(x^2 - 4)^{1/2}}$

Set $f'(x) = 0$.

$x = 0$ (**Note:** $x = 0$ is not in the domain of $f(x)$.)

$f'(x)$ is undefined at $x = \pm 2$.

Critical numbers: $x = \pm 2$

11. $f(x) = 2x - 3$

$f'(x) = 2$

There are no critical numbers. Since the derivative is positive for all x, the function is increasing on $(-\infty, \infty)$.

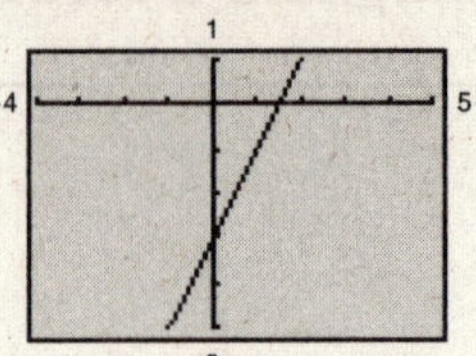

13. $y = x^2 - 6x$

$y' = 2x - 6$

Critical number: $x = 3$

Interval	$-\infty < x < 3$	$3 < x < \infty$
Sign of y'	$y' < 0$	$y' > 0$
Conclusion	Decreasing	Increasing

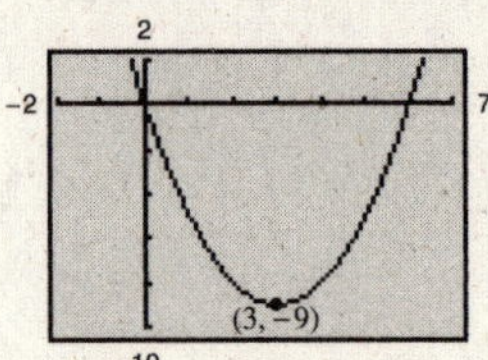

15. $f(x) = -2x^2 + 4x + 3$

$f'(x) = -4x + 4$

Critical number: $x = 1$

Interval	$-\infty < x < 1$	$1 < x < \infty$
Sign of f'	$f' > 0$	$f' < 0$
Conclusion	Increasing	Decreasing

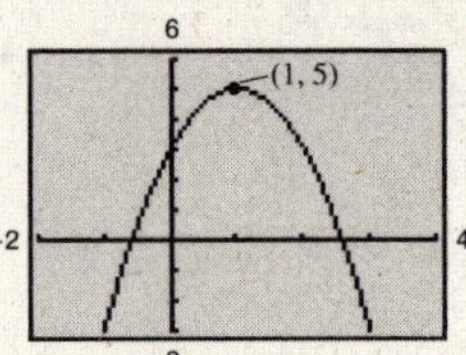

17. $y = 3x^3 + 12x^2 + 15x$

$y' = 9x^2 + 24x + 15 = 3(x + 1)(3x + 5)$

Critical numbers: $x = -1, -\frac{5}{3}$

Interval	$-\infty < x < -\frac{5}{3}$	$-\frac{5}{3} < x < -1$	$-1 < x < \infty$
Sign of y'	$y' > 0$	$y' < 0$	$y' > 0$
Conclusion	Increasing	Decreasing	Increasing

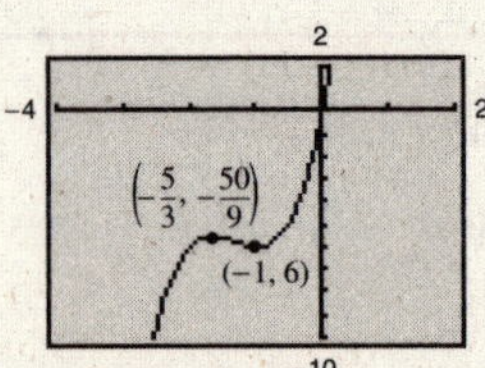

19. $f(x) = x^4 - 2x^3$

$f'(x) = 4x^3 - 6x^2 = 2x^2(2x - 3)$

Critical numbers: $x = 0, x = \frac{3}{2}$

Interval	$-\infty < x < 0$	$0 < x < \frac{3}{2}$	$\frac{3}{2} < x < \infty$
Sign of f'	$f' < 0$	$f' < 0$	$f' > 0$
Conclusion	Decreasing	Decreasing	Increasing

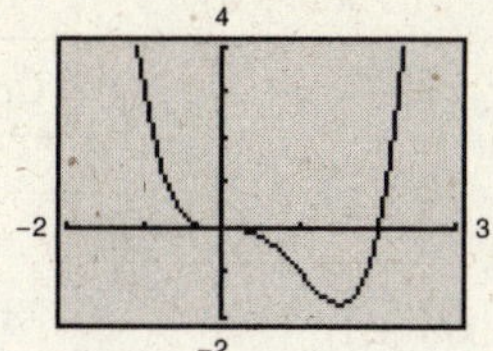

f is decreasing on $\left(-\infty, \frac{3}{2}\right)$ and increasing on $\left(\frac{3}{2}, \infty\right)$.

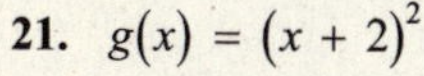

21. $g(x) = (x + 2)^2$

$g'(x) = 2(x + 2)$

Critical number: $x = -2$

Interval	$-\infty < x < -2$	$-2 < x < \infty$
Sign of g'	$g' < 0$	$g' > 0$
Conclusion	Decreasing	Increasing

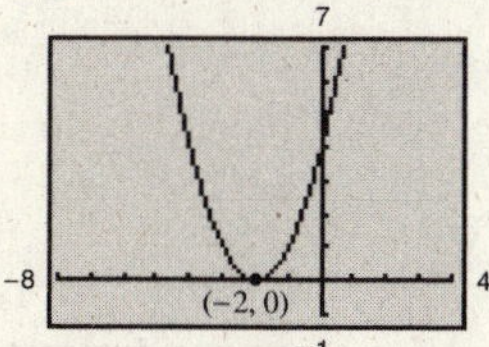

23. $g(x) = -(x - 1)^2$

$g'(x) = -2(x - 1)$

Critical number: $x = 1$

Interval	$-\infty < x < 1$	$1 < x < \infty$
Sign of g'	$g' > 0$	$g' < 0$
Conclusion	Increasing	Decreasing

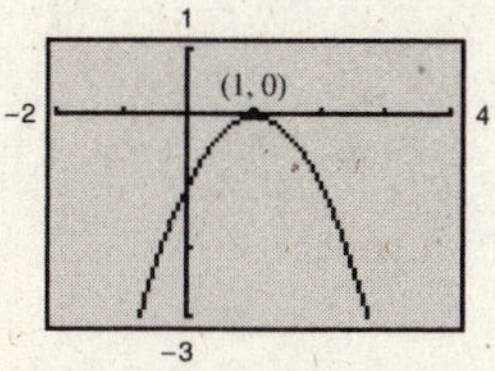

25. $y = x^{1/3} + 1$

$y' = \frac{1}{3}x^{-2/3} = \frac{1}{3x^{2/3}}$

Critical number: $x = 0$

Interval	$-\infty < x < 0$	$0 < x < \infty$
Sign of y'	$y' > 0$	$y' > 0$
Conclusion	Increasing	Increasing

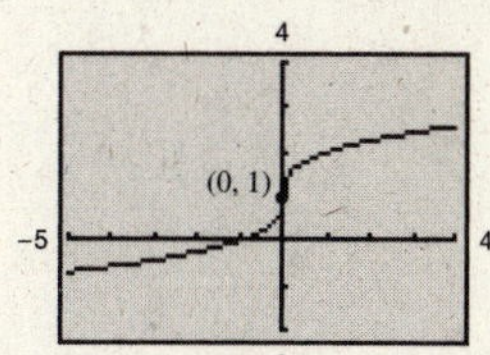

y is increasing on $(-\infty, \infty)$.

27. $f(x) = \sqrt{x^2 - 1} = \left(x^2 - 1\right)^{1/2}$

Domain: $(-\infty, -1] \cup [1, \infty)$

$f'(x) = \frac{1}{2}\left(x^2 - 1\right)^{-1/2}(2x) = \frac{x}{\sqrt{x^2 - 1}}$

Critical numbers: $x = \pm 1$ ($x = 0$ not in domain)

Interval	$-\infty < x < -1$	$1 < x < \infty$
Sign of f'	$f' < 0$	$f' > 0$
Conclusion	Decreasing	Increasing

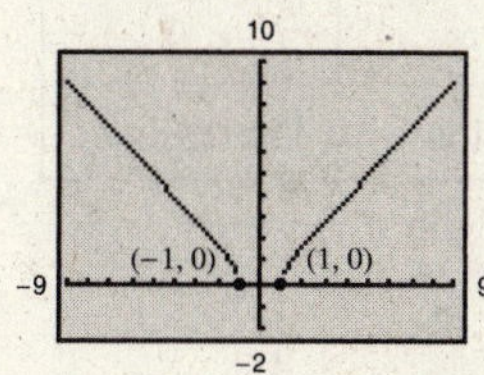

29. $g(x) = (x + 2)^{1/3}$

$$g'(x) = \frac{1}{3}(x + 2)^{-2/3}(1) = \frac{1}{3(x + 2)^{2/3}}$$

$g'(-2)$ is undefined.

Critical number: $x = -2$

Interval	$-\infty < x < -2$	$-2 < x < \infty$
Sign of g'	$g' > 0$	$g' > 0$
Conclusion	Increasing	Increasing

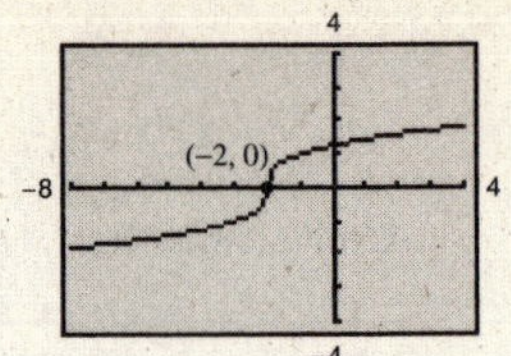

31. $f(x) = x\sqrt{x + 1} = x(x + 1)^{1/2}$

Domain: $[-1, \infty)$

$$f'(x) = x\left[\frac{1}{2}(x + 1)^{-1/2}\right] + (x + 1)^{1/2} = \frac{1}{2}(x + 1)^{-1/2}[x + 2(x + 1)] = \frac{3x + 2}{2\sqrt{x + 1}}$$

Critical numbers: $x = -1, -\frac{2}{3}$

Interval	$-1 < x < -\frac{2}{3}$	$-\frac{2}{3} < x < \infty$
Sign of f'	$f' < 0$	$f' > 0$
Conclusion	Decreasing	Increasing

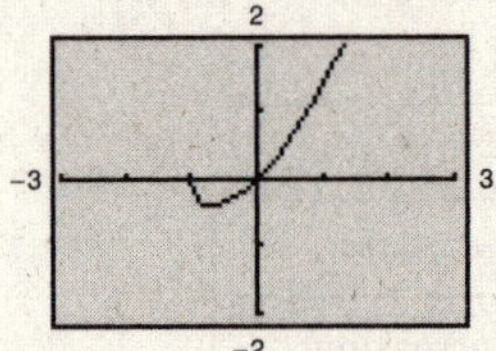

33. $f(x) = \dfrac{x}{x^2 + 9}$

$$f'(x) = \frac{(x^2 + 9)(1) - x(2x)}{(x^2 + 9)^2} = \frac{9 - x^2}{(x^2 + 9)^2}$$

$f'(x) = 0$:

$9 - x^2 = 0$

$x^2 = 9$

$x = \pm 3$

Critical numbers: $x = \pm 3$

Interval	$-\infty < x < -3$	$-3 < x < 3$	$3 < x < \infty$
Sign of f'	$f' < 0$	$f' > 0$	$f' < 0$
Conclusion	Decreasing	Increasing	Decreasing

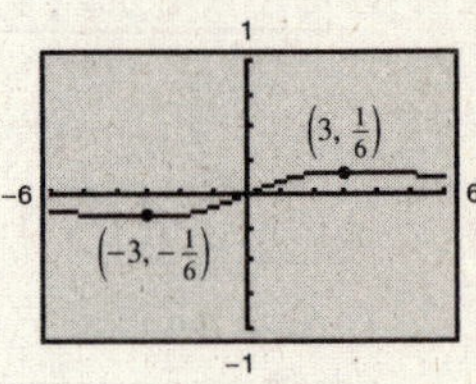

35. $f(x) = \dfrac{x + 4}{x - 5}$

$$f'(x) = \frac{(x - 5)(1) - (x + 4)(1)}{(x - 5)^2} = -\frac{9}{(x - 5)^2}$$

Discontinuity: $x = 5$

Interval	$-\infty < x < 5$	$5 < x < \infty$
Sign of f'	$f' < 0$	$f' < 0$
Conclusion	Decreasing	Decreasing

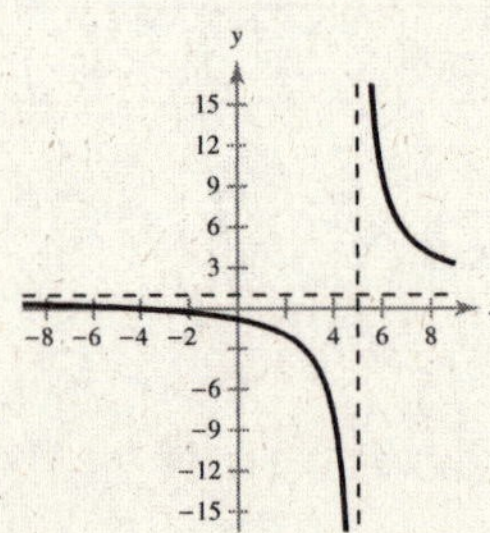

37. $f(x) = \dfrac{2x}{16 - x^2}$

$f'(x) = \dfrac{(16 - x^2)2 - 2x(-2x)}{(16 - x^2)^2} = \dfrac{2x^2 + 32}{(16 - x^2)^2}$

Discontinuities: $x = \pm 4$

Interval	$-\infty < x < -4$	$-4 < x < 4$	$4 < x < \infty$
Sign of f'	$f' > 0$	$f' > 0$	$f' > 0$
Conclusion	Increasing	Increasing	Increasing

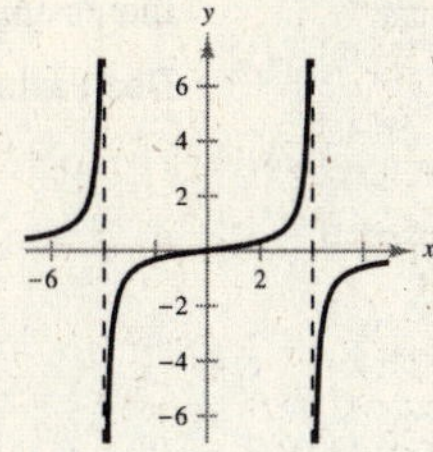

39. $y = \begin{cases} 4 - x^2, & x \le 0 \\ -2x, & x > 0 \end{cases}$

$y' = \begin{cases} -2x, & x < 0 \\ -2, & x > 0 \end{cases}$

$y'(0)$ is undefined.

Critical number: $x = 0$

Interval	$-\infty < x < 0$	$0 < x < \infty$
Sign of y'	$y' > 0$	$y' < 0$
Conclusion	Increasing	Decreasing

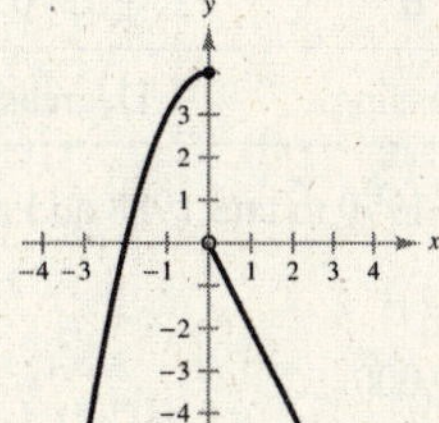

41. $y = \begin{cases} 2x + 1, & x \le -1 \\ x^2 - 2, & x > -1 \end{cases}$

$y' = \begin{cases} 2, & x < -1 \\ 2x, & x > -1 \end{cases}$

$y'(-1)$ is undefined.

Critical numbers: $x = -1, 0$

Interval	$-\infty < x < -1$	$-1 < x < 0$	$0 < x < \infty$
Sign of y'	$y' > 0$	$y' < 0$	$y' > 0$
Conclusion	Increasing	Decreasing	Increasing

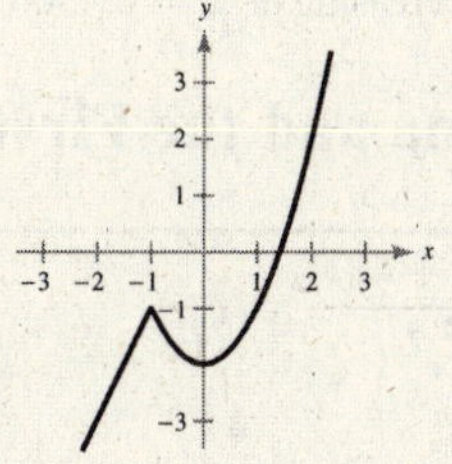

43. $S = -1.598t^2 + 45.61t + 130.2, \ 3 \le t \le 9$

$S' = -3.196t + 45.61$

Since $S' > 0$ when $t \approx 14.3$, there is no critical number in the interval $[3, 9]$.

$S' > 0$ on the interval $[3, 9]$, so S is increasing on $[3, 9]$.

Sales for Wal-Mart are strictly increasing from 2003 to 2009.

45. $y = 0.692t^3 - 50.11t^2 + 1119.7t + 7894$

$y' = 2.076t^2 - 100.22t + 1119.7$

(a)

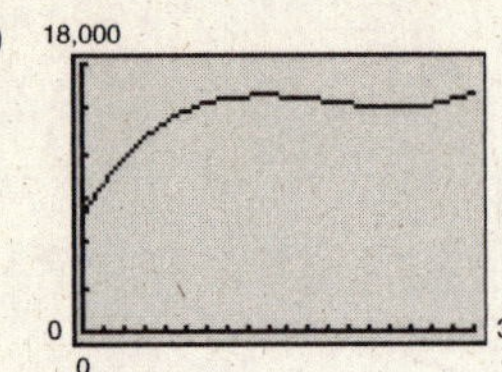

Increasing from 1970 to late 1988 and from late 2001 to 2008

Decreasing from late 1988 to late 2001.

(b) Set $y' = 0$.

$2.076t^2 - 100.22t + 1119.7 = 0$

$$t = \frac{-(-100.22) \pm \sqrt{(-100.22)^2 - 4(2.076)(1119.7)}}{2(2.076)}$$

Critical numbers: $t \approx 17.6, 30.7$

Interval	$0 < t < 17.6$	$17.6 < t < 30.7$	$30.7 < t < 38$
Sign of y'	$y' > 0$	$y' < 0$	$y' > 0$
Conclusion	Increasing	Decreasing	Decreasing

Therefore, the model is increasing from 1970 to late 1988 and from late 2001 to 2008 and decreasing from late 1988 to late 2001.

47. $P = 2.36x - \dfrac{x^2}{25{,}000} - 3500, \ 0 \le x \le 50{,}000$

(a) $P' = 2.36 - \dfrac{1}{12{,}500}x$

Domain: $0 \le x \le 50{,}000$

Critical number: $x = 29{,}500$

Interval	$0 \le x \le 29{,}500$	$29{,}500 < x \le 50{,}000$
Sign of P'	$P' > 0$	$P' < 0$
Conclusion	Increasing	Decreasing

(b) You should charge the price that yields sales of $x = 29{,}500$ bags of popcorn. Since the function changes from increasing to decreasing at $x = 29{,}500$, the maximum profit occurs at this value.

Section 3.2 Extrema and the First-Derivative Test

Skills Warm Up

1. $f(x) = 4x^4 - 2x^2 + 1$

$f'(x) = 16x^3 - 4x = 0$

$4x(4x^2 - 1) = 0$

$4x(2x + 1)(2x - 1) = 0$

$x = 0, \ x = \pm\frac{1}{2}$

2. $f(x) = \frac{1}{3}x^3 - \frac{3}{2}x^2 - 10x$

$f'(x) = x^2 - 3x - 10 = 0$

$(x - 5)(x + 2) = 0$

$x = -2, \ x = 5$

3. $f(x) = 5x^{4/5} - 4x$

$f'(x) = 4x^{-1/5} - 4$

$= \dfrac{4}{x^{1/5}} - 4 = 0$

$x^{1/5} = 1$

$x = 1$

4. $f(x) = \frac{1}{2}x^2 - 3x^{5/3}$

$f'(x) = x - 5x^{2/3} = 0$

$x^{2/3}(x^{1/3} - 5) = 0$

$x = 0, \ x = 125$

Skills Warm Up —*continued*—

5. $f(x) = \dfrac{x+4}{x^2+1}$

$$f'(x) = \frac{-x^2 - 8x + 1}{\left(x^2+1\right)^2} = 0$$

$$-x^2 - 8x + 1 = 0$$

$$x = \frac{8 \pm \sqrt{(-8)^2 - 4(-1)(1)}}{2(-1)} = \frac{8 \pm \sqrt{68}}{-2} = \frac{8 \pm 2\sqrt{17}}{-2} = -4 \pm \sqrt{17}$$

6. $f(x) = \dfrac{x-1}{x^2+4}$

$$f'(x) = \frac{-x^2 + 2x + 4}{\left(x^2+4\right)^2} = 0$$

$$-x^2 + 2x + 4 = 0$$

$$x = \frac{-2 \pm \sqrt{2^2 - 4(-1)(4)}}{2(-1)} = \frac{-2 \pm \sqrt{20}}{-2} = \frac{-2 \pm 2\sqrt{5}}{-2} = 1 \pm \sqrt{5}$$

7. $g'(x) = -5x^4 - 8x^3 + 12x^2 + 2$

$g'(-4) = -574 < 0$

8. $g'(x) = -5x^4 - 8x^3 + 12x^2 + 2$

$g'(0) = 2 > 0$

9. $g'(x) = -5x^4 - 8x^3 + 12x^2 + 2$

$g'(1) = 1 > 0$

10. $g'(x) = -5x^4 - 8x^3 + 12x^2 + 2$

$g'(3) = -511 < 0$

11. $f(x) = 2x^2 - 11x - 6, (3, 6)$

$f'(x) = 4x - 11$

$f'(4) = 5 > 0$

So, f is increasing on $(3, 6)$.

12. $f(x) = x^3 + 2x^2 - 4x - 8, (-2, 0)$

$f'(x) = 3x^2 + 4x - 4$

$f'(-1) = -5 < 0$

So, f is decreasing on $(-2, 0)$.

1. $f(x) = -2x^2 + 4x + 3$

$f'(x) = 4 - 4x = 4(1 - x)$

Critical number: $x = 1$

Interval	$-\infty < x < 1$	$1 < x < \infty$
Sign of f'	$f' > 0$	$f' < 0$
Conclusion	Increasing	Decreasing

Relative maximum: $(1, 5)$

3. $f(x) = x^2 - 6x$

$f'(x) = 2x - 6 = 2(x - 3)$

Critical number: $x = 3$

Interval	$-\infty < x < 3$	$3 < x < \infty$
Sign of f'	$f' < 0$	$f' > 0$
Conclusion	Decreasing	Increasing

Relative minimum: $(3, -9)$

5. $f(x) = x^4 - 12x^3$

$f'(x) = 4x^3 - 36x^2 = 4x^2(x - 9)$

Critical numbers: $x = 0$, $x = 9$

Interval	$-\infty < x < 0$	$0 < x < 9$	$9 < x < \infty$
Sign of f'	$f' < 0$	$f' < 0$	$f' > 0$
Conclusion	Decreasing	Decreasing	Increasing

Relative minimum: $(9, -2187)$

7. $h(x) = -(x + 4)^3$

$h'(x) = -3(x + 4)^2$

Critical number: $x = -4$

No relative extrema

Interval	$-\infty < x < -4$	$-4 < x < \infty$
Sign of h'	$h' < 0$	$h' < 0$
Conclusion	Decreasing	Decreasing

9. $f(x) = x^3 - 6x^2 + 15$

$f'(x) = 3x^2 - 12x = 3x(x - 4)$

Critical numbers: $x = 0$, $x = 4$

Interval	$-\infty < x < 0$	$0 < x < 4$	$4 < x < \infty$
Sign of f'	$f' > 0$	$f' < 0$	$f' > 0$
Conclusion	Increasing	Decreasing	Increasing

Relative maximum: $(0, 15)$

Relative minimum: $(4, -17)$

11. $f(x) = 6x^{2/3} + 4x$

$f'(x) = 4x^{-1/3} + 4 = \dfrac{4}{x^{1/3}} + 4$

$f'(0)$ is undefined.

Set $f'(x) = 0$.

$$\frac{4}{x^{1/3}} + 4 = 0$$

$$4 = -4x^{1/3}$$

$$-1 = x^{1/3}$$

$$-1 = x$$

Critical numbers: $x = -1, 0$

Interval	$-\infty < x < -1$	$-1 < x < 0$	$0 < x < \infty$
Sign of f'	$f' > 0$	$f' < 0$	$f' > 0$
Conclusion	Increasing	Decreasing	Increasing

Relative maximum: $(-1, 2)$

Relative minimum: $(0, 0)$

13. $f(x) = 2x - 6x^{2/3}$

$$f'(x) = 2 - 4x^{-1/3} = 2 - \frac{4}{x^{1/3}} = \frac{2(x^{1/3} - 2)}{x^{1/3}}$$

Critical numbers: $x = 0, 8$

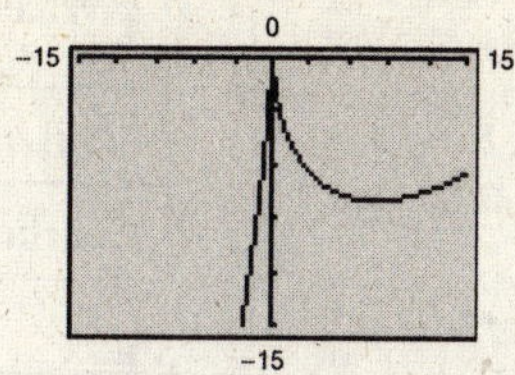

Interval	$-\infty < x < 0$	$0 < x < 8$	$8 < x < \infty$
Sign of f'	$f' > 0$	$f' < 0$	$f' > 0$
Conclusion	Increasing	Decreasing	Increasing

Relative maximum: $(0, 0)$

Relative minimum: $(8, -8)$

15. $g(t) = t - \dfrac{1}{2t^2} = t - \dfrac{1}{2}t^{-2}$

$g'(t) = 1 + t^{-3}$

$= \dfrac{t^3 + 1}{t^3}$

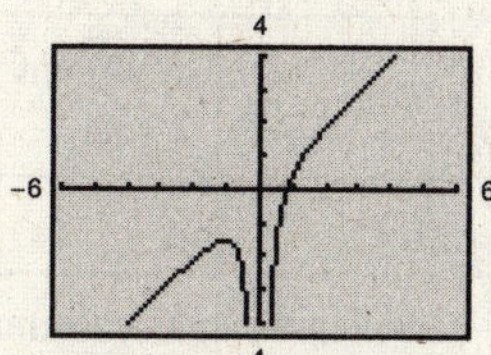

Critical number: $t = -1$

Discontinuity: $t = 0$

Interval	$-\infty < t < -1$	$-1 < t < 0$	$0 < t < \infty$
Sign of g'	$g' > 0$	$g' < 0$	$g' > 0$
Conclusion	Increasing	Decreasing	Increasing

Relative maximum: $\left(-1, -\frac{3}{2}\right)$

17. $f(x) = \dfrac{x}{x + 1}$

$$f'(x) = \frac{(x + 1) - x}{(x + 1)^2} = \frac{1}{(x + 1)^2}$$

Discontinuity: $x = -1$

Interval	$-\infty < x < -1$	$-1 < x < \infty$
Sign of f'	$f' > 0$	$f' > 0$
Conclusion	Increasing	Increasing

No relative extrema

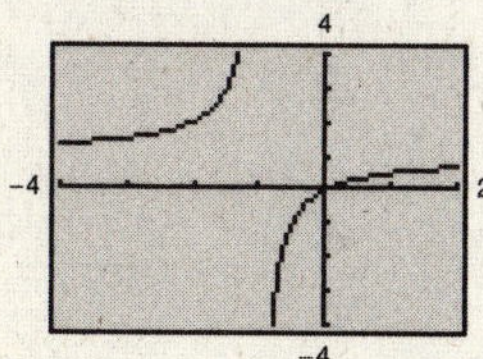

19. $f(x) = 2(3 - x)$, $[-1, 2]$

$f'(x) = -2$

No critical numbers

x-value	Endpoint $x = -1$	Endpoint $x = 2$
$f(x)$	8	2
Conclusion	Maximum	Minimum

21. $f(x) = 5 - 2x^2$, $[0, 3]$

$f'(x) = -4x$

Critical number: $x = 0$ (endpoint)

x-value	Endpoint $x = 0$	Endpoint $x = 3$
$f(x)$	5	−13
Conclusion	Maximum	Minimum

23. $f(x) = x^3 - 3x^2, [-1, 3]$

$f'(x) = 3x^2 - 6x = 3x(x - 2)$

Critical numbers: $x = 0, x = 2$

x-value	Endpoint $x = -1$	Critical $x = 0$	Critical $x = 2$	Endpoint $x = 3$
$f(x)$	–4	0	–4	0
Conclusion	Minimum	Maximum	Minimum	Maximum

25. $h(s) = \dfrac{1}{3 - s} = (3 - s)^{-1}, [0, 2]$

$h'(s) = -(3 - s)^{-2}(-1) = \dfrac{1}{(3 - s)^2}$

No critical numbers

s-value	Endpoint $s = 0$	Endpoint $s = 2$
$h(s)$	$\frac{1}{3}$	1
Conclusion	Minimum	Maximum

27. $g(t) = \dfrac{t^2}{t^2 + 3}, [-1, 1]$

$g'(t) = \dfrac{(t^2 + 3)(2t) - t^2(2t)}{(t^2 + 3)^2} = \dfrac{6t}{(t^2 + 3)^2}$

Critical number: $t = 0$

t-value	Endpoint $t = -1$	Critical $t = 0$	Endpoint $t = 1$
$g(t)$	$\frac{1}{4}$	0	$\frac{1}{4}$
Conclusion	Maximum	Minimum	Maximum

29. $h(t) = (t - 1)^{2/3}, [-7, 2]$

$h'(t) = \dfrac{2}{3}(t - 1)^{-1/3} = \dfrac{2}{3(t - 1)^{1/3}}$

Critical number: $t = 1$

t-value	Endpoint $t = -7$	Critical $t = 1$	Endpoint $t = 2$
$h(t)$	4	0	1
Conclusion	Maximum	Minimum	

31. Critical number: $x = 2$

The function has an absolute maximum at the critical number.

33. Critical number: $x = 1$, absolute maximum and relative maximum

Critical number: $x = 2$, absolute minimum and relative minimum

Critical number: $x = 3$, absolute maximum and relative maximum

35. $f(x) = 0.4x^3 - 1.8x^2 + x - 3, [0, 5]$

Maximum: $(5, 7)$

Minimum: $(2.69, 5.55)$

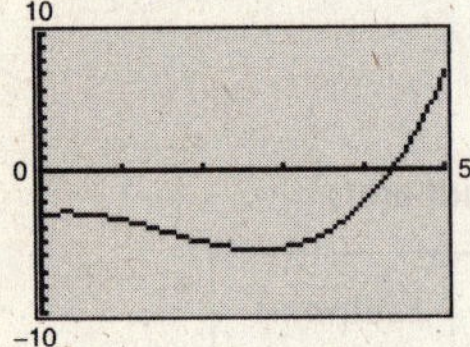

37. $f(x) = \frac{4}{3}x\sqrt{3 - x}, [0, 3]$

Maximum: $(2, 2.67)$

Minimum: $(0, 0), (3, 0)$

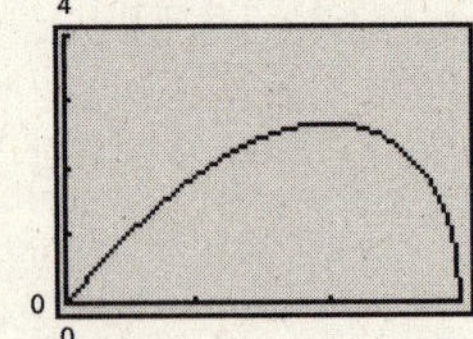

39. $f(x) = x^2 + 16x^{-1}, [0, \infty)$

$$f'(x) = 2x - 16x^{-2} = 2x - \frac{16}{x^2} = \frac{2(x^3 - 8)}{x^2}$$

Critical number: $x = 2$

Absolute minimum: $(2, 12)$

x-value	Critical $x = 2$
$f(x)$	12
Conclusion	Maximum

41. $f(x) = \frac{2x}{x^2 + 4}, [0, \infty)$

$$f'(x) = \frac{(x^2 + 4)(2) - 2x(2x)}{(x^2 + 4)^2} = \frac{8 - 2x^2}{(x^2 + 4)^2} = \frac{2(2 - x)(2 + x)}{(x^2 + 4)^2}$$

Critical number: $x = 2$

x-value	Endpoint $x = 0$	Critical $x = 2$
$f(x)$	0	$\frac{1}{2}$
Conclusion	Minimum	Maximum

43. Answers will vary. Sample answer:

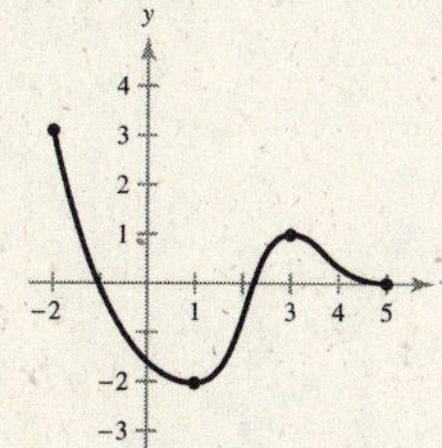

45. (a) Population tends to increase each year, so the minimum population occurred in 1790 and the maximum population occurred in 2010.

(b) $P = 0.000006t^3 + 0.005t^2 + 0.14t + 4.6$

$$\frac{dP}{dt} = P' = 0.000018t^2 + 0.01t + 0.14$$

$$P' = 0: 0.000018t^2 + 0.01t + 0.14 = 0$$

$$t \approx 14.4, \ -541.2$$

No critical numbers in interval $[-10, 210]$.

t-value	Endpoint $t = -10$	Endpoint $t = 210$
$P(t)$	3.69 million	310.07 million
Conclusion	Minimum	Maximum

(c) The minimum population was about 3.69 million in 1790 and the maximum population was about 310.07 million in 2010.

47. $C = 3x + \dfrac{20{,}000}{x} = 3x + 20{,}000x^{-1},\ 0 < x \le 200$

$C' = 3 - 20{,}000x^{-2} = \dfrac{3x^2 - 20{,}000}{x^2}$

Critical numbers: $x = \sqrt{\dfrac{20{,}000}{3}} \approx 82$ units

$C(82) \approx 489.90$, which is the minimum by the First-Derivative Test

49. Demand: (6000, 1.00), (5600, 1.20)

$m = \dfrac{1.20 - 1.00}{5600 - 6000} = \dfrac{0.2}{-400} = -0.0005$

$p - 1 = -0.0005(x - 6000)$

$p = -0.0005x + 4$

$\text{Cost} = C = 5000 + 0.50x$

$\text{Profit} = P = R - C$

$= xp - C$

$= x(-0.0005x + 4) - (5000 + 0.50x)$

$= -0.0005x^2 + 3.50x - 5000$

Selling the cans for \$2.25 will maximize the profit.

Section 3.3 Concavity and the Second-Derivative Test

Skills Warm Up

1. $f(x) = 4x^4 - 9x^3 + 5x - 1$

$f'(x) = 16x^3 - 27x^2 + 5$

$f''(x) = 48x^2 - 54x$

2. $g(s) = (s^2 - 1)(s^2 - 3s + 2)$

$= s^4 - 3s^3 + s^2 + 3s - 2$

$g'(s) = 4s^3 - 9s^2 + 2s + 3$

$g''(s) = 12s^2 - 18s + 2$

3. $g(x) = (x^2 + 1)^4$

$g'(x) = 4(x^2 + 1)^3(2x)$

$= 8x(x^2 + 1)^3$

$g''(x) = 8x\left[3(x^2 + 1)^2(2x)\right] + (x^2 + 1)^3(8)$

$= 8(x^2 + 1)^2(7x^2 + 1)$

4. $f(x) = (x - 3)^{4/3}$

$f'(x) = \frac{4}{3}(x - 3)^{1/3}$

$f''(x) = \frac{4}{9}(x - 3)^{-2/3}$

$= \dfrac{4}{9(x - 3)^{2/3}}$

5. $h(x) = \dfrac{4x + 3}{5x - 1}$

$h'(x) = \dfrac{(5x - 1)(4) - (4x + 3)(5)}{(5x - 1)^2}$

$= \dfrac{20x - 4 - 20x - 15}{(5x - 1)^2}$

$= -19(5x - 1)^{-2}$

$h''(x) = 18(5x - 1)^{-3}(5)$

$= \dfrac{90}{(5x - 1)^3}$

6. $f(x) = \dfrac{2x - 1}{3x + 2}$

$f'(x) = \dfrac{(3x + 2)(2) - (2x - 1)(3)}{(3x + 2)^2}$

$= \dfrac{6x + 4 - 6x + 3}{(3x + 2)^2}$

$= 7(3x + 2)^{-2}$

$f''(x) = -14(3x + 2)^{-3}(3)$

$= -\dfrac{42}{(3x + 2)^3}$

Skills Warm Up —*continued*—

8. $f(x) = x^4 - 4x^3 - 10$

$f'(x) = 4x^3 - 12x^2 = 0$

$4x^2(x - 3) = 0$

$4x^2 = 0 \Rightarrow x = 0$

$x - 3 = 0 \Rightarrow x = 3$

Critical numbers: $x = 0$, $x = 3$

9. $g(t) = \dfrac{16 + t^2}{t} = \dfrac{16}{t} - t$

$g'(t) = -\dfrac{16}{t^2} - 1 = 0$

$t^2 = -16$

No critical numbers

10. $h(x) = \dfrac{x^4 - 50x^2}{8}$

$h'(x) = \dfrac{1}{2}x^3 - \dfrac{25}{2}x = 0$

$\dfrac{1}{2}x(x^2 - 25) = 0$

$\dfrac{1}{2}x(x + 5)(x - 5) = 0$

$\dfrac{1}{2}x = 0 \Rightarrow x = 0$

$x + 5 = 0 \Rightarrow x = -5$

$x - 5 = 0 \Rightarrow x = 5$

Critical numbers: $x = 0$, $x = \pm 5$

1. f is increasing so $f' > 0$.
f is concave upward so $f'' > 0$.

3. f is decreasing so $f' < 0$.
f is concave downward so $f'' < 0$.

5. $f(x) = -3x^2$

$f'(x) = -6x$

$f''(x) = -6$

$f''(x) \neq 0$ for any value of x.

Interval	$-\infty < x < \infty$
Sign of f''	$f'' < 0$
Conclusion	Concave downward

7. $f(x) = -x^3 + 3x^2 - 2$

$f'(x) = -3x^2 + 6x$

$f''(x) = -6x + 6$

$f''(x) = 0$ when $x = 1$.

Interval	$-\infty < x < 1$	$1 < x < \infty$
Sign of f''	$f'' > 0$	$f'' < 0$
Conclusion	Concave upward	Concave downward

9. $f(x) = \dfrac{x^2 - 1}{2x + 1}$

$f'(x) = \dfrac{(2x + 1)(2x) - (x^2 - 1)(2)}{(2x + 1)^2} = \dfrac{2x^2 + 2x + 2}{(2x + 1)^2} = (2x^2 + 2x + 2)(2x + 1)^{-2}$

$f''(x) = (2x^2 + 2x + 2)\left[-2(2x + 1)^{-3}(2)\right] + (2x + 1)^{-2}(4x + 2)$

$= -8(x^2 + x + 1)(2x + 1)^{-3} + 2(2x + 1)^{-2}(2x + 1)$

$= 2(2x + 1)^{-3}\left[-4(x^2 + x + 1) + (2x + 1)(2x + 1)\right]$

$= 2(2x + 1)^{-3}\left[-4x^2 - 4x - 4 + 4x^2 + 4x + 1\right]$

$= \dfrac{-6}{(2x + 1)^3}$

$f''(x) \neq 0$ for any value of x.

$x = -\frac{1}{2}$ is a discontinuity.

Interval	$-\infty < x < -\frac{1}{2}$	$-\frac{1}{2} < x < \infty$
Sign of f''	$f'' > 0$	$f'' < 0$
Conclusion	Concave upward	Concave downward

11. $f(x) = \dfrac{24}{x^2 + 12} = 24(x^2 + 12)^{-1}$

$f'(x) = -24(2x)(x^2 + 12)^{-2} = -48x(x^2 + 12)^{-2}$

$f''(x) = -48x\left[-2(x^2 + 12)^{-3}(2x)\right] + (x^2 + 12)^{-2}(-48) = \dfrac{-48(-4x^2 + x^2 + 12)}{(x^2 + 12)^3} = \dfrac{144(x^2 - 4)}{(x^2 + 12)^3}$

$f''(x) = 0$ when $x = \pm 2$.

Interval	$-\infty < x < -2$	$-2 < x < 2$	$2 < x < \infty$
Sign of f''	$f'' > 0$	$f'' < 0$	$f'' > 0$
Conclusion	Concave upward	Concave downward	Concave upward

13. $f(x) = x^3 - 9x^2 + 24x - 18$

$f'(x) = 3x^2 - 18x + 24$

$f''(x) = 6x - 18 = 6(x - 3)$

$f''(x) = 0$

$6(x - 3) = 0$

$x = 3$

Interval	$-\infty < x < 3$	$3 < x < \infty$
Sign of f''	$f'' < 0$	$f'' > 0$
Conclusion	Concave downward	Concave upward

Point of inflection: $(3, 0)$

15. $f(x) = 2x^3 - 3x^2 - 12x + 5$

$f'(x) = 6x^2 - 6x - 12$

$f''(x) = 12x - 6 = 6(2x - 1)$

$f''(x) = 0$

$6(2x - 1) = 0$

$x = \frac{1}{2}$

Interval	$-\infty < x < \frac{1}{2}$	$\frac{1}{2} < x < \infty$
Sign of f''	$f'' < 0$	$f'' > 0$
Conclusion	Concave downward	Concave upward

Point of inflection: $\left(\frac{1}{2}, -\frac{3}{2}\right)$

17. $g(x) = 2x^4 - 8x^3 + 12x^2 + 12x$

$g'(x) = 8x^3 - 24x^2 + 24x + 12$

$g''(x) = 24x^3 - 48x + 24 = 24(x^2 - 2x + 1) = 24(x - 1)^2$

$g''(x) = 0$

$24(x - 1)^2 = 0$

$x = 1$

Interval	$-\infty < x < -1$	$-1 < x < \infty$
Sign of $g''(x)$	$g'' > 0$	$g'' > 0$
Conclusion	Concave upward	Concave upward

No points of inflection

19. $f(x) = x(6 - x)^2 = x(36 - 12x + x^2) = 36x - 12x^2 + x^3$

$f'(x) = 3x^2 - 24x + 36$

$f''(x) = 6x - 24 = 6(x - 4)$

$f''(x) = 0$

$6(x - 4) = 0$

$x = 4$

Interval	$-\infty < x < 4$	$4 < x < \infty$
Sign of $f''(x)$	$f'' < 0$	$f'' > 0$
Conclusion	Concave downward	Concave upward

Point of inflection: $(4, 16)$

21. $f(x) = 6x - x^2$

$f'(x) = 6 - 2x$

Critical number: $x = 3$

$f''(x) = -2$

$f''(3) = -2 < 0$

(3, 9) is a relative maximum.

23. $f(x) = x^3 - 5x^2 + 7x$

$f'(x) = 3x^2 - 10x + 7 = (3x - 7)(x - 1)$

Critical numbers: $x = 1, x = \frac{7}{3}$

$f''(x) = 6x - 10$

$f''(1) = -4 < 0$

$f''\left(\frac{7}{3}\right) = 4 > 0$

So, (1, 3) is a relative maximum and $\left(\frac{7}{3}, \frac{49}{27}\right)$ is a relative minimum.

25. $f(x) = x^{2/3} - 3$

$f'(x) = \frac{2}{3}x^{-1/3} = \frac{2}{3x^{1/3}}$

Critical number: $x = 0$

The Second-Derivative Test does not apply, so use the First-Derivative Test to conclude that $(0, -3)$ is a relative minimum.

27. $f(x) = \sqrt{x^2 + 1} = (x + 1)^{1/2}$

$f'(x) = \frac{1}{2}(x^2 + 1)^{-1/2}(2x) = \frac{x}{(x^2 + 1)^{1/2}}$

Critical number: $x = 0$

$$f''(x) = \frac{(x^2 + 1)^{1/2} - x\left[\frac{1}{2}(x^2 + 1)^{-1/2}(2x)\right]}{x^2 + 1}$$

$$= \frac{1}{(x^2 + 1)^{3/2}}$$

$f''(0) = 1 > 0$

So, (0, 1) is a relative minimum.

29. $f(x) = \sqrt{9 - x^2} = (9 - x^2)^{1/2}$

$f'(x) = \frac{1}{2}(9 - x^2)^{-1/2}(-2x) = -\frac{x}{(9 - x^2)^{1/2}}$

Critical number: $x = 0$, $x = \pm 3$

$$f''(x) = \frac{(9 - x^2)^{1/2}(-1) - (-x)\left[\frac{1}{2}(9 - x^2)^{-1/2}(-2x)\right]}{9 - x^2}$$

$$= -\frac{9}{(9 - x^2)^{3/2}}$$

$f''(0) = -\frac{1}{3} < 0$

So, $(0, 3)$ is a relative maximum. There are absolute minima at $(\pm 3, 0)$.

31. $f(x) = \frac{8}{x^2 + 2} = 8(x^2 + 2)^{-1}$

$f'(x) = -8(x^2 + 2)^{-2}(2x) = -\frac{16x}{(x^2 + 2)^2}$

Critical number: $x = 0$

$$f''(x) = \frac{(x^2 + 2)^2(-16) - (-16x)\left[(2)(x^2 + 2)(2x)\right]}{(x^2 + 2)^4}$$

$$= \frac{(x^2 + 2)\left[-16(x^2 + 2) + 64x^2\right]}{(x^2 + 2)^4}$$

$$= \frac{48x^2 - 32}{(x^2 + 2)^3}$$

$f''(0) = -4 < 0$

So, $(0, 4)$ is a relative maximum.

33. $f(x) = \frac{x}{x - 1}$

$f'(x) = \frac{(x - 1) - (x)}{(x - 1)^2} = -\frac{1}{(x - 1)^2}$

No critical numbers

No relative extrema

35. $f(x) = 5 + 3x^2 - x^3$

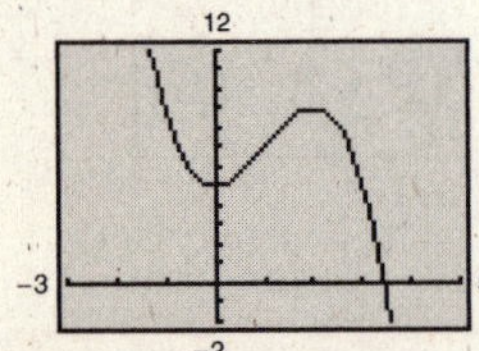

Relative minimum: $(0, 5)$

Relative maximum: $(2, 9)$

37. $f(x) = \frac{1}{2}x^4 - \frac{1}{3}x^3 - \frac{1}{2}x^2$

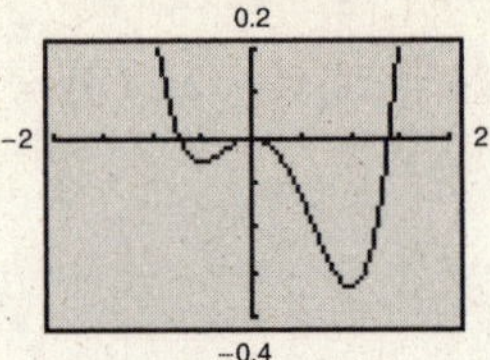

Relative maximum: $(0, 0)$

Relative minima: $(-0.5, -0.052)$, $(1, -0.333)$

39. $f(x) = x^3 - 12x$

$f'(x) = 3x^2 - 12 = 3(x^2 - 4)$

Critical numbers: $x = \pm 2$

$f''(x) = 6x$

$f''(2) = 12 > 0$

$f''(-2) = -12 < 0$

Relative maximum: $(-2, 16)$

Relative minimum: $(2, -16)$

$f''(x) = 0$ when $x = 0$.

$f''(x) < 0$ on $(-\infty, 0)$.

$f''(x) > 0$ on $(0, \infty)$.

Point of inflection: $(0, 0)$

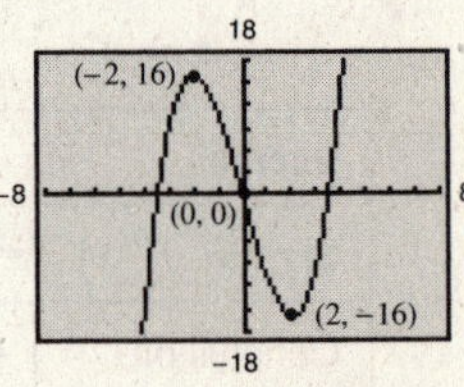

41. $g(x) = x^{1/2} + 4x^{-1/2}$

$g'(x) = \frac{1}{2}x^{-1/2} - 2x^{-3/2}$

$= \frac{1}{2x^{1/2}} - \frac{2}{x^{3/2}} = \frac{x - 4}{2x^{3/2}}$

Critical number: $x = 4$

$g''(x) = -\frac{1}{4}x^{-3/2} + 3x^{-5/2} = -\frac{1}{4x^{3/2}} + \frac{3}{x^{5/2}} = \frac{12 - x}{4x^{5/2}}$

$g''(4) = \frac{1}{16} > 0$

Relative minimum at $(4, 4)$

$g''(x) = 0$ when $x = 12$.

$g''(x) > 0$ on $(0, 12)$

$g''(x) < 0$ on $(12, \infty)$

Point of inflection:

$\left(12, \frac{8\sqrt{3}}{3}\right)$

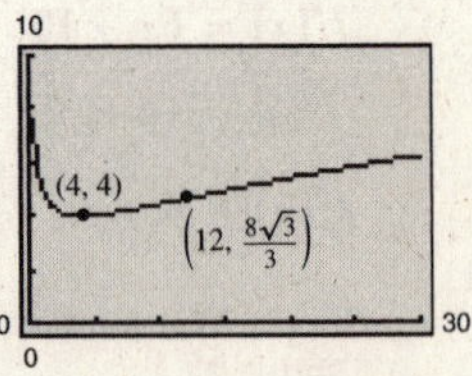

43. $f(x) = \frac{1}{4}x^4 - 2x^2$

$f'(x) = x^3 - 4x = x(x^2 - 4)$

Critical numbers: $x = \pm 2$, $x = 0$

$f''(x) = 3x^2 - 4$

$f''(-2) = 8 > 0$

$f''(0) = -4 < 0$

$f''(2) = 8 > 0$

Relative maximum: $(0, 0)$

Relative minima: $(\pm 2, -4)$

$f''(x) = 3x^2 - 4 = 0$ when $x = \pm\frac{2\sqrt{3}}{3}$.

$f''(x) > 0$ on $\left(-\infty, -\frac{2\sqrt{3}}{3}\right)$.

$f''(x) < 0$ on $\left(-\frac{2\sqrt{3}}{3}, \frac{2\sqrt{3}}{3}\right)$.

$f''(x) > 0$ on $\left(\frac{2\sqrt{3}}{3}, \infty\right)$.

Points of inflection: $\left(-\frac{2\sqrt{3}}{3}, -\frac{20}{9}\right), \left(\frac{2\sqrt{3}}{3}, -\frac{20}{9}\right)$

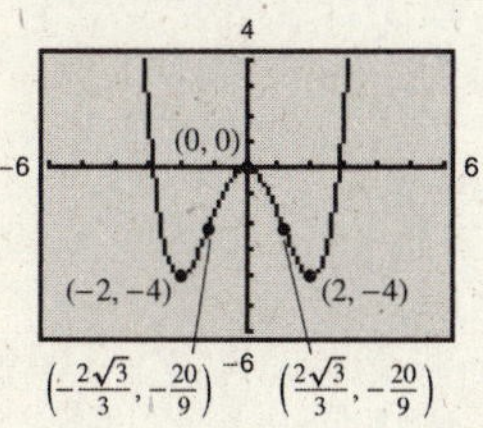

45. $g(x) = (x - 2)(x + 1)^2 = x^3 - 3x - 2$

$g'(x) = (x - 2)[2(x + 1)] + (x + 1)^2 = 3(x^2 - 1)$

Critical numbers: $x = \pm 1$

$g''(x) = 6x$

$g''(-1) = -6 < 0$

$g''(1) = 6 > 0$

Relative maximum: $(-1, 0)$

Relative minimum: $(1, -4)$

$g''(x) = 6x = 0$ when $x = 0$.

$g''(x) < 0$ on $(-\infty, 0)$.

$g''(x) > 0$ on $(0, \infty)$.

Point of inflection: $(0, -2)$

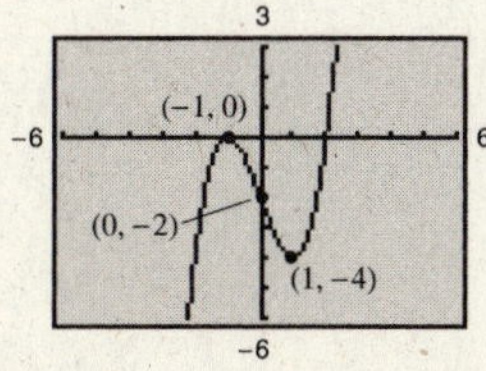

47. $g(x) = x\sqrt{x + 3} = x(x + 3)^{1/2}$

The domain of g is $[-3, \infty)$.

$g'(x) = x\left[\frac{1}{2}(x + 3)^{-1/2}\right] + (x + 3)^{1/2} = \frac{3x + 6}{2(x + 3)^{1/2}}$

Critical numbers: $x = -3$, $x = -2$

$$g''(x) = \frac{2(x + 3)^{1/2}(3) - (3x + 6)\left[2\left(\frac{1}{2}\right)(x + 3)^{-1/2}\right]}{4(x + 3)}$$

$$= \frac{3(x + 4)}{4(x + 3)^{3/2}}$$

$g''(-3)$ is undefined.

$g''(-2) = \frac{3}{2} > 0$

Relative minimum: $(-2, -2)$

$g''(x) > 0$ for all x in the domain, so there are no points of inflection.

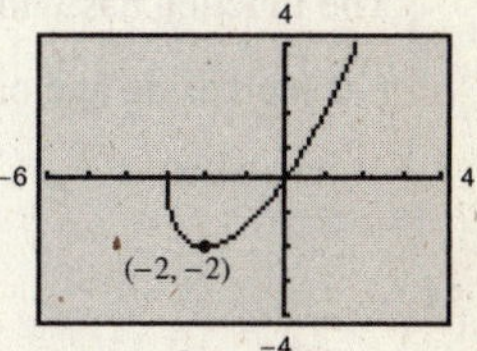

49. $f(x) = \frac{4}{1 + x^2} = 4(1 + x^2)^{-1}$

$f'(x) = -4(1 + x^2)^{-2}(2x) = -\frac{8x}{(1 + x^2)^2}$

Critical number: $x = 0$

$$f''(x) = \frac{(1 + x^2)^2(-8) - (-8x)\left[2(1 + x^2)(2x)\right]}{(1 + x^2)^4}$$

$$= -\frac{8(1 - 3x^2)}{(1 + x^2)^3}$$

$f''(0) = -8 < 0$

Relative maximum: $(0, 4)$

$f''(x) = 1 - 3x^2 = 0$ when $x = \pm\frac{\sqrt{3}}{3}$.

$f''(x) > 0$ on $\left(-\infty, -\frac{\sqrt{3}}{3}\right)$.

$f''(x) < 0$ on $\left(-\frac{\sqrt{3}}{3}, \frac{\sqrt{3}}{3}\right)$.

$f''(x) > 0$ on $\left(\frac{\sqrt{3}}{3}, \infty\right)$.

Points of inflection: $\left(\frac{\sqrt{3}}{3}, 3\right), \left(-\frac{\sqrt{3}}{3}, 3\right)$

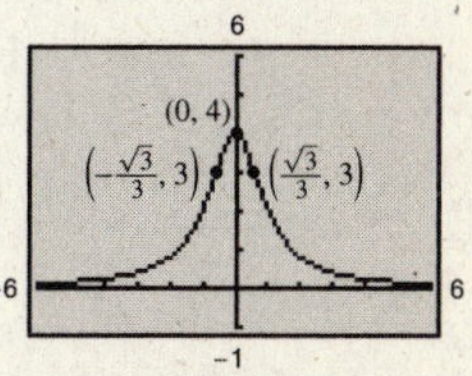

51. *Function* — $f(2) = 0$, $f(4) = 0$

First Derivative — $f'(x) < 0, x < 3$; $f'(3) = 0$; $f'(x) > 0, x > 3$

Second Derivative — $f''(x) > 0$

Answers will vary. Sample answer:

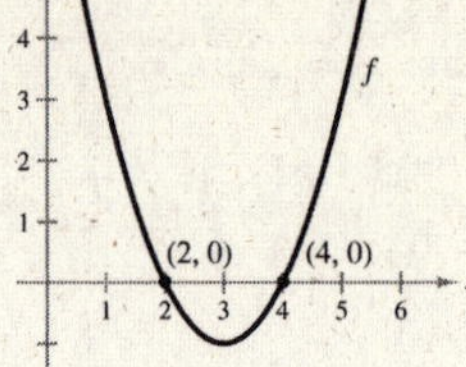

The function has x-intercepts at (2, 0) and (4, 0). On $(-\infty, 3)$, f is decreasing, and on $(3, \infty)$, f is increasing. A relative minimum occurs when $x = 3$. The graph of f is concave upward.

53. *Function* — $f(0) = 0$, $f(2) = 0$

First Derivative — $f'(x) > 0, x < 1$; $f'(1) = 0$; $f'(x) < 0, x > 1$

Second Derivative — $f''(x) < 0$

Answers will vary. Sample answer:

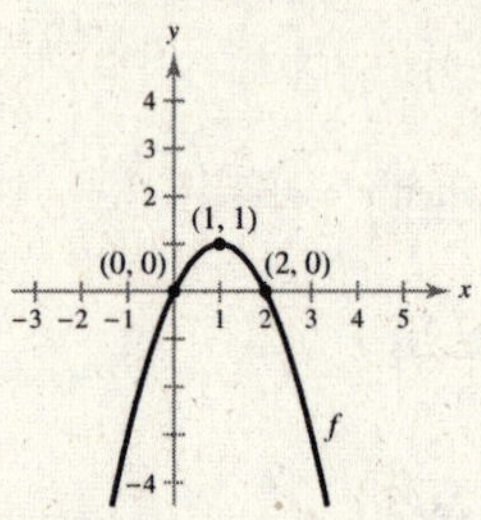

The function has x-intercepts at (0, 0) and (2, 0). On $(-\infty, 1)$, f is increasing and on $(1, \infty)$, f is decreasing. A relative maximum occurs when $x = 1$. The graph of f is concave downward.

55.

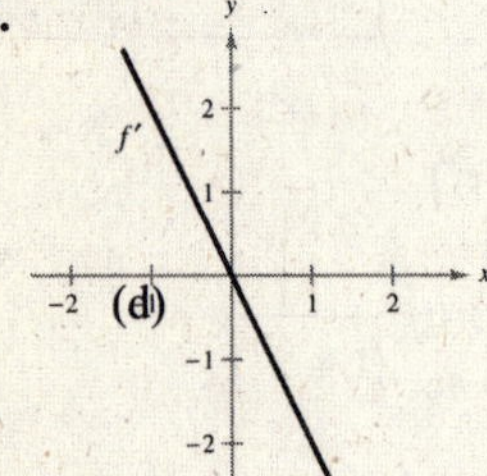

(a) $f'(x) > 0$ on $(-\infty, 0)$ where f is increasing.

(b) $f'(x) < 0$ on $(0, \infty)$ where f is decreasing.

(c) f' is not increasing. f is not concave upward.

f' is decreasing on $(-\infty, \infty)$ where f is concave downward.

57. $f'(x) = 2x + 5$

$f''(x) = 2$

(a) Because the second derivative is positive for all x, f' is increasing on $(-\infty, \infty)$.

(b) Because the second derivative is positive for all x, f is concave upward.

(c) Critical number: $x = -\dfrac{5}{2}$

$$f''\left(-\frac{5}{2}\right) = 2 > 0$$

A relative minimum occurs when $x = -\dfrac{5}{2}$.

$f''(x) > 0$ on $(-\infty, \infty)$

No points of inflection

(d)

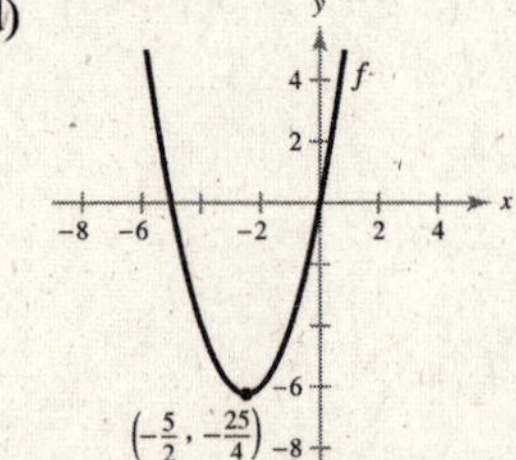

59. $f'(x) = -x^2 + 2x - 1$; Critical number: $x = 1$

$f''(x) = -2x + 2$; $f''(x) = 0$ when $x = 1$

(a) The value of $f''(x)$ is positive on $(-\infty, 1)$ and negative on $(1, \infty)$. So, $f'(x)$ is increasing on $(-\infty, 1)$ and decreasing on $(1, \infty)$.

(b) $f''(x) > 0$ on $(-\infty, 1)$.
$f''(x) < 0$ on $(1, \infty)$.
So, f is concave upward on $(-\infty, 1)$ and concave downward on $(1, \infty)$.

(c) $f''(1) = 0$

The Second-Derivative Test fails. Because $f'(x) \le 0$, f is never increasing and there are no relative extrema. Concavity changes at $x = 1$, so a point of inflection occurs when $x = 1$.

(d)

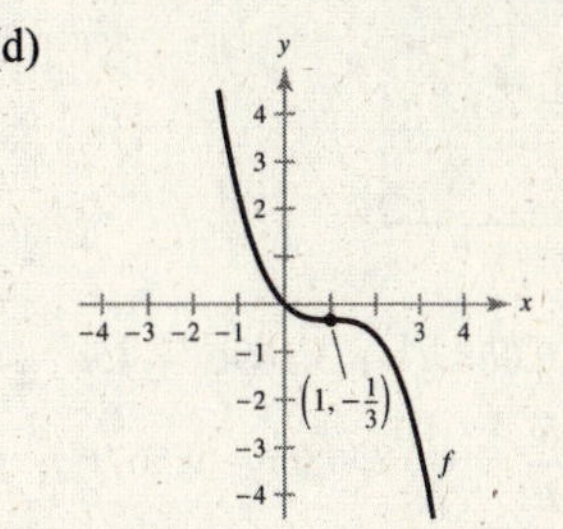

61. $R = \dfrac{1}{50{,}000}\left(600x^2 - x^3\right),\ 0 \le x \le 400$

$R' = \dfrac{1}{50{,}000}\left(1200x - 3x^2\right)$

$R'' = \dfrac{1}{50{,}000}(1200 - 6x) = 0$ when $x = 200$.

$R'' > 0$ on $(0, 200)$.

$R'' < 0$ on $(200, 400)$.

Point of diminishing returns: $(200, 320)$

63. $N(t) = -0.12t^3 + 0.54t^2 + 8.22t,\ 0 \le t \le 4$

$N'(t) = -0.36t^2 + 1.08t + 8.22$

$N''(t) = -0.72t + 1.08 = 0$ when $t = 1.5$.

$N'''(t) = -0.72$

$N'''(1.5) = -0.72 < 0$

The student is assembling components at the greatest rate when $t = 1.5$, or 8:30 P.M.

65. $f(x) = \dfrac{1}{2}x^3 - x^2 + 3x - 5,\ [0, 3]$

$f'(x) = \dfrac{3}{2}x^2 - 2x + 3$

$f''(x) = 3x - 2$

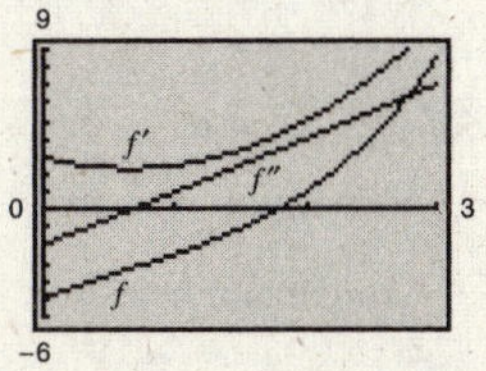

No relative extrema.

Point of inflection: $\left(\frac{2}{3}, -3.30\right)$

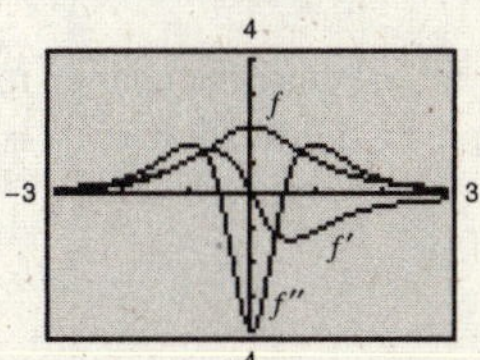

f is increasing when f' is positive. f is concave upward when f'' is positive and concave downward when f'' is negative.

67. $f(x) = \dfrac{2}{x^2 + 1} = 2\left(x^2 + 1\right)^{-1},\ [-3, 3]$

$f'(x) = -2\left(x^2 + 1\right)^{-2}(2x) = -\dfrac{4x}{\left(x^2 + 1\right)^2}$

$f''(x) = \dfrac{\left(x^2 + 1\right)^2(-4) - (-4x)\left[2\left(x^2 + 1\right)(2x)\right]}{\left(x^2 + 1\right)^4}$

$= \dfrac{12x^2 - 4}{\left(x^2 + 1\right)^3}$

Relative maximum: $(0, 2)$

Points of inflection: $(0.58, 1.5)$, $(-0.58, 1.5)$

f is increasing when f' is positive and decreasing when f' is negative. f is concave upward when f'' is positive and concave downward when f'' is negative.

69. $\overline{C} = 0.5x + 10 + \dfrac{7200}{x}$

$\overline{C}' = 0.5 - \dfrac{7200}{x^2}$

Critical number: $x = 120$

$\overline{C}'' = \dfrac{14{,}400}{x^3}$

$\overline{C}''(120) = \dfrac{1}{120} > 0$

So, producing 120 units minimizes the average cost per unit.

71. (a)

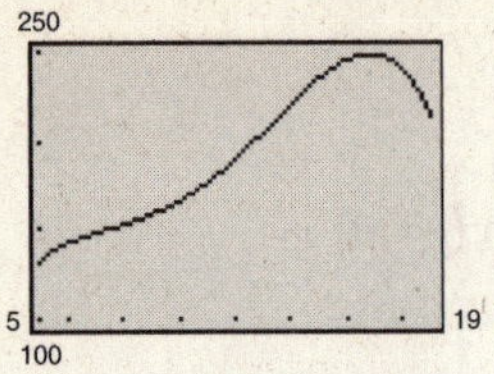

(b) $t \approx 5$: 1995

(c) $t \approx 17$: 2007

(d) $t \approx 13$: 2003; the greatest rate of increase

$t \approx 19$: 2009; the least rate of increase

73. (a)

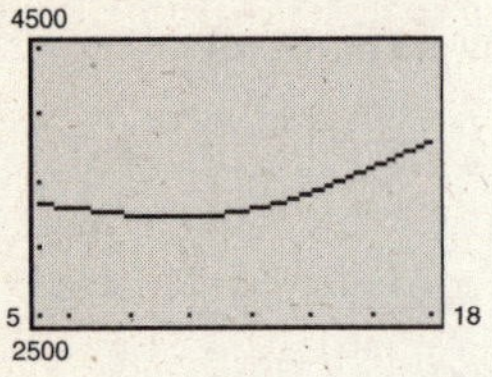

(b) and (c) $v = -0.0687t^4 + 3.169t^3 - 45t^2 + 230.6t + 2950,\ 5 \le t \le 18$

$$v' = \frac{dv}{dt} = -0.2748t^3 + 9.507t^2 - 90t + 230.6$$

$$v'' = \frac{d^2v}{dt^2} = -0.8244t^2 + 19.014t - 90$$

$$v'' = 0$$

$$-8.244t^2 + 19.014t - 90 = 0$$

$$t \approx 16.4, 6.7$$

Interval	$5 < t < 6.7$	$6.7 < t < 16.4$	$16.4 < t < 18$
Sign of v''	$v'' < 0$	$v'' > 0$	$v'' < 0$
Conclusion	Concave downward	Concave upward	Concave downward

Points of inflection: $(6.7, 3289.6)$, $(16.4, 3637.2)$

(d) The first inflection point is where the change in the number of veterans receiving benefits starts to increase after it has been decreasing. The second inflection point is where the change in the number of veterans receiving benefits starts to decrease again.

75. Answers will vary.

Section 3.4 Optimization Problems

Skills Warm Up

1. Let x be the first number and y be the second number.

$$x + \tfrac{1}{2}y = 12$$

2. Let x be the first number and y be the second number.

$$2xy = 24$$

3. Let x be the length of the rectangle and y be the width of the rectangle.

$$xy = 24$$

4. Let (x_1, y_1) be the first point and (x_2, y_2) be the second point.

$$\sqrt{(x_1 - x_2)^2 + (y_1 - y_2)^2} = 10$$

5. $y = x^2 + 6x - 9$

$y' = 2x + 6 = 0$

Critical number: $x = -3$

Skills Warm Up —*continued*—

6. $y = 2x^3 - x^2 - 4x$

$$y' = 6x^2 - 2x - 4 = 0$$
$$2(3x^2 - x - 2) = 0$$
$$2(3x + 2)(x - 1) = 0$$

Critical numbers $x = -\frac{2}{3}, x = 1$

7. $y = 5x + \dfrac{125}{x}$

$$y' = 5 - \frac{125}{x^2} = 0$$
$$\frac{5(x^2 - 25)}{x^2} = 0$$
$$5(x^2 - 25) = 0$$

Critical numbers: $x = \pm 5$

8. $y = 3x + \dfrac{96}{x^2}$

$$y' = 3 - \frac{192}{x^3} = 0$$
$$\frac{3(x^3 - 64)}{x^3} = 0$$
$$3(x^3 - 64) = 0$$

Critical number: $x = 4$

9. $y = \dfrac{x^2 + 1}{x} = x + \dfrac{1}{x}$

$$y' = 1 - \frac{1}{x^2} = 0$$
$$\frac{x^2 - 1}{x^2} = 0$$
$$x^2 - 1 = 0$$

Critical numbers: $x = \pm 1$

10. $y = \dfrac{x}{x^2 + 9}$

$$y' = \frac{x^2 + 9 - x(2x)}{(x^2 + 9)^2} = 0$$
$$\frac{9 - x^2}{(x^2 + 9)^2} = 0$$
$$9 - x^2 = 0$$

Critical numbers: $x = \pm 3$

1. Let l be the length and w be the width of the rectangle. Then $2l + 2w = 100$ and $w = 50 - l$. The area is:

$$A = lw = l(50 - l)$$
$$A' = 50 - 2l$$
$$A'' = -2$$

$A' = 0$ when $x = 25$. Because $A''(25) = -2 < 0$, A is maximum when $l = 25$ meters and $w = 50 - 25 = 25$ meters.

3. Let l and w be the length and width of the rectangle. Then the area is $lw = 64$ and $w = 64/l$. The perimeter is:

$$P = 2l + 2w = 2l + 2\left(\frac{64}{l}\right)$$
$$P' = 2 - \frac{128}{l^2} = \frac{2(l^2 - 64)}{l^2}$$
$$P' = \frac{256}{l^3}$$

$P' = 0$ when $l = 8$. Because $P''(8) = \frac{1}{2} > 0$, P is a minimum when $l = 8$ and $w = 64/8 = 8$ feet.

5. The length of fencing is given by $4x + 3y = 200$ and $y = (200 - 4x)/3$. The area of the corrals is:

$$A = 2xy = 2x\left(\frac{200 - 4x}{3}\right) = \frac{8}{3}(50x - x^2)$$
$$A' = \frac{8}{3}(50 - 2x)$$
$$A'' = -\frac{16}{3}$$

$A' = 0$ when $x = 25$. Because $A''(25) = -\frac{16}{3} < 0$, A is maximum when $x = 25$ feet and $y = \frac{100}{3}$ feet.

7. (a) $9 + 9 + 4(3)(11) = 150 \text{ in.}^2$

$6(25) = 150 \text{ in.}^2$

$36 + 36 + 4(6)(3.25) = 150 \text{ in.}^2$

(b) $V = 3(3)(11) = 99 \text{ in.}^3$

$V = 5(5)(5) = 125 \text{ in.}^3$

$V = 6(6)(3.25) = 117 \text{ in.}^3$

(c) Let the base measure x by x, and the height measure y.

Then the surface area is $2x^2 + 4xy = 150$ and

$y = \frac{1}{2x}(75 - x^2)$. The volume of the solid is:

$$V = x^2y = x^2\left[\frac{1}{2x}(75 - x^2)\right] = \frac{75}{2}x - \frac{x^3}{2}$$

$$V' = \frac{75}{2} - \frac{3x^2}{2}$$

$$V'' = -3x$$

$V' = 0$ when $x = 5$. Because $V''(5) = -15 < 0$, V is maximum when $x = 5$ inches and $y = 5$ inches.

9. (a) Let the base have dimensions x by x and the height have dimension y. Then the volume is given by

$$V = x^2y = 8000 \Rightarrow y = \frac{8000}{x^2}.$$

Surface area: $S = 2x^2 + 4xy$

$$= 2x^2 + 4x\left(\frac{8000}{x^2}\right)$$

$$= 2x^2 + \frac{32{,}000}{x}$$

$$S' = 4x - \frac{32{,}000}{x^2}$$

$$S'' = 4 + \frac{64{,}000}{x^3}$$

$$S' = 0$$

$$4x - \frac{32{,}000}{x^2} = 0$$

$$4x^3 = 32{,}000$$

$$x^3 = 8000$$

$$x = 20$$

$S''(20) > 0 \Rightarrow S$ is a minimum when $x = 20$.

When $x = 20$ in., $y = \frac{8000}{(20)^2} = 20$ in.

(b) $S = 2(20)^2 + \frac{32{,}000}{20} = 2400 \text{ in.}^2$

11. Let x and y be the length and width of the rectangle. The radius of the semicircle is $r = y/2$, and the perimeter is

$200 = 2x + 2\pi r = 2x + 2\pi\left(\frac{y}{2}\right) = 2x + \pi y$ and

$y = (200 - 2x)/\pi$. The area of the rectangle is:

$$A = xy = x\left[\frac{200 - 2x}{\pi}\right] = \frac{2}{\pi}(100x - x^2)$$

$$A' = \frac{2}{\pi}(100 - 2x)$$

$$A'' = -\frac{4}{\pi}$$

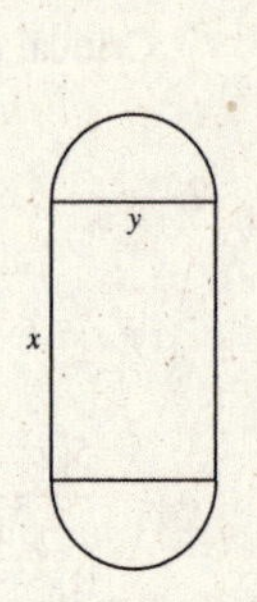

$A' = 0$ when $x = 50$. Because $A'' < 0$, A is maximum when

$x = 50$ meters and

$y = \frac{200 - 2(50)}{\pi} = \frac{100}{\pi}$ meters.

13. Let the base measure x by x and the height measure y. Then the volume is $x^2y = 80$ and $y = \frac{80}{x^2}$. The cost of the box is:

$$C = (0.20)(2)x^2 + (0.10)(4)xy$$

$$= 0.40x^2 + 0.40x\left(\frac{80}{x^2}\right)$$

$$= 0.40x^2 + \frac{32}{x}$$

$$C' = 0.80x - \frac{32}{x^2}$$

$$= \frac{0.80x^3 - 32}{x^2}$$

$$C'' = 0.80 + \frac{64}{x^3}$$

$C' = 0$ when $x = \sqrt[3]{40} = 2\sqrt[3]{5}$. Because $C''(2\sqrt[3]{5}) > 0$, C is minimum when $x = 2\sqrt[3]{5}$ centimeters and $y = 4\sqrt[3]{5}$ centimeters.

15. The volume of the box is:

$$V = x(6 - 2x)^2,\ 0 < x < 3$$

$$V' = 12(x - 1)(x - 3)$$

$V' = 0$ when $x = 3$ and $x = 1$. Because $V = 0$ when $x = 3$ and $V = 16$ when $x = 1$, the volume is maximum when $x = 1$. The corresponding volume is $V = 16$ cubic inches.

17. Let x and y be the lengths shown in the figure.

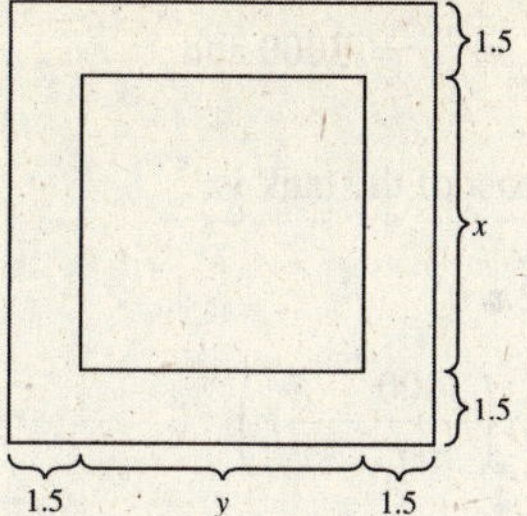

Then $xy = 36$ and $y = \dfrac{36}{x}$.

The area of the page is:

$$A = (x + 3)\left(\frac{36}{x} + 3\right) = 45 + \frac{108}{x} + 3x$$

$$A' = -\frac{108}{x^2} + 3 = \frac{3(x^2 - 36)}{x^2}$$

$$A'' = \frac{216}{x^3}$$

$A' = 0$ when $x = 6$. Because $A''(6) = 1 > 0$, A is minimum when $x = 6$ inches and $y = \dfrac{36}{6} = 6$ inches.

The dimensions of the page are 9 inches by 9 inches.

19. The area of the rectangle is

$$A = xy = x\left(\frac{6 - x}{2}\right) = \frac{1}{2}(6x - x^2)$$

$$A' = \frac{1}{2}(6 - 2x).$$

$$A'' = -1$$

$A' = 0$ when $x = 3$. Because $A'' < 0$, A is maximum when $x = 3$ units and $y = \dfrac{6 - 3}{2} = \dfrac{3}{2}$ units.

21. The area is:

$$A = 2xy = 2x(25 - x^2)^{1/2}$$

$$A' = 2\left[x\left(\frac{1}{2}\right)(25 - x^2)^{-1/2}(-2x) + (25 - x^2)^{1/2}\right] = 2\left[\frac{25 - 2x^2}{(25 - x^2)^{1/2}}\right]$$

$$A'' = 2\left[\frac{(25 - x^2)^{1/2}(-4x) - (25 - 2x^2)\left(\frac{1}{2}\right)(25 - x^2)^{-1/2}(-2x)}{25 - x^2}\right] = 2x\left[\frac{2x^2 - 75}{(25 - x^2)^{3/2}}\right]$$

$A' = 0$ when $x = 5/\sqrt{2}$. Because $A''(5/\sqrt{2}) < 0$, A is maximum when the length is

$2x = \dfrac{10}{\sqrt{2}} \approx 7.07$ units and the width is $y = \sqrt{25 - \left(\dfrac{5}{\sqrt{2}}\right)^2} = \dfrac{5}{\sqrt{2}} \approx 3.54$ units.

23. The volume of the cylinder is $V = \pi r^2 h = (12 \text{ oz})(1.80469 \text{ in.}^3 \text{per oz}) \approx 21.66 \text{ in.}^3$ which implies that $h = 21.66/(\pi r^2)$. The surface area of the container is:

$$S = 2\pi r^2 + 2\pi rh = 2\pi r^2 + 2\pi r\left(\frac{21.66}{\pi r^2}\right) = 2\left(\pi r^2 + \frac{21.66}{r}\right)$$

$$S' = 2\left(2\pi r - \frac{21.66}{r^2}\right) = 2\left(\frac{2\pi r^3 - 21.66}{r^2}\right)$$

$$S'' = 2\left(2\pi + \frac{43.32}{r^3}\right)$$

$S' = 0$ when $r = \sqrt[3]{\dfrac{21.66}{2\pi}} \approx 1.51$ inches. Because $S''(1.51) > 0$, S is minimum when $r \approx 1.51$ inches and

$h = \dfrac{21.66}{\pi(1.51)^2} \approx 3.02$ inches.

25. The distance between a point (x, y) on the graph and the point $(2, 1/2)$ is

$$d = \sqrt{(x-2)^2 + \left(y - \frac{1}{2}\right)^2} = \sqrt{(x-2)^2 + \left(x^2 - \frac{1}{2}\right)^2}$$

and d can be minimized by minimizing its square $L = d^2$.

$$L = (x-2)^2 + \left(x^2 - \frac{1}{2}\right)^2 = x^4 - 4x + \frac{17}{4}$$

$$L' = 4x^3 - 4 = 4(x^3 - 1)$$

$$L'' = 12x^2$$

$L' = 0$ when $x = 1$. Because $L''(1) > 0$, L is minimum when $x = 1$ and $y = (1)^2 = 1$. The point nearest $(2, 1/2)$ is $(1, 1)$.

27. The distance between a point (x, y) on the graph and the point $(4, 0)$ is

$$d = \sqrt{(x-4)^2 + y^2} = \sqrt{(x-4)^2 + x}$$

and d can be minimized by minimizing its square $L = d^2$.

$$L = (x-4)^2 + x = x^2 - 7x + 16$$

$$L' = 2x - 7$$

$$L'' = 2$$

$L' = 0$ when $x = \frac{7}{2}$. Because $L''\left(\frac{7}{2}\right) > 0$, L is minimum when $x = \frac{7}{2}$ and $y = \sqrt{\frac{7}{2}} = \frac{\sqrt{14}}{2}$. The point nearest $(4, 0)$ is $\left(\frac{7}{2}, \frac{\sqrt{14}}{2}\right)$.

29. The length and girth is $4x + y = 108$ and $y = 108 - 4x$. The volume is:

$$V = x^2y = x^2(108 - 4x) = 108x^2 - 4x^3$$

$$V' = 216x - 12x^2 = 12x(18 - x).$$

$$V'' = 216 - 24x$$

$V' = 0$ when $x = 0$ and $x = 18$. Because $V''(18) < 0$, V is minimum when $x = 18$ inches and $y = 36$ inches. The dimensions are 18 inches by 18 inches by 36 inches.

31. Let p represent the cost per square foot of the material. The volume is $\frac{4}{3}\pi r^3 + \pi r^2 h = 3000$ and $h = \frac{3000}{\pi r^2} - \frac{4}{3}r$. The cost of the tank is:

$$C = 2p(4\pi r^2) + p(2\pi rh)$$

$$= 2p(4\pi r^2) + 2\pi rp\left(\frac{3000}{\pi r^2} - \frac{4}{3}r\right)$$

$$= 2p\left(\frac{8}{3}\pi r^2 + \frac{3000}{r}\right)$$

$$C' = 2p\left(\frac{16}{3}\pi r - \frac{3000}{r^2}\right) = 2p\left(\frac{\frac{16}{3}\pi r^3 - 3000}{r^3}\right)$$

$$C'' = 2p\left(\frac{16}{3}\pi + \frac{6000}{r^3}\right)$$

$C' = 0$ when $r = \sqrt[3]{\frac{562.5}{\pi}} \approx 5.64$ feet.

Because $C''(5.64) > 0$, C is minimum when $r \approx 5.64$ feet and $h = \frac{3000}{\pi(5.64)^2} - \frac{4}{3}(5.64) \approx 22.50$ feet.

33. Let x be the length of a side of the square and r be the radius of the circle. Then the combined perimeter is $4x + 2\pi r = 16$ and

$$x = \frac{16 - 2\pi r}{4} = 4 - \frac{\pi r}{2}.$$

The total area is:

$$A = x^2 + \pi r^2 = \left(4 - \frac{\pi r}{2}\right)^2 + \pi r^2$$

$$A' = 2\left(4 - \frac{\pi r}{2}\right)\left(-\frac{\pi}{2}\right) + 2\pi r = \frac{1}{2}(\pi^2 r + 4\pi r - 8\pi)$$

$$A'' = \frac{1}{2}(\pi^2 + 4\pi)$$

$A' = 0$ when $r = \frac{8\pi}{\pi^2 + 4\pi} = \frac{8}{\pi + 4}$. Because $A'' > 0$, A is minimum when $r = \frac{8}{\pi + 4}$ units and

$$x = 4 - \frac{\pi[8/(\pi + 4)]}{2} = \frac{16}{\pi + 4}.$$

35. (a) The perimeter is $4x + 2\pi r = 4$ and $r = \dfrac{2 - 2x}{\pi}$.

The area is: $A(x) = x^2 + \pi r^2 = x^2 + \pi\left(\dfrac{2 - 2x}{\pi}\right)^2 = x^2 + \dfrac{(2 - 2x)^2}{\pi} = \left(1 + \dfrac{4}{\pi}\right)x^2 - \dfrac{8}{\pi}x + \dfrac{4}{\pi}$

(b) Because $x \geq 0$ and $r = \dfrac{2 - 2x}{\pi} \geq 0$, the domain is $0 \leq x \leq 1$

(c)

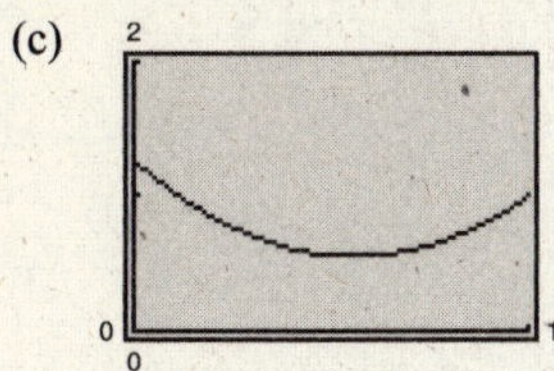

(d) $A'(x) = \left(2 + \dfrac{8}{\pi}\right)x - \dfrac{8}{\pi}$

$A''(x) = 2 + \dfrac{8}{\pi}$

$A'(x) = 0$ when $x = \dfrac{8/\pi}{2 + 8/\pi} = \dfrac{4}{4 + \pi}$.

Because $A''(x) > 0$, $A(x)$ is minimum when $x = \dfrac{4}{4 + \pi} \approx 0.56$ feet and $r = \dfrac{2}{4 + \pi} \approx 0.28$ feet.

So the total area is minimum when 2.24 feet is used for the square and 1.76 feet is used for the circle.

From the graph, $A(x)$ is maximum when $x = 0$ feet and $r = \dfrac{2}{\pi} \approx 0.64$ feet. So the total area is maximum when all 4 feet is used for the circle.

37. Let w be the number of weeks and p be the price per bushel. Use the points $(1, 30)$ and $(2, 29.20)$ to determine the linear equation relating the price per bushel to the number of weeks that pass.

$$p - 30 = \frac{29.20 - 30}{2 - 1}(w - 1)$$
$$p - 30 = -0.80w + 0.80$$
$$p = -0.80w + 30.80$$

Let b be the number of bushels in the field.

Use the points $(1, 120)$ and $(2, 124)$ to determine the linear equation relating the number of bushels in the field to the number of weeks that pass.

$$b - 120 = \frac{124 - 120}{2 - 1}(w - 1)$$
$$b - 120 = 4w - 4$$
$$b = 4w + 116$$

The total value of the crop is:

$$R = pb = (-0.80w + 30.80)(4w + 116) = -3.2w^2 + 30.4w + 3572.8$$
$$R' = -6.4w + 30.4$$
$$R'' = -6.4$$

$R' = 0$ when $w = 4.75$. Because $R'' < 0$, R is maximum when $w = 4.75$. So the farmer should harvest the strawberries after 4 weeks.

The total bushels harvested is $b(4.75) = 135$ bushels.

The maximum value of the strawberries is $R(4.75) = \$3645$.

Chapter 3 Quiz Yourself

1. $f(x) = x^2 - 6x + 1$

   $f'(x) = 2x - 6$

   Critical number: $x = 3$

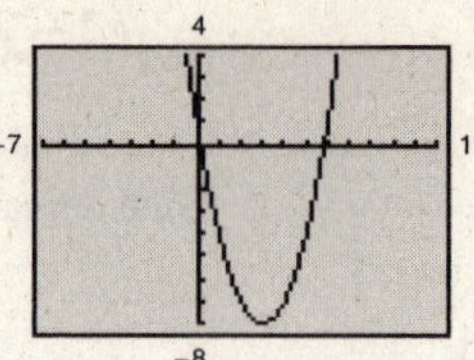

Interval	$-\infty < x < 3$	$3 < x < \infty$
Sign of f'	$f' < 0$	$f' > 0$
Conclusion	Decreasing	Increasing

2. $f(x) = 2x^3 + 12x^2$

   $f'(x) = 6x^2 + 24x = 6x(x + 4)$

   Critical numbers: $x = 0,\ x = -4$

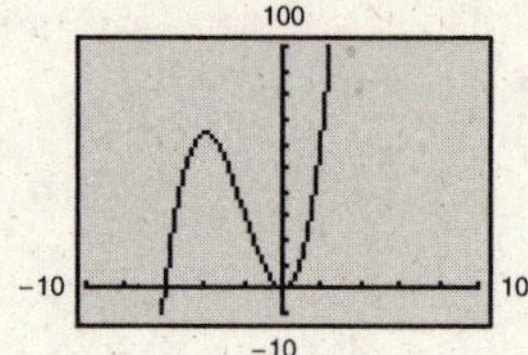

Interval	$-\infty < x < -4$	$-4 < x < 0$	$0 < x < \infty$
Sign of f'	$f' > 0$	$f' < 0$	$f' > 0$
Conclusion	Increasing	Decreasing	Increasing

3. $f(x) = \dfrac{x}{x^2 + 25}$

   $f'(x) = \dfrac{x^2 + 25(1) - x(2x)}{\left(x^2 + 25\right)^2}$

   $f'(x) = \dfrac{25 - x^2}{\left(x^2 + 25\right)^2}$

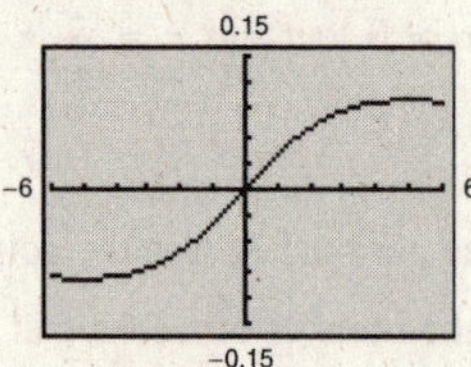

   Critical numbers: $x = \pm 5$

Interval	$-\infty < x < -5$	$-5 < x < 5$	$5 < x < \infty$
Sign of f'	$f' < 0$	$f' > 0$	$f' < 0$
Conclusion	Decreasing	Increasing	Decreasing

4. $f(x) = x^3 + 3x^2 - 5$

   $f'(x) = 3x^2 + 6x = 3x(x + 2)$

   Critical numbers: $x = -2,\ x = 0$

Interval	$-\infty < x < -2$	$-2 < x < 0$	$0 < x < \infty$
Sign of f'	$f' > 0$	$f' < 0$	$f' > 0$
Conclusion	Increasing	Decreasing	Increasing

   Relative maximum: $(-2, -1)$

   Relative minimum: $(0, -5)$

5. $f(x) = x^4 - 8x^2 + 3$

$f'(x) = 4x^3 - 16x = 4x(x^2 - 4)$

Critical numbers: $x = \pm 2,\ x = 0$

Interval	$-\infty < x < -2$	$-2 < x < 0$	$0 < x < 2$	$2 < x < \infty$
Sign of f'	$f' < 0$	$f' > 0$	$f' < 0$	$f' > 0$
Conclusion	Decreasing	Increasing	Decreasing	Increasing

Relative maximum: $(0, 3)$

Relative minima: $(-2, -13), (2, -13)$

6. $f(x) = 2x^{2/3}$

$f'(x) = \frac{4}{3}x^{-1/3} = \frac{4}{3x^{1/3}}$

Critical number: $x = 0$

Interval	$-\infty < x < 0$	$0 < x < \infty$
Sign of f'	$f' < 0$	$f' > 0$
Conclusion	Decreasing	Increasing

Relative minimum: $(0, 0)$

7. $f(x) = x^2 + 2x - 8,\ [-2, 1]$

$f'(x) = 2x + 2$

Critical number: $x = -1$

x-value	Endpoint $x = -2$	Critical $x = -1$	Endpoint $x = 1$
$f(x)$	-8	-9	-5
Conclusion		Minimum	Maximum

8. $f(x) = x^3 - 27x,\ [-4, 4]$

$f'(x) = 3x^2 - 27$

Critical numbers: $x = \pm 3$

x-value	Endpoint $x = -4$	Critical $x = -3$	Critical $x = 3$	Endpoint $x = 4$
$f(x)$	44	54	-54	44
Conclusion		Maximum	Minimum	

9. $f(x) = \frac{x}{x^2 + 1},\ [0, 2]$

$f'(x) = \frac{(x^2 - 1) - x(2x)}{(x^2 + 1)^2} = \frac{1 - x^2}{(x^2 + 1)^2}$

Critical number: $x = 1$

x-value	Endpoint $x = 0$	Critical $x = 1$	Endpoint $x = 2$
$f(x)$	0	$\frac{1}{2}$	$\frac{2}{5}$
Conclusion	Minimum	Maximum	

10. $f(x) = x^3 - 6x^2 + 7x$

$f'(x) = 3x^2 - 12x + 7$

$f''(x) = 6x - 12$

$f''(x) = 0$ when $x = 2$.

Point of inflection: $(2, -2)$

Interval	$-\infty < x < 2$	$2 < x < \infty$
Sign of $f''(x)$	$f''(x) < 0$	$f''(x) > 0$
Conclusion	Concave downward	Concave upward

11. $f(x) = x^4 - 24x^2$

$f'(x) = 4x^3 - 48x$

$f''(x) = 12x^2 - 48$

$f''(x) = 0$ when $x = \pm 2$.

Interval	$-\infty < x < -2$	$-2 < x < 2$	$2 < x < \infty$
Sign of $f''(x)$	$f''(x) > 0$	$f''(x) < 0$	$f''(x) > 0$
Conclusion	Concave upward	Concave downward	Concave upward

Points of inflection: $(-2, -80)$, $(2, -80)$

12. $f(x) = 2x^3 + 3x^2 - 12x + 16$

$f'(x) = 6x^2 + 6x - 12 = 6(x + 2)(x - 1)$

Critical numbers: $x = -2$, $x = 1$

$f''(x) = 12x + 6$

$f''(-2) = -18 < 0$

$f''(1) = 18 > 0$

So, $(-2, 36)$ is a relative maximum and $(1, 9)$ is a relative minimum.

13. $f(x) = 2x + 18x^{-1}$

$f'(x) = 2 - 18x^{-2} = 2 - \dfrac{18}{x^2} = \dfrac{2(x^2 - 9)}{x^2}$

Critical numbers: $x = \pm 3$

$f''(x) = 36x^{-3} = \dfrac{36}{x^3}$

$f''(-3) < 0$ and $f''(3) > 0$

So, $(-3, -12)$ is a relative maximum and $(3, 12)$ is a relative minimum.

14. $S = \dfrac{1}{3600}(360x^2 - x^3)$, $0 \le x \le 240$

$S' = \dfrac{1}{3600}(720x - 3x^2)$

$S'' = \dfrac{1}{3600}(720 - 6x) = 0$ when $x = 120$.

$S'' > 0$ on $(0, 120)$.

$S'' < 0$ on $(120, 240)$.

Since $(120, 960)$ is a point of inflection, it is the point of diminishing returns.

15. The perimeter is $x + 2y = 200$ and

$y = 100 - \dfrac{1}{2}x$. The area is:

$A = xy = x\left(100 - \dfrac{1}{2}x\right) = 100x - \dfrac{1}{2}x^2$

$A' = 100 - x$

$A'' = -1$

$A' = 0$ when $x = 100$. Because $A'' = -1 < 0$, A is maximum when $x = 100$ feet and

$y = 100 - \dfrac{1}{2}(100) = 50$ feet.

16. $P = 0.001t^3 - 0.64t^2 + 10.3t + 1276$, $0 \le t \le 9$

$P' = 0.003t^2 - 1.28t + 10.3$

$P' = 0$

$0.003t^2 - 1.28t + 10.3 = 0$

$t \approx 8.2$ or 418.5 ($t \approx 418.5$ is not in the domain.)

(a)

Interval	$0 < t < 8.2$	$8.2 < t < 9$
Sign of P'	$P' > 0$	$P' < 0$
Conclusion	Increasing	Decreasing

The population was increasing from 2000 to early 2008 $(0 < t < 8.2)$ and decreasing from early 2008 to 2009 $(8.2 < t < 9)$.

(b)

t-value	Endpoint $t = 0$	Critical number $t = 8.2$	Endpoint $t = 9$
$P(t)$	1276	1317.8	1317.4

The maximum population was 1317.8 thousand or 1,317,800 in early 2008 and the minimum population was 1276 thousand or 1,276,000 in 2000.

Section 3.5 Business and Economics Applications

Skills Warm Up

1. $\left|-\frac{300}{150}+3\right| = |-2+3| = 1$

2. $\left|-\frac{600}{5(150)}+2\right| = \left|-\frac{4}{5}+2\right| = \frac{6}{5}$

3. $\left|\frac{20(150)^{-1/2}/150}{-10(150)^{-3/2}}\right| = \left|\frac{20(150)^{-3/2}}{-10(150)^{-3/2}}\right| = 2$

4. $\left|\frac{(4000/150^2)/150}{-8000(150)^{-3}}\right| = \left|\frac{4000(150)^{-3}}{-8000(150)^{-3}}\right| = -\frac{1}{2}$

5. $\frac{dC}{dx} = 1.2 + 0.006x$

6. $\frac{dP}{dx} = 0.02x + 11$

7. $\frac{dP}{dx} = -1.4x + 7$

8. $\frac{dC}{dx} = 4.2 + 0.003x^2$

9. $\frac{dR}{dx} = 14 - \frac{x}{1000}$

10. $\frac{dR}{dx} = 3.4 - \frac{x}{750}$

1. $R = 800x - 0.2x^2$

$R' = 800 - 0.4x$

$R'' = -0.4$

$R' = 0$ when $x = 2000$. Because $R'' < 0$, R is maximum when $x = 2000$ units.

3. $R = 400x - x^2$

$R' = 400 - 2x$

$R'' = -2$

$R' = 0$ when $x = 200$. Because $R'' < 0$, R is maximum when $x = 200$ units.

5. $\overline{C} = 0.125x + 20 + \frac{5000}{x}$

$\overline{C}' = 0.125 - \frac{5000}{x^2}$

$\overline{C}'' = \frac{10,000}{x^3}$

$\overline{C}' = 0$ when $x = 200$.

Because $\overline{C}''(200) > 0$, $\overline{C}$ is minimum when $x = 200$ units.

7. $\overline{C} = 2x + 255 + \frac{5000}{x}$

$\overline{C}' = 2 - \frac{5000}{x^2}$

$\overline{C}'' = \frac{10,000}{x^3}$

$\overline{C}' = 0$ when $x = 50$. Because $\overline{C}''(50) > 0$, $\overline{C}$ is minimum when $x = 50$ units.

9. $P = xp - C = x(90 - x) - (100 + 30x)$

$= -x^2 + 60x - 100$

$P' = -2x + 60$

$P'' = -2$

$P' = 0$ when $x = 30$. Because $P''(30) < 0$, P is maximum when $x = 30$ units and $p = 90 - 30 = \$60$ per unit.

11. $P = xp - C = x\left(50 - \frac{\sqrt{x}}{10}\right) - (35x + 500)$

$= 15x - \frac{1}{10}x^{3/2} - 500$

$P' = 15 - \frac{3}{20}x^{1/2}$

$P'' = -\frac{3}{40}x^{-1/2} = -\frac{3}{40x^{1/2}}$

$P' = 0$ when $x = 10,000$.

Because $P''(10,000) < 0$, P is maximum when $x = 10,000$ units and $p = 50 - \frac{\sqrt{10,000}}{10} = \$40$ per unit.

13. $\overline{C} = 2x + 5 + \dfrac{18}{x}$

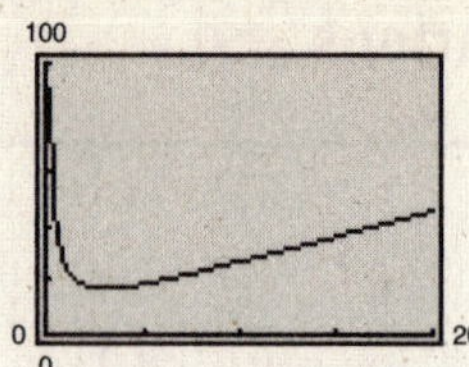

$\overline{C}' = 2 - \dfrac{18}{x^2}$

$\overline{C}'' = \dfrac{36}{x^3}$

$\overline{C}' = 0$ when $x = 3$. Because $\overline{C}''(3) > 0$, $\overline{C}$ is minimum when $x = 3$ units and $\overline{C}(3) = \$17$ per unit.

Marginal cost $= \dfrac{dx}{dc} = 4x + 5$

When $x = 3$, $\dfrac{dx}{dc} = \overline{C} = 17$.

15. (a) $P = xp - C$

$= x(80 - 0.2x) - (30x + 40)$

$= -0.2x^2 + 50x - 40$

$P' = -0.4x + 50$

$P'' = -0.4$

$P' = 0$

$-0.4x + 50 = 0$

$-0.4x = -50$

$x = 125$ units

$P''(125) < 0 \Rightarrow P$ is a maximum when

$x = 125$ units and

$p = 80 - 0.2(125) = \$55$ per unit.

(b) $\overline{C} = \dfrac{C}{x} = \dfrac{30x + 40}{x}$

$\overline{C} = (125) = \$30.32$ per unit

17. $P = -2s^3 + 35s^2 - 100s + 200$

$P' = -6s^2 + 70s - 100$

$= -2(3s^2 - 35s + 50)$

$= -2(3s - 5)(s - 10)$

$P'' = -12x + 70$

$P' = 0$ when $s = \frac{5}{3}$ or $s = 10$.

Because $P''\left(\frac{5}{3}\right) = 50 > 0$, P is minimum when $s = \frac{5}{3}$.

Because $P''(10) = -50 < 0$, P is maximum when $s = 10$.

$P'' = -12s + 70 = 0$ when $s = \frac{35}{6}$. The point of diminishing returns occurs at $s = \frac{35}{6}$.

19. Let x = number of units purchased, p = price per unit, and P = profit.

$p = 150 - (0.10)(x - 100) = 160 - 0.10x, x \geq 100$

$P = xp - C$

$= x(160 - 0.10x) - 90x$

$= -0.10x^2 + 70x$

$P' = -0.2x + 70$

$P'' = -0.2$

$P' = 0$

$-0.02x + 70 = 0$

$-0.2x = -70$

$x = 350$

$P''(350) < 0 \Rightarrow P$ is a maximum when $x = 350$ MP3 players.

21. Let x be the number of units sold per week, p be the price per unit, and R be the revenue. Use the points $(40, 300)$, and $(45, 275)$ to determine the linear equation relating units sold to price per unit.

$x - 300 = \dfrac{275 - 300}{45 - 40}(p - 40)$

$x = -5p + 500$

$R = xp = (-5p + 500)p$

$= -5p^2 + 500p$

$R' = -10p + 500$

$R'' = -10$

$R' = 0$ when $p = 50$. Because $R'' < 0$, R is maximum when $p = \$50$.

23. Let T be the total cost

r = cost under water + cost on land

$T = 25(5280)\sqrt{x^2 + 0.25} + 18(5280)(6 - x)$

$= 132{,}000\sqrt{x^2 + 0.25} + 570{,}240 - 95{,}040x$

$x \approx 0.52$

The line should run from the power station to a point across the river approximately 0.52 mile downstream.

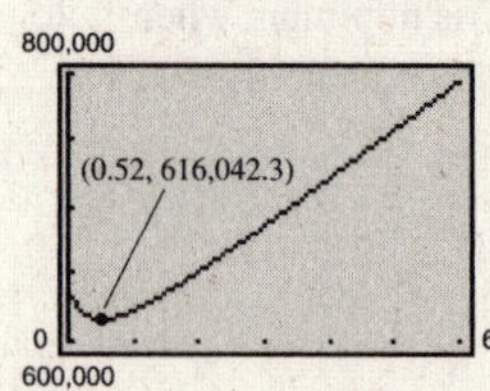

25. distance = (rate)(time)

$$110 = vt$$

$$\frac{110}{v} = t$$

Let T be the total cost.

$$T = \frac{v^2}{300}\left(\frac{110}{v}\right) + 12\left(\frac{110}{v}\right)$$

$$= \frac{11}{30}v + \frac{1320}{v}$$

$$T' = \frac{11}{30} - \frac{1320}{v^2}$$

$$T'' = \frac{2640}{v^3}$$

$T' = 0$ when $v = 60$. Since $T''(60) > 0$, T is minimum when $V = 60$ mi/h.

27. Because $dp/dx = -5$, the price elasticity of demand is

$$\eta = \frac{p/x}{dp/dx} = \frac{\dfrac{600 - 5x}{x}}{-5} = 1 - \frac{600}{5x}.$$

When $x = 60$, you have $\eta = 1 - \dfrac{600}{5(60)} = -1$.

Because $|\eta(60)| = 1$, the demand is unit elastic.

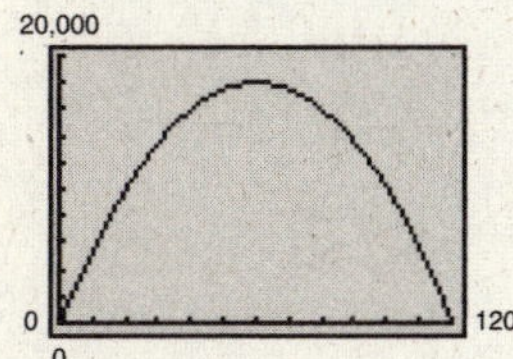

Elastic: $(0, 60)$

Inelastic: $(60, 120)$

29. Because $dp/dx = -0.03$, the price elasticity of demand is

$$\eta = \frac{p/x}{dp/dx} = \frac{(5 - 0.03x)/x}{-0.03} = 1 - \frac{5}{0.03x}.$$

When $x = 100$, you have

$$\eta = 1 - \frac{5}{0.03(100)} = -\frac{2}{3}.$$

Because $|\eta(100)| = \dfrac{2}{3} < 1$, the demand is inelastic.

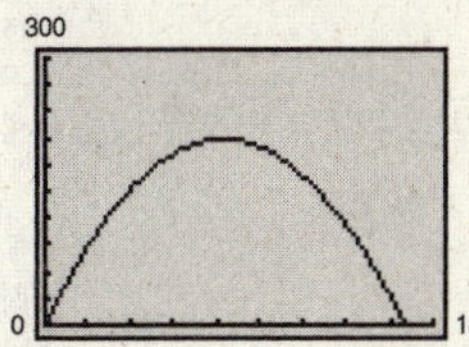

Elastic: $\left(0, \frac{250}{3}\right)$

Inelastic: $\left(\frac{250}{3}, \frac{500}{3}\right)$

31. Because $\dfrac{dp}{dx} = -\dfrac{500}{x(x + 2)^2}$, the price elasticity of demand is

$$\eta = \frac{p/x}{dp/dx} = \frac{500}{x(x + 2)} \cdot \frac{(x + 2)^2}{-500} = -\frac{x + 2}{x}.$$

When $x = 23$, you have

$$\eta = -\frac{23 + 2}{23} = -\frac{25}{23}.$$

Because $|\eta(23)| = \dfrac{25}{23} > 1$, the demand is elastic.

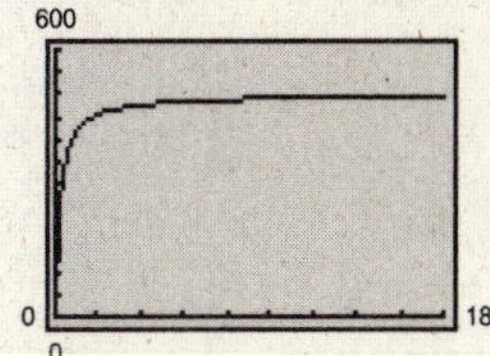

Elastic: $(0, \infty)$

33. (a) $p = 20 - 0.02x,\ 0 \le x \le 1000$

$$\frac{dp}{dx} = -0.02$$

$$\eta = \frac{p/x}{dp/dy} = \frac{\dfrac{20 - 0.02x}{x}}{-0.02} = -\frac{1000}{x} + 1.$$

In the interval $[0, 1000]$, the solution to $|\eta| = \left|-\dfrac{1000}{x} + 1\right| = 1$ is $x = 500$. So the demand is of unit elasticity when $x = 500$. For x-values in $[0, 500)$, $|\eta| > 1$, so demand is elastic. For x-values in $(500, 1000]$, $|\eta| < 1$, so demand is inelastic.

(b) The revenue function increases on the interval $[0, 500)$, then is flat at $x = 500$, and decreases on the interval $(500, 1000]$.

35. $C = 4\left(\dfrac{25}{x^2} - \dfrac{x}{x-10}\right)$

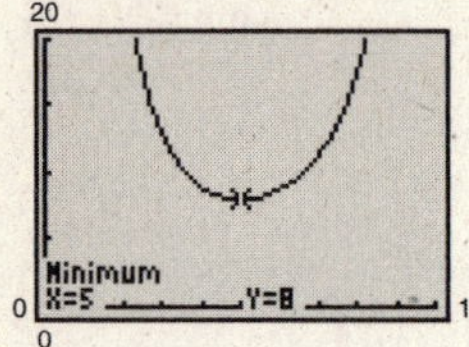

C is minimum when $x = 5$, or 500 units shipped.

37. $x = 900 - 45p \Rightarrow p = 20 - \dfrac{x}{45}$

$\dfrac{dp}{dx} = -\dfrac{1}{45}$

When $p = 8$, $x = 540$ and $\eta = \dfrac{p/x}{dp/dx} = \dfrac{\frac{8}{540}}{-\frac{1}{45}} = -\dfrac{2}{3}$.

Because $|\eta| = \dfrac{2}{3} < 1$, the demand is inelastic.

39. $S = -1.893t^3 + 41.03t^2 - 58.6t + 3972,\ 1 \le t \le 10$

$S' = -5.679t^2 + 82.06t - 58.6$

$S'' = -11.358t + 82.06$

(a)–(c) $S'' = 0$

$-11.358t + 82.06 = 0$

$t \approx 7.2 \Rightarrow t = 7$

t-value	Endpoint $t = 1$	Critical value of S'' $t = 7$	Endpoint $t = 10$
s-value	17.78	237.55	194.1
Conclusion	Minimum	Maximum	

Most rapidly: 2007; rate: \$237.55 million/yr

Slowest rate: 2001; rate: \$17.78 million/yr

(d)

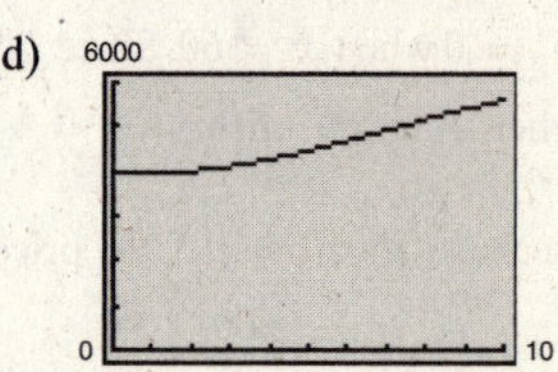

41. $x = \dfrac{a}{p^m},\ m > 1$

$1 = -\dfrac{am}{p^{m+1}}\dfrac{dp}{dx}$

$\dfrac{dp}{dx} = -\dfrac{p^{m+1}}{am}$

$\eta = \dfrac{p/x}{dp/dx} = \dfrac{p}{x} \cdot \dfrac{-am}{p^{m+1}} = \dfrac{p}{a/p^m} \cdot \dfrac{-am}{p^{m+1}} = \dfrac{p^{m+1}}{a} \cdot \dfrac{-am}{p^{m+1}} = -m$

43. Answers will vary.

Section 3.6 Asymptotes

Skills Warm Up

1. $\lim\limits_{x\to 2}(x+1) = 2 + 1 = 3$

2. $\lim\limits_{x\to -1}(3x+4) = 3(-1) + 4 = 1$

3. $\lim\limits_{x\to -3}\dfrac{2x^2 + x - 15}{x+3} = \lim\limits_{x\to -3}\dfrac{(2x-5)(x+3)}{x+3}$
$= \lim\limits_{x\to -3}(2x-5) = 2(-3) - 5$
$= -11$

4. $\lim\limits_{x\to 2}\dfrac{3x^2 - 8x + 4}{x-2} = \lim\limits_{x\to 2}\dfrac{(3x-2)(x-2)}{x-2}$
$= \lim\limits_{x\to 2}(3x-2) = 3(2) - 2 = 4$

5. $\lim\limits_{x\to 2^+}\dfrac{x^2 - 5x + 6}{x^2 - 4} = \lim\limits_{x\to 2^+}\dfrac{(x-3)(x-2)}{(x+2)(x-2)}$
$= \lim\limits_{x\to 2^+}\dfrac{x-3}{x+2} = \dfrac{2-3}{2+2} = -\dfrac{1}{4}$

Skills Warm Up —continued—

6. $\lim_{x\to1^-} \frac{x^2-6x+5}{x^2-1} = \lim_{x\to1^-} \frac{(x-5)(x-1)}{(x+1)(x-1)}$

$= \lim_{x\to1^-} \frac{x-5}{x+1} = \frac{1-5}{1+1} = -2$

7. $\lim_{x\to0^+} \sqrt{x} = \sqrt{0} = 0$

8. $\lim_{x\to1^+} \left(x + \sqrt{x-1}\right) = 1 + \sqrt{1-1} = 1$

9. $\overline{C} = \frac{C}{x} = \frac{150}{x} + 3$

$\frac{dC}{dx} = 3$

10. $\overline{C} = \frac{C}{x} = \frac{1900}{x} + 1.7 + 0.002x$

$\frac{dC}{dx} = 1.7 + 0.004x$

11. $\overline{C} = \frac{C}{x} = 0.005x + 0.5 + \frac{1375}{x}$

$\frac{dC}{dx} = 0.01x + 0.5$

12. $\overline{C} = \frac{C}{x} = \frac{760}{x} + 0.05$

$\frac{dC}{dx} = 0.05$

1. A horizontal asymptote occurs at $y = 1$ because

$\lim_{x\to\infty} \frac{x^2+1}{x^2} = 1, \lim_{x\to-\infty} \frac{x^2+1}{x^2} = 1.$

A vertical asymptote occurs at $x = 0$ because

$\lim_{x\to0^-} \frac{x^2+1}{x^2} = \infty, \lim_{x\to0^+} \frac{x^2+1}{x^2} = \infty.$

3. A horizontal asymptote occurs at $y = 1$ because

$\lim_{x\to\infty} \frac{x^2-2}{x^2-x-2} = 1, \lim_{x\to-\infty} \frac{x^2-2}{x^2-x-2} = 1.$

Vertical asymptotes occur at $x = -1$ and $x = 2$ because

$\lim_{x\to-1^-} \frac{x^2-2}{x^2-x-2} = -\infty, \lim_{x\to-1^+} \frac{x^2-2}{x^2-x-2} = \infty,$

$\lim_{x\to-2^-} \frac{x^2-2}{x^2-x-2} = -\infty, \lim_{x\to-2^+} \frac{x^2-2}{x^2-x-2} = \infty.$

5. A horizontal asymptote occurs at $y = \frac{3}{2}$ because

$\lim_{x\to\infty} \frac{3x^2}{2(x^2+1)} = \frac{3}{2}$ and $\lim_{x\to-\infty} \frac{3x^2}{2(x^2+1)} = \frac{3}{2}.$

The graph has no vertical asymptotes because the denominator is never zero.

7. A horizontal asymptote occurs at $y = \frac{1}{2}$ because

$\lim_{x\to\infty} \frac{x^2-1}{2x^2-8} = \frac{1}{2}$ and $\lim_{x\to-\infty} \frac{x^2-1}{2x^2-8} = \frac{1}{2}.$

Vertical asymptotes occur at $x = \pm2$ because

$\lim_{x\to2^-} \frac{x^2-1}{2x^2-8} = -\infty, \lim_{x\to2^+} \frac{x^2-1}{2x^2-8} = \infty,$

$\lim_{x\to-2^-} \frac{x^2-1}{2x^2-8} = \infty, \lim_{x\to-2^+} \frac{x^2-1}{2x^2-8} = -\infty.$

9. $f(x) = \frac{x-3}{x^2+3x} = \frac{x-3}{x(x+3)}$

$x(x+3) = 0$

$x = 0$ or $x = -3$

Vertical asymptotes: $x = 0, x = -3$

11. $f(x) = \frac{x^2-8x+15}{x^2-9}$

$= \frac{(x-3)(x-5)}{(x-3)(x+3)}$

$= \frac{x-5}{x+3}, x \neq 3$

$x + 3 = 0$

$x = -3$

Vertical asymptote: $x = -3$

13. $f(x) = \frac{2x^2-x-3}{2x^2-11x+12}$

$= \frac{(2x-3)(x+1)}{(2x-3)(x-4)}$

$= \frac{x+1}{x-4}, x \neq \frac{3}{2}$

$x - 4 = 0$

$x = 4$

Vertical asymptote: $x = 4$

15. $\lim_{x\to6^+} \frac{1}{(x-6)^2} = \infty$

17. $\lim_{x\to3^+} \frac{x-4}{x-3} = -\infty$

19. $\lim_{x \to -1^-} \frac{x^2 + 1}{x^2 - 1} = \infty$

21. $\lim_{x \to \infty} \left(1 + \frac{1}{x}\right) = 1 + 0 = 1$

23. $\lim_{x \to -\infty} \left(7 + \frac{4}{x^2}\right) = 7 + 0 = 7$

25. $\lim_{x \to \pm\infty} \frac{4x - 3}{2x + 1} = \frac{4}{2} = 2$

Horizontal asymptote: $y = 2$

27. $\lim_{x \to \pm\infty} \frac{3x}{4x^2 - 1} = 0$

Horizontal asymptote: $y = 0$ or x-axis

29. $\lim_{x \to \pm\infty} \frac{5x^2}{x + 3} = \pm\infty$, does not exist.

No horizontal asymptote

31. $\lim_{x \to \pm\infty} \left(\frac{2x}{x - 1} + \frac{3x}{x + 1}\right) = 2 + 3 = 5$

Horizontal asymptote: $y = 5$

33. The graph of f has a horizontal asymptote at $y = 3$. It matches graph (d).

35. The graph of f has a horizontal asymptote at $y = 2$. It matches graph (a).

37. (a) $h(x) = \frac{5x^3 - 3}{x^2}$

$\lim_{x \to \infty} h(x) = \infty$

(b) $h(x) = \frac{5x^3 - 3}{x^3}$

$\lim_{x \to \infty} h(x) = 5$

(c) $h(x) = \frac{5x^3 - 4}{x^4}$

$\lim_{x \to \infty} h(x) = 0$

39. (a) $\lim_{x \to \infty} \frac{x^2 + 2}{x^3 - 1} = 0$

(b) $\lim_{x \to \infty} \frac{x^2 + 2}{x^2 - 1} = 1$

(c) $\lim_{x \to \infty} \frac{x^2 + 2}{x - 1} = \infty$

41. $f(x) = \sqrt{x^3 + 6} - 2x$

x	10^0	10^1	10^2	10^3	10^4	10^5	10^6
$f(x)$	0.646	11.718	800.003	29,622.777	980,000	31,422,776.6	998,000,000

$\lim_{x \to \infty} \sqrt{x^3 + 6} - 2x = \infty$

43. $f(x) = \frac{x + 1}{x\sqrt{x}}$

x	10^0	10^1	10^2	10^3	10^4	10^5	10^6
$f(x)$	2.000	0.348	0.101	0.032	0.010	0.003	0.001

$\lim_{x \to \infty} f(x) = 0$

45. $y = \frac{3x}{1 - x}$

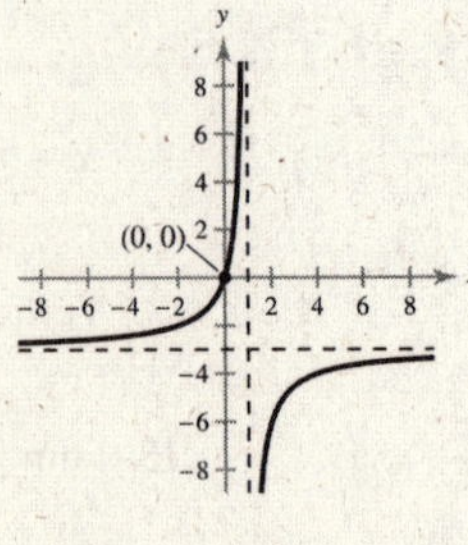

Intercept: $(0, 0)$

Horizontal asymptote: $y = -3$

Vertical asymptote: $x = 1$

$y' = \frac{(1 - x)3 - 3x(-1)}{(1 - x)^2} = \frac{1}{(1 - x)^2}$

$y' \neq 0$ so there are no relative extrema.

47. $f(x) = \dfrac{x^2}{x^2 + 9}$

Intercept: $(0, 0)$

Horizontal asymptote: $y = 1$

$$f'(x) = \frac{(x^2 + 9)(2x) - x^2(2x)}{(x^2 + 9)^2} = \frac{18x}{(x^2 + 9)^2}$$

The critical number is $x = 0$ and by the First-Derivative Test $(0, 0)$ is a relative minimum.

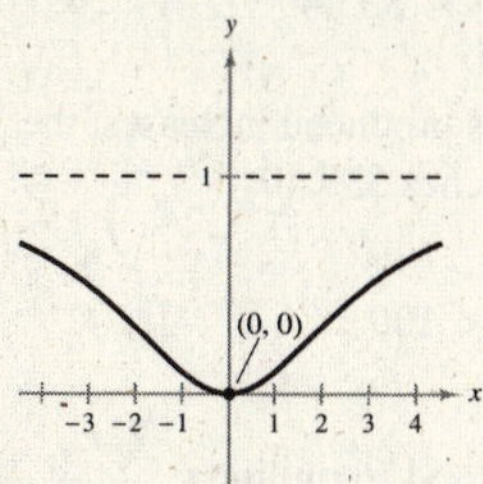

49. $g(x) = \dfrac{x^2}{x^2 - 16}$

Intercept: $(0, 0)$

Horizontal asymptote: $y = 1$

Vertical asymptotes: $x = \pm 4$

$$g'(x) = \frac{(x^2 - 16)(2x) - x^2(2x)}{(x^2 - 16)^2} = \frac{-32x}{(x^2 - 16)^2}$$

The critical number is $x = 0$ and by the First-Derivative Test $(0, 0)$ is a relative maximum.

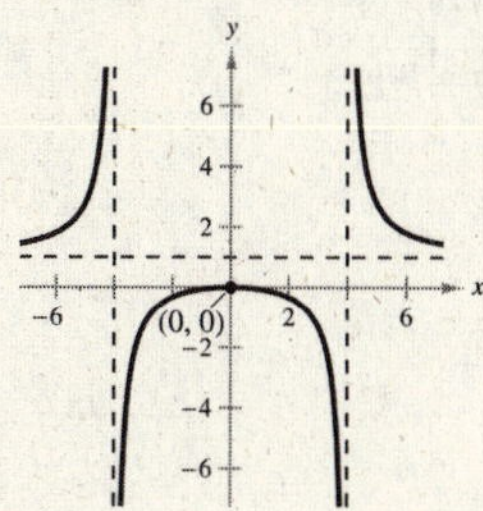

51. $y = 1 - \dfrac{3}{x^2}$

Intercepts: $(\pm\sqrt{3}, 0)$

Horizontal asymptote: $y = 1$

Vertical asymptote: $x = 0$

$$y' = \frac{6}{x^3}$$

No critical numbers because $y' \neq 0$, so there are no relative extrema.

53.

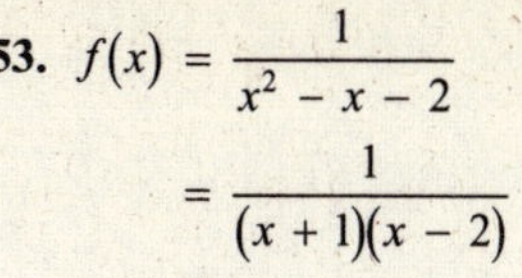

$$f(x) = \frac{1}{x^2 - x - 2} = \frac{1}{(x + 1)(x - 2)}$$

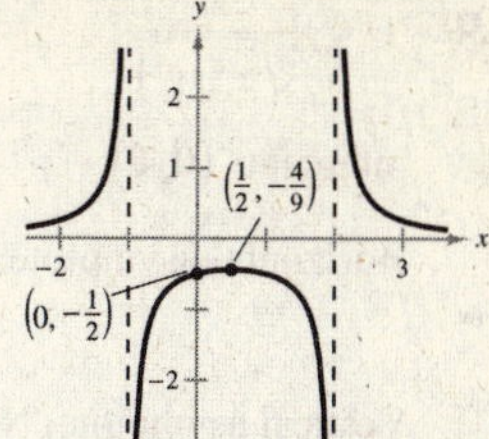

Intercept: $\left(0, -\dfrac{1}{2}\right)$

Horizontal asymptote: $y = 0$

Vertical asymptotes: $x = -1$, $x = 2$

$$f'(x) = -(x^2 - x - 2)^{-2}(2x - 1) = -\frac{2x - 1}{(x^2 - x - 2)^2}$$

The critical number is $x = \dfrac{1}{2}$ and by the First-Derivative Test $\left(\dfrac{1}{2}, -\dfrac{4}{9}\right)$ is a relative maximum.

55. $g(x) = \dfrac{x^2 - x - 2}{x - 2} = \dfrac{(x - 2)(x + 1)}{x - 2}$

$= x + 1$ for $x \neq 2$

Intercepts: $(-1, 0)$, $(0, 1)$

No asymptotes

$g'(x) = 1$ for $x \neq 2$

$g'(x) \neq 0$ so there are no relative extrema.

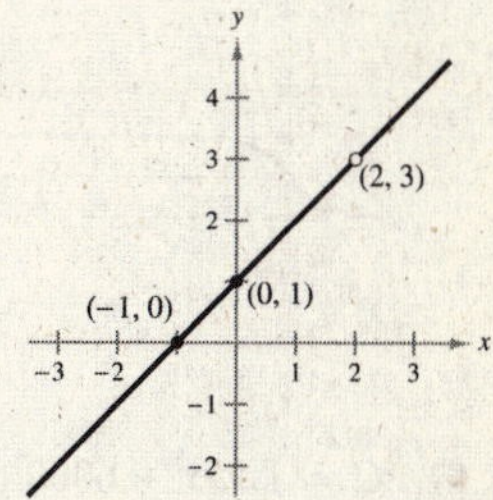

57. $y = \dfrac{2x^2 - 6}{x^2 - 2x + 1} = \dfrac{2(x^2 - 3)}{(x - 1)^2}$

Intercepts: $(\pm\sqrt{3}, 0)$, $(0, -6)$

Vertical asymptote: $x = 1$

Horizontal asymptote: $y = 2$

$$y' = \frac{(x^2 - 2x + 1)(4x) - (2x^2 - 6)(2x - 2)}{(x - 1)^4}$$

$$= \frac{4(3 - x)}{(x - 1)^3}$$

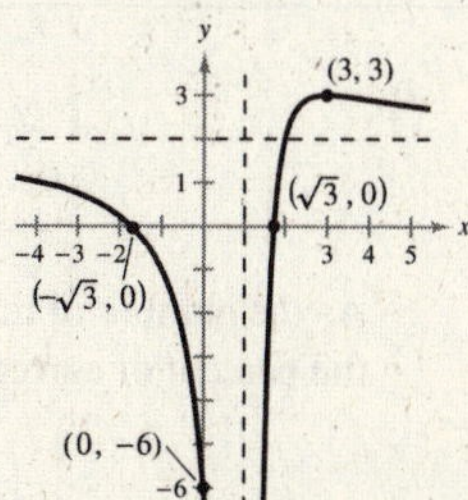

The critical number is $x = 3$ and by the First-Derivative Test $(3, 3)$ is a relative maximum.

59. $y = \dfrac{x}{\sqrt{x^2 + 1}}$

Intercept: $(0, 0)$

Horizontal asymptotes: $y = -1$ as $x \to -\infty$

$y = 1$ as $x \to \infty$

Vertical asymptote: None

$$y' = \frac{(x^2 + 1)^{1/2}(1) - x\left(\frac{1}{2}\right)(x^2 + 1)^{-1/2}(2x)}{\left(\sqrt{x^2 + 1}\right)^2}$$

$$= \frac{(x^2 + 1)^{-1/2}\left[(x^2 + 1) - x^2\right]}{(x^2 + 1)}$$

$$= \frac{1}{(x^2 + 1)^{3/2}}$$

No critical numbers because $y' \neq 0$, so there are no relative extrema.

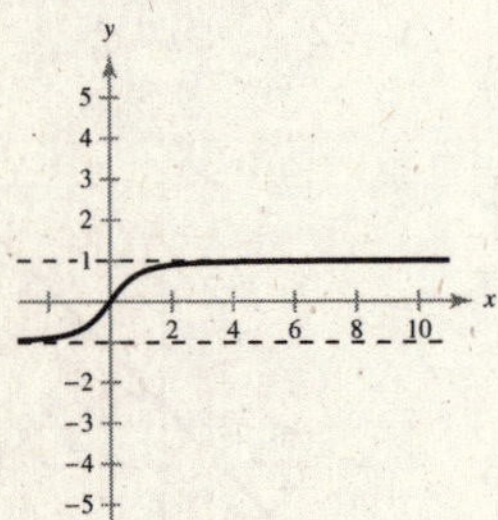

61. $C = 1.15x + 6000$

(a) $\overline{C} = \dfrac{C}{x} = \dfrac{1.15x + 6000}{x}$

(b) $\overline{C}(600) = \$11.15/\text{unit}$

$\overline{C}(6000) = \$2.15/\text{unit}$

(c) $\lim\limits_{x\to\infty} \overline{C} = \lim\limits_{x\to\infty} \dfrac{1.15x + 6000}{x} = \$1.15/\text{unit}$

The cost approaches \$1.15 as the number of units produced increases.

63. $C = 34.5x + 15{,}000$, $R = 69.9x$

(a) $\overline{P} = \dfrac{R - C}{x} = \dfrac{69.9x - (34.5x + 15{,}000)}{x}$

$= 35.4 - \dfrac{15{,}000}{x}$

(b) $\overline{P}(1000) = \$20.40/\text{unit}$

$\overline{P}(10{,}000) = \$33.90/\text{unit}$

$\overline{P}(100{,}000) = \$35.25/\text{unit}$

(c) $\lim\limits_{x\to\infty}\left(35.4 - \dfrac{15{,}000}{x}\right) = \$35.40$

As the number of products produced increases, the average profit approaches \$35.40.

65. $C = \dfrac{528p}{(100 - p)}$, $0 \le p < 100$

(a) $C(25) = \dfrac{528(25)}{100 - 25} = \$176$ million

$C(50) = \dfrac{528(50)}{100 - 50} = \$528$ million

$C(75) = \dfrac{528(75)}{100 - 75} = \$1584$ million

(b) $\lim\limits_{p\to100^-} \dfrac{528p}{100 - p} = \infty$

As the percent of illegal drugs seized approaches 100%, the cost increases without bound.

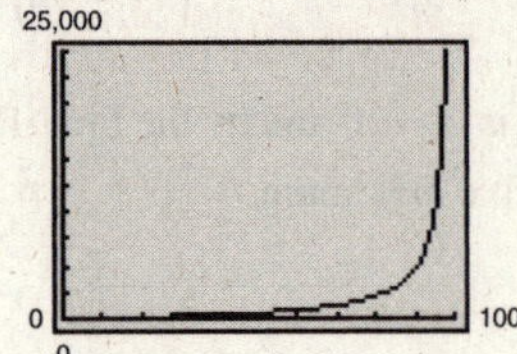

67. (a)

n	1	2	3	4	5	6	7	8	9	10
P	0.50	0.74	0.82	0.86	0.89	0.91	0.92	0.93	0.94	0.95

(b) $\lim\limits_{n\to\infty} \dfrac{0.5 + 0.9(n - 1)}{1 + 0.9(n - 1)} = 1$

(c)

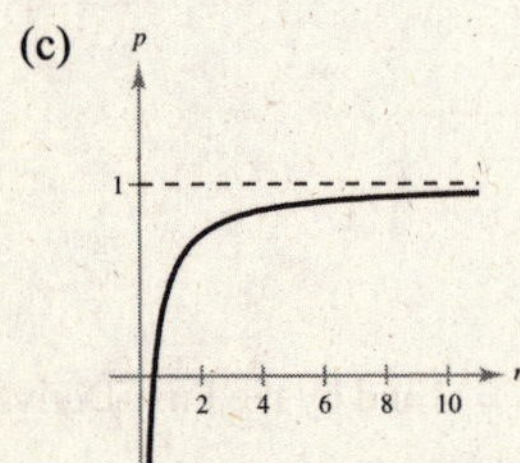

As the number of times the task is performed increases, the percent of correct responses approaches 100%.

Section 3.7 Curve Sketching: A Summary

Skills Warm Up

1. A vertical asymptote occurs at $x = 0$ because

$\lim_{x \to 0^-} \frac{1}{x^2} = \infty$ and $\lim_{x \to 0^+} \frac{1}{x^2} = \infty$.

No horizontal asymptotes.

2. A vertical asymptote occurs at $x = 2$ because

$\lim_{x \to 2^-} \frac{8}{(x-2)^2} = \infty$ and $\lim_{x \to 2^+} \frac{8}{(x-2)^2} = \infty$.

No horizontal asymptotes.

3. A vertical asymptote occurs at $x = -3$ because

$\lim_{x \to -3^-} \frac{40x}{x+3} = \infty$ and $\lim_{x \to -3^+} \frac{40x}{x+3} = -\infty$.

A horizontal asymptote occurs at $y = 40$ because

$\lim_{x \to \infty} \frac{40x}{x+3} = 40$ and $\lim_{x \to -\infty} \frac{40x}{x+3} = 40$.

4. Vertical asymptotes occur at $x = 1$ and $x = 3$ because

$\lim_{x \to 1^-} \frac{x^2 - 3}{x^2 - 4x + 3} = -\infty$, $\lim_{x \to 1^+} \frac{x^2 - 3}{x^2 - 4x + 3} = \infty$,

$\lim_{x \to 3^-} \frac{x^2 - 3}{x^2 - 4x + 3} = -\infty$, and

$\lim_{x \to 3^+} \frac{x^2 - 3}{x^2 - 4x + 3} = \infty$. A horizontal asymptote

occurs at $y = 1$ because $\lim_{x \to \infty} \frac{x^2 - 3}{x^2 - 4x + 3} = 1$ and

$\lim_{x \to -\infty} \frac{x^2 - 3}{x^2 - 4x + 3} = 1$.

5. $f(x) = x^2 + 4x + 2$

$f'(x) = 2x + 4$

Critical number: $x = -2$

Interval	$-\infty < x < -2$	$-2 < x < \infty$
Sign of f'	$f' < 0$	$f' > 0$
Conclusion	Decreasing	Increasing

6. $f(x) = -x^2 - 8x + 1$

$f'(x) = -2x - 8$

Critical number: $x = -4$

Interval	$-\infty < x < -4$	$-4 < x < \infty$
Sign of f'	$f' > 0$	$f' < 0$
Conclusion	Increasing	Decreasing

7. $f(x) = x^3 - 3x + 1$

$f'(x) = 3x^2 - 3$

Critical numbers: $x = \pm 1$

Interval	$-\infty < x < -1$	$-1 < x < 1$	$1 < x < \infty$
Sign of f'	$f' > 0$	$f' < 0$	$f' > 0$
Conclusion	Increasing	Decreasing	Increasing

8. $f(x) = \frac{-x^3 + x^2 - 1}{x^2} = -x + 1 - \frac{1}{x^2}$

$f'(x) = -1 + \frac{2}{x^3} = \frac{-x^3 + 2}{x^3}$

Critical number: $x = \sqrt[3]{2}$

Discontinuity: $x = 0$

Interval	$-\infty < x < 0$	$0 < x < \sqrt[3]{2}$	$\sqrt[3]{2} < x < \infty$
Sign of f'	$f' < 0$	$f' > 0$	$f' < 0$
Conclusion	Decreasing	Increasing	Decreasing

9. $f(x) = \frac{x - 2}{x - 1}$

$f'(x) = \frac{(x-1) - (x-2)}{(x-1)^2} = \frac{1}{(x-1)^2}$

No critical numbers

Discontinuity: $x = 1$

Interval	$-\infty < x < 1$	$1 < x < \infty$
Sign of f'	$f' > 0$	$f' > 0$
Conclusion	Increasing	Increasing

Skills Warm Up —*continued*—

10. $f(x) = -x^3 - 4x^2 + 3x + 2$

$f'(x) = -3x^2 - 8x + 3$

Critical numbers: $x = -3,\ x = \dfrac{1}{3}$

Interval	$-\infty < x < -3$	$-3 < x < \frac{1}{3}$	$\frac{1}{3} < x < \infty$
Sign of f'	$f' < 0$	$f' > 0$	$f' < 0$
Conclusion	Decreasing	Increasing	Decreasing

1. $y = -x^2 - 2x + 3 = -(x + 3)(x - 1)$

$y' = -2x - 2 = -2(x + 1)$

$y'' = -2$

Intercepts: $(0, 3)$, $(1, 0)$, $(-3, 0)$

Relative maximum: $(-1, 4)$

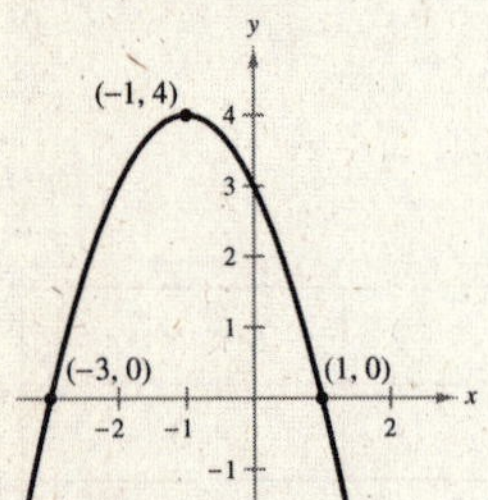

3. $y = x^3 - 4x^2 + 6$

$y' = 3x^2 - 8x = x(3x - 8)$

$y'' = 6x - 8 = 2(3x - 4)$

Relative maximum: $(0, 6)$

Relative minimum: $\left(\dfrac{8}{3}, -\dfrac{94}{27}\right)$

Point of inflection: $\left(\dfrac{4}{3}, \dfrac{34}{27}\right)$

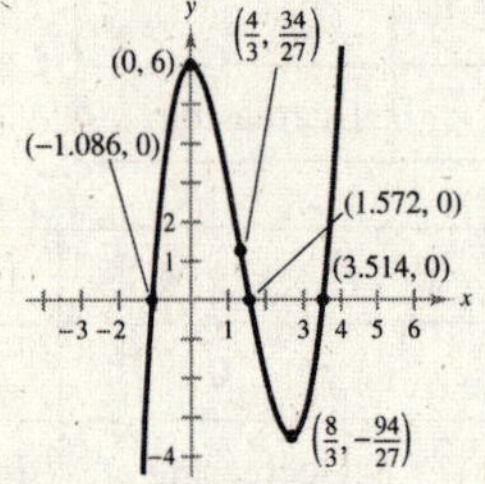

5. $y = 2 - x - x^3$

$y' = -1 - 3x^2$

$y'' = -6x$

No relative extrema

Point of inflection: $(0, 2)$

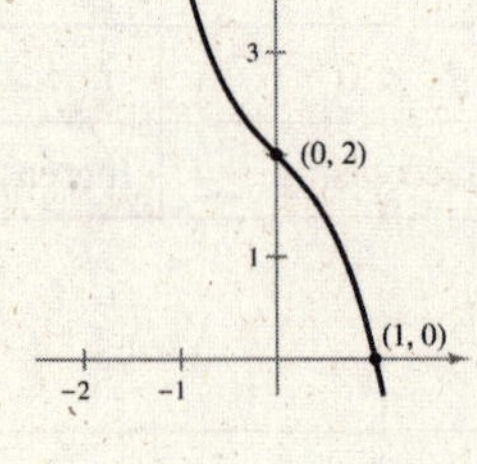

7. $y = 3x^4 + 4x^3 = x^3(3x + 4)$

$y' = 12x^3 + 12x^2 = 12x^2(x + 1)$

$y'' = 36x^2 + 24x = 12x(3x + 2)$

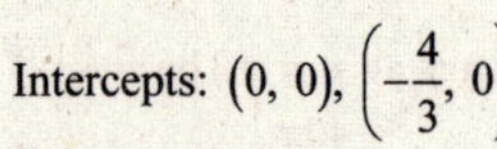

Intercepts: $(0, 0)$, $\left(-\dfrac{4}{3}, 0\right)$

Relative minimum: $(-1, -1)$

Points of inflection:

$(0, 0)$, $\left(-\dfrac{2}{3}, -\dfrac{16}{27}\right)$

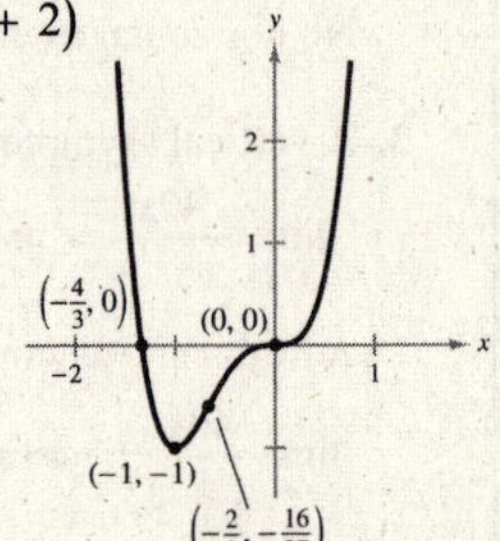

9. $y = x^4 - 8x^3 + 18x^2 - 16x + 5 = (x - 5)(x - 1)^3$

$y' = 4x^3 - 24x^2 + 36x - 16 = 4(x - 4)(x - 1)^2$

$y'' = 12x^2 - 48x + 36 = 12(x - 1)(x - 3)$

Intercepts: $(0, 5)$, $(1, 0)$, $(5, 0)$

Relative minimum: $(4, -27)$

Points of inflection: $(1, 0)$, $(3, -16)$

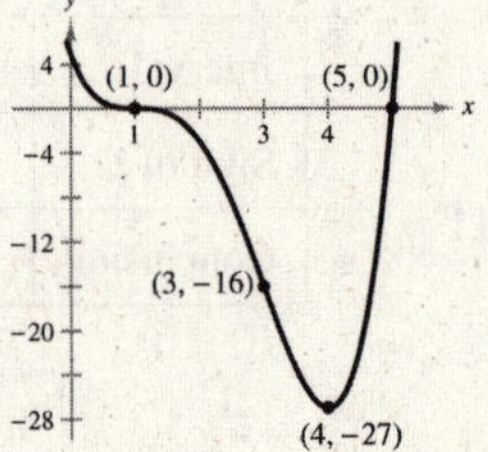

11. $y = \dfrac{x^2 + 1}{x} = x + \dfrac{1}{x}$

$y' = 1 - \dfrac{1}{x^2} = \dfrac{x^2 - 1}{x^2}$

$y'' = \dfrac{2}{x^3}$

No intercepts

Relative maximum: $(-1, -2)$

Relative minimum: $(1, 2)$

No points of inflection

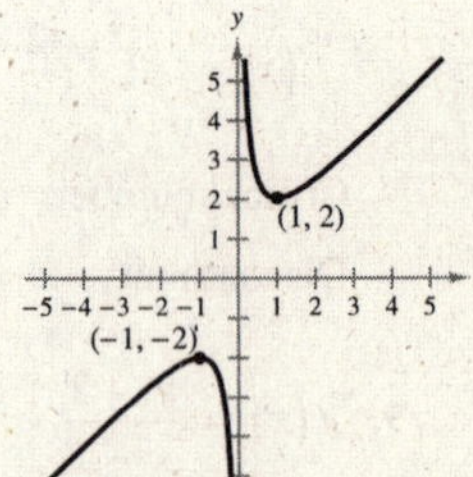

13. $y = \dfrac{x^2 - 6x + 12}{x - 4}$

$$y' = \frac{(x-4)(2x-6) - (x^2 - 6x + 12)}{(x-4)^2}$$
$$= \frac{x^2 - 8x + 12}{(x-4)^2}$$
$$= \frac{(x-2)(x-6)}{(x-4)^2}$$
$$y'' = \frac{(x-4)^2(2x-8) - (x^2 - 8x + 12)[2(x-4)]}{(x-4)^4}$$
$$= \frac{8}{(x-4)^3}$$

Intercept: $(0, -3)$

Relative maximum: $(6, 6)$

Relative minimum: $(2, -2)$

No points of inflection

Vertical asymptote: $x = 4$

Domain: $(-\infty, 4) \cup (4, \infty)$

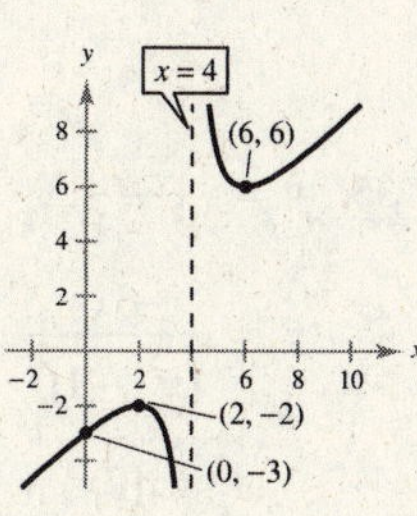

15. $y = \dfrac{x^2 + 1}{x^2 - 9}$

$$y' = \frac{(x^2 - 9)(2x) - (x^2 + 1)(2x)}{(x^2 - 9)^2} = -\frac{20x}{(x^2 - 9)^2}$$
$$y'' = \frac{(x^2 - 9)^2(-20) - (-20x)[2(x^2 - 9)(2x)]}{(x^2 - 9)^4}$$
$$= \frac{60(x^2 + 3)}{(x^2 - 9)^3}$$

No intercepts

Relative maximum: $\left(0, -\dfrac{1}{9}\right)$

No points of inflection

Horizontal asymptote: $y = 1$

Vertical asymptote: $x = \pm 3$

Domain: $(-\infty, -3) \cup (-3, 3) \cup (3, \infty)$

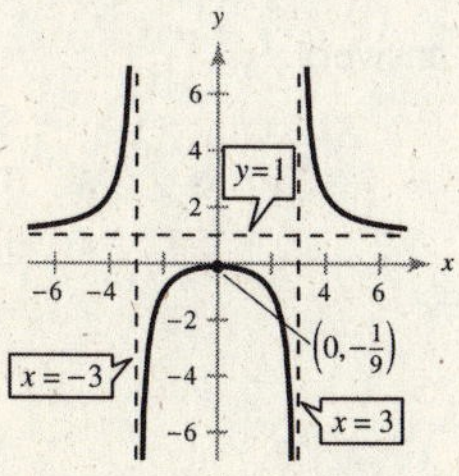

17. $y = 3x^{2/3} - x^2$

$$y' = \frac{2}{x^{1/3}} - 2x$$
$$y'' = -2\left(\frac{1}{3x^{4/3}} + 1\right)$$

Intercepts:

$\left(\pm\sqrt[4]{27}, 0\right), (0, 0)$

Relative maxima: $(\pm 1, 2)$

Relative minimum: $(0, 0)$

No points of inflection

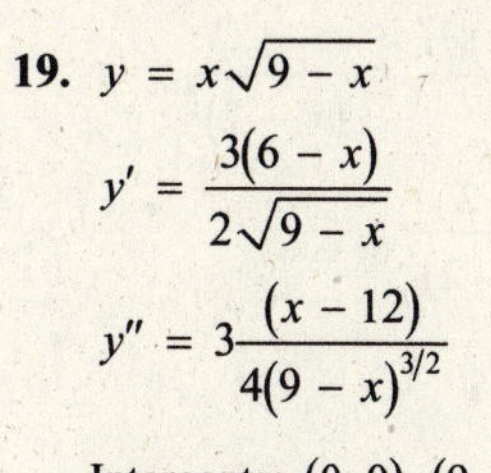
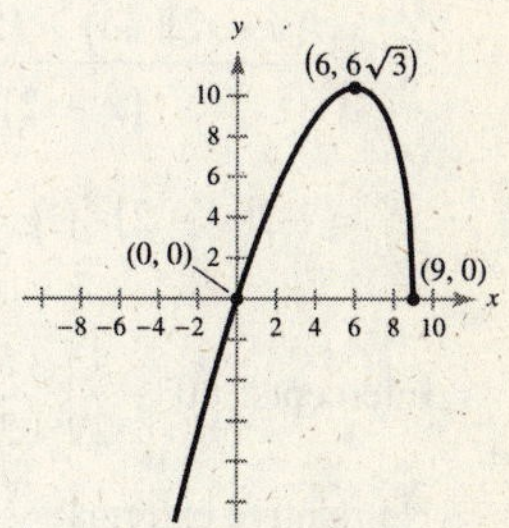

19. $y = x\sqrt{9 - x}$

$$y' = \frac{3(6 - x)}{2\sqrt{9 - x}}$$
$$y'' = 3\frac{(x - 12)}{4(9 - x)^{3/2}}$$

Intercepts: $(0, 0), (9, 0)$

Relative maximum: $\left(6, 6\sqrt{3}\right)$

No points of inflection

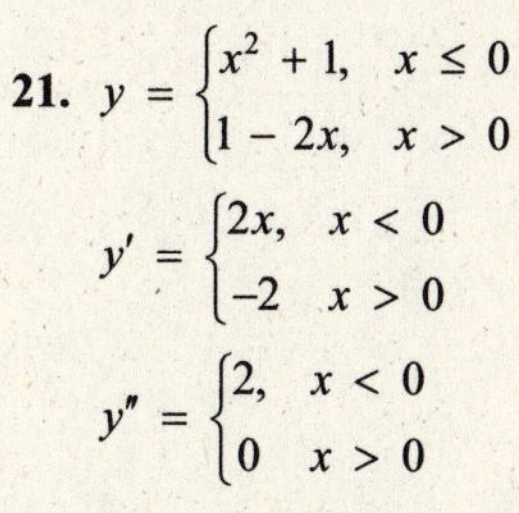
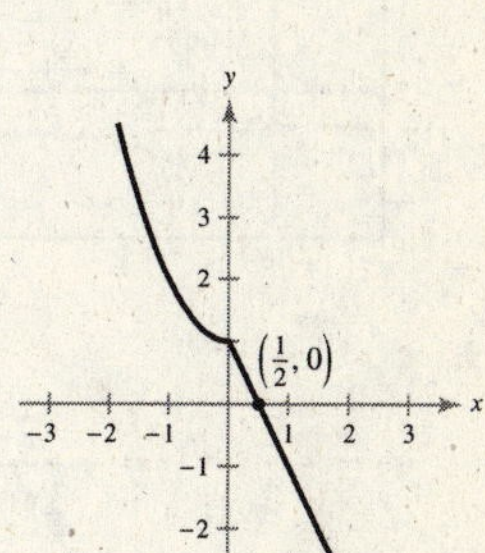

21. $y = \begin{cases} x^2 + 1, & x \le 0 \\ 1 - 2x, & x > 0 \end{cases}$

$$y' = \begin{cases} 2x, & x < 0 \\ -2 & x > 0 \end{cases}$$
$$y'' = \begin{cases} 2, & x < 0 \\ 0 & x > 0 \end{cases}$$

Intercept: $\left(\frac{1}{2}, 0\right)$

No relative extrema

No points of inflection

23. $y = 3x^3 - 9x + 1$

$$y' = 9x^2 - 9 = 9(x^2 - 1)$$
$$y'' = 18x$$

Intercept: $(0, 1)$

Relative maximum: $(-1, 7)$

Relative minimum: $(1, -5)$

Point of inflection: $(0, 1)$

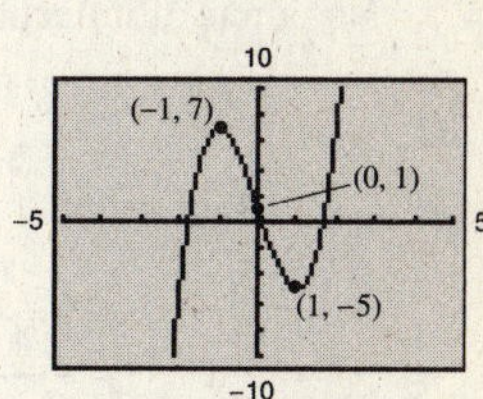

25. $y = x^5 - 5x$

$y' = 5x^4 - 5 = 5(x^4 - 1) = 5(x^2 - 1)(x^2 + 1)$

$y'' = 20x^3$

Intercept: $(0, 0)$

Relative maximum: $(-1, 4)$

Relative minimum: $(1, -4)$

Point of inflection: $(0, 0)$

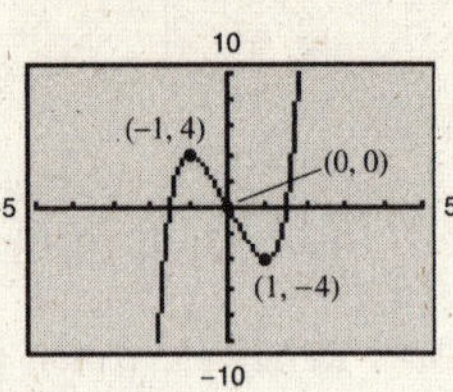

27. $y = \dfrac{5 - 3x}{x - 2}$

$y' = \dfrac{(x - 2)(-3) - (5 - 3x)(1)}{(x - 2)^2} = \dfrac{1}{(x - 2)^2}$

$y'' = -2(x - 2)^{-3}(1) = \dfrac{-2}{(x - 2)^3}$

Intercepts: $\left(0, -\dfrac{5}{2}\right), \left(\dfrac{5}{3}, 0\right)$

No relative extrema

No points of inflection

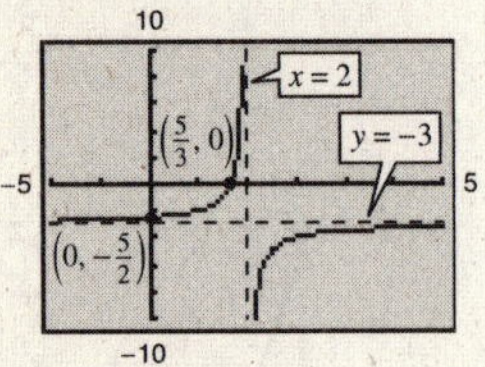

29. $y = 1 - x^{2/3}$

$y' = -\dfrac{2}{3}x^{-1/3} = -\dfrac{2}{3x^{1/3}}$

$y'' = \dfrac{2}{9}x^{-4/3} = \dfrac{2}{9x^{4/3}}$

Intercepts: $(0, 1), (\pm 1, 0)$

Relative maximum: $(0, 1)$

No points of inflection

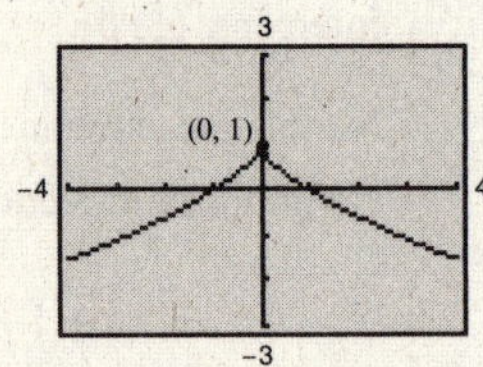

31. $y = x^{4/3}$

$y' = \dfrac{4}{3}x^{1/3}$

$y'' = \dfrac{4}{9}x^{-2/3} = \dfrac{4}{9x^{2/3}}$

Intercept: $(0, 0)$

Relative minimum: $(0, 0)$

No points of inflection

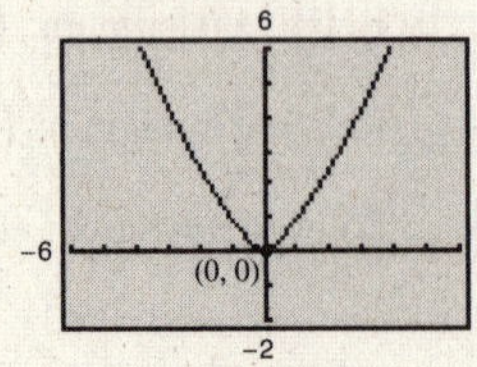

33. $y = \dfrac{x}{\sqrt{x^2 - 4}}, \ |x| > 2$

$y' = \dfrac{(x^2 - 4)^{1/2} - x\left[\frac{1}{2}(x^2 - 4)^{-1/2}(2x)\right]}{x^2 - 4}$

$= -\dfrac{4}{(x^2 - 4)^{3/2}}$

$y'' = 6(x^2 - 4)^{-5/2}(2x) = \dfrac{12x}{(x^2 - 4)^{5/2}}$

No intercepts

No relative extrema

No points of inflection

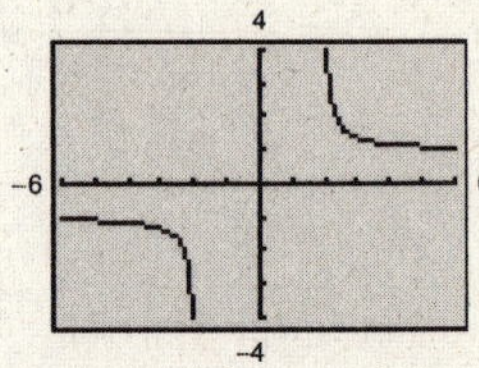

35. $y = \dfrac{x^3}{x^3 - 1}$

$y' = \dfrac{-3x^2}{(x^3 - 1)^2}$

$y'' = \dfrac{6x(2x^3 + 1)}{(x^3 - 1)^3}$

Intercept: $(0, 0)$

No relative extrema

Points of inflection: $\left(-\sqrt[3]{\dfrac{1}{2}}, \dfrac{1}{3}\right), (0, 0)$

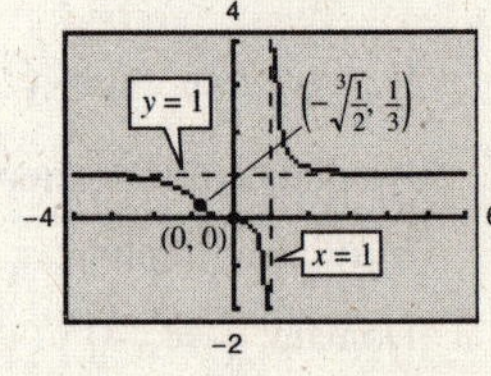

37. Because $f'(x) = 2$, the graph of f is a line with a slope of 2.

Answers will vary. Sample answer:

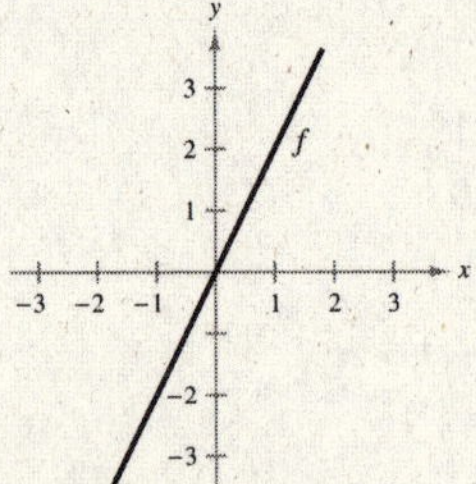

39. $f''(x) = 2 > 0$ so $f(x)$ is a concave upward parabola.

Answers will vary. Sample answer:

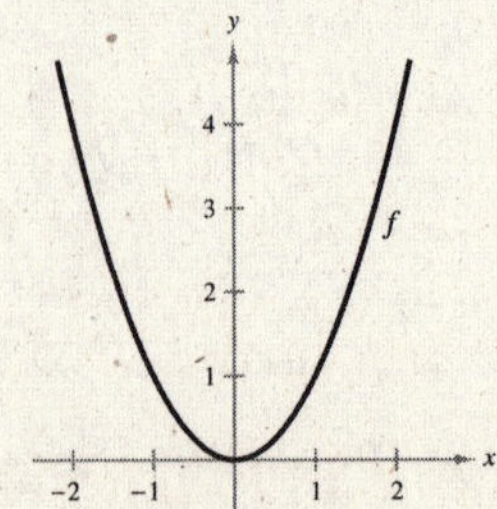

41.

Interval	$-\infty < x < -1$	$-1 < x < 0$	$0 < x < \infty$
Sign of f'	$f' > 0$	$f' < 0$	$f' > 0$
Conclusion	Increasing	Decreasing	Increasing

Intercepts: $(-2, 0)$, $(0, 0)$

Relative maximum: $(-1, f(-1))$

Relative minimum: $(0, f(0)) = (0, 0)$

Answers will vary. Sample answer:

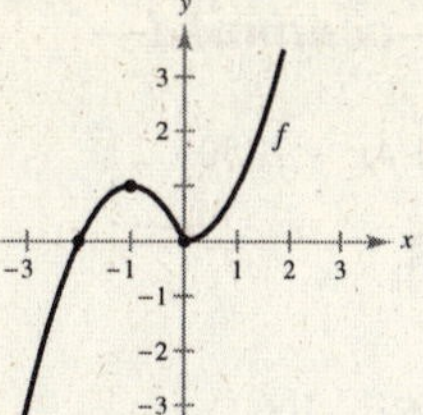

43. Answers will vary. Sample answer:

$$f(x) = \frac{1}{x - 5}$$

45. (a)

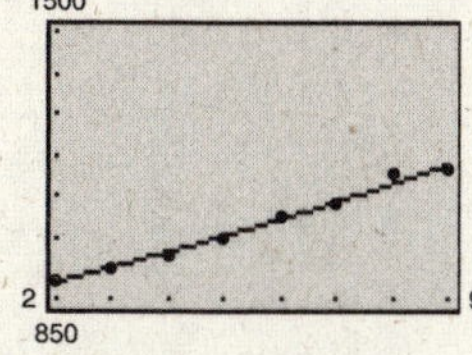

The model fits the data well.

(b) 2014: $B(14) \approx \$1468.54$

(c) No, because the benefits increase without bound as time approaches the year 2035 $(x = 35)$, and the benefits are negative for the years past 2035.

47.

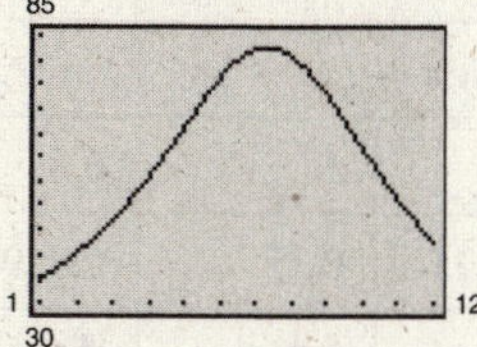

Absolute maximum: $(7, 82.28)$

Absolute minimum: $(1, 34.84)$

The maximum temperature of 82.28°F occurs in July.

The minimum temperature of 34.84°F occurs in January.

49.

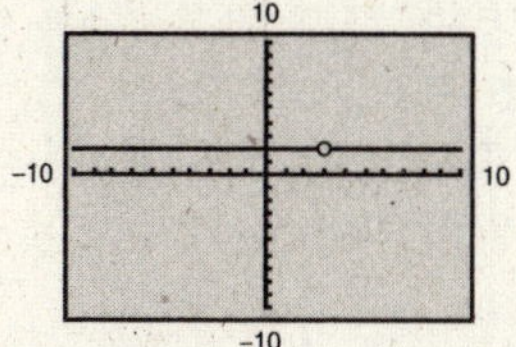

$$h(x) = \frac{6 - 2x}{3 - x} = \frac{2(3 - x)}{3 - x} = 2, \; x \neq 3$$

The rational function simplifies to a constant function that is undefined at $x = 3$.

51. (a) $f(x) = \dfrac{x^2 - 2x + 4}{x - 2} = \dfrac{x^2 - 2x}{x - 2} + \dfrac{4}{x - 2}$

$= \dfrac{x(x - 2)}{x - 2} + \dfrac{4}{x - 2} = x + \dfrac{4}{x - 2}$

(b)

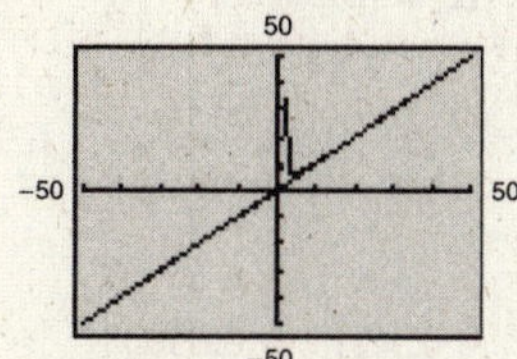

The graphs become almost identical as you zoom out.

(c) A slant asymptote is neither horizontal nor vertical. It is diagonal, following $y = x$.

Section 3.8 Differentials and Marginal Analysis

Skills Warm Up

1. $C = 44 + 0.09x^2$

   $\dfrac{dC}{dx} = 0.18x$

2. $C = 250 + 0.15x$

   $\dfrac{dC}{dx} = 0.15$

3. $R = x(1.25 + 0.02\sqrt{x}) = 1.25x + 0.02x^{3/2}$

   $\dfrac{dR}{dx} = 1.25 + 0.03\sqrt{x}$

4. $R = x(15.5 - 1.55x) = 15.5x - 1.55x^2$

   $\dfrac{dR}{dx} = 15.5 - 3.1x$

Skills Warm Up —*continued*—

5. $P = -0.03x^{1/3} + 1.4x - 2250$

$\frac{dP}{dx} = \frac{-0.01}{x^{2/3}} + 1.4$

6. $P = -0.02x^2 + 25x - 1000$

$\frac{dP}{dx} = -0.04x + 25$

7. $A = \frac{1}{4}\sqrt{3}x^2$

$\frac{dA}{dx} = \frac{1}{2}\sqrt{3}x$

8. $A = 6x^2$

$\frac{dA}{dx} = 12x$

9. $C = 2\pi r$

$\frac{dC}{dr} = 2\pi$

10. $P = 4w$

$\frac{dP}{dw} = 4$

11. $S = 4\pi r^2$

$\frac{dS}{dr} = 8\pi r$

12. $P = 2x + \sqrt{2x}$

$\frac{dP}{dx} = 2 + \sqrt{2}$

13. $A = \pi r^2$

14. $A = x^2$

15. $V = x^3$

16. $V = \frac{4}{3}\pi r^3$

1. $y = 0.5x^3,\ x = 2,\ \Delta x = dx = 0.1$

$$dy = 1.5x^2\ dx$$
$$= 1.5(2)^2(0.1)$$
$$= 0.6$$
$$\Delta y = 0.5(2 + 0.1)^3 - 0.5(2)^3 = 0.6305$$
$$dy \approx \Delta y$$

3. $y = x^4 + 1,\ x = -1,\ \Delta x = dx = 0.01$

$$dy = 4x^3\ dx$$
$$= 4(-1)^3(0.01)$$
$$= -0.04$$
$$\Delta y = (-1 + 0.01)^4 + 1 - \left[(-1)^4 + 1\right] \approx -0.0394$$
$$dy \approx \Delta y$$

5. $y = 3x^{1/2},\ x = 4,\ \Delta x = dx = 0.1$

$$dy = \frac{3}{2x^{1/2}}\ dx$$
$$= \frac{3}{2(4)^{1/2}}(0.1)$$
$$= 0.075$$
$$\Delta y = 3(4 + 0.1)^{1/2} - 3(4)^{1/2} \approx 0.0745$$
$$dy \approx \Delta y$$

7. $dy = 2x\ dx,\ x = 2,\ \Delta y = (x + \Delta x)^2 - x^2$

$dx = \Delta x$	dy	Δy	$\Delta y - dy$	$dy/\Delta y$
1.000	4	5	1	0.8
0.500	2	2.25	0.25	0.889
0.100	0.4	0.41	0.010	0.976
0.010	0.04	0.040	0.000	0.998
0.001	0.004	0.004	0.000	1.000

9. $dy = -\frac{2}{x^3}\ dx,\ x = 2,\ \Delta y = \frac{1}{(x + \Delta x)^2} - \frac{1}{x^2}$

$dx = \Delta x$	dy	Δy	$\Delta y - dy$	$dy/\Delta y$
1.000	−0.25	−0.139	0.111	1.8
0.500	−0.125	−0.09	0.035	1.389
0.100	−0.025	−0.023	0.002	1.076
0.010	−0.003	−0.002	0.000	1.008
0.001	0.000	0.000	0.000	1.001

11. $dy = \frac{1}{4}x^{-3/4}\,dx = \frac{1}{4x^{3/4}}\,dx,\ x = 2,$

$\Delta y = (x + \Delta x)^{1/4} - x^{1/4}$

$dx = \Delta x$	dy	Δy	$\Delta y - dy$	$dy/\Delta y$
1.000	0.149	0.127	−0.022	1.172
0.500	0.074	0.068	−0.006	1.089
0.100	0.015	0.015	0.000	1.019
0.010	0.001	0.001	0.000	1.002
0.001	0.000	0.000	0.000	1.000

13. $x = 12,\ dx = \Delta x = 1$

$$\Delta C \approx dC = (0.10x + 4)\,dx = [0.10(12) + 4](1) = \$5.20$$

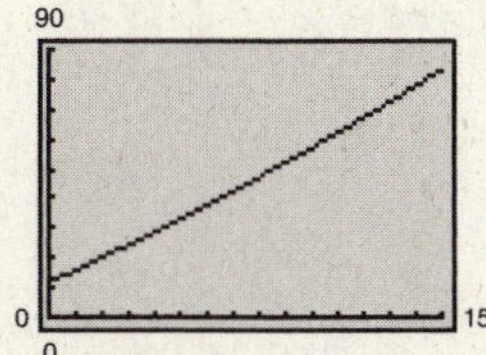

15. $x = 75,\ dx = \Delta x = 1$

$$\Delta R \approx dR = (30 - 0.30x)\,dx = [30 - 0.30(75)](1) = \$7.50$$

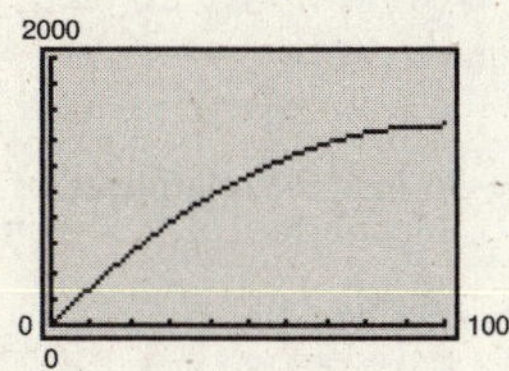

17. $x = 50,\ dx = \Delta x = 1$

$$\Delta P \approx dP = (-1.5x^2 + 2500)\,dx = \left[-1.5(50)^2 + 2500\right](1) = -\$1250$$

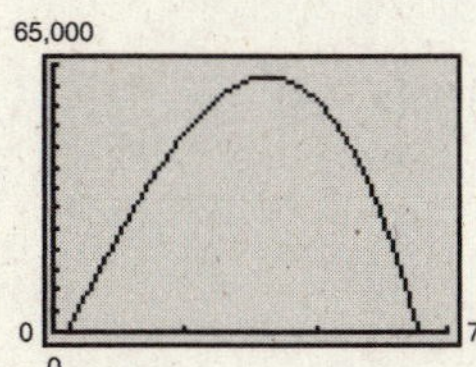

19. $y = 6x^4$

$$\frac{dy}{dx} = 24x^3$$

$$dy = 24x^3\,dx$$

21. $y = 3x^2 - 4$

$$\frac{dy}{dx} = 6x$$

$$dy = 6x\,dx$$

23. $y = (4x - 1)^3$

$$\frac{dy}{dx} = 3(4x - 1)^2(4)$$

$$dy = 12(4x - 1)^2\,dx$$

25. $y = \frac{x + 1}{2x - 1}$

$$\frac{dy}{dx} = \frac{(2x - 1) - (x + 1)(2)}{(2x - 1)^2}$$

$$dy = -\frac{3}{(2x - 1)^2}\,dx$$

27. $y = \sqrt{9 - x^2}$

$$\frac{dy}{dx} = \frac{1}{2}(9 - x^2)^{-1/2}(-2x) = -\frac{x}{\sqrt{9 - x^2}}$$

$$dy = -\frac{x}{\sqrt{9 - x^2}}\,dx$$

29. $f(x) = 2x^3 - x^2 + 1,\ (-2, -19)$

$$f'(x) = 6x^2 - 2x$$

$$f'(-2) = 24 + 4 = 28$$

$$y + 19 = 28(x + 2)$$

$$y = 28x + 37 \quad \text{Tangent line}$$

$$f(-2 + 0.01) \approx -18.72$$

$$y(-2 + 0.01) = -18.72$$

$$f(-2 - 0.01) \approx -19.28$$

$$y(-2 - 0.01) = -19.28$$

31. $f(x) = \frac{x}{x^2 + 1},\ (0, 0)$

$$f'(x) = \frac{(x^2 + 1) - x(2x)}{(x^2 + 1)^2} = \frac{1 - x^2}{(x^2 + 1)^2}$$

$$f'(0) = 1$$

$$y - 0 = 1(x - 0)$$

$$y = x \quad \text{Tangent line}$$

$$f(0 + 0.01) \approx 0.009999$$

$$y(0 + 0.01) = 0.01$$

$$f(0 - 0.01) \approx -0.009999$$

$$y(0 - 0.01) = -0.01$$

33. $P = \left(500x - x^2\right) - \left(\frac{1}{2}x^2 - 77x + 3000\right)$

$= -\frac{3}{2}x^2 + 577x - 3000$

(a) $dP = (-3x + 577)\,dx$

Find dP for $x = 115$ and $dx = 5$.

$dP = \left[-3(115) + 577\right](5) = \$1160$

(b) Find ΔP for $x = 115$ and $\Delta x = 5$.

$$\begin{aligned}\Delta P &= P(x + \Delta x) - P(x)\\ &= P(115 + 5) - P(115)\\ &= \$1122.50\end{aligned}$$

35. $$\begin{aligned}R &= xp\\ &= x(75 - 0.25x)\\ &= 75x - 0.25x^2\end{aligned}$$

(a) $dR = (75 - 0.5x)\,dx$

Find dR for $x = 7$ and $dx = 1$.

$dR = [75 - 0.5(7)](1) = \$71.50$

(b) Find dR for $x = 70$ and $dx = 1$.

$dR = [75 - 0.5(70)](1) = \$40$

37. $N = \dfrac{10(5 + 3t)}{1 + 0.04t}$

$$dN = \frac{(1 + 0.04t)(30) - 10(5 + 3t)(0.04)}{(1 + 0.04t)^2}\,dt$$

$$= \frac{28}{(1 + 0.04t)^2}\,dt$$

When $t = 5$ and $dt = 6 - 5 = 1$, you have the following.

$$dN = \frac{28}{[1 + 0.04(5)]^2}(1) = \frac{28}{1.44} \approx 19.44$$

The change in herd size will be approximately 19 deer.

39. $(150, 50)$, $(120, 60)$

$$m = \frac{60 - 50}{120 - 150} = -\frac{1}{3}$$

$$p - 50 = -\frac{1}{3}(x - 150)$$

$$p = -\frac{1}{3}x + 100$$

$$R = xp = -\frac{1}{3}x^2 + 100x$$

When $x = 141$ and $dx = \Delta x = 1$,

$$\Delta R \approx dR = \left(-\frac{2}{3}x + 100\right)dx$$

$$= \left[-\frac{2}{3}(141) + 100\right](1) = \$6.00.$$

41. $A = x^2$

$dA = 2x\,dx$

When $x = 6$ in. and $dx = \pm\frac{1}{16}$ in.,

$$dA = 2(6)\left(\pm\frac{1}{16}\right) = \pm\frac{3}{4}\text{ in.}^2.$$

When $x = 6$ in. and $A = 36$ in.^{2}, the relative error is

$$\frac{dA}{A} = \frac{\pm\frac{3}{4}}{36} \approx 0.0208 \Rightarrow 2.08\%.$$

43. True; $\dfrac{dy}{dx} = 1$ and $dy = dx$.

Review Exercises for Chapter 3

1. $f(x) = -x^2 + 2x + 4$

$f'(x) = -2x + 2$

Critical number: $x = 1$

3. $y = 4x^3 - 108x$

$y' = 12x^2 - 108 = 12(x^2 - 9)$

Critical numbers: $x = \pm 3$

5. $g(x) = (x - 1)^2(x - 3)$

$$\begin{aligned}g'(x) &= (x - 3)(2)(x - 1) + (x - 1)^2\\ &= (x - 1)(3x - 7)\end{aligned}$$

Critical numbers: $x = 1$, $x = \dfrac{7}{3}$

7. $f(x) = x^2 + x - 2$

$f'(x) = 2x + 1$

Set $f'(x) = 0$.

$2x + 1 = 0$

$x = -\frac{1}{2} \Rightarrow$ Critical number

Interval	$-\infty < x < -\frac{1}{2}$	$-\frac{1}{2} < x < \infty$
Sign of f'	$f' < 0$	$f' > 0$
Conclusion	Decreasing	Increasing

9. $f(x) = x^3 + 6x^2 - 2$

$f'(x) = -3x^2 + 12x = -3x(x - 4)$

Set $f'(x) = 0$.

$-3x(x - 4) = 0$

$x = 0, 4 \Rightarrow$ Critical numbers

Interval	$-\infty < x < 0$	$0 < x < 4$	$4 < x < \infty$
Sign of f'	$f' < 0$	$f' > 0$	$f' < 0$
Conclusion	Decreasing	Increasing	Decreasing

11. $y = (x - 1)^{2/3}$

$y' = \frac{2}{3}(x - 1)^{-1/3}(1) = \frac{2}{3(x - 1)^{1/3}}$

y' is undefined at $x = 1$.

Interval	$-\infty < x < 1$	$1 < x < \infty$
Sign of y'	$y' < 0$	$y' > 0$
Conclusion	Decreasing	Increasing

13. $R = 6.268t^2 + 136.07t - 191.3, 4 \le t \le 9$

$R' = 12.536t + 136.07 = \frac{dR}{dt}$

Set $R' = 0$.

$12.536t + 136.07 = 0$

$t \approx -10.85$

The only critical number is $t \approx -10.85$. Any $t > -10.85$ produces a positive dR/dt, so the sales were increasing from 2004 to 2009.

15. $f(x) = 4x^3 - 6x^2 - 2$

$f'(x) = 12x^2 - 12x = 12x(x - 1)$

Critical numbers: $x = 0, x = 1$

Relative maximum: $(0, -2)$

Relative minimum: $(1, -4)$

Interval	$-\infty < x < 0$	$0 < x < 1$	$1 < x < \infty$
Sign of f'	$f' > 0$	$f' < 0$	$f' > 0$
Conclusion	Increasing	Decreasing	Increasing

17. $g(x) = x^2 - 16x + 12$

$g'(x) = 2x - 16 = 2(x - 8)$

Critical number: $x = 8$

Relative minimum: $(8, -52)$

Interval	$-\infty < x < 8$	$8 < x < \infty$
Sign of g'	$g' < 0$	$g' > 0$
Conclusion	Decreasing	Increasing

19. $h(x) = 2x^2 - x^4$

$h'(x) = 4x - 4x^3$

$= 4x(1 - x)(1 + x)$

Critical numbers: $x = 0, x = \pm 1$

Relative maxima: $(-1, 1), (1, 1)$

Relative minimum: $(0, 0)$

Interval	$-\infty < x < -1$	$-1 < x < 0$	$0 < x < 1$	$1 < x < \infty$
Sign of h'	$h' > 0$	$h' < 0$	$h' > 0$	$h' < 0$
Conclusion	Increasing	Decreasing	Increasing	Decreasing

21. $f(x) = \dfrac{6}{x^2 + 1}$

$f'(x) = -6(x^2 + 1)^{-2}(2x) = -\dfrac{12x}{(x^2 + 1)^2}$

Critical number: $x = 0$

Relative maximum: $(0, 6)$

Interval	$-\infty < x < 0$	$0 < x < \infty$
Sign of f'	$f' > 0$	$f' < 0$
Conclusion	Increasing	Decreasing

23. $h(x) = \dfrac{x^2}{x - 2}$

$h'(x) = \dfrac{(x - 2)(2x) - x^2}{(x - 2)^2} = \dfrac{x(x - 4)}{(x - 2)^2}$

Critical numbers: $x = 0, x = 4$

Discontinuity: $x = 2$

Relative maximum: $(0, 0)$

Relative minimum: $(4, 8)$

Interval	$-\infty < x < 0$	$0 < x < 2$	$2 < x < 4$	$4 < x < \infty$
Sign of h'	$h' > 0$	$h' < 0$	$h' < 0$	$h' > 0$
Conclusion	Increasing	Decreasing	Decreasing	Increasing

25. $f(x) = x^2 + 5x + 6, [-3, 0]$

$f'(x) = 2x + 5$

Critical number: $x = -\dfrac{5}{2}$

x-value	Endpoint $x = -3$	Critical $x = -\frac{5}{2}$	Endpoint $x = 0$
$f(x)$	0	$-\frac{1}{4}$	6
Conclusion		Minimum	Maximum

27. $f(x) = x^3 - 12x + 1, [-4, 4]$

$f'(x) = 3x^2 - 12 = 3(x - 2)(x + 2)$

Critical numbers: $x = \pm 2$

x-value	Endpoint $x = -4$	Critical $x = -2$	Critical $x = 2$	Endpoint $x = 4$
$f(x)$	-15	17	-15	17
Conclusion	Minimum	Maximum	Minimum	Maximum

29. $f(x) = 2\sqrt{x} - x, [0, 9]$

$f'(x) = x^{-1/2} - 1 = \dfrac{1 - \sqrt{x}}{\sqrt{x}}$

Critical numbers: $x = 0$ (endpoint), $x = 1$

x-value	Endpoint $x = 0$	Critical $x = 1$	Endpoint $x = 9$
$f(x)$	0	1	-3
Conclusion		Maximum	Minimum

31. $f(x) = \dfrac{2x}{x^2 + 1}, [-1, 2]$

$f'(x) = \dfrac{(x^2 + 1)(2) - 2x(2x)}{(x^2 + 1)^2} = -\dfrac{2(x^2 - 1)}{(x^2 + 1)^2}$

Critical numbers: $x = 1, x = -1$ (endpoint)

x-value	Endpoint $x = -1$	Critical $x = 1$	Endpoint $x = 2$
$f(x)$	-1	1	$\frac{4}{5}$
Conclusion	Minimum	Maximum	

33. $S = 2\pi r^2 + 50r^{-1}$

$S' = 4\pi r - 50r^{-2} = 4\pi r - \dfrac{50}{r^2}$

Set $S' = 0$.

$$4\pi r - \frac{50}{r^2} = 0$$
$$4\pi r = \frac{50}{r^2}$$
$$4\pi r^3 = 50$$
$$r^3 = \frac{25}{2\pi}$$
$$r = \sqrt[3]{\frac{25}{2\pi}} \approx 1.58 \text{ in.}$$

Interval	$0 < r < 1.58$	$1.58 < r < \infty$
Sign of S'	$S' < 0$	$S' > 0$
Conclusion	Decreasing	Increasing

S is a minimum when $r = 1.58$ inches.

35. $f(x) = (x - 2)^3$

$f'(x) = 3(x - 2)^2$

$f''(x) = 6(x - 2)$

$f''(x) = 0$ when $x = 2$.

Interval	$-\infty < x < 2$	$2 < x < \infty$
Sign of f''	$f'' < 0$	$f'' > 0$
Conclusion	Concave downward	Concave upward

37. $g(x) = \frac{1}{4}(-x^4 + 8x^2 - 12)$

$g'(x) = -x^3 + 4x$

$g''(x) = -3x^2 + 4$

$g''(x) = 0$ when $x = \pm\dfrac{2}{\sqrt{3}}$.

Interval	$-\infty < x < -\dfrac{2}{\sqrt{3}}$	$-\dfrac{2}{\sqrt{3}} < x < \dfrac{2}{\sqrt{3}}$	$\dfrac{2}{\sqrt{3}} < x < \infty$
Sign of g''	$g'' < 0$	$g'' > 0$	$g'' < 0$
Conclusion	Concave downward	Concave upward	Concave downward

39. $f(x) = \frac{1}{2}x^4 - 4x^3$

$f'(x) = 2x^3 - 12x^2$

$f''(x) = 6x^2 - 24x = 6x(x - 4)$

$$f''(x) = 0$$
$$6x(x - 4) = 0$$
$$x = 0 \text{ or } x = 4$$

Interval	$-\infty < x < 0$	$0 < x < 4$	$4 < x < \infty$
Sign of f''	$f'' > 0$	$f'' < 0$	$f'' > 0$
Conclusion	Concave upward	Concave downward	Concave upward

Points of inflection: $(0, 0)$, $(4, -128)$

41. $f(x) = x^3(x-3)^2$

$f'(x) = x^3[2(x-3)(1)] + (x-3)^2(3x^2) = 5x^4 - 24x^3 + 27x^2$

$f''(x) = 20x^3 - 72x^2 + 54x = 2x(10x^2 - 36x + 27)$

$f''(x) = 0$

$2x(10x^2 - 36x + 27) = 0$

$x = 0$

$$x = \frac{36 \pm \sqrt{36^2 - 4(10)(27)}}{2(10)} \approx 1.0652, 2.5348$$

Interval	$-\infty < x < 0$	$0 < x < 1.0652$	$1.0652 < x < 2.5348$	$2.5348 < x < \infty$
Sign of f''	$f'' < 0$	$f'' > 0$	$f'' < 0$	$f'' > 0$
Conclusion	Concave downward	Concave upward	Concave downward	Concave upward

Points of inflection: $(0, 0)$, $(1.6052, 4.5244)$, $(2.5348, 3.5246)$

43. $f(x) = x^3 - 6x^2 + 12x$

$f'(x) = 3x^2 - 12x + 12$

$f''(x) = 6x - 12$

Critical number: $x = 2$

$f''(2) = 0$

By the First-Derivative Test, $(2, 8)$ is not a relative extremum.

45. $f(x) = x^5 - 5x^3$

$f'(x) = 5x^4 - 15x^2 = 5x^2(x^2 - 3)$

$f''(x) = 20x^3 - 30x$

Critical numbers: $x = 0$, $x = \pm\sqrt{3}$

$f''(\sqrt{3}) = 30\sqrt{3} > 0$

Relative minimum: $(\sqrt{3}, -6\sqrt{3})$

$f''(-\sqrt{3}) = -30\sqrt{3} < 0$

Relative maximum: $(-\sqrt{3}, 6\sqrt{3})$

$f''(0) = 0$

By the First-Derivative Test, $(0, 0)$ is not a relative extremum.

47. $f(x) = 2x^2(1 - x^2)$

$f'(x) = (1 - x^2)(4x) + 2x^2(-2x)$

$= 4x - 8x^3$

$= 4x(1 - 2x^2)$

$f''(x) = 4 - 24x^2$

Critical numbers: $x = 0$, $x = \pm\frac{1}{\sqrt{2}}$

$f''(0) = 4 > 0$

Relative minimum: $(0, 0)$

$f''\left(-\frac{1}{\sqrt{2}}\right) = -8 < 0$

$f''\left(\frac{1}{\sqrt{2}}\right) = -8 < 0$

Relative maxima: $\left(\pm\frac{1}{\sqrt{2}}, \frac{1}{2}\right)$

49. $R = \frac{1}{1500}(150x^2 - x^3)$, $0 \le x \le 100$

$R' = \frac{1}{1500}(300x - 3x^2)$

$R'' = \frac{1}{1500}(300 - 6x)$

$R'' = 0$

$300 - 6x = 0$

$x = 50$

Interval	$0 < x < 50$	$50 < x < 100$
Sign of R''	$R'' > 0$	$R'' < 0$
Conclusion	Concave upward	Concave downward

Point of diminishing returns: $\left(50, \frac{500}{3}\right)$

51. 

Let x be the length and y be the width of the rectangle. Then $A = xy = 225$, and $P = 2x + 2y$.

$$P = 2x + 2\left(\frac{225}{x}\right) = 2x + \frac{550}{x} = 2x + 550x^{-1}$$

$$P' = 2 - 550x^{-2} = \frac{2 - 550}{x^2}$$

$$P'' = 1010x^{-3} = \frac{1010}{x^3}$$

$$P' = 0$$

$$\frac{2 - 550}{x^2} = 0$$

$$2x^2 = 550$$

$$x^2 = 225$$

$$x = \pm 15$$

$P''(15) > 0$, so P is a minimum when $x = 15$ meters, and $y = \dfrac{225}{15} = 15$ meters.

53. 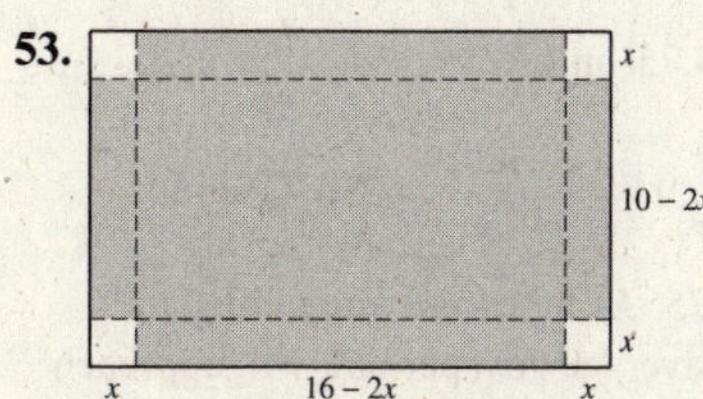

$$V = (16 - 2x)(10 - 2x)x = 4x^3 - 52x^2 + 160x$$

$$V' = 12x^2 - 104x + 160$$

$$V'' = 24x - 104$$

$$V' = 0$$

$$12x^2 - 104x + 160 = 0$$

$$4(3x^2 - 26x + 40) = 0$$

$$4(3x - 20)(x - 2) = 0$$

$$x = \tfrac{20}{3} \text{ or } x = 2$$

(**Note:** $x = \frac{20}{3}$ is not in the domain of V.)

$V''(2) < 0$, so V is a maximum when $x = 2$ in.

$V(2) = 144$ in.^{3}

55. $R = 450x - 0.25x^2$

$$R' = \frac{dR}{dx} = 450 - 0.5x$$

$$R'' = -0.5$$

$$R' = 0$$

$$450 - 0.5x = 0$$

$$-0.5x = -450$$

$$x = 900$$

$R''(900) < 0$, so R is a maximum when $x = 900$, and $R(900) = \$202{,}500$.

57. $\bar{C} = \dfrac{C}{x} = \dfrac{0.2x^2 + 10x + 4500}{x} = 0.2x + 10 + 4500x^{-1}$

$$\bar{C}' = 0.2 - 4500x^{-2} = 0.2 - \frac{4500}{x^2}$$

$$\bar{C}'' = \frac{9000}{x^3}$$

$$\bar{C}' = 0$$

$$0.2 - \frac{4500}{x^2} = 0$$

$$0.2x^2 = 4500$$

$$x^2 = 22{,}500$$

$$x = \pm 150$$

(**Note:** $x = -150$ is not in the domain of $\bar{C}$.)

$\bar{C}'(150) > 0$, so $\bar{C}$ is a minimum when $x = 150$ units, and $\bar{C}(150) = \$70$/unit.

59. $P = R - C$ and $R = xp$

$$R = x(36 - 4x) = 36x - 4x^2$$

$$P = (36x - 4x^2) - (2x^2 + 6)$$

$$P = -6x^2 + 36x - 6$$

$$\frac{dP}{dx} = P' = -12x + 36$$

$$P'' = -12$$

$$P' = 0$$

$$-12x + 36 = 0$$

$$x = 3$$

$P''(3) < 0$, so P is a maximum when $x = 3$ units.

(a) When $x = 3$, $p = 36 - 4(3) = \$24$/unit.

(b) $\bar{C} = \dfrac{C}{x} = \dfrac{2x^2 + 6}{x} = 2x + \dfrac{6}{x}$

When $x = 3$, $\bar{C} = 2(3) + \dfrac{6}{3} = \$8$/unit.

61. (a) $60 - 0.04x,\ 0 \le x \le 1500$

$$\frac{dp}{dx} = -0.04$$

$$\eta = \frac{p/x}{dp/dx} = \frac{\frac{60 - 0.04x}{x}}{-0.04} = \frac{x - 1500}{x}$$

$$|\eta| = 1 = \left|\frac{x - 1500}{x}\right| \Rightarrow x = |x - 1500| \Rightarrow x = 750$$

For $0 < x < 750$, $|\eta| > 1$ and demand is elastic.

For $750 < x < 1500$, $|\eta| < 1$ and demand is inelastic.

For $x = 750$, demand is of unit elasticity.

(b) $R = xp = x(60 - 0.04x) = 60x - 0.04x^2$

$$R' = \frac{dR}{dx} = 60 - 0.08x$$

$$R' = 0$$

$$60 - 0.08x = 0$$

$$x = 750$$

Interval	$0 < x < 750$	$750 < x < 1500$
Sign of R'	$R' > 0$	$R' < 0$
Conclusion	Increasing	Decreasing

From 0 to 750 units, revenue is increasing.

From 750 to 1500 units, revenue does not increase.

63. $f(x) = \dfrac{x + 4}{x^2 + 7x} = \dfrac{x + 4}{x(x + 7)}$

Set denominator equal to 0.

$x(x + 7) = 0$

$x = 0$ or $x + 7 = 0$

$x = -7$

Vertical asymptotes: $x = 0, x = -7$

65. $f(x) = \dfrac{x^2 - 16}{2x^2 + 9x + 4}$

$$= \frac{(x + 4)(x - 4)}{(2x + 1)(x + 4)}$$

$$= \frac{x - 4}{2x + 1},\ x \ne -4$$

Set denominator equal to 0.

$2x + 1 = 0$

$x = -\dfrac{1}{2}$

Vertical asymptote: $x = -\dfrac{1}{2}$

67. $\displaystyle\lim_{x\to 0^+}\left(x - \frac{1}{x^3}\right) = 0 - \infty = -\infty$

69. $\displaystyle\lim_{x\to -1^+}\frac{x^2 - 2x + 1}{x + 1} = \infty$

71. $\displaystyle\lim_{x\to\infty}\frac{2x^2}{3x^2 + 5} = \frac{2}{3}$

$y = \dfrac{2}{3}$ is a horizontal asymptote.

73. $\displaystyle\lim_{x\to\infty}\frac{3x}{x^2 + 1} = 0$

$y = 0$ is a horizontal asymptote.

75. $C = 0.75x + 4000$

(a) $\overline{C} = \dfrac{C}{x} = \dfrac{0.75x + 4000}{x} = 0.75 + \dfrac{4000}{x}$

(b) $\overline{C}(100) = \$40.75/\text{unit.}$

$\overline{C}(1000) = \$4.75/\text{unit.}$

(c) $\displaystyle\lim_{x\to\infty}\overline{C} = \lim_{x\to\infty} 0.75 + \frac{4000}{x} = \$0.75/\text{unit}$

The limit is 0.75. As more and more units are produced, the average cost per unit will approach \$0.75.

77. $C = \dfrac{250p}{100 - p},\ 0 \le p < 100$

(a) $C(20) = \$62.5$ million

$C(50) = \$250$ million

$C(90) = \$2250$ million

(b) $\displaystyle\lim_{p\to 100^-} C = \lim_{p\to 100^-}\frac{250p}{100 - p} = \infty$

The limit is ∞, meaning that as the percent approaches 100%, the cost increases without bound.

79. $f(x) = 4x - x^2$

$f'(x) = 4 - 2x$

$f''(x) = -2$

x-intercepts: $f(x) = 0$

$4x - x^2 = 0$

$x(4 - x) = 0$

$x = 0 \text{ or } x = 4$

y-intercept: $f(0) = 0$

$y = 0$

Critical number: $f'(x) = 0$

$4 - 2x = 0$

$x = 2$

$f''(2) < 0 \Rightarrow$ relative maximum at $(2, 4)$

$f'' \neq 0 \Rightarrow$ no points of inflection

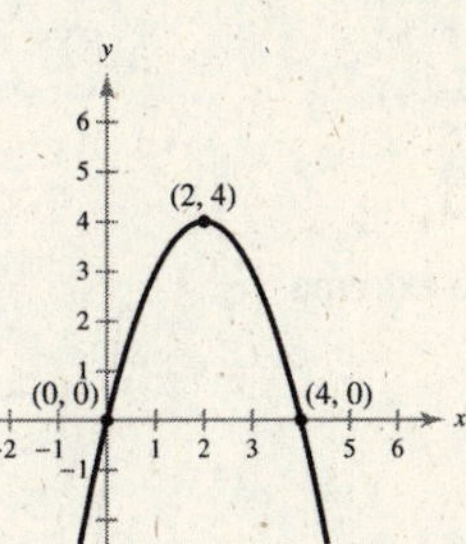

81. $f(x) = x^3 - 6x^2 + 3x + 10$

$f'(x) = 3x^2 - 12x + 3$

$f''(x) = 6x - 12$

x-intercepts: $f(x) = 0$

$x^3 - 6x^2 + 3x + 10 = 0$

$(x - 5)(x - 2)(x + 1) = 0$

$x = 5 \text{ or } x = 2 \text{ or } x = -1$

y-intercept: $f(0) = 10$

$y = 10$

Critical numbers: $f'(x) = 0$

$3x^2 - 12x + 3 = 0$

$3(x^2 - 4x + 1) = 0$

$x = 2 \pm \sqrt{3}$

$f''(2 + \sqrt{3}) > 0 \Rightarrow$ relative minimum at $(2 + \sqrt{3}, -10.39)$

$f''(2 - \sqrt{3}) < 0 \Rightarrow$ relative maximum at $(2 - \sqrt{3}, 10.39)$

$f''(x) = 0$

$6x - 12 = 0$

$x = 2$

Point of inflection: $(2, 0)$

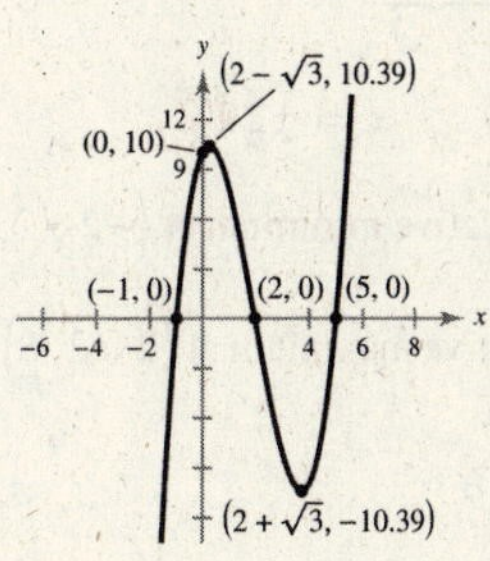

83. $f(x) = x^4 - 4x^3 + 16x - 16$

$f'(x) = 4x^3 - 12x^2 + 16$

$f''(x) = 12x^2 - 24x$

x-intercepts: $f(x) = 0$

$$x^4 - 4x^3 + 16x - 16 = 0$$

$$x = 2 \text{ or } x = -2$$

y-intercept: $f(0) = -16$

$$y = -16$$

Critical numbers: $f'(x) = 0$

$$4x^3 - 12x^2 + 16 = 0$$

$$x = 2 \text{ or } x = -1$$

$f''(-1) > 0 \Rightarrow$ relative minimum at $(-1, -27)$

$f''(2) < 0 \Rightarrow$ Use First-Derivative Test $\Rightarrow$ no extrema

$$f''(x) = 0$$

$$12x^2 - 24x = 0$$

$$12x(x - 2) = 0$$

$$x = 0 \text{ or } x = 2$$

Points of inflection: $(0, -16), (2, 0)$

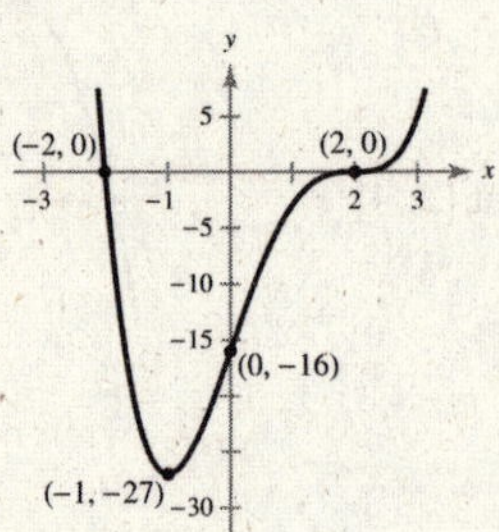

85. $f(x) = x\sqrt{16 - x^2}, \quad -4 \le x \le 4$

$$f'(x) = \frac{16 - 2x^2}{\sqrt{16 - x^2}} = \frac{-2(x^2 - 8)}{\sqrt{16 - x^2}}$$

$$f''(x) = \frac{2x^3 - 48x}{(16 - x^2)^{3/2}} = \frac{2x(x^2 - 24)}{(16 - x^2)^{3/2}}$$

x-intercepts: $f(x) = 0$

$$x\sqrt{16 - x^2} = 0$$

$$x = 0 \text{ or } x = \pm 4$$

y-intercept: $f(0) = 0$

$$y = 0$$

Critical numbers: $f'(x) = 0$

$$\frac{-2(x^2 - 8)}{\sqrt{16 - x^2}} = 0$$

$$x = \pm 2\sqrt{2}$$

$f''(-2\sqrt{2}) > 0 \Rightarrow$ relative minimum at $(-2\sqrt{2}, -8)$

$f''(2\sqrt{2}) < 0 \Rightarrow$ relative maximum at $(2\sqrt{2}, 8)$

$$f''(x) = 0$$

$$\frac{2x(x^2 - 24)}{(16 - x^2)^{3/2}} = 0$$

$$x = 0 \text{ or } x = \pm 2\sqrt{6}$$

(**Note:** $x = \pm 2\sqrt{6}$ not in domain.)

Point of inflection: $(0, 0)$

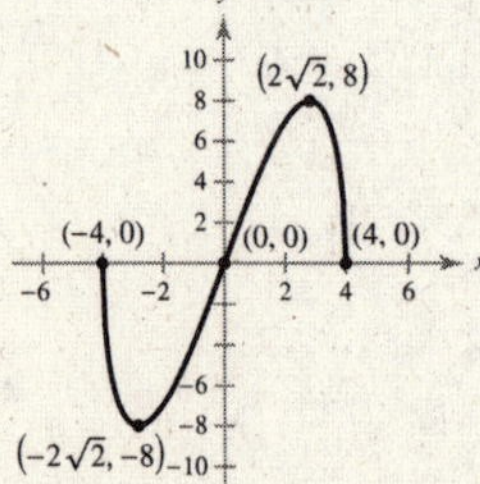

87. $f(x) = \dfrac{x+1}{x-1}$

$f'(x) = -\dfrac{2}{(x-1)^2}$

$f''(x) = \dfrac{4}{(x-1)^3}$

x-intercept: $f(x) = 0$ $\quad$ y-intercept: $f(0) = -1$

$\dfrac{x+1}{x-1} = 0$ $\quad$ $y = -1$

$x = -1$

Vertical asymptote: $x = 1$

Horizontal asymptote: $y = 1$

Critical numbers: $f'(x) \neq 0 \Rightarrow$ no extrema

$f''(x) \neq 0 \Rightarrow$ no points of inflection

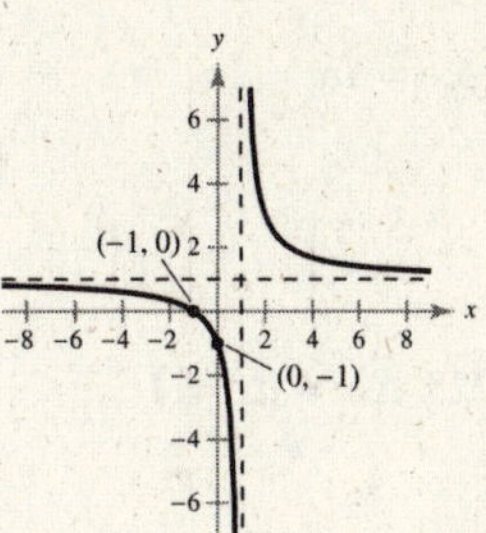

89. $f(x) = 3x^{2/3} - 2x$

$f'(x) = 2x^{-1/3} - 2$

$f''(x) = -\frac{2}{3}x^{-4/3}$

x-intercepts: $f(x) = 0$ $\quad$ y-intercept: $f(0) = 0$

$3x^{2/3} - 2x = 0$ $\quad$ $y = 0$

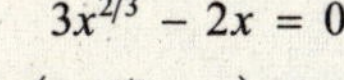

$x(3x^{-1/3} - 2) = 0$

$x = 0 \text{ or } x = \frac{27}{8}$

Critical number: $f'(x) = 0$

$2x^{-1/3} - 2 = 0$

$x = 1$

$f''(1) < 0 \Rightarrow$ relative maximum at $(1, 1)$

$f''(x) = 0 \Rightarrow$ Use First-Derivative Test $\Rightarrow$ relative minimum at $(0, 0)$

No points of inflection

91. (a)

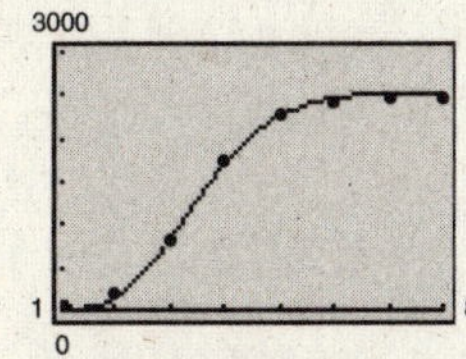

The model fits the data well.

(b) $N(10) \approx 2434$ bacteria

(c) Answers will vary. Sample answer: The limit of N as t approaches infinity is $\dfrac{13{,}250}{7} \approx 1893$ bacteria; however, many different factors can have many different effects on the culture thus affecting the number of bacteria.

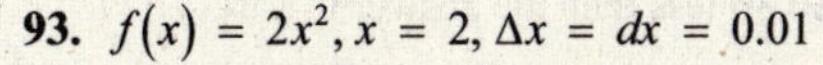

93. $f(x) = 2x^2, x = 2, \Delta x = dx = 0.01$

$\Delta y = f(x + \Delta x) - f(x)$

$= 2(2 + 0.01)^2 - 2(2)^2$

$= 0.0802$

$dy = f'(x)\,dx$

$dy = 4x\,dx$

$dy = 4(2)(0.01) = 0.08$

$dy \approx \Delta y$

95. $f(x) = 6x - x^3, x = 3, \Delta x = dx = 0.1$

$$\Delta y = f(x + \Delta x) - f(x) = \left[6(3 + 0.1) - (3 + 0.1)^3\right] - \left[6(3) - (3)^3\right] = -2.191$$

$$dy = f'(x)\,dx$$

$$dy = \left(6 - 3x^2\right) dx = \left(6 - 3(3)^2\right)(0.1) = -2.1$$

$$dy \approx \Delta y$$

97. $C = 40x^2 + 1225, x = 10, \Delta x = dx = 1$

$$dC = 80x\,dx$$

$$dC = 80(10)(1) = \$800$$

99. $R = 6,25x - 0.4x^{3/2}, x = 225, \Delta x = dx = 1$

$$dR = \left(6.25 + 0.6x^{1/2}\right) dx$$

$$= \left[6.25 + 0.6(225)^{1/2}\right](1)$$

$$= \$15.25$$

101. $P = 0.003x^2 + 0.019x - 1200, x = 750, \Delta x = dx = 1$

$$dP = (0.006x + 0.019)\,dx$$

$$= \left[0.006(750) + 0.019\right](1)$$

$$= 4.519$$

$$\approx \$4.52$$

103. $y = 0.5x^2$

$$dy = 1.5x^2\,dx$$

105. $y = \left(3x^2 - 2\right)^3$

$$\frac{dy}{dx} = 3\left(3x^2 - 2\right)^2(6x)$$

$$dy = 18x\left(3x^2 - 2\right)^2 dx$$

107. $y = \dfrac{2 - x}{x + 5}$

$$\frac{dy}{dx} = \frac{(x + 5)(-1) - (2 - x)}{(x + 5)^2}$$

$$dy = -\frac{7}{(x + 5)^2}\,dx$$

109. $P = -0.8x^2 + 324x - 2000, x = 100, \Delta x = dx = 1$

(a) $dP = (-1.6x + 324)\,dx$

$$dP = \left[-1.6(100) + 324\right](1) = \$164$$

(b) $\Delta P = P(x + \Delta x) = P(x)$

$$= \left[-0.8(100 + 1)^2 + 324(100 + 1) - 2000\right] - \left[-0.8(100)^2 + 324(100) - 2000\right]$$

$$= \$163.20$$

This is \$0.80 less than the answer in part (a).

111. $B = 0.1\sqrt{5w} = 0.1\sqrt{5}w^{1/2}, w = 90, \Delta w = dw = 5$

$$dB = \frac{0.1\sqrt{5}}{2}w^{-1/2}\,dw = \frac{0.05\sqrt{5}}{\sqrt{w}}\,dw$$

$$dB = \frac{0.05\sqrt{5}}{\sqrt{90}}(5) \approx 0.059 \text{ m}^2$$

Chapter 3 Test Yourself

1. $f(x) = 3x^2 - 4$

$$f'(x) = 6x$$

Critical number: $x = 0$

Interval	$-\infty < x < 0$	$0 < x < \infty$
Sign of f'	$f' < 0$	$f' > 0$
Conclusion	Decreasing	Increasing

2. $f(x) = x^3 - 12x$

$f'(x) = 3x^2 - 12 = 3(x - 2)(x + 2)$

Critical numbers: $x = \pm 2$

Interval	$-\infty < x < -2$	$-2 < x < 2$	$2 < x < \infty$
Sign of f'	$f' > 0$	$f' < 0$	$f' > 0$
Conclusion	Increasing	Decreasing	Increasing

3. $f(x) = (x - 5)^4$

$f'(x) = 4(x - 5)^3$

Critical number: $x = 5$

Interval	$-\infty < x < 5$	$5 < x < \infty$
Sign of f'	$f' < 0$	$f' > 0$
Conclusion	Decreasing	Increasing

4. $f(x) = \frac{1}{3}x^3 - 9x + 4$

$f'(x) = x^2 - 9 = (x + 3)(x - 3)$

Critical numbers: $x = \pm 3$

Relative minimum: $(3, -14)$

Relative maximum: $(-3, 22)$

Interval	$-\infty < x < -3$	$-3 < x < 3$	$3 < x < \infty$
Sign of f'	$f' > 0$	$f' < 0$	$f' > 0$
Conclusion	Increasing	Decreasing	Increasing

5. $f(x) = 2x^4 - 4x^2 - 5$

$f'(x) = 8x^3 - 8x = 8x(x + 1)(x - 1)$

Critical numbers: $x = 0$, $x = \pm 1$

Relative minima: $(-1, -7)$, $(1, -7)$

Relative maximum: $(0, -5)$

Interval	$-\infty < x < -1$	$-1 < x < 0$	$0 < x < 1$	$1 < x < \infty$
Sign of f'	$f' < 0$	$f' > 0$	$f' < 0$	$f' > 0$
Conclusion	Decreasing	Increasing	Decreasing	Increasing

6. $f(x) = \dfrac{5}{x^2 + 2}$

$f'(x) = -5(x^2 + 2)^{-2}(2x) = -\dfrac{10x}{(x^2 + 2)^2}$

Critical number: $x = 0$

Relative maximum: $\left(0, \dfrac{5}{2}\right)$

Interval	$-\infty < x < 0$	$0 < x < \infty$
Sign of f'	$f' > 0$	$f' < 0$
Conclusion	Increasing	Decreasing

7. $f(x) = x^2 + 6x + 8$, $[-4, 0]$

$f'(x) = 2x + 6$

Critical number: $x = -3$

x-value	Endpoint $x = -4$	Critical $x = -3$	Endpoint $x = 0$
$f(x)$	0	-1	8
Conclusion		Minimum	Maximum

8. $f(x) = 12\sqrt{x} - 4x$, $[0, 5]$

$f'(x) = 12\left(\dfrac{1}{2}x^{-1/2}\right) - 4 = \dfrac{6}{x^{1/2}} - 4$

Critical number: $x = \dfrac{9}{4}$

x-value	Endpoint $x = 0$	Critical $x = \frac{9}{4}$	Endpoint $x = 5$
$f(x)$	0	9	$12\sqrt{5} - 20$
Conclusion	Minimum	Maximum	

9. $f(x) = \dfrac{6}{x} + \dfrac{x}{2}$, $[1, 6]$

$f'(x) = -\dfrac{6}{x^2} + \dfrac{1}{2}$

Critical number: $x = 2\sqrt{3}$

x-value	Endpoint $x = 1$	Critical $x = 2\sqrt{3}$	Endpoint $x = 6$
$f(x)$	$\frac{13}{2}$	$2\sqrt{3}$	4
Conclusion	Maximum	Minimum	

10. $f(x) = x^5 - 80x^2$

$f'(x) = 5x^4 - 160x$

$f''(x) = 20x^3 - 160 = 20(x^3 - 8)$

$f''(x) = 0$ when $x = 2$

Interval	$-\infty < x < 2$	$2 < x < \infty$
Sign of $f''(x)$	$f''(x) < 0$	$f''(x) > 0$
Conclusion	Concave downward	Concave upward

11. $f(x) = \dfrac{20}{3x^2 + 8}$

$f'(x) = -20(3x^2 + 8)^{-2}(6x) = -\dfrac{120x}{(3x^2 + 8)^2}$

$f''(x) = \dfrac{(3x^2 + 8)^2(-120) - (-120x)[2(3x^2 + 8)(6x)]}{(3x^2 + 8)^4} = \dfrac{120(9x^2 - 8)}{(3x^2 + 8)^3}$

$f''(x) = 0$ when $x = \pm\dfrac{2\sqrt{2}}{3}$

Interval	$-\infty < x < -\dfrac{2\sqrt{2}}{3}$	$-\dfrac{2\sqrt{2}}{3} < x < \dfrac{2\sqrt{2}}{3}$	$\dfrac{2\sqrt{2}}{3} < x < \infty$
Sign of $f''(x)$	$f''(x) > 0$	$f''(x) < 0$	$f''(x) > 0$
Conclusion	Concave upward	Concave downward	Concave upward

12. $f(x) = x^4 + 6$

$f'(x) = 4x^3$

$f''(x) = 12x^2 = 0$ when $x = 0$.

$f''(x) > 0$ on $(-\infty, 0)$.

$f''(x) > 0$ on $(0, \infty)$.

There are no points of inflection.

13. $f(x) = x^4 - 54x^2 + 230$

$f'(x) = 4x^3 - 108x$

$f''(x) = 12x^2 - 108 = 12(x^2 - 9)$

$f''(x) = 0$ when $x = \pm 3$

$f''(x) > 0$ on $(-\infty, -3) \Rightarrow$ concave upward

$f''(x) < 0$ on $(-3, 3) \Rightarrow$ concave downward

$f''(x) > 0$ on $(3, \infty) \Rightarrow$ concave upward

Points of inflection: $(\pm 3, -175)$

14. $f(x) = x^3 - 6x^2 - 36x + 50$

$f'(x) = 3x^2 - 12x - 36$

$= 3(x^2 - 4x - 12)$

$= 3(x - 6)(x + 2)$

Critical numbers: $x = -2, x = 6$

$f''(x) = 6x - 12$

$f''(-2) < 0$

$f''(6) > 0$

Relative maximum: $(-2, 90)$

Relative minimum: $(6, -166)$

15. $f(x) = \frac{3}{5}x^5 - 9x^3$

$f'(x) = 3x^4 - 27x^2 = 3x^2(x + 3)(x - 3)$

Critical numbers: $x = 0,\ x = \pm 3$

$f''(x) = 12x^3 - 54x$

$f''(0) = 0$

$f''(-3) = -162 < 0$

$f''(3) = 162 > 0$

Relative maximum: $\left(-3, \frac{486}{5}\right)$

Relative minimum: $\left(3, -\frac{486}{5}\right)$

By the First-Derivative Test, $(0, 0)$ is not a relative extremum.

16. A vertical asymptote occurs at $x = 5$ because

$\lim_{x\to 5^-} \frac{3x + 2}{x - 5} = -\infty$ and $\lim_{x\to 5^+} \frac{3x + 2}{x - 5} = \infty$.

A horizontal asymptote occurs at $y = 3$ because

$\lim_{x\to \infty} \frac{3x + 2}{x - 5} = 3$ and $\lim_{x\to -\infty} \frac{3x + 2}{x - 5} = 3$.

17. There are no vertical asymptotes because the denominator is never zero. A horizontal asymptote occurs at $y = 2$ because $\lim_{x\to \infty} \frac{2x^2}{x^2 + 3} = 2$ and

$\lim_{x\to -\infty} \frac{2x^2}{x^2 + 3} = 2$.

18. A vertical asymptote occurs at $x = 1$ because

$\lim_{x\to 1^-} \frac{2x^2 - 5}{x - 1} = \infty$ and $\lim_{x\to 1^+} \frac{2x^2 - 5}{x - 1} = -\infty$.

There are no horizontal asymptotes because

$\lim_{x\to \infty} \frac{2x^2 - 5}{x - 1} = \infty$ and $\lim_{x\to -\infty} \frac{2x^2 - 5}{x - 1} = -\infty$.

19. $y = -x^3 + 3x^2 + 9x - 2$

$y' = -3x^2 + 6x + 9$

$y'' = -6x + 6$

x-intercepts: $-x^3 + 3x^2 + 9x - 2 = 0$

$x = -2, \frac{5 \pm \sqrt{21}}{2}$

y-intercept: $f(0) = -2$

$y = -2$

Critical numbers: $y' = 0$

$-3x^2 + 6x + 9 = 0$

$x = -1, 3$

$y''(-1) > 0 \Rightarrow$ relative minimum at $(-1, -7)$

$y''(3) < 0 \Rightarrow$ relative maximum at $(3, 25)$

$y'' = 0$

$-6x + 6 = 0$

$x = 1$

Point of inflection: $(1, 9)$

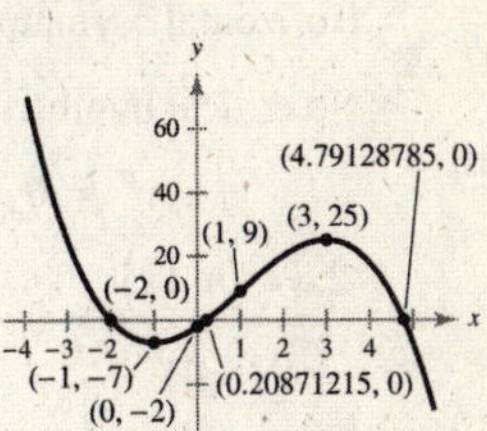

20. $y = x^5 - 5x$

$y' = 5x^4 - 5$

$y'' = 20x^3$

x-intercepts: $x^5 - 5x = 0$

$x(x^4 - 5) = 0$

$x = 0, \pm\sqrt[4]{5}$

y-intercept: $f(0) = 0$

$y = 0$

Critical numbers: $y' = 0$

$5x^4 - 5 = 0$

$5(x^4 - 1) = 0$

$x = \pm 1$

$y''(-1) < 0 \Rightarrow$ relative maximum at $(-1, 4)$

$y''(1) > 0 \Rightarrow$ relative minimum at $(1, -4)$

$y'' = 0$

$20x^3 = 0$

$x = 0$

Point of inflection: $(0, 0)$

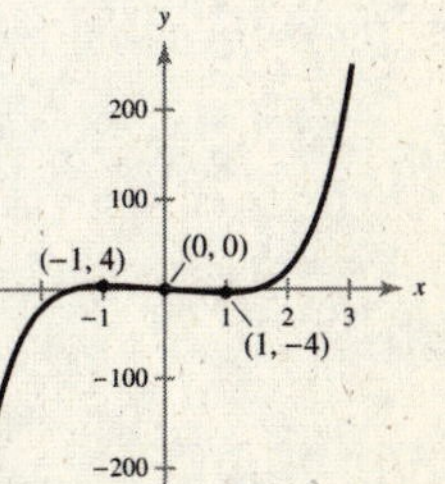

21. $y = \dfrac{x}{x^2 - 4}$

$$y' = -\frac{(x^2 + 4)}{(x^2 - 4)^2}$$

$$y'' = \frac{2x(x^2 + 12)}{(x^2 - 4)^3}$$

x-intercepts: $\dfrac{x}{x^2 - 4} = 0$

$$x = 0$$

y-intercept: $f(0) = 0$

$$y = 0$$

Vertical asymptotes: $x^2 - 4 = 0$

$$x = \pm 2$$

Horizontal asymptote: $y = 0$

No critical numbers: $y' \neq 0$

$$y'' = 0$$

$$\frac{2x(x^2 + 12)}{(x^2 - 4)^3} = 0$$

$$x = 0$$

Point of inflection: $(0, 0)$

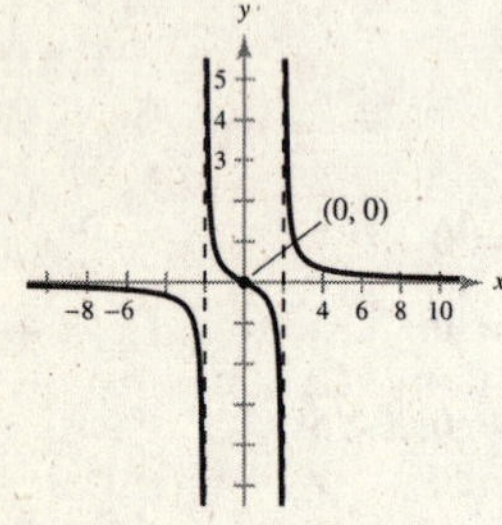

22. $y = 5x^2 - 3$

$$\frac{dy}{dx} = 10x$$

$$dy = 10x\,dx$$

23. $y = \dfrac{1 - x}{x + 3}$

$$\frac{dy}{dx} = \frac{(x + 3)(-1) - (1 - x)}{(x + 3)^2} = -\frac{4}{(x + 3)^2}$$

$$dy = -\frac{4}{(x + 3)^2}\,dx$$

24. $y = (x + 4)^3$

$$\frac{dy}{dx} = 3(x + 4)^2$$

$$dy = 3(x + 4)^2\,dx$$

25. $p = 280 - 0.4x,\ 0 \le x \le 700$

$$\frac{dp}{dx} = -0.4$$

$$\eta = \frac{p/x}{dp/dx} = \frac{\dfrac{280 - 0.4x}{x}}{0.4} = \frac{-700}{x} + 1 = \frac{x - 700}{x}$$

$$|\eta| = 1 = \left|\frac{x - 700}{x}\right| \Rightarrow x = |x - 700| \Rightarrow x = 350$$

For $0 \le x < 350$, $|\eta| > 1$ and demand is elastic.

For $350 < x \le 700$, $|\eta| < 1$ and demand is inelastic.

For $x = 350$, $|\eta| = 1$ and demand is of unit elasticity.

CHAPTER 4
Exponential and Logarithmic Functions

CHAPTER 4
Exponential and Logarithmic Functions

Section 4.1 Exponential Functions

Skills Warm Up

1. The graph of g is the graph of f translated to the left two units.

2. The graph of g is the graph of f reflected in the x-axis.

3. The graph of g is the graph of f translated downward one unit.

4. The graph of g is the graph of f reflected in the y-axis.

5. The graph of g is the graph of f translated to the right one unit.

6. The graph of g is the graph of f translated upward two units.

7. $25^{3/2} = \left(25^{1/2}\right)^3 = 5^3 = 125$

8. $64^{3/4} = \left(64^{1/4}\right)^3 \approx (2.8284)^3 \approx 22.63$

9. $27^{2/3} = \left(27^{1/3}\right)^2 = 3^2 = 9$

10. $\left(\frac{1}{5}\right)^3 = \frac{1}{5^3} = \frac{1}{125}$

11. $\left(\frac{1}{8}\right)^{1/3} = \frac{1}{8^{1/3}} = \frac{1}{2}$

12. $\left(\frac{5}{8}\right)^2 = \frac{5^2}{8^2} = \frac{25}{64}$

13. $2x - 6 = 4$
$$2x = 10$$
$$x = 5$$

14. $3x + 1 = 5$
$$3x = 4$$
$$x = \frac{4}{3}$$

15. $(x + 4)^2 = 25$
$$x + 4 = \pm 5$$
$$x = 1 \text{ or } x = -9$$

16. $(x - 2)^2 = 8$
$$x - 2 = \pm 2\sqrt{2}$$
$$x = 2 \pm 2\sqrt{2}$$

17. $x^2 + 4x - 5 = 0$
$$(x - 1)(x + 5) = 0$$
$$x - 1 = 0 \Rightarrow x = 1$$
$$x + 5 = 0 \Rightarrow x = -5$$

18. $2x^2 - 3x + 1 = 0$
$$(2x - 1)(x - 1) = 0$$
$$2x - 1 = 0 \Rightarrow x = \frac{1}{2}$$
$$x - 1 = 0 \Rightarrow x = 1$$

1. (a) $\left(5^2\right)\left(5^3\right) = 5^5 = 3125$

(b) $\left(5^2\right)\left(5^{-3}\right) = 5^{-1} = \frac{1}{5}$

(c) $\left(5^2\right)^2 = 5^4 = 625$

(d) $5^{-3} = \frac{1}{5^3} = \frac{1}{125}$

3. (a) $\frac{5^3}{25^2} = \frac{5^3}{\left(5^2\right)^2} = \frac{5^3}{5^4} = \frac{1}{5}$

(b) $\left(9^{2/3}\right)(3)\left(3^{2/3}\right) = \left(3^2\right)^{2/3}(3)\left(3^{2/3}\right) = \left(3^{4/3}\right)\left(3^{5/3}\right)$
$$= 3^{9/3} = 3^3 = 27$$

(c) $\left[\left(25^{1/2}\right)5^2\right]^{1/3} = \left[5 \cdot 5^2\right]^{1/3} = \left[5^3\right]^{1/3} = 5$

(d) $\left(8^2\right)\left(4^3\right) = (64)(64) = 4096$

5. $y = 16\left(\frac{1}{2}\right)^{t/30},\ t > 0$

When $t = 90$ years, $y = 16\left(\frac{1}{2}\right)^{90/30} = 16\left(\frac{1}{2}\right)^3 = 2.$

After 90 years, 2 grams of the initial mass remain.

7. $f(x) = 6^x$

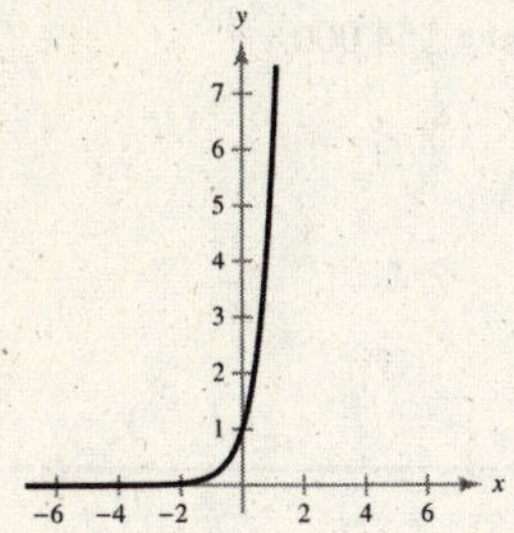

9. $f(x) = 5^{-x}$

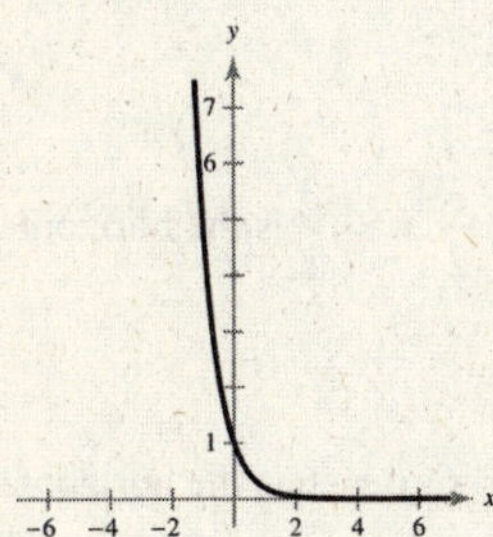

11. $y = 2^{x-1}$

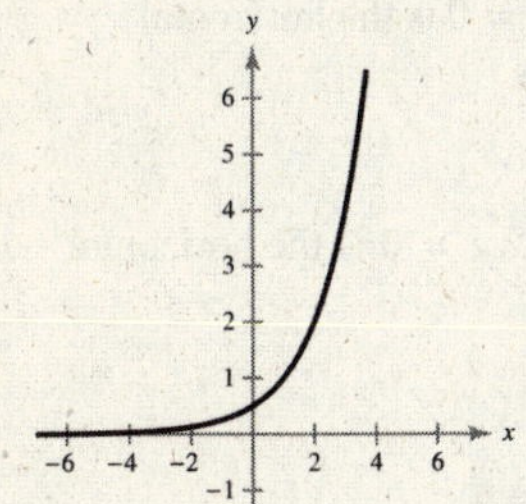

13. $y = -2^x$

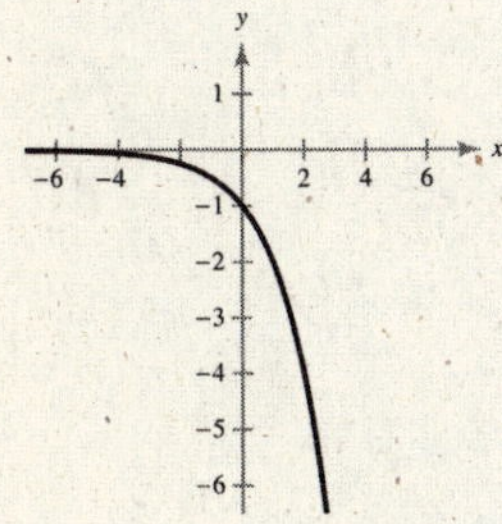

15. $y = 3^{-x^2}$

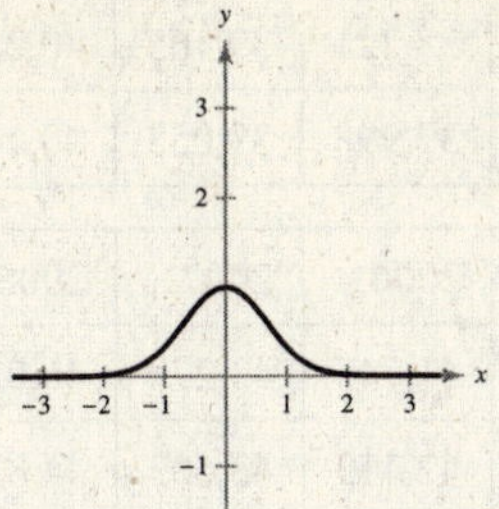

17. $s(t) = \frac{1}{4}\left(3^{-t}\right) = \frac{3^{-t}}{4} = \frac{1}{4\left(3^t\right)}$

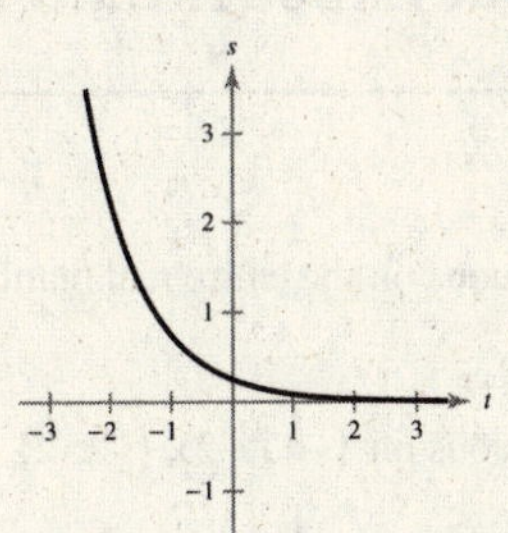

19. $P(t) = 254.75(1.01)^t$, $t = 5$ is 1995.

(a) 2013: $P(23) \approx 320.26 \Rightarrow 320{,}260{,}000$

(b) 2020: $P(30) \approx 343.36 \Rightarrow 343{,}360{,}000$

21. $V(t) = 64{,}000(2)^{t/15}$

(a) $V(5) = 64{,}000(2)^{5/15} \approx \$80{,}634.95$

(b) $V(20) = 64{,}000(2)^{20/15} \approx \$161{,}269.89$

23. $C(t) = P(1.04)^t,\ 0 \le t \le 10$

$C(10) = 24.95(1.04)^{10} \approx \$36.93$

25. $V(t) = 28{,}000\left(\frac{3}{4}\right)^t$

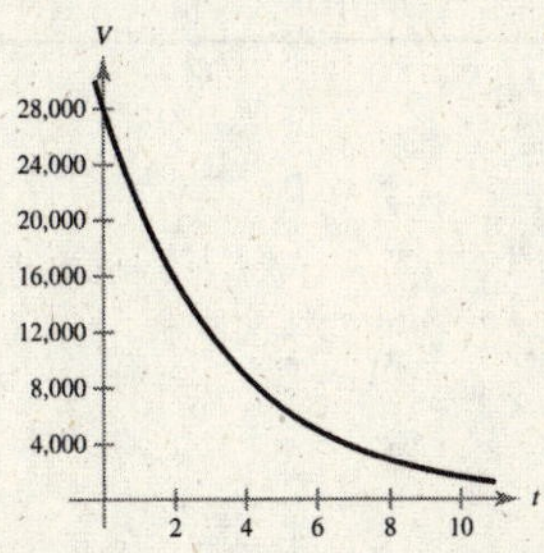

After 4 years: $V(4) \approx \$8859.38$

27. (a)

Year	2001	2002	2003	2004
Actual	37,188	38,221	39,165	40,201
Model	36,966	37,998	39,058	40,148

Year	2005	2006	2007	2008
Actual	40,520	41,746	43,277	46,025
Model	41,268	42,419	43,603	44,819

The model fits the data well. Explanations will vary.

(b) 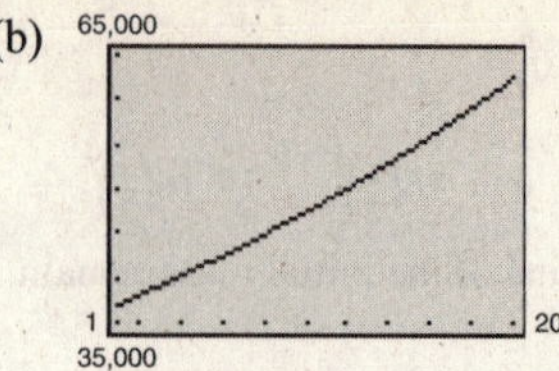

(c) During 2014, $t \approx 14.8$, the average salary of a school nurse will reach $54,000.

Section 4.2 Natural Exponential Functions

Skills Warm Up

1. The function is continuous on the entire real number line.

2. The function is continuous on $(-\infty, -2)$, $(-2, 2)$, and $(2, \infty)$.

3. The function is continuous on $(-\infty, -\sqrt{3})$, $(-\sqrt{3}, \sqrt{3})$, and $(\sqrt{3}, \infty)$.

4. The function is continuous on $(-\infty, 4)$ and $(4, \infty)$.

5. Since $\lim\limits_{x\to\pm\infty} \dfrac{25}{1+4x} = 0$, $y = 0$ is the horizontal asymptote.

6. Since $\lim\limits_{x\to\pm\infty} \dfrac{16x}{3+x^2} = 0$, $y = 0$ is the horizontal asymptote.

7. Since $\lim\limits_{x\to\pm\infty} \dfrac{8x^3+2}{2x^3+x} = 4$, $y = 4$ is the horizontal asymptote.

8. Since $\lim\limits_{x\to\pm\infty} \dfrac{x}{2x} = \dfrac{1}{2}$, $y = \dfrac{1}{2}$ is the horizontal asymptote.

9. Since $\lim\limits_{x\to\pm\infty} \dfrac{3}{2+(1/x)} = \dfrac{3}{2}$, $y = \dfrac{3}{2}$ is the horizontal asymptote.

10. Since $\lim\limits_{x\to\pm\infty} \dfrac{6}{1+(1/x^2)} = 6$, $y = 6$ is the horizontal asymptote.

11. Since $\lim\limits_{x\to\pm\infty} 2^{-x} = 0$, $y = 0$ is the horizontal asymptote.

12. Since $\lim\limits_{x\to\pm\infty} \dfrac{7}{1+5x} = 0$, $y = 0$ is the horizontal asymptote

1. (a) $(e^3)(e^4) = e^{3+4} = e^7$

(b) $(e^3)^4 = e^{3(4)} = e^{12}$

(c) $(e^3)^{-2} = e^{3(-2)} = e^{-6} = \dfrac{1}{e^6}$

(d) $e^0 = 1$

3. (a) $(e^2)^{5/2} = e^{2(5/2)} = e^5$

(b) $(e^2)(e^{1/2}) = e^{2+1/2} = e^{5/2}$

(c) $(e^{-2})^{-3} = e^{-2(-3)} = e^6$

(d) $\dfrac{e^5}{e^{-2}} = e^{5-(-2)} = e^{5+2} = e^7$

5. $f(x) = e^{2x+1}$

Increasing exponential passing through $(0, e)$.

It matches graph (f).

7. $f(x) = e^{x^2}$

Symmetric with respect to y-axis.
It matches graph (d).

9. $f(x) = e^{\sqrt{x}}$

Domain: $x \geq 0$ passing through $(1, e)$.

It matches graph (c).

11. $f(x) = e^{-x/3}$

x	-3	-2	-1	0	1	2	3
$f(x)$	2.72	1.95	1.40	1	0.72	0.51	0.37

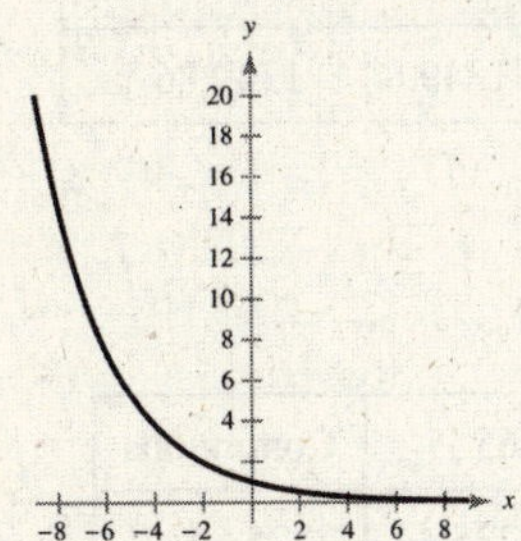

13. $g(x) = e^x - 2$

x	-3	-2	-1	0
$g(x)$	-1.95	-1.87	-1.63	-1

x	1	2	3
$g(x)$	0.72	5.40	18.09

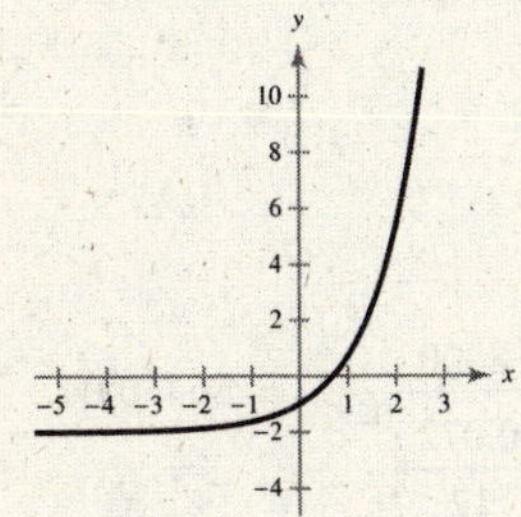

15. $g(x) = e^{1-x}$

x	-1	0	1	2	3
$g(x)$	7.39	2.72	1	0.37	0.14

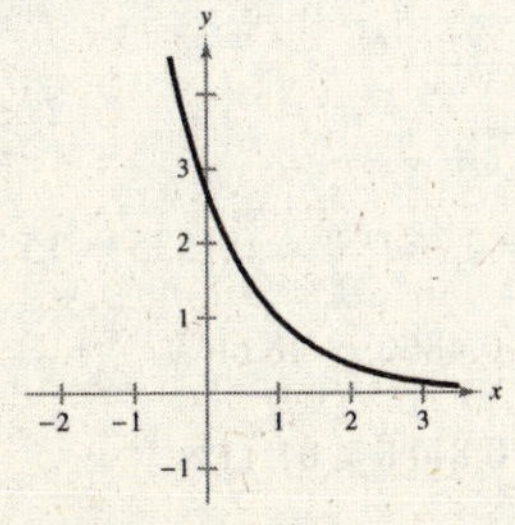

17.

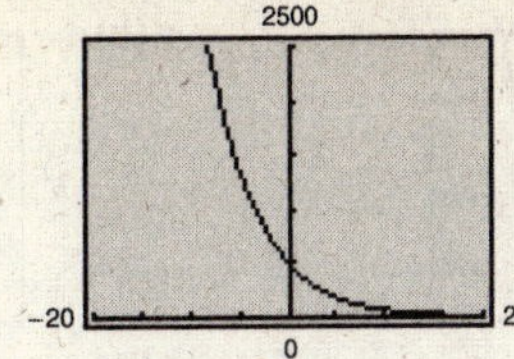

Horizontal asymptote: $N = 0$

Continuous on the entire real number line

19.

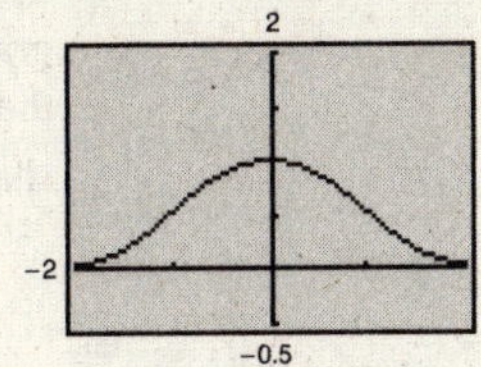

Horizontal asymptote: $g = 0$

Continuous on the entire real number line

21. $f(x) = \dfrac{e^x + e^{-x}}{2}$

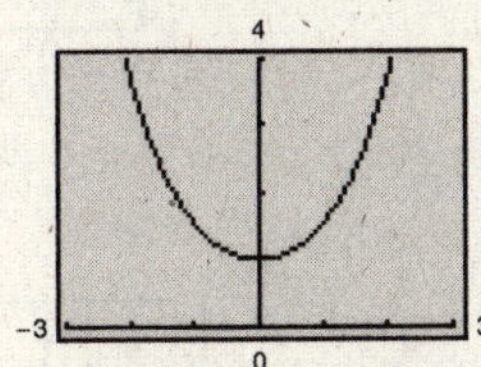

$$\lim_{x\to\infty} \frac{e^x + e^{-x}}{2} = \frac{\infty + 0}{2} = \infty$$

$$\lim_{x\to-\infty} \frac{e^x + e^{-x}}{2} = \frac{0 + \infty}{2} = \infty$$

No horizontal asymptotes

Continuous on the entire real number line

23. $f(x) = \dfrac{2}{1 + e^{1/x}}$

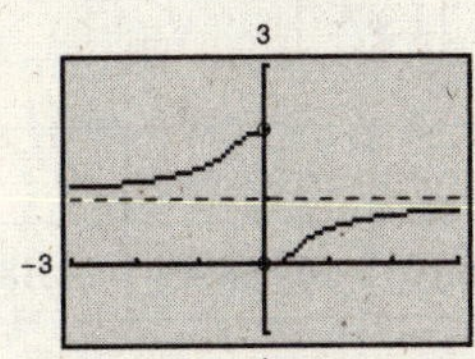

$$\lim_{x\to\infty} \frac{2}{1 + e^{1/x}} = \frac{2}{1 + e^0} = \frac{2}{1 + 1} = 1$$

$$\lim_{x\to-\infty} \frac{2}{1 + e^{1/x}} = \frac{2}{1 + e^0} = \frac{2}{1 + 1} = 1$$

Horizontal asymptote: $y = 1$

Discontinuous at $x = 0$

25. (a)

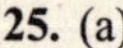

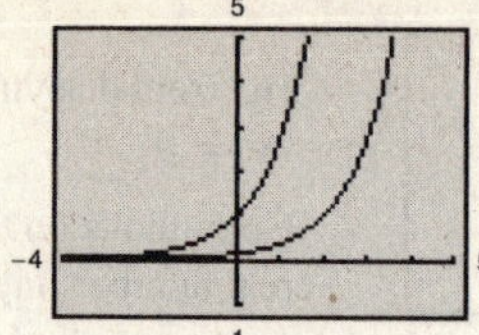

The graph of g is the graph of f shifted to the right two units.

(b)

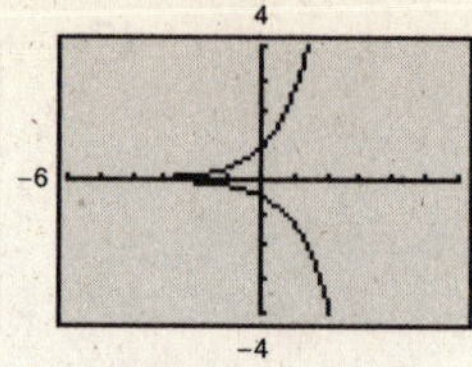

The graph of h is the graph of f reflected in the x-axis and vertically shrunk by a factor of $\frac{1}{2}$.

(c)

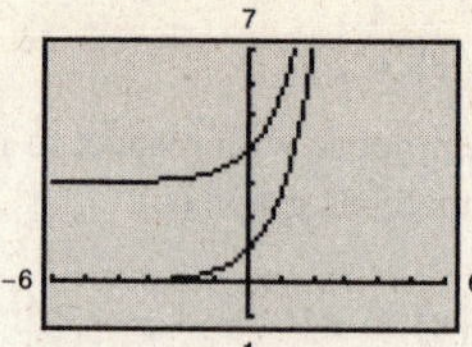

The graph of q is the graph of f shifted upward three units.

27. $P = 1000,\ r = 0.03,\ t = 10$

$$A = P\left(1 + \frac{r}{n}\right)^{nt} = 1000\left(1 + \frac{0.03}{n}\right)^{10n}$$

Continuous compounding:

$$A = Pe^{rt} = 1000e^{(0.03)(10)}$$

n	1	2	4	12	365	Continuous
A	1343.92	1346.86	1348.35	1349.35	1349.84	1349.86

29. $P = 1000,\ r = 0.04,\ t = 20$

$$A = P\left(1 + \frac{r}{n}\right)^{nt} = 1000\left(1 + \frac{0.04}{n}\right)^{20n}$$

Continuous compounding:

$$A = Pe^{rt} = 1000e^{(0.04)(20)}$$

n	1	2	4	12	365	Continuous
A	2191.12	2208.04	2216.72	2222.58	2225.44	2225.54

31. $A = Pe^{rt} \Rightarrow P = Ae^{-rt},\ A = 100{,}000,\ r = 0.04 = 100{,}000e^{-0.04t}$

t	1	10	20	30	40	50
P	96,078.94	67,032.00	44,932.90	30,119.42	20,189.65	13,533.53

33. $A = P\left(1 + \frac{r}{n}\right)^{nt} \Rightarrow P = \dfrac{A}{\left(1 + \frac{r}{n}\right)^{nt}},\ A = 100{,}000,\ r = 0.05,\ n = 12 = \dfrac{100{,}000}{\left(1 + \frac{0.05}{12}\right)^{12t}}$

t	1	10	20	30	40	50
P	95,132.82	60,716.10	36,864.45	22,382.66	13,589.88	8251.24

35. $P = 20{,}000,\ r = 0.08,\ t = 21$

$$A = Pe^{rt} = 20{,}000e^{(0.08)(21)} \approx 107{,}311.12$$

The balance will be \$107,311.12.

37. $r_{eff} = \left(1 + \frac{r}{n}\right)^{n} - 1,\ r = 0.09$

(a) $n = 1: r_{eff} = \left(1 + \frac{0.09}{1}\right)^{1} - 1 = 0.09 = 9\%$

(b) $n = 2: r_{eff} = \left(1 + \frac{0.09}{2}\right)^{2} - 1 \approx 0.0920 = 9.20\%$

(c) $n = 4: r_{eff} = \left(1 + \frac{0.09}{4}\right)^{4} - 1 \approx 0.0931 = 9.31\%$

(d) $n = 12: r_{eff} = \left(1 + \frac{0.09}{12}\right)^{12} - 1 \approx 0.0938 = 9.38\%$

39. $P = \dfrac{A}{\left(1 + \frac{r}{n}\right)^{nt}} = \dfrac{8000}{\left(1 + \frac{0.072}{12}\right)^{(12)(3)}} \approx \$6450.04$

41. (a) $p(100) = 5000\left(1 - \dfrac{4}{4 + e^{-0.002(100)}}\right) \approx \$849.53$

(b) $p(500) = 5000\left(1 - \dfrac{4}{4 - e^{-0.002(500)}}\right) \approx \$421.12$

$\lim_{x\to\infty} p = 0$

(c) $\lim_{x\to\infty} p = 0$

43. (a) $P\left(\frac{1}{2}\right) = 1 - e^{-1/2/3} = 1 - e^{-1/6} \approx 0.1535 = 15.35\%$

(b) $P(2) = 1 - e^{-2/3} \approx 0.4866 = 48.66\%$

(c) $P(5) = 1 - e^{-5/3} \approx 0.8111 = 81.11\%$

45. (a) The model fits the data well.

(b) $y = 637.11x + 502.1$

The linear model fits the data well, but the exponential model fits the data better.

(c) Exponential model: When $x \approx 16$ or 2016, the debt will exceed \$18,000 billion.

Linear model: When $x \approx 21$ or 2021, the debt will exceed \$18,000 billion.

47. (a)

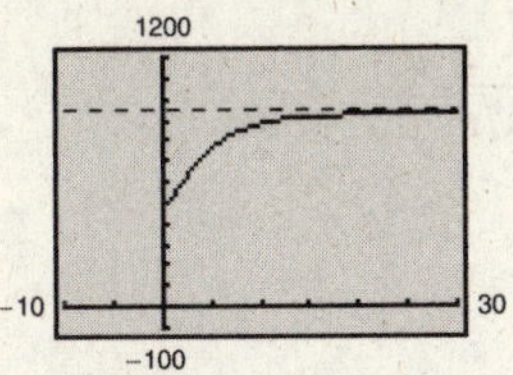

(b) $\lim_{t\to\infty} \dfrac{925}{1 + e^{-0.3t}} = \dfrac{925}{1 + 0} = 925$

As time increases without bound, the population approaches 925.

(c) $\lim_{t\to\infty} \dfrac{1000}{1 + e^{-0.3t}} = \dfrac{1000}{1 + 0} = 1000$

Models of the form $y = \dfrac{a}{1 + e^{-ct}}$, $c > 0$, have a limit of a as $t \to \infty$.

49. $P = \dfrac{0.83}{1 + e^{-0.2n}}$

(a) $P(3) \approx 0.536$

(b) $P(7) \approx 0.666$

(c)

Using the graph, $n = 11$ when $P \approx 0.70$.

(d) The proportion of correct responses does have a limit as n increases without bound because

$$\lim_{n\to\infty} \frac{0.83}{1 + e^{-0.2n}} = 0.83.$$

51. $P = 5000$, $t = 1$

(a) $A = P\left(1 + \dfrac{r}{n}\right)^{nt}$, $r = 0.0525$, and $n = 4$, $A = 5000\left(1 + \dfrac{0.0525}{4}\right)^{4(1)} \approx \$5267.71$

(b) $A = P\left(1 + \dfrac{r}{n}\right)^{nt}$, $r = 0.05$, and $n = 12$, $A = 5000\left(1 + \dfrac{0.05}{12}\right)^{12(1)} \approx \$5255.81$

(c) $A = Pe^{rt}$, $P = 5000$, $r = 0.0475$, and $t = 1$, $A = 5000e^{0.0475(1)} \approx \$5243.23$

You should choose option (a) because it earns the most money.

Section 4.3 Derivatives of Exponential Functions

Skills Warm Up

1. $x^2e^x - \frac{1}{2}e^x = e^x\left(x^2 - \frac{1}{2}\right)$

2. $\left(xe^{-x}\right)^{-1} + e^x = x^{-1}e^x + e^x = e^x\left(x^{-1} + 1\right)$

3. $xe^x - e^{2x} = e^x\left(x - e^x\right)$

4. $e^x - xe^{-x} = e^{-x}\left(e^{2x} - x\right)$

5. $f(x) = \dfrac{3}{7x^2} = \dfrac{3}{7}\left(x^{-2}\right)$

$f'(x) = -2\left(\dfrac{3}{7}\right)x^{-3} = -\dfrac{6}{7x^3}$

6. $g(x) = 3x^2 - \dfrac{x}{6}$

$g'(x) = 6x - \dfrac{1}{6}$

Skills Warm Up —*continued*—

7. $f(x) = (4x - 3)(x^2 + 9)$

$f'(x) = (4x - 3)(2x) + (x^2 + 9)(4) = 8x^2 - 6x + 4x^2 + 36 = 12x^2 - 6x + 36$

8. $f(t) = \dfrac{t - 2}{\sqrt{t}} = \dfrac{t - 2}{t^{1/2}} = t^{1-1/2} - 2t^{-1/2} = t^{1/2} - 2t^{-1/2}$

$f'(t) = \dfrac{1}{2}t^{-1/2} - 2\left(-\dfrac{1}{2}\right)t^{-3/2} = \dfrac{1}{2t^{1/2}} + \dfrac{1}{t^{3/2}} = \dfrac{1}{2\sqrt{t}} + \dfrac{1}{t\sqrt{t}} = \dfrac{t}{2t\sqrt{t}} + \dfrac{2}{2t\sqrt{t}} = \dfrac{t + 2}{2t\sqrt{t}}$

9. $f(x) = \dfrac{1}{8}x^3 - 2x$

$f'(x) = \dfrac{3}{8}x^2 - 2 = 0$

$\dfrac{3}{8}x^2 = 2$

$x^2 = \dfrac{16}{3}$

Critical numbers: $x = \pm\dfrac{4}{\sqrt{3}} = \pm\dfrac{4\sqrt{3}}{3}$

Interval	$-\infty < x < -\frac{4\sqrt{3}}{3}$	$-\frac{4\sqrt{3}}{3} < x < \frac{4\sqrt{3}}{3}$	$\frac{4\sqrt{3}}{3} < x < \infty$
Sign of f'	$f' > 0$	$f' < 0$	$f' > 0$
Conclusion	Increasing	Decreasing	Increasing

Relative maximum: $\left(-\dfrac{4\sqrt{3}}{3}, \dfrac{16\sqrt{3}}{9}\right)$

Relative minimum: $\left(\dfrac{4\sqrt{3}}{3}, -\dfrac{16\sqrt{3}}{9}\right)$

10. $f(x) = x^4 - 2x^2 + 5$

$f'(x) = 4x^3 - 4x = 0$

$4x(x^2 - 1) = 0$

Critical numbers: $x = 0, x = \pm 1$

Interval	$-\infty < x < -1$	$-1 < x < 0$	$0 < x < 1$	$1 < x < \infty$
Sign of f'	$f' < 0$	$f' > 0$	$f' < 0$	$f' > 0$
Conclusion	Decreasing	Increasing	Decreasing	Increasing

Relative maximum: $(0, 5)$

Relative minima: $(-1, 4), (1, 4)$

1. $f(x) = 3e$

$f'(x) = 0$

3. $y = e^{5x}$

$y' = 5e^{5x}$

5. $y = e^{-x^2}$

$$y' = (-2x)e^{-x^2} = -2xe^{-x^2}$$

7. $f(x) = e^{-1/x^2}$

$$f'(x) = e^{-1/x^2} \cdot \frac{d}{dx}\left(-x^{-2}\right) = e^{-1/x^2} \cdot 2x^{-3} = \frac{2}{x^3}e^{-1/x^2}$$

9. $f(x) = \left(x^2 + 1\right)e^{4x}$

$$f'(x) = \left(x^2 + 1\right)4e^{4x} + (2x)e^{4x} = e^{4x}\left(4x^2 + 2x + 4\right)$$

11. $f(x) = \dfrac{2}{\left(e^x + e^{-x}\right)^3} = 2\left(e^x + e^{-x}\right)^{-3}$

$$f'(x) = -6\left(e^x + e^{-x}\right)^{-4}\left(e^x - e^{-x}\right) = -\frac{6\left(e^x - e^{-x}\right)}{\left(e^x + e^{-x}\right)^4}$$

13. $f(x) = \dfrac{e^x + 1}{e^x - 1}$

$$f'(x) = \frac{\left(e^x - 1\right)\left(e^x\right) - \left(e^x + 1\right)\left(e^x\right)}{\left(e^x - 1\right)^2} = \frac{e^{2x} - e^x - e^{2x} - e^x}{\left(e^x - 1\right)^2} = -\frac{2e^x}{\left(e^x - 1\right)^2}$$

15. $y = xe^x - 4e^{-x}$

$y' = xe^x + e^x + 4e^{-x}$

17. $y = e^{4x}$

$y' = e^{4x}(4) = 4e^{4x}$

At $(0, 1)$, $y' = 4e^{4(0)} = 4 = m$.

19. $y = e^{-3x}$

$y' = e^{-3x}(-3) = -3e^{-3x}$

At $(0, 1)$, $y' = -3e^{-3(0)} = -3 = m$.

21.

$$y = e^{-2x+x^2}$$
$$y' = e^{-2x+x^2}(-2 + 2x)$$
$$y'(2) = 1(2) = 2$$
$$y - 1 = 2(x - 2)$$
$$y = 2x - 3$$

23.

$$y = x^2e^{-x}$$
$$y' = x^2\left(-e^{-x}\right) + 2xe^{-x} = xe^{-x}(-x + 2)$$
$$y'(2) = 0$$
$$y - \frac{4}{e^2} = 0(x - 2)$$
$$y = \frac{4}{e^2} = 4e^{-2} \quad \text{(horizontal line)}$$

25.

$$y = \left(e^{2x} + 1\right)^3$$
$$y' = 3\left(e^{2x} + 1\right)^2\left(2e^{2x}\right)$$
$$y'(0) = 3(4)(2) = 24$$
$$y - 8 = 24(x - 0)$$
$$y = 24x + 8$$

27.

$$xe^y - 10x + 3y = 0$$
$$xe^y\frac{dy}{dx} + e^y - 10 + 3\frac{dy}{dx} = 0$$
$$\left(xe^y + 3\right)\frac{dy}{dx} = -e^y + 10$$
$$\frac{dy}{dx} = \frac{-e^y + 10}{xe^y + 3}$$

29.

$$x^2e^{-x} + 2y^2 - xy = 0$$
$$2xe^{-x} - x^2e^{-x} + 4y\frac{dy}{dx} - x\frac{dy}{dx} - y = 0$$
$$(4y - x)\frac{dy}{dx} = y + x^2e^{-x} - 2xe^{-x}$$
$$\frac{dy}{dx} = \frac{y + x^2e^{-x} - 2xe^{-x}}{4y - x}$$

31. $f(x) = 2e^{3x} + 3e^{-2x}$

$f'(x) = 6e^{3x} - 6e^{-2x}$

$f''(x) = 18e^{3x} + 12e^{-2x}$

$= 6(3e^{3x} + 2e^{-2x})$

33. $f(x) = (1 + 2x)e^{4x}$

$f'(x) = (1 + 2x)(4e^{4x}) + 2e^{4x}$

$= 2e^{4x}[(1 + 2x)(2) + 1]$

$= 2e^{4x}(4x + 3)$

$f''(x) = 2e^{4x}(4) + 8e^{4x}(4x + 3)$

$= 8e^{4x}[1 + (4x + 3)]$

$= 8e^{4x}(4x + 4)$

$= 32e^{4x}(x + 1)$

35. $f(x) = \dfrac{1}{2 - e^{-x}}$

$f'(x) = \dfrac{-e^{-x}}{(2 - e^{-x})^2}$

$f''(x) = \dfrac{e^{-x}(2 + e^{-x})}{(2 - e^{-x})^3}$

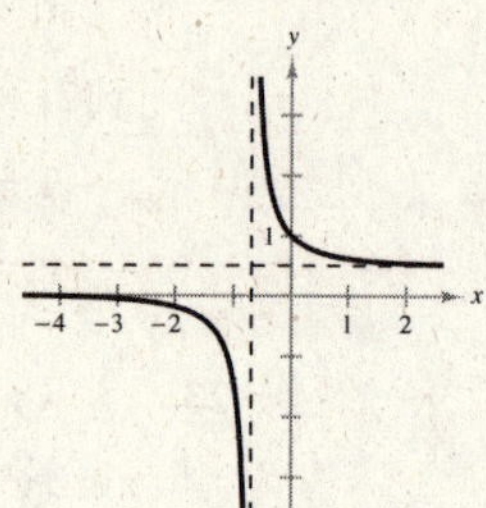

Horizontal asymptote to the right: $y = \dfrac{1}{2}$

Horizontal asymptote to the left: $y = 0$

Vertical asymptote when $2 = e^{-x} \Rightarrow x \approx -0.693$

No relative extrema or inflection points

x	−3	−2	−1	0	1
$f(x)$	−0.055	−0.186	−1.392	1	0.613

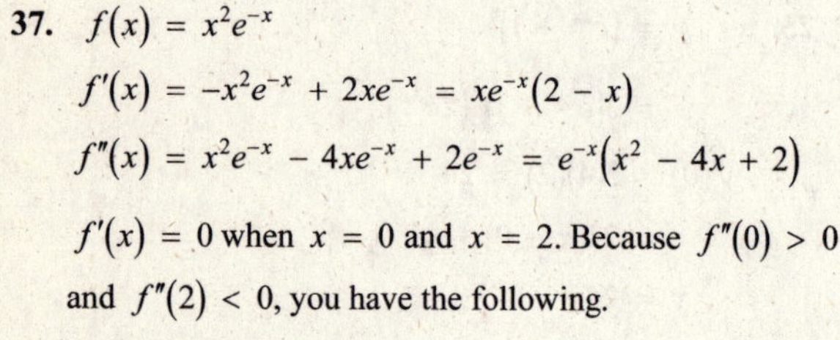

37. $f(x) = x^2e^{-x}$

$f'(x) = -x^2e^{-x} + 2xe^{-x} = xe^{-x}(2 - x)$

$f''(x) = x^2e^{-x} - 4xe^{-x} + 2e^{-x} = e^{-x}(x^2 - 4x + 2)$

$f'(x) = 0$ when $x = 0$ and $x = 2$. Because $f''(0) > 0$ and $f''(2) < 0$, you have the following.

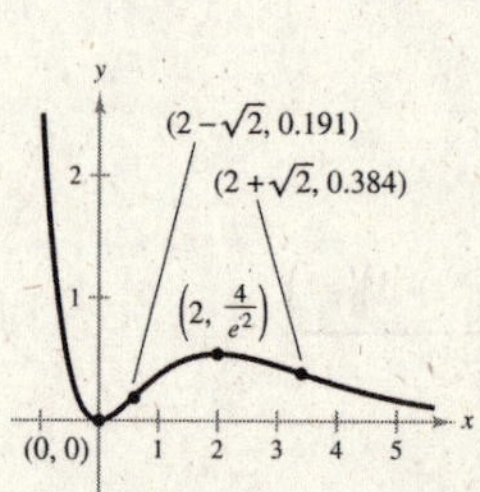

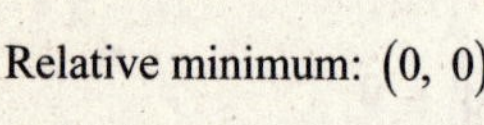

Relative minimum: $(0, 0)$

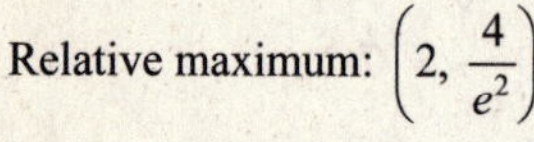

Relative maximum: $\left(2, \dfrac{4}{e^2}\right)$

Because $f''(x) = 0$ when $x = 2 \pm \sqrt{2}$, the inflection points occur at $(2 - \sqrt{2}, 0.191)$ and $(2 + \sqrt{2}, 0.384)$.

Horizontal asymptote: $y = 0$

x	−2	−1	0	1	2	3
$f(x)$	29.56	2.72	0	0.37	0.54	0.45

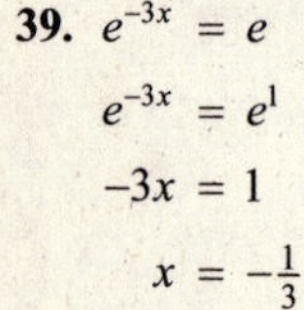

39. $e^{-3x} = e$

$e^{-3x} = e^1$

$-3x = 1$

$x = -\frac{1}{3}$

41. $e^{\sqrt{x}} = e^3$

$\sqrt{x} = 3$

$x = 9$

43. (a) $V = 15{,}000e^{-0.6286t}$

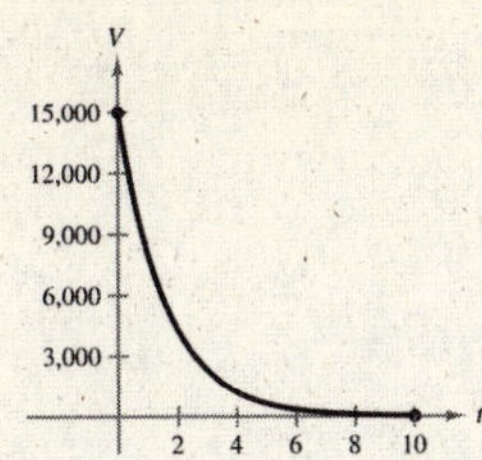

(b) $V' = -9429e^{-0.6286t}$

$V'(1) = -9429e^{-0.6286(1)} \approx -\$5028.84$ per year

(c) $V'(5) = -9429e^{-0.6286(5)} \approx -\$406.89$ per year

(d) $V(0) = 15{,}000,\ V(10) \approx 27.93$

$$\frac{V(10) - V(0)}{10 - 0} = \frac{27.93 - 15{,}000}{10} \approx -1497.21$$

Linear model: $V - 15{,}000 = -1497.21(t - 0)$

$V = -1497.21t + 15{,}000$

(e) Answers will vary.

45. $y = 136.855 - 0.5841t + 0.31664t^2 - 0.002166e^t$

(a)

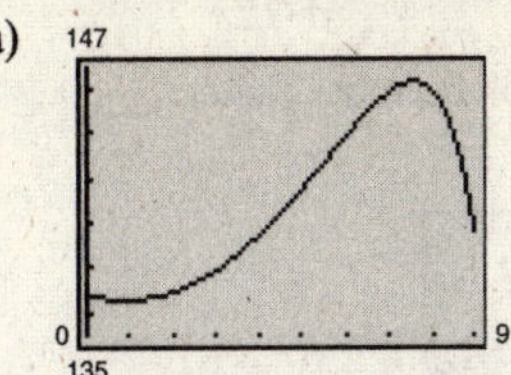

(b) and (c) $y' = -0.5841 + 0.63328t - 0.002166e^t$

2000: $y'(0) \approx -0.59$ million people/yr

2004: $y'(4) \approx 1.83$ million people/yr

2009: $y'(9) \approx -12.44$ million people/yr

47. $A = 5000e^{0.08t},\ A' = 400e^{0.08t}$

(a) When $t = 1$, $A' = 400e^{0.08(1)}$

$\approx \$433.31$ per year.

(b) When $t = 10$, $A' = 400e^{0.08(10)}$

$\approx \$890.22$ per year.

(c) When $t = 50$, $A' = 400e^{0.08(50)}$

$\approx \$21{,}839.26$ per year.

49. (a) Because the mean was $\mu = 516$ and the standard deviation was $\sigma = 116$, $f(x) = \dfrac{1}{116\sqrt{2\pi}}e^{-(x-516)^2/26{,}912}$.

(b)

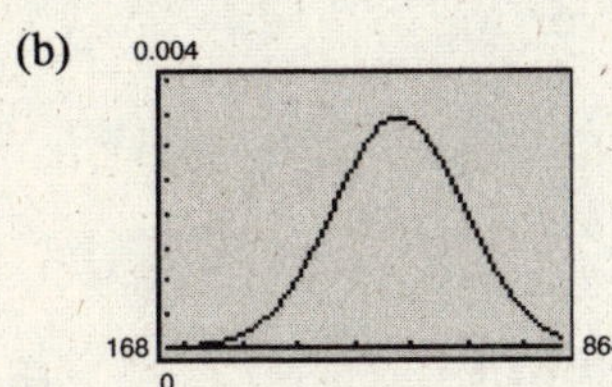

(c) $$f'(x) = \frac{1}{116\sqrt{2\pi}}e^{-(x-516)^2/26{,}912}\left[\frac{-2(x - 516)(1)}{26{,}912}\right] = \frac{-1}{1{,}560{,}896\sqrt{2\pi}}e^{-(x-516)^2/26{,}912}(x - 516)$$

(d) For $x < \mu$, let $x = 515$.

$$f'(515) = \frac{-1}{1{,}560{,}896\sqrt{2\pi}}e^{-(515-516)^2/26{,}912}(515 - 516) \approx 2.46 \times 10^{-7} > 0$$

For $x > \mu$, let $x = 517$.

$$f'(517) = \frac{-1}{1{,}560{,}896\sqrt{2\pi}}e^{-(517-516)^2/26{,}912}(517 - 516) \approx -2.65 \times 10^{-7} < 0$$

51. $f(x) = \frac{1}{\sigma\sqrt{\pi}}e^{-(x-\mu)^2/2\sigma^2} = \frac{1}{\sigma\sqrt{2\pi}}e^{-x^2/2\sigma^2}$

For larger σ, the graph becomes flatter.

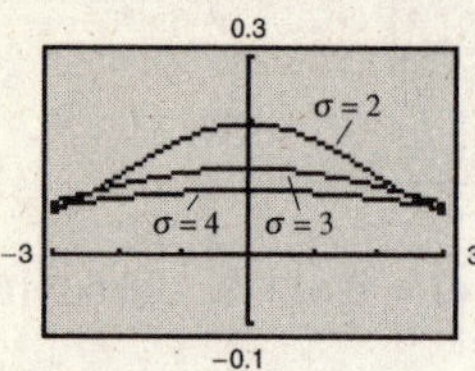

53.
$$f(x) = \frac{1}{\sigma\sqrt{2\pi}}e^{-x^2/2\sigma^2}$$
$$f'(x) = \frac{1}{\sigma\sqrt{2\pi}}\left(-\frac{x}{\sigma^2}\right)e^{-x^2/2\sigma^2} = -\frac{1}{\sigma^3\sqrt{2\pi}}(x)e^{-x^2/2\sigma^2}$$
$$f''(x) = \frac{1}{\sigma\sqrt{2\pi}}\left[\left(-\frac{x}{\sigma^2}\right)\left(-\frac{x}{\sigma^2}\right)e^{-x^2/2\sigma^2} + \left(-\frac{1}{\sigma^2}\right)e^{-x^2/2\sigma^2}\right]$$
$$f''(x) = \frac{1}{\sigma\sqrt{2\pi}}\left(e^{-x^2/2\sigma^2}\right)\left(\frac{x^2}{\sigma^4} - \frac{1}{\sigma^2}\right)$$
$$0 = \frac{1}{\sigma\sqrt{2\pi}}\left(e^{-x^2/2\sigma^2}\right)\left(\frac{x^2}{\sigma^4} - \frac{1}{\sigma^2}\right)$$
$$0 = \left(\frac{x^2}{\sigma^4} - \frac{1}{\sigma^2}\right)$$
$$\frac{1}{\sigma^2} = \frac{x^2}{\sigma^4}$$
$$\sigma^4 = \sigma^2x^2$$
$$\sigma^2 = x^2$$
$$\pm\sigma = x$$

Interval	$-\infty < x < -\sigma$	$-\sigma < x < \sigma$	$\sigma < x < \infty$
Test value	-2σ	$\frac{1}{2}\sigma$	2σ
Sign of f''	$f'' > 0$	$f'' < 0$	$f'' > 0$
Conclusion	Concave upward	Concave downward	Concave upward

Because concavity changes at $x = \pm\sigma$, you know the graph has points of inflection at $x = \pm\sigma$.

Chapter 4 Quiz Yourself

1. $4^3(4^2) = 4^5 = 1024$

2. $\left(\frac{1}{6}\right)^{-3} = 6^3 = 216$

3. $\frac{3^8}{3^5} = 3^3 = 27$

4. $(5^{1/2})(3^{1/2}) = \sqrt{5 \cdot 3} = \sqrt{15}$

5. $(e^2)(e^5) = e^7$

6. $(e^{2/3})(e^3) = e^{11/3}$

7. $\frac{e^2}{e^{-4}} = e^2(e^4) = e^6$

8. $(e^{-1})^{-3} = e^3$

9. $f(x) = 3^x - 2$

x	-2	-1	0	1	2
$f(x)$	$-\frac{17}{9} = -1.8$	$-\frac{5}{3} = -1.\overline{6}$	-1	1	7

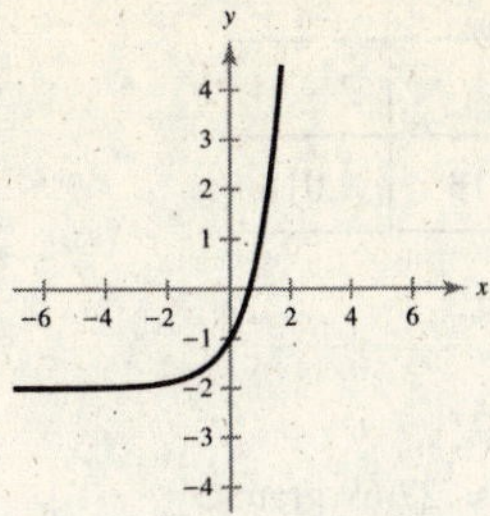

10. $f(x) = 5^{-x} + 2$

x	-2	-1	0	1	2
$f(x)$	27	7	3	$\frac{11}{5} = 2.2$	$\frac{51}{25} = 2.04$

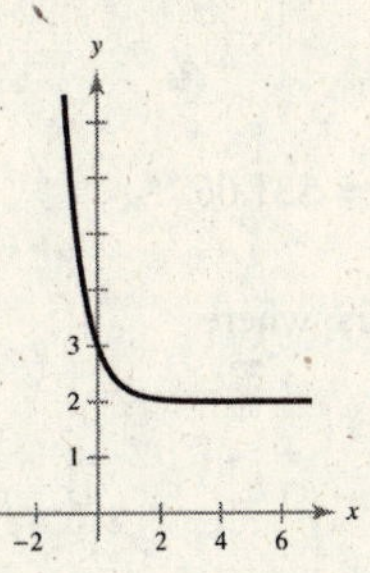

11. $f(x) = 6^{x-3}$

x	0	1	2	3	4
$f(x)$	$\frac{1}{216} \approx 0.005$	$\frac{1}{36} \approx 0.03$	$\frac{1}{6} = 0.1\overline{6}$	1	6

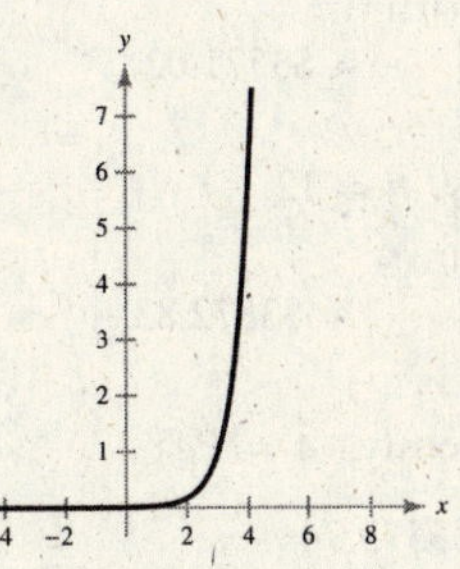
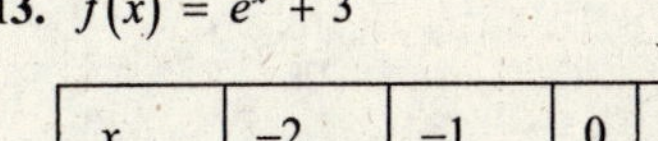

12. $f(x) = e^{x+2}$

x	-2	-1	0	1	2
$f(x)$	1	2.718	7.389	20.086	54.598

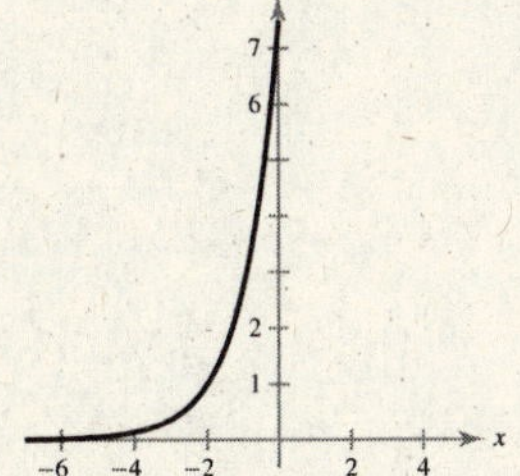

13. $f(x) = e^x + 3$

x	-2	-1	0	1	2
$f(x)$	3.135	3.368	0	5.718	10.389

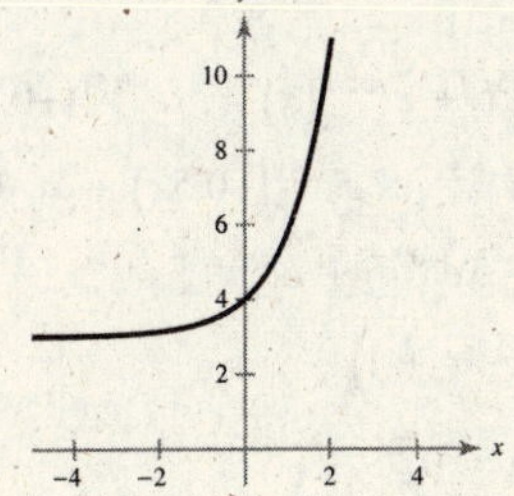

14. $f(x) = e^{-2x} + 1$

x	-2	-1	0	1	2
$f(x)$	55.598	8.389	2	1.135	1.018

15. $y = 35\left(\frac{1}{2}\right)^{t/80}, t \geq 0$

After 50 years, $y(50) = 35\left(\frac{1}{2}\right)^{50/80} \approx 22.69$ grams.

16. $C(t) = P(1.045)^t, 0 \leq t \leq 10$

If $P = 20$, then $C(10) = 20(1.045)^{10} \approx \$31.06$.

17. $P = \$3000$, $r = 3.5\%$, and $t = 5$ years, where

$A = P\left(1 + \frac{r}{n}\right)^{nt}$ or $A = Pe^{rt}$

(a) If compounded quarterly, $n = 4$.

$$A = 3000\left(1 + \frac{0.035}{4}\right)^{(4)(5)} \approx \$3571.02$$

(b) If compounded monthly, $n = 12$.

$$A = 3000\left(1 + \frac{0.035}{12}\right)^{(12)(5)} \approx \$3572.83$$

(c) If compounded continuously, $A = Pe^{rt}$.

$$A = 3000e^{(0.035)(5)} \approx \$3573.74$$

18. $A = P\left(1 + \frac{r}{n}\right)^{nt}$

$$P = \frac{A}{\left(1 + \frac{r}{n}\right)^{nt}}$$

Find P if $A = 14{,}000$, $r = 6\%$, $n = 12$, and $t = 5$ years.

$$P = \frac{14{,}000}{\left(1 + \frac{0.06}{12}\right)^{(12)(5)}} \approx \$10{,}379.21$$

19. $y = e^{5x}$

$y' = 5e^{5x}$

20. $y = e^{x-4}$

$y' = e^{x-4}$

21. $y = 5e^{x+2}$

$y' = 5e^{x+2}$

22. $y = 3e^x - xe^x$

$$\begin{aligned} y' &= 3e^x - xe^x - e^x \\ &= e^x(2 - x) \end{aligned}$$

23.

$$\begin{aligned} y &= e^{-2x} \\ y' &= -2e^{-2x} \\ y'(0) &= -2 \\ y - 1 &= -2(x - 0) \\ y &= -2x + 1 \end{aligned}$$

24. $f(x) = 0.5x^2e^{-0.5x}$

$$\begin{aligned} f'(x) &= 0.5x^2(-0.5)e^{-0.5x} + e^{-0.5x}(x) = xe^{-0.5x}(-0.25x + 1) \\ f''(x) &= -0.25x^2(-0.5)e^{-0.5x} + e^{-0.5x}(-0.5x) + x(-0.5)e^{-0.5x} + e^{-0.5x} \\ &= 0.125x^2e^{-0.5x} - xe^{-0.5x} + e^{-0.5x} \\ &= e^{-0.5x}(0.125x^2 - x + 1) \end{aligned}$$

Horizontal asymptote: $y = 0$

$f'(x) = 0$ when $x = 0$ and $x = 4$. Because $f''(0) > 0$ and $f''(4) < 0$, you have the following.

Relative maximum: $\left(4, \frac{8}{e^2}\right)$

Relative minimum: $(0, 0)$

Because $f''(x) = 0$ when $x = 4 \pm 2\sqrt{2}$, the inflection points occur at $\left(4 + 2\sqrt{2}, 0.77\right)$ and $\left(4 - 2\sqrt{2}, 0.38\right)$.

x	-2	-1	0	1	2
$f(x)$	5.44	0.82	0	0.30	0.74

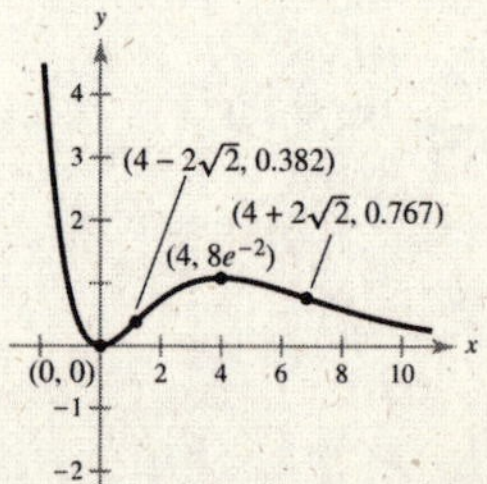

Section 4.4 Logarithmic Functions

Skills Warm Up

1. $f(x) = 5x$
   $y = 5x$
   $x = 5y$
   $\frac{x}{5} = y$
   $f^{-1}(x) = \frac{x}{5}$

2. $f(x) = x - 6$
   $y = x - 6$
   $x = y - 6$
   $x + 6 = y$
   $f^{-1}(x) = x + 6$

3. $f(x) = 3x + 2$
   $y = 3x + 2$
   $x = 3y + 2$
   $\frac{x - 2}{3} = y$
   $f^{-1}(x) = \frac{x - 2}{3}$

4. $f(x) = \frac{3}{4}x - 9$
   $y = \frac{3}{4}x - 9$
   $x = \frac{3}{4}y - 9$
   $x + 9 = \frac{3}{4}y$
   $\frac{4(x + 9)}{3} = y$
   $f^{-1}(x) = \frac{4(x + 9)}{3}$

5. $0 < x + 4$
   $-4 < x$

6. $0 < x^2 + 1$
   x is all real numbers.

7. $0 < \sqrt{x^2 - 1}$
   $0 < x^2 - 1$
   $1 < x^2 \Rightarrow 1 < x,\ -1 < x$
   $(-\infty, -1)$ and $(1, \infty)$

8. $0 < x - 5$
   $5 < x$

9. $A = Pe^{rt}$
   $= 1900e^{0.06(10)}$
   $\approx \$3462.03$

10. $A = Pe^{rt}$
    $= 2500e^{0.03(10)}$
    $\approx \$3374.65$

1. $e^{0.6931\ldots} = 2$

3. $e^{-1.6094\ldots} = 0.2$

5. $\ln 1 = 0$

7. $\ln 0.0498\ldots = -3$

9. $f(x) = 2 + \ln x$
   The graph is a logarithmic curve that passes through the point $(1, 2)$ with a vertical asymptote at $x = 0$.
   It matches graph (c).

11. $f(x) = \ln(x + 2)$
    The graph is a logarithmic curve that passes through the point $(-1, 0)$ with a vertical asymptote at $x = -2$.
    It matches graph (b).

13. $y = \ln(x - 1)$

x	1.5	2	3	4	5
y	−0.69	0	0.69	1.10	1.39

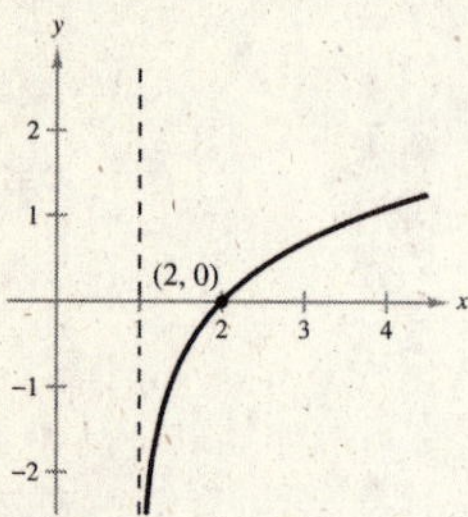

15. $y = \ln 2x$

x	0.25	0.5	1	3	5
y	−0.69	0	0.69	1.79	2.30

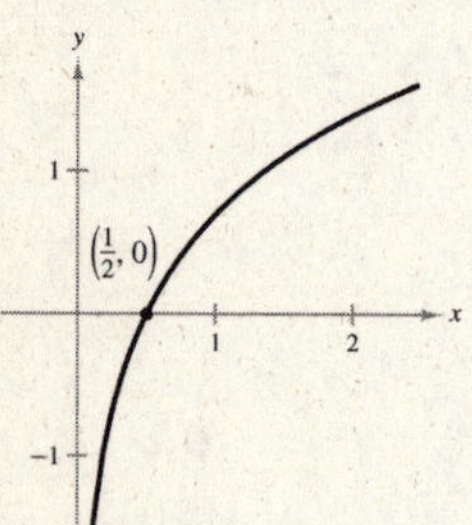

17. $y = 3 \ln x$

x	0.5	1	2	3	4
y	−2.08	0	2.08	3.30	4.16

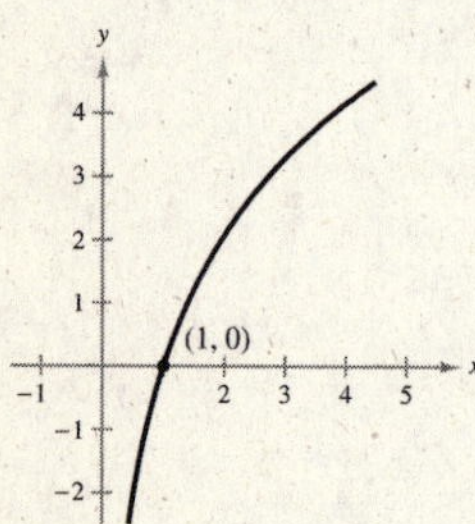

19. $\ln e^{x^2} = x^2$

21. $e^{\ln(5x+2)} = 5x + 2$

23. $-1 + \ln e^{2x} = -1 + 2x = 2x - 1$

25. $\ln \frac{2}{3} = \ln 2 - \ln 3$

27. $\ln xyz = \ln x + \ln y + \ln z$

29. $\ln \sqrt[3]{2x + 7} = \ln(2x + 7)^{1/3} = \frac{1}{3} \ln(2x + 7)$

31. $\ln\left[z(z - 1)^2\right] = \ln z + \ln(z - 1)^2$

$= \ln z + 2 \ln(z - 1)$

33. $\ln \dfrac{3x(x + 1)}{(2x + 1)^2} = \ln\left[3x(x + 1)\right] - \ln(2x + 1)^2$

$= \ln 3 + \ln x + \ln(x + 1) - 2 \ln(2x + 1)$

35. $g(x) = \ln \sqrt{x} = \frac{1}{2} \ln x$

$f(g(x)) = f\left(\frac{1}{2} \ln x\right) = e^{2(1/2 \ln x)} = e^{\ln x} = x$

$g(f(x)) = g\left(e^{2x}\right) = \frac{1}{2} \ln e^{2x} = \frac{1}{2}(2x) \ln e = x$

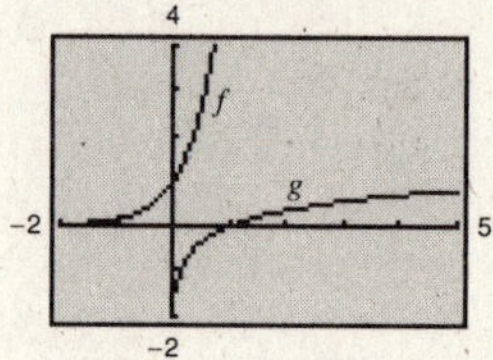

37. $f(g(x)) = f\left(\frac{1}{2} + \ln \sqrt{x}\right)$

$= e^{2\left[(1/2) + \ln \sqrt{x}\right] - 1}$

$= e^{2 \ln x^{1/2}}$

$= e^{\ln x}$

$= x$

$g(f(x)) = g\left(e^{2x-1}\right)$

$= \frac{1}{2} + \ln \sqrt{e^{2x-1}}$

$= \frac{1}{2} + \frac{1}{2} \ln e^{2x-1}$

$= \frac{1}{2} + \frac{1}{2}(2x - 1)$

$= x$

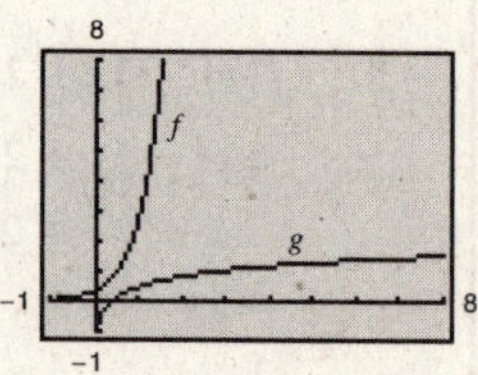

39. (a) $\ln 6 = \ln(2 \cdot 3)$

$= \ln 2 + \ln 3$

$\approx 0.6931 + 1.0986$

$= 1.7917$

(b) $\ln \frac{3}{2} = \ln 3 - \ln 2$

$\approx 1.0986 - 0.6931$

$= 0.4055$

(c) $\ln 81 = \ln 3^4$

$= 4 \ln 3$

$\approx 4(1.0986)$

$= 4.3944$

(d) $\ln \sqrt{3} = \ln 3^{(1/2)}$

$= \left(\frac{1}{2}\right) \ln 3$

$\approx \left(\frac{1}{2}\right)(1.0986)$

$= 0.5493$

41. $\ln(x-2) - \ln(x+2) = \ln\dfrac{x-2}{x+2}$

43. $3\ln x + 2\ln y - 4\ln z = \ln x^3 + \ln y^2 - \ln z^4$

$$= \ln\left(\frac{x^3y^2}{z^4}\right)$$

45. $4\ln(x-6) - \dfrac{1}{2}\ln(3x+1) = \ln(x-6)^4 - \ln(3x+1)^{1/2}$

$$= \ln\left(\frac{(x-6)^4}{\sqrt{3x+1}}\right)$$

47. $3[\ln x + \ln(x+3) - \ln(x+4)] = 3\ln\dfrac{x(x+3)}{x+4}$

$$= \ln\left[\frac{x(x+3)}{x+4}\right]^3$$

49. $\dfrac{3}{2}[\ln x(x^2+1) - \ln(x+1)] = \dfrac{3}{2}\ln\dfrac{x(x^2+1)}{x+1}$

$$= \ln\left[\frac{x(x^2+1)}{x+1}\right]^{3/2}$$

51. $e^{\ln x} = 4$

$x = 4$

53. $e^{x+1} = 4$

$x + 1 = \ln 4$

$x = \ln 4 - 1 \approx 0.39$

55. $300e^{-0.2t} = 700$

$e^{-0.2t} = \dfrac{7}{3}$

$-0.2t = \ln 7 - \ln 3$

$t = \dfrac{\ln 7 - \ln 3}{-0.2}$

≈ -4.24

57. $4e^{2x-1} - 1 = 5$

$4e^{2x-1} = 6$

$e^{2x-1} = \dfrac{3}{2}$

$2x - 1 = \ln 3 - \ln 2$

$x = \dfrac{\ln 3 - \ln 2 + 1}{2}$

$x = \dfrac{\ln 3 - \ln 2 + 1}{2}$

≈ 0.70

59. $\ln x = 0$

$x = e^0$

$x = 1$

61. $\ln 2x = 2.4$

$2x = e^{2.4}$

$x = \dfrac{e^{2.4}}{2}$

≈ 5.51

63. $3 + 4\ln x = 15$

$4\ln x = 12$

$\ln x = 3$

$e^{\ln x} = e^3$

$x = e^3$

65. $\ln x - \ln(x-6) = 3$

$\ln\left(\dfrac{x}{x-6}\right) = 3$

$e^{\ln(x/x-6)} = e^3$

$\dfrac{x}{x-6} = e^3$

$x = e^3x - 6e^3$

$x - e^3x = -6e^3$

$x(1-e^3) = -6e^3$

$x = \dfrac{6e^3}{e^3-1}$

67. $5^{2x} = 15$

$\ln 5^{2x} = \ln 15$

$2x\ln 5 = \ln 15$

$x = \dfrac{\ln 15}{2\ln 5} \approx 0.84$

69. $500(1.07)^t = 1000$

$1.07^t = 2$

$t\ln 1.07 = \ln 2$

$t = \dfrac{\ln 2}{\ln 1.07} \approx 10.24$

71. $\left(1 + \dfrac{0.07}{12}\right)^{12t} = 3$

$12t\ln\left(1 + \dfrac{0.07}{12}\right) = \ln 3$

$t = \dfrac{\ln 3}{12\ln[1 + (0.07/12)]}$

≈ 15.74

73. (a)
$$\begin{aligned} 2P &= Pe^{rt} \\ 2(3000) &= 3000e^{0.085t} \\ 2 &= e^{0.085t} \\ \ln 2 &= 0.085t \\ \frac{\ln 2}{0.085} &= t \\ 8.2 \text{ yr} &\approx t \end{aligned}$$

(b)
$$\begin{aligned} 3P &= Pe^{rt} \\ 3(3000) &= 3000e^{0.085t} \\ 3 &= e^{0.085t} \\ \ln 3 &= 0.085t \\ \frac{\ln 3}{0.085} &= t \\ 12.9 \text{ yr} &\approx t \end{aligned}$$

75. $P = 1000,\ r = 0.05,\ A = 2000$

(a)
$$\begin{aligned} 2000 &= 1000(1 + 0.05)^t \\ 2 &= 1.05^t \\ t &= \frac{\ln 2}{\ln 1.05} \\ &\approx 14.2 \text{ years} \end{aligned}$$

(b)
$$\begin{aligned} 2000 &= 1000\left(1 + \frac{0.05}{12}\right)^{12t} \\ 2 &\approx (1.00417)^{12t} \\ t &= \frac{\ln 2}{12 \ln 1.00417} \\ &\approx 13.9 \text{ years} \end{aligned}$$

(c)
$$\begin{aligned} 2000 &= 1000\left(1 + \frac{0.05}{365}\right)^{365t} \\ 2 &\approx (1.000137)^{365t} \\ t &= \frac{\ln 2}{365 \ln 1.000137} \\ &\approx 13.9 \text{ years} \end{aligned}$$

(d)
$$\begin{aligned} 2000 &= 1000e^{0.05t} \\ 2 &= e^{0.05t} \\ t &= \frac{\ln 2}{0.05} \\ &\approx 13.9 \text{ years} \end{aligned}$$

77. $P = 130e^{0.0205t},\ 0 \le t \le 29$ and $t = 0 \leftrightarrow 1980$

(a) 2009: $P(29) \approx 235.576 \Rightarrow 235{,}576$

(b) Let $P = 300$ and solve for t.
$$\begin{aligned} 300 &= 130e^{0.0205t} \\ \ln\left(\frac{30}{13}\right) &= 0.0205t \\ \frac{\ln\left(\frac{30}{13}\right)}{0.0205} &= t \\ 40.8 &\approx t \end{aligned}$$

During the year 2020

79.
$$\begin{aligned} 0.32 \times 10^{-12} &= 10^{-12}\left(\frac{1}{2}\right)^{t/5715} \\ 0.32 &= \left(\frac{1}{2}\right)^{t/5715} \\ \ln 0.32 &= \frac{t}{5715} \ln\frac{1}{2} \\ t &= \frac{5715 \ln 0.32}{\ln 1/2} \\ &\approx 9394.6 \text{ yr} \end{aligned}$$

81.
$$\begin{aligned} 0.22 \times 10^{-12} &= 10^{-12}\left(\frac{1}{2}\right)^{t/5715} \\ 0.22 &= \left(\frac{1}{2}\right)^{t/5715} \\ \ln 0.22 &= \frac{t}{5715} \ln\left(\frac{1}{2}\right) \\ t &= \frac{5715 \ln 0.22}{\ln 1/2} \\ &\approx 12{,}484.0 \text{ yr} \end{aligned}$$

83. (a) $S(0) = 80 - 14 \ln 1 = 80$

(b) $S(4) = 80 - 14 \ln 5 \approx 57.5$

(c)
$$\begin{aligned} 46 &= 80 - 14 \ln(t + 1) \\ -34 &= -14 \ln(t + 1) \\ \frac{17}{7} &= \ln(t + 1) \\ e^{(17/7)} &= t + 1 \\ -1 + e^{(17/7)} &= t \\ 10 \text{ mo} &\approx t \end{aligned}$$

85. $p = 5000\left(1 - \dfrac{4}{4 + e^{-0.002x}}\right)$

(a) When $p = \$200$, $200 = 5000\left(1 - \dfrac{4}{4 + e^{-0.002x}}\right)$

$$\frac{1}{25} = 1 - \frac{4}{4 + e^{-0.002x}}$$
$$\frac{24}{25} = \frac{4}{4 + e^{-0.002x}}$$
$$4\left(\frac{25}{24}\right) = 4 + e^{-0.002x}$$
$$\frac{25}{6} = 4 + e^{-0.002x}$$
$$\frac{1}{6} = e^{-0.002x}$$
$$\ln\left(\frac{1}{6}\right) = -0.002x$$
$$\frac{0 - \ln 6}{-0.002} = x$$
$$896 \text{ units} \approx x.$$

(b) When $p = \$800$, $800 = 5000\left(1 - \dfrac{4}{4 + e^{-0.002x}}\right)$

$$\frac{4}{25} = 1 - \frac{4}{4 + e^{-0.002x}}$$
$$\frac{21}{25} = \frac{4}{4 + e^{-0.002x}}$$
$$4\left(\frac{25}{21}\right) = 4 + e^{-0.002x}$$
$$\frac{100}{21} = 4 + e^{-0.002x}$$
$$\frac{16}{21} = e^{-0.002x}$$
$$\ln\left(\frac{16}{21}\right) = -0.002x$$
$$\frac{\ln 16 - \ln 21}{-0.002} = x$$
$$136 \text{ units} \approx x.$$

87.

x	y	$\dfrac{\ln x}{\ln y}$	$\ln \dfrac{x}{y}$	$\ln x - \ln y$
1	2	0	−0.6931	−0.6931
3	4	0.7925	−0.2877	−0.2877
10	5	1.4307	0.6931	0.6931
4	0.5	−2	2.0794	2.0794

89.

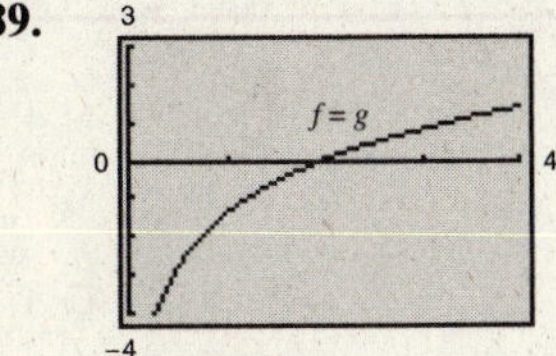

The graphs appear to be identical.

91. False. $\ln(0)$ is undefined.

93. False. $\ln x - \ln 2 = \ln\left(\dfrac{x}{2}\right)$

95. False. $\ln u = 2 \ln v = \ln v^2 \Rightarrow u = v^2$

97. Let $x = rt$ for the following.

(a) $A_1 = 2Pe^{rt}$

$A_1 = 2Pe^{x}$

(b) $A_2 = Pe^{2rt}$

$A_2 = Pe^{2x}$

(c) $A_3 = Pe^{r(2t)}$

$A_3 = Pe^{2rt}$

$A_3 = Pe^{2x}$

$A_2 = A_3$, and for $x < \ln 2$, $A_1 > A_2$. For $x > \ln 2$, $A_1 < A_2$. Option (b) and (c) will give you the same amount, but it makes more sense to double the rate, not the time. So option (b) is better than option (c). If you are looking for a long term investment, choose option (a).

99.

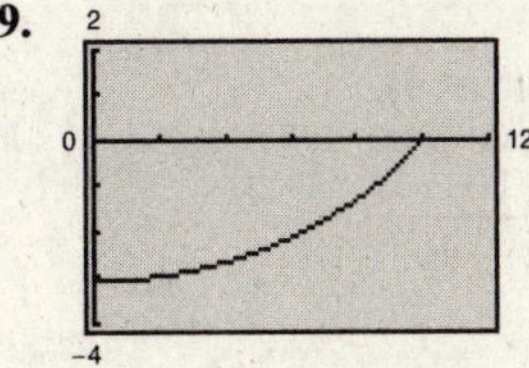

Answers will vary.

Section 4.5 Derivatives of Logarithmic Functions

Skills Warm Up

1. $\ln(x+1)^2 = 2\ln(x+1)$

2. $\ln x(x+1) = \ln x + \ln(x+1)$

3. $\ln \dfrac{x}{x+1} = \ln x - \ln(x+1)$

4. $\ln\left(\dfrac{x}{x-3}\right)^3 = 3\ln\dfrac{x}{x-3} = 3[\ln x - \ln(x-3)]$

5. $$\begin{aligned}\ln\frac{4x(x-7)}{x^2} &= \ln 4x + \ln(x-7) - \ln x^2\\ &= \ln 4x + \ln(x-7) - 2\ln x\\ &= \ln 4 + \ln x + \ln(x-7) - 2\ln x\end{aligned}$$

6. $$\begin{aligned}\ln x^3(x+1) &= \ln x^3 + \ln(x+1)\\ &= 3\ln x + \ln(x+1)\end{aligned}$$

7. $$\begin{aligned}y^2 + xy &= 7\\ 2y\frac{dy}{dx} + x\frac{dy}{dx} + y &= 0\\ \frac{dy}{dx}(2y+x) &= -y\\ \frac{dy}{dx} &= -\frac{y}{2y+x}\end{aligned}$$

8. $$\begin{aligned}x^2y - xy^2 &= 3x\\ x^2\frac{dy}{dx} + 2xy - x2y\frac{dy}{dx} - y^2 &= 3\\ \frac{dy}{dx}(x^2 - 2xy) &= 3 - 2xy + y^2\\ \frac{dy}{dx} &= \frac{3 - 2xy + y^2}{x^2 - 2xy}\end{aligned}$$

9. $$\begin{aligned}f(x) &= x^2(x+1) - 3x^3\\ f'(x) &= x^2 + 2x(x+1) - 9x^2\\ &= x^2 + 2x^2 + 2x - 9x^2\\ &= -6x^2 + 2x\\ f''(x) &= -12x + 2\end{aligned}$$

10. $$\begin{aligned}f(x) &= -\frac{1}{x^2}\\ f'(x) &= \frac{2}{x^3}\\ f''(x) &= -\frac{6}{x^4}\end{aligned}$$

1. $$\begin{aligned}y &= \ln x^2 = 2\ln x\\ y' &= \frac{2}{x}\end{aligned}$$

3. $$\begin{aligned}y &= \ln(x^2+3)\\ y' &= \frac{2x}{x^2+3}\end{aligned}$$

5. $$\begin{aligned}y &= \ln\sqrt{x-4} = \frac{1}{2}\ln(x-4)\\ y' &= \frac{1}{2}\cdot\frac{1}{(x-4)} = \frac{1}{2(x-4)}\end{aligned}$$

7. $$\begin{aligned}y &= (\ln x)^4\\ y' &= \left[4(\ln x)^3\frac{1}{x}\right] = \frac{4(\ln x)^3}{x}\end{aligned}$$

9. $$\begin{aligned}f(x) &= 2x\ln x\\ f'(x) &= 2x\left(\frac{1}{x}\right) + 2\ln x = 2 + 2\ln x\end{aligned}$$

11. $$\begin{aligned}y &= \ln\left(x\sqrt{x^2-1}\right)\\ &= \ln x + \ln(x^2+1)^{1/2}\\ &= \ln x + \frac{1}{2}\ln(x^2-1)\\ y' &= \frac{1}{x} + \left(\frac{1}{2}\right)\frac{1}{x^2-1}(2x)\\ &= \frac{1}{x} + \frac{x}{x^2-1}\\ &= \frac{2x^2-1}{x(x^2-1)}\end{aligned}$$

13. $$\begin{aligned}y &= \ln x - \ln(x+1)\\ y' &= \frac{1}{x} - \frac{1}{x+1} = \frac{1}{x(x+1)}\end{aligned}$$

15. $y = \ln\left(\dfrac{x-1}{x+1}\right)^{1/3} = \dfrac{1}{3}\left[\ln(x-1) - \ln(x+1)\right]$

$y' = \dfrac{1}{3}\left[\dfrac{1}{x-1} - \dfrac{1}{x+1}\right] = \dfrac{1}{3}\left[\dfrac{2}{x^2-1}\right] = \dfrac{2}{3(x^2-1)}$

17. $y = \ln\dfrac{\sqrt{4+x^2}}{x}$

$= \ln\dfrac{(4+x^2)^{1/2}}{x}$

$= \dfrac{1}{2}\ln(4+x^2) - \ln x$

$y' = \dfrac{1}{2}\left(\dfrac{2x}{4+x^2}\right) - \dfrac{1}{x}$

$= \dfrac{x^2 - (4+x^2)}{x(4+x^2)}$

$= -\dfrac{4}{x(4+x^2)}$

19. $g(x) = e^{-x}\ln x$

$g'(x) = e^{-x}\left(\dfrac{1}{x}\right) + \left(-e^{-x}\right)\ln x = e^{-x}\left(\dfrac{1}{x} - \ln x\right)$

21. $g(x) = \ln\dfrac{e^x + e^{-x}}{2} = \ln(e^x + e^{-x}) - \ln 2$

$g'(x) = \dfrac{e^x - e^{-x}}{e^x + e^{-x}}$

23. $\log_5 25 = 2$ because $5^2 = 25$.

25. $\log_3 \frac{1}{27} = -3$ because $3^{-3} = \frac{1}{27}$.

27. $\log_7 49 = 2$ because $7^2 = 49$.

29. $\log_4 7 = \dfrac{\ln 7}{\ln 4} \approx 1.404$

31. $\log_2 48 = \dfrac{\ln 48}{\ln 2} \approx 5.585$

33. $\log_3 \dfrac{1}{2} = \dfrac{\ln(1/2)}{\ln 3} \approx -0.631$

35. $y = 3^x$

$y' = (\ln 3)3^x$

37. $f(x) = \log_2 x$

$f'(x) = \dfrac{1}{\ln 2}\cdot\dfrac{1}{x} = \dfrac{1}{x\ln 2}$

39. $h(x) = 4^{2x-3}$

$h'(x) = (\ln 4)4^{2x-3}(2)$

$= (2\ln 4)4^{2x-3}$

41. $y = \log_{10}(x^2 + 6x)$

$y' = \dfrac{1}{\ln 10}\dfrac{1}{x^2+6x}(2x+6) = \dfrac{2x+6}{(x^2+6x)\ln 10}$

43. $y = x2^x$

$y' = x(\ln 2)2^x + 2^x = 2^x(1 + x\ln 2)$

45. $y = \ln x^3 = 3\ln x$

$y' = 3\left(\dfrac{1}{x}\right) = \dfrac{3}{x}$

$y'(1) = 3$

$y - 0 = 3(x-1)$

$y = 3x - 3$ Tangent line

47. $y = x\ln x$

$y' = x\left(\dfrac{1}{x}\right) + (1)\ln x = 1 + \ln x$

$y'(e) = 1 + \ln e = 2$

$y - e = 2(x - e)$

$y = 2x - e$ Tangent line

49. $f(x) = \ln\dfrac{5(x+2)}{x} = \ln 5 + \ln(x+2) - \ln x$

$f'(x) = \dfrac{1}{x+2} - \dfrac{1}{x} = -\dfrac{2}{x(x+2)}$

$f'\left(-\dfrac{5}{2}\right) = \dfrac{-2}{\left(-\dfrac{5}{2}\right)\left(-\dfrac{5}{2}+2\right)} = -\dfrac{8}{5}$

$y - 0 = -\dfrac{8}{5}\left(x + \dfrac{5}{2}\right)$

$y = -\dfrac{8}{5}x - 4$ Tangent line

51. $y = \log_3 x$

$y' = \dfrac{1}{\ln 3}\dfrac{1}{x}$

$y'(27) = \dfrac{1}{27\ln 3}$

$y - 3 = \dfrac{1}{27\ln 3}(x - 27)$

$y = \dfrac{1}{27\ln 3}x - \dfrac{1}{\ln 3} + 3$ Tangent line

53. $x^2 - 3\ln y + y^2 = 10$

$$2x - 3\left(\frac{1}{y}\right)\frac{dy}{dx} + 2y\frac{dy}{dx} = 0$$

$$2x = \frac{dy}{dx}\left(\frac{3}{y} - 2y\right)$$

$$2x = \frac{dy}{dx}\left(\frac{3 - 2y^2}{y}\right)$$

$$\frac{2xy}{3 - 2y^2} = \frac{dy}{dx}$$

55. $4x^3 + \ln y^2 + 2y = 2x$

$$12x^2 + \frac{2}{y}\frac{dy}{dx} + 2\frac{dy}{dx} = 2$$

$$\left(\frac{2}{y} + 2\right)\frac{dy}{dx} = 2 - 12x^2$$

$$\frac{dy}{dx} = \frac{2 - 12x^2}{(2/y) + 2}$$

$$= \frac{1 - 6x^2}{(1/y) + 1}$$

$$= \frac{(1 - 6x^2)y}{1 + y}$$

57. $x + y - 1 = \ln(x^2 + y^2)$

$$1 + \frac{dy}{dx} = \frac{1}{x^2 + y^2}\left(2x + 2y\frac{dy}{dx}\right)$$

$$1 + \frac{dy}{dx} = \frac{2x}{x^2 + y^2} + \frac{2y}{x^2 + y^2}\frac{dy}{dx}$$

$$\frac{dy}{dx}\left(1 - \frac{2y}{x^2 + y^2}\right) = \frac{2x}{x^2 + y^2} - 1$$

$$\frac{dy}{dx}\left(\frac{x^2 + y^2 - 2y}{x^2 + y^2}\right) = \frac{2x - x^2 - y^2}{x^2 + y^2}$$

$$\frac{dy}{dx} = \frac{(2x - x^2 - y^2)(x^2 + y^2)}{(x^2 + y^2)(x^2 + y^2 - 2y)}$$

$$\frac{dy}{dx} = \frac{2x - x^2 - y^2}{x^2 + y^2 - 2y}$$

For $(1, 0)$: $\dfrac{dy}{dx} = 1$

Tangent line: $y - 0 = 1(x - 1)$

$$y = x - 1$$

59. $f(x) = x\ln\sqrt{x} + 2x = \frac{1}{2}x\ln x + 2x$

$$f'(x) = \frac{1}{2}x\left(\frac{1}{x}\right) + \frac{1}{2}\ln x + 2 = \frac{1}{2}\ln x + \frac{5}{2}$$

$$f''(x) = \frac{1}{2x}$$

61. $f(x) = 2 + x^3\ln x$

$$f'(x) = x^3\left(\frac{1}{x}\right) + 3x^2\ln x$$

$$= x^2 + 3x^2\ln x$$

$$f''(x) = 2x + 3x^2\left(\frac{1}{x}\right) + 6x\ln x$$

$$= 2x + 3x + 6x\ln x$$

$$= 5x + 6x\ln x = x(5 + 6\ln x)$$

63. $f(x) = 5^x$

$$f'(x) = (\ln 5)5^x$$

$$f''(x) = (\ln 5)(\ln 5)5^x = (\ln 5)^2 5^x$$

65. $\beta = 10\log_{10} I - 10\log_{10}(10^{-16})$

$$\frac{d\beta}{dI} = \frac{10}{(\ln 10)I}$$

For $I = 10^{-4}$,

$$\frac{d\beta}{dI} = \frac{10}{(\ln 10)10^{-4}} = \frac{10^5}{\ln 10}$$

$\approx 43{,}429.4$ decibels per watt per cm^2.

67. $y = x - \ln x$

$$y' = 1 - \frac{1}{x} = \frac{x - 1}{x}$$

$y' = 0$ when $x = 1$.

$$y'' = \frac{1}{x^2}$$

Because $y''(1) = 1 > 0$, there is a relative minimum at $(1, 1)$. Because $y'' > 0$ on $(0, \infty)$, it follows that the graph is concave upward on its domain and there are no inflection points.

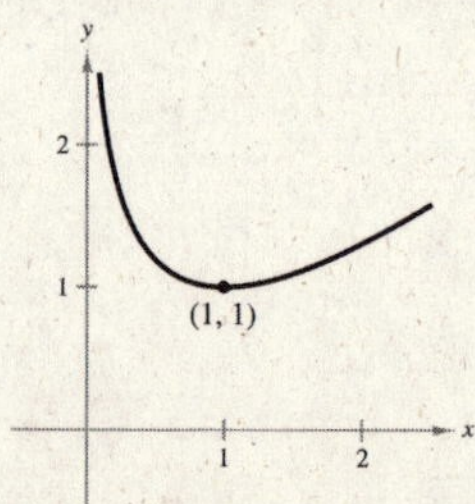

69. $y = \dfrac{x}{\ln x}$

$y' = \dfrac{\ln x - 1}{(\ln x)^2}$

$y' = 0$ when $x = e$.

Discontinuity: $x = 1$

$y'' = \dfrac{-\ln x + 2}{x(\ln x)^3}$

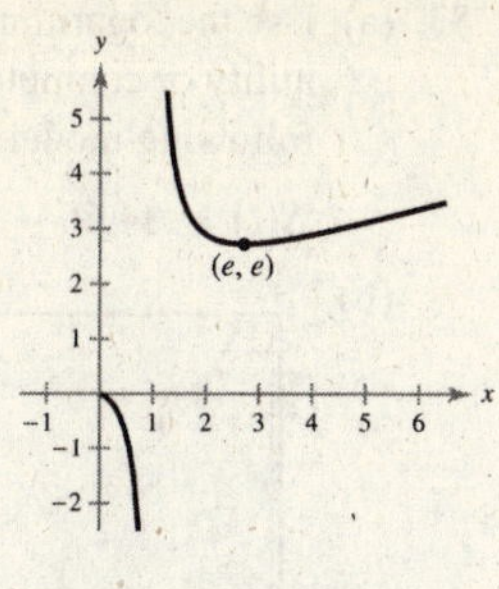

Because $y''(e) = \dfrac{1}{e} > 0$, there is a relative minimum at (e, e). Moreover, because $y'' < 0$ on $(0, 1)$, it follows that the graph is concave downward on $(0, 1)$. Because $y'' > 0$ on $(1, \infty)$, it follows that the graph is concave upward on $(1, \infty)$. There is a vertical asymptote at $x = 1$ and no inflection points.

71. $y = x^2 \ln \dfrac{x}{4}$

$y' = x^2\left(\dfrac{1}{(1/4)x}\right)\left(\dfrac{1}{4}\right) + 2x \ln \dfrac{x}{4}$

$= x + 2x \ln \dfrac{x}{4}$

$y'' = 1 + 2x\left(\dfrac{1}{(1/4)x}\right)\left(\dfrac{1}{4}\right) + 2 \ln \dfrac{x}{4}$

$= 1 + 2 + 2 \ln \dfrac{x}{4}$

$= 3 + 2 \ln \dfrac{x}{4}$

Because $y''\left(4e^{-1/2}\right) > 0$, there is a relative minimum at $\left(4e^{-1/2}, -8e^{-1}\right)$. Moreover, because $y'' > 0$ on $(0, \infty)$, it follows that the graph is concave upward on its domain and there are no inflection points.

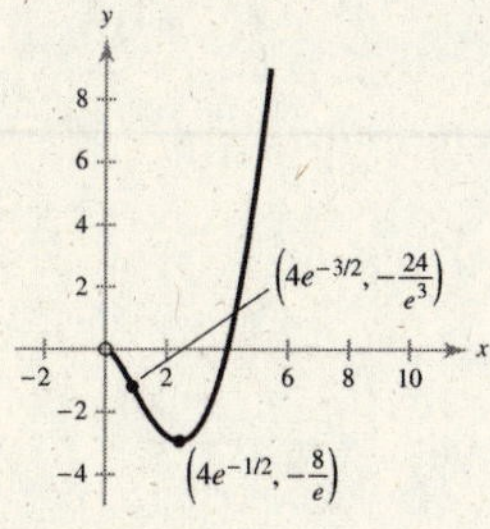

73. $x = \ln \dfrac{1000}{p} = \ln 1000 - \ln p$

$\dfrac{dx}{dp} = -\dfrac{1}{p}$

When $p = 10$, $\dfrac{dy}{dp} = -\dfrac{1}{10}$.

75. $x = \ln \dfrac{1000}{p}$

$e^x = \dfrac{1000}{p}$

$p = 1000e^{-x}$

$\dfrac{dp}{dx} = -1000e^{-x}$

When $p = 10$, $x = \ln \dfrac{1000}{10} = \ln 100$ and

$\dfrac{dp}{dx} = -1000e^{-\ln 100} = \dfrac{-1000}{100} = -10.$

Note that $\dfrac{dp}{dx}$ and $\dfrac{dx}{dp}$ are reciprocals of each other.

77. $C = 500 + 300x - 300 \ln x,\ x \geq 1$

(a) $\overline{C} = \dfrac{C}{x}$

$= \dfrac{500}{x} + 300 - 300\dfrac{\ln x}{x}$

$= \dfrac{500}{x} - 300\left(-1 + \dfrac{\ln x}{x}\right)$

(b) $\overline{C}' = -\dfrac{500}{x^2} - 300\left[\dfrac{x(1/x) - \ln x}{x^2}\right]$

$= -\dfrac{500}{x^2} - \dfrac{300(1 - \ln x)}{x^2}$

Setting $\overline{C}' = 0$,

$500 = -300(1 - \ln x)$

$\dfrac{5}{3} = \ln x - 1$

$\dfrac{8}{3} = \ln x$

$x = e^{8/3} \approx 14.39,\ \overline{C}(14.39) \approx 279.15$

By the First Derivative Test, this is a minimum.

(c)

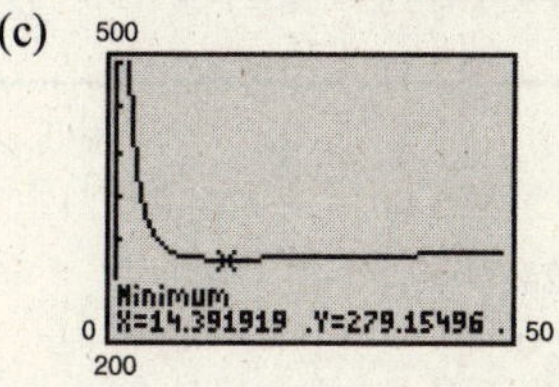

$\overline{C} \approx 279.15$ when $x \approx 14.39$ is the minimum.

79. (a)

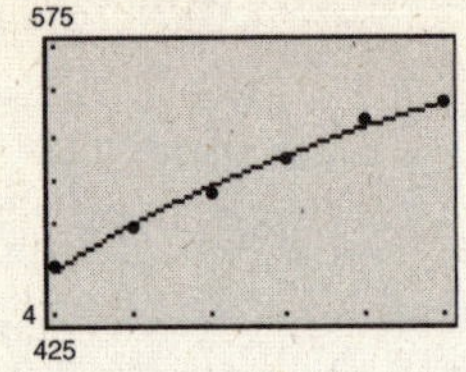

(b) $E = 287 + 116.7 \ln t$

$$E' = 116.7\left(\frac{1}{t}\right) = \frac{116.7}{t}$$

2006: $E'(6) = \dfrac{116.7}{6} = 19.45$ (thousand) per year.

81. $t = -13.375 \ln \dfrac{x - 1250}{x}, \; x > 1250$

(a)

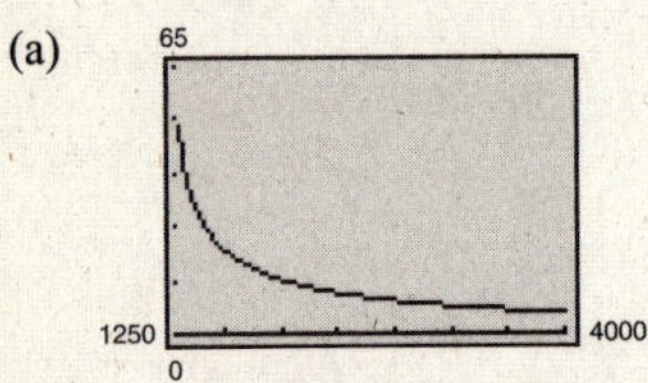

(b) If $x = 1398.43$, $t \approx 30$, and the total amount paid is $(1398.43)(30)(12) = \$503{,}434.80$.

(c) If $x = 1611.19$, $t \approx 20$, and the total amount paid is $(1611.19)(20)(12) = \$386{,}685.60$.

(d) $t' = \dfrac{-13.375x}{x - 1250}\left(\dfrac{1250}{x^2}\right) = \dfrac{-16{,}718.75}{x(x - 1250)}$

When $x = 1398.43$, the instantaneous rate of change is $t'(1398.43) \approx -0.081$.

When $x = 1611.19$, the instantaneous rate of change is $t'(1611.19) \approx -0.03$.

(e) For a higher monthly payment, the term is shorter, and the total amount paid is smaller.

83. (a) Use the logarithmic regression feature of a graphing utility or computer algebra system to obtain the following model.

$$S(t) = 84.66 - 11.00 \ln x$$

(b)

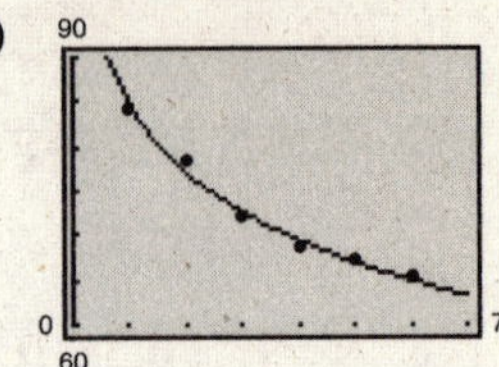

The model fits the data well.

(c) $S(t) = 84.66 - 11.00 \ln x$

$$S'(t) = \frac{-11.00}{x}$$

When $t = 2$, $S'(2) = \dfrac{-11.00}{2} = -5.5$.

The average score is decreasing at a rate of 5.5 points after 2 months.

Section 4.6 Exponential Growth and Decay

Skills Warm Up

1.

$$12 = 24e^{4k}$$
$$\tfrac{1}{2} = e^{4k}$$
$$\ln \tfrac{1}{2} = 4k$$
$$\tfrac{1}{4}(\ln 1 - \ln 2) = k$$
$$\tfrac{1}{4}(-\ln 2) = k$$
$$-0.1733 \approx k$$

2.

$$10 = 3e^{5k}$$
$$\tfrac{10}{3} = e^{5k}$$
$$\ln \tfrac{10}{3} = 5k$$
$$\tfrac{1}{5}(\ln 10 - \ln 3) = k$$
$$0.2408 \approx k$$

Skills Warm Up —*continued*—

3.
$$\begin{aligned} 25 &= 16e^{-0.01k} \\ \frac{25}{16} &= e^{-0.01k} \\ \ln \frac{25}{16} &= -0.01k \\ -100(\ln 25 - \ln 16) &= k \\ -44.6287 &\approx k \end{aligned}$$

4.
$$\begin{aligned} 22 &= 32e^{-0.02k} \\ \frac{11}{16} &= e^{-0.02k} \\ \ln \frac{11}{16} &= -0.02k \\ -50(\ln 11 - \ln 16) &= k \\ 18.7347 &\approx k \end{aligned}$$

5.
$$\begin{aligned} y &= 32e^{0.23t} \\ y' &= 32e^{0.23t}(0.23) \\ &= 7.36e^{0.23t} \end{aligned}$$

6.
$$\begin{aligned} y &= 18e^{0.072t} \\ y' &= 18e^{0.072t}(0.072) \\ &= 1.296e^{0.072t} \end{aligned}$$

7.
$$\begin{aligned} y &= 24e^{-1.4t} \\ y' &= 24e^{-1.4t}(-1.4) \\ &= -33.6e^{-1.4t} \end{aligned}$$

8.
$$\begin{aligned} y &= 25e^{-0.001t} \\ y' &= 25e^{-0.001t}(-0.001) \\ &= -0.025e^{-0.001t} \end{aligned}$$

9. $e^{\ln 4} = 4$

10. $4e^{\ln 3} = 4(3) = 12$

11. $e^{\ln(2x+1)} = 2x + 1$

12. $e^{\ln(x^2+1)} = x^2 + 1$

1. Because $y = 2$ when $t = 0$, it follows that $C = 2$. Moreover, because $y = 3$ when $t = 4$, you have

$$3 = 2e^{4k} \text{ and } k = \frac{\ln(3/2)}{4} \approx 0.1014.$$

So, $y = 2e^{0.1014t}$.

3. Because $y = 4$ when $t = 0$, it follows that $C = 4$. Moreover, because $y = \frac{1}{2}$ when $t = 5$, you have

$$\frac{1}{2} = 4e^{5k} \text{ and } k = \frac{\ln(1/8)}{5} \approx -0.4159.$$

So, $y = 4e^{-0.4159t}$.

5. Because $y = 4$ when $t = 1$ and $y = 2$ when $t = 4$, you have $4 = Ce^k$ and $2 = Ce^{4k}$ or

$$C = \frac{4}{e^k} \text{ and } C = \frac{2}{e^{4k}}.$$

From these two equations you have

$$\begin{aligned} \frac{4}{e^k} &= \frac{2}{e^{4k}} \\ \frac{e^{4k}}{e^k} &= \frac{2}{4} \\ e^{3k} &= \frac{1}{2} \\ 3k &= \ln\left(\frac{1}{2}\right) \end{aligned}$$

So, $k = \frac{1}{3}\ln\left(\frac{1}{2}\right)$ and you have $y = Ce^{t\,1/3 \ln(1/2)}$.

Because $4 = Ce^k \rightarrow 4 = Ce^{(\ln 0.5)/3}$

$$\begin{aligned} \frac{4}{e^{(\ln 0.5)/3}} &= C \\ 4\sqrt[3]{2} &\approx C \end{aligned}$$

So, $y = 4\sqrt[3]{2}e^{(\ln 0.5)/3 \cdot t}$.

7. $\frac{dy}{dt} = 2y,\ y = 10$ when $t = 0$

$y = 10e^{2t}$

$\frac{dy}{dt} = 10(2)e^{2t} = 2(10e^{2t}) = 2y$

Exponential growth

9. $\frac{dy}{dt} = -4y,\ y = 30$ when $t = 0$

$y = 30e^{-4t}$

$\frac{dy}{dt} = 30(-4)e^{-4t} = -4(30e^{-4t}) = -4y$

Exponential decay

11. From Example 1 you have

$y = Ce^{kt} = 10e^{[\ln(1/2)/1599]t}$.

When $t = 1000$,

$y = 10e^{[\ln(1/2)/1599]1000} \approx 6.48$ grams.

When $t = 10{,}000$,

$y = 10e^{[\ln(1/2)/1599]10{,}000} \approx 0.13$ gram.

13. From Example 1 you have

$y = Ce^{kt} = Ce^{[\ln(1/2)/5715]t}$

$2 = Ce^{[\ln(1/2)/5715]10{,}000}$

$C \approx 6.73$.

The initial quantity is 6.73 grams.

When $t = 1000$,

$y = 6.73e^{[\ln(1/2)/5715]1000} \approx 5.96$ grams.

15. From Example 1 you have

$y = Ce^{[\ln(1/2)/24{,}100]t}$

$2.1 = Ce^{[\ln(1/2)/24{,}100]1000}$

$C \approx 2.16$.

The initial quantity is 2.16 grams.

When $t = 10{,}000$,

$y = 2.16e^{[\ln(1/2)/24{,}000]10{,}000} \approx 1.62$ grams.

17. From Example 1 you have

$y = Ce^{[\ln(1/2)/1599]}$.

When $t = 900$,

$y = Ce^{[\ln(1/2)/1620]900} \approx 0.68C$.

After 900 years, approximately 68% of the radioactive radium will remain.

19. $0.15C = Ce^{[\ln(1/2)/5715]t}$

$\ln 0.15 = \frac{\ln(1/2)}{5715}t$

$t = \frac{5715 \ln 0.15}{\ln(1/2)}$

$\approx 15{,}641.8$ years

21. At $(0, 5)$, $y_1 = 5e^{k_1 t}$

At $(12, 20)$, $20 = 5e^{k_1(12)}$

$4 = e^{12k_1}$

$\ln 4 = 12k_1$

$\frac{\ln 4}{12} = k_1$

$0.1155 \approx k_1$

So, $y_1 = 5e^{0.1155t}$.

$k_1 = k_2 \ln 2$

At $(0, 5)$, $y_2 = 5(2)^{k_2 t}$

At $(12, 20)$, $20 = 5(2)^{k_2(12)}$

$4 = 2^{12k_2}$

$\log_2 4 = 12k_2$

$\frac{\log_2 4}{12} = k_2$

$\frac{1}{6} = k_2$

So, $y_2 = 5(2)^{t/6}$.

23. The model is $y = Ce^{kt}$. Because $y = 150$ when $t = 0$, you have $C = 150$. Furthermore,

$450 = 150e^{k(5)}$

$3 = e^{5k}$

$\ln 3 = 5k$

$k = \frac{\ln 3}{5}$.

So, $y = 150e^{[(\ln 3)/5]t} \approx 150e^{0.2197t}$.

(a) When $t = 10$, $y = 150e^{[(\ln 3)/5]10} = 1350$ bacteria.

(b) To find the time required for the population to double, solve for t.

$300 = 150e^{[(\ln 3)/5]t}$

$2 = e^{[(\ln 3)/5]t}$

$\ln 2 = \frac{\ln 3}{5}t$

$t = \frac{5 \ln 2}{\ln 3} \approx 3.2$ hours

(c) No, the doubling time is always 3.2 hours.

25. Because $A = 1000e^{0.12t}$, the time to double is given by $2000 = 1000e^{0.12t}$ and you have

$$2 = e^{0.12t}$$

$$t = \frac{\ln 2}{0.12} \approx 5.8 \text{ years.}$$

Amount after 10 yr: $A = 1000e^{0.12(10)} \approx \$3320.12$

Amount after 25 yr: $A = 1000e^{0.12(25)} \approx \$20,085.54$

27. Because $A = 750e^{rt}$ and $A = 1500$ when $t = 8$, you have

$$1500 = 750e^{8r}$$

$$2 = e^{8r}$$

$$r = \frac{\ln 2}{8} \approx 0.0866 = 8.66\%.$$

Amount after 10 yr: $A = 750e^{0.0866(10)} \approx \$1783.04$

Amount after 25 yr: $A = 750e^{0.0866(25)} \approx \$6535.95$

29. Because $A = 500e^{rt}$ and $A = 1292.85$ when $t = 10$, you have

$$1292.85 = 500e^{10r}$$

$$\frac{1292.85}{500} = e^{10r}$$

$$r = \frac{\ln(1292.85/500)}{10} \approx 0.0950 = 9.5\%.$$

The time to double is given by

$$1000 = 500e^{0.0950t}$$

$$2 = e^{0.0950t}$$

$$t = \frac{\ln 2}{0.0950} \approx 7.3 \text{ years.}$$

Amount after 25 yr: $A = 500e^{0.0950(25)} \approx \$5375.51$

31. Because $A = Pe^{0.045t}$ and $A = 10,000$ when $t = 10$, you have

$$10,000 = Pe^{0.045(10)}$$

$$\frac{10,000}{e^{0.45}} = P$$

$$\$6376.28 \approx P.$$

The time to double is given by

$$12,752.56 = 6376.28e^{0.045t}$$

$$t = \frac{\ln 2}{0.045} \approx 15.4 \text{ years.}$$

Amount after 25 yr:

$A = 6376.28e^{0.045(25)} \approx \$19,640.33$

33. Because $A = Pe^{rt}$, $A = 1,000,000$, $r = 0.075$, and $t = 40$, you have

$$1,000,000 = Pe^{0.075(40)}$$

$$\frac{1,000,000}{e^3} = P$$

$$\$49,787.07 \approx P.$$

35. (a) $$P(1 + r_{eff})^t = P\left(1 + \frac{r}{n}\right)^{nt}$$

$$\sqrt[t]{(1 + r_{eff})^t} = \sqrt[t]{\left(1 + \frac{r}{n}\right)^{nt}}$$

$$1 + r_{eff} = \left(1 + \frac{r}{n}\right)^n$$

$$r_{eff} = \left(1 + \frac{r}{n}\right)^n - 1$$

(b) $$P(1 + r_{eff})^t = Pe^{rt}$$

$$(1 + r_{eff})^t = (e^r)^t$$

$$1 + r_{eff} = e^r$$

$$r_{eff} = e^r - 1$$

37. $$2P = Pe^{rt}$$

$$2 = e^{rt}$$

$$t = \frac{\ln 2}{r} \approx \frac{0.70}{r}$$

If r is entered as a percentage and not a decimal, then

$$t \approx 100\left(\frac{0.70}{r}\right) = \frac{70}{r}.$$

39. (a) Let $V = mt + b$, then use $(0, 21{,}500)$ and $(2, 13{,}600)$ to find the model

$$m = \frac{13{,}600 - 21{,}500}{2 - 0} = -\frac{7900}{2} = -3950.$$

Since $b = 21{,}500$, the linear model is $V = -3950t + 21{,}500.$

(b) Let $V = Ce^{kt}$, then use $(0, 21{,}500)$ and $(2, 13{,}600)$ to find the model

$$21{,}500 = Ce^{k(0)} \Rightarrow C = 21{,}500$$

$$13{,}600 = 21{,}500e^{2k}$$

$$\frac{13{,}600}{21{,}500} = e^{2k}$$

$$\ln\left(\frac{13{,}600}{21{,}500}\right) = 2k$$

$$\frac{1}{2}\ln\left(\frac{13{,}600}{21{,}500}\right) = k$$

$$k \approx -0.229$$

The exponential model is $V = 21{,}500e^{-0.229t}$.

(c) Book value after 1 year:

Linear model: $V = -3950(1) + 21{,}500 = \$17{,}550$

Exponential model:
$V = 21{,}500e^{-0.229(1)} \approx \$17{,}099.56$

Book value after 4 years:

Linear model: $V = -3950(4) + 21{,}500 = \$5700$

Exponential model:
$V = 21{,}500e^{-0.229(4)} \approx \$8602.50$

(d) 22,000

0 5

1700

The exponential model depreciates slightly faster.

(e) After the second year, a buyer would gain an advantage by using the linear model, because it yields a lower value for the vehicle. A seller would want to use the exponential model, because it yields a higher value for the vehicle.

41. $S = Ce^{k/t}$

(a) Because $S = 5$ when $t = 1$, you have

$$5 = Ce^{k} \text{ and } \lim_{t\to\infty} Ce^{k/t} = C = 30.$$

So, $5 = 30e^{k}$, $k = \ln(1/6) \approx -1.7918$,

and $S = 30e^{-1.7918/5}$.

(b) $S(5) = 30e^{-1.7918/5} \approx 20.9646 \approx 20{,}965$ units

(c)

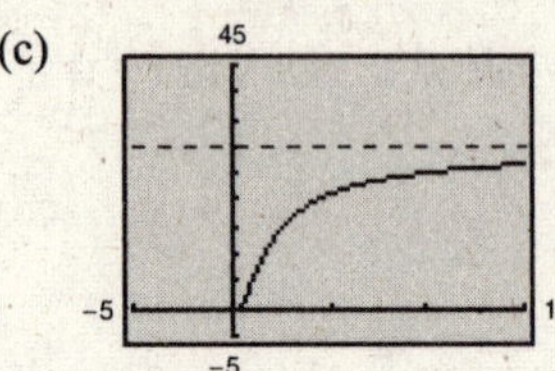

43. $N = 30(1 - e^{kt})$

Because $19 = 30(1 - e^{20k})$, you have

$$30e^{20k} = 11$$

$$e^{20k} = \frac{11}{30}$$

$$k = \frac{\ln(11/30)}{20} \approx -0.0502$$

$$N = 30(1 - e^{-0.0502t})$$

$$25 = 30(1 - e^{-0.0502t})$$

$$\frac{5}{6} = (1 - e^{-0.0502t})$$

$$e^{-0.0502t} = \frac{1}{6}$$

$$t = \frac{\ln(1/6)}{-0.0502} \approx 35.7 \text{ days.}$$

45. (a) Because $p = Ce^{kx}$ where $p = 45$ when $x = 1000$ and $p = 40$ when $x = 1200$, you have the following.

$$45 = Ce^{1000k} \text{ and } 40 = Ce^{1200k}$$
$$\ln 45 = \ln C + 1000k$$
$$\ln 40 = \ln C + 1200k$$
$$\ln 45 - \ln 40 = -200k$$
$$k = \frac{\ln(45/40)}{-200} \approx -0.0005889$$

So, you have $45 = Ce^{1000(-0.0005889)}$ which implies that $C \approx 81.0902$ and $p = 81.0902e^{-0.0005889x}$.

(b) Because $R = xp = 81.0902xe^{-0.0005889x}$, you have the following.

$$R' = 81.0902\left[-0.0005889xe^{-0.0005889x} + e^{-0.0005889x}\right]$$
$$= 81.0902e^{-0.0005889x}[1 - 0.0005889x]$$
$$0 = 81.0902e^{-0.0005889x}[1 - 0.0005889x]$$
$$0 = 1 - 0.0005889x$$
$$x = \frac{1}{0.0005889}$$

Because $R' = 0$ when $x = 1/0.0005889 \approx 1698$ units, you have $p = 81.0902e^{-0.0005889(1698)} \approx \$29.83$.

47. Answers will vary.

Review Exercises for Chapter 4

1. (a) $(4^5)(4^2) = 4^7 = 16{,}384$

(b) $(7^2)^3 = 7^6 = 117{,}649$

(c) $2^{-4} = \frac{1}{2^4} = \frac{1}{16} = 0.0625$

(d) $\frac{3^8}{3^4} = 3^{8-4} = 3^4 = 81$

3. $f(x) = 9^{x/2}$

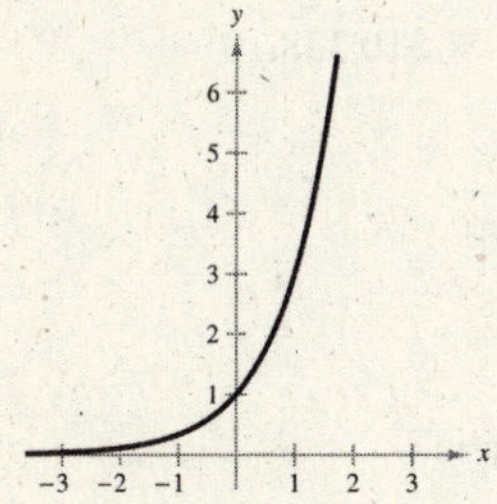

5. $f(t) = \left(\frac{1}{6}\right)^t$

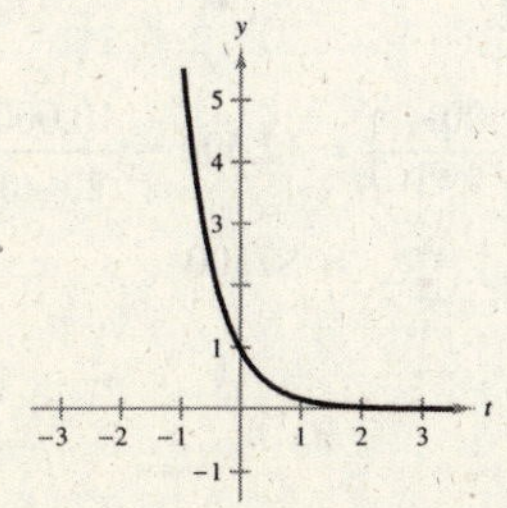

7. $f(x) = \left(\frac{1}{2}\right)^{2x} + 4$

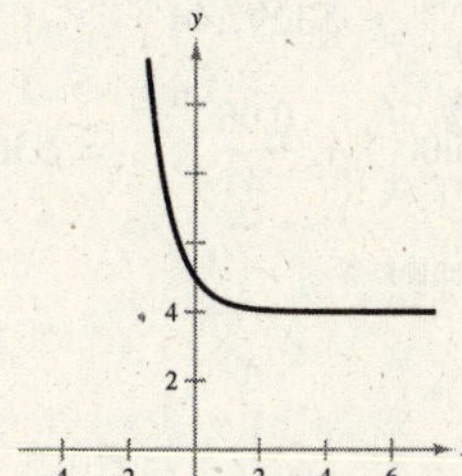

9. $P(t) = 5382(1.0051)^t$, $t = 0 \leftrightarrow 2000$

(a) 2016: $P(16) \approx 5894.39$ thousand

(b) 2025: $P(25) \approx 6203.76$ thousand

11. $V(t) = 55{,}000(2)^{t/12}$

(a) $V(4) = 55{,}000(2)^{4/12} \approx \$69{,}295.66$

(b) $V(25) = 55{,}000(2)^{25/12} \approx \$233{,}081.88$

13. (a) $(e^5)^2 = e^{10}$

(b) $\frac{e^3}{e^5} = \frac{1}{e^2}$ or e^{-2}

(c) $(e^4)(e^{3/2}) = e^{11/2}$

(d) $(e^2)^{-4} = e^{-8} = \frac{1}{e^8}$

15. $f(x) = e^{-x} + 1$

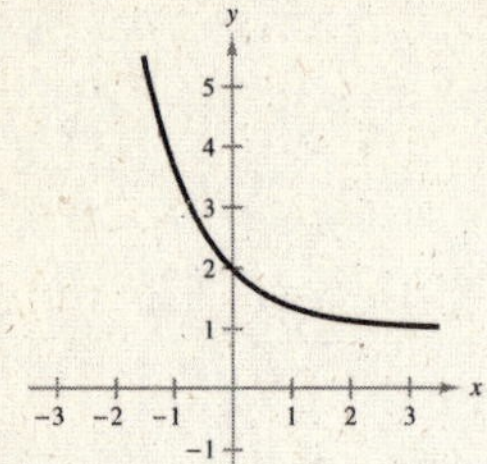

17. $f(x) = 1 - e^x$

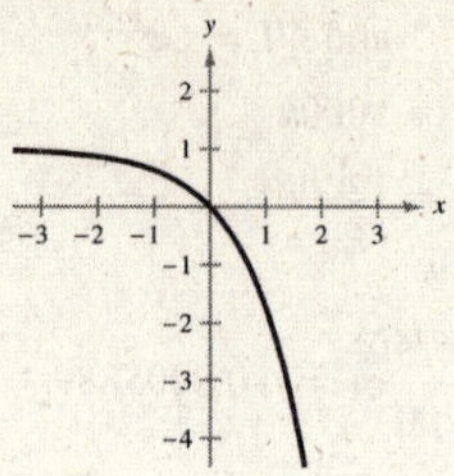

19. $A = P\left(1 + \frac{r}{n}\right)^{nt} = 1000\left(1 + \frac{0.04}{n}\right)^{5n}$

Continuous: $A = Pe^{rt} = 1000e^{(0.04)5}$

n	1	2	4	12	365	Continuous
A	\$1216.65	\$1218.99	\$1220.19	\$1221.00	\$1221.39	\$1221.40

21. $A = P\left(1 + \frac{r}{n}\right)^{nt} = 3000\left(1 + \frac{0.035}{n}\right)^{10n}$

Continuous: $A = Pe^{rt} = 3000e^{(0.035)(10)}$

n	1	2	4	12	365	Continuous
A	\$4231.80	\$4244.33	\$4250.73	\$4255.03	\$4257.13	\$4257.20

23. (a) $A = Pe^{rt} = 2000e^{0.05(10)} \approx \$3297.44$

(b) $A = P\left(1 + \frac{r}{n}\right)^{nt} = 2000\left(1 + \frac{0.06}{4}\right)^{4(10)} \approx \$3628.04$

Account (b) will be greater

25. $r_{eff} = \left(1 + \frac{r}{n}\right)^n - 1$

(a) $r_{eff} = \left(1 + \frac{0.06}{1}\right)^1 - 1 = 0.06 \to 6\%$

(b) $r_{eff} = \left(1 + \frac{0.06}{2}\right)^2 - 1 = 0.0609 \to 6.09\%$

(c) $r_{eff} = \left(1 + \frac{0.06}{4}\right)^4 - 1 \approx 0.0614 \to 6.14\%$

(d) $r_{eff} = \left(1 + \frac{0.06}{12}\right)^{12} - 1 \approx 0.0617 \to 6.17\%$

27. $A = P\left(1 + \frac{r}{4}\right)^{4t}$, $A = 12{,}000$, $r = 0.05$, $t = 3$

$$12{,}000 = P\left(1 + \frac{0.05}{4}\right)^{4(3)}$$

$$P = \frac{12{,}000}{\left(1 + \frac{0.05}{4}\right)^{4(3)}} \approx \$10{,}338.10$$

29. $p = 12{,}500 - \frac{10{,}000}{2 + e^{-0.001x}}$

(a) $p(1000) \approx \$8276.81$

(b) $p(2500) \approx \$7697.12$

(c) $\lim_{x\to\infty}\left(12{,}500 - \frac{10{,}000}{2 + e^{-0.001x}}\right) = 12{,}500 - \frac{10{,}000}{2 + 0}$
$= \$7500$

31. $P = 223.89e^{0.1979t}$, $t = 0 \leftrightarrow 2000$

(a) The model fits the data very well.

(b) $y = 116.85x + 111.1$

The linear model fits the data moderately well.

The exponential model is a better fit.

(c) Exponential: 2015 $P(15) \approx \$4357.50$ million

Linear: 2015 $P(15) \approx \$1863.85$ million

33. $P = \dfrac{10{,}000}{1 + 19e^{-t/5}}$, $t \ge 0$

(a) $P(4) \approx 1049$ fish

(b)

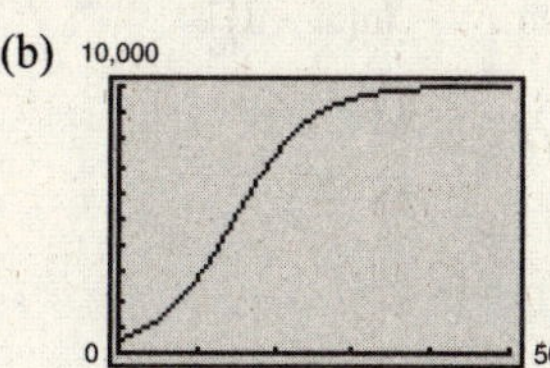

Let $P = 4000$ and solve for t.

$$4000 = \frac{10{,}000}{1 + 19e^{-t/5}}$$
$$1 + 19e^{-t/5} = \frac{5}{2}$$
$$19e^{-t/5} = \frac{3}{2}$$
$$e^{-t/5} = \frac{3}{38}$$
$$-\frac{t}{5} = \ln\left(\frac{3}{38}\right)$$
$$t = -5\ln\left(\frac{3}{38}\right) \approx 12.7 \text{ or } 13 \text{ mo}$$

(c) $\lim\limits_{t\to\infty}\left(\dfrac{10{,}000}{1 + 19e^{-t/5}}\right) = \dfrac{10{,}000}{1 + 0} = 10{,}000$ fish

Yes, P approaches 10,000 fish as t increases without bound.

35. $y = 4e^{x^2}$

$y' = 4e^{x^2}(2x) = 8xe^{x^2}$

37. $y = \dfrac{x}{e^{2x}}$

$$y' = \frac{e^{2x}(1) - x2e^{2x}}{\left(e^{2x}\right)^2} = \frac{1 - 2x}{e^{2x}}$$

39. $y = \dfrac{5}{1 + e^{2x}} = 5\left(1 + e^{2x}\right)^{-1}$

$$y' = -5\left(1 + e^{2x}\right)^{-2}\left(2e^{2x}\right) = -\frac{10e^{2x}}{\left(1 + e^{2x}\right)^2}$$

41. $y = e^{2-x}$

$y' = e^{2-x}(-1) = -e^{2-x}$

$m = y'(2) = -e^0 = -1$

$y - 1 = -(x - 2)$

$y = -x + 3 \Rightarrow$ Tangent line

43. $y = x^2e^{-x}$

$y' = x^2e^{-x}(-1) + e^{-x}(2x) = -x^2e^{-x} + 2xe^{-x}$

$m = y'(1) = -(1)^2e^{-1} + 2(1)e^{-1} = e^{-1}$

$y - e^{-1} = e^{-1}(x - 1)$

$y = e^{-1}x$ or $y = \dfrac{1}{e}x \Rightarrow$ Tangent line

45. $f(x) = x^3e^x$

$f'(x) = x^2e^x(3 + x)$

Critical number: $x = -3$

$f''(x) = xe^x\left(x^2 + 6x + 6\right)$

$f''(-3) = 9e^{-3} \approx 0.45 > 0$

Relative minimum: $(-3, -1.34)$

$f''(x) < 0$ on $\left(-\infty, -3 - \sqrt{3}\right)$

$f''(x) > 0$ on $\left(-3 - \sqrt{3}, -3 + \sqrt{3}\right)$

$f''(x) < 0$ on $\left(-3 + \sqrt{3}, 0\right)$

$f''(x) > 0$ on $(0, \infty)$

Inflection points: $(0, 0)$, $\left(-3 + \sqrt{3}, -0.57\right)$, and $\left(-3 - \sqrt{3}, -0.93\right)$

Horizontal asymptote: $y = 0$

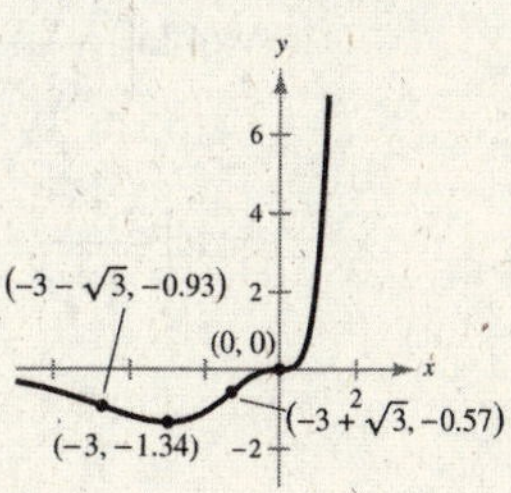

47. $f(x) = \dfrac{1}{xe^x}$

$f'(x) = \dfrac{-x-1}{x^2e^x}$

Critical number: $x = -1$

$f''(x) = \dfrac{x^2+2x+2}{x^3e^x}$

$f''(-1) = -\dfrac{1}{e^{-1}} < 0$

Relative maximum: $(-1, -2.72)$

No inflection points

Horizontal asymptote: $y = 0$

Vertical asymptote: $x = 0$

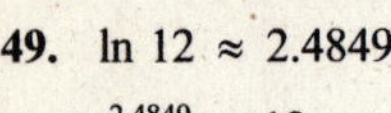

49. $\ln 12 \approx 2.4849$

$e^{2.4849} \approx 12$

51. $e^{1.5} \approx 4.4817$

$\ln 4.4817 \approx 1.5$

53. $y = \ln(4-x)$

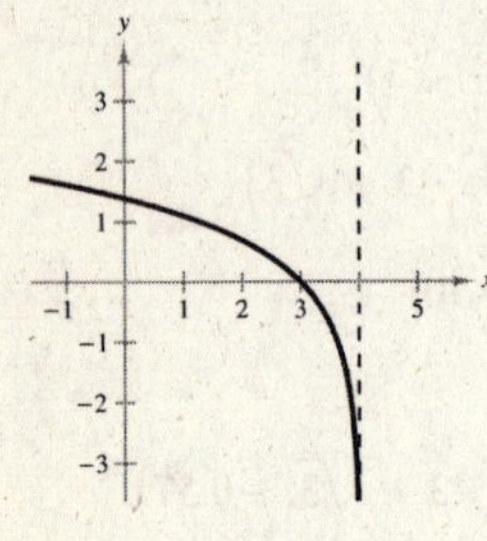

55. $y = \ln\dfrac{x}{3} = \ln x - \ln 3$

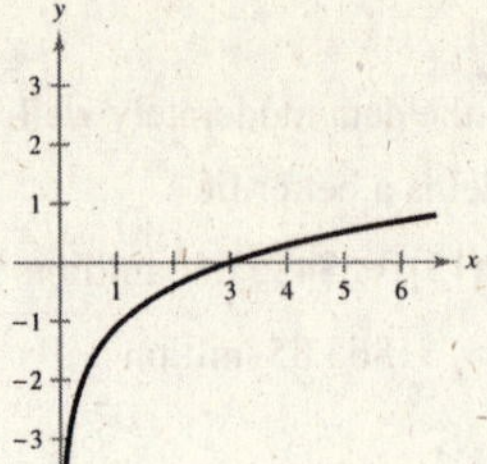

57. $\ln\sqrt{x^2(x-1)} = \frac{1}{2}\ln\left[x^2(x-1)\right]$

$= \frac{1}{2}\left[\ln x^2 + \ln(x-1)\right]$

$= \ln x + \frac{1}{2}\ln(x-1)$

59. $\ln\dfrac{x^2}{(x+1)^3} = \ln x^2 - \ln(x+1)^3$

$= 2\ln x - 3\ln(x+1)$

61. $\ln\left(\dfrac{1-x}{3x}\right)^3 = 3\ln\left(\dfrac{1-x}{3x}\right)$

$= 3\left[\ln(1-x) - \ln 3x\right]$

$= 3\left[\ln(1-x) - \ln 3 - \ln x\right]$

63. $\ln(2x+5) + \ln(x-3) = \ln\left[(2x+5)(x-3)\right]$

$= \ln\left(2x^2 - x - 15\right)$

65. $4\left[\ln\left(x^3-1\right) + 2\ln x - \ln(x-5)\right] = 4\left[\ln\left(x^2\left(x^3-1\right)\right) - \ln(x-5)\right]$

$= 4\ln\left(\dfrac{x^2\left(x^3-1\right)}{(x-5)}\right)$

$= \ln\left(\dfrac{x^2\left(x^3-1\right)}{(x-5)}\right)^4$

67. $e^{\ln x} = 3$

$x = 3$

69. $\ln x = 3$

$x = e^3 \approx 20.09$

71. $\ln 2x - \ln(3x-1) = 0$

$\ln 2x = \ln(3x-1)$

$2x = 3x - 1$

$x = 1$

73. $\ln x + \ln(x-3) = 0$

$\ln\left[x(x-3)\right] = 0$

$x(x-3) = 1$

$x^2 - 3x - 1 = 0$

$x = \dfrac{3 \pm \sqrt{13}}{2}$

$x = \dfrac{3+\sqrt{13}}{2} \approx 3.30$ is the only solution in the domain.

75. $e^{-1.386x} = 0.25$

$$-1.386x = \ln 0.25$$

$$x = \frac{\ln 0.25}{-1.386} \approx 1.00$$

77. $e^{2x-1} - 6 = 0$

$$e^{2x-1} = 6$$

$$2x - 1 = \ln 6$$

$$x = \frac{1 + \ln 6}{2} \approx 1.40$$

79. $100(1.21)^x = 110$

$$1.21^x = \frac{110}{100} = 1.1$$

$$x \ln 1.21 = \ln 1.1$$

$$x = \frac{\ln 1.1}{\ln 1.21} = 0.5$$

81. $A = P\left(1 + \frac{r}{n}\right)^{nt}$

If $P = 400$ and $r = 0.025$, find t when $A = 800$ for given n.

(a) Annually, $n = 1$

$$800 = 400\left(1 + \frac{0.025}{1}\right)^{(1)t}$$

$$2 = 1.025^t$$

$$\ln 2 = \ln 1.025^t$$

$$\frac{\ln 2}{\ln 1.025} = t$$

$$t \approx 28.07 \text{ yr}$$

(b) Monthly, $n = 12$

$$800 = 400\left(1 + \frac{0.025}{12}\right)^{12t}$$

$$2 = \left(1 + \frac{0.025}{12}\right)^{12t}$$

$$\ln 2 = \ln\left(1 + \frac{0.025}{12}\right)^{12t}$$

$$\frac{\ln 2}{\ln\left(1 + \frac{0.025}{12}\right)} = 12t$$

$$\frac{1}{12}\left(\frac{\ln 2}{\ln\left(1 + \frac{0.025}{12}\right)}\right) = t$$

$$t \approx 27.75 \text{ yr}$$

(c) Daily, $n = 365$

$$800 = 400\left(1 + \frac{0.025}{365}\right)^{365t}$$

$$2 = \left(1 + \frac{0.025}{365}\right)^{365t}$$

$$\frac{\ln 2}{\ln\left(1 + \frac{0.025}{365}\right)} = 365t$$

$$\frac{1}{365}\left(\frac{\ln 2}{\ln\left(1 + \frac{0.025}{365}\right)}\right) = t$$

$$t \approx 27.73 \text{ yr}$$

(d) Continuously, $A = Pe^{rt}$

$$800 = 400e^{0.025t}$$

$$2 = e^{0.025t}$$

$$\ln 2 = 0.025t$$

$$\frac{\ln 2}{0.025} = t$$

$$t \approx 27.73 \text{ yr}$$

83. $S = 75 - 6\ln(t+1),\ 0 \le t \le 12$

(a) Original score: $S(0) = 75$

(b) After 4 mo: $S(4) \approx 65.34$

(c) Let $S = 60$ and solve for t.

$$\begin{aligned}60 &= 75 - 6\ln(t+1)\\ -15 &= -6\ln(t+1)\\ \frac{5}{2} &= \ln(t+1)\\ e^{5/2} &= t+1\\ e^{5/2} - 1 &= t\\ t &\approx 11.2 \to 11 \text{ mo}\end{aligned}$$

85. $f(x) = \ln 3x^2 = \ln 3 + 2\ln x$

$$f'(x) = \frac{2}{x}$$

87. $y = \ln\dfrac{x(x-1)}{x-2} = \ln x + \ln(x-1) - \ln(x-2)$

$$y' = \frac{1}{x} + \frac{1}{x-1} - \frac{1}{x-2}$$

89. $f(x) = \ln e^{2x+1} = 2x + 1$

$$f'(x) = 2$$

91. $y = \dfrac{\ln x}{x^3}$

$$y' = \frac{x^3(1/x) - 3x^2 \cdot \ln x}{x^6} = \frac{1 - 3\ln x}{x^4}$$

93. $y = \ln(x^2 - 2)^{2/3} = \dfrac{2}{3}\ln(x^2 - 2)$

$$y' = \frac{2}{3}\cdot\frac{2x}{x^2 - 2} = \frac{4x}{3(x^2 - 2)}$$

95. $f(x) = \ln\left(x^2\sqrt{x+1}\right) = 2\ln x + \dfrac{1}{2}\ln(x+1)$

$$f'(x) = \frac{2}{x} + \frac{1}{2(x+1)}$$

97. $y = \ln\dfrac{e^x}{1+e^x} = \ln e^x - \ln(1+e^x) = x - \ln(1+e^x)$

$$y' = 1 - \frac{e^x}{1+e^x} = \frac{1}{1+e^x}$$

99. $\log_6 36 = \log_6 6^2 = 2\log_6 6 = 2$

101. $\log_{10} 1 = 0$

103. $\log_5 13 = \dfrac{\ln 13}{\ln 5} \approx 1.594$

105. $\log_{16} 64 = \dfrac{\ln 64}{\ln 16} = 1.5$

107. $y = 5^{2x+1}$

$$\begin{aligned}y' &= (\ln 5)\,5^{2x+1}(2)\\ &= (2\ln 5)5^{2x+1}\end{aligned}$$

109. $y = \log_3(2x-1)$

$$y' = \frac{1}{\ln 3}\cdot\frac{2}{2x-1} = \frac{2}{(2x-1)\ln 3}$$

111. $y = \log_{10}\dfrac{3}{x} = \log_{10} 3 - \log_{10} x$

$$y' = \frac{-1}{\ln 10}\cdot\frac{1}{x} = -\frac{1}{x\ln 10}$$

113. $y = \ln(x+3)$

$$y' = \frac{1}{x+3}$$

$$y'' = -\frac{1}{(x+3)^2}$$

No relative extrema

No inflection points

115. $y = \ln\left(\dfrac{10}{x+2}\right) = \ln 10 - \ln(x+2)$

$$y' = -\frac{1}{x+2}$$

$$y'' = \frac{1}{(x+2)^2}$$

No relative extrema

No inflection points

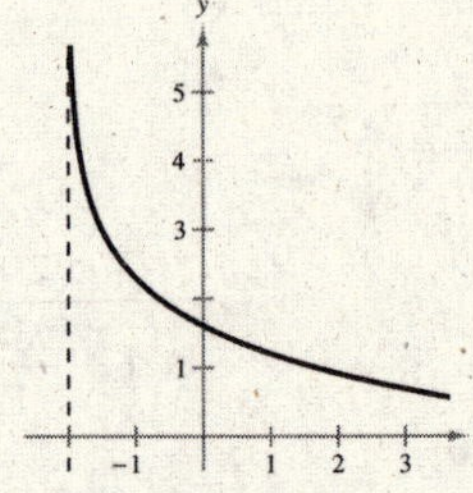

117. $D = -1671.88 + 1282\ln t,\ t = 4 \leftrightarrow 2004$

$$D' = 1282\left(\frac{1}{t}\right) = \frac{1282}{t}$$

2005: $D'(5) = \dfrac{1282}{5} = 256.4$ million/yr

2008: $D'(8) = \dfrac{1282}{8} = 160.25$ million/yr

119. $y = Ce^{kt}$

Since $y = 3$ when $x = 0 \Rightarrow 3 = Ce^{(0)k} \Rightarrow 3 = C.$

Since $y = 1$ when

$x = 4$: $1 = 3e^{4k}$

$\frac{1}{3} = e^{4k}$

$\ln\left(\frac{1}{3}\right) = 4k$

$\frac{1}{4}\ln\left(\frac{1}{3}\right) = k$

So, $y = 3e^{1/4\ \ln(1/3)t}$ or $y = 3e^{-0.27465t}$.

	Isotope	Half-life (in years)	Initial quantity	Amount after 1000 years	Amount after 10,000 years
121.	^{226}Ra	1599	8 g	5.19 g	0.10 g
123.	^{14}C	5715	20.18 g	17.88 g	6 g
125.	^{239}Pa	24,100	2.47 g	2.4 g	1.85 g

	Initial investment	Annual rate	Time to double	Amount after 10 years	Amount after 25 years
127.	\$600	8%	8.66 yr	\$1335.32	\$4433.43
129.	\$15,000	2.0%	34.66 yr	\$18,321.04	\$24,730.82

131. $D = Ce^{kt}$

(a) When $t = 0$, $D = 500$. So, $C = 500$.

$D = 500e^{kt}$

When $t = 6$, $D = 50$.

$50 = 500e^{k(6)}$

$\frac{1}{10} = e^{6k}$

$\ln\left(\frac{1}{10}\right) = 6k$

$\frac{1}{6}\ln\left(\frac{1}{10}\right) = k$

$k \approx -0.38376$

Therefore, $D = 500e^{-0.38376t}$.

(b) Let $t = 4$ and find D.

$D = 500e^{-0.38376(4)} \approx 107.72$ mg/mL

Chapter 4 Test Yourself

1. $3^2(3^{-2}) = \frac{3^2}{3^2} = 3^0 = 1$

2. $\left(\frac{2^3}{2^{-5}}\right)^{-1} = \frac{2^{-3}}{2^5} = \frac{1}{2^3(2^5)} = \frac{1}{2^8} = \frac{1}{256}$

3. $(e^{1/2})(e^4) = e^{9/2}$

4. $(e^3)^4 = e^{12}$

5. $f(x) = 5^{x-2}$

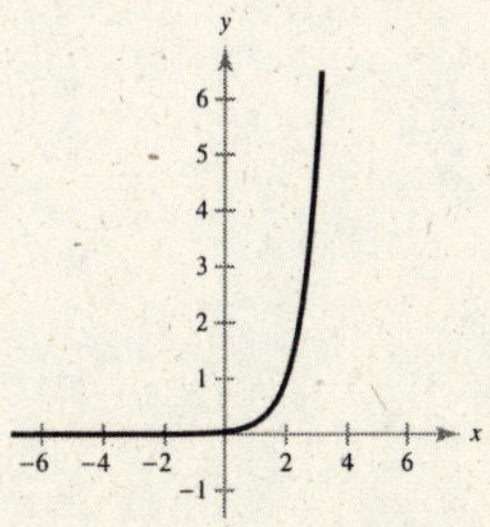

6. $f(x) = 4^{-x}$

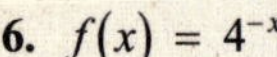

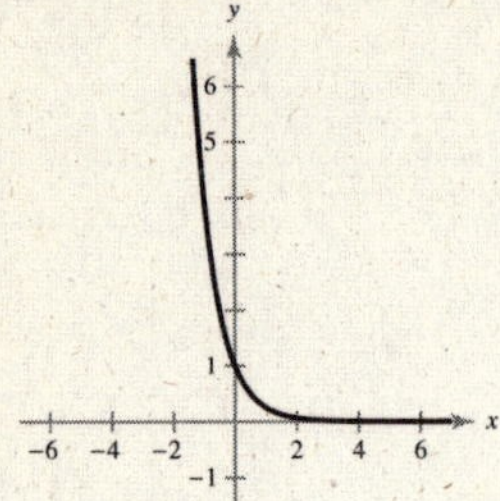

7. $f(x) = e^{x-3}$

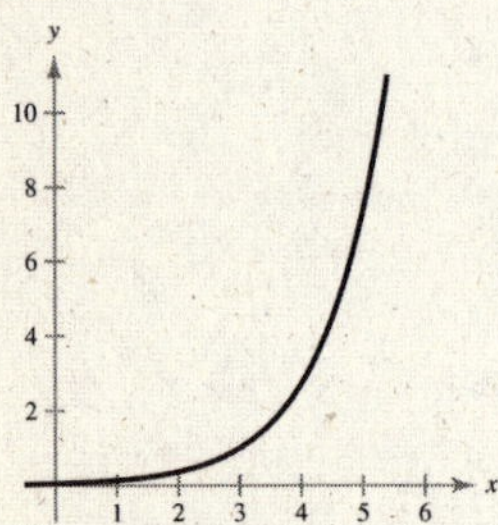

8. $f(x) = 8 + \ln x^2$

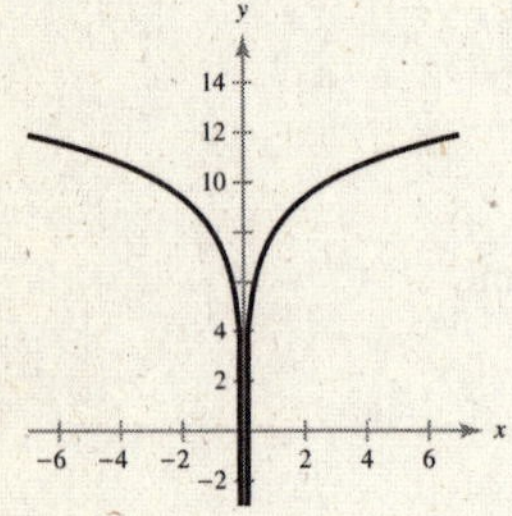

9. $f(x) = \ln(x - 5)$

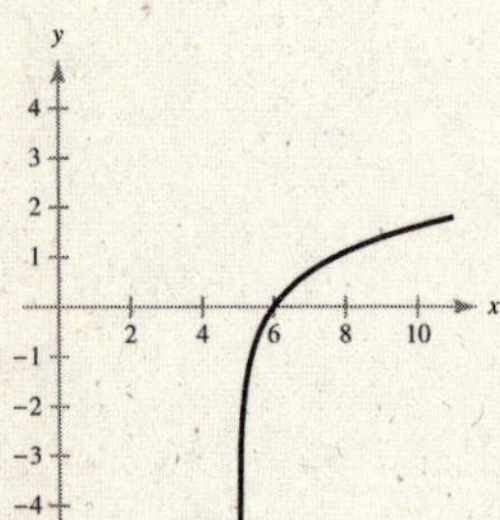

10. $f(x) = 0.5 \ln x$

11. $\ln \frac{3}{2} = \ln 3 - \ln 2$

12. $\ln \sqrt{x + y} = \ln(x + y)^{1/2}$
$= \frac{1}{2} \ln(x + y)$

13. $\ln \dfrac{x + 1}{y} = \ln(x + 1) - \ln y$

14. $\ln y + \ln(x + 1) = \ln[y(x + 1)]$

15. $3 \ln x - 2 \ln(x - 1) = \ln x^3 - \ln(x - 1)^2$
$= \ln \dfrac{x^3}{(x - 1)^2}$

16. $\ln x + 4 \ln y - \frac{1}{2}\ln(z + 4) = \ln x + \ln y^4 - \ln\sqrt{z + 4}$
$= \ln \dfrac{xy^4}{\sqrt{z + 4}}$

17. $e^{x-1} = 9$
$x - 1 = \ln 9$
$x = 1 + \ln 9 \approx 3.20$

18. $10e^{2x+1} = 900$
$e^{2x+1} = 90$
$2x + 1 = \ln 90$
$2x = -1 + \ln 90$
$x = \dfrac{-1 + \ln 90}{2} \approx 1.75$

19. $50(1.06)^x = 1500$
$(1.06)^x = 30$
$x = \log_{1.06} 30$
$x = \dfrac{\ln 30}{\ln 1.06} \approx 58.37$

20. $A = P\left(1 + \frac{r}{n}\right)^{nt}$ when $P = 500$ and $r = 0.04$

(a) $2P = P\left(1 + \frac{0.04}{1}\right)^{(1)t}$

$2 = (1.04)^t$

$\frac{\ln 2}{\ln 1.04} = t$

$17.7 \text{ yr} \approx t$

(b) $2P = P\left(1 + \frac{0.04}{12}\right)^{12t}$

$2 = \left(1 + \frac{0.04}{12}\right)^{12t}$

$\frac{\ln 2}{12 \ln\left(1 + \frac{0.04}{12}\right)} = t$

$17.4 \text{ years} \approx t$

(c) $2P = P\left(1 + \frac{0.04}{365}\right)^{365t}$

$2 = \left(1 + \frac{0.04}{365}\right)^{365t}$

$\frac{\ln 2}{365 \ln\left(1 + \frac{0.04}{365}\right)} = t$

$17.3 \text{ yr} \approx t$

(d) $A = Pe^{rt}$

$2P = Pe^{0.04t}$

$2 = e^{0.04t}$

$\frac{\ln 2}{0.04} = t$

$17.3 \text{ yr} \approx t$

21. $y = e^{-3x} + 5$

$y' = -3e^{-3x}$

22. $y = 7e^{x+2} + 2x$

$y' = 7e^{x+2} + 2$

23. $y = \ln(3 + x^2)$

$y' = \frac{2x}{3 + x^2}$

24. $y = \ln \frac{5x}{x + 2} = \ln 5x - \ln(x + 2)$

$y' = \frac{5}{5x} - \frac{1}{x + 2}$

$= \frac{1}{x} - \frac{1}{x + 2}$

$= \frac{x + 2 - x}{x(x + 2)}$

$= \frac{2}{x(x + 2)}$

25. $R = 1548e^{0.0617t}$, $0 \le t \le 8$ and $t = 0$ is 2000.

(a) 2006: $R(6) \approx \$2241.54$ million

(b) $R' = 1548e^{0.0617t}(0.0617)$

$= 95.5116e^{0.0617t}$

$R'(6) \approx \$138.30$ million

26. $y = Ce^{[\ln(1/2)/1599]t}$

When $t = 1200$, $y = Ce^{[\ln(1/2)/1599]\cdot 1200} \approx 0.59C$. After 1200 years, approximately 59% of the radioactive radium will remain.

27. $y = Ce^{0.0175t}$

$2C = Ce^{0.0175t}$

$2 = e^{0.0175t}$

$\ln 2 = 0.0175t$

$39.6 \text{ yr} \approx t$

CHAPTER 5
Integration and Its Applications

CHAPTER 5
Integration and Its Applications

Section 5.1 Antiderivatives and Indefinite Integrals

Skills Warm Up

1. $\dfrac{\sqrt{x}}{x} = \dfrac{x^{1/2}}{x} = x^{-1/2}$

2. $\sqrt[3]{2x}(2x) = (2x)^{1/3}(2x) = (2x)^{4/3}$

3. $\sqrt{5x^3} + \sqrt{x^5} = (5x^3)^{1/2} + (x^5)^{1/2} = 5^{1/2}x^{3/2} + x^{5/2}$

4. $\dfrac{1}{\sqrt{x}} + \dfrac{1}{\sqrt[3]{x^2}} = \dfrac{1}{x^{1/2}} + \dfrac{1}{x^{2/3}}$
$= x^{-1/2} + x^{-2/3}$

5. $\dfrac{(x+1)^3}{\sqrt{x+1}} = \dfrac{(x+1)^3}{(x+1)^{1/2}} = (x+1)^{5/2}$

6. $\dfrac{\sqrt{x}}{\sqrt[3]{x}} = \dfrac{x^{1/2}}{x^{1/3}} = x^{1/6}$

7. $y = x^2 + 5x + C$
$2 = 2^2 + 5(2) + C$
$-12 = C$

8. $y = 3x^3 - 6x + C$
$2 = 3(2)^3 - 6(2) + C$
$-10 = C$

9. $y = -16x^2 + 26x + C$
$2 = -16(2)^2 + 26(2) + C$
$14 = C$

10. $y = -\frac{1}{4}x^4 - 2x^2 + C$
$2 = -\frac{1}{4}(2)^4 - 2(2)^2 + C$
$14 = C$

1. $\dfrac{d}{dx}\left[2x^2 + C\right] = 2(2x) = 4x$

3. $\dfrac{d}{dx}\left[\dfrac{3}{x^3} + C\right] = \dfrac{d}{dx}\left(3x^{-3} + C\right) = -9x^{-4} = -\dfrac{9}{x^4}$

5. $\dfrac{d}{dx}\left[x^4 + \dfrac{1}{x} + C\right] = 4x^3 - \dfrac{1}{x^2}$

7. $\int du = u + C$
$\dfrac{d}{du}[u + C] = 1$

9. $\int 6\,dx = 6x + C$
$\dfrac{d}{dx}[6x + C] = 6$

11. $\int 7x\,dx = \dfrac{7x^2}{2} + C$
$\dfrac{d}{dx}\left[\dfrac{7x^2}{2} + C\right] = 7x$

13. $\int 5t^2\,dt = \dfrac{5t^3}{3} + C$
$\dfrac{d}{dt}\left[\dfrac{5t^3}{3} + C\right] = 5t^2$

15. $\int 5x^{-3}\,dx = \dfrac{5x^{-2}}{-2} + C = -\dfrac{5}{2x^2} + C$
$\dfrac{d}{dx}\left[-\dfrac{5}{2}x^{-2} + C\right] = 5x^{-3}$

17. $\int y^{3/2}\,dy = \dfrac{y^{5/2}}{5/2} + C = \dfrac{2}{5}y^{5/2} + C$
$\dfrac{d}{dy}\left[\dfrac{2}{5}y^{5/2} + C\right] = y^{3/2}$

	Original Integral	*Rewrite*	*Integrate*	*Simplify*
19.	$\int \sqrt[3]{x^2}\,dx$	$\int x^{2/3}\,dx$	$\dfrac{x^{5/3}}{5/3} + C$	$\dfrac{3}{5}x^{5/3} + C$

	Original Integral	*Rewrite*	*Integrate*	*Simplify*
21.	$\int \frac{1}{x\sqrt{x}}\,dx$	$\int x^{-3/2}\,dx$	$\frac{x^{-1/2}}{-1/2} + C$	$-\frac{2}{\sqrt{x}} + C$
23.	$\int \frac{1}{2x^3}\,dx$	$\frac{1}{2}\int x^{-3}\,dx$	$\frac{1}{2}\left(\frac{x^{-2}}{-2}\right) + C$	$-\frac{1}{4x^2} + C$

25. $\int (x + 3)\,dx = \frac{x^2}{2} + 3x + C$

$\frac{d}{dx}\left[\frac{x^2}{2} + 3x + C\right] = x + 3$

27. $\int (x^3 + 2)\,dx = \frac{x^4}{4} + 2x + C$

$\frac{d}{dx}\left[\frac{x^4}{4} + 2x + C\right] = x^3 + 2$

29. $\int (3x^3 - 6x^2 + 2)\,dx = \frac{3}{4}x^4 - \frac{6}{3}x^3 + 2x + C = \frac{3}{4}x^4 - 2x^3 + 2x + C$

$\frac{d}{dx}\left[\frac{3}{4}x^4 - 2x^3 + 2x + C\right] = x^3 - 6x^2 + 2$

31. $\int (x^2 + 5x + 1)\,dx = \frac{1}{3}x^3 + \frac{5}{2}x^2 + x + C$

$\frac{d}{dx}\left[\frac{1}{3}x^3 + \frac{5}{2}x^2 + x + C\right] = x^2 + 5x + 1$

33. $\int \frac{2x^3 - 1}{x^3}\,dx = \int (2 - x^{-3})\,dx$

$= 2x - \frac{x^{-2}}{-2} + C$

$= 2x + \frac{1}{2x^2} + C$

$\frac{d}{dx}\left[2x + \frac{1}{2x^2} + C\right] = 2 - x^{-3}$

$= 2 - \frac{1}{x^3}$

$= \frac{2x^3 - 1}{x^3}$

35. $\int \left(\frac{5x + 4}{\sqrt[3]{x}}\right) dx = \int x^{-1/3}(5x + 4)\,dx$

$= \int (5x^{2/3} + 4x^{-1/3})\,dx$

$= \frac{5x^{5/3}}{5/3} + \frac{4x^{2/3}}{2/3} + C$

$= 3x^{5/3} + 6x^{2/3} + C$

$= 3x^{2/3}(x + 2) + C$

$\frac{d}{dx}\left[3x^{5/3} + 6x^{2/3} + C\right] = 5x^{2/3} + 4x^{-1/3}$

$= 5x^{2/3} + \frac{4}{x^{1/3}}$

37. If $f'(x) = 2$, then $f(x) = 2x + C$.

Sample answer: $f(x) = 2x$ or $f(x) = 2x + 1$

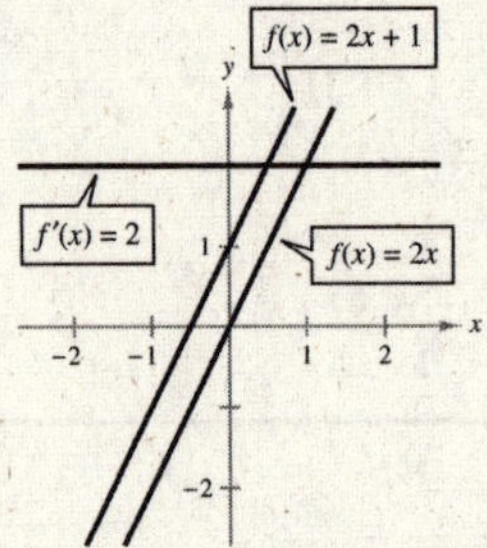

39. If $f'(x) = x$, then $f(x) = \frac{x^2}{2} + C$.

Sample answer: $f(x) = \frac{x^2}{2}$ or $f(x) = \frac{x^2}{2} + 2$

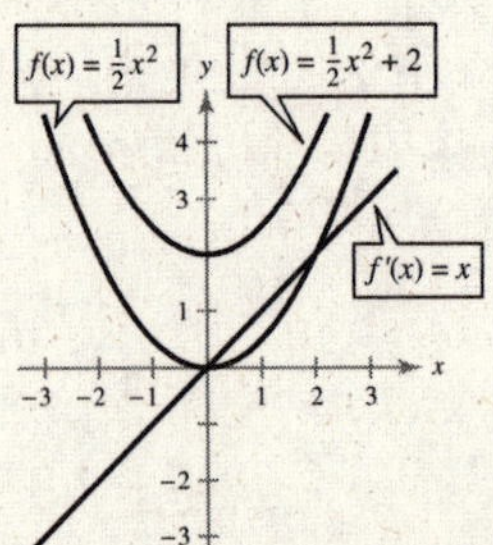

41. $f(x) = \int 4x\,dx = \frac{4}{2}x^2 + C = 2x^2 + C$

$f(0) = 6$

$2(0)^2 + C = 6$

$C = 6$

$f(x) = 2x^2 + 6$

43. $f(x) = \int (2x + 4)\, dx$

$= \dfrac{2x^2}{2} + 4x + C$

$= x^2 + 4x + C$

$f(-2) = 3$

$(-2)^2 + 4(-2) + C = 3$

$4 - 8 + C = 3$

$C = 7$

$f(x) = x^2 + 4x + 7$

45. $f(x) = \int (10x - 12x^3)\, dx$

$= \dfrac{10x^2}{2} - \dfrac{12x^4}{4} + C$

$= 5x^2 - 3x^4 + C$

$f(3) = 2$

$5(3)^2 - 3(3)^4 + C = 2$

$45 - 243 + C = 2$

$C = 200$

$f(x) = 5x^2 - 3x^4 + 200$

47. $f(x) = \int \dfrac{2 - x}{x^3}\, dx$

$= \int (2x^{-3} - x^{-2})\, dx$

$= -x^{-2} + x^{-1} + C$

$= -\dfrac{1}{x^2} + \dfrac{1}{x} + C$

$f(2) = \dfrac{3}{4}$

$-\dfrac{1}{2^2} + \dfrac{1}{2} + C = \dfrac{3}{4}$

$C = \dfrac{1}{2}$

$f(x) = -\dfrac{1}{x^2} + \dfrac{1}{x} + \dfrac{1}{2}$

49. $f'(x) = \int 2\, dx = 2x + C_1$

Because $f'(2) = 2(2) + C_1 = 5$, you know that $C_1 = 1$. So, $f'(x) = 2x + 1$.

$f(x) = \int (2x + 1)\, dx = x^2 + x + C_2$

Because $f(2) = 4 + 2 + C_2 = 10$, you know that $C_2 = 4$. So, $f(x) = x^2 + x + 4$.

51. $f'(x) = \int x^{-2/3}\, dx = 3x^{1/3} + C_1$

Because $f'(8) = 3(8)^{1/3} + C_1 = 6$, you know that $C_1 = 0$. So, $f'(x) = 3x^{1/3}$.

$f(x) = \int 3x^{1/3}\, dx = \frac{9}{4}x^{4/3} + C_2$

Because $f(0) = 0 + C_2 = 0$, you know that $C_2 = 0$. So, $f(x) = \frac{9}{4}x^{4/3}$.

53. $C = \int 85\, dx = 85x + K$

When $x = 0$, $C = 0 + K = 5500$ and $K = 5500$.

So, $C = 85x + 5500$.

55. $C = \int \left(\dfrac{1}{20}x^{-1/2} + 4\right) dx = \dfrac{1}{10}x^{1/2} + 4x + K$

When $x = 0$, $C = 0 + 0 + K = 750$ and $K = 750$.

So, $C = \dfrac{\sqrt{x}}{10} + 4x + 750$.

57. $R = \int (225 - 3x)\, dx = 225x - \dfrac{3}{2}x^2 + C$

When $x = 0$, $R = 0 + 0 + C = 0$ and $C = 0$.

So, $R = 225x - \dfrac{3}{2}x^2$ and the demand function is

$p = \dfrac{R}{x} = 225 - \dfrac{3}{2}x.$

59. $P = \int (-18x + 1650)\, dx = -9x^2 + 1650x + C$

When $x = 15$, $P = -2025 + 24{,}750 + C = 22{,}725$ and $C = 0$. So, $P = -9x^2 + 1650x$.

61. $P = \int (-24x + 805)\, dx = -12x^2 + 805x + C$

When $x = 12$, $P = -1728 + 9660 + C = 8000$ and $C = 68$. So, $P = -12x^2 + 805x + 68$.

63. $a(t) = -32$

$v(t) = \int a(t)\, dt = -32t + C_1$

The rock has no initial velocity, so $v(0) = 0$.

$v(0) = 0 + C_1 = 0$ and $C_1 = 0$, so $v(t) = -32t$.

$s(t) = \int v(t)\, dt = -16t^2 + C_2$

The rock is dropped from a height of 6000 feet, so $s(0) = 6000$. $s(0) = 0 + C_2 = 6000$ and $C_2 = 6000$ so $s(t) = -16t^2 + 6000$. The rock will hit the ground when $s(t) = 0$:

$-16t^2 + 6000 = 0$

$t = \sqrt{375} \approx 19.36$ seconds

65. $v(t) = \int -32\,dt = -32t + C_1$

Letting v_0 be the initial velocity, $v(0) = v_0$ and $C_1 = v_0$, so $v(t) = -32t + v_0$.

$s(t) = \int(-32t + v_0)\,dt = -16t^2 + v_0t + C_2$

Because $s(0) = 0$, $C_2 = 0$. So, the position function is $s(t) = 16t^2 + v_0t$. At the highest point, the velocity is zero. This occurs when $v(t) = -32t + v_0 = 0$, and $t = v_0/32$ seconds. Substituting this value into the position function, $s\left(\frac{v_0}{32}\right) = -16\left(\frac{v_0}{32}\right)^2 + v_0\left(\frac{v_0}{32}\right) = 550$ which implies that $v_0^2 = 35{,}200$ and the initial velocity should be $v_0 = 40\sqrt{22} \approx 187.62$ feet per second.

67. (a) $C(x) = \int(2x - 12)\,dx = x^2 - 12x + C_1$

Because $C(0) = 0 + 0 + C_1 = 125$, you know that $C_1 = 125$. So, $C(x) = x^2 - 12x + 125$ and the average cost is $C/x = x - 12 + (125/x)$.

(b) $C(50) = 50^2 - 12(50) + 125 = \$2025$

(c) C_1 represents the fixed cost, so \$125 is fixed cost and \$1900 is variable cost. Answers will vary.

69. (a) $\frac{dP}{dt} = 158.80t + 1758.6$, $t = 0$ is 1970.

$P(t) = \int(158.80t + 1758.6)\,dt$

$P(t) = 79.4t^2 + 1758.6t + C$

Since the population was 263,868 in 2009, when $t = 39$, $P = 263{,}868$.

$$79.4(39)^2 + 1758.6(39) + C = 263{,}868$$
$$C = 74{,}515.2$$

So, $P(t) = 79.4t^2 + 1758.6t + 74{,}515.2$.

(b) 2015: Let $t = 45$ and find P.

$P \approx 314{,}437.2$

Yes, the prediction seems reasonable if the population continues to increase at the rate dP/dt.

71. (a) $\frac{dM}{dt} = -0.105t^2 + 14.02t + 217.8$, $t = 0$ is 1980.

$M = \int(-0.105t^2 + 14.02t + 217.8)\,dt$

$M = -0.035t^3 + 7.01t^2 + 217.8t + C$

Since the number of married couples was 60,844 thousand in 2009, $M = 60{,}844$ when $t = 29$.

$$-0.035(29)^3 + 7.01(29)^2 + 217.8(29) + C = 60{,}844$$
$$C = 49{,}486.005$$

So, $M(t) = -0.035t^3 + 7.01t^2 + 217.8t + 49{,}486.005$.

(b) 2015: Let $t = 35$ and find M.

$M = 64{,}195.63$ thousand

Yes, the number seems reasonable if the number of married couples continues to increase at the rate of dM/dt.

73. (a)

$$\frac{dB}{dx} = -19.9x + 351$$

$$\frac{dC}{dx} = 5.38x^2 - 40.6x + 182$$

(b) $B(x) = \int(-19.9x + 351)\,dx$

$B(x) = -9.95x^2 + 351x + C_1$

$B(0) = 0 \rightarrow C_1 = 0$

$B(x) = -9.95x + 351x$

$C(x) = \int(5.38x^2 - 40.6x + 182)\,dx$

$C(x) = 1.79x^3 - 20.3x^2 + 182x + C_2$

$C(0) = 425 \rightarrow C_2 = 425$

$C(x) = 1.79x^3 - 20.3x^2 + 182x + 425$

(c) Graphing the benefit and cost equations together, you see that benefit exceeds cost on the interval $(2.32, 12.00)$.

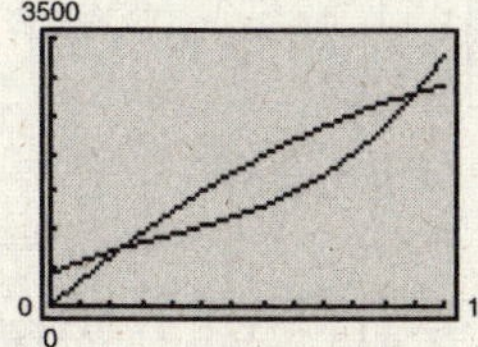

The company should produce from 3 to 11 units.

Section 5.2 Integration by Substitution and the General Power Rule

Skills Warm Up

1. $\int(2x^3 + 1)\,dx = \frac{1}{2}x^4 + x + C$

2. $\int(x^{1/2} + 3x - 4)\,dx = \frac{2}{3}x^{3/2} + \frac{3}{2}x^2 - 4x + C$

3. $\int\frac{1}{x^2}\,dx = \int x^{-2}\,dx = -\frac{1}{x} + C$

4. $\int\frac{1}{3t^3}\,dt = \frac{1}{3}\int t^{-3}\,dt = -\frac{1}{6t^2} + C$

5. $\int(1 + 2t)t^{3/2}\,dt = \int(t^{3/2} + 2t^{5/2})\,dt$

$= \frac{2}{5}t^{5/2} + \frac{4}{7}t^{7/2} + C$

6. $\int\sqrt{x}(2x - 1)\,dx = \int(2x^{3/2} - x^{1/2})\,dx$

$= \frac{4}{5}x^{5/2} - \frac{2}{3}x^{3/2} + C$

7. $\int\frac{5x^3 + 2}{x^2}\,dx = \int(5x + 2x^{-2})\,dx$

$= \frac{5}{2}x^2 - \frac{2}{x} + C$

$= \frac{5x^3 - 4}{2x} + C$

8. $\int\frac{2x^2 - 5}{x^4}\,dx = \int(2x^{-2} - 5x^{-4})\,dx$

$= -\frac{2}{x} + \frac{5}{3x^3} + C$

$= \frac{-6x^2 + 5}{3x^3} + C$

9. $\int\left(\frac{8x^2 + 3}{\sqrt{x}}\right)dx = \int x^{-1/2}(8x^2 + 3)\,dx$

$= \int(8x^{3/2} + 3x^{-1/2})\,dx$

$= \frac{8x^{5/2}}{5/2} + \frac{3x^{1/2}}{1/2} + C$

$= \frac{16}{5}x^{5/2} + 6x^{1/2} + C$

$\int u^n \frac{du}{dx}\,dx$	u	$\frac{du}{dx}$
1. $\int (5x^2+1)^2(10x)\,dx$	$5x^2+1$	$10x$
3. $\int \sqrt{1-x^2}(-2x)\,dx$	$1-x^2$	$-2x$
5. $\int \left(4+\frac{1}{x^2}\right)^5\left(\frac{-2}{x^3}\right)dx$	$4+\frac{1}{x^2}$	$-\frac{2}{x^3}$
7. $\int (1+\sqrt{x})^3 \frac{1}{2\sqrt{x}}\,dx$	$1+\sqrt{x}$	$\frac{1}{2\sqrt{x}}$

9. $$\int (x-1)^4\,dx = \int (x-1)^4(1)\,dx = \frac{(x-1)^5}{5} + C$$

$$\frac{d}{dx}\left[\frac{(x-1)^5}{5} + C\right] = (x-1)^4$$

11. $$\int (1+2x)^4(2)\,dx = \frac{(1+2x)^5}{5} + C$$

$$\frac{d}{dx}\left[\frac{(1+2x)^5}{5} + C\right] = (1+2x)^4(2)$$

13. $$\int (x^2+3x)(2x+3)\,dx = \frac{(x^2+3x)^2}{2} + C$$

$$\frac{d}{dx}\left[\frac{(x^2+3x)^2}{2} + C\right] = (x^2+3x)(2x+3)$$

15. $$\int \sqrt{4x^2-5}(8x)\,dx = \frac{(4x^2-5)^{3/2}}{3/2} + C$$

$$\frac{d}{dx}\left[\frac{(4x^2-5)^{3/2}}{3/2} + C\right] = (4x^2-5)^{1/2}(8x) = \sqrt{4x^2-5}(8x)$$

17. $$\int \frac{6x}{(3x^2-5)^4}\,dx = \int (3x^2-5)^{-4}(6x)\,dx = \frac{(3x^2-5)^{-3}}{-3} + C$$

$$\frac{d}{dx}\left[\frac{(3x^2-5)^{-3}}{-3} + C\right] = (3x^2-5)^{-4}(6x) = \frac{6x}{(3x^2-5)^4}$$

19. $$\int x^2(2x^3-1)^4\,dx = \frac{1}{6}\int (2x^3-1)^4(6x^2)\,dx = \frac{1}{6}\left[\frac{(2x^3-1)^5}{5}\right] + C = \frac{1}{30}(2x^3-1)^5 + C$$

$$\frac{d}{dx}\left[\frac{1}{30}(2x^3-1)^5 + C\right] = \frac{1}{30}\left(5(2x^3-1)^4(6x^2)\right) = x^2(2x^3-1)^4$$

21. $$\int t(t^2+6)^{1/2}\,dt = \frac{1}{2}\int (t^2+6)^{1/2}(2t)\,dt = \frac{1}{2}\left[\frac{(t^2+6)^{3/2}}{3/2}\right] + C = \frac{1}{3}(t^2+6)^{3/2} + C$$

$$\frac{d}{dt}\left[\frac{1}{3}(t^2+6)^{3/2} + C\right] = \frac{1}{3}\left(\frac{3}{2}(t^2+6)^{1/2}(2t)\right) = t(t^2+6)^{1/2}$$

23. $$\int x^5(4-x^6)^{-3}\,dx = -\frac{1}{6}\int (4-x^6)^{-3}(-6x^5)\,dx = -\frac{1}{6}\left[\frac{(4-x^6)^{-2}}{-2}\right] + C = \frac{1}{12(4-x^6)^2} + C$$

$$\frac{d}{dx}\left[\frac{1}{12}(4-x^6)^{-2} + C\right] = \frac{1}{12}\left(-2(4-x^6)^{-3}(-6x^5)\right) = x^5(4-x^6)^{-3} = \frac{x^5}{(4-x^6)^3}$$

25. $$\int (x^2-6x)^4(x-3)\,dx = \frac{1}{2}\int (x^2-6x)^4(2(x-3))\,dx = \frac{1}{2}\left[\frac{(x^2-6x)^5}{5}\right] + C = \frac{1}{10}(x^2-6x)^5 + C$$

$$\frac{d}{dx}\left[\frac{1}{10}(x^2-6x)^5 + C\right] = \frac{1}{10}\left(5(x^2-6x)^4(2x-6)\right) = \frac{1}{10}\left(5(x^2-6x)^4(2(x-3))\right) = (x^2-6x)^4(x-3)$$

27. $\displaystyle\int \frac{x+1}{\left(x^2+2x-3\right)^2}\,dx = \frac{1}{2}\int\left(x^2+2x-3\right)^{-2}(2x+2)\,dx = \frac{1}{2}\frac{\left(x^2+2x-3\right)^{-1}}{-1} + C = -\frac{1}{2\left(x^2+2x-3\right)} + C$

$$\frac{d}{dx}\left[-\frac{1}{2\left(x^2+2x-3\right)} + C\right] = \frac{1}{2}\left(x^2+2x-3\right)^{-2}(2x+2) = \frac{x+1}{\left(x^2+2x-3\right)^2}$$

29. $\displaystyle\int 5x\left(1-x^2\right)^{1/3}\,dx = -\frac{5}{2}\int\left(1-x^2\right)^{1/3}(-2x)\,dx$

$$= -\frac{5}{2}\left[\frac{\left(1-x^2\right)^{4/3}}{4/3}\right] + C$$

$$= -\frac{15}{8}\left(1-x^2\right)^{4/3} + C$$

$$\frac{d}{dx}\left[-\frac{15}{8}\left(1-x^2\right)^{4/3} + C\right] = -\frac{15}{8}\left(\frac{4}{3}\left(1-x^2\right)^{1/3}(-2x)\right)$$

$$= 5x\left(1-x^2\right)^{1/3}$$

31. $\displaystyle\int \frac{6x}{\left(1+x^2\right)^3}\,dx = 3\int\left(1+x^2\right)^{-3}(2x)\,dx$

$$= 3\left[\frac{\left(1+x^2\right)^{-2}}{-2}\right] + C$$

$$= -\frac{3}{2\left(1+x^2\right)^2} + C$$

$$\frac{d}{dx}\left[-\frac{3}{2\left(1+x^2\right)^2} + C\right] = 3\left(1+x^2\right)^{-3}(2x)$$

$$= \frac{6x}{\left(1+x^2\right)^3}$$

33. $\displaystyle\int \frac{-3}{\sqrt{2t+3}}\,dt = -\frac{3}{2}\int(2t+3)^{-1/2}(2)\,dt$

$$= -\frac{3}{2}\left[\frac{(2t+3)^{1/2}}{1/2}\right] + C$$

$$= -3\sqrt{2t+3} + C$$

$$\frac{d}{dt}\left[-3\sqrt{2t+3} + C\right] = -\frac{3}{2}(2t+3)^{-1/2}(2)$$

$$= \frac{-3}{\sqrt{2t+3}}$$

35. Let $u = 6x^2 - 1$, then $du = 12x\,dx$.

$$\int 12x\left(6x^2-1\right)^3\,dx = \int\left(6x^2-1\right)^3(12x\,dx)$$

$$= \int u^3\,du$$

$$= \frac{u^4}{4} + C$$

$$= \frac{1}{4}\left(6x^2-1\right)^4 + C$$

37. Let $u = 4x + 3$, then $du = 4\,dx$, which implies that $\frac{1}{4}\,du = dx$.

$$\int(4x+3)^{1/3}\,dx \to \frac{1}{4}\int u^{1/3}\,du = \frac{1}{4}\left[\frac{u^{4/3}}{4/3}\right] + C$$

$$= \frac{3}{16}u^{4/3} + C$$

$$= \frac{3}{16}(4x+3)^{4/3} + C$$

39. Let $u = x^2 + 25$, then $du = 2x\,dx$, which implies that $x\,dx = \frac{1}{2}\,du$.

$$\int\frac{x}{\sqrt{x^2+25}}\,dx = \int\left(x^2+25\right)^{-1/2}(x)\,dx$$

$$= \int u^{-1/2}\left(\frac{1}{2}\right)du$$

$$= \frac{1}{2}\left[\frac{u^{1/2}}{1/2}\right] + C$$

$$= \sqrt{u} + C$$

$$= \sqrt{x^2+25} + C$$

41. Let $u = x^3 + 3x + 4$, then

$$du = \left(3x^2+3\right)dx = 3\left(x^2+1\right)dx,$$

which implies that $\left(x^2+1\right)dx = \frac{1}{3}\,du$.

$$\int\frac{x^2+1}{\sqrt{x^3+3x+4}}\,dx = \int\left(x^3+3x+4\right)^{-1/2}\left(x^2+1\right)dx$$

$$= \int u^{-1/2}\left(\frac{1}{3}\right)du$$

$$= \frac{1}{3}\left[\frac{u^{1/2}}{1/2}\right] + C$$

$$= \frac{2}{3}\sqrt{u} + C$$

$$= \frac{2}{3}\sqrt{x^3+3x+4} + C$$

43. (a) $\int (x-1)^2\,dx = \int (x^2 - 2x + 1)\,dx$

$$= \frac{1}{3}x^3 - x^2 + x + C_1$$

$$\int (x-1)^2\,dx = \frac{(x-1)^3}{3} + C_2$$

$$= \frac{1}{3}(x^3 - 3x^2 + 3x - 1) + C_2$$

$$= \frac{1}{3}x^3 - x^2 + x - \frac{1}{3} + C_2$$

(b) The two answers differ by a constant:

$$C_1 = -\frac{1}{3} + C_2$$

(c) Answers will vary.

45. (a) $\int x(x^2-1)^2\,dx = \int (x^5 - 2x^3 + x)\,dx$

$$= \frac{1}{6}x^6 - \frac{1}{2}x^4 + \frac{1}{2}x^2 + C_1$$

$$\int x(x^2-1)^2\,dx = \frac{1}{2}\int (x^2-1)^2(2x)\,dx$$

$$= \frac{1}{2}\frac{(x^2-1)^3}{3} + C_2$$

$$= \frac{1}{6}(x^6 - 3x^4 + 3x^2 - 1) + C_2$$

$$= \frac{1}{6}x^6 - \frac{1}{2}x^4 + \frac{1}{2}x^2 - \frac{1}{6} + C_2$$

(b) The two answers differ by a constant:

$$C_1 = -\frac{1}{6} + C_2$$

(c) Answers will vary.

47. $f(x) = \int 2x(4x^2 - 10)^2\,dx$

$$= \frac{1}{4}\int (4x^2-10)^2(8x)\,dx$$

$$= \frac{1}{4}\left[\frac{(4x^2-10)^3}{3}\right] + C$$

$$= \frac{1}{12}(4x^2-10)^3 + C$$

Because $f(2) = 10$, it follows that $C = -8$, and you have $f(x) = \frac{1}{12}(4x^2-10)^3 - 8$

$$= \frac{16}{3}x^6 - 40x^4 + 100x^2 - \frac{226}{3}.$$

49. $\frac{dC}{dx} = \frac{4}{\sqrt{x+1}}$

(a) $C = \int 4(x+1)^{-1/2}\,dx = 4\left[\frac{(x+1)^{1/2}}{1/2}\right] + k$

$$= 8(x+1)^{1/2} + k$$

$$C(15) = 50$$

$$8(15+1)^{1/2} + k = 50$$

$$k = 18$$

$$C = 8\sqrt{x+1} + 18$$

(b) Let $x = 50$ and find C.

$$C(50) = 8\sqrt{51} + 18$$

$$\approx \$75.13$$

51. $x = \int p\sqrt{p^2 - 25}\,dp$

$$= \frac{1}{2}\int (p^2-25)^{1/2}(2p)\,dp$$

$$= \frac{1}{2}\left[\frac{(p^2-25)^{3/2}}{3/2}\right] + C$$

$$= \frac{1}{3}(p^2-25)^{3/2} + C$$

Because $x = 600$ when $p = 13$, it follows that $C = 24$, and you have $x = \frac{1}{3}(p^2-25)^{3/2} + 24$.

53. $x = \int -\frac{6000p}{(p^2-16)^{3/2}}\,dp$

$$= -\frac{6000}{2}\int (p^2-16)^{-3/2}(2p)\,dp$$

$$= -6000\left[\frac{(p^2-16)^{-1/2}}{-1/2}\right] + C$$

$$= \frac{6000}{\sqrt{p^2-16}} + C$$

Because $x = 5000$ when $p = 5$, it follows that $C = 3000$, and you have $x = \frac{6000}{\sqrt{p^2-16}} + 3000$.

55. (a) $h = \int \frac{17.6t}{\sqrt{17.6t^2+1}}\,dt = \frac{1}{2}\int (17.6t^2+1)^{-1/2}(35.2t)\,dt = (17.6t^2+1)^{1/2} + C$

Because $h(0) = 6$, it follows that $C = 5$, and you have $h(t) = \sqrt{17.6t^2+1} + 5$.

(b) $h(5) = 26$ inches

57. (a) $Q = \int \frac{0.95}{(x - 24{,}999)^{0.05}}\, dx = (x - 24{,}999)^{0.95} + C$

Because $Q = 25{,}000$ when $x = 25{,}000$, it follows that

$$25{,}000 = (25{,}000 - 24{,}999)^{0.95} + C = 1 + C$$
$$C = 24{,}999,$$

and you have $Q = (x - 24{,}999)^{0.95} + 24{,}999$.

(b)

x	25,000	50,000	100,000	150,000
Q	25,000	40,067.14	67,786.18	94,512.29
$x - Q$	0	9932.86	32,213.82	55,487.71

(c)

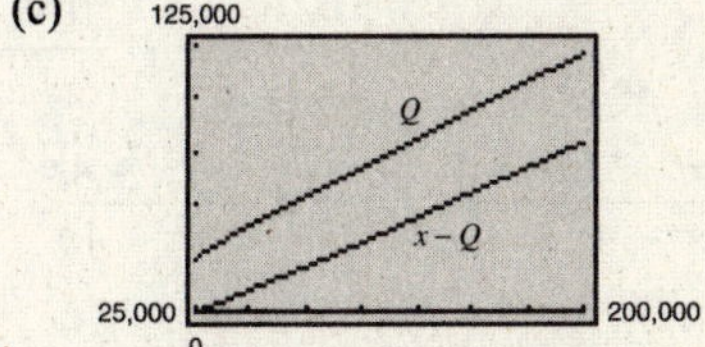

59. $\int \frac{1}{\sqrt{x} + \sqrt{x+1}}\, dx = \int \frac{1}{\sqrt{x} + \sqrt{x+1}} \cdot \frac{\sqrt{x} - \sqrt{x+1}}{\sqrt{x} - \sqrt{x+1}}\, dx$

$$= \int \frac{\sqrt{x} - \sqrt{x+1}}{x - (x+1)}\, dx$$
$$= \int -\sqrt{x} + \sqrt{x+1}\, dx$$
$$= \frac{-x^{3/2}}{3/2} + \frac{(x+1)^{3/2}}{3/2} + C$$
$$= -\frac{2}{3}x^{3/2} + \frac{2}{3}(x+1)^{3/2} + C$$

$$\frac{d}{dx}\left[-\frac{2}{3}x^{3/2} + \frac{2}{3}(x+1)^{3/2} + C\right] = -x^{1/2} + (x+1)^{1/2}$$
$$= \frac{-\sqrt{x} + \sqrt{x+1}}{1} \cdot \frac{\sqrt{x} + \sqrt{x+1}}{\sqrt{x} + \sqrt{x+1}}$$
$$= \frac{-x + x + 1}{\sqrt{x} + \sqrt{x+1}}$$
$$= \frac{1}{\sqrt{x} + \sqrt{x+1}}$$

Section 5.3 Exponential and Logarithmic Integrals

Skills Warm Up

1. $\frac{x^2 + 4x + 2}{x + 2} = x + 2\overline{\big)\, x^2 + 4x + 2}$ with quotient $x + 2 - \frac{2}{x+2}$

$$\begin{array}{r} x^2 + 4x + 2 \\ -(x^2 + 2x) \\ \hline 2x + 2 \\ -(2x + 4) \\ \hline -2 \end{array}$$

$$\frac{x^2 + 4x + 2}{x + 2} = x + 2 - \frac{2}{x + 2}$$

2. $\frac{x^2 - 6x + 9}{x - 4} = x - 4\overline{\big)\, x^2 - 6x + 9}$ with quotient $x - 2 + \frac{1}{x-4}$

$$\begin{array}{r} x^2 - 6x + 9 \\ -(x^2 + 4x) \\ \hline -2x + 9 \\ -(-2x + 8) \\ \hline 1 \end{array}$$

$$\frac{x^2 - 6x + 9}{x - 4} = x - 2 + \frac{1}{x - 4}$$

Skills Warm Up —continued—

3. $\dfrac{x^3 + 4x^2 - 30x - 4}{x^2 - 4x} = $

$$\begin{array}{r}
x + 8 + \dfrac{2x - 4}{x^2 - 4x} \\
x^2 - 4x \,\big)\, \overline{x^3 + 4x^2 - 30x - 4} \\
-(x^3 - 4x^2) \\
\hline
8x^2 - 30x \\
-(8x^2 - 32x) \\
\hline
2x - 4
\end{array}$$

$$\frac{x^3 + 4x^2 - 30x - 4}{x^2 - 4x} = x + 8 + \frac{2x - 4}{x^2 - 4x}$$

4. $\dfrac{x^4 - x^3 + x^2 + 15x + 2}{x^2 + 5} = $

$$\begin{array}{r}
x^2 - x - 4 + \dfrac{20x + 22}{x^2 + 5} \\
x^2 + 5 \,\big)\, \overline{x^4 - x^3 + x^2 + 15x + 2} \\
-(x^4 + 5x^2) \\
\hline
-x^3 - 4x^2 + 15x \\
-(-x^3 - 5x) \\
\hline
-4x^2 + 20x + 2 \\
-(-4x^2 - 20) \\
\hline
20x + 22
\end{array}$$

$$\frac{x^4 - x^3 + x^2 + 15x + 2}{x^2 + 5} = x^2 - x - 4 + \frac{20x + 22}{x^2 + 5}$$

5. $\displaystyle\int\left(x^3 + \frac{1}{x^2}\right) dx = \int x^3\, dx + \int \frac{1}{x^2}\, dx$

$\displaystyle = \frac{1}{4}x^4 - \frac{1}{x} + C$

6. $\displaystyle\int \frac{x^2 + 2x}{x}\, dx = \int \frac{x^2}{x}\, dx + \int \frac{2x}{x}\, dx$

$\displaystyle = \int x\, dx + \int 2\, dx$

$\displaystyle = \frac{1}{2}x^2 + 2x + C$

7. $\displaystyle\int \frac{x^3 + 4}{x^2}\, dx = \int \frac{x^3}{x^2}\, dx + \int \frac{4}{x^2}\, dx$

$\displaystyle = \int x\, dx + \int \frac{4}{x^2}\, dx$

$\displaystyle = \frac{1}{2}x^2 - \frac{4}{x} + C$

8. $\displaystyle\int \frac{x + 3}{x^3}\, dx = \int \frac{x}{x^3}\, dx + \int \frac{3}{x^3}\, dx$

$\displaystyle = \int \frac{1}{x^2}\, dx + \int \frac{3}{x^3}\, dx$

$\displaystyle = -\frac{1}{x} - \frac{3}{2x^2} + C$

1. $\int 2e^{2x}\, dx = \int e^{2x}(2)\, dx = e^{2x} + C$

3. $\int e^{4x}\, dx = \frac{1}{4}\int e^{4x}(4)\, dx = \frac{1}{4}e^{4x} + C$

5. $\int e^{5x-3}\, dx = \frac{1}{5}\int e^{5x-3}(5)\, dx = \frac{1}{5}e^{5x-3} + C$

7. $\int 9xe^{-x^2}\, dx = -\frac{9}{2}\int e^{-x^2}(-2x)\, dx = -\frac{9}{2}e^{-x^2} + C$

9. $\int 5x^2e^{x^3}\, dx = \frac{5}{3}\int e^{x^3}(3x^2)\, dx = \frac{5}{3}e^{x^3} + C$

11. $\int (2x + 1)e^{x^2+x}\, dx = \int e^{x^2+x}(2x + 1)\, dx = e^{x^2+x} + C$

13. $\displaystyle\int \frac{1}{x + 1}\, dx = \ln|x + 1| + C$

15. $\displaystyle\int \frac{5}{5x + 2}\, dx = \ln|5x + 2| + C$

17. $\displaystyle\int \frac{1}{3 - 2x}\, dx = -\frac{1}{2}\int \frac{1(-2)}{3 - 2x}\, dx = -\frac{1}{2}\ln|3 - 2x| + C$

19. $\int \frac{2}{3x+5}\,dx = \frac{2}{3}\int \frac{3}{3x+5}\,dx = \frac{2}{3}\ln|3x+5| + C$

21. $\int \frac{x}{x^2+1}\,dx = \frac{1}{2}\int \frac{x(2)}{x^2+1}\,dx$

$= \frac{1}{2}\ln(x^2+1) + C$

$= \ln(x^2+1)^{1/2} + C$

$= \ln\sqrt{x^2+1} + C$

23. $\int \frac{x^2}{x^3+1}\,dx = \frac{1}{3}\int \frac{3x^2}{x^3+1}\,dx = \frac{1}{3}\ln|x^3+1| + C$

25. $\int \frac{x+3}{x^2+6x+7}\,dx = \frac{1}{2}\int \frac{2(x+3)}{x^2+6x+7}\,dx$

$= \frac{1}{2}\ln|x^2+6x+7| + C$

27. $\int \frac{1}{x\ln x}\,dx = \int \frac{1}{\ln x}\left(\frac{1}{x}\right)dx = \ln|\ln x| + C$

29. $\int \frac{e^{-x}}{1-e^{-x}}\,dx = \int \frac{1}{1-e^{-x}}\left(e^{-x}\,dx\right) = \ln|1-e^{-x}| + C$

31. $\int \frac{x^3-8x}{2x^2}\,dx = \int\left(\frac{x}{2} - \frac{4}{x}\right)dx = \frac{1}{4}x^2 - 4\ln|x| + C$

General Power Rule, Logarithmic Rule

33. $\int \frac{8x^3+3x^2+6}{x^3}\,dx = \int\left(\frac{8x^3}{x^3} + \frac{3x^2}{x^3} + \frac{6}{x^3}\right)dx$

$= \int\left(8 + \frac{3}{x} + 6x^{-3}\right)dx$

$= 8x + 3\ln|x| - \frac{3}{x^2} + C$

General Power Rule, Logarithmic Rule

35. $\int \frac{e^{2x}+2e^x+1}{e^x}\,dx = \int\left(e^x + 2 + e^{-x}\right)dx$

$= e^x + 2x - e^{-x} + C$

General Power Rule, Exponential Rule

37. $\int e^x\sqrt{1-e^x}\,dx = -\int(1-e^x)^{1/2}(-e^x)\,dx$

$= -\frac{2}{3}(1-e^x)^{3/2} + C$

Exponential Rule

39. $\int \frac{1+e^{-x}}{1+xe^{-x}}\,dx = \int \frac{e^x+1}{e^x+x}\,dx = \ln|e^x+x| + C$

Logarithmic Rule

41. $\int \frac{5}{e^{-5x}+7}\,dx = \frac{1}{7}\int \frac{1}{7e^{5x}+1}35e^{5x}\,dx$

$= \frac{1}{7}\ln(7e^{5x}+1) + C$

Logarithmic Rule

43. $\int \frac{x^2+2x+5}{x-1}\,dx = \int\left(x+3+\frac{8}{x-1}\right)dx$

$= \frac{1}{2}x^2 + 3x + 8\ln|x-1| + C$

General Power Rule, Logarithmic Rule

45. $\int \frac{x-3}{x+3}\,dx = \int\left(1 - \frac{6}{x+3}\right)dx$

$= x - 6\ln|x+3| + C$

General Power Rule, General Logarithmic Rule

47. $f(x) = \int \frac{1}{x^2}e^{2/x}\,dx$

$= \int x^{-2}e^{2x^{-1}}\,dx$

$= -\frac{1}{2}\int e^{2x^{-1}}\left(-2x^{-2}\right)dx$

$= -\frac{1}{2}e^{2/x} + C$

At $(4, 6)$, $-\frac{1}{2}e^{2/4} + C = 6$

$C = 6 + \frac{1}{2}e^{1/2}.$

So, $f(x) = -\frac{1}{2}e^{2/x} + \frac{1}{2}e^{1/2} + 6.$

49. $f(x) = \int \frac{x^2+4x+3}{x-1}\,dx$

$= \int\left(x+5+\frac{8}{x-1}\right)dx$

$= \frac{x^2}{2} + 5x + 8\ln|x-1| + C$

At $(2, 4)$, $4 = \frac{2^2}{2} + 5(2) + 8\ln|2-1| + C$

$-8 = C.$

So, $f(x) = \frac{1}{2}x^2 + 5x + 8\ln|x-1| - 8.$

51. (a) $P = \int \frac{3000}{1 + 0.25t}\, dt$

$= \frac{3000}{0.25} \int \frac{0.25}{1 + 0.25t}\, dt$

$= 12{,}000 \ln|1 + 0.25t| + C$

Because $P(0) = 12{,}000 \ln 1 + C = 1000$, it follows that $C = 1000$. So,

$P(t) = 12{,}000 \ln|1 + 0.25t| + 1000$

$= 1000\left[12 \ln|1 + 0.25t| + 1\right]$

$= 1000\left[1 + \ln(1 + 0.25t)^{12}\right].$

(b) $P(3) = 1000\left[1 + \ln(1 + 0.75)^{12}\right] \approx 7715$ bacteria

(c) $12{,}000 = 1000\left[1 + \ln(1 + 0.25t)^{12}\right]$

$12 = 1 + \ln(1 + 0.25t)^{12}$

$11 = 12 \ln(1 + 0.25t)$

$\frac{11}{12} = \ln(1 + 0.25t)$

$e^{11/12} = 1 + 0.25t$

$\frac{e^{11/12} - 1}{0.25} = t$

$t \approx 6$ days

53. (a) $p = \int 0.1e^{-x/500}\, dx = -50e^{-x/500} + C$

Because $x = 600$ when $p = 30$, it follows that

$30 = -50e^{-600/500} + C$, and $C \approx 45.06$.

So, $p = -50e^{-x/500} + 45.06$.

(b)

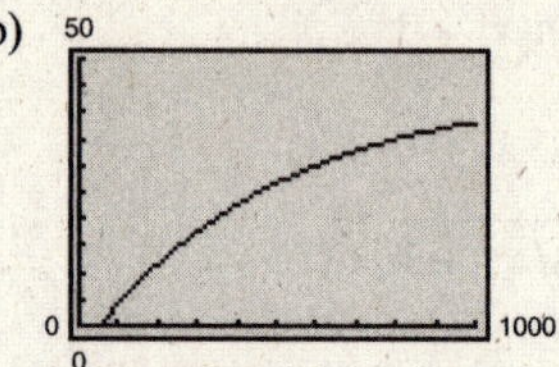

Price increases as demand increases.

(c) When $p = 22$, $x \approx 387$.

55. (a) $\frac{dR}{dt} = 320.1e^{0.0993t}$, $2 \le t \le 9$, $t = 2 \leftrightarrow 2002$

$R(t) = \int 320.1e^{0.0993t}\, dt$

$= \frac{320.1}{0.0993} \int e^{0.0993t}(0.0993)\, dt$

$= 3223.5e^{0.0993t} + C$

$R(7) = 6484.5$

$3223.5e^{0.0993(7)} + C = 6484.5$

$C \approx 24.9$

$R(t) = 3223.5e^{0.0993t} + 24.9$

(b) 2009: Let $t = 9$ and find R.

$R \approx \$7903.64$

57. False. $(\ln 5)^{1/2} \approx 1.27 \ne \frac{1}{2} \ln 5 \approx 0.80$

$\ln x^{1/2} = \frac{1}{2} \ln x$

Chapter 5 Quiz Yourself

1. $\int 3\, dx = 3\int dx = 3x + C$

$\frac{d}{dx}[3x + C] = 3$

2. $\int 10x\, dx = 10\int x\, dx = 10\left(\frac{1}{2}x^2 + C\right) = 5x^2 + C$

$\frac{d}{dx}\left[5x^2 + C\right] = 10x$

3. $\int \frac{1}{x^5}\, dx = -\frac{1}{4}x^{-4} + C = -\frac{1}{4x^4} + C$

$\frac{d}{dx}\left[-\frac{1}{4x^4} + C\right] = \frac{1}{x^5}$

4. $\int \left(x^2 - 2x + 15\right) dx = \frac{1}{3}x^3 - x^2 + 15x + C$

$\frac{d}{dx}\left[\frac{1}{3}x^3 - x^2 + 15x + C\right] = x^2 - 2x + 15$

5. $\int (6x + 1)^3(6)\, dx = \frac{1}{4}(6x + 1)^4 + C$

$\frac{d}{dx}\left[\frac{1}{4}(6x + 1)^4 + C\right] = (6x + 1)^3(6)$

6. $\int x\left(5x^2 - 2\right)^4 dx = \frac{1}{10}\int \left(5x^2 - 2\right)^4(10x)\, dx$

$= \frac{1}{10}\left[\frac{\left(5x^2 - 2\right)^5}{5}\right] + C$

$= \frac{1}{50}\left(5x^2 - 2\right)^5 + C$

$\frac{d}{dx}\left[\frac{1}{50}\left(5x^2 - 2\right)^5 + C\right] = \frac{1}{50}(5)\left(5x^2 - 2\right)^4(10x)$

$= x\left(5x^2 - 2\right)^4$

7. $\int (x^2 - 5x)(2x - 5)\,dx = \int (2x^3 - 15x^2 + 25x)\,dx$

$$= \frac{1}{2}x^4 - 5x^3 + \frac{25}{2}x^2 + C$$

$$\frac{d}{dx}\left[\frac{1}{2}x^4 - 5x^3 + \frac{25}{2}x^2 + C\right] = 2x^3 - 15x^2 + 25x = (x^2 - 5x)(2x - 5)$$

8. $\int \frac{3x^2}{(x^3 + 3)^3} = -\frac{1}{2}(x^3 + 3)^{-2} + C$

$$\frac{d}{dx}\left[-\frac{1}{2}(x^3 + 3)^{-2} + C\right] = (x^3 + 3)^{-3}(3x^2) = \frac{3x^2}{(x^3 + 3)^3}$$

9. $\int \sqrt{5x + 2}\,dx = \frac{1}{5}\int 5\sqrt{5x + 2}\,dx$

$$= \frac{2}{15}(5x + 2)^{3/2} + C$$

$$\frac{d}{dx}\left[\frac{2}{15}(5x + 2)^{3/2} + C\right] = \frac{1}{5}(5x + 2)^{1/2}(5) = \sqrt{5x + 2}$$

10. $f'(x) = 16x;\ f(0) = 1$

$$f(x) = \int 16x\,dx = 8x^2 + C$$

$$1 = 8(0)^2 + C$$

$$C = 1$$

$$f(x) = 8x^2 + 1$$

11. $f'(x) = 9x^2 + 4;\ f(1) = 5$

$$f(x) = \int (9x^2 + 4)\,dx = 3x^3 + 4x + C$$

$$5 = 3(1)^3 + 4(1) + C$$

$$5 = 7 + C$$

$$C = -2$$

$$f(x) = 3x^3 + 4x - 2$$

12. $\frac{dC}{dx} = 16 - 0.06x$

(a) $C = \int (16 - 0.06x)\,dx = 16x - 0.03x^2 + k$

$$C(1) = 25$$

$$16(1) - 0.03(1)^2 + k = 25$$

$$k = 9.03$$

$$C = 16x - 0.03x^2 + 9.03$$

(b) $C(0) = \$9.03$

(c) $C(500) = 16(500) - 0.03(500)^2 + 9.03 = \$509.03$

13. $f'(x) = 2x^2 + 1$

$$f(x) = \int (2x^2 + 1)\,dx = \tfrac{2}{3}x^3 + x + C$$

$$1 = \tfrac{2}{3}(0)^3 + 0 + C$$

$$C = 1$$

$$f(x) = \tfrac{2}{3}x^3 + x + 1$$

14. $B = \int \frac{250t}{\sqrt{t^2 + 36}}\,dt$

$$= 2\int \frac{125t}{\sqrt{t^2 + 36}} = 250\sqrt{t^2 + 36} + C$$

(a) $\int_0^8 \frac{250t}{\sqrt{t^2 + 36}}\,dt = 250\left(\sqrt{t^2 + 36}\right)\Big]_0^8 = 2500 - 1500 = 1000$ bolts

(b) $\int_0^{40} \frac{250t}{\sqrt{t^2 + 36}}\,dt = 250\left(\sqrt{t^2 + 36}\right)\Big]_0^{40} = 250\sqrt{1636} - 1500 \approx 8612$ bolts

15. $\int 5e^{5x+4}\,dx = e^{5x+4} + C$

$$\frac{d}{dx}\left[e^{5x+4} + C\right] = 5e^{5x+4}$$

16. $\int 3x^2e^{x^5}\,dx = e^{x^3} + C$

$$\frac{d}{dx}\left[e^{x^3} + C\right] = 3x^2e^{x^3}$$

17. $\int (x - 3)e^{x^2-6x}\,dx = \frac{1}{2}\int e^{x^2-6x}(2(x - 3))\,dx$

$$= \frac{1}{2}e^{x^2-6x} + C$$

$$\frac{d}{dx}\left[\frac{1}{2}e^{x^2-6x} + C\right] = \frac{1}{2}e^{x^2-6x}(2x - 6) = (x - 3)e^{x^2-6x}$$

18. $\int \frac{2}{2x-1}\,dx = \ln|2x-1| + C$

$\frac{d}{dx}\left[\ln|2x-1| + C\right] = \frac{1}{2x-1}(2) = \frac{2}{2x-1}$

19. $\int \frac{1}{3-8x}\,dx = -\frac{1}{8}\int \frac{(-8)}{3-8x}\,dx$

$= -\frac{1}{8}\ln|3-8x| + C$

20. $\int \frac{x}{3x^2+4}\,dx = \frac{1}{6}\int \frac{6x}{3x^2+4}\,dx$

$= \frac{1}{6}\ln(3x^2+4) + C$

21. (a) $\frac{ds}{dt} = 26.32t + \frac{848.99}{t}, 1 \le t \le 9, t = 1 \leftrightarrow 2001$

$S(t) = \int\left(26.32t + \frac{848.99}{t}\right) dt$

$= 13.16t^2 + 848.99 \ln|t| + C$

$S(1) = 2517.6$

$13.16(1)^2 + 848.99 \ln(1) + C = 2517.6$

$C = 2504.44$

$S(t) = 13.16t^2 + 848.99 \ln t + 2504.44$

(b) 2008: Let $t = 8$ and find S.

$S = 13.16(8)^2 + 848.99 \ln 8 + 2504.44$

$= \$5112.11$ million

Section 5.4 Area and the Fundamental Theorem of Calculus

Skills Warm Up

1. $\int(3x+7)\,dx = \frac{3}{2}x^2 + 7x + C$

2. $\int\left(x^{3/2} + 2\sqrt{x}\right) dx = \frac{2}{5}x^{5/2} + \frac{4}{3}x^{3/2} + C$

3. $\int \frac{1}{5x}\,dx = \frac{1}{5}\ln|x| + C$

4. $\int e^{-6x}\,dx = -\frac{1}{6}e^{-6x} + C$

5. $C = \int\left(0.02x^{3/2} + 29{,}500\right) dx$

$= 0.008x^{5/2} + 29{,}500x + C$

6. $R = \int(9000 + 2x)\,dx = 9000x + x^2 + C$

7. $P = \int(25{,}000 - 0.01x)\,dx = 25{,}000x - 0.005x^2 + C$

8. $C = \int\left(0.03x^2 + 4600\right) dx = 0.01x^3 + 4600x + C$

1. $\int_0^2 3\,dx$

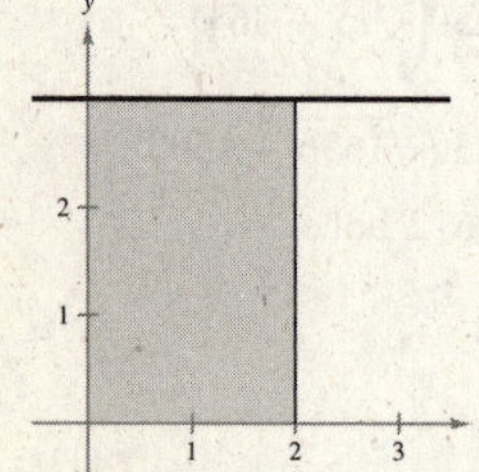

Area = (base)(height)

$= (2)(3)$

$= 6$

3. $\int_0^4 x\,dx$

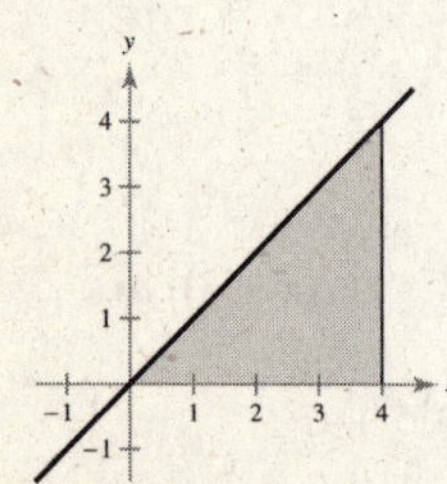

Area = $\frac{1}{2}$(base)(height)

$= \frac{1}{2}(4)(4)$

$= 8$

5. $\int_{-3}^{3} \sqrt{9-x^2}\,dx$

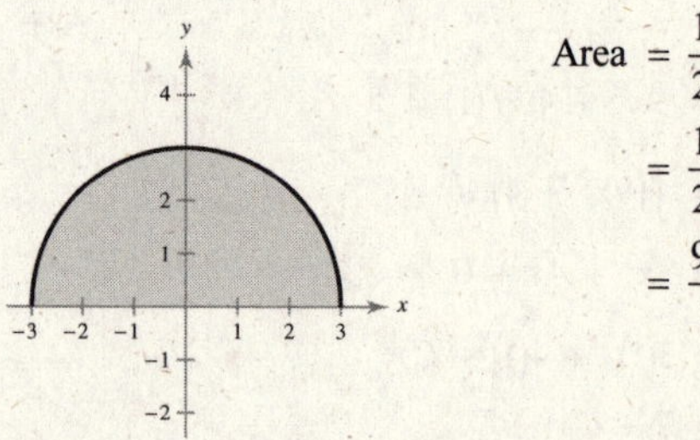

Area $= \frac{1}{2}\pi r^2$

$= \frac{1}{2}\pi(3)^2$

$= \frac{9\pi}{2}$

7. (a) $\int_0^5 [f(x) + g(x)]\,dx = \int_0^5 f(x)\,dx + \int_0^5 g(x)\,dx$
$= 6 + 2$
$= 8$

(b) $\int_0^5 [f(x) - g(x)]\,dx = \int_0^5 f(x)\,dx - \int_0^5 g(x)\,dx$
$= 6 - 2$
$= 4$

(c) $\int_0^5 -4f(x)\,dx = -4\int_0^5 f(x)\,dx$
$= -4(6)$
$= -24$

(d) $\int_0^5 [f(x) - 3g(x)] = \int_0^5 f(x)\,dx - 3\int_0^5 g(x)\,dx$
$= 6 - 3(2)$
$= 0$

9. $A = \int_0^1 (1 - x^2)\,dx = \left[\frac{x^2}{2} - \frac{x^3}{3}\right]_0^1 = \frac{1}{2} - \frac{1}{3} = \frac{1}{6}$

11. $A = \int_1^2 \frac{1}{x^2}\,dx$
$= \int_1^2 x^{-2}\,dx$
$= \left[\frac{x^{-1}}{-1}\right]_1^2$
$= -\frac{1}{x}\Big]_1^2$
$= -\frac{1}{2} + 1$
$= \frac{1}{2}$

13. $A = \int_0^4 3e^{-x/2}\,dx$
$= 3(-2)\int_0^4 e^{-x/2}\left(-\frac{1}{2}\right)dx$
$= -6e^{-x/2}\Big]_0^4$
$= -6(e^{-2} - 1)$
$= 6(1 - e^{-2})$
$= 6\left(1 - \frac{1}{e^2}\right)$

15. $A = \int_1^4 \frac{x^2 + 4}{x}\,dx$
$= \int_1^4 \left[x + 4\left(\frac{1}{x}\right)\right]dx$
$= \left[\frac{x^2}{2} + 4\ln|x|\right]_1^4$
$= (8 + 4\ln 4) - \left(\frac{1}{2} + 4\ln 1\right)$
$= \frac{15}{2} + 8\ln 2$

17. $\int_0^1 2x\,dx = x^2\Big]_0^1 = 1 - 0 = 1$

19. $\int_{-1}^0 (x - 2)\,dx = \left[\frac{x^2}{2} - 2x\right]_{-1}^0 = 0 - \left(\frac{1}{2} + 2\right) = -\frac{5}{2}$

21. $\int_{-1}^1 (3t + 4)^2\,dt = \frac{1}{3}\int_{-1}^1 (3t + 4)^2(3)\,dt$
$= \frac{1}{3}\left[\frac{(3t + 4)^3}{3}\right]_{-1}^1$
$= \frac{1}{9}\left[7^3 - 1^3\right] = 38$

23. $\int_0^3 (x - 2)^3\,dx = \frac{(x - 2)^4}{4}\Big]_0^3 = \frac{1}{4} - 4 = -\frac{15}{4}$

25. $\int_{-1}^1 (\sqrt[3]{t} - 2)\,dt = \left[\frac{3}{4}t^{4/3} - 2t\right]_{-1}^1$
$= \left(\frac{3}{4} - 2\right) - \left(\frac{3}{4} + 2\right)$
$= -\frac{5}{4} - \frac{11}{4}$
$= -4$

27. $\int_{-1}^0 (t^{1/3} - t^{2/3})\,dt = \left[\frac{3}{4}t^{4/3} - \frac{3}{5}t^{5/3}\right]_{-1}^0$
$= 0 - \left(\frac{3}{4} + \frac{3}{5}\right)$
$= -\frac{27}{20}$

29. $\int_2^8 \frac{3}{x}\,dx = \left[3\ln|x|\right]_2^8 = 3\ln 8 - 3\ln 2 = 3(\ln 8 - \ln 2)$ or $3\ln 4$
or $3\ln 2^2$
or $6\ln 2 \approx 4.16$

31. $\displaystyle\int_0^4 \frac{1}{\sqrt{2x+1}}\,dx = \frac{1}{2}\int_0^4 (2x+1)^{-1/2}(2)\,dx$

$\displaystyle = \frac{1}{2}(2)(2x+1)^{1/2}\Big]_0^4$

$\displaystyle = \sqrt{2x+1}\Big]_0^4$

$= 3 - 1$

$= 2$

33. $\displaystyle\int_1^2 e^{1-x}\,dx = -\int_1^2 e^{1-x}(-1)\,dx$

$\displaystyle = -e^{1-x}\Big]_1^2$

$= -e^{-1} + 1$

≈ 0.63

35. $\displaystyle\int_0^1 e^{2x}\sqrt{e^{2x}+1}\,dx = \frac{1}{2}\int_0^1 \left(e^{2x}+1\right)^{1/2} 2e^{2x}\,dx$

$\displaystyle = \frac{1}{3}\left(e^{2x}+1\right)^{3/2}\Big]_0^1$

$\displaystyle = \frac{1}{3}\left[\left(e^2+1\right)^{3/2} - 2\sqrt{2}\right]$

≈ 7.16

37. $\displaystyle\int_0^2 \frac{x}{1+4x^2}\,dx = \frac{1}{8}\int_0^2 \frac{1}{1+4x^2}(8x)\,dx$

$\displaystyle = \frac{1}{8}\ln\left(1+4x^2\right)\Big]_0^2$

$\displaystyle = \frac{1}{8}(\ln 17 - 0)$

$\displaystyle = \frac{1}{8}\ln 17 \approx 0.35$

39. $\displaystyle\int_{-2}^1 |4x|\,dx = \int_{-2}^0 -4x\,dx + \int_0^1 4x\,dx$

$\displaystyle = \left[-2x^2\right]_{-2}^0 + \left[2x^2\right]_0^1$

$= 0 - \left(-2(-2)^2\right) + \left(2(1)^2 - 0\right)$

$= 8 + 2 = 10$

41. $\displaystyle\int_2^8 |3x-9|\,dx = \int_2^3 -(3x-9)\,dx + \int_3^8 (3x-9)\,dx$

$\displaystyle = \left[-\tfrac{3}{2}x^2 + 9x\right]_2^3 + \left[\tfrac{3}{2}x^2 - 9x\right]_3^8$

$\displaystyle = \left(-\tfrac{3}{2}(3)^2 + 9(3)\right) - \left(-\tfrac{3}{2}(2)^2 + 9(2)\right) + \left(\tfrac{3}{2}(8)^2 - 9(8)\right) - \left(\tfrac{3}{2}(3)^2 - 9(3)\right)$

$= 39$

43. $\displaystyle A = \int_0^2 \left(3x^2+1\right)dx$

$\displaystyle = \left[x^3 + x\right]_0^2$

$= 10$

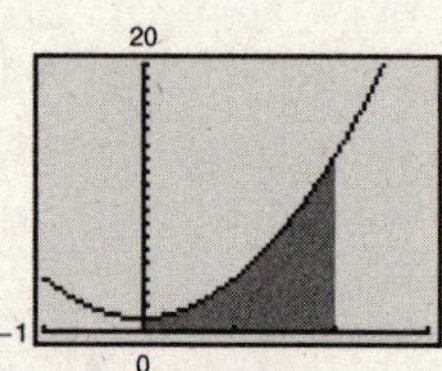

45. $\displaystyle\int_1^3 \frac{4}{x}\,dx = 4\ln x\Big]_1^3$

$= 4\ln 3 - 4\ln 1$

$= 4\ln 3$

≈ 4.39

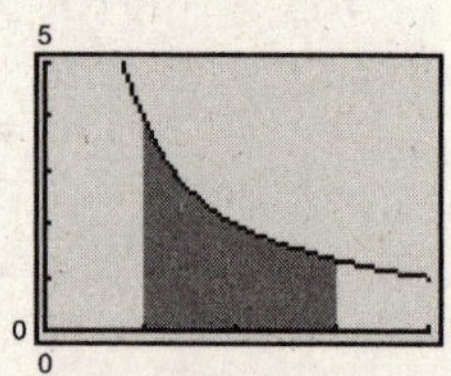

47. $\displaystyle\Delta C = \int_{100}^{103} 2.25\,dx = 2.25x\Big]_{100}^{103}$

$= 231.75 - 225 = \$6.75$

49. $\displaystyle\Delta R = \int_{12}^{15} (48 - 3x)\,dx$

$\displaystyle = \left[48x - \frac{3}{2}x^2\right]_{12}^{15}$

$= 382.5 - 360 = \$22.50$

51. $\displaystyle\Delta P = \int_{200}^{203} \frac{400 - x}{150}\,dx$

$\displaystyle = \frac{1}{150}\left[400x - \frac{x^2}{2}\right]_{200}^{203} = \$3.97$

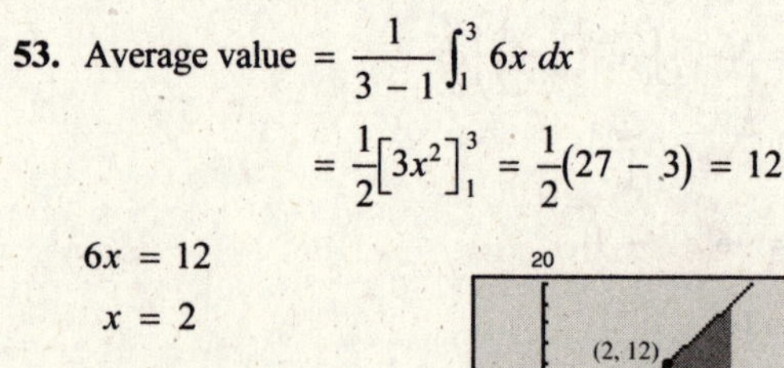

53. $\displaystyle\text{Average value} = \frac{1}{3-1}\int_1^3 6x\,dx$

$\displaystyle = \frac{1}{2}\left[3x^2\right]_1^3 = \frac{1}{2}(27 - 3) = 12$

$6x = 12$

$x = 2$

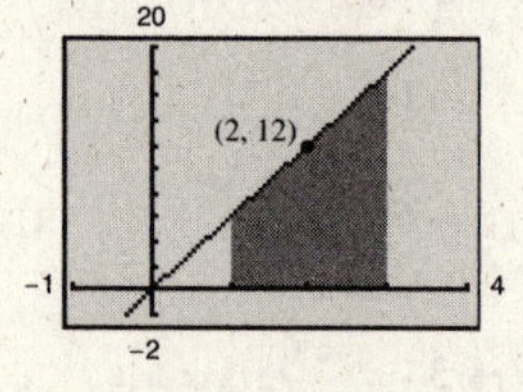

55. Average value $= \dfrac{1}{2-(-2)}\displaystyle\int_{-2}^{2}\left(4-x^2\right)dx$

$$= \frac{1}{4}\left[4x - \frac{x^3}{3}\right]_{-2}^{2}$$

$$= \frac{1}{4}\left[8 - \frac{8}{3} - \left(-8 + \frac{8}{3}\right)\right]$$

$$= \frac{8}{3}$$

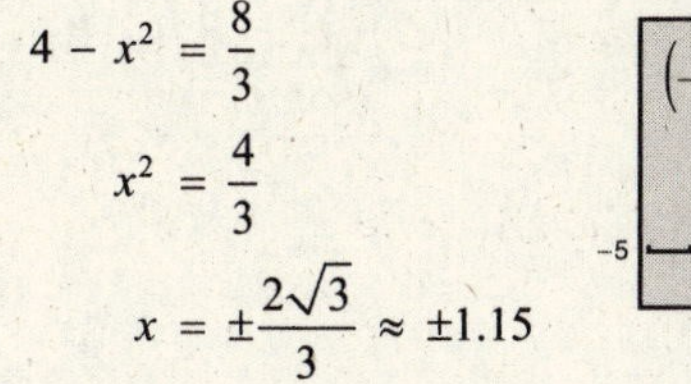

$$4 - x^2 = \frac{8}{3}$$

$$x^2 = \frac{4}{3}$$

$$x = \pm\frac{2\sqrt{3}}{3} \approx \pm 1.15$$

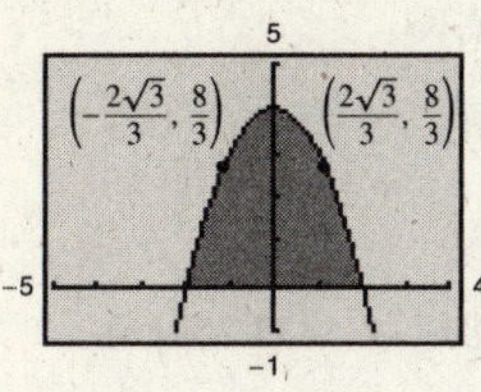

57. Average value $= \dfrac{1}{1-(-1)}\displaystyle\int_{-1}^{1} 2e^x\,dx$

$$= \frac{1}{2}\left[2e^x\right]_{-1}^{1}$$

$$= \frac{1}{2}\left[2e - 2e^{-1}\right]$$

$$\approx 2.35$$

$$2e^x = 2.35$$

$$e^x = 1.175$$

$$x = \ln 1.175 \approx 0.16$$

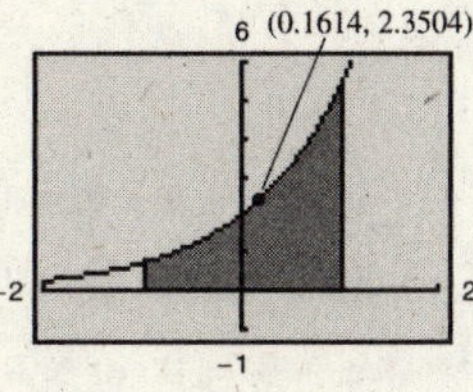

59. Average value $= \dfrac{1}{5-1}\displaystyle\int_{1}^{5}\frac{3}{x+2}\,dx$

$$= \frac{1}{4}\Big[3\ln|x+2|\Big]_{1}^{5}$$

$$= \frac{3}{4}(\ln 7 - \ln 3)$$

$$= \frac{3}{4}\ln\left(\frac{7}{3}\right) \approx 0.6355$$

$$\frac{3}{x+2} = \frac{3}{4}\ln\left(\frac{7}{3}\right)$$

$$4 = (x+2)\ln\left(\frac{7}{3}\right)$$

$$\frac{4}{\ln(7/3)} = x + 2$$

$$-2 + \frac{4}{\ln(7/3)} = x$$

$$x \approx 2.7209$$

(2.7209, 0.6355)

61. $\displaystyle\int_{-1}^{1} 3x^4\,dx = 2\int_0^1 3x^4\,dx = \tfrac{6}{5}\left[x^5\right]_0^1 = \tfrac{6}{5}$

because $f(x) = 3x^4$ is an even function.

63. $\displaystyle\int_{-1}^{1}\left(2t^5 - 2t\right)dt = 0$ because $f(t) = 2t^5 - 2t$ is an odd function.

65. Given $\displaystyle\int_0^1 x^2\,dx = \tfrac{1}{3}$.

(a) $\displaystyle\int_{-1}^{0} x^2\,dx = \tfrac{1}{3}$ because $\displaystyle\int_0^1 x^2\,dx = \tfrac{1}{3}$ and the function is even.

(b) $\displaystyle\int_{-1}^{1} x^2\,dx = \tfrac{2}{3}$ because $\displaystyle\int_0^1 x^2\,dx = \tfrac{1}{3}$ and the function is even.

(c) $\displaystyle\int_0^1 -x^2\,dx = -\tfrac{1}{3}$ because $\displaystyle\int_0^1 -x^2\,dx = -\int_0^1 x^2\,dx.$

67. $A = e^{rT}\displaystyle\int_0^T c(t)e^{-rt}\,dt$

$$= e^{0.08(6)}\int_0^6 250e^{-0.08t}\,dt$$

$$= \$1925.23$$

69. $A = e^{rT}\displaystyle\int_0^T c(t)e^{-rt}\,dt$

$$= e^{0.02(10)}\int_0^{10} 1500e^{-0.02t}\,dt$$

$$= \$16{,}605.21$$

71. $\displaystyle\int_0^5 500\,dt = 500t\Big]_0^5 = 2500 - 0 = \$2500$

73. $\displaystyle\int_0^5 500\sqrt{t+1}\,dt = \frac{1000}{3}(t+1)^{3/2}\Big]_0^5$

$$= \frac{1000}{3}(6)^{3/2} - \frac{1000}{3}$$

$$= \$4565.65$$

75. $C(x) = 5000\left(25 + 3\displaystyle\int_0^x t^{1/4}\,dt\right)$

$$= 5000\left(25 + \tfrac{12}{5}t^{5/4}\Big]_0^x\right)$$

$$= 5000\left(25 + \tfrac{12}{5}x^{5/4}\right)$$

(a) $C(1) = 5000\left[25 + \left(\tfrac{12}{5}\right)(1)^{5/4}\right] = \$137{,}000.00$

(b) $C(5) = 5000\left[25 + \left(\tfrac{12}{5}\right)(5)^{5/4}\right] = \$214{,}720.93$

(c) $C(10) = 5000\left[25 + \left(\tfrac{12}{5}\right)(10)^{5/4}\right] = \$338{,}393.53$

77. Average balance $= \dfrac{1}{5-0}\int_0^5 2250e^{0.06t}\,dt = 450\int_0^5 e^{0.06t}\,dt = 450\left(\dfrac{1}{0.06}\right)e^{0.06t}\Big]_0^5 = 7500\left(e^{0.3}-1\right) = \$2623.94$

79. $\dfrac{dM}{dt} = 547.56t - 69.459t^2 + 331.258e^{-t},\ 0 \le t \le 9,\ t = 0 \leftrightarrow 2000$

(a) $M(t) = \int\left(547.56t - 69.459t^2 + 331.258e^{-t}\right)dt = 273.78t^2 - 23.153t^3 - 331.258e^{-t} + C$

$$M(0) = 5107$$
$$273.78(0)^2 - 23.153(0)^3 - 331.258e^0 + C = 5107$$
$$C = 5438.258$$
$$M(t) = 273.78t^2 - 23.153t^3 - 331.258e^{-t} + 5438.258$$

(b) Average value $= \dfrac{1}{9-0}\int_0^9\left(273.78t^2 - 23.153t^3 - 331.258e^{-t} + 5438.258\right)dt$

$$= \frac{1}{9}\left[91.26t^3 - 5.78825t^4 + 331.258e^{-t} + 5438.258t\right]_0^9$$
$$\approx \frac{1}{9}(77{,}496.1946 - 331.258) = \$8573.88 \text{ billion}$$

Section 5.5 The Area of a Region Bounded by Two Graphs

Skills Warm Up

1. $\left(-x^2 + 4x + 3\right) - (x + 1) = -x^2 + 3x + 2$

2. $\left(-2x^2 + 3x + 9\right) - (-x + 5) = -2x^2 + 4x + 4$

3. $\left(-x^3 + 3x^2 - 1\right) - \left(x^2 - 4x + 4\right) = -x^3 + 2x^2 + 4x - 5$

4. $(3x + 1) - \left(-x^3 + 9x + 2\right) = x^3 - 6x - 1$

5.
$$x^2 - 4x + 4 = 4$$
$$x(x - 4) = 0$$
$$x = 0 \text{ or } x - 4 = 0$$
$$x = 4$$
$$g(0) = 4 \text{ and } g(4) = 4$$

The graphs intersect at $(0, 4)$ and $(4, 4)$.

6.
$$-3x^2 = 6 - 9x$$
$$0 = 3x^2 - 9x + 6$$
$$0 = x^2 - 3x + 2$$
$$0 = (x - 1)(x - 2)$$
$$x - 1 = 0 \text{ or } x - 2 = 0$$
$$x = 1 \qquad x = 2$$
$$f(1) = -3(1)^2 = -3 \text{ and } f(2) = -3(2)^2 = -12$$

The graphs intersect at $(1, -3)$ and $(2, -12)$.

7.
$$x^2 = -x + 6$$
$$x^2 + x - 6 = 0$$
$$(x + 3)(x - 2) = 0$$
$$x + 3 = 0 \text{ or } x - 2 = 0$$
$$x = -3 \qquad x = 2$$
$$f(-3) = (-3)^2 = 9 \text{ and } f(2) = (2)^2 = 4$$

The graphs intersect at $(-3, 9)$ and $(2, 4)$.

8.
$$\frac{1}{2}x^3 = 2x$$
$$\frac{1}{2}x^3 - 2x = 0$$
$$x\left(\frac{1}{2}x^2 - 2\right) = 0$$
$$x = 0 \text{ or } \frac{1}{2}x^2 - 2 = 0$$
$$x^2 = 4$$
$$x = \pm 2$$
$$g(0) = 2(0) = 0,\ g(-2) = 2(-2) = -4, \text{ and}$$
$$g(2) = 2(2) = 4$$

The graphs intersect at $(0, 0)$, $(-2, -4)$, and $(2, 4)$.

1. $A = \int_0^6 \left[0 - \left(x^2 - 6x\right)\right] dx$

$= \left(-\frac{x^3}{3} + 3x^2\right)\Big]_0^6$

$= 36$

3. $A = \int_0^3 \left[\left(-x^2 + 2x + 3\right) - \left(x^2 - 4x + 3\right)\right] dx$

$= \int_0^3 \left(-2x^2 + 6x\right) dx$

$= \left[-\frac{2x^3}{3} + 3x^2\right]_0^3$

$= 9$

5. $A = \int_0^1 \left[\left(e^x - 1\right) - 0\right] dx$

$= \int_0^1 \left(e^x - 1\right) dx$

$= \left[e^x - x\right]_0^1$

$= (e - 1) - \left(e^0 - 0\right)$

$= e - 2$

7. $A = 2\int_0^1 \left[0 - 3\left(x^3 - x\right)\right] dx$

$= -6\left(\frac{x^4}{4} - \frac{x^2}{2}\right)\Big]_0^1$

$= \frac{3}{2}$

9. The region is bounded by the graphs of $y = x + 1$, $y = x/2$, $x = 0$, and $x = 4$, as shown in the figure.

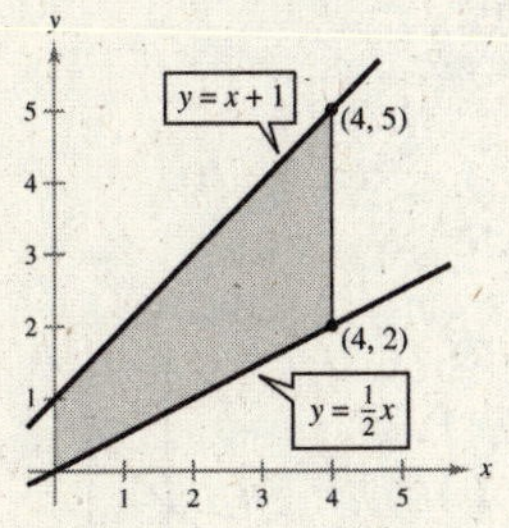

11. The region is bounded by the graphs of $y = 2x^2$ and $y = x^4 - 2x^2$ from $x = -2$ to $x = 2$, as shown in the figure.

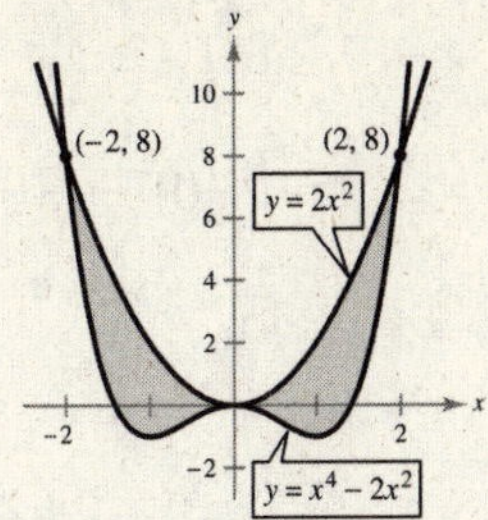

13. $f(x) = x + 1$

$g(x) = (x - 1)^2$

$A \approx 4$

Matches d

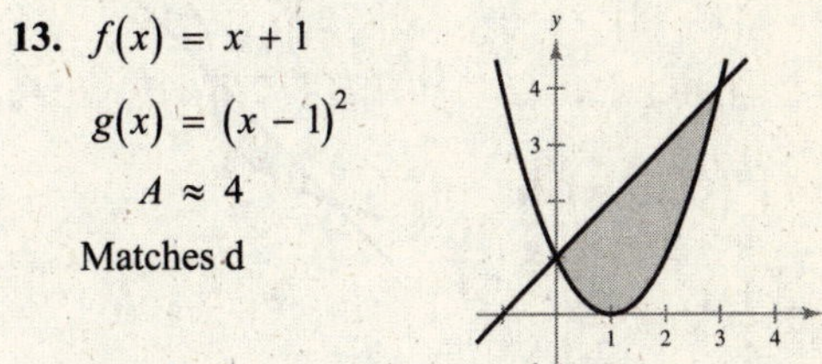

15. $A = \int_1^5 \frac{1}{x^2}\, dx = -\frac{1}{x}\Big]_1^5 = \frac{4}{5}$

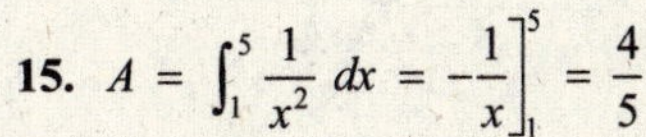

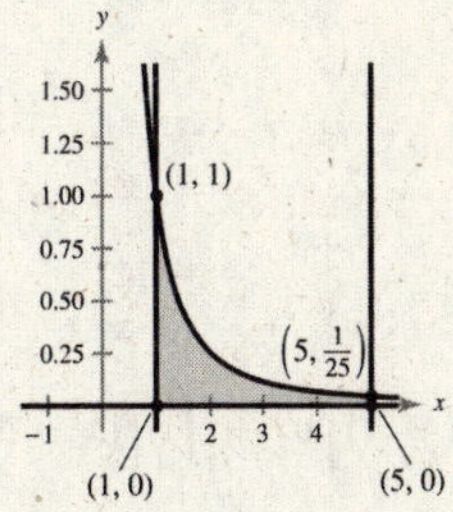

17. $x^2 - 4x + 3 = 3 + 4x - x^2$

$2x^2 - 8x = 0$

$2x(x - 4) = 0$

$x = 0, 4$

$A = \int_0^4 \left[\left(3 + 4x - x^2\right) - \left(x^2 - 4x + 3\right)\right] dx$

$= \int_0^4 \left(-2x^2 + 8x\right) dx$

$= \left[-\frac{2x^3}{3} + 4x^2\right]_0^4$

$= \frac{64}{3}$

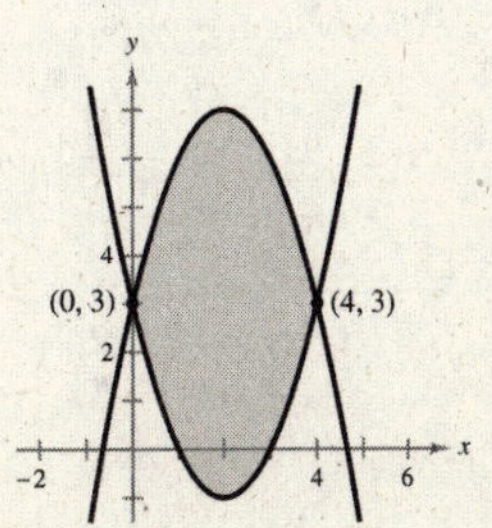

19.

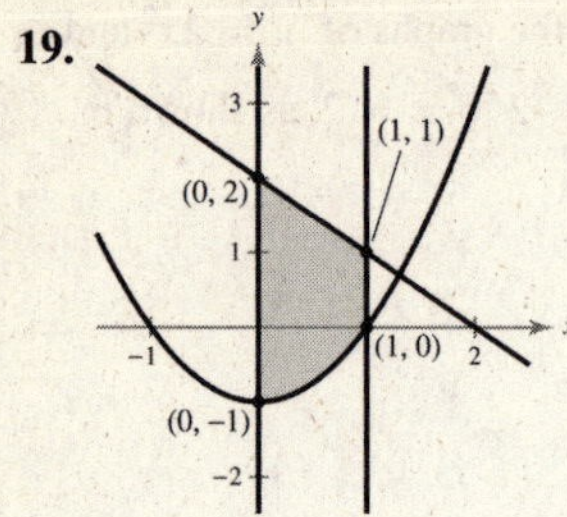

$$A = \int_0^1 \left[(-x + 2) - (x^2 - 1)\right] dx$$
$$= \int_0^1 (-x^2 - x + 3)\, dx$$
$$= \left[-\tfrac{1}{3}x^3 - \tfrac{1}{2}x^2 + 3x\right]_0^1$$
$$= \left(-\tfrac{1}{3}(1)^3 - \tfrac{1}{2}(1)^2 + 3(1)\right) - 0$$
$$= \tfrac{13}{6}$$

21. $\sqrt[3]{x} = x$

$$x = x^3$$
$$0 = x(x^2 - 1)$$
$$x = -1, 0, 1$$
$$A = 2\int_0^1 \left(\sqrt[3]{x} - x\right) dx$$
$$= 2\left[\frac{3}{4}x^{4/3} - \frac{x^2}{2}\right]_0^1$$
$$= \frac{1}{2}$$

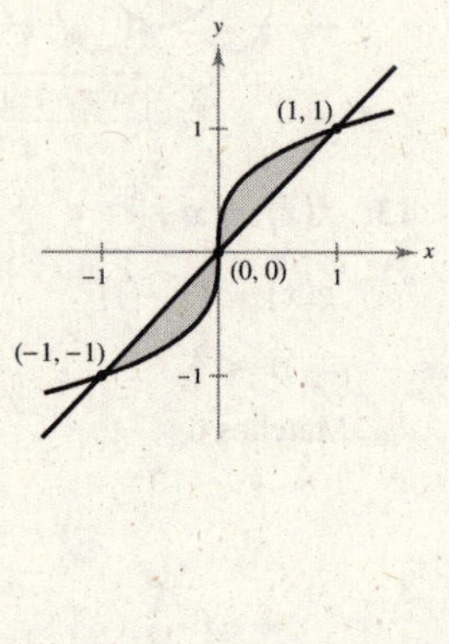

23.

$$x^3 + 4x^2 = x + 4$$
$$x^3 + 4x^2 - x - 4 = 0$$
$$x^2(x + 4) - (x + 4) = 0$$
$$(x + 4)(x^2 - 1) = 0$$
$$x = -4, -1, 1$$
$$A = \int_{-4}^{-1} \left[(x^3 + 4x^2) - (x + 4)\right] dx + \int_{-1}^{1} \left[(x + 4) - (x^3 + 4x^2)\right] dx$$
$$= \int_{-4}^{-1} (x^3 + 4x^2 - x - 4)\, dx + \int_{-1}^{1} (-x^3 - 4x^2 + x + 4)\, dx$$
$$= \left[\tfrac{1}{4}x^4 + \tfrac{4}{3}x^3 - \tfrac{1}{2}x^2 - 4x\right]_{-4}^{-1} + \left[-\tfrac{1}{4}x^4 - \tfrac{4}{3}x^3 + \tfrac{1}{2}x^2 + 4x\right]_{-1}^{1}$$
$$= \tfrac{63}{4} + \tfrac{16}{3} = \tfrac{253}{12}$$

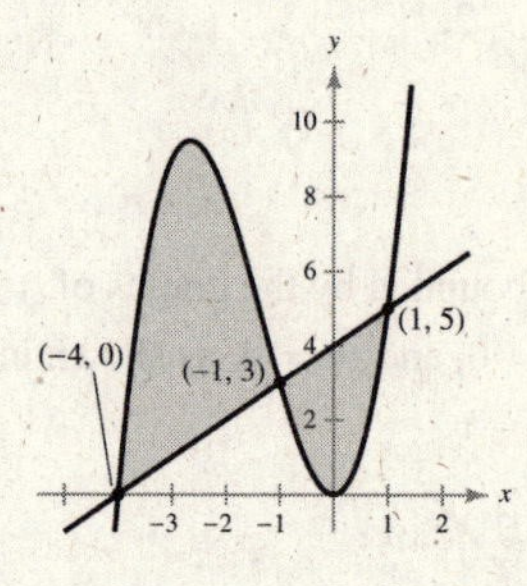

25. $A = \displaystyle\int_0^1 xe^{-x^2}\, dx = -\frac{1}{2}e^{-x^2}\Big]_0^1$

$$= -\frac{1}{2}e^{-1} + \frac{1}{2} \approx 0.316$$

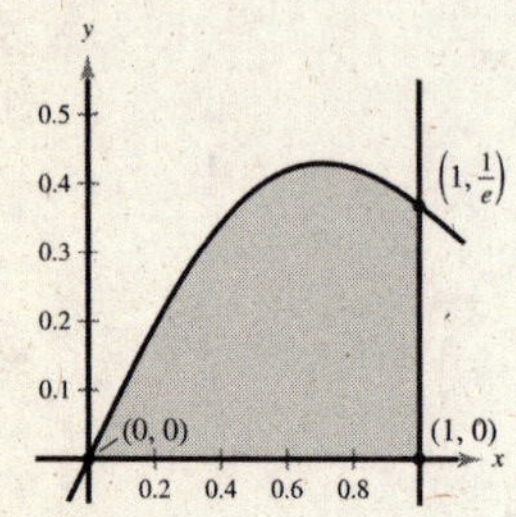

27. $A = \displaystyle\int_1^2 \left[e^{0.5x} - \left(-\frac{1}{x}\right)\right] dx$

$$= 2e^{0.5x} + \ln x\Big]_1^2$$
$$= (2e + \ln 2) - 2e^{0.5}$$
$$\approx 2.832$$

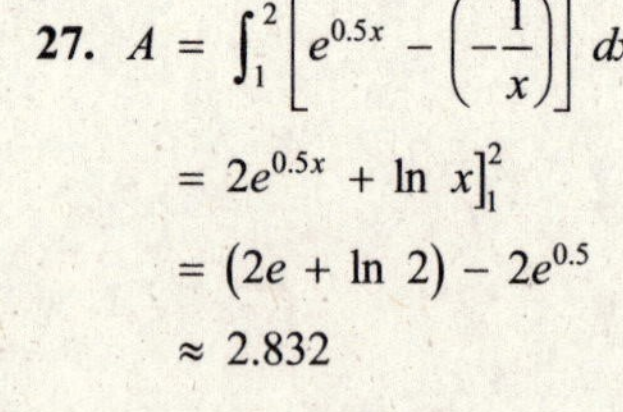

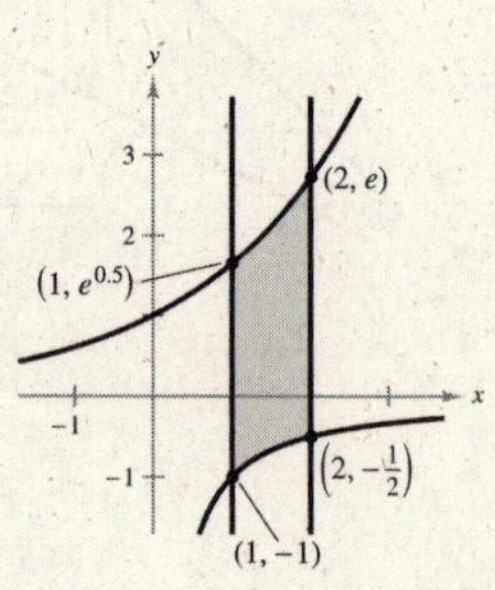

29. $A = \int_1^2 x^2\,dx + \int_2^4 \frac{8}{x}\,dx = \left[\frac{x^3}{3}\right]_1^2 + 8\ln x\Big]_2^4$

$= \left(\frac{8}{3} - \frac{1}{3}\right) + 8(\ln 4 - \ln 2)$

$= \frac{7}{3} + 8\ln 2$

≈ 7.879

31. $2x = 4 - 2x$

$4x = 4$

$x = 1$

$A = \int_0^1 2x\,dx + \int_1^2 (4 - 2x)\,dx$

33. $\frac{4}{x} = x$

$4 = x^2$

$x = -2, 2$

$A = \int_1^2 \left(\frac{4}{x} - x\right) dx + \int_2^4 \left(x - \frac{4}{x}\right) dx$

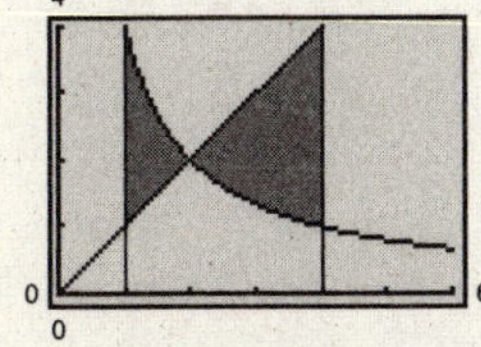

35.

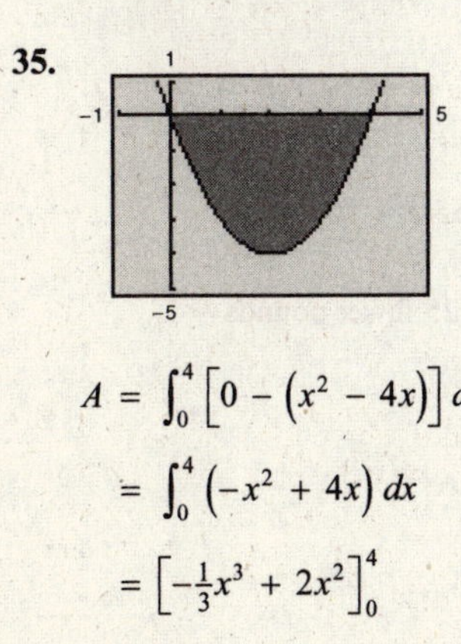

$A = \int_0^4 \left[0 - \left(x^2 - 4x\right)\right] dx$

$= \int_0^4 \left(-x^2 + 4x\right) dx$

$= \left[-\frac{1}{3}x^3 + 2x^2\right]_0^4$

$= \frac{32}{3}$

37.

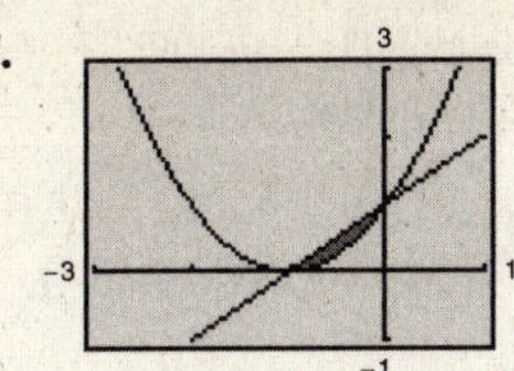

$A = \int_{-1}^0 \left[(x + 1) - \left(x^2 + 2x + 1\right)\right] dx$

$= \int_{-1}^0 \left(-x^2 - x\right) dx$

$= \left[-\frac{1}{3}x^3 - \frac{1}{2}x^2\right]_{-1}^0$

$= \frac{1}{6}$

39. The equation of the line passing through $(0, 0)$ and $(4, 4)$ is $y = x$. So, the area is given by

$A = \int_0^4 x\,dx = \left[\frac{x^2}{2}\right]_0^4 = 8.$

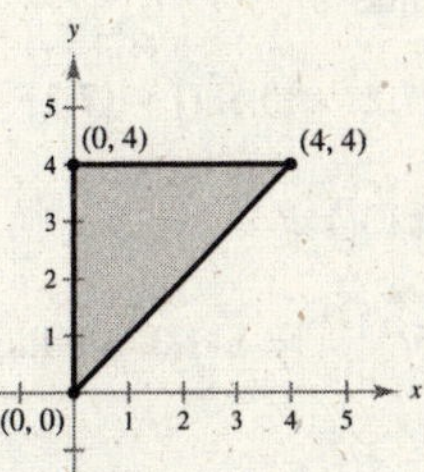

41. The point of equilibrium is found by equating $50 - 0.5x$ and $0.125x$ to obtain $x = 80$ and $p = 10$.

$CS = \int_0^{80} \left[(50 - 0.5x) - 10\right] dx$

$= \int_0^{80} (40 - 0.5x)\,dx$

$= \left[40x - \frac{0.5x^2}{2}\right]_0^{80}$

$= \$1600.00$

$PS = \int_0^{80} (10 - 0.125x)\,dx$

$= \left[10x - 0.0625x^2\right]_0^{80}$

$= \$400.00$

43. The point of equilibrium is found by equating $200 - 0.4x$ and $100 + 1.6x$ to obtain $x = 50$ and $p = 180$.

$$\text{Consumer surplus} = \int_0^{50} \left[(200 - 0.4x) - 180\right] dx = \int_0^{50} (20 - 0.4x)\, dx = \left[20x - 0.2x^2\right]_0^{50} = \$500.00$$

$$\text{Producer surplus} = \int_0^{50} 180 - (100 + 1.6x)\, dx = \int_0^{50} (80 - 1.6x)\, dx = \left[80x - 0.8x^2\right]_0^{50} = \$2000.00$$

45. The point of equilibrium is found by equating $42 - 0.015x^2$ and $0.01x^2 + 2$ to obtain $x = 40$ and $p = 18$.

$$\text{Consumer surplus} = \int_0^{40} \left[(42 - 0.015x^2) - 18\right] dx = \int_0^{40} (-0.015x^2 + 24)\, dx = \left[-0.005x^3 + 24x\right]_0^{40} = \$640.00$$

$$\text{Producer surplus} = \int_0^{40} \left[18 - (0.01x^2 + 2)\right] dx = \int_0^{40} (-0.01x^2 + 16)\, dx = \left[-\frac{0.01}{3}x^3 + 16x\right]_0^{40} = \$426.67$$

47. The model R_1 projects greater revenue than R_2. The difference in total revenue is

$$\int_{15}^{20} (R_1 - R_2)\, dt = \int_{15}^{20} \left[(7.21 + 0.58t) - (7.21 + 0.45t)\right] dt = \int_{15}^{20} (0.13t)\, dt = \left[0.065t^2\right]_{15}^{20} = \$11.375 \text{ billion.}$$

49. $$\int_{15}^{25} (C_1 - C_2)\, dt = \int_{15}^{25} \left[(568.5 + 7.15t) - (525.6 + 6.43t)\right] dt = \int_{15}^{25} (42.9 + 0.72t)\, dt = \left[42.9t + 0.36t^2\right]_{15}^{25} = \$573 \text{ million}$$

Explanations will vary.

51. (a)

300

8 18

200

(b) $$\int_{13}^{18} \left[(-0.443t^2 + 5.02t + 277.7) - (-0.775t^2 + 18.73t + 170.5)\right] dt = \int_{13}^{18} (-0.332t^2 - 13.71t + 107.2)\, dt$$

$$= \left[\frac{0.332t^3}{3} - 6.855t^2 + 107.2t\right]_{13}^{18}$$

$$\approx -124.25 \text{ or } 124.25 \text{ fewer pounds}$$

53. $$\int_0^{10} \left[R(t) - C(t)\right] dt = \int_0^{10} \left[100 - (60 + 0.2t^2)\right] dt = \int_0^{10} (40 - 0.2t^2)\, dt = \left[40t - \frac{0.2}{3}t^3\right]_0^{10} = \$333.33 \text{ million}$$

55. Demand: $(6000, 331), (8000, 303)$

$$m = \frac{303 - 331}{8000 - 6000} = -0.014$$

$$p - 331 = -0.014(x - 6000)$$

$$p = -0.014x + 415$$

Point of equilibrium: $-0.014x + 415 = 0.0275x$

$$x = 10{,}000 \text{ and } p = 275$$

$$\begin{aligned}\text{Consumer surplus} &= \int_0^{10{,}000} \left[(-0.014x + 415) - 275\right] dx \\ &= \int_0^{10{,}000} (-0.014x + 140)\, dx \\ &= \left[-0.007x^2 + 140x\right]_0^{10{,}000} \\ &= \$700{,}000\end{aligned}$$

$$\begin{aligned}\text{Producer surplus} &= \int_0^{10{,}000} (275 - 0.0275x)\, dx \\ &= \left[275x - 0.01375x^2\right]_0^{10{,}000} \\ &= \$1{,}375{,}000\end{aligned}$$

57.

$$\begin{aligned}\text{Income inequality} &= \int_0^{3.03} \left[(0.00061x^2 + 0.0224x + 1.666)^2 - x\right] dx \\ &\quad + \int_{3.03}^{99.94} \left[x - (0.00061x^2 + 0.0224x + 1.666)^2\right] dx \\ &\quad + \int_{99.94}^{100} \left[(0.00061x^2 + 0.0224x + 1.666)^2 - x\right] dx \\ &= \$2085.48\end{aligned}$$

59. Answers will vary.

Section 5.6 The Definite Integral as the Limit of a Sum

Skills Warm Up

1. $\left[0, \frac{1}{3}\right]$ midpoint: $\dfrac{0 + \frac{1}{3}}{2} = \dfrac{1}{6}$

2. $\left[\frac{1}{10}, \frac{2}{10}\right]$ midpoint: $\dfrac{\frac{1}{10} + \frac{2}{10}}{2} = \dfrac{3}{20}$

3. $\left[\frac{3}{20}, \frac{4}{20}\right]$ midpoint: $\dfrac{\frac{3}{20} + \frac{4}{20}}{2} = \dfrac{7}{40}$

4. $\left[1, \frac{7}{6}\right]$ midpoint: $\dfrac{1 + \frac{7}{6}}{2} = \dfrac{13}{12}$

5. $\left[2, \frac{31}{15}\right]$ midpoint: $\dfrac{2 + \frac{31}{15}}{2} = \dfrac{61}{30}$

6. $\left[\frac{26}{9}, 3\right]$ midpoint: $\dfrac{\frac{26}{9} + 3}{2} = \dfrac{53}{18}$

7. $\displaystyle \lim_{x\to\infty} \frac{2x^2 + 4x - 1}{3x^2 - 2x} = \lim_{x\to\infty} \frac{2 + \frac{4}{x} - \frac{1}{x^2}}{3 - \frac{2}{x}} = \frac{2}{3}$

8. $\displaystyle \lim_{x\to\infty} \frac{4x + 5}{7x - 5} = \lim_{x\to\infty} \frac{4 + \frac{5}{x}}{7 - \frac{5}{x}} = \frac{4}{7}$

9. $\displaystyle \lim_{x\to\infty} \frac{x - 7}{x^2 + 1} = \lim_{x\to\infty} \frac{\frac{1}{x} - \frac{7}{x^2}}{1 + \frac{1}{x^2}} = 0$

10. $\displaystyle \lim_{x\to\infty} \frac{5x^3 + 1}{x^3 + x^2 + 4} = \lim_{x\to\infty} \frac{5 + \frac{1}{x^3}}{1 + \frac{1}{x} + \frac{4}{x^3}} = 5$

1. The midpoints of the four intervals are $\frac{1}{8}$, $\frac{3}{8}$, $\frac{5}{8}$, and $\frac{7}{8}$. The approximate area is

$$A \approx \frac{1-0}{4}\left[f\left(\frac{1}{8}\right) + f\left(\frac{3}{8}\right) + f\left(\frac{5}{8}\right) + f\left(\frac{7}{8}\right)\right] = \frac{1}{4}\left[\frac{11}{4} + \frac{9}{4} + \frac{7}{4} + \frac{5}{4}\right] = 2.$$

The exact area is $A = \int_0^1 (-2x + 3)\,dx = \left[-x^2 + 3x\right]_0^1 = 2.$

3. The midpoints of the four intervals are $\frac{1}{8}$, $\frac{3}{8}$, $\frac{5}{8}$, and $\frac{7}{8}$. The approximate area is

$$A \approx \frac{1-0}{4}\left[f\left(\frac{1}{8}\right) + f\left(\frac{3}{8}\right) + f\left(\frac{5}{8}\right) + f\left(\frac{7}{8}\right)\right]$$
$$= \frac{1}{4}\left[\sqrt{\frac{1}{8}} + \sqrt{\frac{3}{8}} + \sqrt{\frac{5}{8}} + \sqrt{\frac{7}{8}}\right]$$
$$= \frac{1}{4}\left[\frac{\sqrt{2}}{4} + \frac{\sqrt{6}}{4} + \frac{\sqrt{10}}{4} + \frac{\sqrt{14}}{4}\right] \approx 0.6730.$$

The exact area is $A = \int_0^1 \sqrt{x}\,dx = \frac{2}{3}x^{3/2}\Big]_0^1 = \frac{2}{3} \approx 0.6667.$

5. The midpoints of the five intervals are $\frac{1}{10}$, $\frac{3}{10}$, $\frac{1}{2}$, $\frac{7}{10}$, and $\frac{9}{10}$.

The approximate area is

$$A = \frac{1-0}{5}\left[f\left(\frac{1}{10}\right) + f\left(\frac{3}{10}\right) + f\left(\frac{1}{2}\right) + f\left(\frac{7}{10}\right) + f\left(\frac{9}{10}\right)\right] = \frac{1}{5}\left[\frac{1001}{1000} + \frac{1027}{1000} + \frac{9}{8} + \frac{1343}{1000} + \frac{1729}{1000}\right] = \frac{249}{200} = 1.245.$$

The exact area is $A = \int_0^1 \left(x^3 + 1\right)dx = \left[\frac{1}{4}x^4 + x\right]_0^1 = \frac{5}{4} = 1.25.$

7. Using the Midpoint Rule with $n = 5$,

$$\Delta x = \frac{6-1}{5} = \frac{5}{5} = 1.$$

The midpoints are 1.5, 2.5, 3.5, 4.5, and 5.5.

The approximate area is

$$A \approx (1)\left[f(1.5) + f(2.5) + f(3.5) + f(4.5) + f(5.5)\right]$$
$$= 71.25.$$

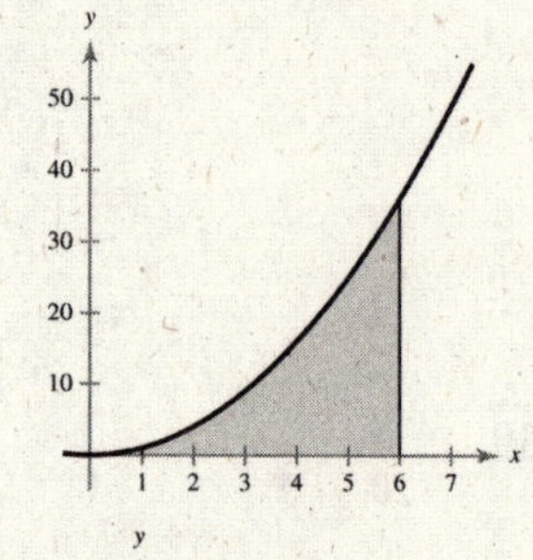

9. Using the Midpoint Rule with $n = 5$,

$$\Delta x = \frac{1-0}{5} = \frac{1}{5}.$$

The midpoints are 0.1, 0.3, 0.5, 0.7, and 0.9.

The approximate area is

$$A \approx \frac{1}{5}\left[f(0.1) + f(0.3) + f(0.5) + f(0.7) + f(0.9)\right]$$
$$\approx 1.07901.$$

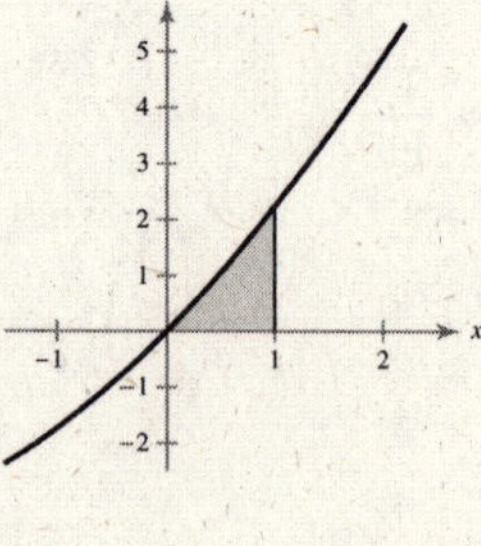

11. Using the Midpoint Rule with $n = 5$,

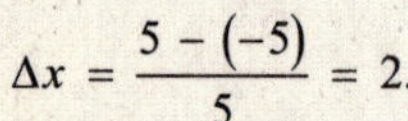

$$\Delta x = \frac{5-(-5)}{5} = 2.$$

The midpoints are -4, -2, 0, 2, and 4.

The approximate area is

$$A \approx (2)\left[f(-4) + f(-2) + f(0) + f(2) + f(4)\right]$$
$$\approx 24.28.$$

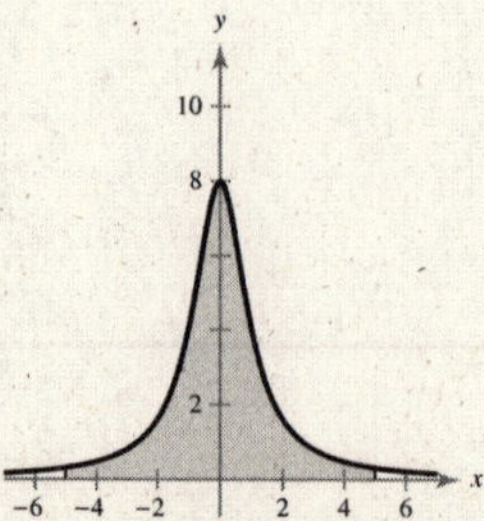

13. The midpoints are $\frac{5}{4}$, $\frac{7}{4}$, $\frac{9}{4}$, and $\frac{11}{4}$.

The approximate area is

$$A \approx \frac{3-1}{4}\left[f\left(\frac{5}{4}\right) + f\left(\frac{7}{4}\right) + f\left(\frac{9}{4}\right) + f\left(\frac{11}{4}\right)\right]$$

$$\approx \frac{1}{2}\left[\frac{25}{8} + \frac{49}{8} + \frac{81}{8} + \frac{121}{8}\right] = \frac{69}{4} = 17.25.$$

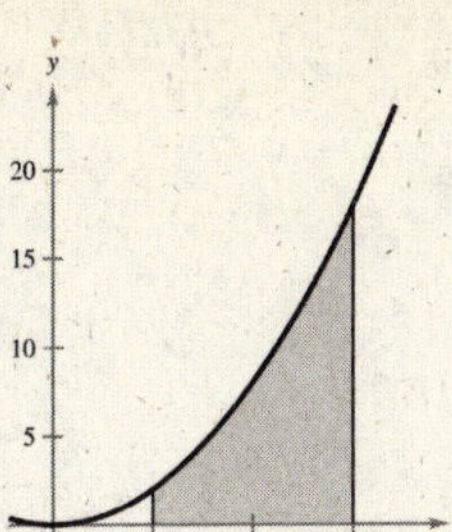

15. The midpoints are -1.5, -0.5, 0.5, and 1.5.

The approximate area is

$$A \approx \left(\frac{2-(-2)}{4}\right)\left[f(-1.5) + f(-0.5) + f(0.5) + f(1.5)\right]$$

$$= 34.25.$$

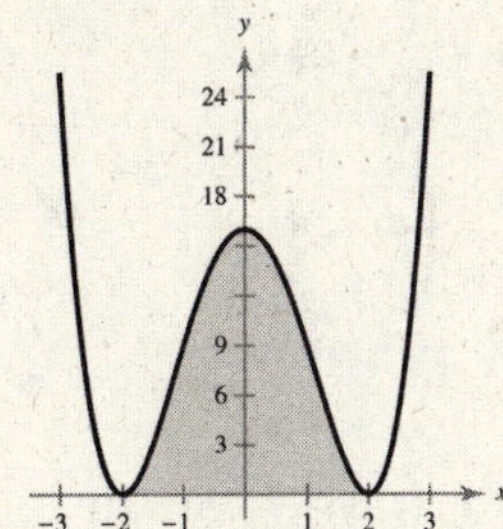

17. The midpoints are -0.5, 0.5, 1.5, and 2.5.

The approximate area is

$$A \approx \frac{3-(-1)}{4}\left[f(-0.5) + f(0.5) + f(1.5) + f(2.5)\right]$$

$$\approx 1.39.$$

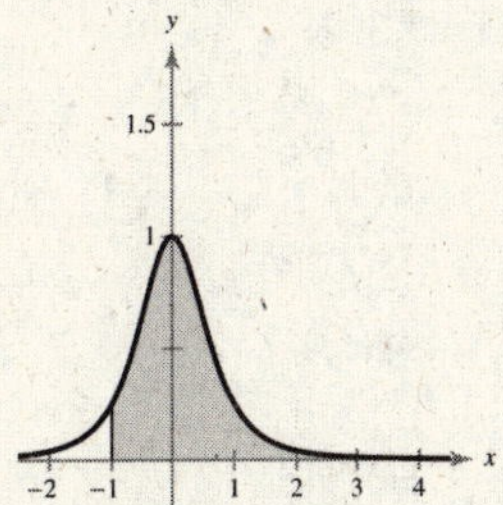

19.

n	10	20	30	40	50
Approximation	15.4543	15.4628	15.4644	15.4650	15.4652

21.

n	10	20	30	40	50
Approximation	5.8520	5.8526	5.8528	5.8528	5.8528

23. $\int_0^4 \left(2x^2 + 3\right) dx = \left[\frac{2x^3}{3} + 3x\right]_0^4 = \frac{164}{3} \approx 54.6667$

Using the Midpoint Rule with $n = 31$, you get 54.66.

25. $\int_1^4 \frac{1}{x+1}\, dx = \ln(x+1)\Big]_1^4 = \ln 5 - \ln 2 \approx 0.916$

Using the Midpoint Rule with $n = 5$, you get 0.913.

27. $A \approx \frac{160}{8}(25 + 52 + 68 + 82 + 77.5 + 74 + 77.5 + 40) \approx 9920 \text{ ft}^2$

29. $A \approx \frac{32}{8}(5.5 + 12.25 + 13.85 + 14.1 + 14.1 + 14.6 + 14.25 + 6.75) \approx 381.6 \text{ mi}^2$

31. Midpoint Rule: $\int_0^1 \frac{4}{1+x^2}\, dx \approx 3.146801$

Graphing utility: $\int_0^1 \frac{4}{1+x^2}\, dx \approx 3.141593$

Review Exercises for Chapter 5

1. $\int 16\, dx = 16x + C$

3. $\int \frac{3}{5}x\, dx = \frac{3}{10}x^2 + C$

5. $\int 3x^2\,dx = x^3 + C$

7. $\int\left(2x^2 + 5x\right)dx = \frac{2}{3}x^3 + \frac{5}{2}x^2 + C$

9. $\int \frac{2}{3\sqrt[3]{x}}\,dx = \int \frac{2}{3}x^{-1/3}\,dx = x^{2/3} + C$

11. $\int\left(3\sqrt{x^4} + 3x\right)dx = \int\left(x^{4/3} + 3x\right)dx$

$= \frac{3}{7}x^{7/3} + \frac{3}{2}x^2 + C$

13. $\int \frac{2x^4 - 1}{\sqrt{x}}\,dx = \int\left(2x^{7/2} - x^{-1/2}\right)dx$

$= \frac{4}{9}x^{9/2} - 2\sqrt{x} + C$

15. $f(x) = \int 12x\,dx = 6x^2 + C$

When $f(0) = -3$, $6(0)^2 + C = -3$

$C = -3.$

So, $f(x) = 6x^2 - 3.$

17. $f(x) = \int\left(3x^2 - 8x\right)dx = x^3 - 4x^2 + C$

When $f(1) = 12$, $(1)^3 - 4(1)^2 + C = 12$

$C = 15.$

So, $f(x) = x^3 - 4x^2 + 15.$

19. $v(t) = \int -32\,dt = -32t + C_1$

$v(0) = 80$

$-32(0) + C_1 = 80$

$C_1 = 80$

$v(t) = -32t + 80$

$s(t) = \int(-32t + 80)\,dt = -16t^2 + 80t + C_2$

$s(0) = 0$

$-16(0)^2 + 80(0) + C_2 = 0$

$C_2 = 0$

$s(t) = -16t^2 + 80t$

To find how long the object will be in the air, set $s(t)$ equal to 0 and solve for t.

$-16t^2 + 80t = 0$

$-16t(t - 5) = 0$

$t = 0$ and $t = 5$

The object will be in the air for 5 seconds.

21. $\int(x + 4)^3\,dx = \frac{1}{4}(x + 4)^4 + C$

23. $\int(5x + 1)^4(5)\,dx = \frac{1}{5}(5x + 1)^5 + C$

25. $\int(1 + 5x)^2\,dx = \int\left(1 + 10x + 25x^2\right)dx$

$= x + 5x^2 + \frac{25}{3}x^3 + C$

or

$\int(1 + 5x)^2\,dx = \frac{1}{5}\frac{(1 + 5x)^3}{3} + C_1$

$= \frac{1}{15}(1 + 5x)^3 + C_1$

27. $\int x^2\left(3x^3 + 1\right)^2 dx = \frac{1}{9}\int\left(3x^3 + 1\right)^2\left(9x^2\right)dx$

$= \frac{1}{9}\left[\frac{\left(3x^3 + 1\right)^3}{3}\right] + C$

$= \frac{1}{27}\left(3x^3 + 1\right)^3 + C$

29. $\int x^2\left(2x^3 - 5\right)^{-3} dx = \frac{1}{6}\int\left(2x^3 - 5\right)^{-3}\left(6x^2\right)dx$

$= \frac{1}{6}\left[\frac{\left(2x^3 - 5\right)^{-2}}{-2}\right] + C$

$= -\frac{1}{12\left(2x^3 - 5\right)^2} + C$

31. $\int \frac{1}{\sqrt{5x - 1}}\,dx = \frac{1}{5}\int(5x - 1)^{-1/2}(5)\,dx$

$= \frac{1}{5}(2)(5x - 1)^{1/2} + C$

$= \frac{2}{5}\sqrt{5x - 1} + C$

33. $P = \int 2t\left(0.001t^2 + 0.5\right)^{1/4} dt$

$= 800\left(0.001t^2 + 0.5\right)^{5/4} + C$

When $P(0) = 0$, $800(0.5)^{5/4} + C = 0$ and $C \approx -336.36.$

So, $P = 800\left(0.001t^2 + 0.5\right)^{5/4} - 336.36.$

(a) $P(6) \approx 30.5$ board feet

(b) $P(12) \approx 125.2$ board feet

35. $\int 4e^{4x}\,dx = e^{4x} + C$

37. $\int e^{-5x}\,dx = -\frac{1}{5}\int e^{-5x}(-5)\,dx = -\frac{1}{5}e^{-5x} + C$

39. $\int 7xe^{3x^2}\,dx = \frac{7}{6}\int e^{3x^2}(6x)\,dx = \frac{7}{6}e^{3x^2} + C$

41. $\int \frac{1}{x-6}\,dx = \ln|x-6| + C$

43. $\int \frac{4}{6x-1}\,dx = \frac{2}{3}\int \frac{1}{6x-1}(6)\,dx = \frac{2}{3}\ln|6x-1| + C$

45. $\int \frac{x^2}{1-x^3}\,dx = -\frac{1}{3}\int \frac{1}{1-x^3}\left(-3x^2\right)dx$

$= -\frac{1}{3}\ln\left|1-x^3\right| + C$

47. $\int_0^3 2\,dx$

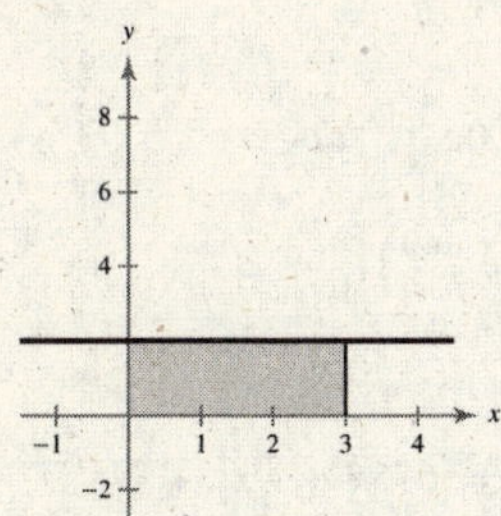

$A = \ell \cdot w = (3)(2) = 6$

49. $\int_0^4 (4-x)\,dx$

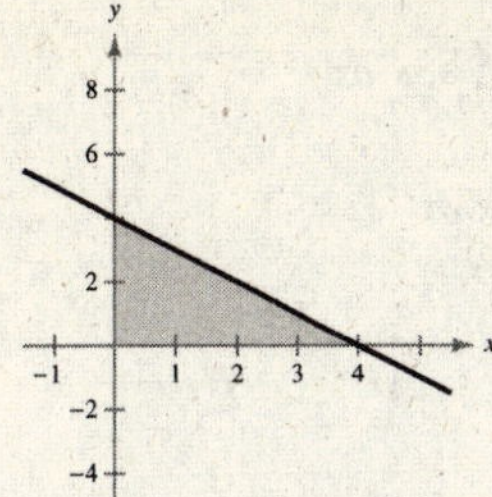

$A = \frac{1}{2}bh = \frac{1}{2}(4)(4) = 8$

51. (a) $\int_2^6 \left[f(x) + g(x)\right]dx = \int_2^6 f(x)\,dx + \int_2^6 g(x)\,dx = 10 + 3 = 13$

(b) $\int_2^6 \left[f(x) - g(x)\right]dx = \int_2^6 f(x)\,dx - \int_2^6 g(x)\,dx = 10 - 3 = 7$

(c) $\int_2^6 \left[2f(x) - 3g(x)\right]dx = 2\int_2^6 f(x)\,dx - 3\int_2^6 g(x)\,dx = 2(10) - 3(3) = 11$

(d) $\int_2^6 5f(x)\,dx = 5\int_2^6 f(x)\,dx = 5(10) = 50$

53. $A = \int_{-2}^2 \left(4 - x^2\right)dx$

$= 2\int_0^2 \left(4 - x^2\right)dx$

$= 2\left[4x - \frac{1}{3}x^3\right]_0^2$

$= \frac{32}{3}$

55. $A = \int_0^1 \frac{2}{x+1}\,dx$

$= 2\left[\ln(x+1)\right]_0^1$

$= 2\ln 2$

≈ 1.39

57. $A = \int_0^1 2e^{x/2}\,dx = 4\int_0^1 e^{x/2}\left(\frac{1}{2}\right)dx$

$= \left[4e^{x/2}\right]_0^1 = 4e^{1/2} - 4$

59. $\int_0^4 (2 + x)\,dx = \left[2x + \frac{1}{2}x^2\right]_0^4 = 16$

61. $\int_{-1}^1 \left(4t^3 - 2t\right)dt = \left[t^4 - t^2\right]_{-1}^1 = 0$

63. $A = \int_{-2}^0 (x+2)^3\,dx = \left[\frac{1}{4}(x+2)^4\right]_{-2}^0$

$= \frac{1}{4}(2)^4 - \frac{1}{4}(0)^4$

$= 4$

65. $\int_0^3 \frac{1}{\sqrt{1+x}}\,dx = \int_0^3 (1+x)^{-1/2}\,dx$

$= \left[2\sqrt{1+x}\right]_0^3$

$= 2$

67. $\int_3^9 \frac{5}{x}\,dx = \left[5\ln|x|\right]_3^9$

$= 5\ln 9 - 5\ln 3$

$= 5(\ln 9 - \ln 3)$

$= 5\ln 3 \approx 5.49$

69. $\int_0^{\ln 5} e^{x/5}\,dx = \left[5e^{x/5}\right]_0^{\ln 5} = 5^{6/5} - 5 \approx 1.899$

71. Average value $= \dfrac{1}{2-0}\int_0^2 3x\,dx$

$= \dfrac{1}{2}\int_0^2 3x\,dx$

$= \left[\dfrac{3}{4}x^2\right]_0^2$

$= 3 - 0$

$= 3$

To find the value of x for which $f(x) = 3$, you solve $3x = 3$ for x.

$3x = 3$

$x = 1$

73. Average value $= \dfrac{1}{3-0}\int_0^3 -2e^x\,dx$

$= \dfrac{1}{3}\int -2e^x\,dx$

$= \left[-\dfrac{2}{3}e^x\right]_0^3$

$= -\dfrac{2}{3}e^3 - \left(-\dfrac{2}{3}e^0\right)$

$= -\dfrac{2}{3}e^3 + \dfrac{2}{3}$ or $-\dfrac{2}{3}(e^3 - 1)$

To find the value of x for which $f(x) = -\dfrac{2}{3}(e^3 - 1)$, you solve $-2e^x = -\dfrac{2}{3}(e^3 - 1)$ for x.

$-2e^x = -\dfrac{2}{3}(e^3 - 1)$

$e^x = \dfrac{1}{3}(e^3 - 1)$

$x = \ln\left(\dfrac{1}{3}(e^3 - 1)\right) \approx 1.850$

75. Average value: $\dfrac{1}{9-4}\int_4^9 \dfrac{1}{\sqrt{x}}\,dx = \dfrac{1}{5}\left[2\sqrt{x}\right]_4^9 = \dfrac{2}{5}$

To find the value of x for which $f(x) = \dfrac{2}{5}$, you solve $\dfrac{1}{\sqrt{x}} = \dfrac{2}{5}$ for x.

$\dfrac{1}{\sqrt{x}} = \dfrac{2}{5}$

$2\sqrt{x} = 5$

$x = \dfrac{25}{4}$

77. $\int_{-2}^{2} 6x^5\,dx = 0$ since $f(x) = 6x^5$ is an odd function.

79. $\int_{-3}^{3}\left(x^4 + x^2\right)dx = 2\int_0^3\left(x^4 + x^2\right)dx$

$= 2\left[\frac{1}{5}x^5 + \frac{1}{3}x^3\right]_0^3$

$= 2\left(\frac{288}{5}\right)$

$= \frac{576}{5}$

since $f(x) = x^4 + x^2$ is an even function.

81. Amount $= e^{rT}\int_0^T c(t)e^{-rt}\,dt$

$= e^{0.06(5)}\int_0^5 3000e^{-0.06t}\,dt$

$= e^{0.3}\left(-50{,}000e^{-0.06t}\right)\Big]_0^5$

$= \$17{,}492.94$

83. $C = \int(675 + 0.5x)\,dx = 675x + \frac{1}{4}x^2 + C$

$C(51) - C(50) = \$700.25$

85. Average value $= \dfrac{1}{b-a}\int_a^b f(t)\,dt$

$= \dfrac{1}{2-0}\int_0^2 500e^{0.04t}\,dt$

$= 250\left[\dfrac{e^{0.04t}}{0.04}\right]_0^2$

$= 6250\left[e^{0.08} - 1\right]$

$= \$520.54$

87.

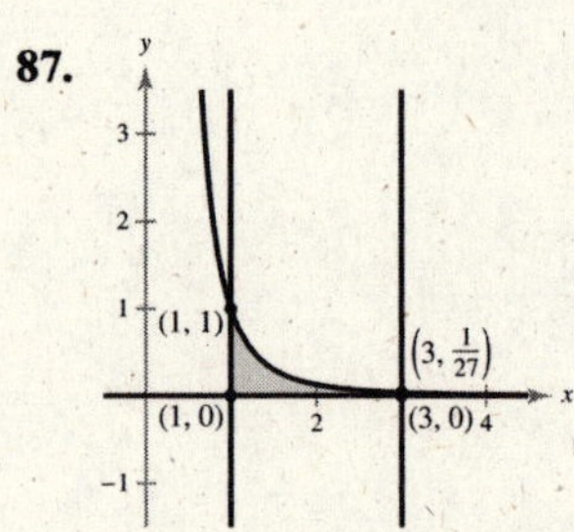

$A = \int_1^3 \dfrac{1}{x^3}\,dx = \int_1^3 x^{-3}\,dx = \left[-\dfrac{1}{2x^2}\right]_1^3 = \dfrac{4}{9}$

89. $(x-3)^2 = 8-(x-3)^2$

$$2(x-3)^2 = 8$$
$$(x-3)^2 = 4$$
$$x-3 = \pm 2$$
$$x = 1, 5$$

$$A = \int_1^5 \left\{\left[8-(x-3)^2\right]-(x-3)^2\right\} dx$$
$$= \int_1^5 \left[8-2(x-3)^2\right] dx$$
$$= \left[8x - \tfrac{2}{3}(x-3)^3\right]_1^5$$
$$= \tfrac{64}{3}$$

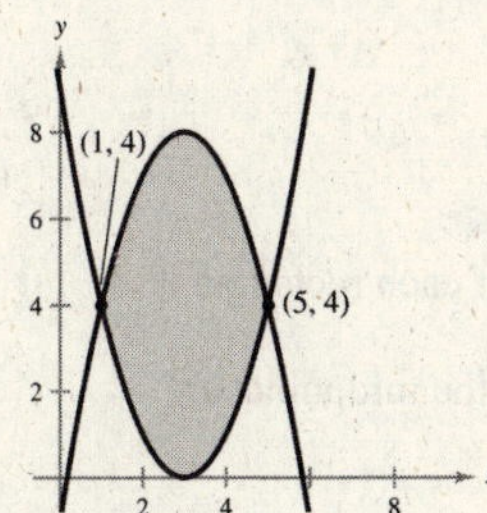

91.

$$A = \int_0^8 \frac{4}{\sqrt{x+1}}\, dx$$
$$= 4\int_0^8 (x+1)^{-1/2}\, dx$$
$$= 4(2)(x+1)^{1/2}\Big]_0^8$$
$$= 8\sqrt{x+1}\Big]_0^8 = 16$$

93. $x^3 = x$

$$x^3 - x = 0$$
$$x(x^2-1) = 0$$
$$x = 0, \pm 1$$

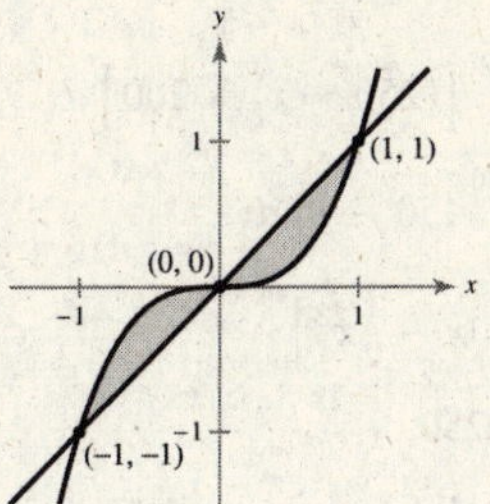

$$2\int_0^1 (x - x^3)\, dx = 2\left[\tfrac{1}{2}x^2 - \tfrac{1}{4}x^4\right]_0^1 = \tfrac{1}{2}$$

95. Demand function = Supply function

$$36 - 0.35x = 0.05x$$
$$36 = 0.4x$$
$$x = 90 \Rightarrow \text{price} = \$4.50$$

$$\text{Consumer surplus} = \int_0^{90} (\text{Demand function} - \text{Price})\, dx$$
$$= \int_0^{90} \left[(36 - 0.35x) - 4.50\right] dx$$
$$= \int_0^{90} (31.5 - 0.35x)\, dx$$
$$= \left[31.5x - 0.175x^2\right]_0^{90}$$
$$= \$1417.50$$

$$\text{Producer surplus} = \int_0^{90} (\text{Price} - \text{Supply function})\, dx$$
$$= \int_0^{90} \left[4.50 - (0.05x)\right] dx$$
$$= \left[4.50x - 0.025x^2\right]_0^{90}$$
$$= \$202.50$$

97. Demand function = Supply function

$$250 - x = 150 + x$$
$$100 = 2x$$
$$x = 50 \Rightarrow \text{price} = \$200$$

$$\text{Consumer surplus} = \int_0^{50} (\text{Demand function} - \text{Price})\, dx = \int_0^{50} \left[(250 - x) - 200\right] dx = \int_0^{50} (50 - x)\, dx = \left[50x - \tfrac{1}{2}x^2\right]_0^{50} = \$1250$$

$$\text{Producer surplus} = \int_0^{50} (\text{Price} - \text{Supply function})\, dx = \int_0^{50} \left[200 - (150 + x)\right] dx = \int_0^{50} (50 - x)\, dx = \left[50x - \tfrac{1}{2}x^2\right]_0^{50} = \$1250$$

99. Total revenue for $R_1 = \int_{15}^{21} (24.3 + 8.24t)\, dt = \left[24.3t + 4.12t^2\right]_{15}^{21} = \$1{,}035.72$ million

Total revenue for $R_2 = \int_{15}^{21} (21.6 + 9.36t)\, dt = \left[21.6t + 4.68t^2\right]_{15}^{21} = \$1{,}140.48$ million

The total revenue for R_2 is greater by \$104.76 million.

101. Profit = Revenue − Cost

$$P = 70 - (30 + 0.3t^2) = 40 - 0.3t^2$$

$$\text{Total profit} = \int_0^{10} (40 - 0.3t^2)\, dt = \left[40t - 0.1t^3\right]_0^{10} = \$300 \text{ million}$$

103. The width of each rectangle is $\Delta x = \dfrac{3 - 0}{3} = 1$ and the midpoints are $\dfrac{1}{2}, \dfrac{3}{2}$, and $\dfrac{5}{2}$.

The approximate area is

$$\text{Area} \approx (1)\left[f\left(\frac{1}{2}\right) + f\left(\frac{3}{2}\right) + f\left(\frac{5}{2}\right)\right] = 1.5.$$

The exact area is

$$A = \int_0^3 \frac{1}{3}x\, dx = \left[\frac{1}{6}x^2\right]_0^3 = 1.5.$$

105. Using $n = 4$, the width of each rectangle is $\Delta x = \dfrac{2 - 0}{4} = \dfrac{1}{2}$ and the midpoints are $\dfrac{1}{4}, \dfrac{3}{4}, \dfrac{5}{4}$, and $\dfrac{7}{4}$.

The approximate area is

$$A \approx \left(\frac{1}{2}\right)\left[f\left(\frac{1}{4}\right) + f\left(\frac{3}{4}\right) + f\left(\frac{5}{4}\right) + f\left(\frac{7}{4}\right)\right] = 2.625.$$

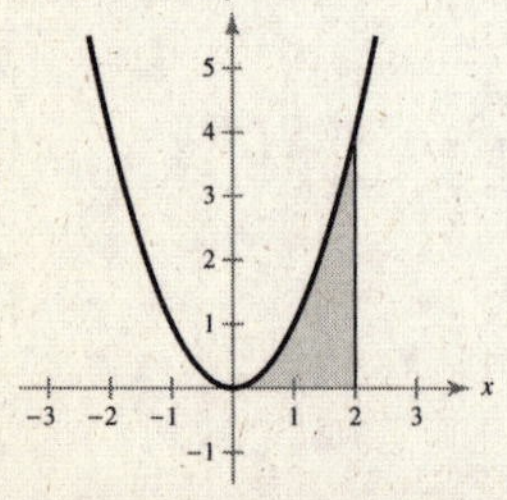

107. Using $n = 4$, the width of each rectangle is $\Delta x = \dfrac{1 - (-1)}{4} = \dfrac{1}{2}$ and the midpoints are $-\dfrac{3}{4}, -\dfrac{1}{4}, \dfrac{1}{4}$, and $\dfrac{3}{4}$.

The approximate area is

$$A \approx \left(\frac{1}{2}\right)\left[f\left(-\frac{3}{4}\right) + f\left(-\frac{1}{4}\right) + f\left(\frac{1}{4}\right) + f\left(\frac{3}{4}\right)\right] = 1.0703125.$$

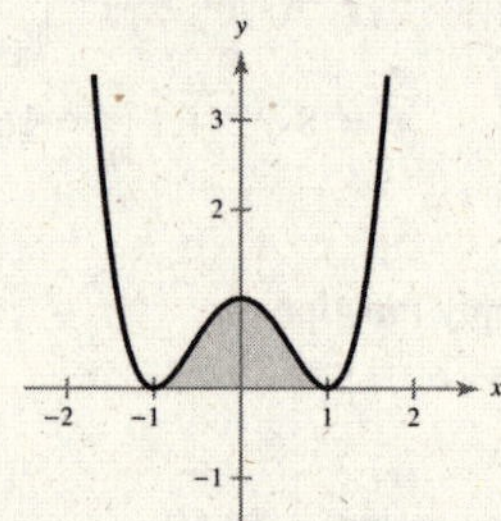

109. Using $n = 6$, the width of each rectangle is $\Delta x = \dfrac{3-0}{6} = \dfrac{1}{2}$ and the midpoints are $\dfrac{1}{4}, \dfrac{3}{4}, \dfrac{5}{4}, \dfrac{7}{4}, \dfrac{9}{4}$, and $\dfrac{11}{4}$.

The approximate area is

$$A \approx \left(\frac{1}{2}\right)\left[f\left(\frac{1}{4}\right) + f\left(\frac{3}{4}\right) + f\left(\frac{5}{4}\right) + f\left(\frac{7}{4}\right) + f\left(\frac{9}{4}\right) + f\left(\frac{11}{4}\right)\right]$$

$$= 13.5.$$

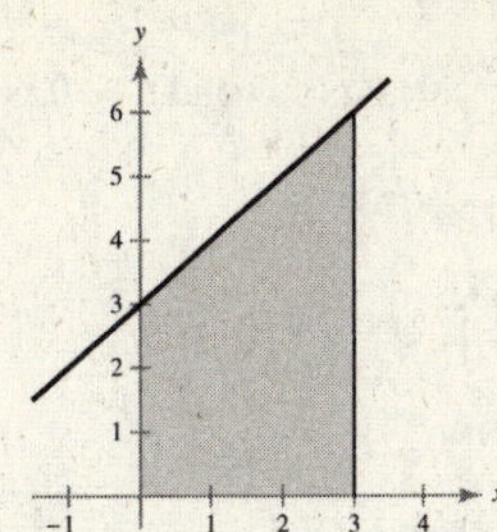

111. Using $n = 6$, the width of each rectangle is $\Delta x = \dfrac{2-0}{6} = \dfrac{1}{3}$ and the midpoints are $\dfrac{1}{6}, \dfrac{1}{2}, \dfrac{5}{6}, \dfrac{7}{6}, \dfrac{3}{2}$, and $\dfrac{11}{6}$.

The approximate area is

$$A \approx \left(\frac{1}{3}\right)\left[f\left(\frac{1}{6}\right) + f\left(\frac{1}{2}\right) + f\left(\frac{5}{6}\right) + f\left(\frac{7}{6}\right) + f\left(\frac{3}{2}\right) + f\left(\frac{11}{6}\right)\right]$$

$$\approx 3.0319.$$

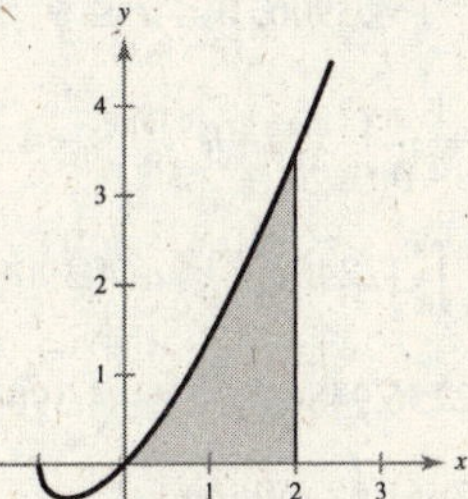

113. Area $\approx \frac{180}{9}(27.5 + 62.5 + 65 + 60 + 66 + 68 + 60.5 + 55.5 + 27) = 9840 \text{ ft}^2$

Chapter 5 Test Yourself

1. $\int(9x^2 - 4x + 13)\,dx = 3x^3 - 2x^2 + 13x + C$

2. $\int(x+1)^2\,dx = \frac{1}{3}(x+1)^3 + C$

3. $\int 4x^3\sqrt{x^4 - 7}\,dx = \frac{2}{3}(x^4 - 7)^{3/2} + C$

4. $\int \dfrac{5x-6}{\sqrt{x}}\,dx = \int \dfrac{5x}{\sqrt{x}}\,dx - \int \dfrac{6}{\sqrt{x}}\,dx$

$= \dfrac{10}{3}x^{3/2} - 12x^{1/2} + C$

5. $\int 15e^{3x}\,dx = 5e^{3x} + C$

6. $\int \dfrac{3}{4x-1}\,dx = \dfrac{3}{4}\int \dfrac{1}{4x-1}(4)\,dx = \dfrac{3}{4}\ln|4x-1| + C$

7. $f(x) = \int(6x-5)\,dx = 3x^2 - 5x + C$

$f(-1) = 6$

$3(-1)^2 - 5(-1) + C = 6$

$C = -2$

$f(x) = 3x^2 - 5x - 2$

8. $\int(e^x + 1)\,dx = e^x + x + C$

$1 = e^0 + 0 + C$

$C = 0$

$f(x) = e^2 + x$

9. $\int_0^1 16x\,dx = 8x^2\Big]_0^1 = 8$

10. $\int_{-3}^3 (3 - 2x)\,dx = 3x - x^2\Big]_{-3}^3 = 18$

11. $\int_{-1}^1 (x^3 + x^2)\,dx = \frac{1}{4}x^4 + \frac{1}{3}x^3\Big]_{-1}^1 = \frac{2}{3}$

12. $\int_{-1}^2 \dfrac{2x}{\sqrt{x^2+1}}\,dx = 2(x^2+1)^{1/2}\Big]_{-1}^2$

$= 2\sqrt{5} - 2\sqrt{2}$

≈ 1.64

13. $\int_0^3 e^{4e}\,dx = \dfrac{1}{4}e^{4x}\Big]_0^3$

$= \dfrac{e^{12}}{4} - \dfrac{1}{4}$

$\approx 40{,}688.45$

14. $\int_{-2}^{3} \frac{1}{x+3}\,dx = \ln(x+3)\Big]_{-2}^{3} = \ln(6) \approx 1.79$

15. $\frac{dS}{dt} = 226.912e^{0.1013t}$, $0 \le t \le 9$ and $t = 0$ is 2000.

(a) $S(t) = \int 226.912e^{0.1013t}\,dt$

$$= \frac{226.912}{0.1013}\int e^{0.1013t}(0.1013)\,dt$$

$$= 2240e^{0.1013t} + C$$

$$S(4) = 3363.5$$

$$2240e^{0.1013(4)} + C = 3363.5$$

$$C \approx 4.3906$$

$$S(t) = 2240e^{0.1013t} + 4.3906,\ 0 \le t \le 9$$

(b) Average sales $= \frac{1}{9-0}\int_0^9 \left(2240e^{0.1013t} + 4.3906\right)dt$

$$= \frac{1}{9}\int_0^9 \left(2240e^{0.1013t} + 4.3906\right)dt$$

$$= \Big[2256.9486e^{0.1013t} + 4.3906t\Big]_0^9$$

$$\approx \$3661.68 \text{ million}$$

16.

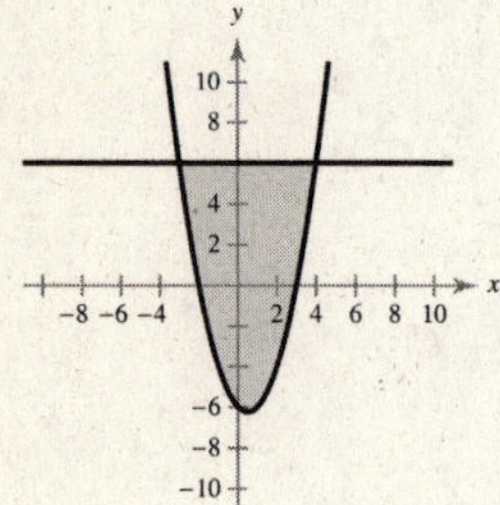

$$A = \int_{-3}^{4}\left[6 - \left(x^2 - x - 6\right)\right]dx$$

$$= \int_{-3}^{4}\left(-x^2 + x + 12\right)dx$$

$$= \left[-\tfrac{1}{3}x^3 + \tfrac{1}{2}x^2 + 12x\right]_{-3}^{4}$$

$$= \frac{343}{6}$$

17.

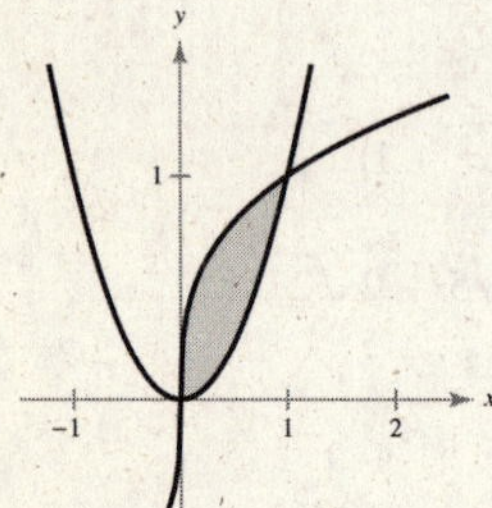

$$A = \int_0^1 \left(x^{1/3} - x^2\right)dx$$

$$= \left[\tfrac{3}{4}x^{4/3} - \tfrac{1}{3}x^3\right]_0^1$$

$$= \frac{5}{12}$$

18. $-0.625x + 10 = 0.25x + 3$

$$\tfrac{7}{8}x = 7$$

$$x = 8;\ \text{price} = 5$$

Consumer surplus $= \int_0^8 \left[(-0.625x + 10) - 5\right]dx$

$$= -0.3125x^2 + 5x\Big]_0^8$$

$$= 20 \text{ million}$$

Producer surplus $= \int_0^8 \left[5 - (0.25x + 3)\right]dx$

$$= -0.125x^2 + 2x\Big]_0^8$$

$$= 8 \text{ million}$$

19. The midpoints are $\frac{1}{8}, \frac{3}{8}, \frac{5}{8}$, and $\frac{7}{8}$.

The approximate area is

$$A \approx \frac{1-0}{4}\left[f\left(\frac{1}{8}\right) + f\left(\frac{3}{8}\right) + f\left(\frac{5}{8}\right) + f\left(\frac{7}{8}\right)\right]$$

$$= \frac{1}{4}\left(\frac{3}{64} + \frac{27}{64} + \frac{75}{64} + \frac{147}{64}\right)$$

$$= \frac{63}{64}$$

$$\approx 0.984.$$

The exact area is

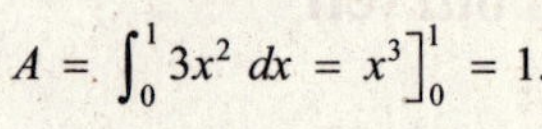

$$A = \int_0^1 3x^2\,dx = x^3\Big]_0^1 = 1.$$

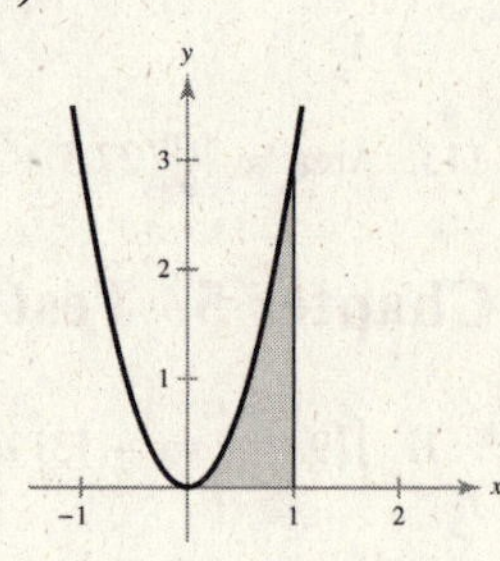

20. The midpoints are $-\frac{3}{4}, -\frac{1}{4}, \frac{1}{4}$ and $\frac{3}{4}$.

The approximate area is

$$A \approx \frac{1-(-1)}{4}\left[f\left(-\frac{3}{4}\right) + f\left(-\frac{1}{4}\right) + f\left(\frac{1}{4}\right) + f\left(\frac{3}{4}\right)\right]$$

$$= \frac{1}{2}\left(\frac{25}{16} + \frac{17}{16} + \frac{17}{16} + \frac{25}{16}\right)$$

$$= \frac{42}{16}$$

$$= 2.625.$$

The exact area is

$$A = \int_{-1}^{1}\left(x^2 + 1\right)dx = \frac{1}{3}x^3 + x\Big]_{-1}^{1} = 2.67.$$

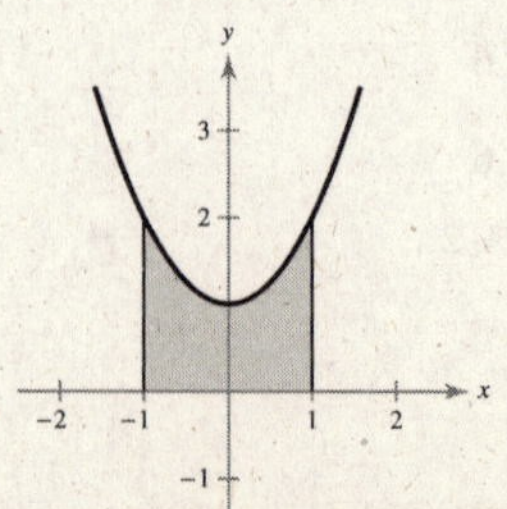

CHAPTER 6
Techniques of Integration

CHAPTER 6
Techniques of Integration

Section 6.1 Integration by Parts and Present Value

Skills Warm Up

1. $f(x) = \ln(x + 1)$

   $f'(x) = \dfrac{1}{x + 1}$

2. $f(x) = \ln(x^2 - 1)$

   $f'(x) = \dfrac{2x}{x^2 - 1}$

3. $f(x) = e^{x^3}$

   $f'(x) = 3x^2 e^{x^3}$

4. $f(x) = e^{-x^2}$

   $f'(x) = -2xe^{-x^2}$

5. $f(x) = x^2 e^x$

   $f'(x) = x^2 e^x + 2xe^x$

6. $f(x) = xe^{-2x}$

   $f'(x) = -2xe^{-2x} + e^{-2x}$

7. $$\begin{aligned} A &= \int_a^b \left[f(x) - g(x)\right] dx \\ &= \int_{-2}^{2} \left[(-x^2 + 4) - (x^2 - 4)\right] dx \\ &= \int_{-2}^{2} (-2x^2 + 8)\, dx \\ &= \left[-\tfrac{2}{3}x^3 + 8x\right]_{-2}^{2} \\ &= \left(-\tfrac{16}{3} + 16\right) - \left(\tfrac{16}{3} - 16\right) = \tfrac{64}{3} \end{aligned}$$

8. $$\begin{aligned} A &= \int_a^b \left[f(x) - g(x)\right] dx \\ &= \int_{-1}^{1} \left[(-x^2 + 2) - 1\right] dx \\ &= \int_{-1}^{1} (-x^2 + 1)\, dx \\ &= \left[-\tfrac{1}{3}x^3 + x\right]_{-1}^{1} \\ &= \left(-\tfrac{1}{3} + 1\right) - \left(\tfrac{1}{3} - 1\right) \\ &= \tfrac{4}{3} \end{aligned}$$

9. $$\begin{aligned} A &= \int_a^b \left[f(x) - g(x)\right] dx \\ &= \int_{-1}^{5} \left[4x - (x^2 - 5)\right] dx \\ &= \int_{-1}^{5} (-x^2 + 4x + 5)\, dx \\ &= \left[-\tfrac{1}{3}x^3 + 2x^2 + 5x\right]_{-1}^{5} \\ &= \left(-\tfrac{125}{3} + 50 + 25\right) - \left(\tfrac{1}{3} + 2 - 5\right) \\ &= 36 \end{aligned}$$

10. $$\begin{aligned} A &= \int_a^b \left[f(x) - g(x)\right] dx \\ &= \int_{-1}^{1} \left[f(x) - g(x)\right] dx + \int_1^3 \left[g(x) - f(x)\right] dx \\ &= \int_{-1}^{1} \left[(x^3 - 3x^2 + 2) - (x - 1)\right] dx + \int_1^3 \left[(x - 1) - (x^3 - 3x^2 + 2)\right] dx \\ &= \int_{-1}^{1} (x^3 - 3x^2 - x + 3)\, dx + \int_1^3 (-x^3 + 3x^2 + x - 3)\, dx \\ &= \left[\tfrac{1}{4}x^4 - x^3 - \tfrac{1}{2}x^2 + 3x\right]_{-1}^{1} + \left[-\tfrac{1}{4}x^4 + x^3 + \tfrac{1}{2}x^2 - 3x\right]_1^3 \\ &= \left(\tfrac{1}{4} - 1 - \tfrac{1}{2} + 3\right) - \left(\tfrac{1}{4} + 1 - \tfrac{1}{2} - 3\right) + \left(-\tfrac{81}{4} + 27 + \tfrac{9}{2} - 9\right) - \left(-\tfrac{1}{4} + 1 + \tfrac{1}{2} - 3\right) = 8 \end{aligned}$$

1. $\int xe^{3x}\,dx$

$u = x,\ dv = e^{3x}\,dx$

3. $\int x \ln 2x\,dx$

$u = \ln 2x,\ dv = x\,dx$

5. Let $u = x$ and $dv = e^{3x}\,dx$. Then $du = dx$ and $v = \frac{1}{3}e^{3x}$.

$$\begin{aligned}\int xe^{3x}\,dx &= \tfrac{1}{3}xe^{3x} - \int \tfrac{1}{3}e^{3x}\,dx\\ &= \tfrac{1}{3}xe^{3x} - \tfrac{1}{9}e^{3x} + C\\ &= \tfrac{1}{9}e^{3x}(3x-1) + C\end{aligned}$$

7. Let $u = \ln x$ and $dv = x^3\,dx$. Then $du = \frac{1}{x}\,dx$ and $v = \frac{x^4}{4}$.

$$\begin{aligned}\int x^3 \ln x\,dx &= \frac{x^4}{4}\ln x - \int \frac{x^4}{4}\left(\frac{1}{x}\right)dx\\ &= \frac{x^4}{4}\ln x - \int \frac{x^3}{4}\,dx\\ &= \frac{x^4}{4}\ln x - \frac{x^4}{16} + C\\ &= \frac{x^4}{16}(4\ln x - 1) + C\end{aligned}$$

9. Let $u = \ln 2x$ and $dv = dx$. Then $du = \frac{1}{x}\,dx$ and $v = x$.

$$\begin{aligned}\int \ln 2x\,dx &= x\ln 2x - \int x\left(\frac{1}{x}\right)dx\\ &= x\ln 2x - \int dx\\ &= x\ln 2x - x + C\\ &= x(\ln 2x - 1) + C\end{aligned}$$

11. Let $u = x^2$ and $dv = e^{-x}\,dx$. Then $du = 2x\,dx$ and $v = -e^{-x}$.

$$\int x^2e^{-x}\,dx = -x^2e^{-x} + 2\int xe^{-x}\,dx$$

Let $u = x$ and $dv = e^{-x}\,dx$. Then $du = dx$ and $v = -e^{-x}$.

$$\begin{aligned}\int x^2e^{-x}\,dx &= -x^2e^{-x} + 2\left[-xe^{-x} + \int e^{-x}\,dx\right]\\ &= -x^2e^{-x} - 2xe^{-x} - 2e^{-x} + C\\ &= -e^{-x}\left(x^2 + 2x + 2\right) + C\end{aligned}$$

13. Let $u = \ln x$ and $dv = x^{1/2}\,dx$.

Then $du = \frac{1}{x}\,dx$ and $v = \frac{2}{3}x^{3/2}$.

$$\begin{aligned}\int x^{1/2}\ln x\,dx &= \frac{2}{3}x^{3/2}\ln x - \int \frac{2}{3}x^{3/2}\left(\frac{1}{x}\right)dx\\ &= \frac{2}{3}x^{3/2}\ln x - \frac{2}{3}\int x^{1/2}\,dx\\ &= \frac{2}{3}x^{3/2}\ln x - \frac{4}{9}x^{3/2} + C\end{aligned}$$

15. Let $u = 2x^2$ and $dv = e^x\,dx$. Then $du = 4x\,dx$ and $v = e^x$.

$$\int 2x^2e^x\,dx = 2x^2e^x - \int 4xe^x\,dx$$

Let $u = 4x$ and $dv = e^x\,dx$. Then $du = 4\,dx$ and $v = e^x$.

$$\begin{aligned}\int 2x^2e^x\,dx &= 2x^2e^x - \left[4xe^x - \int e^x(4\,dx)\right]\\ &= 2x^2e^x - 4xe^x + 4e^x + C\\ &= 2e^x\left(x^2 - 2x + 2\right) + C\end{aligned}$$

17. $\int e^{4x}\,dx = \frac{1}{4}\int e^{4x}(4)\,dx = \frac{1}{4}e^{4x} + C$

19. Let $u = x$ and $dv = e^{4x}\,dx$. Then $du = dx$ and $v = \frac{1}{4}e^{4x}$.

$$\begin{aligned}\int xe^{4x}\,dx &= \frac{1}{4}xe^{4x} - \frac{1}{4}\int e^{4x}\,dx\\ &= \frac{1}{4}xe^{4x} - \frac{1}{16}e^{4x} + C\\ &= \frac{e^{4x}}{16}(4x-1) + C\end{aligned}$$

21. Let $u = x$ and $dv = e^{-x/4}\,dx$.

Then $du = dx$ and $v = -4e^{-x/4}$.

$$\begin{aligned}\int xe^{-x/4}\,dx &= -4xe^{-x/4} - \int -4e^{-x/4}\,dx\\ &= -4xe^{-x/4} + 4\int e^{-x/4}\,dx\\ &= -4xe^{-x/4} - 16e^{-x/4} + C\end{aligned}$$

23. Let $u = \ln(t+1)$ and $dv = t\,dt$. Then $du = \dfrac{1}{(t+1)}dt$ and $v = \dfrac{t^2}{2}$.

$$\begin{aligned}\int t\ln(t+1)\,dt &= \frac{t^2}{2}\ln(t+1) - \frac{1}{2}\int\frac{t^2}{t+1}\,dt\\ &= \frac{t^2}{2}\ln(t+1) - \frac{1}{2}\int\left(t - 1 + \frac{1}{t+1}\right)dt\\ &= \frac{t^2}{2}\ln(t+1) - \frac{1}{2}\left[\frac{t^2}{2} - t + \ln|t+1|\right] + C\\ &= \frac{1}{4}\left[2t^2\ln(t+1) + t(2-t) - 2\ln|t+1|\right] + C\end{aligned}$$

25. Let $u = \dfrac{1}{t}$. Then $du = \left(-\dfrac{1}{t^2}\right)dt$.

$$\int\frac{e^{1/t}}{t^2}\,dt = -\int e^u\,du = -e^u + C = -e^{1/t} + C$$

27. Let $u = (\ln x)^2$ and $dv = x\,dx$. Then

$du = \dfrac{(2\ln x)}{x}dx$ and $v = \dfrac{x^2}{2}$.

$$\int x(\ln x)^2\,dx = \frac{x^2}{2}(\ln x)^2 - \int x\ln x\,dx$$

Let $u = \ln x$ and $v = x\,dx$. Then $du = \dfrac{1}{x}dx$ and

$v = \dfrac{x^2}{2}$.

$$\begin{aligned}\int x(\ln x)^2\,dx &= \frac{x^2}{2}(\ln x)^2 - \left[\frac{x^2}{2}\ln x - \int\frac{x}{2}\,dx\right]\\ &= \frac{x^2}{2}(\ln x)^2 - \frac{x^2}{2}\ln x + \frac{x^2}{4} + C\\ &= \frac{x^2}{4}\left[2(\ln x)^2 - 2\ln x + 1\right] + C\end{aligned}$$

29. $$\int\frac{(\ln x)^2}{x}\,dx = \int(\ln x)^2\left(\frac{1}{x}\right)dx = \frac{(\ln x)^3}{3} + C$$

31. Let $u = \ln x$ and $dv = \dfrac{1}{x^2}dx$. Then $du = \dfrac{1}{x}dx$ and

$v = -\dfrac{1}{x}$.

$$\begin{aligned}\int\frac{\ln x}{x^2}\,dx &= -\frac{\ln x}{x} - \int -\frac{1}{x}\cdot\frac{1}{x}\,dx\\ &= -\frac{\ln x}{x} + \int\frac{1}{x^2}\,dx\\ &= \frac{\ln x}{x} - \frac{1}{x} + C\\ &= -\frac{\ln x + 1}{x} + C\end{aligned}$$

33. Let $u = x$ and $dv = \sqrt{x-1}\,dx$. Then $du = dx$ and $v = \frac{2}{3}(x-1)^{3/2}$.

$$\begin{aligned}\int x\sqrt{x-1}\,dx &= \tfrac{2}{3}x(x-1)^{3/2} - \int\tfrac{2}{3}(x-1)^{3/2}\,dx\\ &= \tfrac{2}{3}x(x-1)^{3/2} - \tfrac{4}{15}(x-1)^{5/2} + C\\ &= \tfrac{2}{15}(x-1)^{3/2}(3x+2) + C\end{aligned}$$

35. $$\begin{aligned}\int x(x+1)^2\,dx &= \int x(x^2 + 2x + 1)\,dx\\ &= \int\left(x^3 + 2x^2 + x\right)dx\\ &= \frac{x^4}{4} + \frac{2x^3}{3} + \frac{x^2}{2} + C\end{aligned}$$

37. Let $u = xe^{2x}$ and $dv = (2x+1)^{-2}\,dx$. Then

$du = e^{2x}(2x+1)\,dx$ and $v = -\dfrac{1}{2(2x+1)}$.

$$\begin{aligned}\int\frac{xe^{2x}}{(2x+1)^2}\,dx &= -\frac{xe^{2x}}{2(2x+1)} + \frac{1}{2}\int e^{2x}\\ &= -\frac{xe^{2x}}{2(2x+1)} + \frac{1}{4}e^{2x} + C\\ &= \frac{e^{2x}}{4(2x+1)} + C\end{aligned}$$

39. Let $u = \ln x$ and $dv = x^5\,dx$. Then $du = \left(\dfrac{1}{x}\right)dx$ and

$v = \dfrac{x^6}{6}$.

$$\begin{aligned}\int_1^e x^5\ln x\,dx &= \frac{x^6}{6}\ln x\Big]_1^e - \int_1^e\frac{x^5}{6}\,dx\\ &= \frac{e^6}{6} - \left[\frac{x^6}{36}\right]_1^e\\ &= \frac{e^6}{6} - \frac{e^6}{36} + \frac{1}{36}\\ &= \frac{5}{36}e^6 + \frac{1}{36}\\ &\approx 56.060\end{aligned}$$

41. Let $u = \ln(1 + 2x)$ and $dv = dx$. Then

$du = \dfrac{2}{1 + 2x}\,dx$ and $v = x$.

$$\int_0^1 \ln(1 + 2x)\,dx = x\ln(1 + 2x)\Big]_0^1 - \int_0^1 \frac{2x}{1 + 2x}\,dx$$
$$= \ln 3 - \int_0^1\left[1 - \frac{1}{2x + 1}\right]dx$$
$$= \ln 3 - \left[x - \frac{1}{2}\ln(2x + 1)\right]_0^1$$
$$= \ln 3 - 1 + \frac{1}{2}\ln 3$$
$$= \frac{3}{2}\ln 3 - 1$$
$$\approx 0.648$$

43. Let $u = x$ and $dv = (x + 1)^{1/2}\,dx$.

Then $du = dx$ and $v = \frac{2}{3}(x + 1)^{3/2}$.

$$\int_0^8 x\sqrt{x + 1}\,dx = \tfrac{2}{3}x(x + 1)^{3/2} - \int \tfrac{2}{3}(x + 1)^{3/2}\,dx$$
$$= \left[\tfrac{2}{3}x(x + 1)^{3/2} - \tfrac{4}{15}(x + 1)^{5/2}\right]_0^8$$
$$= \frac{1192}{15}$$

45. Let $u = x^2$ and $dv = e^x\,dx$. Then $du = 2x\,dx$ and $v = e^x$.

$$\int_1^2 x^2e^x\,dx = x^2e^x\Big]_1^2 - 2\int_1^2 xe^x\,dx$$

Let $u = x$ and $dv = e^x\,dx$. Then $du = dx$ and $v = e^x$.

$$\int_1^2 x^2e^x\,dx = 4e^2 - e - 2\left(\left[xe^x\right]_1^2 - \int_1^2 e^x\,dx\right)$$
$$= 4e^2 - e - 2\left(2e^2 - e - \left[e^x\right]_1^2\right)$$
$$= 4e^2 - e - 4e^2 + 2e + 2\left(e^2 - e\right)$$
$$= 2e^2 - e$$
$$\approx 12.060$$

47. Area $= \int_0^2 x^3e^x\,dx$

Let $u = x^3$ and $dv = e^x\,dx$. Then $du = 3x^2\,dx$ and $v = e^x$.

$$\int x^3e^x\,dx = x^3e^x - 3\int x^2e^x\,dx$$

Let $u = x^2$ and $dv = e^x\,dx$. Then $du = 2x\,dx$ and $v = e^x$.

$$\int x^3e^x\,dx = x^3e^x - 3\left[x^2e^x - \int 2xe^x\,dx\right]$$
$$= x^3e^x - 3x^2e^x + 6\int xe^x\,dx$$

Let $u = x$ and $dv = e^x\,dx$. Then $du = dx$ and $v = e^x$.

$$\int x^3e^x\,dx = x^3e^x - 3x^2e^x + 6\left[xe^x - \int e^x\,dx\right]$$
$$= \left(x^3 - 3x^2 + 6x - 6\right)e^x + C$$

$$\text{Area} = \int_0^2 x^3e^x\,dx = \left[\left(x^3 - 3x^2 + 6x - 6\right)e^x\right]_0^2$$
$$= 2e^2 + 6 \approx 20.778$$

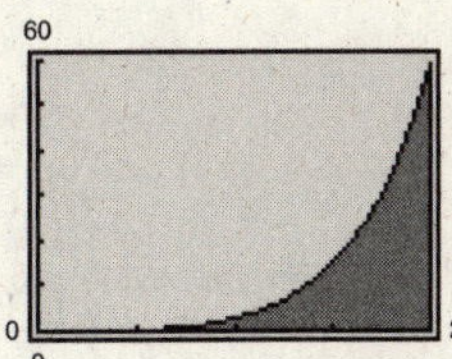

49. Area $= \int_0^3 \frac{1}{9}xe^{-x/3}\,dx = \frac{1}{9}\int_0^3 xe^{-x/3}\,dx$

Let $u = x$ and $dv = e^{-x/3}\,dx$.

Then $du = dx$ and $v = -3e^{-x/3}$.

$$\frac{1}{9}\int xe^{-x/3}\,dx = \frac{1}{9}\left[-3xe^{-x/3} + 3\int e^{-x/3}\,dx\right]$$
$$= \frac{1}{9}\left[-3xe^{-x/3} - 9e^{-x/3}\right] + C$$

$$\text{Area} = \frac{1}{9}\left[-3xe^{-x/3} - 9e^{-x/3}\right]_0^3$$
$$= \left[-\frac{1}{3}e^{-x/3}(x + 3)\right]_0^3$$
$$= \frac{e - 2}{e}$$
$$\approx 0.2642$$

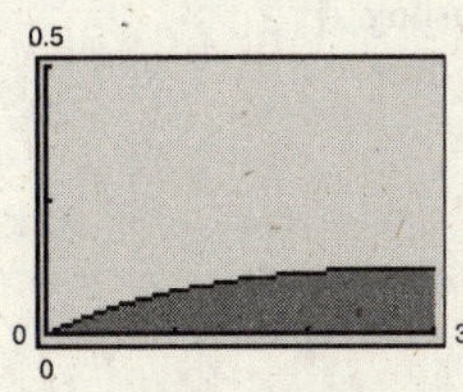

51. Area $= \int_1^e x^2 \ln x \, dx$

Let $u = \ln x$ and $dv = x^2$. Then $du = \frac{1}{x}\,dx$ and

$v = \frac{x^3}{3}$.

$$\int x^2 \ln x \, dx = \frac{x^3}{3}\ln x - \int \frac{x^2}{3}\,dx$$
$$= \frac{x^3}{3}\ln x - \frac{x^3}{9} + C$$

$$\text{Area} = \int_1^e x^2 \ln x \, dx = \left[\frac{x^3}{3}\ln x - \frac{x^3}{9}\right]_1^e$$
$$= \left(\frac{e^3}{3} - \frac{e^3}{9}\right) - \left(-\frac{1}{9}\right) = \frac{2e^3}{9} + \frac{1}{9}$$
$$= 4.575$$

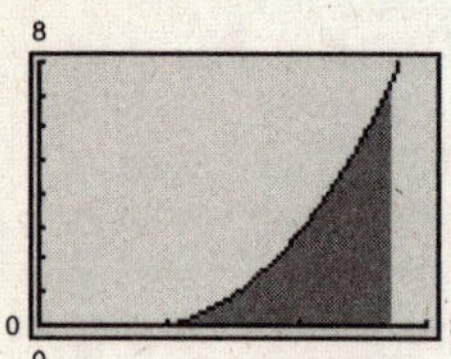

53. Let $u = \ln x$ and $dv = x^n\,dx$. Then $du = \left(\frac{1}{x}\right)dx$ and

$v = \frac{x^{n+1}}{(n+1)}$.

$$\int x^n \ln x \, dx = \frac{x^{n+1}}{n+1}\ln x - \int \frac{1}{x}\cdot\frac{x^{n+1}}{n+1}\,dx$$
$$= \frac{x^{n+1}}{n+1}\ln x - \frac{1}{n+1}\int x^n\,dx$$
$$= \frac{x^{n+1}}{n+1}\ln x - \frac{1}{n+1}\cdot\frac{x^{n+1}}{n+1} + C$$
$$= \frac{x^{n+1}}{(n+1)^2}\left[-1 + (n+1)\ln x\right] + C$$

55. Using $n = 2$ and $a = 5$,

$$\int x^2 e^{5x}\,dx = \frac{x^2e^{5x}}{5} - \frac{2}{5}\int xe^{5x}\,dx.$$

Now, using $n = 1$ and $a = 5$,

$$\int x^2 e^{5x}\,dx = \frac{x^2e^{5x}}{5} - \frac{2}{5}\left[\frac{xe^{5x}}{5} - \frac{1}{5}\int e^{5x}\,dx\right]$$
$$= \frac{x^2e^{5x}}{5} - \frac{2xe^{5x}}{25} + \frac{2e^{5x}}{125} + C$$
$$= \frac{e^{5x}}{125}\left(25x^2 - 10x + 2\right) + C.$$

57. Using $n = -4$,

$$\int x^{-4}\ln x\,dx = \frac{x^{-4+1}}{(-4+1)^2}\left[-1 + (-4+1)\ln x\right] + C$$
$$= \frac{x^{-3}}{9}(-1 - 3\ln x) + C$$
$$= \frac{-1}{9x^3}(1 + 3\ln x) + C$$

59. $\int_0^2 t^3 e^{-4t}\,dt = \frac{3}{128} - \frac{379}{128}e^{-8} \approx 0.022$

61. $\int_0^5 x^4\left(25 - x^2\right)^{3/2}\,dx = \frac{1{,}171{,}875}{256}\pi \approx 14{,}381.070$

63. (a)

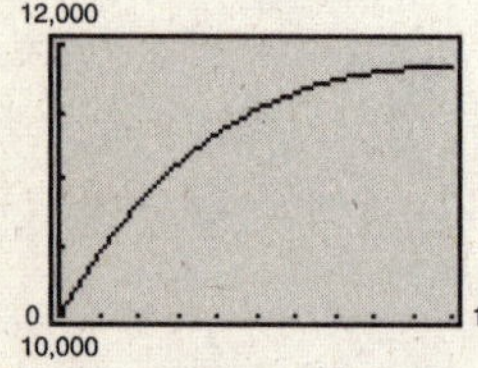

The company is forecasting an increase in demand over the next decade.

(b) $\int_0^{10} 500\left(20 + te^{-0.1t}\right)dt = 500\left[\int_0^{10} 20\,dt + \int_0^{10} te^{-0.1t}\,dt\right]$

$= 500(200) + 500\int_0^{10} te^{-0.1t}\,dt$

Let $u = t$ and $dv = e^{-0.1t}\,dt$. Then $du = dt$ and $v = -10e^{-0.1t}$.

$$= 100{,}000 + 500\left[-10te^{-0.1t}\right]_0^{10} + 10\int_0^{10} e^{-0.1t}\,dt$$
$$= 100{,}000 + 500\left\{-100e^{-1} - \left[100e^{-0.1t}\right]_0^{10}\right\}$$
$$= 100{,}000 + 500\left(-100e^{-1} - 100e^{-1} + 100\right)$$
$$= 100{,}000 + 50{,}000\left(1 - 2e^{-1}\right)$$
$$\approx 113{,}212 \text{ units}$$

(c) Average $= \frac{113{,}212}{10} \approx 11{,}321$ per year

65. $\int (1.6t \ln t + 1)\,dt = \int dt + \int 1.6t \ln t\,dt$

$= t + 1.6\int t \ln t\,dt$

Let $u = \ln t$ and $dv = t\,dt$. Then $du = \frac{1}{t}\,dt$ and $v = \frac{t^2}{2}$.

$= t + 1.6\left[\frac{t^2}{2}\ln t - \int \frac{t}{2}\,dt\right]$

$= t + 1.6\left[\frac{t^2}{2}\ln t - \frac{t^2}{4}\right]$

$= t + 0.8t^2 \ln t - 0.4t^2$

(a) Average value $= \frac{1}{2-1}\int_1^2 (1.6t \ln t + 1)\,dt$

$= \left[t + 0.8t^2 \ln t - 0.4t^2\right]_1^2$

$= 3.2 \ln 2 - 0.2$

≈ 2.018

(b) Average value $= \frac{1}{4-3}\int_3^4 (1.6t \ln t + 1)\,dt$

$= \left[t + 0.8t^2 \ln t - 0.4t^2\right]_3^4$

$= 12.8 \ln 4 - 7.2 \ln 3 - 1.8$

≈ 8.035

67. $V = \int_0^{t_1} c(t)e^{-rt}\,dt = \int_0^4 5000e^{-0.04t}\,dt = \frac{5000}{-0.04}e^{-0.04t}\Big]_0^4 = \$18{,}482.03$

69. Present value $= \int_0^{10} (100{,}000 + 4000t)e^{-0.05t}\,dt$

$= \int_0^{10} 100{,}000e^{-0.05t}\,dt + 4000\int_0^{10} te^{-0.05t}\,dt$

$= -2{,}000{,}000e^{-0.05t}\Big]_0^{10} + 4000\int_0^{10} te^{-0.05t}\,dt$

$= 2{,}000{,}000\left(1 - e^{-0.5}\right) + 4000\int_0^{10} te^{-0.05t}\,dt$

Let $u = t$ and $dv = e^{-0.05t}\,dt$. Then $du = dt$ and $v = -20e^{-0.05t}$.

$= 2{,}000{,}000\left(1 - e^{-0.5}\right) + 4000\left\{\left[-20te^{-0.05t}\right]_0^{10} + 20\int_0^{10} e^{-0.05t}\,dt\right\}$

$= 2{,}000{,}000\left(1 - e^{-0.5}\right) + 4000\left\{-200e^{-0.5} + \left[-400e^{-0.05t}\right]_0^{10}\right\}$

$= 2{,}000{,}000\left(1 - e^{-0.5}\right) + 4000\left(-600e^{-0.5} + 400\right)$

$= \$931{,}265.0973$

71. Present value $= \int_0^4 \left(1000 + 50e^{t/2}\right)e^{-0.06t}\,dt$

$= \int_0^4 \left(1000e^{-0.06t} + 50e^{0.44t}\right)dt$

$= \left[-\frac{1000}{0.06}e^{-0.06t} + \frac{50}{0.44}e^{0.44t}\right]_0^4$

$= \left(-\frac{1000}{0.06}e^{-0.24} + \frac{50}{0.44}e^{1.76}\right) - \left(-\frac{1000}{0.06} + \frac{50}{0.44}\right)$

$= \$4103.07$

73. Present value $= \int_0^{15} 100{,}000e^{-0.05t}\,dt$

$= \frac{100{,}000}{-0.05}\int_0^{15} e^{-0.05t}(-0.05)\,dt$

$= \frac{100{,}000}{-0.05}\left[e^{-0.05t}\right]_0^{15}$

$= \$1{,}055{,}266.89$

75. (a) Actual income $= \int_0^4 (150{,}000 + 75{,}000t)\,dt = 150{,}000t + 37{,}500t^2\Big]_0^4 = \$1{,}200{,}000$

(b) Present value $= \int_0^4 (150{,}000 + 75{,}000t)e^{-0.04t}\,dt$

$$= \int_0^4 150{,}000e^{-0.04t}\,dt + \int_0^4 75{,}000te^{-0.04t}\,dt$$

$$= -3{,}750{,}000e^{-0.04t}\Big]_0^4 + 75{,}000\int_0^4 te^{-0.04t}\,dt$$

$$= -3{,}750{,}000\left(e^{-0.16} - 1\right) + 75{,}000\int_0^4 te^{-0.04t}\,dt$$

Let $u = t$ and $dv = e^{-0.04t}\,dt$. Then $du = dt$ and $v = -25e^{-0.04t}$.

$$= -3{,}750{,}000\left(e^{-0.16} - 1\right) + 75{,}000\left\{\left[-25te^{-0.04t}\right]_0^4 + 25\int_0^4 e^{-0.04t}\,dt\right\}$$

$$= -3{,}750{,}000\left(e^{-0.16} - 1\right) + 75{,}000\left\{-100e^{-0.16} - 625\left[e^{-0.04t}\right]_0^4\right\}$$

$$= -3{,}750{,}000\left(e^{-0.16} - 1\right) + 75{,}000\left(-725e^{-0.16} + 625\right) = \$1{,}094{,}142.27$$

77. (a) Actual income $= \int_0^4 (3{,}000{,}000 + 750{,}000t)\,dt$

$$= \left[3{,}000{,}000t + 375{,}000t^2\right]_0^4$$

$$= \$18{,}000{,}000$$

(b) Present value $= \int_0^4 (3{,}000{,}000 + 750{,}000t)e^{-0.05t}\,dt$

$$= \int_0^4 3{,}000{,}000e^{-0.05t}\,dt + \int_0^4 750{,}000te^{-0.05t}\,dt$$

$$= \left[-60{,}000{,}000e^{-0.05t}\right]_0^4 + 750{,}000\int_0^4 te^{-0.05t}\,dt$$

$$= -60{,}000{,}000\left(e^{-0.2} - 1\right) + 750{,}000\int_0^4 te^{-0.05t}\,dt$$

Let $u = t$ and $dv = e^{-0.05t}\,dt$. Then $du = dt$ and $v = -20e^{-0.05t}$.

$$= -60{,}000{,}000\left(e^{-0.2} - 1\right) + 750{,}000\left\{\left[-20te^{-0.05t}\right]_0^4 + 20\int_0^4 e^{-0.05t}\,dt\right\}$$

$$= -60{,}000{,}000\left(e^{-0.2} - 1\right) + 750{,}000\left\{-80e^{-0.2} - 400\left[e^{-0.05t}\right]_0^4\right\}$$

$$= -60{,}000{,}000\left(e^{-0.2} - 1\right) + 750{,}000\left\{-80e^{-0.2} - 400e^{-0.2} + 400\right\}$$

$$\approx \$16{,}133{,}083.71$$

79. Future value $= e^{rt_1}\int_0^{t_1} f(t)e^{-rt}\,dt$

$$= e^{(0.08)10}\int_0^{10} 3000e^{-0.08t}\,dt$$

$$= e^{0.8}\left[\frac{3000}{-0.08}e^{-0.08t}\right]_0^{10}$$

$$\approx \$45{,}957.78$$

81. (a) Future value $= e^{(0.07)(10)} \int_0^{10} 1200e^{-0.07t}\,dt$

$$= e^{0.7} \int_0^{10} 1200e^{-0.07t}\,dt$$

$$= e^{0.7}\left[\frac{1200}{-0.07}e^{-0.07t}\right]_0^{10}$$

$$\approx \$17{,}378.62$$

(b) Difference $= \left[e^{(0.10)(15)} \int_0^{15} 1200e^{-0.10t}\,dt\right] - \left[e^{(0.09)(15)} \int_0^{15} 1200e^{-0.09t}\,dt\right]$

$$= \left[e^{1.5} \int_0^{15} 1200e^{-0.1t}\,dt\right] - \left[e^{1.35} \int_0^{15} 1200e^{-0.09t}\,dt\right]$$

$$= e^{1.5}\left[\frac{1200}{-0.1}e^{-0.1t}\right]_0^{15} - e^{1.35}\left[\frac{1200}{-0.09}e^{-0.09t}\right]_0^{15}$$

$$\approx \$41{,}780.27 - \$38{,}099.01$$

$$= \$3681.26$$

83. $\int_1^4 \frac{4}{\sqrt{x} + \sqrt[3]{x}}\,dx = 4\int_1^4 \frac{4}{\sqrt{x} + \sqrt[3]{x}}\,dx$

$$\approx 4\left(\frac{4-1}{10}\right)\left[f\left(\frac{23}{20}\right) + f\left(\frac{29}{20}\right) + f\left(\frac{35}{20}\right) + f\left(\frac{41}{20}\right) + f\left(\frac{47}{20}\right) + f\left(\frac{53}{20}\right)\right.$$

$$\left. + f\left(\frac{59}{20}\right) + f\left(\frac{65}{20}\right) + f\left(\frac{71}{20}\right) + f\left(\frac{77}{20}\right)\right] \approx 4.254$$

Section 6.2 Integration Tables

Skills Warm Up

1. $(x + 4)^2 = x^2 + 8x + 16$

2. $(x - 1)^2 = x^2 - 2x + 1$

3. $\left(x + \frac{1}{2}\right)^2 = x^2 + x + \frac{1}{4}$

4. $\left(x - \frac{1}{3}\right)^2 = x^2 - \frac{2}{3}x + \frac{1}{9}$

5. Let $u = 2x$ and $dv = e^x\,dx$. Then $du = 2\,dx$ and $v = e^x$.

$$\int 2xe^x\,dx = 2xe^x - \int 2e^x\,dx = 2xe^x - 2e^x + C = 2e^x(x - 1) + C$$

6. Let $u = \ln x$ and $dv = 3x^2\,dx$. Then $du = \frac{1}{x}\,dx$ and $v = x^3$.

$$\int 3x^2 \ln x\,dx = x^3 \ln x - \int x^2\,dx = x^3 \ln x - \frac{1}{3}x^3 + C = x^3\left(\ln x - \frac{1}{3}\right) + C$$

1. Formula 4: $u = x$, $du = dx$, $a = 2$, $b = 3$

$$\int \frac{x}{(2 + 3x)^2}\,dx = \frac{1}{9}\left[\frac{2}{2 + 3x} + \ln|2 + 3x|\right] + C$$

3. Formula 19: $u = x$, $du = dx$, $a = 2$, $b = 3$

$$\int \frac{x}{\sqrt{2 + 3x}}\,dx = -\frac{2(4 - 3x)}{27}\sqrt{2 + 3x} + C$$

$$= \frac{2(3x - 4)}{27}\sqrt{2 + 3x} + C$$

5. Formula 25: $u = x^2$, $du = 2x\,dx$, $a = 3$

$$\int \frac{2x}{\sqrt{x^4 - 9}}\,dx = \ln\left|x^2 + \sqrt{x^4 - 9}\right| + C$$

7. Formula 35: $u = x^2$, $du = 2x\,dx$

$$\int x^3 e^{x^2}\,dx = \frac{1}{2}\int x^2 e^{x^2}\,2x\,dx = \frac{1}{2}(x^2 - 1)e^{x^2} + C$$

9. Formula 10: $u = x$, $du = dx$, $a = b = 1$

$$\int \frac{1}{x(1+x)}\,dx = \ln\left|\frac{x}{1+x}\right| + C$$

11. Formula 26: $u = x$, $du = dx$, $a = 3$

$$\int \frac{1}{x\sqrt{x^2+9}}\,dx = -\frac{1}{3}\ln\left|\frac{3+\sqrt{x^2+9}}{x}\right| + C$$

13. Formula 32: $u = x$, $du = dx$, $a = 2$

$$\int \frac{1}{x\sqrt{4-x^2}}\,dx = -\frac{1}{2}\ln\left|\frac{2+\sqrt{4-x^2}}{x}\right| + C$$

15. Formula 42: $u = 3x$, $du = 3\,dx$

$$\int 3x\ln(3x)\,dx = \frac{1}{3}\int 3x\ln(3x)(3)\,dx$$

$$= \frac{1}{3}\left[\frac{(3x)^2}{4}\left(-1 + 2\ln(3x)\right)\right] + C$$

$$= \frac{3x^2}{4}\left[-1 + 2\ln(3x)\right] + C$$

17. Formula 37: $u = 3x^2$, $du = 6x\,dx$

$$\int \frac{6x}{1+e^{3x^2}}\,dx = 3x^2 - \ln\left(1 + e^{3x^2}\right) + C$$

19. Formula 14: $u = x$, $du = dx$, $a = 3$, $b = 1$

$$\int x^2\sqrt{3+x}\,dx = \frac{2}{(1)(2(2)+3)}\left[x^2(3+x)^{3/2} - 2(3)\int x\sqrt{3+x}\,dx\right]$$

$$= \frac{2}{7}\left[x^2(3+x)^{3/2} - 6\int x(3+x)^{1/2}\,dx\right]$$

Formula 14: $u = x$, $du = dx$, $a = 3$, $b = 1$

$$= \frac{2}{7}\left\{x^2(3+x)^{3/2} - 6\left[\frac{2}{(1)(2(1)+3)}\left(x(3+x)^{3/2} - 3\int\sqrt{3+x}\,dx\right)\right]\right\}$$

$$= \frac{2}{7}\left\{x^2(3+x)^{3/2} - \frac{12}{5}x(3+x)^{3/2} + \frac{36}{5}\int(3+x)^{1/2}\,dx\right\}$$

$$= \frac{2}{7}\left(x^2(3+x)^{3/2} - \frac{12}{5}x(3+x)^{3/2} + \frac{24}{5}(3+x)^{3/2}\right) + C$$

$$= \frac{2}{35}(3+x)^{3/2}\left(5x^2 - 12x + 24\right) + C$$

21. Formula 8: $u = t$, $du = dt$, $a = 2$, $b = 3$

$$\int \frac{t^2}{(2+3t)^3}\,dt = \frac{1}{27}\left[\frac{4}{2+3t} - \frac{4}{2(2+3t)^2} + \ln|2+3t|\right] + C = \frac{1}{27}\left[\frac{4}{2+3t} - \frac{2}{(2+3t)^2} + \ln|2+3t|\right] + C$$

23. Formula 15: $u = x$, $du = dx$, $a = 3$, $b = 4$

$$\int \frac{1}{x\sqrt{3+4x}}\,dx = \frac{1}{\sqrt{3}}\ln\left|\frac{\sqrt{3+4x}-\sqrt{3}}{\sqrt{3+4x}+\sqrt{3}}\right| + C$$

25. Formula 6: $u = x$, $du = dx$, $a = 1$, $b = 1$

$$\int \frac{x^2}{1+x}\,dx = -\frac{x}{2}(2-x) + \ln|1+x| + C$$

27. Formula 9: $u = x$, $du = dx$, $a = 3$, $b = 2$, $n = 5$

$$\int \frac{x^2}{(3+2x)^5}\,dx = \frac{1}{8}\left[\frac{-1}{2(3+2x)^2} + \frac{6}{3(3+2x)^3} - \frac{9}{4(3+2x)^4}\right] + C$$

$$= \frac{1}{8}\left[\frac{-1}{2(3+2x)^2} + \frac{2}{(3+2x)^3} - \frac{9}{4(3+2x^4)}\right] + C$$

29. Formula 33: $u = x$, $du = dx$, $a = 1$

$$\int \frac{1}{x^2\sqrt{1-x^2}}\,dx = -\frac{\sqrt{1-x^2}}{x} + C$$

31. Formula 43: $u = 2x, du = 2\,dx, n = 2$

$$\begin{aligned}\int 4x^2 \ln(2x)\,dx &= \frac{1}{2}\int (2x)^2 \ln(2x)(2)\,dx\\ &= \frac{1}{2}\left(\frac{(2x)^{2+1}}{(2+1)^2}\left[-1 + (2+1)\ln(2x)\right]\right) + C\\ &= \frac{1}{2}\left(\frac{8x^3}{9}\left[-1 + 3\ln(2x)\right]\right) + C\\ &= \frac{4x^3}{9}\left(-1 + 3\ln(2x)\right) + C\end{aligned}$$

33. Formula 7: $u = x, du = dx, a = -5, b = 3$

$$\int \frac{x^2}{(3x-5)^2}\,dx = \frac{1}{27}\left[3x - \frac{25}{3x-5} + 10\ln|3x-5|\right] + C$$

35. Formula 3: $u = \ln x, du = (1/x)\,dx, a = 4, b = 3$

$$\int \frac{\ln x}{x(4 + 3\ln x)}\,dx = \int \frac{\ln x}{4 + 3\ln x}\left(\frac{1}{x}\,dx\right) = \frac{1}{9}\left[3\ln x - 4\ln|4 + 3\ln x|\right] + C$$

37. Formula 19: $u = x, du = dx, a = 1, b = 1$

$$\int_0^1 \frac{x}{\sqrt{1+x}}\,dx = -\frac{2(2-x)}{3}\sqrt{1+x}\Bigg]_0^1 = -\frac{2}{3}\sqrt{2} - \left(-\frac{4}{3}\right) = \frac{4 - 2\sqrt{2}}{3} \approx 0.391$$

39. Formula 4: $u = x, du = dx, a = 4, b = 1$

$$\int_0^5 \frac{x}{(4+x)^2}\,dx = \left[\frac{4}{4+x} + \ln|4+x|\right]_0^5 = \left(\frac{4}{9} + \ln 9\right) - (1 + \ln 4) = -\frac{5}{9} + \ln\frac{9}{4} \approx 0.255$$

41. Formula 38: $u = x, du = dx, n = 0.5$

$$6\int_0^4 \frac{1}{1 + e^{0.5x}}\,dx = 6\left[x - \frac{1}{0.5}\ln\left(1 + e^{0.5x}\right)\right]_0^4 = 6\left\{\left[4 - 2\ln\left(1 + e^2\right)\right] - (0 - 2\ln 2)\right\} \approx 6.795$$

43. Formula 43: $u = x, du = dx, n = 3$

$$\begin{aligned}\int_1^2 x^3 \ln\left(x^2\right)dx &= 2\int_1^2 x^3 \ln x\,dx\\ &= 2\left[\left(\frac{x^{3+1}}{(3+1)^2}\left[-1 + (3+1)\ln x\right]\right)\right]_1^2\\ &= 2\left[\frac{x^4}{16}(4\ln x - 1)\right]_1^2\\ &= 2\left[(4\ln 2 - 1) - \left(\frac{1}{16}(-1)\right)\right]\\ &= 2\left(4\ln 2 - \frac{15}{16}\right)\\ &= 8\ln 2 - \frac{15}{8} \approx 3.6702\end{aligned}$$

45. Formula 35: $u = x, du = dx, a = 4$

$$\begin{aligned}\int_{-2}^2 \frac{1}{\left(16 - x^2\right)^{3/2}}\,dx &= \left[\frac{x}{(4)^2\sqrt{(4)^2 - (x)^2}}\right]_{-2}^2\\ &= \left[\frac{x}{16\sqrt{16 - x^2}}\right]_{-2}^2\\ &= \frac{2}{16\sqrt{12}} - \frac{(-2)}{16\sqrt{12}}\\ &= \frac{1}{4\sqrt{12}} = \frac{1}{8\sqrt{3}}\end{aligned}$$

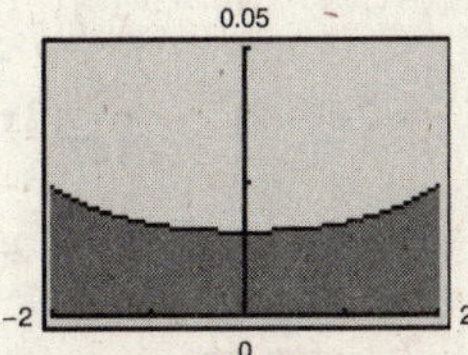

Approximate area: 0.0722

47. Formula 12: $u = x, du = dx, a = 2, b = 3$

$$\int_1^2 \frac{1}{9x^2(2+3x)}\,dx = \frac{1}{9}\left[-\frac{1}{2}\left(\frac{1}{x} + \frac{3}{2}\ln\left|\frac{x}{2+3x}\right|\right)\right]_1^2$$

$$= -\frac{1}{18}\left[\left(\frac{1}{2} + \frac{3}{2}\ln\left(\frac{1}{4}\right)\right) - \left(1 + \frac{3}{2}\ln\left(\frac{1}{5}\right)\right)\right]$$

$$= -\frac{1}{18}\left[-\frac{1}{2} + \frac{3}{2}\left(\ln\frac{1}{4} - \ln\frac{1}{5}\right)\right]$$

$$= -\frac{1}{36}\left[-1 + 3\left(\ln\frac{(1/4)}{(1/5)}\right)\right]$$

$$= -\frac{1}{36}\left[-1 + 3\ln\frac{5}{4}\right]$$

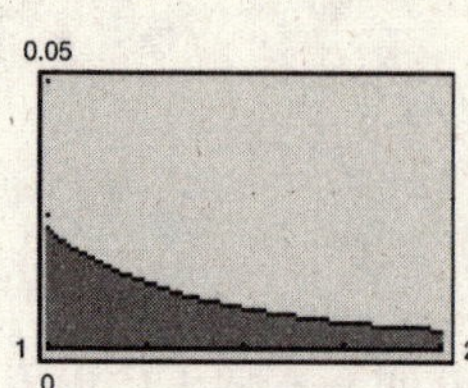

Approximate area: 0.0092

49. Formula 22: $u = x, a = 2, du = dx$

$$A = \int_0^{\sqrt{5}} x^2\sqrt{x^2+4}$$

$$= \frac{1}{8}\left[x\left(2x^2+4\right)\sqrt{x^2+4} - 16\ln\left|x + \sqrt{x^2+4}\right|\right]_0^{\sqrt{5}}$$

$$= \frac{1}{8}\left[42\sqrt{5} - 16\ln\left(\sqrt{5}+3\right) + 16\ln 2\right]$$

$$= \frac{1}{8}\left[42\sqrt{5} + 16\ln\left(\frac{2}{\sqrt{5}+3}\right)\right]$$

$$= \frac{1}{4}\left[21\sqrt{5} + 8\ln\left(\frac{2}{\sqrt{5}+3}\right)\right]$$

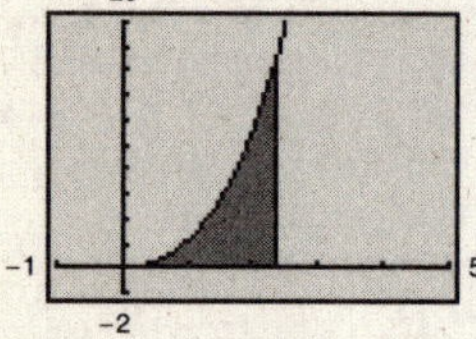

Approximate area: 9.8145

51. (a) Formula 41: $u = \frac{x}{3}, du = \frac{1}{3}\,dx$

$$\int \ln\left(\frac{x}{3}\right)dx = 3\int \ln\left(\frac{x}{3}\right)\left(\frac{1}{3}\right)dx$$

$$= 3\left[\frac{x}{3}\left(-1 + \ln\left(\frac{x}{3}\right)\right)\right] + C$$

$$= x\left(-1 + \ln\left(\frac{x}{3}\right)\right) + C$$

(b) Let $u = \ln\left(\frac{x}{3}\right)$ and $dv = dx$.

Then $du = \frac{1}{x/3}\left(\frac{1}{3}\right)dx = \frac{1}{x}\,dx$ and $v = x$.

$$\int \ln\left(\frac{x}{3}\right)dx = x\ln\left(\frac{x}{3}\right) - \int x\left(\frac{1}{x}\right)dx$$

$$= x\ln\left(\frac{x}{3}\right) - \int dx$$

$$= x\ln\left(\frac{x}{3}\right) - x + C$$

$$= x\left(\ln\left(\frac{x}{3}\right) - 1\right) + C$$

53. (a) Formula 19: $u = x, du = dx, a = -3, b = 7$

$$\int \frac{x}{\sqrt{-3+7x}}\,dx = -\frac{2\left(2(-3) - 7x\right)}{3(7)^2}\sqrt{-3+7x} + C = \frac{2}{147}(7x+6)\sqrt{7x-3} + C$$

(b) Let $u = x$ and $dv = (7x-3)^{-1/2}\,dx$.

Then $du = dx$ and $v = \frac{2}{7}(7x-3)^{1/2}$.

$$\int x(7x-3)^{-1/2}\,dx = \frac{2}{7}x(7x-3)^{1/2} - \int \frac{2}{7}(7x-3)^{1/2}\,dx$$

$$= \frac{2}{7}x(7x-3)^{1/2} - \frac{4}{147}(7x-3)^{3/2} + C$$

$$= \frac{2}{147}(7x-3)^{1/2}\left[21x - 2(7x-3)\right] + C$$

$$= \frac{2}{147}\sqrt{7x-3}(7x+6) + C$$

55. Formula 19: $u = x$, $du = dx$, $a = 4$, $b = 5$

$$P(a \le x \le b) = \int_a^b \frac{75}{14}\left(\frac{x}{\sqrt{4+5x}}\right) dx$$
$$= \frac{75}{14}\left[-\frac{2(8-5x)}{75}\sqrt{4+5x}\right]_a^b$$
$$= -\frac{1}{7}\left[(8-5x)\sqrt{4+5x}\right]_a^b$$

(a) $P(0.4 \le x \le 0.8) = -\frac{1}{7}\left[(8-5x)\sqrt{4+5x}\right]_{0.4}^{0.8}$
$$= -\frac{1}{7}\left(4\sqrt{8} - 6\sqrt{6}\right)$$
$$\approx 0.483$$

(b) $P(0 \le x \le 0.5) = -\frac{1}{7}\left[(8-5x)\sqrt{4+5x}\right]_0^{0.5}$
$$= -\frac{1}{7}\left(5.5\sqrt{6.5} - 16\right)$$
$$\approx 0.283$$

57. Formula 37: $u = 4.8 - 1.9t$, $du = -1.9\, dt$

$$\text{Average value} = \frac{1}{2-0}\int_0^2 \frac{5000}{1+e^{4.8-1.9t}}\, dt$$
$$= -\frac{2500}{1.9}\int_0^2 \frac{-1.9}{1+e^{4.8-1.9t}}\, dt$$
$$= -\frac{2500}{1.9}\left[4.8 - 1.9t - \ln\left(1+e^{4.8-1.9t}\right)\right]_0^2$$
$$= -\frac{2500}{1.9}\left[\left(1 - \ln(1+e)\right) - \left(4.8 - \ln\left(1+e^{4.8}\right)\right)\right]$$
$$\approx 401.402$$

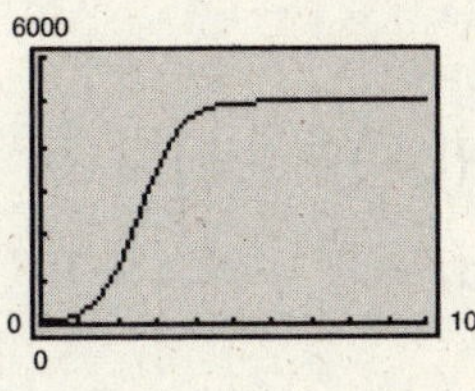

59. $R = \int_0^2 10{,}000\left[1 - \frac{1}{\left(1+0.1t^2\right)^{1/2}}\right] dt = 10{,}000t\Big]_0^2 - 10{,}000\int_0^2 \frac{1}{\left(1+0.1t^2\right)^{1/2}}\, dt$

Formula 25: $u = t\sqrt{0.1}$, $du = \sqrt{0.1}\, dt$, $a = 1$

$$R = 20{,}000 - \frac{10{,}000}{\sqrt{0.1}}\left[\ln\left(t\sqrt{0.1} + \sqrt{0.1t^2+1}\right)\right]_0^2 \approx \$1138.43$$

61. Formula 21: $u = \sqrt{0.00645}t$, $du = \sqrt{0.00645}\, dt$, $a = \sqrt{0.1673}$

$$\text{Average net profit} = \frac{1}{5-2}\int_2^5 \sqrt{0.00645t^2 - 0.1673}\, dt$$
$$= \frac{1}{3\sqrt{0.00645}}\int_2^5 \sqrt{0.00645t^2 + 0.1673}\sqrt{0.00645}\, dt$$
$$= \frac{1}{3\sqrt{0.00645}}\left(\frac{1}{2}\right)\left[\sqrt{0.00645}t\sqrt{0.00645t^2+0.1673} + 0.1673\ln\left(\sqrt{0.00645}t + \sqrt{0.00645t^2+0.1673}\right)\right]_2^5$$
$$\approx \$0.50 \text{ billion per year}$$

Chapter 6 Quiz Yourself

1. Let $u = x$ and $dv = e^{5x}\, dx$. Then $du = dx$ and $v = \frac{1}{5}e^{5x}$.

$$\int xe^{5x}\, dx = \tfrac{1}{5}xe^{5x} - \tfrac{1}{5}\int e^{5x}\, dx$$
$$= \tfrac{1}{5}xe^{5x} - \tfrac{1}{25}e^{5x} + C$$
$$= \tfrac{1}{5}e^{5x}\left(x - \tfrac{1}{5}\right) + C$$

2. $\int \ln x^3\, dx = 3\int \ln x\, dx$

Let $u = \ln x$ and $dv = dx$. Then $du = \frac{1}{x}\, dx$ and $v = x$.

$$\int \ln x^3\, dx = 3\left[x\ln x - \int dx\right]$$
$$= 3x\ln x - 3x + C$$
$$= 3x(\ln x - 1) + C$$

3. Let $u = \ln x$ and $dv = (x + 1)\,dx$. Then $du = \frac{1}{x}\,dx$ and $v = \frac{1}{2}x^2 + x$.

$$\begin{aligned}\int (x + 1)\ln x\,dx &= \ln x\left(\frac{1}{2}x^2 + x\right) - \int\left(\frac{1}{2}x^2 + x\right)\left(\frac{1}{x}\right)dx \\ &= \ln x\left(\frac{1}{2}x^2 + x\right) - \int\left(\frac{x}{2} + 1\right)dx \\ &= \ln x\left(\frac{1}{2}x^2 + x\right) - \frac{1}{4}x^2 - x + C \\ &= \frac{1}{2}x^2\ln x + x\ln x - \frac{1}{4}x^2 - x + C\end{aligned}$$

4. Let $u = x$ and $dv = \sqrt{x + 3}\,dx$. Then $du = dx$ and $v = \frac{2}{3}(x + 3)^{3/2}$.

$$\begin{aligned}\int x\sqrt{x + 3}\,dx &= \tfrac{2}{3}x(x + 3)^{3/2} - \tfrac{2}{3}\int (x + 3)^{3/2}\,dx \\ &= \tfrac{2}{3}x(x + 3)^{3/2} - \tfrac{4}{15}(x + 3)^{5/2} + C\end{aligned}$$

5. $\int x\ln\sqrt{x}\,dx = \int x\ln x^{1/2}\,dx = \frac{1}{2}\int x\ln x\,dx$

Let $u = \ln x$ and $dv = x\,dx$. Then $du = \frac{1}{x}\,dx$ and $v = \frac{1}{2}x^2$.

$$\begin{aligned}\int x\ln\sqrt{x}\,dx &= \frac{1}{2}\left[\frac{1}{2}x^2\ln x - \frac{1}{2}\int x\,dx\right] \\ &= \frac{1}{4}x^2\ln x - \frac{1}{8}x^2 + C\end{aligned}$$

6. Let $u = x^2$ and $du = e^{-2x}\,dx$. Then $du = 2x\,dx$ and $v = -\frac{1}{2}e^{-2x}$.

$$\int x^2e^{-2x}\,dx = -\frac{x^2}{2}e^{-2x} + \int xe^{-2x}\,dx$$

Let $u = x$ and $dv = e^{-2x}\,dx$. Then $du = dx$ and $v = -\frac{1}{2}e^{-2x}$.

$$\int x^2e^{-2x}\,dx = -\frac{x^2}{2}e^{-2x} - \frac{x}{2}e^{-2x} + \frac{1}{2}\int e^{-2x}\,dx = -\frac{x^2}{2}e^{-2x} - \frac{x}{2}e^{-2x} - \frac{1}{4}e^{-2x} + C = -\frac{1}{4}e^{-2x}\left(2x^2 + 2x + 1\right) + C$$

7. $x = 1000\left(45 + 20te^{-0.5t}\right)$

(a)
$$\begin{aligned}\text{Total demand} &= \int_0^5 1000\left(45 + 20te^{-0.5t}\right)dt \\ &= 1000\int_0^5 45\,dt + 1000\int_0^5 20te^{-0.5t}\,dt \\ &= 45{,}000\int_0^5 dt + 20{,}000\int_0^5 te^{-0.5t}\,dt \\ &= 45{,}000\big[t\big]_0^5 + 20{,}000\int_0^5 te^{-0.5t}\,dt \\ &= 225{,}000 + 20{,}000\int_0^5 te^{-0.5t}\,dt\end{aligned}$$

Let $u = t$ and $dv = e^{-0.5t}\,dt$. Then $du = dt$ and $v = -2e^{-0.5t}$.

$$\begin{aligned}&= 225{,}000 + 20{,}000\left\{\left[-2te^{-0.5t}\right]_0^5 + 2\int_0^5 e^{-0.5t}\,dt\right\} \\ &= 225{,}000 + 20{,}000\left\{-10e^{-2.5} - 4\left[e^{-0.5t}\right]_0^5\right\} \\ &= 225{,}000 - 200{,}000e^{-2.5} - 80{,}000e^{-2.5} + 80{,}000 \approx 282{,}016 \text{ units}\end{aligned}$$

(b) $\text{Average annual demand} = \frac{1}{5 - 0}\int_0^5 1000\left(45 + 20te^{-0.5t}\right)dt = \frac{1}{5}\int_0^5 1000\left(45 + 20te^{-0.5t}\right)dt \approx \frac{1}{5}(282{,}016) \approx 56{,}403 \text{ units}$

8. (a) Actual income $= \int_0^7 32{,}000t\,dt = \left[16{,}000t^2\right]_0^7 = \$784{,}000$

(b) Present value $= \int_0^7 32{,}000te^{-0.033t}\,dt = 32{,}000\int_0^7 te^{-0.033t}\,dt$

Let $u = t$ and $dv = e^{-0.033t}\,dt$. Then $du = dt$ and $v = -\dfrac{1}{0.033}e^{-0.033t}$.

$$= 32{,}000\left\{\left[-\frac{1}{0.033}te^{-0.033t}\right]_0^7 + \frac{1}{0.033}\int_0^7 e^{-0.033t}\,dt\right\}$$

$$= 32{,}000\left\{-\frac{1}{0.033}(7)e^{-0.033(7)} - \left[\frac{1}{(0.033)^2}e^{-0.033t}\right]_0^7\right\}$$

$$= 32{,}000\left\{-\frac{7}{0.033}e^{-0.231} - \frac{1}{(0.033)^2}e^{-0.231} + \frac{1}{(0.033)^2}\right\}$$

$$\approx \$673{,}108.31$$

9. Formula 3: $u = x$, $du = dx$, $a = 1$, $b = 2$

$$\int \frac{x}{1 + 2x}\,dx = \frac{1}{4}\left(2x - \ln|1 + 2x|\right) + C$$

10. Formula 10: $u = x$, $du = dx$, $a = 0.1$, $b = 0.2$

$$\int \frac{1}{x(0.1 + 0.2x)}\,dx = \frac{1}{0.1}\ln\left|\frac{x}{0.1 + 0.2x}\right| + C$$

$$= 10\ln\left|\frac{x}{0.1 + 0.2x}\right| + C$$

11. Formula 24: $u = x$, $du = dx$, $a = 4$

$$\int \frac{\sqrt{x^2 - 16}}{x^2}\,dx = -\frac{\sqrt{x^2 - 16}}{x} + \ln\left|x + \sqrt{x^2 - 16}\right| + C$$

12. Formula 15: $u = x$, $du = dx$, $a = 4$, $b = 9$

$$\int \frac{1}{x\sqrt{4 + 9x}}\,dx = \frac{1}{2}\ln\left|\frac{\sqrt{4 + 9x} - 2}{\sqrt{4 + 9x} + 2}\right| + C$$

13. Formula 38: $u = x^2$, $du = 2x\,dx$, $n = 4$

$$\int \frac{2x}{1 + e^{4x^2}}\,dx = x^2 - \frac{1}{4}\ln\left(1 + e^{4x^2}\right) + C$$

14. Formula 35: $u = x^2 + 1$, $du = 2x\,dx$

$$\int 2x\left(x^2 + 1\right)e^{x^2+1}\,dx = x^2e^{x^2+1} + C$$

15. $R = \sqrt{144t^2 + 400}$

$\int \sqrt{144t^2 + 400}\,dt$

Formula 23: $u = 12t$, $du = 12\,dt$, $a = 20$

$$\int \sqrt{144t^2 + 400}\,dt = \tfrac{1}{12}\int \sqrt{144t^2 + 400}(12)\,dt$$

$$= \tfrac{1}{12}\left[\tfrac{1}{2}\left(12t\sqrt{144t^2 + 400} + 400\ln\left|12t + \sqrt{144t^2 + 400}\right|\right)\right] + C$$

$$= \tfrac{1}{2}t\sqrt{144t^2 + 400} + \tfrac{50}{3}\ln\left(12t + \sqrt{144t^2 + 400}\right) + C$$

(a) Total revenue over first three years:

$$\int_0^3 \sqrt{144t^2 + 400}\,dt = \left[\tfrac{1}{2}t\sqrt{144t^2 + 400} + \tfrac{50}{3}\ln\left(12t + \sqrt{144t^2 + 400}\right)\right]_0^3$$

$$\approx \$84.28112652 \text{ million or } \$84{,}281{,}126.52$$

(b) Total revenue over first six years:

$$\int_0^6 \sqrt{144t^2 + 400}\,dt = \left[\tfrac{1}{2}t\sqrt{144t^2 + 400} + \tfrac{50}{3}\ln\left(12t + \sqrt{144t^2 + 400}\right)\right]_0^6$$

$$\approx \$257.39242972 \text{ million or } \$257{,}392{,}429.72$$

16. Let $u = x$ and $dv = e^{x/2}\,dx$. Then $du = dx$ and $v = 2e^{x/2}$.

$$\int_{-2}^{0} xe^{x/2}\,dx = 2xe^{x/2}\Big]_{-2}^{0} - \int_{-2}^{0} 2e^{x/2}\,dx = 4e^{-1} - \Big[4e^{x/2}\Big]_{-2}^{0} = 4e^{-1} - \left(4 - 4e^{-1}\right) = \frac{8}{e} - 4 \approx -1.057$$

17. Formula 42: $u = x,\ du = dx$

$$\int_1^2 5x\ln x\,dx = 5\int_1^2 x\ln x\,dx$$
$$= 5\left[\frac{x^2}{4}(-1 + 2\ln x)\right]_1^2$$
$$= 5\left[(-1 + 2\ln 2) - \frac{1}{4}(-1)\right]$$
$$= 5\left(-\frac{3}{4} + 2\ln 2\right)$$
$$= 10\ln 2 - \frac{15}{4} \approx 3.1815$$

18. Formula 19: $u = x,\ du = dx,\ a = 8,\ b = 1$

$$\int_0^8 \frac{x}{\sqrt{8 + x}}\,dx = \left[-\frac{2(16 - x)}{3}\sqrt{x + 8}\right]_0^8$$
$$= \left(\frac{-16}{3}(4)\right) - \left(\frac{-32}{3}\left(2\sqrt{2}\right)\right)$$
$$= \frac{-64}{3} + \frac{64}{3}\sqrt{2}$$
$$= \frac{64}{3}\left(\sqrt{2} - 1\right)$$
$$\approx 8.8366$$

19. Formula 42: $u = x,\ du = dx$

$$\int_1^e (\ln x)^2\,dx = x\left[2 - 2\ln x + (\ln x)^2\right]_1^e$$
$$= e - 2$$
$$\approx 0.718$$

20. Formula 33: $u = x,\ du = dx,\ a = 3$

$$\int_2^3 \frac{1}{x^2\sqrt{9 - x^2}}\,dx = -\frac{\sqrt{9 - x^2}}{9x}\Bigg]_2^3$$
$$= \frac{\sqrt{5}}{18}$$
$$\approx 0.124$$

21. Formula 29: $u = x^2,\ du = 2x\,dx,\ a = 2$

$$\int_4^6 \frac{2x}{x^4 - 4}\,dx = \left[\frac{1}{4}\ln\left|\frac{x^2 - 2}{x^2 + 2}\right|\right]_4^6$$
$$= \frac{1}{4}\left[\ln\frac{17}{19} - \ln\frac{7}{9}\right]$$
$$\approx 0.035$$

Section 6.3 Numerical Integration

Skills Warm Up

1. $f(x) = \dfrac{1}{x}$

$f'(x) = -\dfrac{1}{x^2}$

$f''(x) = \dfrac{2}{x^3}$

2. $f(x) = \ln(2x + 1)$

$f'(x) = \dfrac{2}{2x + 1}$

$f''(x) = -\dfrac{4}{(2x + 1)^2}$

$f'''(x) = \dfrac{16}{(2x + 1)^3}$

$f^{(4)}(x) = -\dfrac{96}{(2x + 1)^4}$

3. $f(x) = 2\ln x$

$f'(x) = \dfrac{2}{x}$

$f''(x) = -\dfrac{2}{x^2}$

$f'''(x) = \dfrac{4}{x^3}$

$f^{(4)}(x) = -\dfrac{12}{x^4}$

4. $f(x) = x^3 - 2x^2 + 7x - 12$

$f'(x) = 3x^2 - 4x + 7$

$f''(x) = 6x - 4$

Skills Warm Up —*continued*—

5. $f(x) = e^{2x}$

$f'(x) = 2e^{2x}$

$f''(x) = 4e^{2x}$

$f'''(x) = 8e^{2x}$

$f^{(4)}(x) = 16e^{2x}$

6. $f(x) = e^{x^2}$

$f'(x) = 2xe^{x^2}$

$f''(x) = 4x^2e^{x^2} + 2e^{x^2}$

7. $f(x) = -x^2 + 6x + 9, \quad [0, 4]$

$f'(x) = -2x + 6 = 0$ when $x = 3$

Interval	$\infty < x < 3$	$3 < x < \infty$
Test value	$x = 0$	$x = 4$
Sign of $f'(x)$	$f'(x) > 0$	$f'(x) < 0$
Conclusion	Increasing	Decreasing

$f(0) = 9$, $f(3) = 18$, and $f(4) = 17$, so $(3, 18)$ is the absolute maximum.

8. $f(x) = \dfrac{8}{x^3}, \quad [1, 2]$

$f'(x) = -\dfrac{24}{x^4}$

Because $-\dfrac{24}{x^4} \neq 0$ for any value of x, there are no critical numbers. $f(1) = 8$ and $f(2) = 1$, so $(1, 8)$ is the absolute maximum.

9.

$$\frac{1}{4n^2} < 0.001$$
$$\frac{1}{0.001} < 4n^2$$
$$1000 < 4n^2$$
$$250 < n^2$$
$$n^2 - 250 = 0$$
$$n^2 = 250$$
$$n = \pm 5\sqrt{10}$$

So, $n < -5\sqrt{10}$ or $n > 5\sqrt{10}$.

Test interval	$(-\infty, -5\sqrt{10})$	$(-5\sqrt{10}, 5\sqrt{10})$	$(5\sqrt{10}, \infty)$
n-value	–16	1	16
Function value	0.00098	0.25	0.00098
Conclusion	Does satisfy	Does not satisfy	Does satisfy

10.

$$\frac{1}{16n^4} < 0.0001$$
$$\frac{1}{0.0001} < 16n^4$$
$$10{,}000 < 16n^4$$
$$625 < n^4$$
$$n^4 = 625$$
$$n = \pm 5$$

So, $n < -5$ or $n > 5$.

Test interval	$(-\infty, -5)$	$(-5, 5)$	$(5, \infty)$
n-value	–6	1	6
Function value	0.000045	0.0625	0.000045
Conclusion	Does satisfy	Does not satisfy	Does satisfy

1. Exact: $\int_0^2 x^2\,dx = \frac{1}{3}x^3\Big]_0^2 = \frac{8}{3} \approx 2.6667$

Trapezoidal Rule: $\int_0^2 x^2\,dx \approx \frac{1}{4}\left[0 + 2\left(\frac{1}{2}\right)^2 + 2(1)^2 + 2\left(\frac{3}{2}\right)^2 + (2)^2\right] = \frac{11}{4} = 2.75$

Simpson's Rule: $\int_0^2 x^2\,dx \approx \frac{1}{6}\left[0 + 4\left(\frac{1}{2}\right)^2 + 2(1)^2 + 4\left(\frac{3}{2}\right)^2 + (2)^2\right] = \frac{8}{3} \approx 2.6667$

3. Exact: $\int_0^2 e^{-4x}\,dx = \left[-\frac{1}{4}e^{-4x}\right]_0^2 = -\frac{1}{4}e^{-8} + \frac{1}{4} \approx 0.2499$

Trapezoidal Rule: $\int_0^2 e^{-4x}\,dx \approx \frac{1}{8}\left[e^0 + 2e^{-1} + 2e^{-2} + 2e^{-3} + 2e^{-4} + 2e^{-5} + 2e^{-6} + 2e^{-7} + e^{-8}\right] \approx 0.2704$

Simpson's Rule: $\int_0^2 e^{-4x}\,dx \approx \frac{1}{12}\left[e^0 + 4e^{-1} + 2e^{-2} + 4e^{-3} + 2e^{-4} + 4e^{-5} + 2e^{-6} + 4e^{-7} + e^{-8}\right] \approx 0.2512$

5. Exact: $\int_0^2 x^3\,dx = \frac{1}{4}x^4\Big]_0^2 = 4$

Trapezoidal Rule: $\int_0^2 x^3\,dx \approx \frac{1}{8}\left[0 + 2\left(\frac{1}{4}\right)^3 + 2\left(\frac{1}{2}\right)^3 + 2\left(\frac{3}{4}\right)^3 + 2(1)^3 + 2\left(\frac{5}{4}\right)^3 + 2\left(\frac{3}{2}\right)^3 + 2\left(\frac{7}{4}\right)^3 + 8\right] = \frac{65}{16} = 4.0625$

Simpson's Rule: $\int_0^2 x^3\,dx \approx \frac{1}{12}\left[0 + 4\left(\frac{1}{4}\right)^3 + 2\left(\frac{1}{2}\right)^3 + 4\left(\frac{3}{4}\right)^3 + 2(1)^3 + 4\left(\frac{5}{4}\right)^3 + 2\left(\frac{3}{2}\right)^3 + 4\left(\frac{7}{4}\right)^3 + 8\right] = 4$

7. Exact: $\int_1^2 \frac{1}{x}\,dx = \ln|x|\Big]_1^2 = \ln 2 \approx 0.6931$

Trapezoidal Rule: $\int_1^2 \frac{1}{x}\,dx \approx \frac{1}{16}\left[1 + 2\left(\frac{8}{9}\right) + 2\left(\frac{4}{5}\right) + 2\left(\frac{8}{11}\right) + 2\left(\frac{2}{3}\right) + 2\left(\frac{8}{13}\right) + 2\left(\frac{4}{7}\right) + 2\left(\frac{8}{15}\right) + \frac{1}{2}\right] \approx 0.6941$

Simpson's Rule: $\int_1^2 \frac{1}{x}\,dx \approx \frac{1}{24}\left[1 + 4\left(\frac{8}{9}\right) + 2\left(\frac{4}{5}\right) + 4\left(\frac{8}{11}\right) + 2\left(\frac{2}{3}\right) + 4\left(\frac{8}{13}\right) + 2\left(\frac{4}{7}\right) + 4\left(\frac{8}{15}\right) + \frac{1}{2}\right] \approx 0.6932$

9. Exact: $\int_0^1 xe^{3x^2}\,dx = \left[\frac{1}{6}e^{3x^2}\right]_0^1 = \frac{1}{6}e^3 - \frac{1}{6} \approx 3.1809$

Trapezoidal Rule: $\int_0^1 xe^{3x^2}\,dx \approx \frac{1}{8}\left[0 + 2\left(\frac{1}{4}e^{3/16}\right) + 2\left(\frac{1}{2}e^{3/4}\right) + 2\left(\frac{3}{4}e^{27/16}\right) + e^3\right] \approx 3.8643$

Simpson's Rule: $\int_0^1 xe^{3x^2}\,dx \approx \frac{1}{12}\left[0 + 4\left(\frac{1}{4}e^{3/16}\right) + 2\left(\frac{1}{2}e^{3/4}\right) + 4\left(\frac{3}{4}e^{27/16}\right) + e^3\right] \approx 3.3022$

11. Exact: $\int_4^9 \sqrt{x}\,dx = \frac{2}{3}x^{3/2}\Big]_4^9 = \frac{38}{3} \approx 12.6667$

Trapezoidal Rule: $\int_4^9 \sqrt{x}\,dx \approx \frac{5}{16}\left[2 + 2\sqrt{\frac{37}{8}} + 2\sqrt{\frac{21}{4}} + 2\sqrt{\frac{47}{8}} + 2\sqrt{\frac{13}{2}} + 2\sqrt{\frac{57}{8}} + 2\sqrt{\frac{31}{4}} + 2\sqrt{\frac{67}{8}} + 3\right] \approx 12.6640$

Simpson's Rule: $\int_4^9 \sqrt{x}\,dx \approx \frac{5}{24}\left[2 + 4\sqrt{\frac{37}{8}} + 2\sqrt{\frac{21}{4}} + 4\sqrt{\frac{47}{8}} + 2\sqrt{\frac{13}{2}} + 4\sqrt{\frac{57}{8}} + 2\sqrt{\frac{31}{4}} + 4\sqrt{\frac{67}{8}} + 3\right] \approx 12.6667$

13. (a) Trapezoidal Rule:

$$\int_0^4 \frac{1}{\sqrt[3]{x^2+1}}\,dx \approx \frac{1}{4}\left[0 + 2\left(\frac{1}{\sqrt[3]{(1/2)^2+1}}\right) + 2\left(\frac{1}{\sqrt[3]{1^2+1}}\right) + 2\left(\frac{1}{\sqrt[3]{(3/2)^2+1}}\right) + 2\left(\frac{1}{\sqrt[3]{2^2+1}}\right)\right.$$
$$\left. + 2\left(\frac{1}{\sqrt[3]{(5/2)^2+1}}\right) + 2\left(\frac{1}{\sqrt[3]{3^2+1}}\right) + 2\left(\frac{1}{\sqrt[3]{(7/2)^2+1}}\right) + \frac{1}{\sqrt[3]{4^2+1}}\right] \approx 2.540$$

(b) Simpson's Rule:

$$\int_0^4 \frac{1}{\sqrt[3]{x^2+1}}\,dx \approx \frac{1}{6}\left[0 + 4\left(\frac{1}{\sqrt[3]{(1/2)^2+1}}\right) + 2\left(\frac{1}{\sqrt[3]{1^2+1}}\right) + 4\left(\frac{1}{\sqrt[3]{(3/2)^2+1}}\right) + 2\left(\frac{1}{\sqrt[3]{2^2+1}}\right)\right.$$
$$\left. + 4\left(\frac{1}{\sqrt[3]{(5/2)^2+1}}\right) + 2\left(\frac{1}{\sqrt[3]{3^2+1}}\right) + 4\left(\frac{1}{\sqrt[3]{(7/2)^2+1}}\right) + \frac{1}{\sqrt[3]{4^2+1}}\right] \approx 2.541$$

15. (a) Trapezoidal Rule: $\int_0^2 \sqrt{1+x^3}\,dx \approx \frac{1}{4}\left[1 + 2\sqrt{\frac{9}{8}} + 2\sqrt{2} + 2\sqrt{\frac{35}{8}} + 3\right] \approx 3.283$

(b) Simpson's Rule: $\int_0^2 \sqrt{1+x^3}\,dx \approx \frac{1}{6}\left[1 + 4\sqrt{\frac{9}{8}} + 2\sqrt{2} + 4\sqrt{\frac{35}{8}} + 3\right] \approx 3.240$

17. (a) Trapezoidal Rule:

$$\int_0^1 e^{x^2}\,dx \approx \frac{1}{16}\left[e^{(0)^2} + 2e^{(1/8)^2} + 2e^{(1/4)^2} + 2e^{(3/8)^2} + 2e^{(1/2)^2} + 2e^{(5/8)^2} + 2e^{(3/4)^2} + 2e^{(7/8)^2} + e^{(1)^2}\right] \approx 1.470$$

(b) Simpson's Rule:

$$\int_0^1 e^{x^2}\,dx \approx \frac{1}{24}\left[e^{(0)^2} + 4e^{(1/8)^2} + 2e^{(1/4)^2} + 4e^{(3/8)^2} + 2e^{(1/2)^2} + 4e^{(5/8)^2} + 2e^{(3/4)^2} + 4e^{(7/8)^2} + e^{(1)^2}\right] \approx 1.463$$

19. (a) Trapezoidal Rule: $\int_0^3 \frac{1}{2-2x+x^2}\,dx \approx \frac{1}{4}\left[\frac{1}{2} + 2\left(\frac{4}{5}\right) + 2(1) + 2\left(\frac{4}{5}\right) + 2\left(\frac{1}{2}\right) + 2\left(\frac{4}{13}\right) + \frac{1}{5}\right] \approx 1.879$

(b) Simpson's Rule: $\int_0^3 \frac{1}{2-2x+x^2}\,dx \approx \frac{1}{6}\left[\frac{1}{2} + 4\left(\frac{4}{5}\right) + 2(1) + 4\left(\frac{4}{5}\right) + 2\left(\frac{1}{2}\right) + 4\left(\frac{4}{13}\right) + \frac{1}{5}\right] \approx 1.888$

21. $V = \int_0^{t_1} c(t)e^{-rt}\,dt$

$= \int_0^4 \left(6000 + 200\sqrt{t}\right)e^{-0.07t}\,dt$

Using a program similar to the Simpson's Rule program, when $n = 8$, $V \approx \$21{,}831.20$.

23. $\Delta R = \int_{14}^{16} 5\sqrt{8000 - x^3}\,dx$

Using a program similar to the Simpson's Rule program, when $n = 4$, $\Delta R \approx \$678.36$.

25. $P(0 \le x \le 1) = \int_0^1 \frac{1}{\sqrt{2\pi}}e^{-x^2/2}\,dx$

Using a program similar to the Simpson's Rule program, when $n = 6$, $P \approx 0.3413 = 34.13\%$.

27. $P(0 \le x \le 4) = \int_0^4 \frac{1}{\sqrt{2\pi}}e^{-x^2/2}\,dx$

Using a program similar to the Simpson's Rule program, when $n = 6$, $P \approx 0.499958 = 49.996\%$.

29. $A \approx \frac{1000}{3(10)}\left[125 + 4(125) + 2(120) + 4(112) + 2(90) + 4(90) + 2(95) + 4(88) + 2(75) + 4(35) + 0\right]$

$= 89{,}500$ square feet

31. $f(x) = x^2 + 2x$

$f'(x) = 2x + 2$

$f''(x) = 2$

$f'''(x) = 0$

$f^{(4)}(x) = 0$

(a) Trapezoidal Rule: Because $|f''(x)|$ is maximum for all x in $[0, 2]$ and $|f''(x)| = 2$, you have

$$|\text{Error}| \le \frac{(2-0)^3}{12(4)^2}(2) = \frac{1}{12} \approx 0.0833.$$

(b) Simpson's Rule: Because $|f^{(4)}(x)|$ is maximum for all x in $[0, 2]$ and $f^{(4)}(x) = 0$, you have

$$|\text{Error}| \le \frac{(2-0)^5}{180(4)^4}(0) = 0.$$

33. $f(x) = e^{x^3}$

$f'(x) = 3x^2e^{x^3}$

$f''(x) = 3(3x^4 + 2x)e^{x^3}$

$f'''(x) = 3(9x^6 + 18x^3 + 2)e^{x^3}$

$f^{(4)}(x) = 9(9x^8 + 36x^5 + 20x^2)e^{x^3}$

(a) Trapezoidal Rule: Because $|f''(x)|$ is maximum in $[0, 1]$ when $x = 1$ and $|f''(1)| = 15e$, you have

$$|\text{Error}| \le \frac{(1-0)^3}{12(4)^2}(15e) = \frac{5e}{64} \approx 0.212.$$

(b) Simpson's Rule: Because $|f^{(4)}(x)|$ is maximum in $[0, 1]$ when $x = 1$ and $|f^{(4)}(1)| = 585e$, you have

$$|\text{Error}| \le \frac{(1-0)^5}{180(4)^4}(585e) = \frac{13e}{1024} \approx 0.035.$$

35. $f(x) = x^4$

$f'(x) = 4x^3$

$f''(x) = 12x^2$

$f'''(x) = 24x$

$f^{(4)}(x) = 24$

(a) Trapezoidal Rule: Because $f''(x)$ is maximum in $[0, 2]$ when $x = 2$ and $|f''(2)| = 48$, you have

$$|\text{Error}| \le \frac{(2-0)^3(48)}{12n^2} < 0.0001$$

$$\frac{32}{n^2} < 0.0001$$

$$320{,}000 < n^2$$

$$565.69 < n.$$

Let $n = 566$.

(b) Simpson's Rule: Because $f^{(4)}(x) = 24$ for all x in $[0, 2]$, you have

$$|\text{Error}| \le \frac{(2-0)^5}{180n^4}(24) < 0.0001$$

$$\frac{64}{15n^4} < 0.0001$$

$$n^4 > 42{,}666.67$$

$$n > 14.37.$$

Let $n = 16$.

37. $f(x) = e^{2x}$

$f'(x) = 2e^{2x}$

$f''(x) = 4e^{2x}$

$f'''(x) = 8e^{2x}$

$f^{(4)}(x) = 16e^{2x}$

(a) Trapezoidal Rule: Because $|f''(x)|$ is maximum in $[1, 3]$ when $x = 3$ and $|f''(3)| = 4e^6$, you have

$$|\text{Error}| \le \frac{(3-1)^3}{12n^2}(4e^6) < 0.0001$$

$$\frac{8e^6}{3n^2} < 0.0001$$

$$n^2 > 10{,}758{,}101.16$$

$$n > 3279.95.$$

Let $n = 3280$.

(b) Simpson's Rule: Because $|f^{(4)}(x)|$ is maximum in $[1, 3]$ when $x = 3$ and $|f^{(4)}(3)| = 16e^6$, you have

$$|\text{Error}| \le \frac{(3-1)^5}{180n^4}(16e^6) < 0.0001$$

$$\frac{128e^6}{45n^4} < 0.0001$$

$$n^4 > 11{,}475{,}307.90$$

$$n > 58.2.$$

Let $n = 60$. (n must be even.)

39. Using a program similar to the Simpson's Rule program, when $n = 100$,

$$\int_1^4 x\sqrt{x+4}\,dx \approx 19.5215.$$

41. Using a program similar to the Simpson's Rule program, when $n = 100$,

$$\int_2^5 10xe^{-x}\,dx \approx 3.6558.$$

43. (a) Average median age $= \dfrac{1}{9-1}\displaystyle\int_1^9 f(x)\,dx \approx \frac{1}{8}\left\{\frac{1}{3}\left[35.5 + 4(35.7) + 2(35.9) + 4(36.0) + 2(36.2) + 4(36.3) + 2(36.5) + 4(36.7) + 36.8\right]\right\} \approx 36.2$ years

(b) Average median age $= \dfrac{1}{9-1}\displaystyle\int_1^9 \left(35.4 + 0.16t - 0.000004e^t\right)dt = \frac{1}{8}\left[35.4t + 0.08t^2 - 0.000004e^t\right]_1^9 \approx 36.2$ years

(c) The results are approximately equal.

45. $C = \int_0^{12} \left[8 - \ln\left(t^2 - 2t + 4\right)\right] dt$

$\approx \frac{1}{2}\left[6.61 + 4(6.82) + 2(6.05) + 4(5.28) + 2(4.67) + 4(4.19) + 2(3.80) + 4(3.46) + 3.18\right]$

≈ 58.915 mg

47. $S = \int_0^6 1000t^2e^{-t}\, dt \approx \frac{1}{6}[0 + 4(151.63) + 2(367.88) + 4(502.04) + 2(541.34) + 4(513.03) + 2(448.08)$

$+ 4(369.92) + 2(293.05) + 4(224.96) + 2(168.45) + 4(123.62) + 89.24]$

≈ 1878 subscribers

49. Answers will vary.

Section 6.4 Improper Integrals

Skills Warm Up

1. $\lim_{x\to 2} (2x + 5) = 2(2) + 5 = 9$

2. $\lim_{x\to 1}\left(\frac{1}{x} + 2x^2\right) = \frac{1}{1} + 2(1^2) = 3$

3. $\lim_{x\to -4} \frac{x + 4}{x^2 - 16} = \lim_{x\to -4} \frac{x + 4}{(x + 4)(x - 4)}$

$= \lim_{x\to -4} \frac{1}{x - 4}$

$= \frac{1}{-4 - 4}$

$= -\frac{1}{8}$

4. $\lim_{x\to 0} \frac{x^2 - 2x}{x^3 + 3x^2} = \lim_{x\to 0} \frac{x(x - 2)}{x^2(x + 3)} = \lim_{x\to 0} \frac{x - 2}{x(x + 3)}$

$\lim_{x\to 0^-} \frac{x - 2}{x(x + 3)} = \infty$

$\lim_{x\to 0^+} \frac{x - 2}{x(x + 3)} = -\infty$

Limit does not exist.

5. $\lim_{x\to 1} \frac{1}{\sqrt{x - 1}} = \infty$

Limit does not exist.

6. $\lim_{x\to -3} \frac{x^2 + 2x - 3}{x + 3} = \lim_{x\to -3} \frac{(x + 3)(x - 1)}{x + 3}$

$= \lim_{x\to -3} x - 1$

$= -3 - 1$

$= -4$

7. $\frac{4}{3}(2x - 1)^3$

(a) $\frac{4}{3}(2b - 1)^3$

(b) $\frac{4}{3}(2 \cdot 0 - 1)^3 = -\frac{4}{3}$

8. $\frac{1}{x - 5} + \frac{3}{(x - 2)^2}$

(a) $\frac{1}{b - 5} + \frac{3}{(b - 2)^2}$

(b) $\frac{1}{0 - 5} + \frac{3}{(0 - 2)^2} = -\frac{1}{5} + \frac{3}{4} = \frac{11}{20}$

9. $\ln\left(5 - 3x^2\right) - \ln(x + 1)$

(a) $\ln\left(5 - 3b^2\right) - \ln(b + 1)$

(b) $\ln\left(5 - 3 \cdot 0^2\right) - \ln(0 + 1) = \ln 5 - \ln 1$

$= \ln 5$

≈ 1.609

10. $e^{3x^2} + e^{-3x^2}$

(a) $e^{3b^2} + e^{-3b^2} = e^{-3b^2}\left(e^{6b^2} + 1\right)$

(b) $e^{3(0)^2} + e^{-3(0)^2} = e^0 + e^0 = 1 + 1 = 2$

1. The integral is improper because the function has an infinite discontinuity in $[0, 1]$.

3. The integral is proper.

5. The integral is proper.

7. This integral converges because

$$\int_1^{\infty} \frac{1}{x^2}\,dx = \lim_{b\to\infty}\left[-\frac{1}{x}\right]_1^b = 0 + 1 = 1.$$

9. This integral diverges because

$$\int_0^{\infty} e^{x/3}\,dx = \lim_{b\to\infty}\left[3e^{x/3}\right]_0^b = \infty.$$

11. This integral diverges because

$$\int_5^{\infty} \frac{x}{\sqrt{x^2-16}}\,dx = \lim_{b\to\infty}\left[\sqrt{x^2-16}\right]_5^b = \infty - 3 = \infty.$$

13. This integral diverges because

$$\int_{-\infty}^{0} e^{-x}\,dx = \lim_{a\to-\infty}\left[-e^{-x}\right]_a^0 = -1 + \infty = \infty.$$

15. This integral diverges because

$$\begin{aligned}\int_1^{\infty} \frac{e^{\sqrt{x}}}{\sqrt{x}}\,dx &= \lim_{b\to\infty}\left[2e^{\sqrt{x}}\right]_1^b\\ &= \lim_{b\to\infty}\left[2e^{\sqrt{b}} - 2e\right]\\ &= \infty.\end{aligned}$$

17. This integral converges because

$$\begin{aligned}\int_{-\infty}^{\infty} 2xe^{-3x^2}\,dx &= \int_{-\infty}^{0} 2xe^{-3x^2}\,dx + \int_0^{\infty} 2xe^{-3x^2}\,dx\\ &= \lim_{a\to-\infty}\left[-\tfrac{1}{3}e^{-3x^2}\right]_a^0 + \lim_{b\to\infty}\left[-\tfrac{1}{3}e^{-3x^2}\right]_0^b\\ &= \left(-\tfrac{1}{3} + 0\right) + \left(0 + \tfrac{1}{3}\right)\\ &= 0.\end{aligned}$$

19. This integral converges because

$$\int_4^{\infty} \frac{1}{x(\ln x)^3}\,dx = \lim_{b\to\infty}\left[-\frac{1}{2(\ln x)^2}\right]_4^b = \frac{1}{2(\ln 4)^2}.$$

21. $A = \int_0^{\infty} e^{-x}\,dx = \lim_{b\to\infty}\left[-e^{-x}\right]_0^b = 1$

23. $A = \int_{-\infty}^{-1}\left(-\frac{1}{x^3}\right)dx = \lim_{a\to-\infty}\left[\frac{1}{2x^2}\right]_a^{-1} = \frac{1}{2}$

25. $A = \int_0^{\infty} \frac{6x}{x^2+1}\,dx = \lim_{b\to\infty}\left[3\ln\left(x^2+1\right)\right]_0^b = \infty$

27. $\mu = 63.8,\ \sigma = 2.9$

$$\begin{aligned}f(x) &= \frac{1}{\sigma\sqrt{2\pi}}e^{-(x-\mu)^2/2\sigma^2}\\ &= 0.137566e^{-(x-63.8)^2/16.82}\end{aligned}$$

Using a graphing utility:

(a) $\int_{60}^{72} f(x)\,dx \approx 0.9026$

(b) $\int_{68}^{\infty} f(x)\,dx \approx 0.0738$

(c) $\int_{72}^{\infty} f(x)\,dx \approx 0.00235$

29. $\mu = 36,\ \sigma = 0.2$

$$\begin{aligned}f(x) &= \frac{1}{\sigma\sqrt{2\pi}}e^{-(x-\mu)^2/2\sigma^2}\\ &= 1.99471e^{-(x-36)^2/0.08}\end{aligned}$$

Using a graphing utility:

(a) $\int_{35.5}^{\infty} f(x)\,dx \approx 0.9938$

(b) $\int_{35.9}^{\infty} f(x)\,dx \approx 0.6915$

31. Present value $= \frac{P}{r} = \frac{5000}{0.075} \approx \$66,666.67$

33. Present value $= \frac{P}{r} = \frac{18,000}{0.05} = \$360,000$

The amount you need to start the scholarship fund is \$360,000. Yes, you have enough money to start the scholarship fund.

35. (a) Present value $= \int_0^{20} 500,000e^{-0.09t}\,dt$

$$\begin{aligned}&= -\frac{500,000}{0.09}e^{-0.09t}\Big]_0^{20}\\ &\approx \$4,637,228.40\end{aligned}$$

(b) Present value $= \int_0^{\infty} 500,000e^{-0.09t}\,dt$

$$\begin{aligned}&= \lim_{b\to\infty}\left[-\frac{500,000}{0.09}e^{-0.09t}\right]_0^b\\ &= \$5,555,555.56\end{aligned}$$

37. $C = 650{,}000 + \int_0^n 25{,}000e^{-0.10t}\, dt$

$= 650{,}000 - \left[250{,}000e^{-0.10t}\right]_0^n$

(a) For $n = 5$, you have

$C = 650{,}000 - \left[250{,}000\left(e^{-0.50} - 1\right)\right]$

$\approx \$748{,}367.34.$

(b) For $n = 10$, you have

$C = 650{,}000 - \left[250{,}000\left(e^{-1} - 1\right)\right] \approx \$808{,}030.14.$

(c) For $n = \infty$, you have

$C = 650{,}000 - \lim_{n\to\infty} \left[250{,}000e^{-0.10t}\right]_0^n$

$= 650{,}000 - 250{,}000(0 - 1) = \$900{,}000.$

39. $C = 300{,}000 + \int_0^n 15{,}000te^{-0.06t}\, dt$

$= 300{,}000 - 15{,}000\left[\left(\frac{50}{3}t + \frac{2500}{9}\right)e^{-0.06t}\right]_0^n$

(a) For $n = 5$, you have

$C = 300{,}000 - 15{,}000\left(\frac{3250}{9}e^{-0.3} - \frac{2500}{9}\right)$

$\approx \$453{,}901.30.$

(b) For $n = 10$, you have

$C = 300{,}000 - 15{,}000\left(\frac{4000}{9}e^{-0.6} - \frac{2500}{9}\right)$

$\approx \$807{,}922.43.$

(c) For $n = \infty$, you have

$C = 300{,}000 - 15{,}000 \lim_{n\to\infty} \left[\left(\frac{50}{3}t + \frac{2500}{9}\right)e^{-0.06t}\right]_0^n$

$= 300{,}000 - 15{,}000\left(0 - \frac{2500}{9}\right)$

$\approx \$4{,}466{,}666.67.$

Review Exercises for Chapter 6

1. Let $u = \ln x$ and $dv = \left(1/\sqrt{x}\right) dx$. Then $du = (1/x)\, dx$ and $v = 2\sqrt{x}$.

$$\int \frac{\ln x}{\sqrt{x}}\, dx = 2\sqrt{x}\ln x - \int 2\sqrt{x}\left(\frac{1}{x}\right) dx$$
$$= 2\sqrt{x}\ln x - \int 2x^{-1/2}\, dx$$
$$= 2\sqrt{x}\ln x - 4\sqrt{x} + C$$

3. Let $u = x - 1$ and $dv = e^x\, dx$. Then $du = dx$ and $v = e^x$.

$$\int (x-1)e^x\, dx = (x-1)e^x - \int e^x\, dx$$
$$= (x-1)e^x - e^x + C$$
$$= (x-2)e^x + C$$

5. Let $u = x$ and $dv = (x-5)^{1/2}\, dx$.

Then $du = dx$ and $v = \frac{2}{3}(x-5)$.

$$\int x\sqrt{x-5}\, dx = \frac{2}{3}x(x-5)^{3/2} - \int \frac{2}{3}(x-5)^{3/2}\, dx$$
$$= \frac{2}{3}x(x-5)^{3/2} - \frac{4}{15}(x-5)^{5/2} + C$$

7. Let $u = 2x^2$ and $dv = e^{2x}$. Then $du = 4x\, dx$ and $v = \frac{1}{2}e^{2x}$.

$$\int 2x^2e^{2x}\, dx = \left(2x^2\right)\left(\tfrac{1}{2}e^{2x}\right) - \int \tfrac{1}{2}e^{2x}(4x\, dx) = x^2e^{2x} - \int 2xe^{2x}\, dx$$

Let $u = 2x$ and $dv = e^{2x}$. Then $du = 2\, dx$ and $v = \frac{1}{2}e^{2x}$.

$$\int 2x^2e^{2x}\, dx = x^2e^{2x} - \left[(2x)\left(\tfrac{1}{2}e^{2x}\right) - \int \tfrac{1}{2}e^{2x}(2)\, dx\right]$$
$$= x^2e^{2x} - xe^{2x} + \int e^{2x}\, dx$$
$$= x^2e^{2x} - xe^{2x} + \tfrac{1}{2}e^{2x} + C$$

9. Let $u = \ln x$ and $dv = 6x\,dx$.

Then $du = \frac{1}{x}\,dx$ and $v = 3x^2$.

$$\int_1^e 6x \ln x\,dx = 3x^2 \ln x - \int 3x^2\left(\frac{1}{x}\right)dx$$
$$= \left[3x^2 \ln x - \frac{3}{2}x^2\right]_1^e$$
$$= \left(3e^2 - \frac{3}{2}e^2\right) - \left(0 - \frac{3}{2}\right)$$
$$= \frac{3}{2}e^2 + \frac{3}{2}$$
$$\approx 12.584$$

11. Let $u = x$ and $dv = e^{-x/4}\,dx$.

Then $du = dx$ and $v = -4e^{-x/4}$.

$$\int_0^1 xe^{-x/4}\,dx = -4xe^{-x/4} + 4\int e^{-x/4}\,dx$$
$$= \left[-4xe^{-x/4} - 16e^{-x/4}\right]_0^1$$
$$= \left(-4e^{-1/4} - 16e^{-1/4}\right) - (0 - 16)$$
$$= 16 - 20e^{-1/4}$$
$$\approx 0.4240$$

13. Present value $= \int_0^5 20{,}000e^{-0.04t}\,dt = -500{,}000e^{-0.04t}\Big]_0^5 = \$90{,}634.62$

15. Present value $= \int_0^{10} 24{,}000te^{-0.05t}\,dt = 24{,}000\int_0^{10} e^{-0.05t}\,dt$

Let $u = t$ and $dv = e^{-0.05t}\,dt$. Then $du = dt$ and $v = -20e^{-0.05t}$.

$$= 24{,}000\left\{\left[-20te^{-0.05t}\right]_0^{10} + 20\int_0^{10} e^{-0.05t}\,dt\right\}$$
$$= 24{,}000\left\{-200e^{-0.5} - 400\left[e^{-0.05t}\right]_0^{10}\right\}$$
$$= 24{,}000\left[-200e^{-0.5} - 400\left(e^{-0.5} - 1\right)\right]$$
$$= 24{,}000\left(400 - 600e^{-0.5}\right)$$
$$= \$865{,}958.50$$

17. (a) Actual income $= \int_0^4 (200{,}000 + 50{,}000t)\,dt$
$$= \left[200{,}000t + 25{,}000t^2\right]_0^4$$
$$= \$1{,}200{,}000$$

(b) Present value $= \int_0^4 (200{,}000 + 50{,}000t)e^{-0.06t}\,dt$
$$= \int_0^4 200{,}000e^{-0.06t}\,dt + \int_0^4 50{,}000te^{-0.06t}\,dt$$
$$= \left[\frac{200{,}000}{-0.06}e^{-0.06t}\right]_0^4 + 50{,}000\int_0^4 te^{-0.06t}\,dt$$
$$= 711{,}240.46 + 50{,}000\int_0^4 te^{-0.06t}\,dt$$

Let $u = t$ and $dv = e^{-0.06t}\,dt$. Then $du = dt$ and $v = -\frac{1}{0.06}e^{-0.06t}$.

$$= 711{,}240.46 + 50{,}000\left[-\frac{1}{0.06}te^{-0.06t} + \frac{1}{0.06}\int e^{-0.06t}\,dt\right]$$
$$= 711{,}240.46 + 50{,}000\left[-\frac{1}{0.06}te^{-0.06t} - \frac{1}{0.0036}e^{-0.06t}\right]_0^4$$
$$\approx \$1{,}052{,}649.52$$

19. Formula 6: $u = x$, $du = dx$, $a = 2$, $b = 3$

$$\int \frac{x^2}{2+3x}\,dx = \frac{1}{(3)^3}\left[-\frac{3x}{2}(4-3x) + 4\ln|2+3x|\right] + C$$
$$= \frac{1}{27}\left[-\frac{1}{2}\left(12x - 9x^2\right) + 4\ln|2+3x|\right] + C$$
$$= \frac{1}{54}\left(9x^2 - 12x + 8\ln|2+3x|\right) + C$$

21. Formula 23: $u = x$, $du = dx$, $a = 4$

$$\int \sqrt{x^2-16}\,dx = \tfrac{1}{2}\left(x\sqrt{x^2-16} - 16\ln\left|x + \sqrt{x^2-16}\right|\right) + C$$

23. Formula 4: $u = x$, $du = dx$, $a = 2$, $b = 3$

$$\int \frac{x}{(2+3x)^2}\,dx = \frac{1}{9}\left(\frac{2}{2+3x} + \ln|2+3x|\right) + C$$

25. Formula 23: $u = x$, $du = dx$, $a = 5$

$$\int \frac{\sqrt{x^2+25}}{x}\,dx = \sqrt{x^2+25} - 5\ln\left|\frac{5+\sqrt{x^2+25}}{x}\right| + C$$

27. Formula 29: $u = x$, $du = dx$, $a = 2$

$$\int \frac{1}{x^2-4}\,dx = \frac{1}{4}\ln\left|\frac{x-2}{x+2}\right| + C$$

29. Formula 19: $u = x$, $du = dx$, $a = 1$, $b = 1$

$$\int \frac{x}{\sqrt{1+x}}\,dx = -\frac{2(2(1)-x)}{3(1)^2}\sqrt{1+x} + C$$
$$= -\frac{2(2-x)}{3}\sqrt{1+x} + C$$

31. Formula 17: $u = x$, $du = dx$, $a = b = 1$

$$\int \frac{\sqrt{1+x}}{x}\,dx = 2\sqrt{1+x} + \int \frac{1}{x\sqrt{1+x}}\,dx$$

Formula 15: $u = x$, $du = dx$, $a = b = 1$

$$\int \frac{\sqrt{1+x}}{x}\,dx = 2\sqrt{1+x} + \ln\left|\frac{\sqrt{1+x}-1}{\sqrt{1+x}+1}\right| + C$$

33. Formula 19: $u = x$, $du = dx$, $a = 9$, $b = 16$

$$P(a \le x \le b) = \int_a^b \frac{96}{11}\left(\frac{x}{\sqrt{9-16x}}\right)dx$$
$$= \frac{96}{11}\left[-\frac{2(18-16x)}{768}\sqrt{9+16x}\right]_a^b$$
$$= -\frac{1}{22}\left[(9-8x)\sqrt{9+16x}\right]_a^b$$

(a) $P(0 \le x \le 0.8) = -\frac{1}{22}\left[(9-8x)\sqrt{9+16x}\right]_0^{0.8}$
$= -\frac{1}{22}\left(2.6\sqrt{21.8} - 27\right)$
≈ 0.675

(b) $P(0 \le x \le 0.5) = -\frac{1}{22}\left[(9-8x)\sqrt{9+16x}\right]_0^{0.5}$
$= -\frac{1}{22}\left(5\sqrt{17} - 27\right)$
≈ 0.290

35. Trapezoidal Rule: $\int_1^3 \frac{1}{x^2}\,dx \approx \frac{1}{4}\left[\frac{1}{(1)^2} + 2\left(\frac{1}{(3/2)^2}\right) + 2\left(\frac{1}{(2)^2}\right) + 2\left(\frac{1}{(5/2)^2}\right) + \left(\frac{1}{(3)^2}\right)\right] = 0.705$

Simpson's Rule: $\int_1^3 \frac{1}{x^2}\,dx \approx \frac{1}{6}\left[\frac{1}{(1)^2} + 4\left(\frac{1}{(3/2)^2}\right) + 2\left(\frac{1}{(2)^2}\right) + 4\left(\frac{1}{(5/2)^2}\right) + \left(\frac{1}{(3)^2}\right)\right] \approx 0.6715$

Exact value: $\int_1^3 \frac{1}{x^2}\,dx = \left[-\frac{1}{x}\right]_1^3 = \frac{2}{3}$

37. Trapezoidal Rule: $\displaystyle\int_1^2 \frac{1}{x^3}\,dx \approx \frac{1}{16}\left[\frac{1}{(1)^3} + 2\left(\frac{1}{(9/8)^3}\right) + 2\left(\frac{1}{(5/4)^3}\right) + 2\left(\frac{1}{(11/8)^3}\right) + 2\left(\frac{1}{(3/2)^3}\right) + 2\left(\frac{1}{(13/8)^3}\right)\right.$

$\displaystyle\left. + 2\left(\frac{1}{(7/4)^3}\right) + 2\left(\frac{1}{(15/8)^3}\right) + \left(\frac{1}{(2)^3}\right)\right] \approx 0.3786$

Simpson's Rule: $\displaystyle\int_1^2 \frac{1}{x^3}\,dx \approx \frac{1}{24}\left[\frac{1}{(1)^3} + 4\left(\frac{1}{(9/8)^3}\right) + 2\left(\frac{1}{(5/4)^3}\right) + 4\left(\frac{1}{(11/8)^3}\right) + 2\left(\frac{1}{(3/2)^3}\right) + 4\left(\frac{1}{(7/4)^3}\right)\right.$

$\displaystyle\left. + 2\left(\frac{1}{(15/8)^3}\right) + 4\left(\frac{1}{(15/8)^3}\right) + \left(\frac{1}{(2)^3}\right)\right] \approx 0.3751$

Exact value: $\displaystyle\int_1^2 \frac{1}{x^3}\,dx = \left[-\frac{1}{2x^2}\right]_1^2 = \frac{3}{8} = 0.375$

39. Trapezoidal Rule: $\displaystyle\int_0^4 e^{-x/2}\,dx \approx \tfrac{1}{2}\left[e^0 + 2e^{-1/2} + 2e^{-2/2} + 2e^{-3/2} + e^{-4/2}\right] \approx 1.7652$

Simpson's Rule: $\displaystyle\int_0^4 e^{-x/2}\,dx \approx \tfrac{1}{3}\left[e^0 + 4e^{-1/2} + 2e^{-2/2} + 4e^{-3/2} + e^{-4/2}\right] \approx 1.7299$

Exact value: $\displaystyle\int_0^4 e^{-x/2}\,dx = \left[-2e^{-x/2}\right]_0^4 = 2 - 2e^{-2} \approx 1.7293$

41. (a) Trapezoidal Rule:

$$\int_1^2 \frac{1}{1+\ln x}\,dx \approx \frac{1}{8}\left[\left(\frac{1}{1+\ln(1)}\right) + 2\left(\frac{1}{1+\ln(5/4)}\right) + 2\left(\frac{1}{1+\ln(3/2)}\right) + 2\left(\frac{1}{1+\ln(7/4)}\right) + \left(\frac{1}{1+\ln(2)}\right)\right] \approx 0.741$$

(b) Simpson's Rule:

$$\int_1^2 \frac{1}{1+\ln x}\,dx \approx \frac{1}{12}\left[\left(\frac{1}{1+\ln(1)}\right) + 4\left(\frac{1}{1+\ln(5/4)}\right) + 2\left(\frac{1}{1+\ln(3/2)}\right) + 4\left(\frac{1}{1+\ln(7/4)}\right) + \left(\frac{1}{1+\ln(2)}\right)\right] \approx 0.737$$

43. (a) Trapezoidal Rule: $\displaystyle\int_0^1 \frac{x^{3/2}}{2-x^2}\,dx \approx \frac{1}{8}\left[0 + 2\left(\frac{(1/4)^{3/2}}{2-(1/4)^2}\right) + 2\left(\frac{(1/2)^{3/2}}{2-(1/2)^2}\right) + 2\left(\frac{(3/4)^{3/2}}{2-(3/4)^2}\right) + \left(\frac{(1)^{3/2}}{2-(1)^2}\right)\right] \approx 0.305$

(b) Simpson's Rule: $\displaystyle\int_0^1 \frac{x^{3/2}}{2-x^2}\,dx \approx \frac{1}{12}\left[0 + 4\left(\frac{(1/4)^{3/2}}{2-(1/4)^2}\right) + 2\left(\frac{(1/2)^{3/2}}{2-(1/2)^2}\right) + 4\left(\frac{(3/4)^{3/2}}{2-(3/4)^2}\right) + \left(\frac{(1)^{3/2}}{2-(1)^2}\right)\right] \approx 0.289$

45. (a) Trapezoidal Rule: $\displaystyle\int_0^8 \frac{3}{x^2+2}\,dx \approx \frac{1}{2}\left[\left(\frac{3}{(0)^2+2}\right) + 2\left(\frac{3}{(1)^2+2}\right) + 2\left(\frac{3}{(2)^2+2}\right) + 2\left(\frac{3}{(3)^2+2}\right) + 2\left(\frac{3}{(4)^2+2}\right)\right.$

$\displaystyle\left. + 2\left(\frac{3}{(5)^2+2}\right) + 2\left(\frac{3}{(6)^2+2}\right) + 2\left(\frac{3}{(7)^2+2}\right) + \left(\frac{3}{(8)^2+2}\right)\right] \approx 2.961$

(b) Simpson's Rule: $\displaystyle\int_0^8 \frac{3}{x^2+2}\,dx \approx \frac{1}{3}\left[\left(\frac{3}{(0)^2+2}\right) + 4\left(\frac{3}{(1)^2+2}\right) + 2\left(\frac{3}{(2)^2+2}\right) + 4\left(\frac{3}{(3)^2+2}\right) + 2\left(\frac{3}{(4)^2+2}\right)\right.$

$\displaystyle\left. + 4\left(\frac{3}{(5)^2+2}\right) + 2\left(\frac{3}{(6)^2+2}\right) + 4\left(\frac{3}{(7)^2+2}\right) + \left(\frac{3}{(8)^2+2}\right)\right] \approx 2.936$

47. $f(x) = e^{2x}$

$f'(x) = 2e^{2x}$

$f''(x) = 4e^{2x}$

$f'''(x) = 8e^{2x}$

$f^{(4)}(x) = 16e^{2x}$

(a) Trapezoidal Rule: Because $|f''(x)|$ is maximum in $[0, 2]$ when $x = 2$ and $|f''(x)| = 4e^4$, you have

$$|\text{Error}| \le \frac{(2-0)^3}{12(4)^2}4e^4 = \frac{e^4}{6} \approx 9.0997.$$

(b) Simpson's Rule: Because $|f^{(4)}(x)|$ is maximum in $[0, 2]$ when $x = 2$ and $|f^{(4)}(x)| = 16e^4$, you have

$$|\text{Error}| \le \frac{(2-0)^5}{180(4)^4}16e^4 = \frac{e^4}{90} \approx 0.6066.$$

49. $f(x) = x^2$

$f'(x) = 2x$

$f''(x) = 2$

$f'''(x) = 0$

$f^{(4)}(x) = 0$

(a) Trapezoidal Rule: Because $|f''(x)| = 2$ for all x, you have

$$|\text{Error}| \le \frac{(3-0)^3}{12n^2}(2) \le 0.0001$$

$$\frac{9}{2n^2} \le 0.0001$$

$$45{,}000 \le n^2$$

$$n > 212.13.$$

Let $n = 214$.

(b) Simpson's Rule: Because $|f^{(4)}(x)| = 0$ for all x, n can be any even value and will result in zero error. Let $n = 2$.

51. This integral converges because $\int_{-\infty}^{-1} \frac{1}{x^5}\,dx = \lim_{a\to-\infty}\left[-\frac{1}{4x^4}\right]_a^{-1} = -\frac{1}{4} + 0 = -\frac{1}{4}.$

53. This integral diverges because $\int_{-\infty}^{0} \frac{1}{\sqrt[3]{8-x}}\,dx = \lim_{a\to-\infty}\left[-\frac{3(8-x)^{2/3}}{2}\right]_a^0 = -6 - \infty = -\infty.$

55. This integral diverges because $\int_1^{\infty} \frac{\ln x}{x}\,dx = \lim_{b\to\infty}\left[\frac{(\ln x)^2}{2}\right]_1^b = \infty - 0 = \infty.$

57. $A = \int_0^{\infty} e^{-x/4}\,dx = \lim_{b\to\infty}\left[-4e^{-x/4}\right]_0^b = 0 - (-4) = 4$

59. $A = \int_0^{\infty} 4xe^{-2x^2}\,dx = \lim_{b\to\infty} -\int_0^b e^{-2x^2}(-4x\,dx) = \lim_{b\to\infty}\left[-e^{2x^2}\right]_0^b = 1$

61. Present value $= \int_0^{\infty} 8000e^{-0.03t}\,dt$

$$= \lim_{b\to\infty}\left[-\frac{8000}{0.03}e^{-0.03t}\right]_0^b$$

$$= 0 - (-266{,}666.67)$$

$$= \$266{,}666.67$$

63. No, you do not have enough money to start the scholarship fund because

Present value $= \int_0^{\infty} 21{,}000e^{-0.07t}\,dt$

$$= \lim_{b\to\infty}\left[-300{,}000e^{-0.07t}\right]_0^b$$

$$= 0 - (-300{,}000)$$

$$= \$300{,}000.$$

Chapter 6 Test Yourself

1. Let $u = x$ and $dv = e^{x+1}\,dx$. Then $du = dx$ and $v = e^{x+1}$.

$$\int xe^{x+1}\,dx = xe^{x+1} - \int e^{x+1}\,dx$$

$$= xe^{x+1} - e^{x+1} + C$$

$$= (x-1)e^{x+1} + C$$

2. Let $u = \ln x$ and $dv = 9x^2\,dx$. Then $du = \frac{1}{x}\,dx$ and $v = 3x^3$.

$$\int 9x^2 \ln x\,dx = 3x^3 \ln x - 3\int x^2\,dx$$
$$= 3x^3 \ln x - x^3 + C$$
$$= x^3(3 \ln x - 1) + C$$

3. Let $u = x^2$ and $dv = e^{-x/3}\,dx$. Then $du = 2x\,dx$ and $v = -3e^{-x/3}$.

$$\int x^2 e^{-x/3}\,dx = -3x^2 e^{-x/3} + \int 6xe^{-x/3}\,dx$$

Let $u = 6x$ and $dv = e^{-x/3}\,dx$. Then $du = 6\,dx$ and $v = -3e^{-x/3}$.

$$\int x^2 e^{-x/3}\,dx = -3x^2 e^{-x/3} - 18xe^{-x/3} + \int 18e^{-x/3}\,dx$$
$$= -3x^2 e^{-x/3} - 18xe^{-x/3} - 54e^{-x/3} + C$$
$$= -3e^{-x/3}(x^2 + 6x + 18) + C$$

4. $R = 295.1 + 147.66\sqrt{t}\ln t,\ 1 \le t \le 9$ and $t = 1 \to 2001$

(a) Total revenue $= \int_1^9 \left(295.1 + 147.66\sqrt{t}\ln t\right) dt = \int_1^9 295.1\,dt + 147.66\int_1^9 t^{1/2}\ln t\,dt$

Let $u = \ln t$ and $dv = t^{1/2}\,dt$. Then $du = \frac{1}{t}\,dt$ and $v = \frac{2}{3}t^{2/3}$.

$$= \left[295.1t\right]_1^9 + 147.66\left\{\frac{2}{3}t^{3/2}\ln t - \int \frac{2}{3}t^{1/2}\,dt\right\}$$
$$= 2360.8 + 147.66\left[\frac{2}{3}t^{3/2}\ln t - \frac{4}{9}t^{3/2}\right]_1^9$$
$$\approx \$6494.47 \text{ million}$$

(b) Average revenue $= \frac{1}{9-1}\int_1^9 R(t)\,dt = \frac{1}{8}(6494.47) \approx \$811.81$ million

5. Formula 4: $u = x,\ du = dx,\ a = 7,\ b = 2$

$$\int \frac{x}{(7+2x)^2}\,dx = \frac{1}{4}\left(\frac{7}{7+2x} + \ln|7+2x|\right) + C$$

6. Formula 37: $u = x^3,\ du = 3x^2\,dx$

$$\int \frac{3x^2}{1+e^{x^3}}\,dx = x^3 - \ln\left(1 + e^{x^3}\right) + C$$

7. Formula 19: $u = x^2,\ du = 2x\,dx,\ a = 1,\ b = 5$

$$\int \frac{2x^3}{\sqrt{1+5x^2}}\,dx = \int \frac{x^2}{\sqrt{1+5x^2}}(2x\,dx)$$
$$= -\frac{2(2-5x^2)}{75}\sqrt{1+5x^2} + C$$

8. Formula 39: $u = 3 - 2x,\ du = -2\,dx$

$$\int_0^1 \ln(3-2x)\,dx = \left[-\tfrac{1}{2}(3-2x)\{-1 + \ln(3-2x)\}\right]_0^1$$
$$= -\tfrac{1}{2}\left[-1 + \ln 1 - 3(-1 + \ln 3)\right]$$
$$= -\tfrac{1}{2}(2 - 3\ln 3)$$
$$\approx 0.6479$$

9. Let $u = x$ and $dv = (x-2)^{-1/2}\,dx$.

Then $du = dx$ and $v = 2(x-2)^{1/2}$.

$$\int_3^6 x(x-2)^{-1/2}\,dx = 2x(x-2)^{1/2} - \int 2(x-2)^{1/2}\,dx$$
$$= \left[2x(x-2)^{1/2} - \tfrac{4}{3}(x-2)^{3/2}\right]_3^6$$
$$= \left(24 - \tfrac{4}{3}(8)\right) - \left(6 - \tfrac{4}{3}\right)$$
$$= \tfrac{26}{3}$$

10. Formula 23: $u = x$, $du = dx$, $a = 4$

$$\int_{-3}^{-1} \frac{\sqrt{x^2 + 16}}{x}\,dx = \left[\sqrt{x^2 + 16} - 4\ln\left|\frac{4 + \sqrt{x^2 + 16}}{x}\right|\right]_{-3}^{-1}$$

$$= \left(\sqrt{17} - 4\ln\left|-4 - \sqrt{17}\right|\right) - \left(5 - 4\ln|-3|\right)$$

$$\approx -4.8613$$

11. Trapezoidal Rule: $\int_2^5 \left(x^2 - 2x\right) dx \approx \frac{3}{8}\left[\left((2)^2 - 2(2)\right) + 2\left(\left(\frac{11}{4}\right)^2 - 2\left(\frac{11}{4}\right)\right) + 2\left(\left(\frac{7}{2}\right)^2 - 2\left(\frac{7}{2}\right)\right)\right.$

$$\left. + 2\left(\left(\frac{17}{4}\right)^2 - 2\left(\frac{17}{4}\right)\right) + \left((5)^2 - 2(5)\right)\right]$$

$$= \frac{585}{32} \approx 18.28$$

Exact value: $\int_2^5 \left(x^2 - 2x\right) dx = \left[\frac{1}{3}x^3 - x^2\right]_2^5 = 18$

12. Exact: Let $u = 9x$ and $dv = e^{3x}\,dx$. Then $du = 9\,dx$ and $v = \frac{1}{3}e^{3x}$.

$$\int_0^1 9xe^{3x}\,dx = 3xe^{3x}\Big]_0^1 - \int 3e^{3x}\,dx$$

$$= 3e^3 - \left[e^{3x}\right]_0^1$$

$$= 3e^3 - \left(e^3 - 1\right)$$

$$= 2e^3 + 1$$

$$\approx 41.1711$$

Trapezoidal Rule: $\int_0^1 9xe^{3x}\,dx = \frac{1}{12}\left[0 + 4\left(\frac{9}{4}e^{3/4}\right) + 2\left(\frac{9}{2}e^{3/2}\right) + 4\left(\frac{27}{4}e^{9/4}\right) + 9e^3\right]$

$$\approx 41.3606$$

13. This integral converges because

$$\int_0^\infty e^{-3x}\,dx = \lim_{b\to\infty}\left[-\frac{1}{3}e^{-3x}\right]_0^b = 0 + \frac{1}{3} = \frac{1}{3}.$$

14. This integral converges because

$$\int_0^9 \frac{2}{\sqrt{x}}\,dx = \lim_{a\to 0^+}\left[4\sqrt{x}\right]_b^9 = 12.$$

15. This integral diverges because

$$\int_{-\infty}^0 \frac{1}{(4x - 1)^{2/3}}\,dx = \frac{3}{4}\lim_{a\to-\infty}\left[(4x - 1)^{1/3}\right]_a^0 = -\frac{3}{4} + \infty = \infty.$$

16. (a) Present value $= \int_0^\infty 19.95e^{-0.04t}\,dt = \lim_{b\to\infty}\left[-498.75e^{-0.04t}\right]_0^b = \$498.75$

(b) The subscriber should choose plan B because the lifetime subscription for \$149 is less than \$498.75, the present value of plan A.

CHAPTER 7
Functions of Several Variables

CHAPTER 7
Functions of Several Variables

Section 7.1 The Three-Dimensional Coordinate System

Skills Warm Up

1. $(5, 1), (3, 5)$

$$d = \sqrt{(3-5)^2 + (5-1)^2} = \sqrt{4+16} = \sqrt{20} = 2\sqrt{5}$$

2. $(2, 3), (-1, -1)$

$$d = \sqrt{(-1-2)^2 + (-1-3)^2} = \sqrt{9+16} = \sqrt{25} = 5$$

3. $(-5, 4), (-5, -4)$

$$d = \sqrt{(-5-(-5))^2 + (-4-4)^2} = \sqrt{64} = 8$$

4. $(-3, 6), (-3, -2)$

$$d = \sqrt{(-3-(-3))^2 + (-2-6)^2} = \sqrt{64} = 8$$

5. $(2, 5), (6, 9)$

$$\text{Midpoint} = \left(\frac{2+6}{2}, \frac{5+9}{2}\right) = (4, 7)$$

6. $(-1, -2), (3, 2)$

$$\text{Midpoint} = \left(\frac{-1+3}{2}, \frac{-2+2}{2}\right) = (1, 0)$$

7. $(-6, 0), (6, 6)$

$$\text{Midpoint} = \left(\frac{-6+6}{2}, \frac{0+6}{2}\right) = (0, 3)$$

8. $(-4, 3), (2, -1)$

$$\text{Midpoint} = \left(\frac{-4+2}{2}, \frac{3+(-1)}{2}\right) = (-1, 1)$$

9. $c: (2, 3), r = 2$

$$(x-2)^2 + (y-3)^2 = 2^2$$

$$(x-2)^2 + (y-3)^2 = 4$$

10. $C = \left(\frac{4+(-2)}{2}, \frac{0+8}{2}\right) = (1, 4)$

$$r = \frac{1}{2}\sqrt{(-2-4)^2 + (8-0)^2} = \frac{1}{2}\sqrt{36+64} = \frac{1}{2}\sqrt{100} = 5$$

$$(x-1)^2 + (y-4)^2 = 5^2$$

$$(x-1)^2 + (y-4)^2 = 25$$

1.

(3, −2, 5)

(2, 1, 3)

(−1, 2, 1)

$\left(\frac{3}{2}, 4, -2\right)$

3.

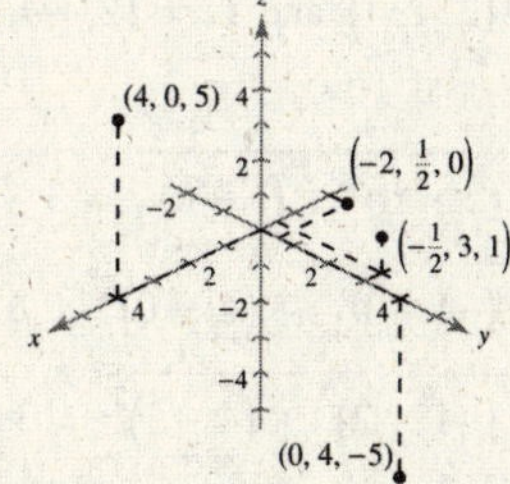

5. $x = -3, y = 4, z = 5: (-3, 4, 5)$

7. $y = z = 0, x = 10: (10, 0, 0)$

9. The z-coordinate is 0.

11. $(4, 1, 5), (8, 2, 6)$

$$d = \sqrt{(8-4)^2 + (2-1)^2 + (6-5)^2} = \sqrt{18} = 3\sqrt{2}$$

13. $(-1, -5, 7), (-3, 4, -4)$

$$d = \sqrt{(-3+1)^2 + (4+5)^2 + (-4-7)^2} = \sqrt{206}$$

15. $(6, -4, 2), (-2, 1, 3)$

$$\text{Midpoint} = \left(\frac{6+(-2)}{2}, \frac{-4+1}{2}, \frac{2+3}{2}\right) = \left(2, -\frac{3}{2}, \frac{5}{2}\right)$$

17. $(-5, -2, 5), (6, 3, -7)$

$$\text{Midpoint} = \left(\frac{-5+6}{2}, \frac{-2+3}{2}, \frac{5+(-7)}{2}\right) = \left(\frac{1}{2}, \frac{1}{2}, -1\right)$$

19. $(2, -1, 3) = \left(\frac{x+(-2)}{2}, \frac{y+1}{2}, \frac{z+1}{2}\right)$

$$2 = \frac{x-2}{2} \qquad -1 = \frac{y+1}{2} \qquad 3 = \frac{z+1}{2}$$

$$4 = x-2 \qquad -2 = y+1 \qquad 6 = z+1$$

$$x = 6 \qquad y = -3 \qquad z = 5$$

$$(x, y, z) = (6, -3, 5)$$

21. $\left(\frac{3}{2}, 1, 2\right) = \left(\frac{x+2}{2}, \frac{y+0}{2}, \frac{z+3}{2}\right)$

$$\frac{3}{2} = \frac{x+2}{2} \qquad 1 = \frac{y}{2} \qquad 2 = \frac{z+3}{2}$$

$$3 = x+2 \qquad 4 = z+3$$

$$x = 1 \qquad y = 2 \qquad z = 1$$

$$(x, y, z) = (1, 2, 1)$$

23. Let $A = (0, 0, 0)$, $B = (2, 2, 1)$, and $C = (2, -4, 4)$.

Then you have

$$d(AB) = \sqrt{(2-0)^2 + (2-0)^2 + (1-0)^2} = 3$$

$$d(AC) = \sqrt{(2-0)^2 + (-4-0)^2 + (4-0)^2} = 6$$

$$d(BC) = \sqrt{(2-2)^2 + (-4-2)^2 + (4-1)^2} = 3\sqrt{5}.$$

The triangle is a right triangle because

$$d^2(AB) + d^2(AC) = (3)^2 + (6)^2 = 45 = \left(3\sqrt{5}\right)^2 = d^2(BC).$$

25. Let $A = (-1, 0, -2)$, $B = (-1, 5, 2)$, and $C = (-3, -1, 1)$. Then you have

$$d(AB) = \sqrt{[-1-(-1)]^2 + (0-5)^2 + (-2-2)^2} = \sqrt{41}$$

$$d(AC) = \sqrt{[-1-(-3)]^2 + [0-(-1)]^2 + (-2-1)^2} = \sqrt{14}$$

$$d(BC) = \sqrt{[-1-(-3)]^2 + [5-(-1)]^2 + (2-1)^2} = \sqrt{41}.$$

Because $d(AB) = d(BC)$, the triangle is isosceles.

The triangle is not a right triangle because

$$d^2(AB) + d^2(BC) = \left(\sqrt{41}\right)^2 + \left(\sqrt{41}\right)^2 = 82 \neq \left(\sqrt{14}\right)^2 = 14 = d^2(AC).$$

27. Each z-coordinate is decreased by 5 units:

$(0, 0, -5), (2, 2, -4), (2, -4, -1)$

29. $x^2 + (y-2)^2 + (z-2)^2 = 4$

31. The midpoint of the diameter is the center.

$$\text{Center} = \left(\frac{2+1}{2}, \frac{1+3}{2}, \frac{3+(-1)}{2}\right) = \left(\frac{3}{2}, 2, 1\right)$$

The radius is the distance between the center and either endpoint.

$$\text{Radius} = \sqrt{\left(2-\frac{3}{2}\right)^2 + (1-2)^2 + (3-1)^2} = \sqrt{\frac{1}{4} + 1 + 4} = \frac{\sqrt{21}}{2}$$

$$\left(x-\frac{3}{2}\right)^2 + (y-2)^2 + (z-1)^2 = \frac{21}{4}$$

33. $(x-3)^2 + (y+2)^2 + (z+3)^2 = 16$

35. The midpoint of the diameter is the center.

$$\text{Center} = \left(\frac{2+0}{2}, \frac{0+6}{2}, \frac{0+0}{2}\right) = (1, 3, 0)$$

The radius is the distance from the center to either endpoint.

$$\text{Radius} = \sqrt{(1-2)^2 + (3-0)^2 + (0-0)^2} = \sqrt{10}$$

$$(x-1)^2 + (y-3)^2 + z^2 = 10$$

37. The distance from $(-4, 3, 2)$ to the xy-plane is the radius $r = 2$.

$$(x+4)^2 + (y-3)^2 + (z-2)^2 = 4$$

39.
$$\left(x^2 - 5x + \tfrac{25}{4}\right) + y^2 + z^2 = \tfrac{25}{4}$$
$$x^2 + y^2 + z^2 - 5x = 0$$
$$\left(x - \tfrac{5}{2}\right)^2 + (y-0)^2 + (z-0)^2 = \tfrac{25}{4}$$

Center: $\left(\frac{5}{2}, 0, 0\right)$

Radius: $\frac{5}{2}$

41.
$$x^2 + y^2 + z^2 + 4x - 2y + 8z - 4 = 0$$
$$\left(x^2 + 4x + 4\right) + \left(y^2 - 2y + 1\right) + \left(z^2 + 8z + 16\right) = 4 + 4 + 1 + 16$$
$$(x+2)^2 + (y-1)^2 + (z+4)^2 = 25$$

Center: $(-2, 1, -4)$

Radius: 5

43.
$$2x^2 + 2y^2 + 2z^2 - 4x - 12y - 8z + 3 = 0$$
$$\left(x^2 - 2x + 1\right) + \left(y^2 - 6y + 9\right) + \left(z^2 - 4z + 4\right) = -\frac{3}{2} + 1 + 9 + 4$$
$$(x-1)^2 + (y-3)^2 + (z-2)^2 = \frac{25}{2}$$

Center: $(1, 3, 2)$

Radius: $\dfrac{5}{\sqrt{2}} = \dfrac{5\sqrt{2}}{2}$

45. $(x-1)^2 + (y-3)^2 + (z-2)^2 = 25$

To find the xy-trace, let $z = 0$.

$$(x-1)^2 + (y-3)^2 + (0-2)^2 = 25$$
$$(x-1)^2 + (y-3)^2 = 21$$

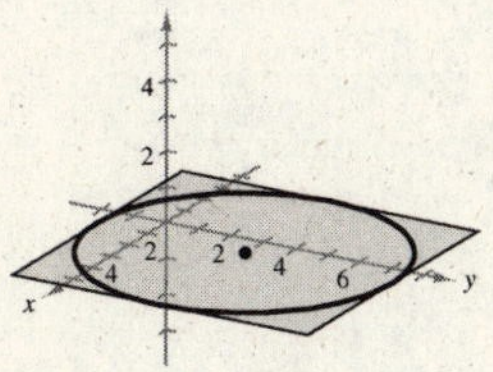

47. $x^2 + y^2 + z^2 - 6x - 10y + 6z + 30 = 0$

To find the xy-trace, let $z = 0$.

$$x^2 + y^2 + (0)^2 - 6x - 10y + 6(0) + 30 = 0$$
$$\left(x^2 - 6x + 9\right) + \left(y^2 - 10y + 25\right) = -30 + 9 + 25$$
$$(x-3)^2 + (y-5)^2 = 4$$

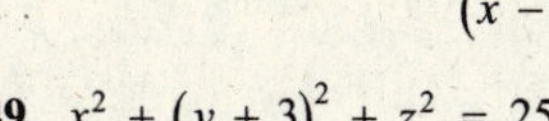

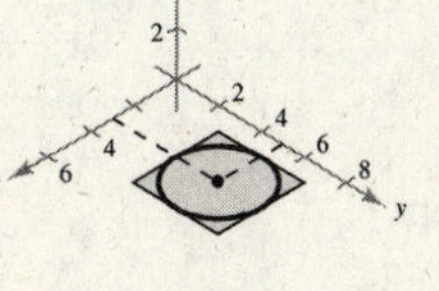

49. $x^2 + (y+3)^2 + z^2 = 25$

To find the yz-trace, let $x = 0$.

$$0^2 + (y+3)^2 + z^2 = 25$$
$$(y+3)^2 + z^2 = 25$$

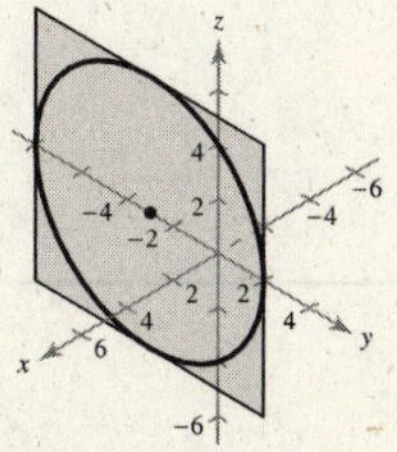

51. $x^2 + y^2 + z^2 - 4x - 4y - 6z - 12 = 0$

To find the yz-trace, let $x = 0$.

$$(0)^2 + y^2 + z^2 - 4(0) - 4y - 6z - 12 = 0$$
$$\left(y^2 - 4y + 4\right) + \left(z^2 - 6z + 9\right) = 12 + 4 + 9$$
$$(y-2)^2 + (z-3)^2 = 25$$

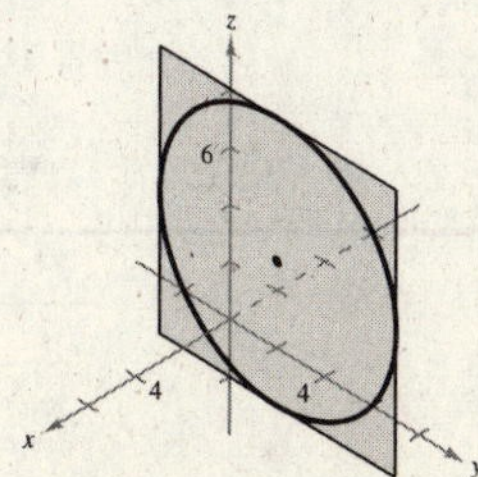

53. $x^2 + y^2 + z^2 = 25$

(a) To find the trace, let $z = 3$.

$$x^2 + y^2 + 3^2 = 25$$
$$x^2 + y^2 = 16$$

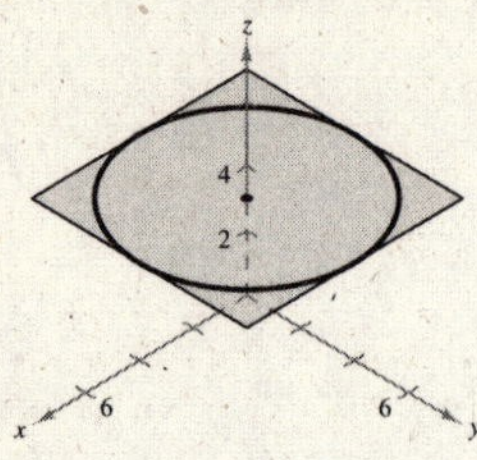

(b) To find the trace, let $x = 4$.

$$4^2 + y^2 + z^2 = 25$$
$$y^2 + z^2 = 9$$

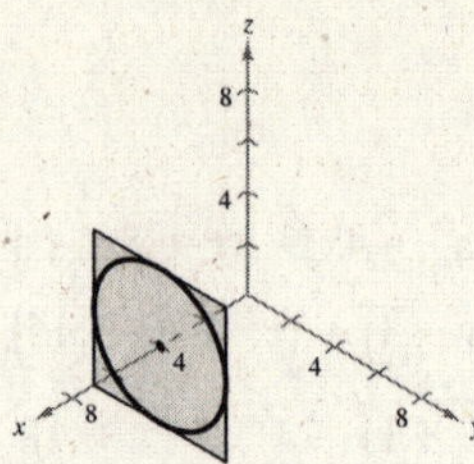

55. $x^2 + y^2 + z^2 - 4x - 6y + 9 = 0$

(a) To find the trace, let $x = 2$.

$$2^2 + y^2 + z^2 - 4(2) - 6y + 9 = 0$$
$$(y^2 - 6y + 9) + z^2 = -9 - 4 + 8 + 9$$
$$(y - 3)^2 + z^2 = 2^2$$

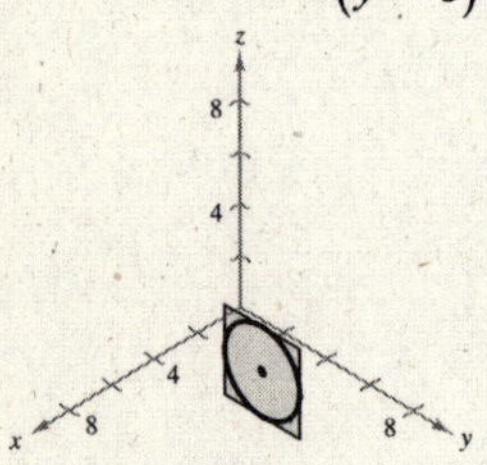

(b) To find the trace, let $y = 3$.

$$x^2 + 3^2 + z^2 - 4x - 6(3) + 9 = 0$$
$$(x^2 - 4x + 4) + z^2 = -9 + 18 - 9 + 4$$
$$(x - 2)^2 + z^2 = 2^2$$

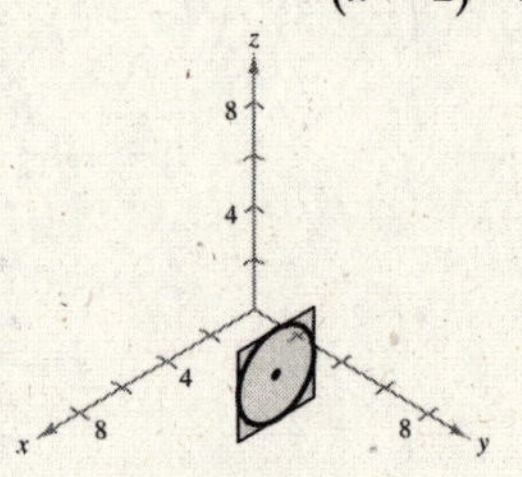

57. $x^2 + y^2 + z^2 = \left(\frac{165}{2}\right)^2$

$$x^2 + y^2 + z^2 = 6806.25$$

Section 7.2 Surfaces in Space

Skills Warm Up

1. $3x + 4y = 12$

Let $x = 0$ to find the y-intercept.

$$3(0) + 4y = 12$$
$$y = 3$$

y-intercept: $(0, 3)$

Let $y = 0$ to find the x-intercept.

$$3x + 4(0) = 12$$
$$x = 4$$

x-intercept: $(4, 0)$

2. $6x + y = -8$

Let $x = 0$ to find the y-intercept.

$$6(0) + y = -8$$
$$y = -8$$

y-intercept: $(0, -8)$

Let $y = 0$ to find the x-intercept.

$$6x + 0 = -8$$
$$x = -\frac{4}{3}$$

x-intercept: $\left(-\frac{4}{3}, 0\right)$

Skills Warm Up —*continued*—

3. $-2x + y = -2$

Let $x = 0$ to find the y-intercept.

$-2(0) + y = -2$

$y = -2$

y-intercept: $(0, -2)$

Let $y = 0$ to find the x-intercept.

$-2x + 0 = -2$

$x = 1$

x-intercept: $(1, 0)$

4. $-x - y = 5$

Let $x = 0$ to find the y-intercept.

$-0 - y = 5$

$y = -5$

y-intercept: $(0, -5)$

Let $y = 0$ to find the x-intercept.

$-x - 0 = 5$

$x = -5$

x-intercept: $(-5, 0)$

5. $16x^2 + 16y^2 + 16z^2 = 4$

$x^2 + y^2 + z^2 = \frac{1}{4}$

6. $9x^2 + 9y^2 + 9z^2 = 36$

$x^2 + y^2 + z^2 = 4$

1. $4x + 2y + 6z = 12$

To find the x-intercept, let $y = 0$ and $z = 0$.

$4x = 12 \Rightarrow x = 3$

To find the y-intercept, let $x = 0$ and $z = 0$.

$2y = 12 \Rightarrow y = 6$

To find the z-intercept, let $x = 0$ and $y = 0$.

$6z = 12 \Rightarrow z = 2$

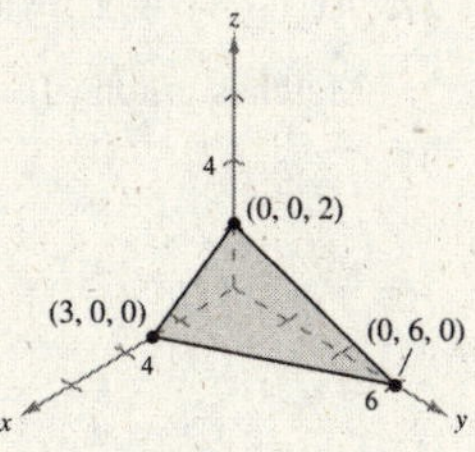

3. $3x + 3y + 5z = 15$

To find the x-intercept, let $y = 0$ and $z = 0$.

$3x = 15 \Rightarrow x = 5$

To find the y-intercept, let $x = 0$ and $z = 0$.

$3y = 15 \Rightarrow y = 5$

To find the z-intercept, let $x = 0$ and $y = 0$.

$5z = 15 \Rightarrow z = 3$

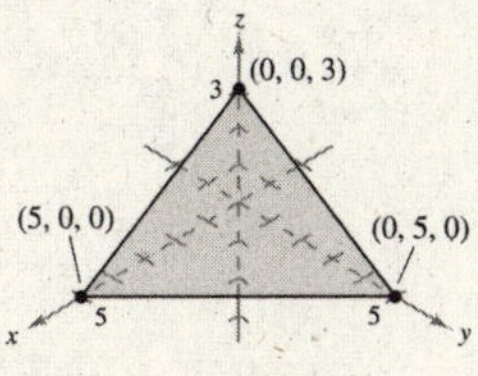

5. $2x - y + 3z = 4$

To find the x-intercept, let $y = 0$ and $z = 0$.

$2x = 4 \Rightarrow x = 2$

To find the y-intercept, let $x = 0$ and $z = 0$.

$-y = 4 \Rightarrow y = -4$

To find the z-intercept, let $x = 0$ and $y = 0$.

$3z = 4 \Rightarrow z = \frac{4}{3}$

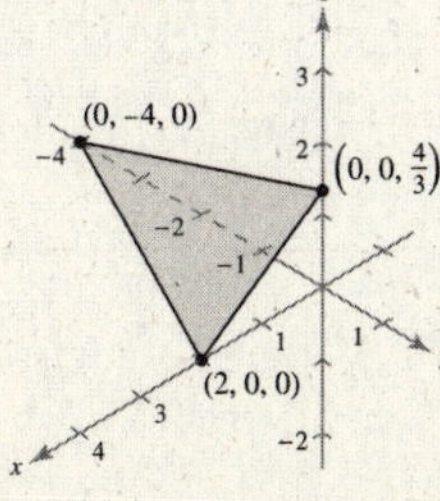

7. $z = 8$

Because the coefficients of x and y are zero, the only intercept is the z-intercept of 8. The plane is parallel to the xy-plane.

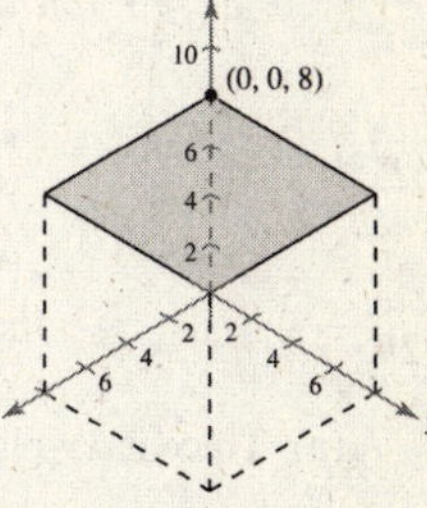

9. $y + z = 5$

Because the coefficient of x is zero, there is no x-intercept.

To find the y-intercept, let $z = 0$.

$y = 5$

To find the z-intercept, let $y = 0$.

$z = 5$

The plane is parallel to the x-axis.

11. $x + z = 6$

To find the x-intercept, let $z = 0$, so $x = 6$.

Because the coefficient of y is zero, there is no y-intercept.

To find the z-intercept, let $x = 0$, so $z = 6$.

The plane is parallel to the y-axis.

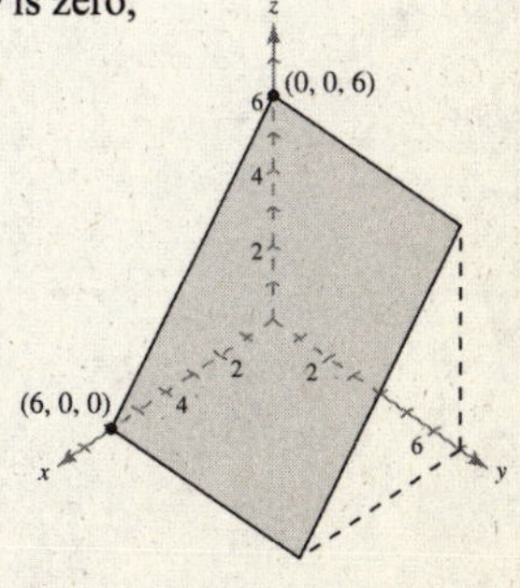

13. For the first plane, $5x - 3y + z = 4$, $a_1 = 5$, $b_1 = -3$, and $c_1 = 1$. For the second plane, $x + 4y + 7z = 1$, $a_2 = 1$, $b_2 = 4$, and $c_2 = 7$. So you have

$$\begin{aligned} a_1a_2 + b_1b_2 + c_1c_2 &= (5)(1) + (-3)(4) + (1)(7) \\ &= 5 - 12 + 7 \\ &= 0. \end{aligned}$$

The planes are perpendicular.

15. For the first plane, $x - 5y - z = 1$, $a_1 = 1$, $b_1 = -5$, and $c_1 = -1$. For the second plane, $5x - 25y - 5z = -3$, $a_2 = 5$, $b_2 = -25$, and $c_2 = -5$. So you have $a_2 = 5a_1$, $b_2 = 5b_1$, and $c_2 = 5c_1$.

The planes are parallel.

17. For the first plane, $x + 2y = 3$, $a_1 = 1$, $b_1 = 2$, and $c_1 = 0$. For the second plane, $4x + 8y = 5$, $a_2 = 4$, $b_2 = 8$, and $c_2 = 0$. So you have $a_2 = 4a_1$, $b_2 = 4b_1$, and $c_2 = 4c_1$. The planes are parallel.

19. For the first plane, $2x + y = 3$, $a_1 = 2$, $b_1 = 1$, and $c_1 = 0$. For the second plane, $3x - 5z = 0$, $a_2 = 3$, $b_2 = 0$, and $c_2 = -5$. The planes are not parallel because $3a_1 = 2a_2$ and $3b_1 \neq 2b_2$. The planes are not perpendicular because

$$\begin{aligned} a_1a_2 + b_1b_2 + c_1c_2 &= (2)(3) + (1)(0) + (0)(-5) \\ &= 6 \neq 0. \end{aligned}$$

21. For the first plane, $x = 3$, $a_1 = 1$, $b_1 = 0$, and $c_1 = 0$. For the second plane, $z = -1$, $a_2 = 0$, $b_2 = 0$, and $c_2 = 1$. So you have

$$\begin{aligned} a_1a_2 + b_1b_2 + c_1c_2 &= (1)(0) + (0)(0) + (0)(1) \\ &= 0. \end{aligned}$$

The planes are perpendicular.

23. $\dfrac{x^2}{9} + \dfrac{y^2}{16} + \dfrac{z^2}{9} = 1$ is an ellipsoid.

Matches graph (c).

25. $4x^2 - y^2 + 4z^2 = 4$ is a hyperboloid of one sheet.

Matches graph (f).

27. $4x^2 - 4y + z^2 = 0$ is an elliptic paraboloid.

Matches graph (d).

29. $z = x^2 - y^2$

(a) Trace in xy-plane $(z = 0)$: $0 = x^2 - y^2$

$\pm x = y$ Lines

(b) Trace in plane $x = 3$: $z = 9 - y^2$ Parabola

(c) Trace in xz-plane $(y = 0)$: $z = x^2$ Parabola

The graph is a hyperbolic parabola.

31. $\dfrac{x^2}{4} + y^2 + z^2 = 1$

(a) Trace in xy-plane $(z = 0)$: $\dfrac{x^2}{4} + y^2 = 1$ Ellipse

(b) Trace in xz-plane $(y = 0)$: $\dfrac{x^2}{4} + z^2 = 1$ Ellipse

(c) Trace in yz-plane $(x = 0)$: $y^2 + z^2 = 1$ Circle

The graph is an ellipsoid.

33. $z^2 - \frac{x^2}{9} - \frac{y^2}{16} = 1$

(a) Trace in xz-plane $(y = 0)$: $z^2 - \frac{x^2}{9} = 1$ Hyperbola

(b) Trace in plane $x = 2$: $z^2 - \frac{y^2}{16} = \frac{13}{9}$

$\frac{z^2}{13/9} - \frac{y^2}{208/9} = 1$ Hyperbola

(c) Trace in plane $z = 4$: $\frac{x^2}{9} + \frac{y^2}{16} = 15$

$\frac{x^2}{135} + \frac{y^2}{240} = 1$ Ellipse

The graph is a hyperboloid of two sheets.

35. The graph of $x^2 + \frac{y^2}{4} + z^2 = 1$ is an ellipsoid.

37. $25x^2 + 25y^2 - z^2 = 5$

Standard form: $\frac{x^2}{1/5} + \frac{y^2}{1/5} - \frac{z^2}{5} = 1$

The graph is a hyperboloid of one sheet.

39. $x^2 - y^2 + z = 0$

Standard form: $z = y^2 - x^2$

The graph is a hyperbolic paraboloid.

41. $x^2 - y + z^2 = 0$

Standard form: $y = x^2 + z^2$

The graph is an elliptic paraboloid.

43. $z^2 = 9x^2 + y^2$

Standard form: $x^2 + \frac{y^2}{9} - \frac{z^2}{9} = 0$

The graph is an elliptic cone.

45. $2x^2 - y^2 + 2z^2 = -4$

Standard form: $-\frac{x^2}{2} + \frac{y^2}{4} - \frac{z^2}{2} = 1$

The graph is a hyperboloid of two sheets.

47. $3z = -y^2 + x^2$

Standard form: $z = \frac{x^2}{3} - \frac{y^2}{3}$

The graph is a hyperbolic paraboloid.

49. $\frac{x^2}{3963^2} + \frac{y^2}{3963^2} + \frac{z^2}{3950^2} = 1$

51. $z = 0.62x - 0.41y + 0.38$

(a)

Year	2004	2005	2006	2007	2008	2009
x	33.1	34.9	37.4	40.6	43.0	41.8
y	13.2	13.8	14.9	15.0	15.4	14.5
z (actual)	15.5	16.3	17.8	19.5	20.5	20.7
z (model)	15.5	16.4	17.5	19.4	20.7	20.4

The approximated values of z are very close to the actual values.

(b) According to the model, increases in expenditures of recreation types y and z will correspond to an increase in expenditures of recreation type x.

Section 7.3 Functions of Several Variables

Skills Warm Up

1. $f(x) = 5 - 2x, x = -3$

$f(-3) = 5 - 2(-3) = 11$

2. $f(x) = -x^2 + 4x + 5, x = -3$

$f(-3) = -(-3)^2 + 4(-3) + 5 = -16$

Skills Warm Up —continued—

3. $y = \sqrt{4x^2 - 3x + 4},\ x = -3$

$y = \sqrt{4(-3)^2 - 3(-3) + 4} = \sqrt{49} = 7$

4. $y = \sqrt[3]{34 - 4x + 2x^2},\ x = -3$

$y = \sqrt[3]{34 - 4(-3) + 2(-3)^2} = \sqrt[3]{64} = 4$

5. $f(x) = 5x^2 + 3x - 2$

Domain: $(-\infty, \infty)$

6. $g(x) = \dfrac{1}{2x} - \dfrac{2}{x + 3}$

Domain: $(-\infty, -3) \cup (-3, 0) \cup (0, \infty)$

7. $h(y) = \sqrt{y - 5}$

Domain: $[5, \infty)$

8. $f(y) = \sqrt{y^2 - 5}$

Domain: $\left(-\infty, -\sqrt{5}\right] \cup \left[\sqrt{5}, \infty\right)$

9. $(476)^{0.65} \approx 55.0104$

10. $(251)^{0.35} \approx 6.9165$

1. $f(x, y) = \dfrac{x}{y}$

(a) $f(3, 2) = \dfrac{3}{2}$ (b) $f(-1, 4) = -\dfrac{1}{4}$

(c) $f(30, 5) = \dfrac{30}{5} = 6$ (d) $f(5, y) = \dfrac{5}{y}$

(e) $f(x, 2) = \dfrac{x}{2}$ (f) $f(5, t) = \dfrac{5}{t}$

3. $f(x, y) = xe^y$

(a) $f(5, 0) = 5e^0 = 5$ (b) $f(3, 2) = 3e^2$

(c) $f(2, -1) = 2e^{-1} = \dfrac{2}{e}$ (d) $f(5, y) = 5e^y$

(e) $f(x, 2) = xe^2$ (f) $f(t, t) = te^t$

5. $h(x, y, z) = \dfrac{xy}{z}$

(a) $h(2, 3, 9) = \dfrac{(2)(3)}{9} = \dfrac{2}{3}$

(b) $h(1, 0, 1) = \dfrac{(1)(0)}{1} = 0$

7. $V(r, h) = \pi r^2 h$

(a) $V(3, 10) = \pi(3)^2(10) = 90\pi$

(b) $V(5, 2) = \pi(5)^2(2) = 50\pi$

9. $A(P, r, t) = P\left[\left(1 + \dfrac{r}{12}\right)^{12t} - 1\right]\left(1 + \dfrac{12}{r}\right)$

(a) $A(100, 0.10, 10) = 100\left[\left(1 + \dfrac{0.10}{12}\right)^{120} - 1\right]\left(1 + \dfrac{12}{0.10}\right) = \$20,655.20$

(b) $A(275, 0.0925, 40) = 275\left[\left(1 + \dfrac{0.0925}{12}\right)^{480} - 1\right]\left(1 + \dfrac{12}{0.0925}\right) = \$1,397,672.67$

11. $f(x, y) = \int_x^y (2t - 3)\, dt$

(a) $f(1, 2) = \int_1^2 (2t - 3)\, dt = \left[(t^2 - 3t)\right]_1^2 = (-2) - (-2) = 0$

(b) $f(1, 4) = \int_1^4 (2t - 3)\, dt = \left[(t^2 - 3t)\right]_1^4 = 4 - (-2) = 6$

13. $f(x, y) = x^2 - 2y$

(a) $f(x + \Delta x, y) = (x + \Delta x)^2 - 2y = x^2 + 2x\,\Delta x + (\Delta x)^2 - 2y$

(b) $\dfrac{f(x, y + \Delta y) - f(x, y)}{\Delta y} = \dfrac{\left[x^2 - 2(y + \Delta y)\right] - (x^2 - 2y)}{\Delta y} = \dfrac{x^2 - 2y - 2\,\Delta y - x^2 + 2y}{\Delta y} = -\dfrac{2\,\Delta y}{\Delta y} = -2, \Delta y \neq 0$

15. $f(x, y) = \sqrt{16 - x^2 - y^2}$

The domain is the set of all points inside and on the circle $x^2 + y^2 = 16$ because $16 - x^2 - y^2 \geq 0$. The range is $[0, 4]$.

17. $f(x, y) = x^2 + y^2$

The domain is the set of all points in the xy-plane. The range is $[0, \infty)$.

19. $f(x, y) = e^{x/y}$

The domain is the set of all points above or below the x-axis because $y \neq 0$. The range is $(0, \infty)$.

21. $g(x, y) = \ln(4 - x - y)$

The domain is the half-plane below the line $y = -x + 4$ because $4 - x - y > 0$. The range is $(-\infty, \infty)$.

23. $z = \sqrt{9 - 3x^2 - y^2}$

The domain is the set of all points inside and on the ellipse $3x^2 + y^2 = 9$ because $9 - 3x^2 - y^2 \geq 0$. The range is $[0, 3]$.

25. $z = \dfrac{y}{x}$

The domain is the set of all points in the xy-plane above or below the y-axis because $x \neq 0$. The range is $(-\infty, \infty)$.

27. $f(x, y) = \dfrac{1}{xy}$

The domain is the set of all points in the xy-plane except those on the x-axis and y-axis because $x \neq y \neq 0$. The range is all $z \neq 0$, or $(-\infty, 0) \cup (0, \infty)$.

29. $h(x, y) = x\sqrt{y}$

The domain is the set of all points in the xy-plane such that $y \geq 0$. The range is $(-\infty, \infty)$.

31. $f(x, y) = x^2 + \dfrac{y^2}{4}$

The contour map consists of ellipses

$$x^2 + \frac{y^2}{4} = C.$$

Matches (b).

33. $f(x, y) = e^{1-x^2-y^2}$

The contour map consists of curves $e^{1-x^2-y^2} = C$, or

$1 - x^2 - y^2 = \ln C \Rightarrow x^2 + y^2 = 1 - \ln C$, circles.

Matches (a).

35. $c = -1, \quad -1 = x + y, \quad y = -x - 1$

$c = 0, \quad 0 = x + y, \quad y = -x$

$c = 2, \quad 2 = x + y, \quad y = -x + 2$

$c = 4, \quad 4 = x + y, \quad y = -x + 4$

The level curves are parallel lines.

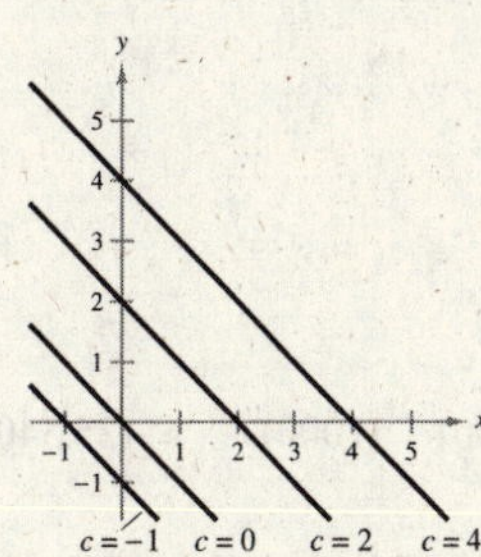

37. $c = 0, \quad 0 = \sqrt{25 - x^2 - y^2}, \quad x^2 + y^2 = 25$

$c = 1, \quad 1 = \sqrt{25 - x^2 - y^2}, \quad x^2 + y^2 = 24$

$c = 2, \quad 2 = \sqrt{25 - x^2 - y^2}, \quad x^2 + y^2 = 21$

$c = 3, \quad 3 = \sqrt{25 - x^2 - y^2}, \quad x^2 + y^2 = 16$

$c = 4, \quad 4 = \sqrt{25 - x^2 - y^2}, \quad x^2 + y^2 = 9$

$c = 5, \quad 5 = \sqrt{25 - x^2 - y^2}, \quad x^2 + y^2 = 0$

The level curves are circles.

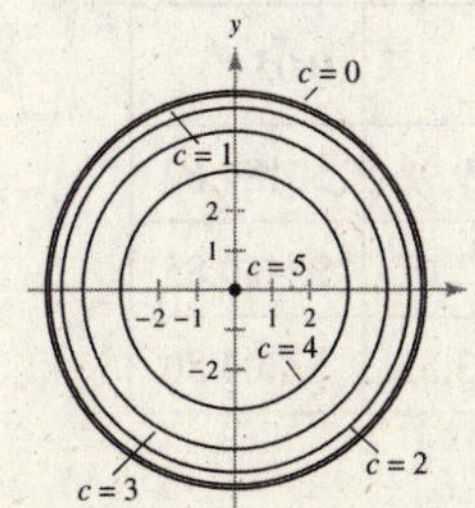

39. $c = \pm 1, \quad xy = \pm 1$

$c = \pm 2, \quad xy = \pm 2$

$c = \pm 3, \quad xy = \pm 3$

$c = \pm 4, \quad xy = \pm 4$

$c = \pm 5, \quad xy = \pm 5$

$c = \pm 6, \quad xy = \pm 6$

The level curves are hyperbolas.

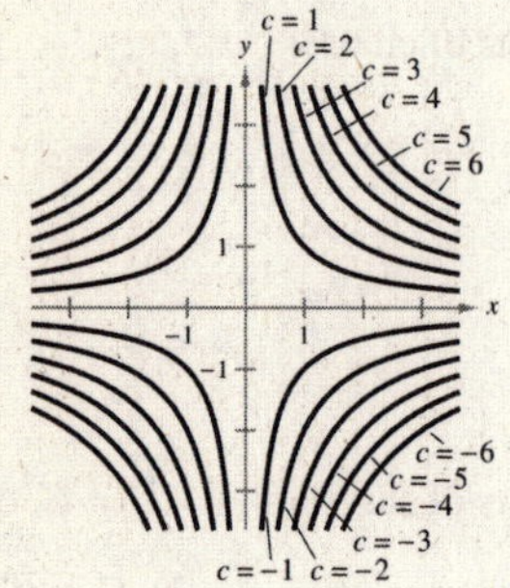

41. $c = \frac{1}{2}, \quad \frac{1}{2} = \frac{x}{x^2 + y^2}, \quad (x - 1)^2 + y^2 = 1$

$c = -\frac{1}{2}, \quad -\frac{1}{2} = \frac{x}{x^2 + y^2}, \quad (x + 1)^2 + y^2 = 1$

$c = 1, \quad 1 = \frac{x}{x^2 + y^2}, \quad \left(x - \frac{1}{2}\right)^2 + y^2 = \frac{1}{4}$

$c = -1, \quad -1 = \frac{x}{x^2 + y^2}, \quad \left(x + \frac{1}{2}\right)^2 + y^2 = \frac{1}{4}$

$c = \frac{3}{2}, \quad \frac{3}{2} = \frac{x}{x^2 + y^2}, \quad \left(x - \frac{1}{3}\right)^2 + y^2 = \frac{1}{9}$

$c = -\frac{3}{2}, \quad -\frac{3}{2} = \frac{x}{x^2 + y^2}, \quad \left(x + \frac{1}{3}\right)^2 + y^2 = \frac{1}{9}$

$c = 2, \quad 2 = \frac{x}{x^2 + y^2}, \quad \left(x - \frac{1}{4}\right)^2 + y^2 = \frac{1}{16}$

$c = -2, \quad -2 = \frac{x}{x^2 + y^2}, \quad \left(x + \frac{1}{4}\right)^2 + y^2 = \frac{1}{16}$

The level curves are circles.

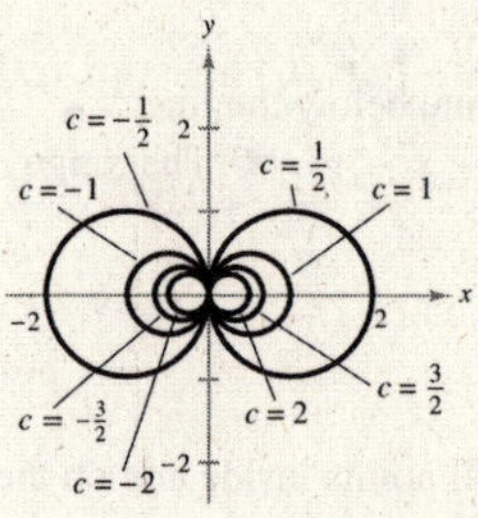

43. $f(x, y) = 100x^{0.75}y^{0.25}$

$f(1500, 1000) = 100(1500)^{0.75}(1000)^{0.25} \approx 135{,}540$ units

45. $P(x_1, x_2) = 50(x_1 + x_2) - C_1(x_1) - C_2(x_2) = 50(x_1 + x_2) - \left(0.02x_1^2 + 4x_1 + 500\right) - \left(0.05x_2^2 + 4x_2 + 275\right)$

(a) $P(250, 150) = 50(250 + 150) - \left[0.02(250)^2 + 4(250) + 500\right] - \left[0.05(150)^2 + 4(150) + 275\right] = \$15{,}250$

(b) $P(300, 200) = 50(300 + 200) - \left[0.02(300)^2 + 4(300) + 500\right] - \left[0.05(200)^2 + 4(200) + 275\right] = \$18{,}425$

(c) $P(600, 400) = 50(600 + 400) - \left[0.02(600)^2 + 4(600) + 500\right] - \left[0.05(400)^2 + 4(400) + 275\right] = \$30{,}025$

47. $V(I, R) = 2000\left[\frac{1 + 0.10(1 - R)}{1 + I}\right]^{10}$

R \ I	0	0.03	0.05
0	\$5187.48	\$3859.98	\$3184.67
0.28	\$4008.46	\$2982.67	\$2460.85
0.35	\$3754.27	\$2793.53	\$2304.80

49. (a) C, highest pressure

(b) A, lowest pressure

(c) B, highest wind velocity

51. $z = 0.379x - 0.135y - 3.45$

(a) $z = 0.379(20) - 0.135(10) - 3.45 = \$2.78$ earnings per share

(b) The x-variable

Explanations will vary. Sample answer: The x-variable has a greater influence on the earnings per share because the absolute value of its coefficient is larger than the absolute value of the coefficient of the y-term.

53. $$M = \frac{\left(\dfrac{P \cdot r}{12}\right)}{1 - \left[\dfrac{1}{1 + (r/12)}\right]^{12t}}$$

(a) $$M = \frac{\left[\dfrac{120{,}000(0.08)}{12}\right]}{1 - \left[\dfrac{1}{1 + (0.08/12)}\right]^{12(20)}} = \$1003.73$$

$\$1003.73 \times 240$ payments $= \$240{,}895.20$

(b) $$M = \frac{\left[\dfrac{120{,}000(0.07)}{12}\right]}{1 - \left[\dfrac{1}{1 + (0.07/12)}\right]^{12(30)}} = \$798.36$$

$\$798.36 \times 360$ payments $= \$287{,}409.60$

(c) $$M = \frac{\left[\dfrac{120{,}000(0.07)}{12}\right]}{1 - \left[\dfrac{1}{1 + (0.07/12)}\right]^{12(15)}} = \$1078.59$$

$\$1078.59 \times 180$ payments $= \$194{,}146.20$

Choices will vary, as well as explanations.

Section 7.4 Partial Derivatives

Skills Warm Up

1. $f(x) = \sqrt{x^2 + 3}$

$$f'(x) = \frac{1}{2}(x^2 + 3)^{-1/2}(2x) = \frac{x}{\sqrt{x^2 + 3}}$$

2. $g(x) = (3 - x^2)^3$

$$g'(x) = 3(3 - x^2)^2(-2x) = -6x(3 - x^2)^2$$

3. $g(t) = te^{2t+1}$

$$g'(t) = te^{2t+1}(2) + e^{2t+1}(1) = e^{2t+1}(2t + 1)$$

4. $f(x) = e^{2x}\sqrt{1 - e^{2x}}$

$$f'(x) = e^{2x}\left(\frac{1}{2}\right)(1 - e^{2x})^{-1/2}(-e^{2x})(2) + \sqrt{1 - e^{2x}}e^{2x}(2)$$
$$= \frac{e^{2x}(-e^{2x})}{\sqrt{1 - e^{2x}}} + 2e^{2x}\sqrt{1 - e^{2x}}$$
$$= \frac{e^{2x}}{\sqrt{1 - e^{2x}}}\left(-e^{2x} + 2(1 - e^{2x})\right)$$
$$= \frac{e^{2x}}{\sqrt{1 - e^{2x}}}(2 - 3e^{2x})$$

5. $f(x) = \ln(3 - 2x)$

$$f'(x) = \frac{-2}{3 - 2x}$$

6. $u(t) = \ln\sqrt{t^3 - 6t}$

$$u'(t) = \frac{1}{\sqrt{t^3 - 6t}}\left(\frac{1}{2}\right)(t^3 - 6t)^{-1/2}(3t^2 - 6)$$
$$= \frac{3(t^2 - 2)}{2t(t^2 - 6)}$$

7. $g(x) = \dfrac{5x^2}{(4x - 1)^2}$

$$g'(x) = \frac{(4x - 1)^2(10x) - 5x^2(2)(4x - 1)(4)}{(4x - 1)^4}$$
$$= \frac{(4x - 1)10x - 40x^2}{(4x - 1)^3}$$
$$= -\frac{10x}{(4x - 1)^3}$$

Skills Warm Up —continued—

8. $f(x) = \dfrac{(x+2)^3}{(x^2-9)^2}$

$$f'(x) = \frac{(x^2-9)^2(3)(x+2)^2 - (x+2)^3(2)(x^2-9)(2x)}{(x^2-9)^4}$$

$$= \frac{3(x+2)^2(x^2-9) - 4x(x+2)^3}{(x^2-9)^3} = \frac{(x+2)^2\left[3(x^2-9) - 4x(x+2)\right]}{(x^2-9)^3} = -\frac{(x+2)^2(x^2+8x+27)}{(x^2-9)^3}$$

9. $f(x) = x^2e^{x-2}$

$f'(x) = x^2e^{x-2} + e^{x-2}(2x)$

$f'(2) = (2)^2e^{2-2} + e^{2-2}(2(2)) = 4 + 4 = 8$

10. $g(x) = x\sqrt{x^2 - x + 2}$

$$g'(x) = x\left(\frac{1}{2}\right)(x^2 - x + 2)^{-1/2}(2x-1) + \sqrt{x^2 - x + 2} = \frac{x^2 - x/2}{\sqrt{x^2 - x + 2}} + \sqrt{x^2 - x + 2}$$

$$g'(2) = \frac{2^2 - 2/2}{\sqrt{2^2 - 2 + 2}} + \sqrt{2^2 - 2 + 2} = \frac{3}{2} + 2 = \frac{7}{2}$$

1. $\dfrac{\partial z}{\partial x} = 3$

$\dfrac{\partial z}{\partial y} = 5$

3. $f_x(x, y) = 3$

$f_y(x, y) = -12y$

5. $f_x(x, y) = \dfrac{1}{y}$

$f_y(x, y) = -xy^{-2} = -\dfrac{x}{y^2}$

7. $f_x(x, y) = \dfrac{1}{2}(x^2 + y^2)^{-1/2}(2x) = \dfrac{x}{\sqrt{x^2 + y^2}}$

$f_y(x, y) = \dfrac{1}{2}(x^2 + y^2)^{-1/2}(2y) = \dfrac{y}{\sqrt{x^2 + y^2}}$

9. $\dfrac{\partial z}{\partial x} = 2xe^{2y}$

$\dfrac{\partial z}{\partial y} = 2x^2e^{2y}$

11. $h_x(x, y) = -2xe^{-(x^2+y^2)}$

$h_y(x, y) = -2ye^{-(x^2+y^2)}$

13. $z = \ln\dfrac{x+y}{x-y} = \ln(x+y) - \ln(x-y)$

$\dfrac{\partial z}{\partial x} = \dfrac{1}{x+y} - \dfrac{1}{x-y} = -\dfrac{2y}{x^2 - y^2}$

$\dfrac{\partial z}{\partial y} = \dfrac{1}{x+y} + \dfrac{1}{x-y} = \dfrac{2x}{x^2 - y^2}$

15. $f_x(x, y) = 6x + y,\ f_x(2, 1) = 13$

$f_y(x, y) = x - 2y,\ f_y(2, 1) = 0$

17. $f_x(x, y) = 3ye^{3xy},\ f_x(0, 4) = 12$

$f_y(x, y) = 3xe^{3xy},\ f_y(0, 4) = 0$

19. $f_x(x, y) = \dfrac{(x-y)y - xy(1)}{(x-y)^2} = -\dfrac{y^2}{(x-y)^2},\ f_x(2, -2) = -\dfrac{1}{4}$

$f_y(x, y) = \dfrac{(x-y)x - xy(-1)}{(x-y)^2} = \dfrac{x^2}{(x-y)^2},\ f_y(2, -2) = \dfrac{1}{4}$

21. $f_x(x, y) = \dfrac{3}{3x + 5y}$, $f_x(1, 0) = 1$

$f_y(x, y) = \dfrac{5}{3x + 5y}$, $f_y(1, 0) = \dfrac{5}{3}$

23. (a) $\dfrac{\partial z}{\partial x} = y \qquad \dfrac{\partial z}{\partial x}(1, 2, 2) = 2$

(b) $\dfrac{\partial z}{\partial y} = x \qquad \dfrac{\partial z}{\partial y}(1, 2, 2) = 1$

25. (a) $\dfrac{\partial z}{\partial x} = -2x \qquad \dfrac{\partial z}{\partial x}(1, 1, 2) = -2$

(b) $\dfrac{\partial z}{\partial y} = -2y \qquad \dfrac{\partial z}{\partial y}(1, 1, 2) = -2$

27. $w_x = y^2z^4$, $w_y = 2xyz^4$, $w_z = 4xy^2z^3$

29. $w_x = \dfrac{(x + y)(0) - 2z(1)}{(x + y)^2} = -\dfrac{2z}{(x + y)^2}$

$w_y = \dfrac{(x + y)(0) - 2z(1)}{(x + y)^2} = -\dfrac{2z}{(x + y)^2}$

$w_z = \dfrac{2}{x + y}$

31. $w_x = 2z^2 + 3yz$, $w_x(1, -1, 2) = 2$

$w_y = 3xz - 12yz$, $w_y(1, -1, 2) = 30$

$w_z = 4xz + 3xy - 6y^2$, $w_z(1, -1, 2) = -1$

33. $w_x = \dfrac{x}{\sqrt{x^2 + y^2 + z^2}}$, $w_x(2, -1, 2) = \dfrac{2}{3}$

$w_y = \dfrac{y}{\sqrt{x^2 + y^2 + z^2}}$, $w_y(2, -1, 2) = -\dfrac{1}{3}$

$w_z = \dfrac{z}{\sqrt{x^2 + y^2 + z^2}}$, $w_z(2, -1, 2) = \dfrac{2}{3}$

35. $w_x = 4xy^3z^2e^{2x^2}$, $w_x\left(\frac{1}{2}, -1, 2\right) = -8\sqrt{e}$

$w_y = 3y^2z^2e^{2x^2}$, $w_y\left(\frac{1}{2}, -1, 2\right) = 12\sqrt{e}$

$w_z = 2y^3ze^{2x^2}$, $w_z\left(\frac{1}{2}, -1, 2\right) = -4\sqrt{e}$

37. $w_x = \dfrac{5}{5x + 2y^3 - 3z}$, $w_x(4, 1, -1) = \dfrac{1}{5}$

$w_y = \dfrac{6y^2}{5x + 2y^3 - 3z}$, $w_y(4, 1, -1) = \dfrac{6}{25}$

$w_z = -\dfrac{3}{5x + 2y^3 - 3z}$, $w_z(4, 1, -1) = -\dfrac{3}{25}$

39.
$$\begin{aligned} f_x(x, y) &= 2x + 4y - 4 = 0 \Rightarrow & -4x - 8y &= -8 \\ f_y(x, y) &= 4x + 2y + 16 = 0 \Rightarrow & 4x + 2y &= -16 \\ & & -6y &= -24 \\ & & y &= 4 \\ & & x &= -6 \end{aligned}$$

Solution: $(-6, 4)$

41.
$$\left.\begin{aligned} f_x(x, y) &= -\frac{1}{x^2} + y = 0 \Rightarrow x^2y = 1 \\ f_y(x, y) &= -\frac{1}{y^2} + x = 0 \Rightarrow y^2x = 1 \end{aligned}\right\} x = y = 1$$

Solution: $(1, 1)$

43. $\dfrac{\partial z}{\partial x} = 3x^2$

$\dfrac{\partial z}{\partial y} = -8y$

$\dfrac{\partial^2 z}{\partial x^2} = 6x$

$\dfrac{\partial^2 z}{\partial x \partial y} = 0$

$\dfrac{\partial^2 z}{\partial y \partial x} = 0$

$\dfrac{\partial^2 z}{\partial y^2} = -8$

45. $\dfrac{\partial z}{\partial x} = 2x - 2y$

$\dfrac{\partial z}{\partial y} = -2x + 6y$

$\dfrac{\partial^2 z}{\partial x^2} = 2$

$\dfrac{\partial^2 z}{\partial x \partial y} = -2$

$\dfrac{\partial^2 z}{\partial y \partial x} = -2$

$\dfrac{\partial^2 z}{\partial y^2} = 6$

47. $\dfrac{\partial z}{\partial x} = 3(3x^4 - 2y^3)^2(12x^3) = 36x^3(3x^4 - 2y^3)^2$

$\dfrac{\partial z}{\partial y} = 3(3x^4 - 2y^3)^2(-6y^2) = -18y^2(3x^4 - 2y^3)^2$

$\dfrac{\partial^2 z}{\partial x^2} = 36x^3\left[2(3x^4 - 2y^3)(12x^3)\right] + (3x^4 - 2y^3)^2(108x^2) = 108x^2(3x^4 - 2y^3)(11x^4 - 2y^3)$

$\dfrac{\partial^2 z}{\partial y \partial y} = 36x^3\left[2(3x^4 - 2y^3)(-6y^2)\right] = -432x^3y^2(3x^4 - 2y^3)$

$\dfrac{\partial^2 z}{\partial y^2} = -18y^2\left[2(3x^4 - 2y^3)(-6y^2)\right] + (3x^4 - 2y^3)^2(-36y) = -36y(3x^4 - 2y^3)(3x^4 - 8y^3)$

$\dfrac{\partial^2 z}{\partial y \partial x} = -18y^2\left[2(3x^4 - 2y^3)(12x^3)\right] = -432x^3y^2(3x^4 - 2y^3)$

49. $\dfrac{\partial z}{\partial x} = \dfrac{2xy(2x) - (x^2 - y^2)(2y)}{4x^2y^2} = \dfrac{x^2 + y^2}{2x^2y}$

$\dfrac{\partial z}{\partial y} = \dfrac{2xy(-2y) - (x^2 - y^2)(2x)}{4x^2y^2} = -\dfrac{x^2 + y^2}{2xy^2}$

$\dfrac{\partial^2 z}{\partial x^2} = \dfrac{2x^2y(2x) - (x^2 + y^2)(4xy)}{4x^4y^2} = -\dfrac{y}{x^3}$

$\dfrac{\partial^2 z}{\partial x \partial y} = \dfrac{2x^2y(2y) - (x^2 + y^2)(2x^2)}{4x^4y^2} = -\dfrac{x^2 - y^2}{2x^2y^2}$

$\dfrac{\partial^2 z}{\partial y \partial x} = -\dfrac{2xy^2(2x) - (x^2 + y^2)(2y^2)}{4x^2y^2} = -\dfrac{x^2 - y^2}{2x^2y^2}$

$\dfrac{\partial^2 z}{\partial y^2} = -\dfrac{2xy^2(2y) - (x^2 + y^2)(4xy)}{4x^2y^4} = \dfrac{x}{y^3}$

51. $f_x(x, y) = 4x^3 - 6xy^2, \quad f_y(x, y) = -6x^2y + 2y$

$f_{xx}(x, y) = 12x^2 - 6y^2, \quad f_{xx}(1, 0) = 12$

$f_{xy}(x, y) = -12xy, \quad f_{xy}(1, 0) = 0$

$f_{yx}(x, y) = -12xy, \quad f_{yx}(1, 0) = 0$

$f_{yy}(x, y) = -6x^2 + 2, \quad f_{yy}(1, 0) = -4$

53. $f_x(x, y) = 2xy^3e^{x^2}, \quad f_y(x, y) = 3y^2e^{x^2}$

$f_{xx}(x, y) = e^{x^2}(4x^2y^3 + 2y^3), \quad f_{xx}(1, -1) = -6e$

$f_{xy}(x, y) = 6xy^2e^{x^2}, \quad f_{xy}(1, -1) = 6e$

$f_{yx}(x, y) = 6xy^2e^{x^2}, \quad f_{yx}(1, -1) = 6e$

$f_{yy}(x, y) = 6ye^{x^2}, \quad f_{yy}(1, -1) = -6e$

55. $w_x = 2x - 3y, \quad w_y = -3x + 4z, \quad w_z = 4y + 3z^2$

$w_{xx} = 2, \quad w_{yx} = -3, \quad w_{zx} = 0$

$w_{xy} = -3, \quad w_{yy} = 0, \quad w_{zy} = 4$

$w_{xz} = 0, \quad w_{yz} = 4, \quad w_{zz} = 6z$

57. $w_x = \dfrac{(x + y)(4z) - 4xz(1)}{(x + y)^2} = \dfrac{4yz}{(x + y)^2}$

$w_y = \dfrac{(x + y)(0) - 4xz(1)}{(x + y)^2} = \dfrac{-4xz}{(x + y)^2}$

$w_z = \dfrac{4x}{x + y}$

$w_{xx} = 4yz\left[-2(x + y)^{-3}(1)\right] = -\dfrac{8yz}{(x + y)^3}$

$w_{xy} = 4yz\left[-2(x + y)^{-3}(1)\right] + (x + y)^{-2}(4z)$

$= \dfrac{4z(x - y)}{(x + y)^3}$

$w_{xz} = \dfrac{4y}{(x + y)^2}$

$w_{yx} = -4xz\left[-2(x + y)^{-3}(1)\right] + (x + y)^{-2}(-4z)$

$= \dfrac{4z(x - y)}{(x + y)^3}$

$w_{yy} = -4xz\left[-2(x + y)^{-3}(1)\right] = \dfrac{8xz}{(x + y)^3}$

$w_{yz} = -\dfrac{4x}{(x + y)^2}$

$w_{zx} = \dfrac{(x + y)(4) - 4x(1)}{(x + y)^2} = \dfrac{4y}{(x + y)^2}$

$w_{zy} = 4\left[(-1)(x + y)^{-2}(1)\right] = -\dfrac{4x}{(x + y)^2}$

$w_{zz} = 0$

59. (a) $\frac{\partial C}{\partial x} = \frac{5y}{\sqrt{xy}} + 149, \quad \frac{\partial C}{\partial x}(120, 160) \approx 154.77$

$\frac{\partial C}{\partial y} = \frac{5x}{\sqrt{xy}} + 189, \quad \frac{\partial C}{\partial y}(120, 160) \approx 193.33$

(b) Increasing the production of racing bikes increases the cost at a higher rate than increasing the production of mountain bikes. This is determined by comparing the marginal cost for mountain bikes $\frac{\partial C}{\partial x} = \$154.77$ to that for racing bikes $\frac{\partial C}{\partial y} = \$193.33$ at the production level $(120, 160)$.

61. (a) $\frac{\partial f}{\partial x} = 140x^{-0.3}y^{0.3} = 140\left(\frac{y}{x}\right)^{0.3}$

When $x = 1000$ and $y = 500$,

$\frac{\partial f}{\partial x} = 140\left(\frac{500}{1000}\right)^{0.3} = 140\left(\frac{1}{2}\right)^{0.3} \approx 113.72.$

(b) $\frac{\partial f}{\partial y} = 60x^{0.7}y^{-0.7} = 60\left(\frac{x}{y}\right)^{0.7}$

When $x = 1000$ and $y = 500$,

$\frac{\partial f}{\partial y} = 60\left(\frac{1000}{500}\right)^{0.7} = 60(2)^{0.7} \approx 97.47.$

63. Complementary because $\frac{\partial x_1}{\partial p_2} = -\frac{5}{2} < 0$ and $\frac{\partial x_2}{\partial p_1} = -\frac{3}{2} < 0.$

65. $z = 0.62x - 0.41y + 0.38$

(a) $\frac{\partial z}{\partial x} = 0.62; \frac{\partial z}{\partial y} = -0.41$

(b) For every increase of \$1 billion in expenditures on amusement parks and campgrounds, the expenditures for spectator sports will increase by \$0.62 billion. For every increase of \$1 billion in expenditures on live entertainment (excluding sports), the expenditures for spectator sports will decrease by \$0.41 billion.

67. $IQ_M = \frac{100}{C}$

$IQ_C = \frac{-100M}{C^2}$

$IQ_M(12, 10) = \frac{100}{10} = 10$

$IQ_C(12, 10) = \frac{-100(12)}{10^2} = -12$

For a child who has a current mental age of 12 years and a chronological age of 10 years, the IQ is increasing at a rate of 10 IQ points for every increase of 1 year in the child's mental age. For a child who has a current mental age of 12 years and a chronological age of 10 years, the IQ is decreasing at a rate of 12 IQ points for every increase of 1 year in the child's chronological age.

69. $V(I, R) = 1000\left[\frac{1 + 0.10(1 - R)}{1 + I}\right]^{10}$

$V_I(I, R) = 10{,}000\left[\frac{1 + 0.10(1 - R)}{1 + I}\right]^9\left[-\frac{1 + 0.10(1 - R)}{(1 + I)^2}\right] = -10{,}000\frac{[1 + 0.10(1 - R)]^{10}}{(1 + I)^{11}}$

$V_I(0.03, 0.28) \approx -14{,}478.99$

$V_R(I, R) = 10{,}000\left[\frac{1 + 0.10(1 - R)}{1 + I}\right]^9\left[-\frac{0.10}{1 + I}\right] = -1000\frac{[1 + 0.10(1 - R)]^9}{(1 + I)^{10}}$

$V_R(0.03, 0.28) \approx -1391.17$

The rate of inflation has the greater negative influence on the growth of the investment because $|-14{,}478.99| > |-1391.17|$.

71. (a) $U_x = -10x + y$

(b) $U_y = x - 6y$

(c) When $x = 2$ and $y = 3$, $U_x = -17$ and $U_y = -16$.

The person should consume one more unit of good y, since the rate of decrease of satisfaction is less for y.

The slope of U in the x-direction is 0 when $y = 10x$ and negative when $y < 10x$.

The slope of U in the y-direction is 0 when $x = 6y$ and negative when $x < 6y$.

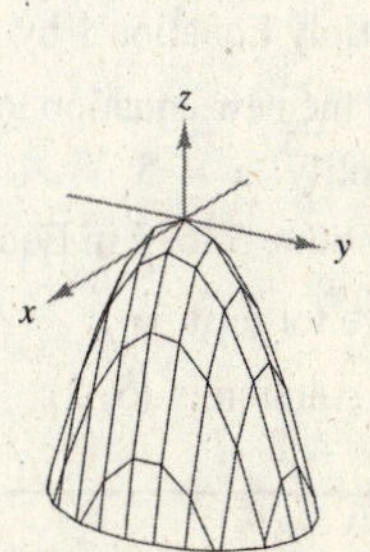

Section 7.5 Extrema of Functions of Two Variables

Skills Warm Up

1. $\begin{cases} 5x = 15 \\ 3x - 2y = 5 \end{cases}$

$5x = 15$

$x = 3$

Substitute in the other equation.

$3(3) - 2y = 5$

$-2y = -4$

$y = 2$

The solution is $(3, 2)$.

2. $\begin{cases} \frac{1}{2}y = 3 \\ -x + 5y = 19 \end{cases}$

$\frac{1}{2}y = 3$

$y = 6$

Substitute in the other equation.

$-x + 5(6) = 19$

$-x = -11$

$x = 11$

The solution is $(11, 6)$.

3. $\begin{cases} x + y = 5 \\ x - y = -3 \end{cases}$

Adding the two equations gives $2x = 2$, so $x = 1$.

Substitute.

$1 + y = 5$

$y = 4$

The solution is $(1, 4)$.

4. $\begin{cases} x + y = 8 \\ 2x - y = 4 \end{cases}$

Adding the two equations gives $3x = 12$, so $x = 4$.

Substitute.

$4 + y = 8$

$y = 4$

The solution is $(4, 4)$.

5. $\begin{cases} 2x - y = 8 & \text{Equation 1} \\ 3x - 4y = 7 & \text{Equation 2} \end{cases}$

Multiply Equation 1 by -4: $-8x + 4y = -32$

Add the new equation to Equation 2: $-5x = -25$

Simplify: $x = 5$

Substitute 5 for x in Equation 1: $2(5) - y = 8$

Solve for y: $y = 2$

The solution is $(5, 2)$.

6. $\begin{cases} 2x - 4y = 14 & \text{Equation 1} \\ 3x + y = 7 & \text{Equation 2} \end{cases}$

Multiply equation 2 by 4: $12x + 4y = 28$

Add new equation to Equation 1: $14x = 42$

Simplify: $x = 3$

Substitute 3 for x in Equation 2: $3(3) + y = 7$

Simplify: $y = -2$

The solution is $(3, -2)$.

7. $\begin{cases} x^2 + x = 0 & \text{Equation 1} \\ 2yx + y = 0 & \text{Equation 2} \end{cases}$

Factor Equation 1: $x(x + 1) = 0$

Solve equation 1 for x: $x = -1$ or $x = 0$

Substitute -1 for x in Equation 2: $2y(-1) + y = 0$

Solve for y: $y = 0$

Substitute 0 for x in Equation 2: $2y(0) + y = 0$

Solve for y: $y = 0$

The solutions are $(-1, 0)$ and $(0, 0)$.

8. $\begin{cases} 3y^2 + 6y = 0 & \text{Equation 1} \\ xy + x + 2 = 0 & \text{Equation 2} \end{cases}$

Factor Equation 1: $y(3y + 6) = 0$

Solve for y: $y = 0$

$3y + 6 = 0$

$y = -2$

Substitute 0 for y in Equation 2: $x(0) + x + 2 = 0$

Solve for x: $x = -2$

Substitute -2 for y in Equation 2: $x(-2) + x + 2 = 0$

Solve for x: $x = 2$

The solutions are $(-2, 0)$ and $(2, -2)$.

9. $z = 4x^3 - 3y^2$

$\frac{\partial z}{\partial x} = 12x^2$, $\quad \frac{\partial z}{\partial y} = -6y$

$\frac{\partial^2 z}{\partial x^2} = 24x$, $\quad \frac{\partial^2 z}{\partial y^2} = -6$

$\frac{\partial^2 z}{\partial x \partial y} = 0$, $\quad \frac{\partial^2 z}{\partial y \partial x} = 0$

Skills Warm Up —*continued*—

10. $z = 2x^5 - y^3$

$$\frac{\partial z}{\partial x} = 10x^4, \qquad \frac{\partial z}{\partial y} = -3y^2$$

$$\frac{\partial^2 z}{\partial x^2} = 40x^3, \qquad \frac{\partial^2 z}{\partial y^2} = -6y$$

$$\frac{\partial^2 z}{\partial x \partial y} = 0, \qquad \frac{\partial^2 z}{\partial y \partial x} = 0$$

11. $z = x^4 - \sqrt{xy} + 2y$

$$\frac{\partial z}{\partial x} = 4x^3 - \frac{y}{2\sqrt{xy}} = 4x^3 - \frac{\sqrt{xy}}{2x}$$

$$\frac{\partial z}{\partial x} = -\frac{x}{2\sqrt{xy}} + 2 = -\frac{\sqrt{xy}}{2y} + 2$$

$$\frac{\partial^2 z}{\partial x^2} = 12x^2 - \frac{y}{4x\sqrt{xy}} + \frac{2\sqrt{xy}}{4x^2} = 12x^2 + \frac{\sqrt{xy}}{4x^2}$$

$$\frac{\partial^2 z}{\partial y^2} = -\frac{x}{4y\sqrt{xy}} + \frac{xy}{2y^2} = \frac{\sqrt{xy}}{4y^2}$$

$$\frac{\partial^2 z}{\partial x \partial y} = -\frac{x}{4x\sqrt{xy}} = -\frac{xy}{4xy}$$

$$\frac{\partial^2 z}{\partial y \partial y} = -\frac{x}{4x\sqrt{xy}} = -\frac{xy}{4xy}$$

12. $z = 2x^2 - 3xy + y^2$

$$\frac{\partial z}{\partial x} = 4x - 3y, \qquad \frac{\partial z}{\partial y} = -3x + 2y$$

$$\frac{\partial^2 z}{\partial x^2} = 4, \qquad \frac{\partial z}{\partial y^2} = 2$$

$$\frac{\partial^2 z}{\partial x \partial y} = -3, \qquad \frac{\partial^2 z}{\partial y \partial x} = -3$$

13. $z = ye^{xy^2}$

$$\frac{\partial z}{\partial x} = y(y^2)e^{xy^2} = y^3e^{xy^2}$$

$$\frac{\partial z}{\partial y} = y(2xy)e^{xy^2} + e^{xy^2}(1) = 2xy^2e^{xy^2} + e^{xy^2}$$

$$\frac{\partial^2 z}{\partial x^2} = y^3(y^2)e^{xy^2} = y^5e^{xy^2}$$

$$\frac{\partial^2 z}{\partial y^2} = 2xy^2(2xy)e^{xy^2} + e^{xy^2}(4xy) + 2xye^{xy^2}$$

$$= 4x^2y^3e^{xy^2} + 6xye^{xy^2}$$

$$\frac{\partial^2 z}{\partial x \partial y} = 2xy^4e^{xy^2} + 3y^2e^{xy^2}$$

$$\frac{\partial^2 z}{\partial y \partial x} = y^3(2xy)e^{xy^2} + e^{xy^2}3y^2 = 2xy^4e^{xy^2} + 3y^2e^{xy^2}$$

14. $z = xe^{xy}$

$$\frac{\partial z}{\partial x} = xye^{xy} + e^{xy} = e^{xy}(xy + 1)$$

$$\frac{\partial z}{\partial y} = x^2e^{xy}$$

$$\frac{\partial^2 z}{\partial x^2} = e^{xy}(y) + (xy + 1)ye^{xy} = ye^{xy}(xy + 2)$$

$$\frac{\partial^2 z}{\partial y^2} = x^3e^{xy}$$

$$\frac{\partial^2 z}{\partial x \partial y} = e^{xy}(x) + (xy + 1)xe^{xy} = xe^{xy}(xy + 2)$$

$$\frac{\partial^2 z}{\partial y \partial x} = x^2ye^{xy} + e^{xy}(2x) = xe^{xy}(xy + 2)$$

1. $f(x, y) = x^2 - y^2 + 4x - 8y - 11$

The first partial derivatives of f, $f_x(x, y) = 2x + 4$ and $f_y(x, y) = -2y - 8$, are zero at the critical point $(-2, -4)$. Because

$f_{xx}(x, y) = 2$, $f_{yy}(x, y) = -2$, and $f_{xy}(x, y) = 0$,

it follows that $f_{xx}(-2, -4) > 0$ and $f_{xx}(-2, -4)f_{yy}(-2, -4) - [f_{xy}(-2, -4)]^2 = -4 < 0$.

So, $(-2, -4, 1)$ is a saddle point. There are no relative extrema.

3. $f(x, y) = \sqrt{x^2 + y^2 + 1}$

The first partial derivatives of f,

$f_x(x, y) = \dfrac{x}{\sqrt{x^2 + y^2 + 1}}$ and

$f_y(x, y) = \dfrac{y}{\sqrt{x^2 + y^2 + 1}}$, are zero at the critical point $(0, 0)$. Because

$f_{xx}(x, y) = \dfrac{y^2 + 1}{(x^2 + y^2 + 1)^{3/2}}$,

$f_{yy}(x, y) = \dfrac{x^2 + 1}{(x^2 + y^2 + 1)^{3/2}}$, and

$f_{xy}(x, y) = \dfrac{-xy}{(x^2 + y^2 + 1)^{3/2}}$,

it follows that $f_{xx}(0, 0) = 1 > 0$ and

$f_{xx}(0, 0)f_{yy}(0, 0) - [f_{xy}(0, 0)]^2 = 1 > 0$.

So, $(0, 0, 1)$ is a relative minimum.

5. $f(x, y) = (x - 1)^2 + (y - 3)^2$

The first partial derivatives of f, $f_x(x, y) = 2(x - 1)$ and $f_y(x, y) = 2(y - 3)$, are zero at the critical point $(1, 3)$. Because

$f_{xx}(x, y) = 2$, $f_{yy}(x, y) = 2$, and $f_{xy}(x, y) = 0$,

it follows that $f_{xx}(1, 3) > 0$ and

$f_{xx}(1, 3)f_{yy}(1, 3) - [f_{xy}(1, 3)]^2 = 4 > 0$.

So, $(1, 3, 0)$ is a relative minimum.

7. $f(x, y) = 2x^2 + 2xy + y^2 + 2x - 3$

The first partial derivatives of f,

$f_x(x, y) = 4x + 2y + 2$ and $f_y(x, y) = 2x + 2y$, are zero at the critical point $(-1, 1)$.

Because $f_{xx}(x, y) = 4$, $f_{yy}(x, y) = 2$, and

$f_{xy}(x, y) = 2$, it follows that $f_{xx}(-1, 1) > 0$ and

$f_{xx}(-1, 1)f_{yy}(-1, 1) - [f_{xy}(-1, 0)]^2 = 4 > 0$.

So, $(-1, 1, -4)$ is a relative minimum.

9. $f(x, y) = -5x^2 + 4xy - y^2 + 16x + 10$

The first partial derivatives of f,

$f_x(x, y) = -10x + 4y + 16$ and $f_y(x, y) = 4x - 2y$, are zero at the critical point $(8, 16)$.

Because $f_{xx}(x, y) = -10$, $f_{yy}(x, y) = -2$, and

$f_{xy}(x, y) = 4$, it follows that $f_{xx}(8, 16) < 0$ and

$f_{xx}(8, 16)f_{yy}(8, 16) - [f_{xy}(8, 16)]^2 = 4 > 0$.

So, (8, 16, 74) is a relative maximum.

11. $f(x, y) = 3x^2 + 2y^2 - 6x - 4y + 16$

The first partial derivatives of f,

$f_x(x, y) = 6x - 6 = 6(x - 1)$ and

$f_y(x, y) = 4y - 4 = 4(y - 1)$,

are zero at the critical point $(1, 1)$.

Because $f_{xx}(x, y) = 6$, $f_{yy}(x, y) = 4$, and

$f_{xy}(x, y) = 0$, it follows that $f_{xx}(1, 1) > 0$ and

$f_{xx}(1, 1)f_{yy}(1, 1) - [f_{xy}(1, 1)]^2 = 24 > 0$.

So, $(1, 1, 11)$ is a relative minimum.

13. $f(x, y) = -x^3 + 4xy - 2y^2 + 1$

The first partial derivatives of f, $f_x(x, y) = -3x^2 + 4y$ and $f_y(x, y) = 4x - 4y$, are zero at the critical points $(0, 0)$ and $\left(\frac{4}{3}, \frac{4}{3}\right)$. Because

$f_{xx}(x, y) = -6x$, $f_{yy}(x, y) = -4$, and

$f_{xy}(x, y) = 4$, it follows that $f_{xx}(0, 0) = 0$

and $f_{xx}(0, 0)f_{yy}(0, 0) - [f_{xy}(0, 0)]^2 = -16 < 0$.

So, $(0, 0, 1)$ is a saddle point.

Because $f_{xx}(x, y) = -6x$, $f_{yy}(x, y) = -4$, and

$f_{xy}(x, y) = 4$, it follows that $f_{xx}\left(\frac{4}{3}, \frac{4}{3}\right) < 0$ and

$f_{xx}\left(\frac{4}{3}, \frac{4}{3}\right)f_{yy}\left(\frac{4}{3}, \frac{4}{3}\right) - \left[f_{xy}\left(\frac{4}{3}, \frac{4}{3}\right)\right]^2 = 16 > 0$.

So, $\left(\frac{4}{3}, \frac{4}{3}, \frac{59}{27}\right)$ is a relative maximum.

15. $f(x, y) = \frac{1}{2}xy$

The first partial derivatives of f, $f_x(x, y) = \frac{1}{2}y$ and $f_y(x, y) = \frac{1}{2}x$, are zero at the critical point $(0, 0)$.

Because $f_{xx}(x, y) = 0$, $f_{yy}(x, y) = 0$, and

$f_{xy}(x, y) = \frac{1}{2}$, it follows that $f_{xx}(0, 0) = 0$ and

$f_{xx}(0, 0)f_{yy}(0, 0) - [f_{xy}(0, 0)]^2 = -\frac{1}{4} < 0$.

So, (0, 0, 0) is a saddle point.

17. $f(x, y) = (x + y)e^{1-x^2-y^2}$

The first partial derivatives of f,

$f_x(x, y) = (-2x^2 - 2xy + 1)e^{1-x^2-y^2}$ and

$f_y(x, y) = (-2y^2 - 2xy + 1)e^{1-x^2-y^2}$, are zero at

the critical points $\left(\frac{1}{2}, \frac{1}{2}\right)$ and $\left(-\frac{1}{2}, -\frac{1}{2}\right)$.

Because

$f_{xx}(x, y) = (4x^3 + 4x^2y - 6x - 2y)e^{1-x^2-y^2}$,

$f_{yy}(x, y) = (4y^3 + 4xy^2 - 6y - 2x)e^{1-x^2-y^2}$, and

$f_{xy}(x, y) = (4x^2y + 4xy^2 - 2y - 2x)e^{1-x^2-y^2}$,

it follows that $f_{xx}\left(\frac{1}{2}, \frac{1}{2}\right) = -3e^{1/2} < 0$,

$f_{xx}\left(\frac{1}{2}, \frac{1}{2}\right)f_{yy}\left(\frac{1}{2}, \frac{1}{2}\right) - \left[f_{xy}\left(\frac{1}{2}, \frac{1}{2}\right)\right]^2 = 0$,

$f_{xx}\left(-\frac{1}{2}, -\frac{1}{2}\right) = 3e^{1/2} > 0$, and

$f_{xx}\left(-\frac{1}{2}, -\frac{1}{2}\right)f_{yy}\left(-\frac{1}{2}, -\frac{1}{2}\right) - \left[f_{xy}\left(-\frac{1}{2}, -\frac{1}{2}\right)\right]^2 = 0$.

So, $\left(\frac{1}{2}, \frac{1}{2}, e^{1/2}\right)$ is a relative maximum and

$\left(-\frac{1}{2}, -\frac{1}{2}, -e^{1/2}\right)$ is a relative minimum.

19. $f_{xx} > 0$ and $f_{xx}f_{yy} - (f_{xy})^2 = (9)(4) - 6^2 = 0$

Insufficient information

21. $f_{xx} < 0$ and $f_{xx}f_{yy} - (f_{xy})^2 = (-9)(6) - 10^2 < 0$

f has a saddle point at (x_0, y_0).

23. $f_{xx} > 0$ and $f_{xx}f_{yy} - (f_{xy})^2 = (5)(5) - 3^2 > 0$

f has a relative minimum at (x_0, y_0).

25. $f(x, y) = (xy)^2$

The first partial derivatives of f, $f_x(x, y) = 2xy^2$ and

$f_y(x, y) = 2x^2y$, are zero at the critical points $(a, 0)$ and

$(0, b)$ where a and b are any real numbers. Because

$f_{xx}(x, y) = 2y^2$, $f_{yy}(x, y) = 2x^2$, and

$f_{xy}(x, y) = 4xy$, it follows that

$f_{xx}(a, 0)f_{yy}(a, 0) - [f_{xy}(a, 0)]^2 = 0$ and

$f_{xx}(0, b)f_{yy}(0, b) - [f_{xy}(0, b)]^2 = 0$ and the

Second-Derivative Test fails. Note that $f(x, y) = (xy)^2$

is nonnegative for all $(a, 0, 0)$ and $(0, b, 0)$ where a and

b are real numbers.

So, $(a, 0, 0)$ and $(0, b, 0)$ are relative minima.

27. $f(x, y) = x^3 + y^3$

The first partial derivatives of f, $f_x(x, y) = 3x^2$ and

$f_y(x, y) = 3y^2$, are zero at the critical point $(0, 0)$.

Because

$f_{xx}(x, y) = 6x$, $f_{yy}(x, y) = 6y$, $f_{xy}(x, y) = 0$,

and $f_{xx}(0, 0)f_{yy}(0, 0) - [f_{xy}(0, 0)]^2 = 0$, the

Second-Partials Test fails. By testing "nearby" points,

you can conclude that (0, 0, 0) is a saddle point.

29. $f(x, y) = x^{2/3} + y^{2/3}$

The first partial derivatives of f,

$f_x(x, y) = \dfrac{2}{3\sqrt[3]{x}}$ and $f_y(x, y) = \dfrac{2}{3\sqrt[3]{y}}$, are undefined

at the point (0, 0). Because $f_{xx}(x, y) = -\dfrac{2}{9x^{4/3}}$,

$f_{yy}(x, y) = -\dfrac{2}{9y^{4/3}}$, $f_{xy}(x, y) = 0$ and $f_{xx}(0, 0)$ is

undefined, the Second-Derivative Test fails. Because

$f(x, y) \ge 0$ for all points in the xy-coordinate plane,

(0, 0, 0) is a relative minimum.

31. $f(x, y, z) = (x - 1)^2 + (y + 3)^2 + z^2$

Critical point: $(x, y, z) = (1, -3, 0)$

Relative minimum

33. The sum is $x + y + z = 45$, or $z = 45 - x - y$,

and the product is $P = xyz$, or

$P = xy(45 - x - y) = 45xy - x^2y - xy^2$.

The first partial derivatives of P are

$P_x(x, y) = 45y - 2xy - y^2 = y(45 - 2x - y)$

$P_y(x, y) = 45x - x^2 - 2xy = x(45 - x - 2y)$.

Setting these equal to zero produces the system

$2x + y = 45$

$x + 2y = 45$.

Solving the system, you have $x = 15$, $y = 15$, and

$z = 45 - 15 - 15 = 15$.

35. The sum is $x + y + z = 60$, or $z = 60 - x - y$,

and the sum of the squares is

$S = x^2 + y^2 + z^2 = x^2 + y^2 + (60 - x - y)^2$.

The first partial derivatives of S are

$S_x(x, y) = 2x + 2(60 - x - y)(-1) = 4x + 2y - 120$

and

$S_y(x, y) = 2y + 2(60 - x - y)(-1) = 2x + 4y - 120$.

Setting these equal to zero produces the system

$4x + 2y = 120$

$2x + 4y = 120$.

Solving this system, you have $x = 20$, $y = 20$, and

$z = 60 - 20 - 20 = 20$.

37. The revenue function is

$R(x_1, x_2) = -5x_1^2 - 8x_2^2 - 2x_1x_2 + 42x_1 + 102x_2$ and

the first partial derivatives of R are

$R_{x_1} = -10x_1 - 2x_2 + 42$ and

$R_{x_2} = -16x_2 - 2x_1 + 102.$

Setting these equal to zero produces the system

$$5x_1 + x_2 = 21$$
$$x_1 + 8x_2 = 51.$$

Solving the system, you have $x_1 = 3$ and $x_2 = 6$. Because $R_{x_1x_1} = -10$, $R_{x_1x_2} = -2$, and $R_{x_2x_2} = -16$, it follows that $R_{x_1x_1} < 0$ and

$$R_{x_1x_1}R_{x_2x_2} - \left(R_{x_1x_2}\right)^2 > 0.$$

So, the revenue is maximized when $x_1 = 3$ and $x_2 = 6$.

39. The revenue function is

$$R = x_1p_1 + x_2p_2$$
$$= 1000p_1 + 1500p_2 + 3p_1p_2 - 2p_1^2 - 1.5p_2^2$$

and the first partial derivatives of R are

$R_{p_1} = 1000 + 3p_2 - 4p_1$ and

$R_{p_2} = 1500 + 3p_1 - 3p_2.$

Setting these equal to zero produces the system

$$4p_1 + 3p_2 = 1000$$
$$-3p_1 + 3p_2 = 1500.$$

Solving this system, you have $p_1 = 2500$ and $p_2 = 3000$, and by the Second-Partials Test you can conclude that the revenue is maximized when $p_1 = 2500$ and $p_2 = 3000$.

41. The profit is

$$P = R - C_1 - C_2$$
$$= \left[225 - 0.4(x_1 + x_2)\right](x_1 + x_2) - \left(0.05x_1^2 + 15x_1 + 5400\right) - \left(0.03x_2^2 + 15x_2 + 6100\right)$$
$$= -0.45x_1^2 - 0.43x_2^2 - 0.8x_1x_2 + 210x_1 + 210x_2 - 11{,}500$$

and the first partial derivatives of P are $P_{x_1} = -0.9x_1 - 0.8x_2 + 210$ and $P_{x_2} = -0.86x_2 - 0.8x_1 + 210$.

Setting these equal to zero produces the system

$$0.9x_1 + 0.8x_2 = 210$$
$$0.8x_1 + 0.86x_2 = 210.$$

Solving this system, you have $x_1 \approx 94$ and $x_2 \approx 157$, and by the Second-Partials Test you can conclude that the profit is maximized when $x_1 \approx 94$ and $x_2 \approx 157$.

43. Let $x =$ length, $y =$ width, and $z =$ height.

The sum of length and girth is

$$x + (2y + 2z) = 96$$
$$x = 96 - 2y - 2z.$$

The volume is $V = xyz = 96yz - 2zy^2 - 2yz^2$ and the first partial derivatives are

$V_y = 96z - 4zy - 2z^2 = 2z(48 - 2y - z)$ and $V_z = 96y - 2y^2 - 4yz = 2y(48 - y - 2z)$.

Setting these equal to zero produces the system

$$2y + z = 48$$
$$y + 2z = 48.$$

So, $x = 32$.

Solving this system, you have $y = 16$ and $z = 16$. The volume is a maximum when the length is 32 inches and the width and height are each 16 inches.

45. Let x = length y = width, h = height and C = cost.

The volume is $xyz = 18$ or $z = \dfrac{18}{xy}$.

$$\begin{aligned}\text{The cost is } C &= 0.2xy + 0.15(2)xz + 0.15(2)yz \\ &= 0.2xy + 0.3xz + 0.3yz \\ &= 0.2xy + 0.3x\left(\frac{18}{xy}\right) + 0.3y\left(\frac{18}{xy}\right) \\ &= 0.2xy + \frac{5.4}{y} + \frac{5.4}{x}.\end{aligned}$$

The first partial derivatives of C are

$C_x = (x, y) = 0.2y - \dfrac{5.4}{x^2}$ and $C_y = (x, y) = 0.2x - \dfrac{5.4}{y^2}$.

Setting these equal to zero produces the system

$$-\frac{5.4}{x^2} + 0.2y = 0$$

$$0.2x - \frac{5.4}{y^2} = 0.$$

Solving this system, you have $x = 3, y = 3$, and $z = \dfrac{18}{(3)(3)} = 2$.

The cost is a minimum when $x = 3$ feet, $y = 3$ feet, and $z = 2$ feet.

The minimum cost is $C = 0.2(3)(3) + \dfrac{5.4}{3} + \dfrac{5.4}{3} = \$5.40$.

47. The total cost function is $C(x, y) = 2x^2 + 3y^2 - 15x - 20y + 4xy + 39$ and the first partial derivatives are $C_x = 4x - 15 + 4y$ and $C_y = 6y - 20 + 4x$.

Setting these equal to zero produces the system

$$4x + 4y = 15$$

$$4x + 6y = 20.$$

Solving this system, you have $x = 1.25$ and $y = 2.5$. So, the minimum total cost is

$2(1.25)^2 + 3(2.5)^2 - 15(1.25) - 20(2.5) + 4(1.25)(2.5) + 39 = \$4.625$ million.

49. The total weight function is

$$\begin{aligned}T &= xW_1 + yW_2 \\ &= x(3 - 0.002x - 0.001y) + y(4.5 - 0.004x - 0.005y) \\ &= 3x - 0.002x^2 + 4.5y - 0.005y^2 - 0.005xy\end{aligned}$$

and the first partial derivatives are $T_x = 3 - 0.004x - 0.005y$ and $T_y = 4.5 - 0.010y - 0.005x$.

Setting these equal to zero produces the system

$$0.004x + 0.005y = 3$$

$$0.005x + 0.01y = 4.5.$$

Solving this system, you have $x = 500$ and $y = 200$.

The lake should be stocked with approximately 500 smallmouth bass and 200 largemouth bass.

51. The population function is $P(p, q, r) = 2pq + 2pr + 2qr$.

Because $p + q + r = 1$, $r = 1 - p - q$.

So, $P = 2pq + 2p(1 - p - q) + 2q(1 - p - q)$

$= -2p^2 + 2p - 2q^2 + 2q - 2pq$

and the first partial derivatives are

$P_p = -4p + 2 - 2q$

$P_q = -4q + 2 - 2p.$

Setting these equal to zero produces the system

$4p + 2q = 2$

$2p + 4q = 2.$

Solving this system, you have $p = \frac{1}{3}$ and $q = \frac{1}{3}$. So, $r = \frac{1}{3}$.

The proportion is a maximum when $p = \frac{1}{3}$, $q = \frac{1}{3}$, and $r = \frac{1}{3}$.

The maximum proportion is $P = 2\left(\frac{1}{3}\right)\left(\frac{1}{3}\right) + 2\left(\frac{1}{3}\right)\left(\frac{1}{3}\right) + 2\left(\frac{1}{3}\right)\left(\frac{1}{3}\right) = \frac{6}{9} = \frac{2}{3}$.

53. True

Chapter 7 Quiz Yourself

1. (a)

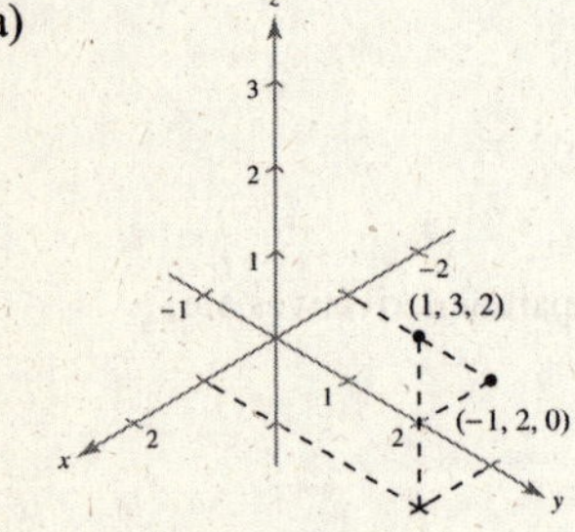

(b) $d = \sqrt{(-1-1)^2 + (2-3)^2 + (0-2)^2} = 3$

(c) Midpoint $= \left(\frac{1+(-1)}{2}, \frac{3+2}{2}, \frac{2+0}{2}\right)$

$= \left(0, \frac{5}{2}, 1\right)$

2. (a)

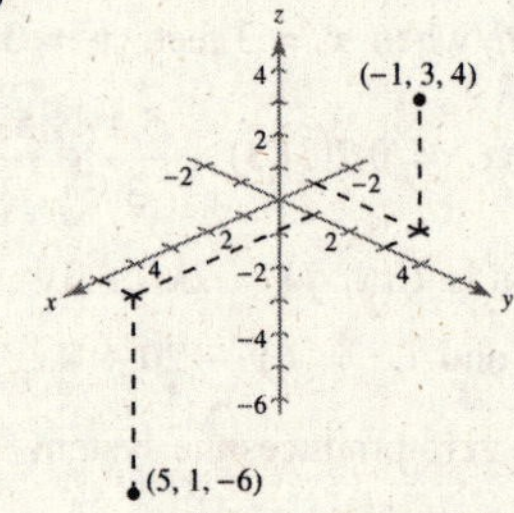

(b) $d = \sqrt{(5-(-1))^2 + (1-3)^2 + (-6-4)^2}$

$= \sqrt{140} = 2\sqrt{35}$

(c) Midpoint $= \left(\frac{5+(-1)}{2}, \frac{1+3}{2}, \frac{-6+4}{2}\right) = (2, 2, -1)$

3. (a)

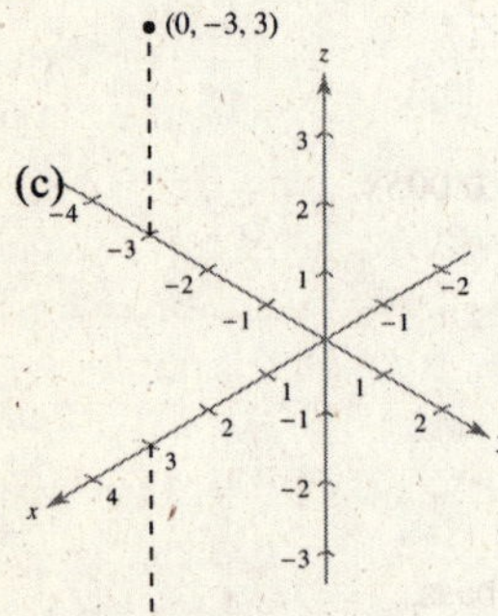

(b) $d = \sqrt{(3-0)^2 + (0+3)^2 + (-3-3)^2} = \sqrt{54} = 3\sqrt{6}$

(c) Midpoint $= \left(\frac{0+3}{2}, \frac{-3+0}{2}, \frac{3+-3}{2}\right) = \left(\frac{3}{2}, -\frac{3}{2}, 0\right)$

4. $(x-2)^2 + (y+1)^2 + (z-3)^2 = 16$

5. Center: $\left(\dfrac{0+2}{2}, \dfrac{3+5}{2}, \dfrac{1+(-5)}{2}\right) = (1, 4, -2)$

Radius $= \sqrt{(1-0)^2 + (4-3)^2 + (-2-1)^2} = \sqrt{11}$

Standard form: $(x-1)^2 + (y-4)^2 + (z+2)^2 = 11$

6. $x^2 + y^2 + z^2 - 8x - 2y - 6z - 23 = 0$

$(x^2 - 8x + 16) + (y^2 - 2y + 1) + (z^2 - 6z + 9) = 23 + 16 + 1 + 9$

$(x-4)^2 + (y-1)^2 + (z-3)^2 = 49$

Center: $(4, 1, 3)$

Radius: $\sqrt{49} = 7$

7. $2x + 3y + z = 6$

To find the x-intercept, let $y = 0$ and $z = 0$.

$2x = 6 \Rightarrow x = 3$

To find the y-intercept, let $x = 0$ and $z = 0$.

$3y = 6 \Rightarrow y = 2$

To find the z-intercept, let $x = 0$ and $y = 0$.

$z = 6$

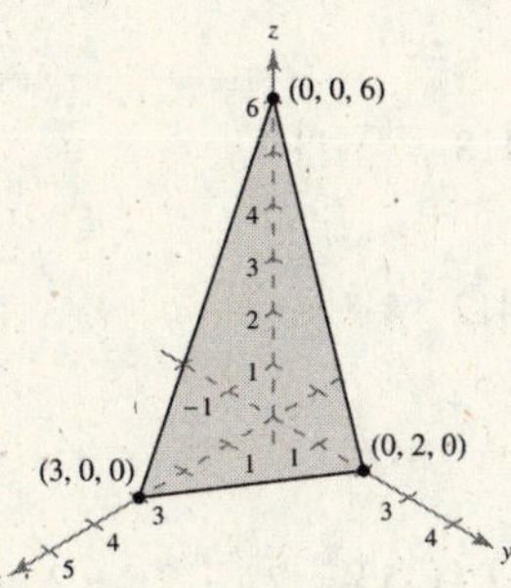

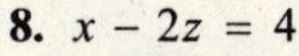

8. $x - 2z = 4$

To find the x-intercept, let $z = 0$.

$x = 4$

Because the y-coefficient is 0, there is no y-intercept. The plane is parallel to the y-axis.

To find the z-intercept,

let $x = 0$.

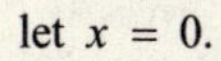

$-2z = 4 \Rightarrow z = -2$

9.

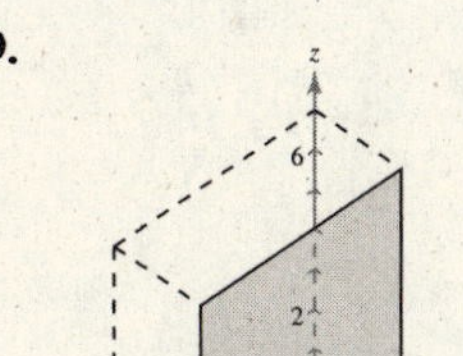

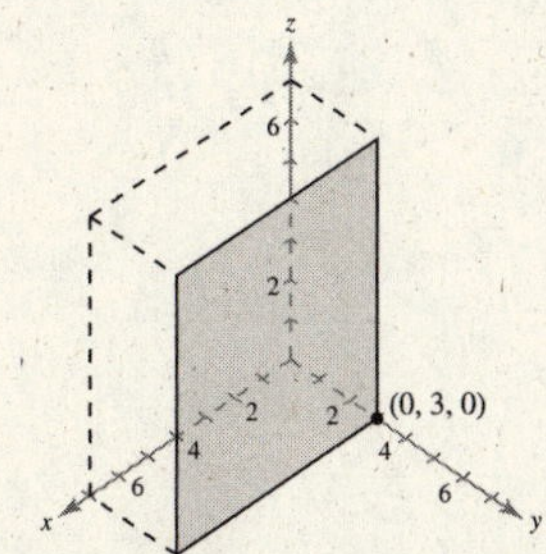

The only intercept is $y = 3$. The plane is parallel to the xz-plane.

10. The graph of $\dfrac{x^2}{4} + \dfrac{y^2}{9} + \dfrac{z^2}{16} = 1$ is an ellipsoid.

11. The graph of $z^2 - x^2 - y^2 = 25$ or

$\dfrac{z^2}{25} - \dfrac{x^2}{25} - \dfrac{y^2}{25} = 1$ is a hyperboloid of two sheets.

12. The graph of $81z - 9x^2 - y^2 = 0$ or $z = \dfrac{x^2}{9} + \dfrac{y^2}{81}$

is an elliptic paraboloid.

13. $f(x, y) = x - 9y^2$

$f(1, 0) = 1 - 9(0)^2 = 1$

$f(4, -1) = 4 - 9(-1)^2 = -5$

14. $f(x, y) = \sqrt{4x^2 + y}$

$f(1, 0) = \sqrt{4(1)^2 + 0} = 2$

$f(4, -1) = \sqrt{4(4)^2 + (-1)} = \sqrt{63} = 3\sqrt{7}$

15. $f(x, y) = \ln(x - 2y)$

$f(1, 0) = \ln(1) = 0$

$f(4, -1) = \ln(4 - 2(-1)) = \ln 6 \approx 1.79$

16. (a) The temperatures in the Great Lakes region range from about 30° to about 50°.

(b) The temperatures in the United States range from 40° to 80°.

(c) The temperatures in Mexico range from about 70° to almost 90°.

17. $f(x, y) = x^2 + 2y^2 - 3x - y + 1$

$f_x(x, y) = 2x - 3 \qquad f_x(-2, 3) = 2(-2) - 3 = -7$

$f_y(x, y) = 4y - 1 \qquad f_y(-2, 3) = 4(3) - 1 = 11$

18. $f(x, y) = \dfrac{3x - y^2}{x + y}$

$f_x(x, y) = \dfrac{(x + y)(3) - (3x - y^2)}{(x + y)^2} = \dfrac{y^2 + 3y}{(x + y)^2} = \dfrac{y(y + 3)}{(x + y)^2}$, $\quad f_x(-2, 3) = \dfrac{3^2 + 3(3)}{(-2 + 3)^2} = 18$

$f_y(x, y) = \dfrac{(x + y)(-2y) - (3x - y^2)}{(x + y)^2} = \dfrac{-y^2 - 2xy - 3x}{(x + y)^2}$, $\quad f_y(-2, 3) = \dfrac{-3^2 - 2(-2)(3) - 3(-2)}{(-2 + 3)^2} = 9$

19. $f(x, y) = x^3e^{2y}$

$f_x(x, y) = 3x^2e^{2y}, \quad f_x(-2, 3) = 3(-2)^2e^{2(3)} = 12e^6 \approx 4841.15$

$f_y(x, y) = 2x^3e^{2y}, \quad f_y(-2, 3) = 2(-2)^3e^{2(3)} = -16e^6 \approx -6454.86$

20. $f(x, y) = \ln(2x + 7y)$

$f_x(x, y) = \dfrac{2}{2x + 7y}, \quad f_x(-2, 3) = \dfrac{2}{2(-2) + 7(3)} = \dfrac{2}{17} \approx 0.118$

$f_y(x, y) = \dfrac{7}{2x + 7y}, \quad f_y(-2, 3) = \dfrac{7}{2(-2) + 7(3)} = \dfrac{7}{17} \approx 0.412$

21. $f(x, y) = 3x^2 + y^2 - 2xy - 6x + 2y$

The first partial derivatives of f, $f_x(x, y) = 6x - 2y - 6$ and $f_y(x, y) = 2y - 2x + 2$, are zero at the point $(1, 0)$. Moreover, because $f_{xx}(x, y) = 6$, $f_{yy}(x, y) = 2$, and $f_{xy}(x, y) = -2$, it follows that $f_{xx}(1, 0) > 0$ and $f_{xx}(1, 0)f_{yy}(1, 0) - [f_{xy}(1, 0)]^2 = 8 > 0$. So, $(1, 0, -3)$ is a relative minimum.

22. $f(x, y) = -x^3 + 4xy - 2y^2 + 1$

The first partial derivatives of f, $f_x(x, y) = -3x^2 + 4y$ and $f_y(x, y) = 4x - 4y$, are zero at the points $(0, 0)$ and $\left(\frac{4}{3}, \frac{4}{3}\right)$. Moreover, because $f_{xx}(x, y) = -6x$, $f_{yy}(x, y) = -4$, and $f_{xy} = 4$, it follows that $f_{xx}(0, 0) = 0$ and $f_{xx}(0, 0)f_{yy}(0, 0) - [f_{xy}(0, 0)]^2 = -16 < 0$, $f_{xx}\left(\frac{4}{3}, \frac{4}{3}\right) = -8$ and $f_{xx}\left(\frac{4}{3}, \frac{4}{3}\right)f_{yy}\left(\frac{4}{3}, \frac{4}{3}\right) - \left[f_{xy}\left(\frac{4}{3}, \frac{4}{3}\right)\right]^2 = 16 > 0$.

So, $(0, 0, 1)$ is a saddle point and $\left(\frac{4}{3}, \frac{4}{3}, \frac{59}{27}\right)$ is a relative maximum.

23. The cost function is $C(x, y) = \frac{1}{16}x^2 + y^2 - 10x - 40y + 820$ and the first partial derivatives are $C_x = \frac{1}{8}x - 10$ and $C_y = 2y - 40$. Setting these equal to zero and solving for x and y gives you $x = 80$ and $y = 20$.

The minimum combined cost is $\frac{1}{16}(80)^2 + 20^2 - 10(80) - 40(20) + 820 = 20$, or \$20,000.

Section 7.6 Lagrange Multipliers

Skills Warm Up

1. $\begin{cases} 4x - 6y = 3 & \text{Equation 1} \\ 2x + 3y = 2 & \text{Equation 2} \end{cases}$

Multiply Equation 2 by 2: $4x + 6y = 4$

Add to Equation 1: $8x = 7$

Simplify: $x = \frac{7}{8}$

Substitute $\frac{7}{8}$ for x in Equation 2: $2\left(\frac{7}{8}\right) + 3y = 2$

Solve for y: $y = \frac{1}{12}$

The solution is $\left(\frac{7}{8}, \frac{1}{12}\right)$.

2. $\begin{cases} 6x - 6y = 5 & \text{Equation 1} \\ -3x - y = 1 & \text{Equation 2} \end{cases}$

Multiply Equation 2 by 2: $-6x - 2y = 2$

Add to Equation 1: $-8y = 7$

Simplify: $y = -\frac{7}{8}$

Substitute $-\frac{7}{8}$ for x in Equation 2: $-3x - \left(-\frac{7}{8}\right) = 1$

Solve for x: $x = -\frac{1}{24}$

The solution is $\left(-\frac{1}{24}, -\frac{7}{8}\right)$.

3. $\begin{cases} 5x - y = 25 & \text{Equation 1} \\ x - 5y = 15 & \text{Equation 2} \end{cases}$

Multiply Equation 2 by -5: $-5x + 25y = -75$

Add to Equation 1: $24y = -50$

Simplify: $y = -\frac{25}{12}$

Substitute $-\frac{25}{12}$ for y in Equation 1:

$5x - \left(-\frac{25}{12}\right) = 25$

Solve for x: $x = \frac{55}{12}$

The solution is $\left(\frac{55}{12}, -\frac{25}{12}\right)$.

4. $\begin{cases} 4x - 9y = 5 & \text{Equation 1} \\ -x + 8y = -2 & \text{Equation 2} \end{cases}$

Multiply Equation 2 by 4: $-4x + 32y = -8$

Add to Equation 1: $23y = -3$

Simplify: $y = -\frac{3}{23}$

Substitute $-\frac{3}{23}$ for y in Equation 2:

$-x + 8\left(-\frac{3}{23}\right) = -2$

Solve for x: $x = \frac{22}{23}$

The solution is $\left(\frac{22}{23}, -\frac{3}{23}\right)$.

5. $\begin{cases} 2x - y + z = 3 & \text{Equation 1} \\ 2x + 2y + z = 4 & \text{Equation 2} \\ -x + 2y + 3z = -1 & \text{Equation 3} \end{cases}$

Multiply Equation 3 by 2: $-2x + 4y + 6z = -2$

Add to Equation 2: $6y + 7z = 2$ New Equation 1

Multiply Equation 3 by 2: $-2x + 4y + 6z = -2$

Add to Equation 1: $3y + 7z = 1$ New Equation 2

Subtract New Equation 2 from New Equation 1: $3y = 1$

Simplify: $y = \frac{1}{3}$

Substitute $\frac{1}{3}$ for y in New Equation 2: $3\left(\frac{1}{3}\right) + 7z = 1$

Solve for z: $z = 0$

Substitute $\frac{1}{3}$ for y and 0 for z in Equation 3:

$-x + 2\left(\frac{1}{3}\right) + 3(0) = -1$

Solve for x: $x = \frac{5}{3}$

The solution is $\left(\frac{5}{3}, \frac{1}{3}, 0\right)$.

Skills Warm Up —continued—

6. $\begin{cases} -x - 4y + 6z = -2 & \text{Equation 1} \\ x - 3y - 3z = 4 & \text{Equation 2} \\ 3x + y + 3z = 0 & \text{Equation 3} \end{cases}$

Add Equation 2 and Equation 3: $4x - 2y = 4$ New Equation 1

Multiply Equation 2 by 2: $2x - 6y - 6z = 8$

Add to Equation 1: $x - 10y = 6$ New Equation 2

Multiply New Equation 1 by 5: $20x - 10y = 20$ New Equation 3

Subtract New Equation 2 from New Equation 3: $19x = 14$

Simplify: $x = \frac{14}{19}$

Substitute $\frac{14}{19}$ for x in New Equation 2: $\frac{14}{19} - 10y = 6$

Solve for y: $y = -\frac{10}{19}$

Substitute $\frac{14}{19}$ for x and $-\frac{10}{19}$ for y in Equation 3: $3\left(\frac{14}{19}\right) + \left(-\frac{10}{19}\right) + 3z = 0$

Solve for z: $z = -\frac{32}{57}$

The solution is $\left(\frac{14}{19}, -\frac{10}{19}, -\frac{32}{57}\right)$.

7. $f(x, y) = x^2y + xy^2$

$f_x(x, y) = 2xy + y^2$

$f_y(x, y) = x^2 + 2xy$

8. $f(x, y) = 25(xy + y^2)^2$

$f_x(x, y) = 50(xy + y^2)(y)$

$= 50y^2(x + y)$

$f_y(x, y) = 50(xy + y^2)(x + 2y)$

$= 50y(x + y)(x + 2y)$

9. $f(x, y, z) = x(x^2 - 2xy + yz)$

$f_x(x, y, z) = x(2x - 2y) + (x^2 - 2xy + yz)(1)$

$= 3x^2 - 4xy + yz$

$f_y(x, y, z) = x(-2x + z) + (x^2 - 2xy + yz)(0)$

$= -2x^2 + xz$

$f_z(x, y, z) = x(y) + (x^2 - 2xy + yz)(0)$

$= xy$

10. $f(x, y, z) = z(xy + xz + yz)$

$f_x(x, y, z) = z(y + z) + (xy + xz + yz)(0)$

$= z^2 + yz$

$f_y(x, y, z) = z(x + z) + (xy + xz + yz)(0)$

$= z^2 + xz$

$f_z(x, y, z) = z(x + y) + (xy + xz + yz)(1)$

$= xy + 2xz + 2yz$

1. $F(x, y, \lambda) = xy - \lambda(x + y - 10)$

$F_x = y - \lambda = 0,\ y = \lambda$

$F_y = x - \lambda = 0,\ x = \lambda$

$F_\lambda = -(x + y - 10) = 0,\ 2\lambda = 10,\ \lambda = 5$

So, $\lambda = 5$, $x = 5$, $y = 5$, and $f(x, y)$ has a maximum at $(5, 5)$. The maximum is $f(5, 5) = 25$.

3. $F(x, y, \lambda) = x^2 + y^2 - \lambda(x + y - 8)$

$F_x = 2x - \lambda = 0,\ x = \frac{1}{2}\lambda$

$F_y = 2y - \lambda = 0,\ y = \frac{1}{2}\lambda$

$F_\lambda = -(x + y - 8) = 0,\ -\lambda = -8,\ \lambda = 8$

So, $\lambda = 8$, $x = 4$, and $y = 4$, and $f(x, y)$ has a minimum at $(4, 4)$. The minimum is $f(4, 4) = 32$.

5. $F(x, y, \lambda) = x^2 - y^2 - \lambda(2y - x^2)$

$F_x = 2x + 2\lambda x = 0,\ 2x(1 + \lambda) = 0,\ \lambda = -1$

$F_y = -2y - 2\lambda = 0,\ y = -\lambda$

$F_\lambda = -(2y - x^2) = 0,\ x = \sqrt{2y}$

So, $\lambda = -1$, $x = \sqrt{2}$, $y = 1$, and $f(x, y)$ has a maximum at $(\sqrt{2}, 1)$. The maximum is $f(\sqrt{2}, 1) = 1$.

7. $F(x, y, \lambda) = 2x + 2xy + y - \lambda(2x + y - 100)$

$F_x = 2 + 2y - 2\lambda = 0,\ y = \lambda - 1$

$F_y = 2x + 1 - \lambda = 0,\ x = \dfrac{\lambda - 1}{2}$

$F_\lambda = -(2x + y - 100) = 0,$

$-2\left(\dfrac{\lambda - 1}{2}\right) - (\lambda - 1) = -100,\ \lambda = 51$

So, $\lambda = 51$, $x = 25$, $y = 50$, and $f(x, y)$ has a maximum at $(25, 50)$. The maximum is $f(25, 50) = 2600$.

9. Note: $f(x, y)$ has a maximum value when $g(x, y) = 6 - x^2 - y^2$ is maximum.

$F(x, y, \lambda) = 6 - x^2 - y^2 - \lambda(x + y - 2)$

$$\left.\begin{aligned} F_x &= -2x - \lambda = 0, \quad -2x = \lambda \\ F_y &= -2y - \lambda = 0, \quad -2y = \lambda \end{aligned}\right\} x = y$$

$F_\lambda = -(x + y - 2) = 0,\ \ 2x = 2,\ x = 1$

So, $x = 1$, $y = 1$, and $f(x, y)$ has a maximum at $(1, 1)$. The maximum is $f(1, 1) = 2$.

11. $F(x, y, \lambda) = e^{xy} - \lambda(x^2 + y^2 - 8)$

$$\left.\begin{aligned} F_x &= ye^{xy} - 2x\lambda = 0,\ e^{xy} = \frac{2x\lambda}{y} \\ F_y &= xe^{xy} - 2y\lambda = 0,\ e^{xy} = \frac{2y\lambda}{x} \end{aligned}\right\} x = y$$

$F_\lambda = -(x^2 + y^2 - 8) = 0,\ 2x^2 = 8,\ x = 2$

So, $x = 2$, $y = 2$, and $f(x, y)$ has a maximum at (2, 2). The maximum is $f(2, 2) = e^4$.

13. $F(x, y, z, \lambda) = 2x^2 + 3y^2 + 2z^2 - \lambda(x + y + z - 24)$

$F_x = 4x - \lambda = 0, \quad \lambda = 4x$

$F_y = 6y - \lambda = 0, \quad \lambda = 6y$

$F_z = 4z - \lambda = 0, \quad \lambda = 4z$

$F_\lambda = -(x + y + z - 24) = 0$

$$\frac{\lambda}{4} + \frac{\lambda}{6} + \frac{\lambda}{4} = 24$$

$$8\lambda = 288$$

$$\lambda = 36$$

So, $\lambda = 36$, $x = 9$, $y = 6$, $z = 9$, and $f(x, y, z)$ has a minimum at $(9, 6, 9)$.

The minimum is $f(9, 6, 9) = 432$.

15. $F(x, y, z, \lambda) = x^2 + y^2 + z^2 - \lambda(x + y + z - 1)$

$$\left.\begin{aligned} F_x &= 2x - \lambda = 0 \\ F_y &= 2y - \lambda = 0 \\ F_z &= 2z - \lambda = 0 \end{aligned}\right\} x = y = z$$

$F_\lambda = -(x + y + z - 1) = 0,\ 3x = 1,\ x = \frac{1}{3}$

So, $x = \frac{1}{3}$, $y = \frac{1}{3}$, $z = \frac{1}{3}$, and $f(x, y, z)$ has a minimum at $\left(\frac{1}{3}, \frac{1}{3}, \frac{1}{3}\right)$. The minimum is $f\left(\frac{1}{3}, \frac{1}{3}, \frac{1}{3}\right) = \frac{1}{3}$.

17. $F(x, y, z, \lambda) = x + y + z - \lambda(x^2 + y^2 + z^2 - 1)$

$$\left.\begin{aligned} F_x &= 1 - 2x\lambda = 0 \\ F_y &= 1 - 2y\lambda = 0 \\ F_z &= 1 - 2z\lambda = 0 \end{aligned}\right\} x = y = z$$

$F_\lambda = -(x^2 + y^2 + z^2 - 1) = 0,\ 3x^2 = 1,\ x = \dfrac{\sqrt{3}}{3}$

So, $x = \dfrac{\sqrt{3}}{3}$, $y = \dfrac{\sqrt{3}}{3}$, $z = \dfrac{\sqrt{3}}{3}$, and $f(x, y, z)$ has a maximum at $\left(\dfrac{\sqrt{3}}{3}, \dfrac{\sqrt{3}}{3}, \dfrac{\sqrt{3}}{3}\right)$.

The maximum is $f\left(\dfrac{\sqrt{3}}{3}, \dfrac{\sqrt{3}}{3}, \dfrac{\sqrt{3}}{3}\right) = \sqrt{3}$.

19. Maximize $f(x, y, z) = xyz$ subject to the constraint $x + y + z = 60$.

$$F(x, y z, \lambda) = xyz - \lambda(x + y + z - 60)$$

$$\left.\begin{aligned} F_x &= yz - \lambda = 0, \quad yz = \lambda \\ F_y &= xz - \lambda = 0, \quad xz = \lambda \\ F_z &= xy - \lambda = 0, \quad xy = \lambda \end{aligned}\right\} yz = xz = xy \Rightarrow x = y = z$$

$$F_\lambda = -(x + y + z - 60) = 0, -3x = -60, x = 20$$

So, $x = 20$, $y = 20$, and $z = 20$.

21. Minimize $f(x, y, z) = x^2 + y^2 + z^2$ subject to the constraint $x + y + z = 120$.

$$F(x, y, z, \lambda) = x^2 + y^2 + z^2 - \lambda(x + y + z - 120)$$

$$\left.\begin{aligned} F_x &= 2x - \lambda = 0 \\ F_y &= 2y - \lambda = 0 \\ F_z &= 2z - \lambda = 0 \end{aligned}\right\} x = y = z$$

$$F_\lambda = -(x + y + z - 120) = 0, 3x = 120, x = 40$$

So, $x = 40$, $y = 40$, and $z = 40$.

23. $F(x, y, \lambda) = x^2 + y^2 - \lambda(x + y - 6)$

$$F_x = 2x - \lambda = 0, \ x = \frac{\lambda}{2}$$

$$F_y = 2y - \lambda = 0, \ y = \frac{\lambda}{2} = x$$

$$F_\lambda = -(x + y - 6) = 0, x + x = 6 \Rightarrow x = 3$$

So, $x = 3$, $y = 3$, and $d = \sqrt{x^2 + y^2} = \sqrt{9 + 9} = \sqrt{18} = 3\sqrt{2}$.

25. $F(x, y, z, \lambda) = (x - 2)^2 + (y - 1)^2 + (z - 1)^2 - \lambda(x + y + z - 1)$

$$\left.\begin{aligned} F_x &= 2(x - 2) - \lambda = 0 \\ F_y &= 2(y - 1) - \lambda = 0 \\ F_z &= 2(z - 1) - \lambda = 0 \end{aligned}\right\} \begin{aligned} x - 2 &= y - 1 = z - 1 \\ x - 1 &= y = z \end{aligned}$$

$$F_\lambda = -(x + y + z - 1) = 0$$

So, $y = 0$, $z = 0$, $x = 1$, and $d = \sqrt{(1 - 2)^2 + (0 - 1)^2 + (0 - 1)^2} = \sqrt{3}$.

27. Maximize $f(x, y, z) = xyz$ subject to the constraint $2x + 3y + 5z - 90 = 0$.

$$F(x, y z, \lambda) = xyz - \lambda(2x + 3y + 5z - 90)$$

$$\left.\begin{aligned} F_x &= yz - 2\lambda = 0, \quad \lambda = \tfrac{1}{2}yz \\ F_y &= xz - 3\lambda = 0, \quad \lambda = \tfrac{1}{3}xz \\ F_z &= xy - 5\lambda = 0, \quad \lambda = \tfrac{1}{5}xy \end{aligned}\right\} \begin{aligned} \tfrac{1}{2}y &= \tfrac{1}{3}x \Rightarrow y = \tfrac{2}{3}x \\ \tfrac{1}{3}z &= \tfrac{1}{5}y \Rightarrow z = \tfrac{3}{5}y \end{aligned}$$

$$F_\lambda = -(2x + 3y + 5z - 90) = 0$$

Using F_λ, $-\left(2x + 3\left(\tfrac{2}{3}x\right) + 5\left[\tfrac{3}{5}\left(\tfrac{2}{3}x\right)\right] - 90\right) = 0$

$$-2x - 2x - 2x + 90 = 0 \Rightarrow x = 15.$$

So, $x = 15$, $y = 10$, and $z = 6$.

The rectangular box has dimensions 15 units by 10 units by 6 units.

29. $F(x, y, z, \lambda) = xyz - \lambda(3xy + 2xz + 2yz - 1296)$

$$\left.\begin{aligned} F_x &= yz - (3\lambda y + 2\lambda z) = 0 \\ F_y &= xz - (3\lambda x + 2\lambda z) = 0 \end{aligned}\right\} x = y \text{ or } z = 3\lambda$$

$F_z = xy - (2\lambda x + 2\lambda y) = 0,\ x = 4\lambda$ (if $x = y$)

$F_\lambda = -(3xy + 2xz + 2yz - 1296) = 0$

From F_y, $z = 6\lambda = \frac{3}{2}x$ and F_λ gives $x = 12$.

So, $x = 12$, $y = 12$, and $z = \frac{3}{2}(12) = 18$.

(The case $z = 3\lambda$ results in $\lambda = 0$ or $z = 0$, which are impossible.)

The volume is maximized when the dimensions are $12 \times 12 \times 18$ feet.

31. $F(x_1, x_2, \lambda) = 0.25x_1^2 + 25x_1 + 0.05x_2^2 + 12x_2 - \lambda(x_1 + x_2 - 1000)$

$F_{x_1} = 0.5x_1 + 25 - \lambda = 0,\ x_1 = 2\lambda - 50$

$F_{x_2} = 0.1x_2 + 12 - \lambda = 0,\ x_2 = 10\lambda - 120$

$F_\lambda = -(x_1 + x_2 - 1000) = 0$

$$\begin{aligned} (2\lambda - 50) + (10\lambda - 120) &= 1000 \\ 12\lambda &= 1170 \\ \lambda &= 97.5 \end{aligned}$$

So, $x_1 = 145$ and $x_2 = 855$.

To minimize cost, let $x_1 = 145$ units and $x_2 = 855$ units.

33. (a) Maximize $f(x, y) = 100x^{0.25}y^{0.75}$ subject to the constraint $48x + 36y = 100{,}000$.

$F(x, y, \lambda) = 100x^{0.25}y^{0.75} - \lambda(48x + 36y - 100{,}000)$

$F_x = 25x^{-0.75}y^{0.75} - 48\lambda = 0$

$F_y = 75x^{0.25}y^{-0.25} - 36\lambda = 0$

$F_\lambda = -(48x + 36y - 100{,}000) = 0$

Using F_x, $\lambda = \dfrac{25x^{-0.75}y^{0.75}}{48}$ and F_y, $\lambda = \dfrac{75x^{0.25}y^{-0.25}}{36}$, so

$$\begin{aligned} \frac{25x^{-0.75}y^{0.75}}{48} &= \frac{75x^{0.25}y^{-0.25}}{36} \\ \frac{25y^{0.75}}{48x^{0.75}} &= \frac{75x^{0.25}}{36y^{0.25}} \\ \frac{3}{4}y &= 3x \\ y &= 4x. \end{aligned}$$

Then using F_λ, $-(48x + 36(4x) - 100{,}000) = 0 \Rightarrow x = \dfrac{3125}{6}$.

So, $x = \dfrac{3125}{6}$ and $y = \dfrac{6250}{3}$, and $f(x, y)$ has a maximum at $\left(\dfrac{3125}{6}, \dfrac{6250}{3}\right)$.

The maximum production level is $f\left(\dfrac{3125}{6}, \dfrac{6250}{3}\right) \approx 147{,}314$ units.

(b) Using F_x, $25x^{-0.75}y^{0.75} - 48\lambda = 0$, so

$$\lambda = \frac{25x^{-0.75}y^{0.75}}{48} = \frac{25}{48}\left(\frac{3125}{6}\right)^{-0.75}\left(\frac{6250}{3}\right)^{0.75} \approx 1.4731.$$

(c) $147{,}314 + (125{,}000 - 100{,}000)\lambda \approx 147{,}314 + 25{,}000(1.4731) \approx 181{,}142$ units

(d) $147{,}314 + (350{,}000 - 100{,}000)\lambda \approx 147{,}314 + 250{,}000(1.4731) \approx 515{,}589$ units

35. (a) Minimize $f(x, y) = 50x + 100y$ subject to $100x^{0.7}y^{0.3} - 20{,}000 = 0$.

$$F(x, y, \lambda) = 50x + 100y - \lambda(100x^{0.7}y^{0.3} - 20{,}000)$$

$$F_x = 50 - 70\lambda x^{-0.3}y^{0.3} = 0$$
$$F_y = 100 - 30\lambda x^{0.7}y^{-0.7} = 0$$
$$F_\lambda = -(100x^{0.7}y^{0.3} - 20{,}000) = 0$$

Using F_x, $\left(\frac{y}{x}\right)^{0.3} = \frac{50}{70\lambda}$ and

using F_y, $\left(\frac{y}{x}\right)^{0.7} = \frac{30\lambda}{100}$.

So, $\left(\frac{y}{x}\right)^{0.3}\left(\frac{y}{x}\right)^{0.7} = \frac{y}{x} = \left(\frac{50}{70\lambda}\right)\left(\frac{30\lambda}{100}\right)$

$$\frac{y}{x} = \frac{3}{14} \Rightarrow y = \frac{3}{14}x.$$

Since $100x^{0.7}y^{0.3} = 20{,}000$,

$$100x^{0.7}\left(\frac{3}{14}x\right)^{0.3} = 20{,}000$$

$$\left(\frac{3}{14}\right)^{0.3} x = 200$$

$$x = \frac{200}{\left(\frac{3}{14}\right)^{0.3}} \approx 317 \text{ units.}$$

So, $x \approx 317$ units and $y = \frac{3}{14}(317) \approx 68$ units.

(b) $f(x, y) = 100x^{0.7}y^{0.3}$

$$f_x(x, y) = 70x^{-0.3}y^{0.3}$$
$$f_y(x, y) = 30x^{0.7}y^{-0.7}$$

$$\frac{f_x(x, y)}{f_y(x, y)} = \frac{70x^{-0.3}y^{0.3}}{30x^{0.7}y^{-0.7}} = \frac{50}{100}$$

$$\frac{70y}{30x} = \frac{50}{100}$$

$$\frac{y}{x} = \frac{3}{14}$$

So, $y = \frac{3}{14}x$ and the conditions are met.

37. (a)

Barn

x x x

y y

Minimize $f(x, y) = 10(2x + 2y) + 4x = 24x + 20y$ subject to the constraint $g(x, y) = 2xy - 6000 = 0$.

$$F(x, y, \lambda) = 24x + 20y - \lambda(2xy - 6000)$$

$$\left.\begin{aligned} F_x &= 24 - 2\lambda y = 0 \\ F_y &= 20 - 2\lambda x = 0 \end{aligned}\right\} \begin{aligned} y &= 12/\lambda \\ x &= 10/\lambda \end{aligned}$$

$$F_\lambda = -2xy + 6000 = 0$$
$$2xy = 6000$$
$$2\left(\frac{10}{\lambda}\right)\left(\frac{12}{\lambda}\right) = 6000$$
$$\frac{1}{\lambda^2} = 25$$
$$\lambda = \frac{1}{5}$$

So, $x = 50$ and $y = 60$.

To minimize the cost of the fencing, make the fence 50 feet by 120 feet.

(b) $f(50, 60) = \$2400$

The minimum cost is \$2400.

39. Minimize $C(x, y, z) = x + 2y + 3z$ subject to the constraint $12xyz = 0.13$.

$F(x, y, z, \lambda) = x + 2y + 3z - \lambda(12xyz - 0.13)$

$$\left.\begin{array}{l} F_x = 1 - 12\lambda yz = 0, \ 12\lambda yz = 1 \\ F_y = 2 - 12\lambda xz = 0, \ 12\lambda xz = 2 \\ F_z = 3 - 12\lambda xy = 0, \ 12\lambda xy = 3 \end{array}\right\} \begin{array}{l} x = 2y \\ x = 3z \end{array}$$

$$F_\lambda = -(12xyz - 0.13) = 0, \ 12x\left(\frac{x}{2}\right)\left(\frac{x}{3}\right) = 0.13$$

$$2x^3 = 0.13$$

$$x = \sqrt[3]{0.065} \approx 0.402$$

So, $x = \sqrt[3]{0.065} \approx 0.402$, $y = \frac{1}{2}\sqrt[3]{0.065} \approx 0.201$, $z = \frac{1}{3}\sqrt[3]{0.065} \approx 0.134$, and $f(x, y, z)$ is a minimum at about $(0.402, 0.201, 0.134)$. To minimize the cost, use 0.402 liter of solution x, 0.201 liter of solution y, and 0.134 liter of solution z.

41. Maximize $G(x, y, z) = 0.05x^2 + 0.16xy + 0.25z^2$ subject to the constraint $9x + 4y + 4z = 400$.

$F(x, y, z, \lambda) = 0.05x^2 + 0.16xy + 0.25z^2 - \lambda(9x + 4y + 4z - 400)$

$F_x = 0.1x + 0.16y - 9\lambda = 0$

$$\left.\begin{array}{l} F_y = 0.16x - 4\lambda = 0 \\ F_z = 0.5z - 4\lambda = 0 \end{array}\right\} \begin{array}{l} x = 25\lambda \\ z = 8\lambda \end{array}$$

$F_\lambda = -(9x + 4y + 4z - 400) = 0$

From F_λ, you have $9(25\lambda) + 4y + 4(8\lambda) = 400$. So, $y = 100 - 64.25\lambda$. From F_x, you have $0.1(25\lambda) + 0.16(100 - 64.25\lambda) - 9\lambda = 0$. So, $\lambda = \frac{800}{839} \approx 0.953516$. So, $x \approx 23.8$, $y \approx 38.7$, $z \approx 7.6$, and $f(x, y, z)$ has a maximum at about $(23.8, 38.7, 7.6)$. To maximize the amount of ice cream you can eat, have $f(23.8, 38.7, 7.6) \approx 190.1$ grams.

43. (a) Maximize $A = 0.0001t^2pr^{1.5}$ subject to the constraint $30t + 12p + 15r = 2700$.

$F(A) = 0.0001t^2pr^{1.5} - \lambda(30t + 12p + 15r - 2700)$

$F_t = 0.0002tpr^{1.5} - 30\lambda = 0$

$F_p = 0.0001t^2r^{1.5} - 12\lambda = 0, \ \lambda = \dfrac{0.0001t^2r^{1.5}}{12}$

$F_r = 0.00015t^2pr^{0.5} - 15\lambda = 0$

$F_\lambda = -(30t + 12p + 15r - 2700) = 0$

From F_t, you obtain $t = 0.8p$.

From F_r, you obtain $r = 1.2p$.

From F_λ, you obtain $p = 50$.

So, $t = 40$ and $r = 60$. To maximize the number of responses, spend $30(40) = \$1200$ on a cable television ad, $12(50) = \$600$ on a newspaper ad, and $\$15(60) = \$900$ on a radio ad.

(b) $A = 0.0001(40)^2(50)(60)^{1.5} \approx 3718$

The maximum number of responses is about 3718.

Section 7.7 Least Squares Regression Analysis

Skills Warm Up

1. $(2.5-1)^2+(3.25-2)^2+(4.1-3)^2=(1.5)^2+(1.25)^2+(1.1)^2=2.25+1.5625+1.21=5.0225$

2. $(1.1-1)^2+(2.08-2)^2+(2.95-3)^2=(0.1)^2+(0.08)^2+(-0.05)^2=0.01+0.0064+0.0025=0.0189$

3. $S=a^2+6b^2-4a-8b-4ab+6$

$$\frac{\partial S}{\partial a}=2a-4-4b$$

$$\frac{\partial S}{\partial b}=12b-8-4a$$

4. $S=4a^2+9b^2-6a-4b-2ab+8$

$$\frac{\partial S}{\partial a}=8a-6-2b$$

$$\frac{\partial S}{\partial b}=18b-4-2a$$

5. $\sum_{i=1}^{5} i=1+2+3+4+5=15$

6. $\sum_{i=1}^{6} 2i=2(1)+2(2)+2(3)+2(4)+2(5)+2(6)$

$=42$

7. $\sum_{i=1}^{4}\frac{1}{i}=\frac{1}{1}+\frac{1}{2}+\frac{1}{3}+\frac{1}{4}=\frac{25}{12}$

8. $\sum_{i=1}^{3} i^2=1^2+2^2+3^2=14$

9. $\sum_{i=1}^{6}(2-i)^2=(2-1)^2+(2-2)^2+(2-3)^2+(2-4)^2+(2-5)^2+(2-6)^2$

$=1^2+0^2+(-1)^2+(-2)^2+(-3)^2+(-4)^2=31$

10. $\sum_{i=1}^{5}(30-i^2)=(30-1)+(30-(2^2))+(30-(3^2))+(30-(4^2))+(30-(5^2))$

$=29+26+21+14+5=95$

1.

x-values	-3	-2	-1	0	1
Actual y-values	2	2	4	6	8
Linear model, $f(x)=1.6x+6$	1.2	2.8	4.4	6	7.6
Quadratic model, $g(x)=0.29x^2+2.2x+6$	2.01	2.76	4.09	6	8.49

Linear model sum of the squared errors:

$S=(1.2-2)^2+(2.8-2)^2+(4.4-4)^2+(6-6)^2+(7.6-8)^2=1.6$

Quadratic model sum of the squared errors:

$S=(2.01-2)^2+(2.76-2)^2+(4.09-4)^2+(6-6)^2+(8.49-8)^2=0.8259$

3.

x-values	0	1	2	3
Actual y-values	10	9	6	0
Linear model, $f(x) = -3.3x + 11$	11	7.7	4.4	1.1
Quadratic model, $g(x) = -1.25x^2 + 0.5x + 10$	10	9.25	6	0.25

Linear model sum of the squared errors:

$$S = (11 - 10)^2 + (7.7 - 9)^2 + (4.4 - 6)^2 + (1.1 - 0)^2 = 6.46$$

Quadratic model sum of the squared errors:

$$S = (10 - 10)^2 + (9.25 - 9)^2 + (6 - 6)^2 + (0.25 - 0)^2 = 0.125$$

5. The sum of the squared errors is as follows.

$$S = (-2a + b + 1)^2 + (0a + b)^2 + (2a + b - 3)^2$$

$$\frac{\partial S}{\partial a} = 2(-2a + b + 1)(-2) + 2(2a + b - 3)(2) = 16a - 16$$

$$\frac{\partial S}{\partial b} = 2(-2a + b + 1) + 2b + 2(2a + b - 3) = 6b - 4$$

Setting these partial derivatives equal to zero produces $a = 1$ and $b = \frac{2}{3}$. So, $y = x + \frac{2}{3}$.

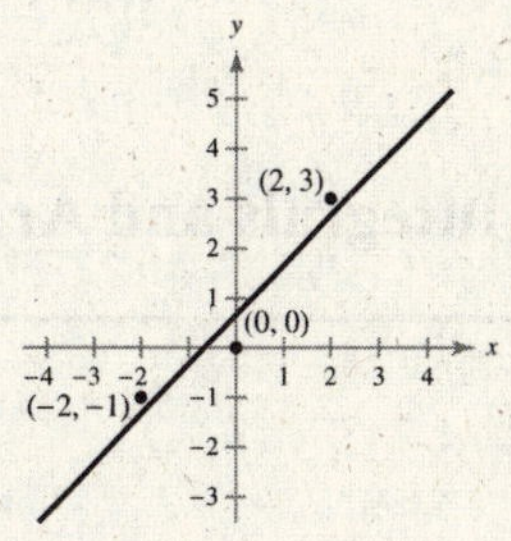

7. The sum of the squared errors is as follows.

$$S = (-2a + b - 4)^2 + (-a + b - 1)^2 + (b + 1)^2 + (a + b + 3)^2$$

$$\frac{\partial S}{\partial a} = -4(-2a + b - 4) - 2(-a + b - 1) + 2(a + b + 3) = 12a - 4b + 24$$

$$\frac{\partial S}{\partial b} = 2(-2a + b - 4) + 2(-a + b - 1) + 2(b + 1) + 2(a + b + 3) = -4a + 8b - 2$$

Setting these partial derivatives equal to zero produces:

$$12a - 4b = -24$$

$$-4a + 8b = 2$$

So, $a = -2.3$ and $b = -0.9$, and $y = -2.3x - 0.9$.

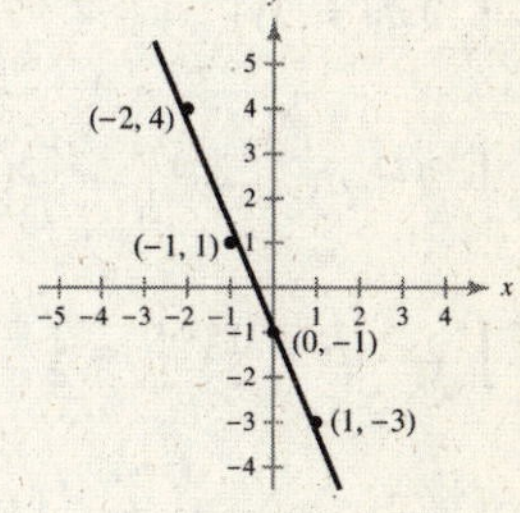

9. $y = 0.8x + 2$

11. $y = -1.1824x + 6.3851$

13. (a) $y = 4.13t + 11.6$, $t = 2$ is 2002.

(b) 2014: Let $t = 14$.

$y = 4.13(14) + 11.6 = 69.42$ or \$69.42 billion

(c) Let $y = 85$ and find t.

$85 = 4.13t + 11.6$

$73.4 = 4.13t$

$t \approx 17.77$ or 2018

15. (a) $y = 0.138x + 22.1$

(b) Let $x = 160$.

$y = 0.138(160) + 22.1 = 44.18$ bushels/acre

17. Positive correlation

$r \approx 0.9981$

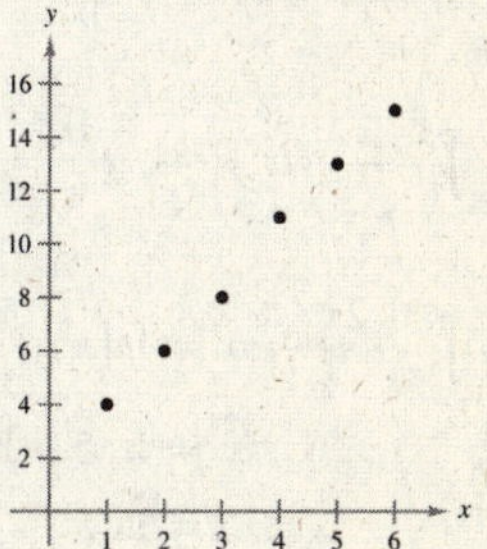

19. No correlation

$r = 0$

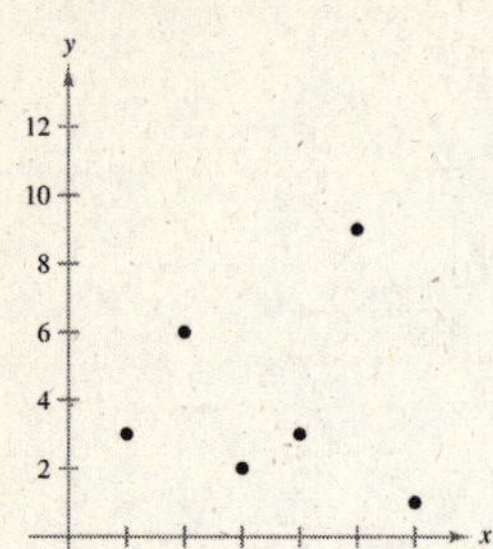

21. No linear correlation

$r \approx 0.0750$

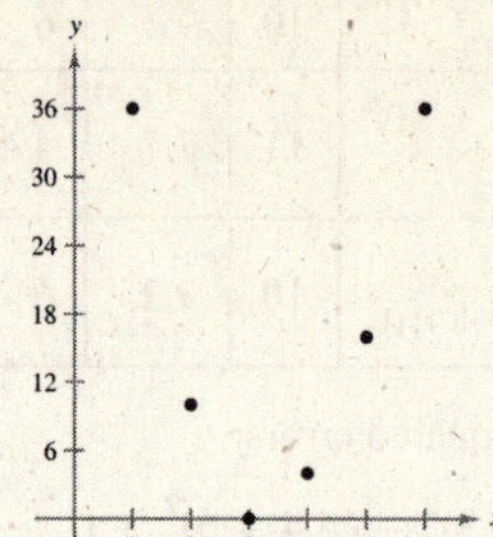

23. False, the slope is positive, which means there is a positive correlation.

25. True

27. True

29. Answers will vary.

Section 7.8 Double Integrals and Area in the Plane

Skills Warm Up

1. $\int_0^1 dx = x\Big]_0^1 = 1$

2. $\int_0^2 3\,dy = 3y\Big]_0^2 = 6$

3. $\int_1^4 2x^2\,dx = \frac{2}{3}x^3\Big]_1^4 = \left[\frac{2}{3}(4^3) - \frac{2}{3}(1^3)\right] = 42$

4. $\int_0^1 2x^3\,dx = \frac{1}{2}x^4\Big]_0^1 = \frac{1}{2}$

5. $\int_1^2 (x^3 - 2x + 4)\,dx = \left[\frac{1}{4}x^4 - x^2 + 4x\right]_1^2$

$= (4 - 4 + 8) - \left(\frac{1}{4} - 1 + 4\right)$

$= \frac{19}{4}$

6. $\int_0^2 (4 - y^2)\,dy = \left[4y - \frac{1}{3}y^3\right]_0^2 = \frac{16}{3}$

7. $\int_1^2 \frac{2}{7x^2}\,dx = -\frac{2}{7x}\Big]_1^2 = -\frac{2}{14} + \frac{2}{7} = \frac{1}{7}$

8. $\int_1^4 \frac{2}{\sqrt{x}}\,dx = 4\sqrt{x}\Big]_1^4 = 8 - 4 = 4$

9. $\int_0^2 \frac{2x}{x^2 + 1}\,dx = \ln(x^2 + 1)\Big]_0^2$

$= \ln 5 - \ln 1$

$= \ln 5$

≈ 1.609

10. $\int_2^e \frac{1}{y - 1}\,dy = \ln(y - 1)\Big]_2^e$

$= \ln(e - 1) - \ln(2 - 1)$

$= \ln(e - 1) \approx 0.541$

11. $\int_0^2 xe^{x^2+1}\,dx = \frac{1}{2}e^{x^2+1}\Big]_0^2 = \frac{1}{2}e^5 - \frac{1}{2}e \approx 72.847$

12. $\int_0^1 e^{-2y}\,dy = -\frac{1}{2}e^{-2y}\Big]_0^1 = -\frac{1}{2}e^{-2} + \frac{1}{2} \approx 0.432$

13. $y = x,\ y = 0,\ x = 3$

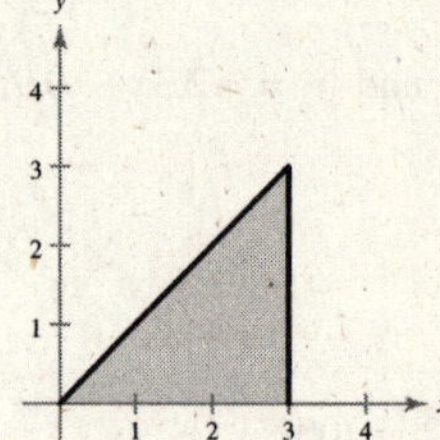

14. $y = x,\ y = 3,\ x = 0$

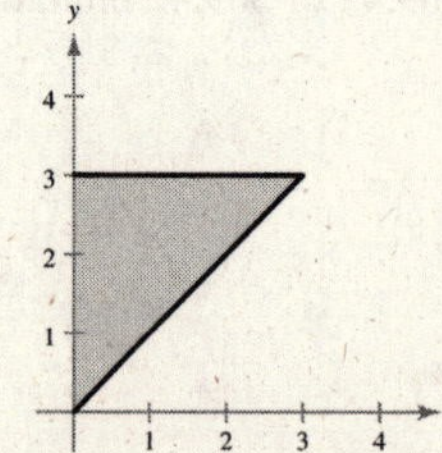

Skills Warm Up —continued—

15. $y = 4 - x^2,\ y = 0,\ x = 0$

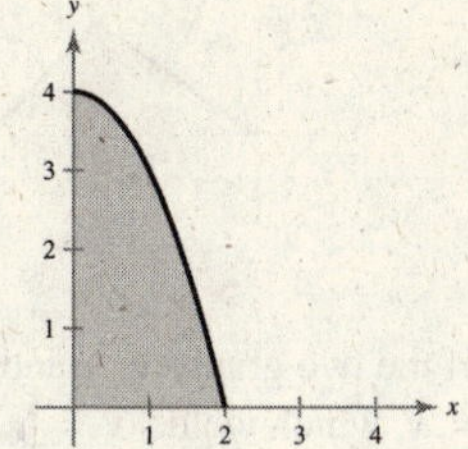

16. $y = x^2,\ y = 4x$

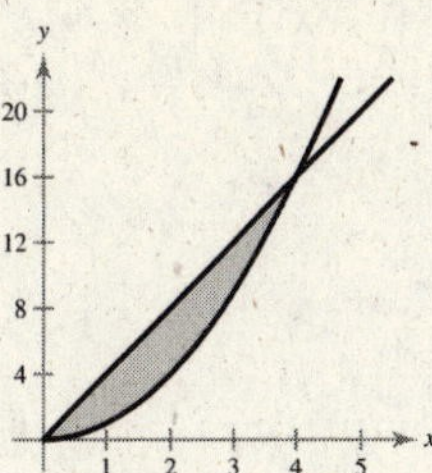

1. $\int_0^x (2x - y)\,dy = \left[2xy - \frac{y^2}{2}\right]_0^x = \frac{3x^2}{2}$

3. $\int_x^{x^2} \frac{y}{x}\,dy = \left.\frac{y^2}{2x}\right]_x^{x^2} = \frac{x^3}{2} - \frac{x}{2} = \frac{x}{2}(x^2 - 1)$

5. $\int_2^y (6x^2y + y^2)\,dx = \left[2x^3y + xy^2\right]_2^y$

$= (2y^4 + y^3) - (16y + 2y^2)$

$= 2y^4 + y^3 - 2y^2 - 16y$

7. $\int_{x^3}^{\sqrt{x}} (x^2 + 3y^2)\,dy = \left[x^2y + y^3\right]_{x^3}^{\sqrt{x}}$

$= (x^{5/2} + x^{3/2}) - (x^5 + x^9)$

$= -x^9 - x^5 + x^{5/2} + x^{3/2}$

9. $\int_1^{e^y} \frac{y \ln x}{x}\,dx = \left.\frac{y(\ln x)^2}{2}\right]_1^{e^y}$

$= \frac{y(\ln e^y)^2}{2} - \frac{y(\ln 1)^2}{2}$

$= \frac{y^3}{2}$

11. $\int_0^1 \int_0^2 (x + y)\,dy\,dx = \int_0^1 \left[xy + \frac{y^2}{2}\right]_0^2 dx$

$= \int_0^1 (2x + 2)\,dx$

$= \left[x^2 + 2x\right]_0^1$

$= 3$

13. $\int_0^3 \int_0^4 xy\,dx\,dy = \int_0^3 \left[\frac{1}{2}x^2y\right]_0^4 dy$

$= \int_0^3 8y\,dy$

$= \left[4y^2\right]_0^3$

$= 36$

15. $\int_0^2 \int_0^{6x^2} x^3\,dy\,dx = \int_0^2 \left.x^3y\right]_0^{6x^2} dx$

$= \int_0^2 6x^5\,dx$

$= \left.x^6\right]_0^2$

$= 64$

17. $\int_0^1 \int_0^y (x + y)\,dx\,dy = \int_0^1 \left[\frac{x^2}{2} + xy\right]_0^y dy = \int_0^1 \frac{3y^2}{2}\,dy = \left[\frac{y^3}{2}\right]_0^1 = \frac{1}{2}$

19. $\int_0^1 \int_0^{3x} (3x^2 + 3y^2 + 1)\,dy\,dx = \int_0^1 \left[3x^2y + y^3 + y\right]_0^{3x} dx$

$= \int_0^1 (9x^3 + 27x^3 + 3x)\,dx$

$= \int_0^1 (36x^3 + 3x)\,dx$

$= \left[9x^4 + \frac{3}{2}x^2\right]_0^1 = \frac{21}{2}$

21. $\int_0^1 \int_0^x \sqrt{1 - x^2}\, dy\, dx = \int_0^1 \sqrt{1 - x^2}\, y \Big]_0^x dx$

$= -\frac{1}{2}\int_0^1 \sqrt{1 - x^2}(-2x)\, dx$

$= -\frac{1}{3}(1 - x^2)^{3/2}\Big]_0^1$

$= \frac{1}{3}$

23. Because (for a fixed x)

$$\lim_{b\to\infty} -2e^{-(x+y)/2}\Big]_0^b = 2e^{-x/2},$$

you have

$$\int_0^\infty \int_0^\infty e^{-(x+y)/2}\, dy\, dx = \int_0^\infty 2e^{-x/2}\, dx$$

$$= \lim_{b\to\infty} -4e^{-x/2}\Big]_0^b = 4.$$

25. $A = \int_0^8 \int_0^3 dy\, dx$

$= \int_0^8 [y]_0^3\, dx$

$= \int_0^8 3\, dx$

$= [3x]_0^8$

$= 24$

27. $A = \int_0^4 \int_0^x dy\, dx$

$= \int_0^4 [y]_0^x\, dx$

$= \int_0^4 x\, dx$

$= \left[\frac{1}{2}x^2\right]_0^4$

$= 8$

29. $A = \int_0^2 \int_0^{4-x^2} dy\, dx$

$= \int_0^2 (4 - x^2)\, dx$

$= \left[4x - \frac{x^3}{3}\right]_0^2$

$= \frac{16}{3}$

31. $A = \int_{-3}^3 \int_0^{9-x^2} dy\, dx$

$= \int_{-3}^3 (9 - x^2)\, dx$

$= \left[9x - \frac{x^3}{3}\right]_{-3}^3$

$= 36$

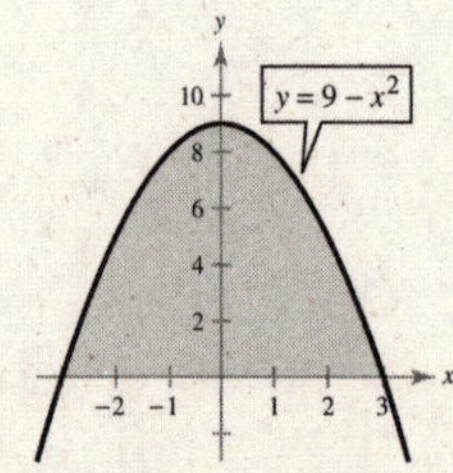

33. $A = \int_0^2 \int_{\frac{3}{2}y}^{-y+5} dx\, dy$

$= \int_0^2 \left(-y + 5 - \frac{3}{2}y\right) dy$

$= \int_0^2 \left(-\frac{5}{2}y + 5\right) dy$

$= \left[-\frac{5y^2}{4} + 5y\right]_0^2$

$= 5$

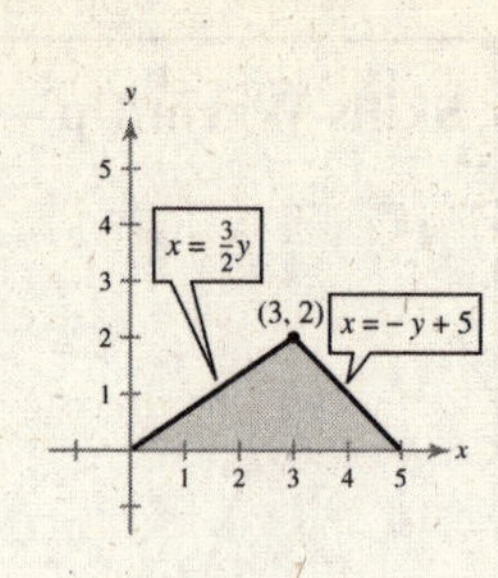

35. The point of intersection of the two graphs is found by equating $y = 2x$ and $y = x$, which yields $x = y = 0$.

$A = \int_0^2 \int_x^{2x} dy\, dx$

$= \int_0^2 (2x - x)\, dx$

$= \int_0^2 x\, dx$

$= \frac{x^2}{2}\Big]_0^2 = 2$

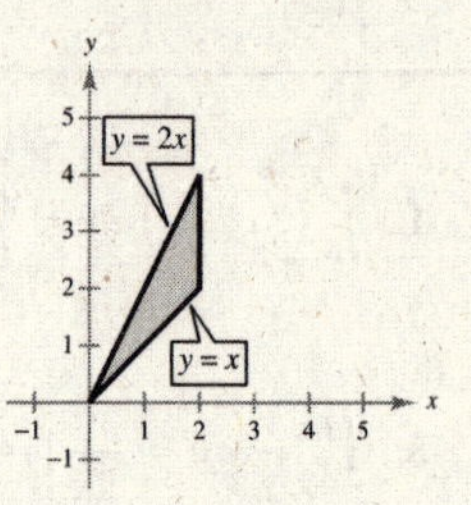

37. $\int_0^1 \int_0^2 dy\, dx = \int_0^1 2\, dx = 2$

$\int_0^2 \int_0^1 dx\, dy = \int_0^2 dy = 2$

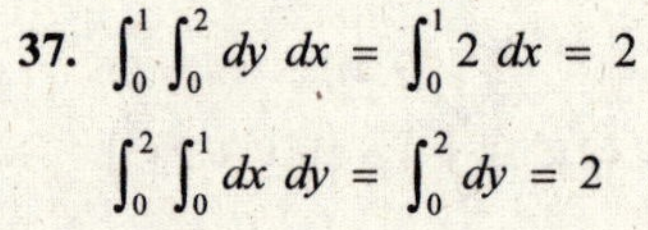

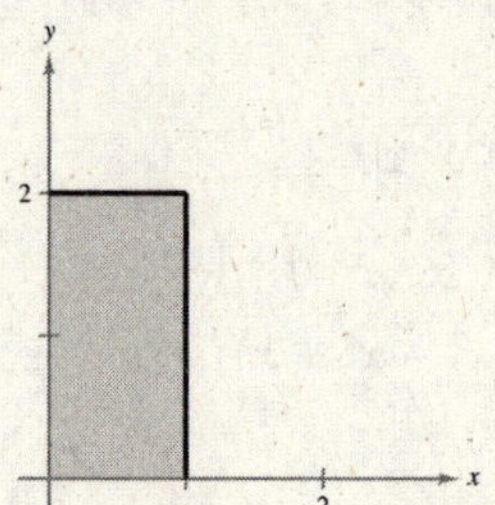

39. $\int_0^1 \int_{2y}^2 dx\, dy = \int_0^1 (2 - 2y)\, dy = \left[2y - y^2\right]_0^1 = 1$

$\int_0^2 \int_0^{x/2} dy\, dx = \int_0^2 \frac{x}{2}\, dx = \frac{x^2}{4}\Big]_0^2 = 1$

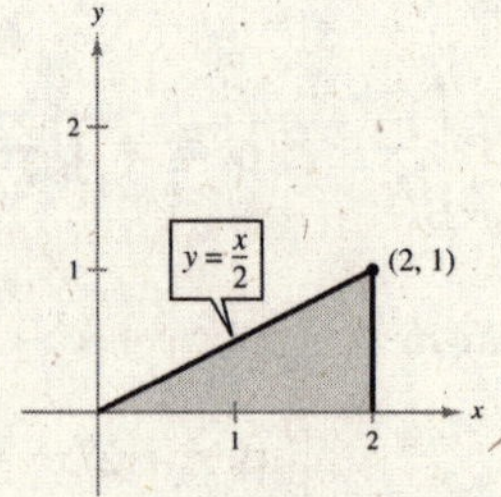

41. $\int_0^2 \int_{x/2}^1 dy\, dx = \int_0^2 \left(1 - \frac{x}{2}\right) dx = \left[x - \frac{x^2}{4}\right]_0^2 = 1$

$\int_0^1 \int_0^{2y} dx\, dy = \int_0^1 2y\, dy = y^2\Big]_0^1 = 1$

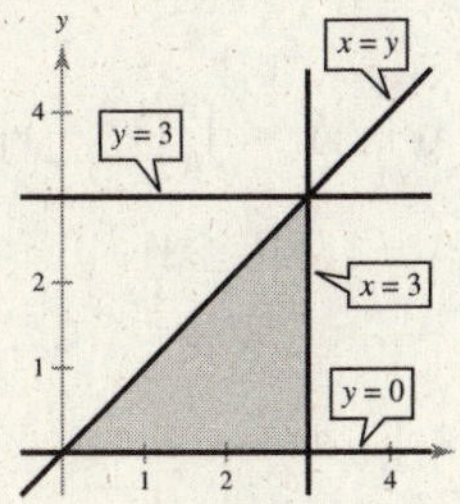

43. $\int_0^1 \int_{y^2}^{\sqrt[3]{y}} dx\, dy = \int_0^1 \left(\sqrt[3]{y} - y^2\right) dy$

$= \left[\frac{3}{4}y^{4/3} - \frac{y^3}{3}\right]_0^1$

$= \frac{5}{12}$

$\int_0^1 \int_{x^3}^{\sqrt{x}} dy\, dx = \int_0^1 \left(\sqrt{x} - x^3\right) dx$

$= \left[\frac{2}{3}x^{3/2} - \frac{x^4}{4}\right]_0^1$

$= \frac{5}{12}$

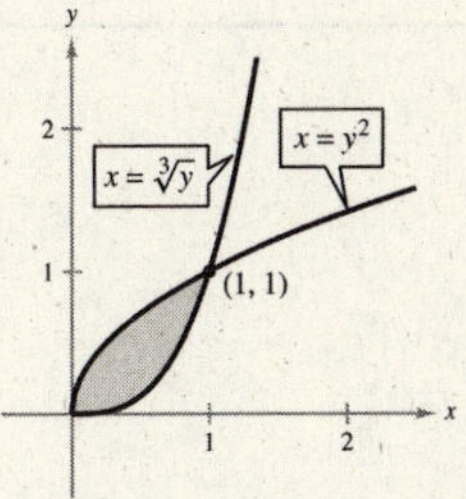

45. (a) $\int_0^3 \int_y^3 e^{x^2}\, dx\, dy$ cannot be evaluated in the order as given since no antiderivative for e^{x^2} can be found.

The region bounded by $y \le x \le 3$ and $0 \le y \le 3$

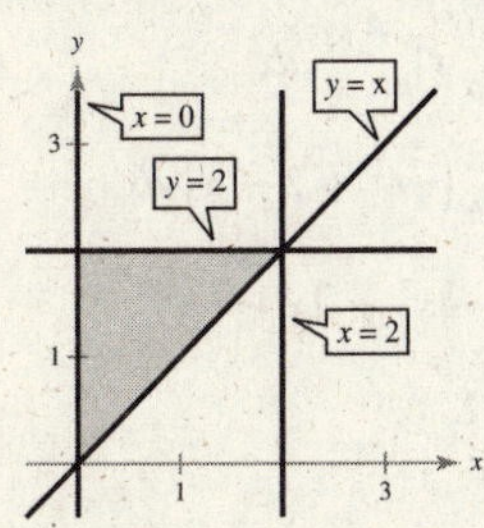

is rewritten as $0 \le y \le x$ and $0 \le x \le 3$ so that the integral can be changed to $\int_0^3 \int_0^x e^{x^2}\, dy\, dx.$

$\int_0^3 \int_0^x e^{x^2}\, dy\, dx = \int_0^3 \left[e^{x^2} y\right]_0^x dx = \int_0^3 xe^{x^2}\, dx = \left[\frac{1}{2}e^{x^2}\right]_0^3 = \frac{1}{2}\left(e^9 - 1\right) \approx 4051.042$

(b) $\int_0^2 \int_x^2 e^{-y^2}\, dy\, dx$ cannot be evaluated in the order as given since no antiderivative for e^{-y^2} can be found.

The region bounded by $x \le y \le 2$ and $0 \le x \le 2$

x = 0
y = x
y = 2
x = 2

is rewritten as $0 \le x \le y$ and $0 \le y \le 2$ so that the integral can be changed to $\int_0^2 \int_0^y e^{-y^2}\, dx\, dy.$

$\int_0^2 \int_0^y e^{-y^2}\, dx\, dy = \int_0^2 \left[xe^{-y^2} y\right]_0^y dy = \int_0^2 ye^{-y^2}\, dy = \left[-\frac{1}{2}e^{-y^2}\right]_0^2 = -\frac{1}{2}\left(e^{-4} - 1\right) \approx 0.491$

47. $\int_0^1 \int_0^2 e^{-x^2-y^2}\,dx\,dy \approx 0.6588$

49. $\int_1^2 \int_0^x e^{xy}\,dy\,dx \approx 8.1747$

51. $\int_0^1 \int_x^1 \sqrt{1-x^2}\,dy\,dx \approx 0.4521$

53. $\int_0^2 \int_{\sqrt{4-x^2}}^{4-x^2/4} \dfrac{xy}{x^2+y^2+1}\,dy\,dx \approx 1.1190$

55. True

$$\int_{-1}^1 \int_{-2}^2 y\,dy\,dx = \int_{-1}^1 \frac{y^2}{2}\Big]_{-2}^2\,dx = \int_{-1}^1 0\,dx = 0$$

$$\int_{-1}^1 \int_{-2}^2 y\,dx\,dy = \int_{-1}^1 xy\Big]_{-2}^2\,dy = \int_{-1}^1 4y\,dy = 2y^2\Big]_{-1}^1 = 0$$

Section 7.9 Applications of Double Integrals

Skills Warm Up

1.

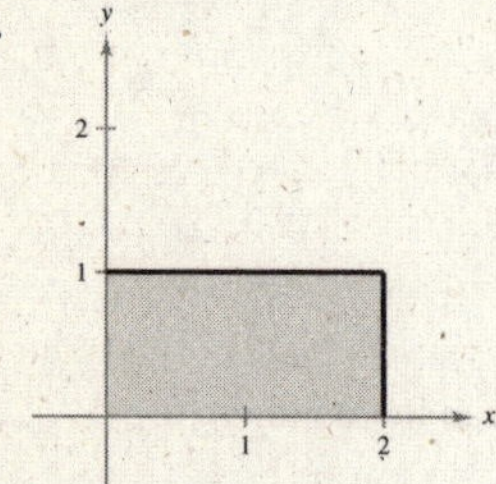

2.

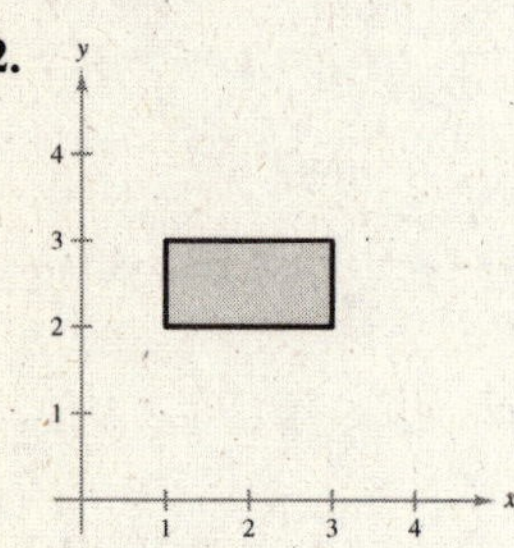

3.

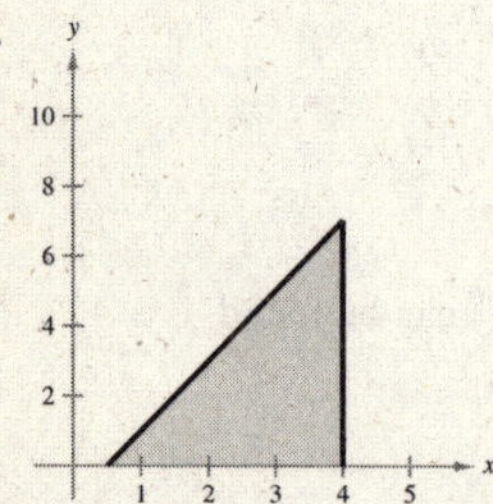

4.

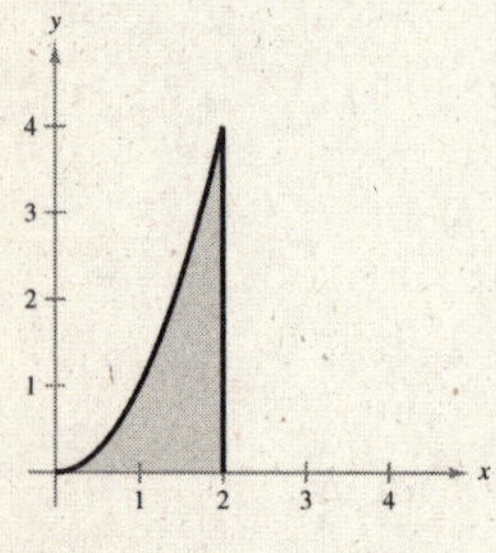

5. $\int_0^1 \int_1^2 dy\,dx = \int_0^1 y\Big]_1^2\,dx = \int_0^1 dx = x\Big]_0^1 = 1$

6. $\int_0^3 \int_1^3 dx\,dy = \int_0^3 x\Big]_1^3\,dy = \int_0^3 2\,dy = 2y\Big]_0^3 = 6$

7.
$$\begin{aligned}\int_0^1 \int_0^x x\,dy\,dx &= \int_0^1 xy\Big]_0^x\,dx \\ &= \int_0^1 x^2\,dx = \tfrac{1}{3}x^3\Big]_0^1 = \tfrac{1}{3}\end{aligned}$$

8.
$$\begin{aligned}\int_0^4 \int_1^y y\,dx\,dy &= \int_0^4 xy\Big]_1^y\,dy = \int_0^4 (y^2 - y)\,dy \\ &= \left[\tfrac{1}{3}y^3 - \tfrac{1}{2}y^2\right]_0^4 = \tfrac{40}{3}\end{aligned}$$

9.
$$\begin{aligned}\int_1^3 \int_x^{x^2} 2\,dy\,dx &= \int_1^3 2y\Big]_x^{x^2}\,dx \\ &= \int_1^3 (2x^2 - 2x)\,dx \\ &= \left[\tfrac{2}{3}x^3 - x^2\right]_1^3 \\ &= 9 + \tfrac{1}{3} \\ &= \tfrac{28}{3}\end{aligned}$$

10.
$$\begin{aligned}\int_0^1 \int_x^{-x^2+2} dy\,dx &= \int_0^1 y\Big]_x^{-x^2+2}\,dx \\ &= \int_0^1 (-x^2 + 2 - x)\,dx \\ &= \left[-\tfrac{1}{3}x^3 + 2x - \tfrac{1}{2}x^2\right]_0^1 \\ &= \tfrac{7}{6}\end{aligned}$$

1. $\int_0^2 \int_0^1 (3x + 4y)\,dy\,dx = \int_0^2 \Big[3xy + 2y^2\Big]_0^1 dx$

$= \int_0^2 (3x + 2)\,dx$

$= \Big[\frac{3}{2}x^2 + 2x\Big]_0^2$

$= 10$

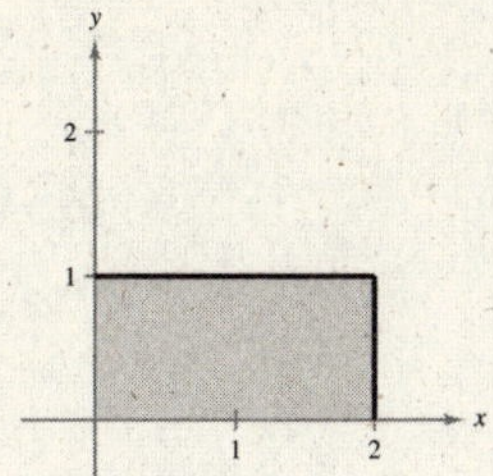

3. $\int_{-1}^1 \int_0^{\sqrt{1-x^2}} x^2 y\,dy\,dx = \int_{-1}^1 \Big[\frac{1}{2}x^2y^2\Big]_0^{\sqrt{1-x^2}} dx$

$= \frac{1}{2}\int_{-1}^1 x^2(1 - x^2)\,dx$

$= \frac{1}{2}\int_{-1}^1 (x^2 - x^4)\,dx$

$= \frac{1}{2}\Big[\frac{1}{3}x^3 - \frac{1}{5}x^5\Big]_{-1}^1$

$= \frac{2}{15}$

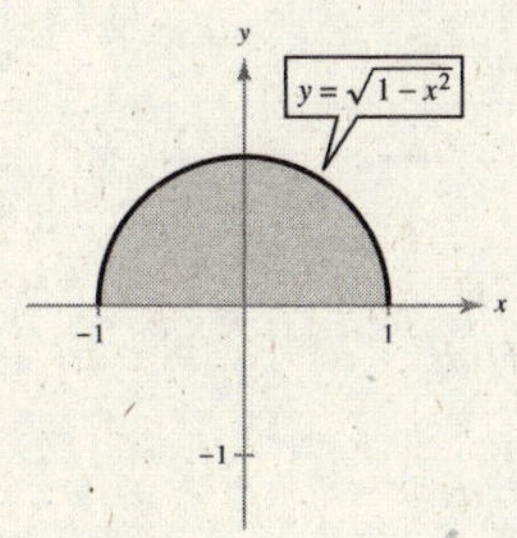

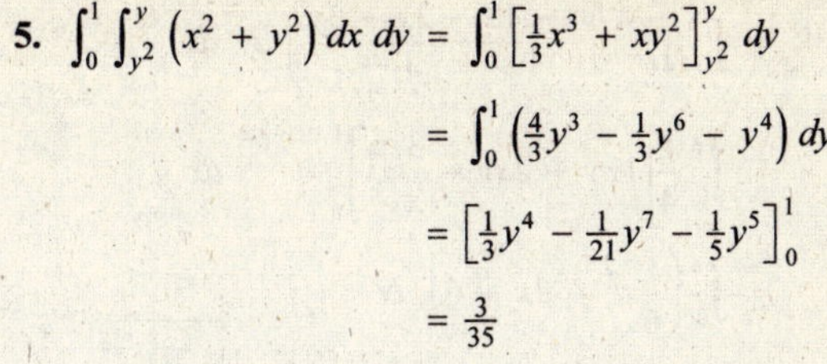

5. $\int_0^1 \int_{y^2}^y (x^2 + y^2)\,dx\,dy = \int_0^1 \Big[\frac{1}{3}x^3 + xy^2\Big]_{y^2}^y dy$

$= \int_0^1 \left(\frac{4}{3}y^3 - \frac{1}{3}y^6 - y^4\right) dy$

$= \Big[\frac{1}{3}y^4 - \frac{1}{21}y^7 - \frac{1}{5}y^5\Big]_0^1$

$= \frac{3}{35}$

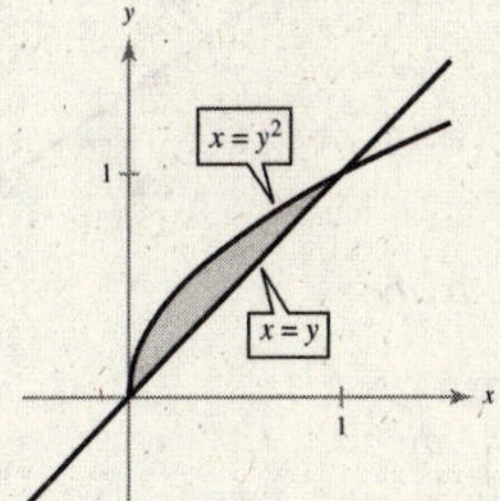

7. $\int_{-a}^a \int_{-\sqrt{a^2-x^2}}^{\sqrt{a^2-x^2}} dy\,dx = \int_{-a}^a y\Big]_{-\sqrt{a^2-x^2}}^{\sqrt{a^2-x^2}} dx$

$= \int_{-a}^a 2\sqrt{a^2 - x^2}\,dx$

$= \pi a^2$ (area of circle)

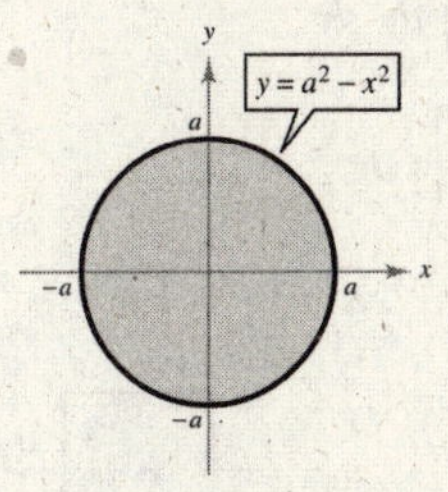

9. $\int_0^3 \int_0^5 xy\,dy\,dx = \int_0^5 \int_0^3 xy\,dx\,dy$

$\int_0^3 \int_0^5 xy\,dy\,dx = \int_0^3 \frac{xy^2}{2}\Big]_0^5 dx = \int_0^3 \frac{25}{2}x\,dx = \frac{25}{4}x^2\Big]_0^3 = \frac{225}{4}$

11. $\int_0^2 \int_x^{2x} \frac{y}{x^2 + y^2}\,dy\,dx = \int_0^2 \int_{y/2}^y \frac{y}{x^2 + y^2}\,dx\,dy + \int_2^4 \int_{y/2}^2 \frac{y}{x^2 + y^2}\,dx\,dy$

$\int_0^2 \int_x^{2x} \frac{y}{x^2 + y^2}\,dy\,dx = \int_0^2 \frac{1}{2}\ln(x^2 + y^2)\Big]_x^{2x} dx$

$= \frac{1}{2}\int_0^2 \Big[\ln(5x^2) - \ln(2x^2)\Big] dx = \frac{1}{2}\int_0^2 \ln\frac{5}{2}\,dx = \left(\frac{1}{2}\ln\frac{5}{2}\right)x\Big]_0^2 = \ln\frac{5}{2} \approx 0.916$

13. $V = \int_0^2 \int_0^4 \frac{y}{2}\,dx\,dy = \int_0^2 \frac{xy}{2}\Big]_0^4 dy = \int_0^2 2y\,dy = y^2\Big]_0^2 = 4$

15. $V = \int_0^6 \int_0^{(-2/3)x+4} \left(\dfrac{12 - 2x - 3y}{4}\right) dy\, dx$

$= \int_0^6 \dfrac{1}{4}\left[12y - 2xy - \dfrac{3}{2}y^2\right]_0^{(-2/3)x+4} dx$

$= \int_0^6 \left(\dfrac{1}{6}x^2 - 2x + 6\right) dx$

$= \left[\dfrac{1}{18}x^3 - x^2 + 6x\right]_0^6$

$= 12$

17. $V = \int_0^2 \int_0^y (4 - x - y)\, dx\, dy$

$= \int_0^2 \left[4x - \dfrac{x^2}{2} - xy\right]_0^y dy$

$= \int_0^2 \left(4y - \dfrac{3y^2}{2}\right) dy$

$= \left[2y^2 - \dfrac{y^3}{2}\right]_0^2 = 4$

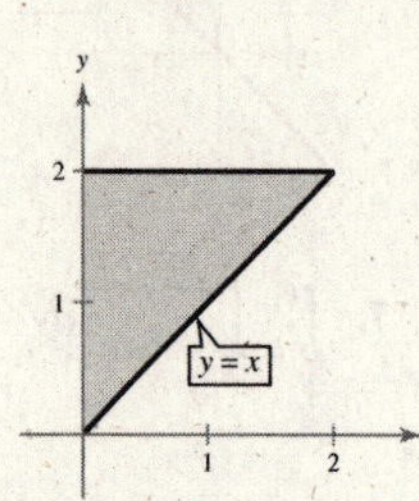

19. $V = 4\int_0^1 \int_0^1 \left(4 - x^2 - y^2\right) dy\, dx$

$= 4\int_0^1 \left[4y - x^2y - \dfrac{y^3}{3}\right]_0^1 dx$

$= 4\int_0^1 \left[4 - x^2 - \dfrac{1}{3}\right] dx$

$= 4\int_0^1 \left(\dfrac{11}{3} - x^2\right) dx$

$= 4\left[\dfrac{11x}{3} - \dfrac{x^3}{3}\right]_0^1 = \dfrac{40}{3}$

21. $V = \int_0^3 \int_0^{2x} (xy)\, dy\, dx$

$= \int_0^3 \left[\frac{1}{2}xy^2\right]_0^{2x} dx$

$= \frac{1}{2}\int_0^3 4x^3\, dx$

$= \frac{1}{2}\left[x^4\right]_0^3$

$= \frac{81}{2}$

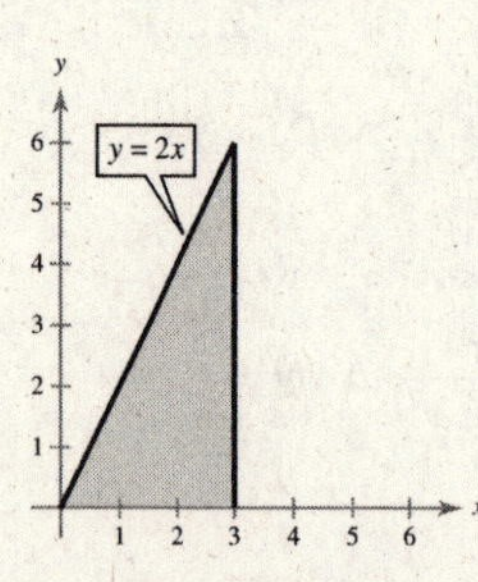

23. $V = \int_0^2 \int_0^{x+2} \left(9 - x^2\right) dy\, dx$

$= \int_0^2 \left[9y - x^2y\right]_0^{x+2} dx$

$= \int_0^2 \left(-x^3 - 2x^2 + 9x + 18\right) dx$

$= \left[-\frac{1}{4}x^4 - \frac{2}{3}x^3 + \frac{9}{2}x^2 + 18x\right]_0^2$

$= \frac{134}{3}$

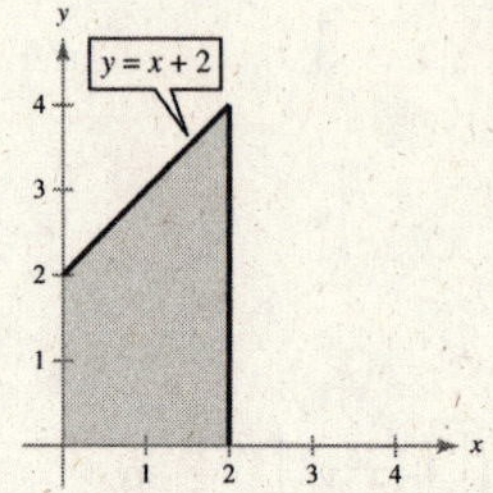

25. $P = \int_0^2 \int_0^2 \dfrac{120{,}000}{(2 + x + y)^3}\, dy\, dx$

$= \int_0^2 -60{,}000(2 + x + y)^{-2}\Big]_0^2 dx$

$= -60{,}000\int_0^2 \left(\dfrac{1}{(4 + x)^2} - \dfrac{1}{(2 + x)^2}\right) dx$

$= 60{,}000\left[\dfrac{1}{4 + x} - \dfrac{1}{2 + x}\right]_0^2$

$= 10{,}000$ people

27. $\text{Average} = \frac{1}{15}\int_0^5 \int_0^3 y\, dy\, dx$

$= \frac{1}{15}\int_0^5 \left[\frac{1}{2}y^2\right]_0^3 dx$

$= \frac{1}{30}\int_0^5 9\, dx$

$= \frac{1}{30}\left[9x\right]_0^5$

$= \frac{3}{2}$

29. $\text{Average} = \dfrac{1}{4}\int_0^2 \int_0^2 \left(x^2 + y^2\right) dx\, dy$

$= \dfrac{1}{4}\int_0^2 \left[\dfrac{x^3}{3} + xy^2\right]_0^2 dy$

$= \dfrac{1}{4}\int_0^2 \left(\dfrac{8}{3} + 2y^2\right) dy$

$= \dfrac{1}{4}\left[\dfrac{8}{3}y + \dfrac{2}{3}y^3\right]_0^2$

$= \dfrac{8}{3}$

31. $$\text{Average} = \frac{1}{50}\int_{45}^{50}\int_{40}^{50}\left(192x_1 + 576x_2 - x_1^2 - 5x_2^2 - 2x_1x_2 - 5000\right)dx_1\,dx_2$$

$$= \frac{1}{50}\int_{45}^{50}\left[96x_1^2 + 576x_1x_2 - \frac{x_1^3}{3} - 5x_1x_2^2 - x_1^2x_2 - 5000x_1\right]_{40}^{50}dx_2$$

$$= \frac{1}{50}\int_{45}^{50}\left(\frac{48,200}{3} + 4860x_2 - 50x_2^2\right)dx_2$$

$$= \frac{1}{50}\left[\frac{48,200x}{3} + 2430x_2^2 - \frac{50x_2^3}{3}\right]_{45}^{50} = \$13,400$$

33. $$\text{Average} = \frac{1}{1250}\int_{100}^{150}\int_{50}^{75}\left[(500 - 3p_1)p_1 + (750 - 2.4p_2)p_2\right]dp_1\,dp_2$$

$$= \frac{1}{1250}\int_{100}^{150}\int_{50}^{75}\left[-3p_1^2 + 500p_1 - 2.4p_2^2 + 750p_2\right]dp_1\,dp_2$$

$$= \frac{1}{1250}\int_{100}^{150}\left[-p_1^3 + 250p_1^2 - 2.4p_1p_2^2 + 750p_1p_2\right]_{50}^{75}dp_2$$

$$= \frac{1}{1250}\int_{100}^{150}\left[484,375 - 60p_2^2 + 18,750p_2\right]dp_2$$

$$= \frac{1}{1250}\left[484,375p_2 - 20p_2^3 + 9375p_2^2\right]_{100}^{150} = \$75,125$$

35. $$\text{Average} = \frac{1}{1250}\int_{300}^{325}\int_{200}^{250}100x^{0.6}y^{0.4}\,dx\,dy = \frac{1}{1250}\int_{300}^{325}\left(100y^{0.4}\right)\frac{x^{1.6}}{1.6}\Bigg]_{200}^{250}dy$$

$$= \frac{128,844.1}{1250}\int_{300}^{325}y^{0.4}\,dy = 103.0753\left[\frac{y^{1.4}}{1.4}\right]_{300}^{325} \approx 25,645.24$$

Review Exercises for Chapter 7

1.

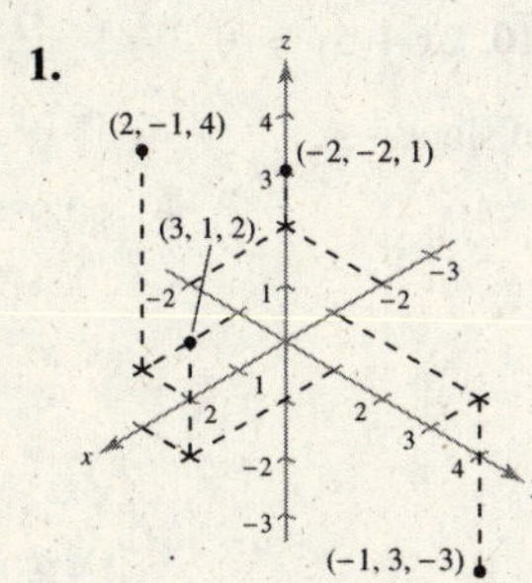

3. $$d = \sqrt{(3-1)^2 + (5-0)^2 + (8-2)^2} = \sqrt{4 + 25 + 36} = \sqrt{65}$$

5. $$\text{Midpoint} = \left(\frac{2 + (-4)}{2}, \frac{6+2}{2}, \frac{4+8}{2}\right) = (-1, 4, 6)$$

7. $$(x-0)^2 + (y-1)^2 + (z-0)^2 = 5^2$$
$$x^2 + (y-1)^2 + z^2 = 25$$

9. $$\text{Center} = \left(\frac{3+1}{2}, \frac{-4+0}{2}, \frac{-1+(-5)}{2}\right) = (2, -2, -3)$$

$$\text{Radius} = \sqrt{(2-3)^2 + (-2+4)^2 + (-3+1)^2} = \sqrt{1+4+4} = 3$$

$$\text{Sphere: } (x-2)^2 + (y+2)^2 + (z+3)^2 = 9$$

11. $$x^2 + y^2 + z^2 - 8x + 4y - 6z - 20 = 0$$
$$\left(x^2 - 8x + 16\right) + \left(y^2 + 4y + 4\right) + \left(z^2 - 6z + 9\right) = 20 + 16 + 4 + 9$$
$$(x-4)^2 + (y+2)^2 + (z-3)^2 = 49$$

Center: $(4, -2, 3)$

Radius: 7

13. Let $z = 0$.

$$(x + 2)^2 + (y - 1)^2 + (0 - 3)^2 = 25$$

$$(x + 2)^2 + (y - 1)^2 = 16$$

Circle of radius 4

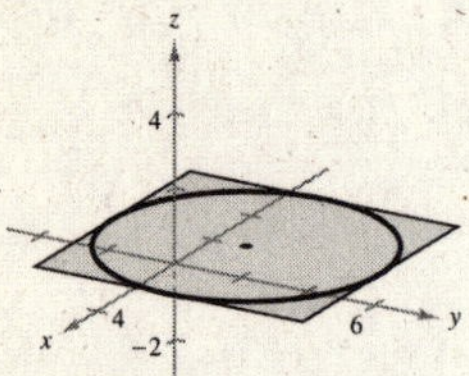

15. $x + 2y + 3z = 6$

To find the x-intercept, let $y = 0$ and $z = 0$.

$x = 6$

To find the y-intercept, let $x = 0$ and $z = 0$.

$2y = 6 \Rightarrow y = 3$

To find the z-intercept, let $x = 0$ and $y = 0$.

$3z = 6 \Rightarrow z = 2$

x-intercept: (6, 0, 0)

y-intercept: (0, 3, 0)

z-intercept: (0, 0, 2)

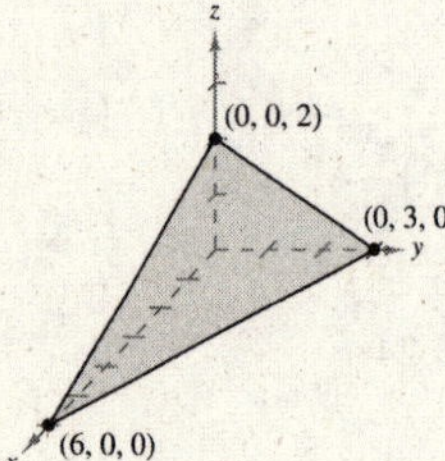

17. $3x - 6z = 12$

To find the x-intercept, let $z = 0$.

$3x = 12 \Rightarrow x = 4$

Because the coefficient of y is zero, there is no y-intercept.

To find the z-intercept, let $x = 0$.

$-6z = 12 \Rightarrow z = -2$

x-intercept: (4, 0, 0)

z-intercept: (0, 0, −2)

The plane is parallel to the y-axis.

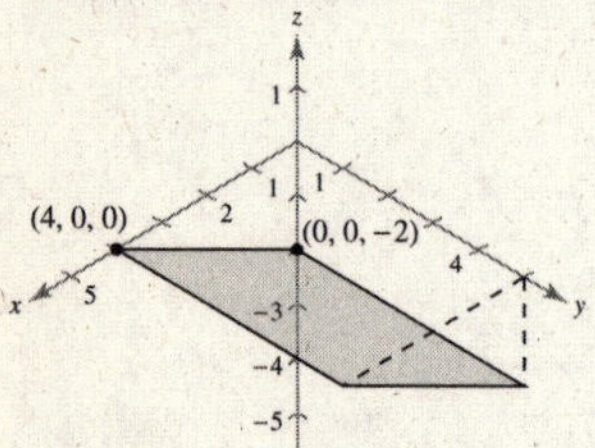

19. The graph of $x^2 + y^2 + z^2 - 2x + 4y - 6z + 5 = 0$ is a sphere whose standard equation is

$$(x - 1)^2 + (y + 2)^2 + (z - 3)^2 = 9.$$

21. The graph of $x^2 + \frac{y^2}{16} + \frac{z^2}{9} = 1$ is an ellipsoid.

23. The graph of $z = \frac{x^2}{9} + y^2$ is an elliptic paraboloid.

25. The graph of $z = \sqrt{x^2 + y^2}$ is the top half of a circular cone whose standard equation is $x^2 + y^2 - z^2 = 0$.

27. $f(x, y) = xy^2$

(a) $f(2, 3) = 2(3)^2 = 18$

(b) $f(0, 1) = 0(1)^2 = 0$

(c) $f(-5, 7) = -5(7)^2 = -245$

(d) $f(-2, -4) = -2(-4)^2 = -32$

29. The domain of $f(x, y) = \sqrt{1 - x^2 - y^2}$ is the set of all points inside or on the circle $x^2 + y^2 = 1$. The range is $[0, 1]$.

31. The domain of $f(x, y) = e^{xy}$ is the set of all points in the xy-plane. The range is $(0, \infty)$.

33. $z = 10 - 2x - 5y$

$c = 0: 10 - 2x - 5y = 0, 2x + 5y = 10$

$c = 2: 10 - 2x - 5y = 2, 2x + 5y = 8$

$c = 4: 10 - 2x - 5y = 4, 2x + 5y = 6$

$c = 5: 10 - 2x - 5y = 5, 2x + 5y = 5$

$c = 10: 10 - 2x - 5y = 10, 2x + 5y = 0$

The level curves are lines of slope $-\frac{2}{5}$.

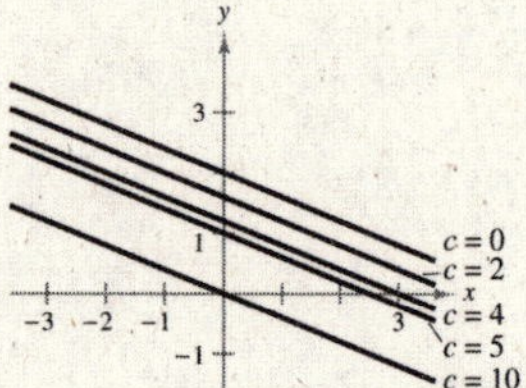

35. $z = (xy)^2$

$c = 1: (xy)^2 = 1, \ y = \pm\frac{1}{x}$

$c = 4: (xy)^2 = 4, \ y = \pm\frac{2}{x}$

$c = 9: (xy)^2 = 9, \ y = \pm\frac{3}{x}$

$c = 12: (xy)^2 = 12, \ y = \pm\frac{2\sqrt{3}}{x}$

$c = 16: (xy)^2 = 16, \ y = \pm\frac{4}{x}$

The level curves are hyperbolas.

37. (a) No; the precipitation increments are 7.99 inches, 9.99 inches, 9.99 inches, 9.99 inches, and 19.99 inches.

(b) You could increase the number of level curves to correspond to smaller increments of precipitation.

39. $z = -4.51 + 0.046x + 0.060y$

(a) $z(100, 40) = -4.51 + 0.046(100) + 0.060(40)$

$= \$2.49$

(b) Because $\dfrac{\partial z}{\partial y} = 0.060 > \dfrac{\partial z}{\partial x} = 0.46$, y has the greater influence on earnings per share.

41. $f(x, y) = x^2y + 3xy + 2x - 5y$

$f_x = 2xy + 3y + 2$

$f_y = x^2 + 3x - 5$

43. $z = \dfrac{x^2}{y^2}$

$\dfrac{\partial z}{\partial x} = \dfrac{2x}{y^2}$

$\dfrac{\partial z}{\partial y} = -\dfrac{2x^2}{y^3}$

45. $f(x, y) = \ln(5x + 4y)$

$f_x(x, y) = \dfrac{5}{5x + 4y}$

$f_y(x, y) = \dfrac{4}{5x + 4y}$

47. $f(x, y) = xe^y + ye^x$

$f_x = ye^x + e^y$

$f_y = xe^y + e^x$

49. $w = xyz^2$

$\dfrac{\partial w}{\partial x} = yz^2$

$\dfrac{\partial w}{\partial y} = xz^2$

$\dfrac{\partial w}{\partial z} = 2xyz$

51. $z = 3xy$

(a) $\dfrac{\partial z}{\partial x} = 3y;\ \dfrac{\partial z}{\partial x}(-2, -3, 18) = -9$

(b) $\dfrac{\partial z}{\partial y} = 3x;\ \dfrac{\partial z}{\partial y}(-2, -3, 18) = -6$

53. $z = 8 - x^2 - y^2$

(a) $\dfrac{\partial z}{\partial x} = -2x;\ \dfrac{\partial z}{\partial x}(1, 1, 6) = -2$

(b) $\dfrac{\partial z}{\partial y} = -2y;\ \dfrac{\partial z}{\partial y}(1, 1, 6) = -2$

55. $f(x, y) = 3x^2 - xy + 2y^3$

$f_x = 6x - y \qquad f_y = -x + 6y^2$

$f_{xx} = 6 \qquad f_{yy} = 12y \qquad f_{xy} = f_{yx} = -1$

57. $f(x, y) = \sqrt{1 + x + y}$

$f_x = \dfrac{1}{2\sqrt{1 + x + y}} \qquad f_y = \dfrac{1}{2\sqrt{1 + x + y}}$

$f_{xx} = f_{yy} = f_{xy} = f_{yx} = -\dfrac{1}{4(1 + x + y)^{3/2}}$

59. $f(x, y, z) = xy + 5x^2yz^3 - 3y^3z$

$f_x = y + 10xyz^3 \qquad f_y = x + 5x^2z^3 - 9y^2z \qquad f_z = 15x^2yz^2 - 3y^3$

$f_{xx} = 10yz^3 \qquad f_{yx} = 1 + 10xz^3 \qquad f_{zx} = 30xyz^2$

$f_{xy} = 1 + 10xz^3 \qquad f_{yy} = -18yz \qquad f_{zy} = 15x^2z^2 - 9y^2$

$f_{xz} = 30xyz^2 \qquad f_{yz} = 15x^2z^2 - 9y^2 \qquad f_{zz} = 30x^2yz$

61. $C = 15(xy)^{1/3} + 99x + 139y + 2293$

(a) $\dfrac{\partial C}{\partial x} = 5x^{-2/3}y^{1/3} + 99$; At $(500, 250)$, $\dfrac{\partial C}{\partial x} = \$99.50$.

$\dfrac{\partial C}{\partial y} = 5x^{1/3}y^{-2/3} + 139$;

At $(500, 250)$, $\dfrac{\partial C}{\partial y} = \$140.00$.

(b) Downhill skis; this is determined by comparing the marginal costs for the two models of skis at the production level $(500, 250)$.

63. $f(x, y) = x^2 + 2y^2$

The first partial derivatives of f, $f_x(x, y) = 2x$ and $f_y(x, y) = 4y$, are zero at the critical point $(0, 0)$. Because $f_{xx}(xy) = 2$, $f_{yy}(xy) = 4$, and $f_{xy}(x, y) = 0$, it follows that $f_{xx}(0, 0) > 0$ and $f_{xx}(0, 0)f_{yy}(0, 0) - \left[f_{xy}(x, y)\right]^2 = 8 > 0$.

So, $(0, 0, 0)$ is a relative minimum.

65. $f(x, y) = 1 - (x + 2)^2 + (y - 3)^2$

The first partial derivatives of f, $f_x(x, y) = -2(x + 2)$ and $f_y(x, y) = 2(y - 3)$, are zero at the critical point $(-2, 3)$. Because $f_{xx}(x, y) = -2$, $f_{yy}(x, y) = 2$, and $f_{xy}(x, y) = 0$, it follows that $f_{xx}(-2, 3) > 0$ and

$f_{xx}(-2, 3)f_{yy}(-2, 3) - [f_{xy}(-2, 3)]^2 = -4 < 0.$

So, $(-2, 3, 1)$ is a saddle point.

67. $f(x, y) = x^3 + y^2 - xy$

The first partial derivatives of f, $f_x(x, y) = 3x^2 - y$ and $f_y(x, y) = 2y - x$, are zero at the critical points $(0, 0)$ and $\left(\frac{1}{6}, \frac{1}{12}\right)$. Because $f_{xx}(x, y) = 6x$, $f_{yy}(x, y) = 2$, and $f_{xy}(x, y) = -1$, it follows that $f_{xx}(0, 0) = 0$,

$f_{xx}(0, 0)f_{yy}(0, 0) - [f_{xy}(0, 0)]^2 = -1 < 0,$

$f_{xx}\left(\frac{1}{6}, \frac{1}{12}\right) = 1 > 0$, and

$f_{xx}\left(\frac{1}{6}, \frac{1}{12}\right)f_{yy}\left(\frac{1}{6}, \frac{1}{12}\right) - \left[f_{xy}\left(\frac{1}{6}, \frac{1}{12}\right)\right]^2 = 1 > 0.$

So, $(0, 0, 0)$ is a saddle point and $\left(\frac{1}{6}, \frac{1}{12}, -\frac{1}{432}\right)$ is a relative minimum.

69. $f(x, y) = x^3 + y^3 - 3x - 3y + 2$

The first partial derivatives of f, $f_x(x, y) = 3x^2 - 3$ and $f_y(x, y) = 3y^2 - 3$, are zero at the critical points $(1, 1)$, $(-1, -1)$, $(1, -1)$, and $(-1, 1)$. Because $f_{xx}(x, y) = 6x$, $f_{yy}(x, y) = 6y$, and $f_{xy}(x, y) = 0$, it follows that $f_{xx}(1, 1) = 6 > 0$,

$f_{xx}(1, 1)f_{yy}(1, 1) - [f_{xy}(1, 1)]^2 = 36 > 0,$

$f_{xx}(-1, -1) = -6 < 0,$

$f_{xx}(-1, -1)f_{yy}(-1, -1) - [f_{xy}(-1, -1)]^2 = 36 > 0,$

$f_{xx}(1, -1) = 6 > 0,$

$f_{xx}(1, -1)f_{yy}(1, -1) - [f_{xy}(1, -1)]^2 = -36 < 0,$

$f_{xx}(-1, 1) = -6 < 0$, and

$f_{xx}(-1, 1)f_{yy}(-1, 1) - [f_{xy}(-1, 1)]^2 = -36 < 0.$

So, $(1, 1, -2)$ is a relative minimum, $(-1, -1, 6)$ is a relative maximum, $(1, -1, 2)$ is a saddle point, and $(-1, 1, 2)$ is a saddle point.

71. $R = x_1p_1 + x_2p_2$

$= x_1(100 - x_1) + x_2(200 - 0.5x_2)$

$= -x_1^2 - 0.5x_2^2 + 100x_1 + 200x_2$

$R_{x_1} = -2x_1 + 100 = 0 \Rightarrow x_1 = 50$

$R_{x_2} = -x_2 + 200 = 0 \Rightarrow x_2 = 200$

$R_{x_1x_1} = -2$

$R_{x_2x_2} = -1$

$R_{x_1x_2} = 0$

$d = R_{x_1x_1}R_{x_2x_2} - [R_{x_1x_2}]^2$

$d = (-2)(-1) - (0)^2 = 2 > 0$

Since $R_{x_1x_1} < 0$, R is maximum at $(50, 200)$, or when $x_1 = 50$ and $x_2 = 200$.

The maximum revenue is $22,500.

73. $F(x, y, \lambda) = 2xy - \lambda(2x + y - 12)$

$$\left.\begin{aligned} F_x(x, y, \lambda) &= 2y - 2\lambda = 0 \\ F_y(x, y, \lambda) &= 2x - \lambda = 0 \end{aligned}\right\} y = 2x$$

$F_\lambda(x, y, \lambda) = -(2x + y - 12) = 0$

Using F_x, $\quad -(2x + 2x - 12) = 0$

$-4x = -12$

$x = 3.$

So, $x = 3$ and $y = 6$, and $f(x, y)$ has a maximum at $(3, 6)$. The relative maximum is $f(3, 6) = 36$.

75. $F(x, y, \lambda) = x^2 + y^2 - \lambda(x + y - 4)$

$$\left.\begin{aligned} F_x(x, y, \lambda) &= 2x - \lambda = 0 \\ F_y(x, y, \lambda) &= 2y - \lambda = 0 \end{aligned}\right\} x = y$$

$F_\lambda(x, y, \lambda) = -(x + y - 4) = 0, x = y = 2$

Minimum $f(2, 2) = 8$.

So, $x = 2$, $y = 2$, and $f(x, y)$ has a minimum at $(2, 2)$. The relative minimum is $f(2, 2) = 8$.

77. $F(x, y, z, \lambda) = xyz - \lambda(x + 2y + z - 4)$

$$\left.\begin{aligned} F_x &= yz - \lambda = 0 \\ F_y &= xz - 2\lambda = 0 \\ F_z &= xy - \lambda = 0 \end{aligned}\right\} \begin{aligned} xz = 2yz = 2xy \Rightarrow x &= 2y \\ z &= 2y \end{aligned}$$

$$F_\lambda = -(x + 2y + z - 4) = 0$$

$$2y + 2y + 2y - 4 = 0$$

So, $y = \frac{2}{3}$, $x = \frac{4}{3}$, $z = \frac{4}{3}$, and $f(x, y, z)$ has a maximum at $\left(\frac{4}{3}, \frac{2}{3}, \frac{4}{3}\right)$. The relative maximum is $f\left(\frac{4}{3}, \frac{2}{3}, \frac{4}{3}\right) = \frac{32}{27}$.

79. Minimize $C = 0.25x_1^2 + 10x_1 + 0.15x_2^2 + 12x_2$ subject to the constraint $x_1 + x_2 = 1000$.

$$F(x_1, x_2, \lambda) = 0.25x_1^2 + 10x_1 + 0.15x_2^2 + 12x_2 - \lambda(x_1 + x_2 - 1000)$$

$$\left.\begin{aligned} F_{x_1}(x_1, x_2, \lambda) &= 0.50x_1 + 10 - \lambda = 0 \\ F_{x_2}(x_1, x_2, \lambda) &= 0.30x_2 + 12 - \lambda = 0 \end{aligned}\right\} 5x_1 - 3x_2 = 20$$

$$F_\lambda(x_1, x_2, \lambda) = -(x_1 + x_2 - 1000) = 0, \quad x_2 = 1000 - x_1$$

So, $x_1 = 377.5$ and $x_2 = 622.5$, and $f(x_1, x_2)$ has a minimum at $(377.5, 622.5)$.

The cost is minimized when 378 units of x_1 and 623 units of x_2 are ordered.

81. $\sum x_i = 1$

$\sum y_i = 0$

$\sum x_i^2 = 15$

$\sum x_i y_i = 15$

$$a = \frac{4(15) - (1)(0)}{4(15) - (1)^2} = \frac{60}{59}$$

$$b = \frac{1}{4}\left(0 - \frac{60}{59}(1)\right) = -\frac{15}{59}$$

$$y = \frac{60}{59}x - \frac{15}{59}$$

83. (a) $y = -2.6x + 347$

(b) Let $x = 85$.

$y = -2.6(85) + 347 = 126$ cameras

(c) Let $y = 200$ and find x.

$$200 = -2.6x + 347$$
$$-147 = -2.6x$$
$$x \approx \$56.54$$

85.
$$\int_0^1 \int_0^{1+x} (4x - 2y)\,dy\,dx = \int_0^1 \left[4xy - y^2\right]_0^{1+x} dx$$
$$= \int_0^1 \left[4x(1 + x) - (1 + x)^2\right] dx$$
$$= \int_0^1 \left(3x^2 + 2x - 1\right) dx$$
$$= \left[x^3 + x^2 - x\right]_0^1$$
$$= 1$$

87.
$$\int_1^2 \int_1^{2y} \frac{x}{y^2}\,dx\,dy = \int_1^2 \left[\frac{x^2}{2y^2}\right]_1^{2y} dy$$
$$= \int_1^2 \left(\frac{4y^2}{2y^2} - \frac{1}{2y^2}\right) dy$$
$$= \int_1^2 \left(2 - \frac{1}{2}y^{-2}\right) dy$$
$$= \left[2y + \frac{1}{2y}\right]_1^2$$
$$= \left(4 + \frac{1}{4}\right) - \left(2 + \frac{1}{2}\right)$$
$$= \frac{7}{4}$$

89.
$$A = \int_{-2}^2 \int_5^{9-x^2} dy\,dx$$
$$= \int_{-2}^2 [y]_5^{9-x^2}\,dx$$
$$= \int_{-2}^2 (4 - x^2)\,dx$$
$$= \left[4x - \tfrac{1}{3}x^3\right]_{-2}^2$$
$$= \frac{32}{3}$$

91. $A = \int_{-3}^{6} \int_{(1/3)x+1}^{\sqrt{x+3}} dy\, dx$

$= \int_{-3}^{6} \left[y\right]_{(1/3)x+1}^{\sqrt{x+3}} dx$

$= \int_{-3}^{6} \left(\sqrt{x+3} - \frac{1}{3}x - 1\right) dx$

$= \left[\frac{2}{3}(x+3)^{3/2} - \frac{1}{6}x^2 - x\right]_{-3}^{6}$

$= \frac{9}{2}$

93. $V = \int_0^4 \int_0^2 \left(3 - \frac{1}{2}y\right) dy\, dx$

$= \int_0^4 \left[3y - \frac{1}{4}y^2\right]_0^2 dx$

$= \int_0^4 5\, dx$

$= \left[5x\right]_0^4$

$= 20$

95. $V = \int_0^2 \int_0^x 4\, dy\, dx$

$= \int_0^2 \left[4y\right]_0^x dx$

$= \int_0^2 4x\, dx$

$= \left[2x^2\right]_0^2 = 8$

97. $V = \int_0^4 \int_0^4 (xy)^2\, dy\, dx$

$= \int_0^4 \int_0^4 x^2y^2\, dy\, dx$

$= \int_0^4 \left[\frac{x^2y^3}{3}\right]_0^4 dx$

$= \int_0^4 \frac{64x^2}{3}\, dx$

$= \left[\frac{64x^3}{9}\right]_0^4$

$= \frac{4096}{9}$

99. $\text{Average} = \frac{1}{12} \int_0^4 \int_0^3 xy\, dy\, dx$

$= \frac{1}{12} \int_0^4 \left[\frac{1}{2}xy^2\right]_0^3 dx$

$= \frac{1}{24} \int_0^4 9x\, dx$

$= \frac{1}{24}\left[\frac{9}{2}x^2\right]_0^4$

$= 3$

101. $\text{Average} = \frac{1}{100} \int_{40}^{50} \int_{30}^{40} \left(150x_1 + 400x_2 - x_1^2 - 5x_2^2 - 2x_1x_2 - 3000\right) dx_1\, dx_2$

$= \frac{1}{100} \int_{40}^{50} \left[75x_1^2 + 400x_1x_2 - \frac{1}{3}x_1^3 - 5x_1x_2^2 - x_1^2x_2 - 3000x_1\right]_{30}^{40} dx_2$

$= \frac{1}{100} \int_{40}^{50} \left(\frac{30{,}500}{3} + 3300x_2 - 50x_2^2\right) dx_2$

$= \frac{1}{100}\left[\frac{30{,}500}{3}x_2 + 1650x_2^2 - \frac{50}{3}x_2^3\right]_{40}^{50}$

$= \$5700$

103. $\text{Average} = \frac{1}{20{,}908{,}800} \int_0^{5280} \int_0^{3960} 0.003x^{2/3}y^{3/4}\, dy\, dx$

$= \frac{1}{20{,}908{,}800} \int_0^{5280} \left[\frac{0.012}{7}x^{2/3}y^{7/4}\right]_0^{3960} dx$

$= \frac{1}{20{,}908{,}800} \int_0^{5280} \left(3388.830x^{2/3}\right) dx$

$= \frac{1}{20{,}908{,}800}\left[2033.298x^{5/3}\right]_0^{5280}$

$\approx \$155.69/\text{ft}^2$

Chapter 7 Test Yourself

1. (a)

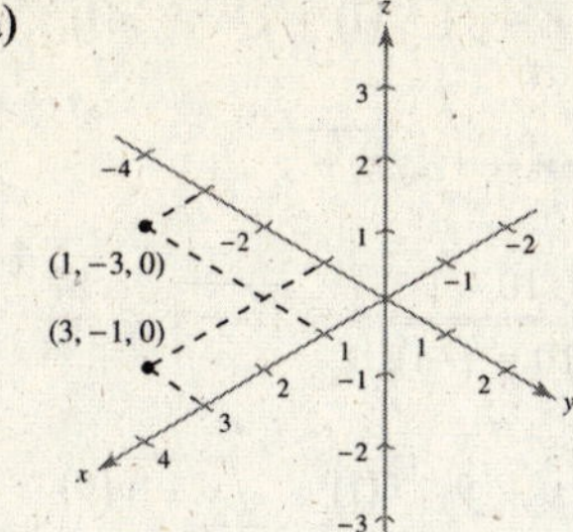

(b) $d = \sqrt{(3-1)^2 + (-1+3)^2 + (0-0)^2}$

$= \sqrt{4+4+0} = 2\sqrt{2}$

(c) Midpoint $= \left(\dfrac{1+3}{2}, \dfrac{-3-1}{2}, \dfrac{0+0}{2}\right) = (2, -2, 0)$

2. (a)

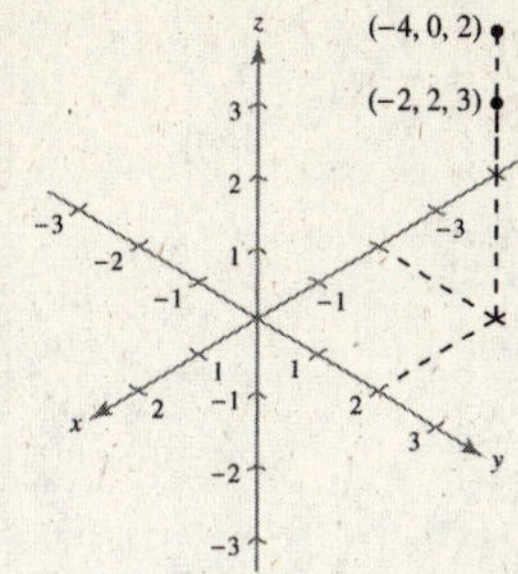

(b) $d = \sqrt{(-4+2)^2 + (0-2)^2 + (2-3)^2}$

$= \sqrt{4+4+1} = 3$

(c) Midpoint $= \left(\dfrac{-2-4}{2}, \dfrac{2+0}{2}, \dfrac{3+2}{2}\right) = \left(-3, 1, \dfrac{5}{2}\right)$

3. (a)

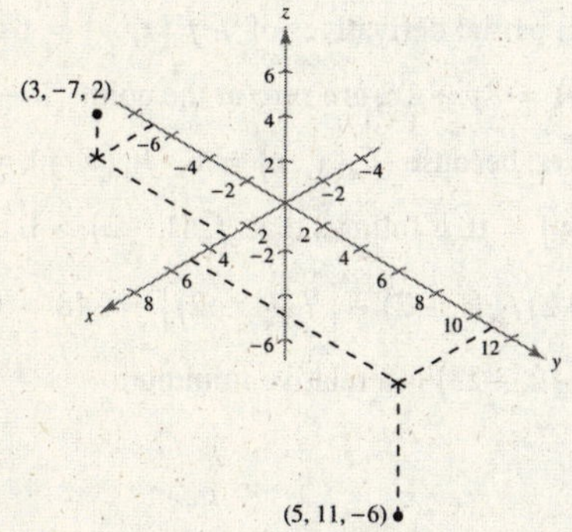

(b) $d = \sqrt{(5-3)^2 + (11+7)^2 + (-6-2)^2} = \sqrt{4+324+64} = 14\sqrt{2}$

(c) Midpoint $= \left(\dfrac{3+5}{2}, \dfrac{-7+11}{2}, \dfrac{2-6}{2}\right) = (4, 2, -2)$

4.

$$x^2 + y^2 + z^2 - 20x + 10y - 10z + 125 = 0$$

$$(x^2 - 20x + 100) + (y^2 + 10y + 25) + (z^2 - 10z + 25) = -125 + 100 + 25 + 25$$

$$(x-10)^2 + (y+5)^2 + (z-5)^2 = 25$$

Center: $(10, -5, 5)$

Radius: 5

5. The graph of $4x^2 + 2y^2 - z^2 = 16$ is a hyperboloid of one sheet whose standard equation is $\dfrac{x^2}{4} + \dfrac{y^2}{8} - \dfrac{z^2}{16} = 1$.

6. The graph of $36x^2 + 9y^2 - 4z^2 = 0$ is an elliptic cone whose standard equation is $x^2 + \dfrac{y^2}{4} - \dfrac{z^2}{9} = 0$.

7. The graph of $4x^2 - y^2 - 16z = 0$ is a hyperbolic paraboloid whose standard equation is $z = \dfrac{x^2}{4} - \dfrac{y^2}{16}$.

8. $f(x, y) = x^2 + xy + 1$

$f(3, 3) = 3^2 + 3(3) + 1 = 19$

$f(1, 4) = 1^2 + 1(4) + 1 = 6$

9. $f(x, y) = \dfrac{x + 2y}{3x - y}$

$f(3, 3) = \dfrac{3 + 2(3)}{3(3) - 3} = \dfrac{9}{6} = \dfrac{3}{2}$

$f(1, 4) = \dfrac{1 + 2(4)}{3(1) - 4} = -9$

10. $f(x, y) = xy \ln\left(\dfrac{x}{y}\right)$

$f(3, 3) = 3(3)\ln\left(\dfrac{3}{3}\right) = 9\ln(1) = 0$

$f(1, 4) = 1(4)\ln\left(\dfrac{1}{4}\right) = 4\ln\left(\dfrac{1}{4}\right) \approx -5.5$

11. $f(x, y) = 3x^2 + 9xy^2 - 2$

$f_x(x, y) = 6x + 9y^2$

$f_x(10, -1) = 6(10) + 9(-1)^2 = 69$

$f_y(x, y) = 18xy$

$f_y(10, -1) = 18(10)(-1) = -180$

12. $f(x, y) = x\sqrt{x + y}$

$f_x(x, y) = x\left(\dfrac{1}{2}\right)(x + y)^{-1/2}(1) + \sqrt{x + y}(1)$

$= \dfrac{x}{2\sqrt{x + y}} + \sqrt{x + y}$

$f_x(10, -1) = \dfrac{10}{2\sqrt{10 + (-1)}} + \sqrt{10 - 1} = \dfrac{14}{3}$

$f_y(x, y) = x\left(\dfrac{1}{2}\right)(x + y)^{-1/2}(1) + \sqrt{x + y}(0)$

$= \dfrac{x}{2\sqrt{x + y}}$

$f_y(10, -1) = \dfrac{10}{2\sqrt{10 + (-1)}} = \dfrac{5}{3}$

13. $f(x, y) = 3x^2 + 4y^2 - 6x + 16y - 4$

The first partial derivatives of f, $f_x(x, y) = 6x - 6$ and $f_y(x, y) = 8y + 16$, are zero at the point $(1, -2)$. Moreover, because $f_{xx}(x, y) = 6$, $f_{yy}(x, y) = 8$, and $f_{xy}(x, y) = 0$, it follows that $f_{xx}(1, -2) > 0$ and $f_{xx}(1, -2)f_{yy}(1, -2) - \left[f_{xy}(1, -2)\right]^2 = 48 > 0$.

So, $(1, -2, -23)$ is a relative minimum.

14. $f(x, y) = 4xy - x^4 - y^4$

The first partial derivatives of f, $f_x(x, y) = 4y - 4x^3$ and $f_y(x, y) = 4x - 4y^3$, are zero at the points $(0, 0)$, $(1, 1)$, and $(-1, -1)$.

Moreover, because $f_{xx}(x, y) = -12x^2$, $f_{yy}(x, y) = -12y^2$, and $f_{xy}(x, y) = 4$, it follows that $f_{xx}(0, 0) = 0$, $f_{xx}(0, 0)f_{yy}(0, 0) - \left[f_{xy}(0, 0)\right]^2 = -16 < 0$, $f_{xx}(1, 1) = -12 < 0$, $f_{xx}(1, 1)f_{yy}(1, 1) - \left[f_{xy}(1, 1)\right]^2 = 128 > 0$, $f_{xx}(-1, -1) = -12 < 0$, and $f_{xx}(-1, -1)f_{yy}(-1, -1) - \left[f_{xy}(-1, -1)\right]^2 = 128 > 0$.

So, $(0, 0, 0)$ is a saddle point, $(1, 1, 2)$ is a relative minimum, and $(-1, -1, 2)$ is a relative minimum.

15. Maximize $f(x, y) = 60x^{0.7}y^{0.3}$ subject to the constraint $42x + 144y = 240{,}000$.

$F(x, y, \lambda) = 60x^{0.7}y^{0.3} - \lambda(42x + 144y - 240{,}000)$

$F_x = 42x^{-0.3}y^{0.3} - 42\lambda = 0,\ \lambda = \dfrac{42x^{-0.3}y^{0.3}}{42},\ \lambda = x^{-0.3}y^{0.3}$

$F_y = 18x^{0.7}y^{-0.7} - 144\lambda = 0$

$F_\lambda = -42x - 144y + 240{,}000 = 0$

Substituting $\lambda = x^{-0.3}y^{0.3}$ into F_y yields $18x^{0.7}y^{-0.7} - 144x^{-0.3}y^{0.3} = 0 \Rightarrow x = 8y$.

From F_λ, you can obtain $y = 500$. So, $x = 4000$. To maximize production, the company should use 4000 units of labor and 500 units of capital.

16. $\sum x_i = 29$

$\sum y_i = 22$

$\sum x_i^2 = 231$

$\sum x_i y_i = 160$

$$a = \frac{5(160) - (29)(22)}{5(231) - (29)^2} \approx 0.52$$

$$b = \frac{1}{5}(22 - 0.52(29)) \approx 1.4$$

$$y = 0.52x + 1.4$$

17. $$\int_0^1 \int_x^1 \left(30x^2y - 1\right) dy\, dx = \int_0^1 \left[15x^2y^2 - y\right]_x^1 dx$$
$$= \int_0^1 \left[\left(15x^2 - 1\right) - \left(15x^2(x)^2 - x\right)\right] dx$$
$$= \int_0^1 \left(-15x^4 + 15x^2 + x - 1\right) dx$$
$$= \left[-3x^5 + 5x^3 + \frac{x}{2} - x\right]_0^1 = \frac{3}{2}$$

18. $$\int_0^{\sqrt{e-1}} \int_0^{2y} \frac{1}{y^2 + 1}\, dx\, dy = \int_0^{\sqrt{e-1}} \left[\frac{x}{y^2 + 1}\right]_0^{2y} dy = \int_0^{\sqrt{e-1}} \frac{2y}{y^2 + 1}\, dy = \left[\ln\left|y^2 + 1\right|\right]_0^{\sqrt{e-1}} = \ln e = 1$$

19. $$\int_0^2 \int_{x^2-2x+3}^3 dy\, dx = \int_0^2 \left[3 - \left(x^2 - 2x + 3\right)\right] dx = \int_0^2 \left(-x^2 + 2x\right) dx = \left[-\frac{x^3}{3} + x^2\right]_0^2 = \frac{4}{3} \text{ square units}$$

20. $$V = \int_0^4 \int_0^3 (8 - 2x)\, dy\, dx = \int_0^4 \left[8y - 2xy\right]_0^3 dx = \int_0^4 (24 - 6x)\, dx = \left[24x - 3x^2\right]_0^4 = 48$$

21. $$\text{Average} = \frac{\int_0^1 \int_0^3 \left(x^2 + y\right) dy\, dx}{\text{area}} = \frac{\int_0^1 \left[x^2y + \frac{y^2}{2}\right]_0^3 dx}{3} = \frac{\int_0^1 \left(3x^2 + \frac{9}{2}\right) dx}{3} = \frac{\left[x^3 + \frac{9}{2}x\right]_0^1}{3} = \frac{11}{6}$$

CHAPTER 8
Trigonometric Functions

CHAPTER 8

Trigonometric Functions

Section 8.1 Radian Measure of Angles

Skills Warm Up

1. $A = \frac{1}{2}bh$

$= \frac{1}{2}(10)(7)$

$= 35 \text{ cm}^2$

2. $A = \frac{1}{2}bh$

$= \frac{1}{2}(4)(6)$

$= 12 \text{ in.}^2$

3. $a^2 + b^2 = c^2$

$5^2 + 12^2 = c^2$

$169 = c^2$

$13 = c$

4. $a^2 + b^2 = c^2$

$3^2 + b^2 = 5^2$

$9 + b^2 = 25$

$b^2 = 16$

$b = 4$

5. $a^2 + b^2 = c^2$

$8^2 + b^2 = 17^2$

$64 + b^2 = 289$

$b^2 = 225$

$b = 15$

6. $a^2 + b^2 = c^2$

$a^2 + 8^2 = 10^2$

$a^2 + 64 = 100$

$a^2 = 36$

$a = 6$

7. Because all side lengths are 4, the triangle is an equilateral triangle.

8. Because $a = b$, the triangle is an isosceles triangle.

9. Because $a^2 + b^2 = c^2$, the triangle is a right triangle.

10. Because $a = b$ and $a^2 + b^2 = c^2$, the triangle is an isosceles right triangle.

1. Positive: $45° + 360° = 405°$

Negative: $45° - 360° = -315°$

3. Positive: $-120° + 360° = 240°$

Negative: $-120° - 360° = -480°$

5. Positive: $-420° + 2(360°) = 300°$

Negative: $-420° + 360° = -60°$

7. Positive: $\frac{\pi}{9} + 2\pi = \frac{19\pi}{9}$

Negative: $\frac{\pi}{9} - 2\pi = -\frac{17\pi}{9}$

9. Positive: $-\frac{2\pi}{15} + 2\pi = \frac{28\pi}{15}$

Negative: $-\frac{2\pi}{15} - 2\pi = -\frac{32\pi}{15}$

11. $30°\left(\frac{\pi \text{ radians}}{180°}\right) = \frac{\pi}{6} \text{ radians}$

13. $270°\left(\frac{\pi \text{ radians}}{180°}\right) = \frac{3\pi}{2} \text{ radians}$

15. $675°\left(\frac{\pi \text{ radians}}{180°}\right) = \frac{15\pi}{4} \text{ radians}$

17. $-24°\left(\frac{\pi \text{ radians}}{180°}\right) = -\frac{2\pi}{15} \text{ radians}$

19. $-144°\left(\frac{\pi \text{ radians}}{180°}\right) = -\frac{4\pi}{5} \text{ radians}$

21. $330°\left(\frac{\pi \text{ radians}}{180°}\right) = \frac{11\pi}{6} \text{ radians}$

23. $\frac{5\pi}{2}\left(\frac{180°}{\pi}\right) = 450°$

25. $\frac{7\pi}{3}\left(\frac{180^\circ}{\pi}\right) = 420^\circ$

27. $-\frac{\pi}{12}\left(\frac{180^\circ}{\pi}\right) = -15^\circ$

29. $\frac{4\pi}{15}\left(\frac{180^\circ}{\pi}\right) = 48^\circ$

31. $\frac{19\pi}{6}\left(\frac{180^\circ}{\pi}\right) = 570^\circ$

33. The angle θ is $\theta = 90^\circ - 30^\circ = 60^\circ$. By the Pythagorean Theorem, the value of c is

$$C = \sqrt{(5\sqrt{3})^2 + 5^2} = \sqrt{75 + 25} = \sqrt{100} = 10.$$

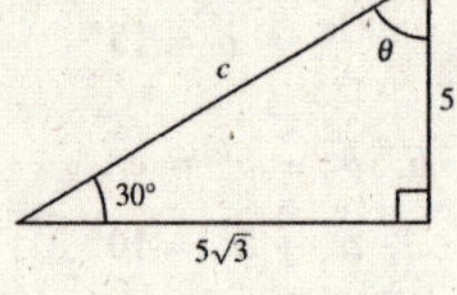

35. The angle θ is $\theta = 90^\circ - 60^\circ = 30^\circ$. By the Pythagorean Theorem, the value of a is

$$a = \sqrt{8^2 - 4^2} = \sqrt{64 - 16} = \sqrt{48} = 4\sqrt{3}.$$

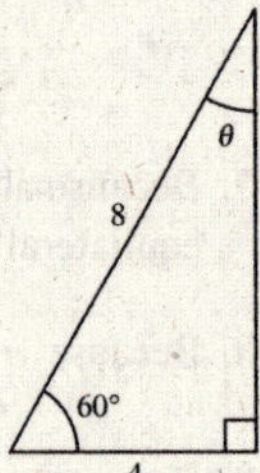

37. Because the triangle is isosceles, you have $\theta = 40^\circ$.

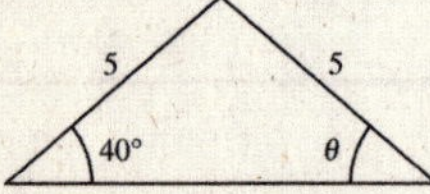

39. Because the largest triangle is similar to the two smaller triangles, $\theta = 60^\circ$. The hypotenuse of the large triangle is

$$\sqrt{2^2 + (2\sqrt{3})^2} = \sqrt{16} = 4.$$

By similar triangles you have

$$\frac{s}{2} = \frac{2\sqrt{3}}{4}$$
$$s = \sqrt{3}.$$

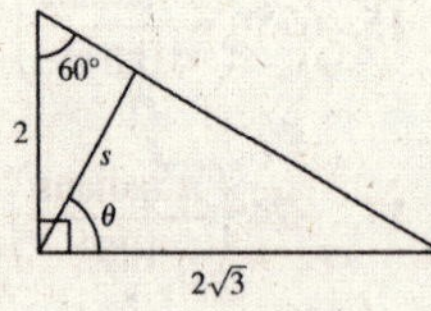

41. $h^2 + 2^2 = 4^2$

$h^2 = 12$

$h = 2\sqrt{3}$

$A = \frac{1}{2}bh$

$= \frac{1}{2}(4)(2\sqrt{3})$

$= 4\sqrt{3}$ sq in.

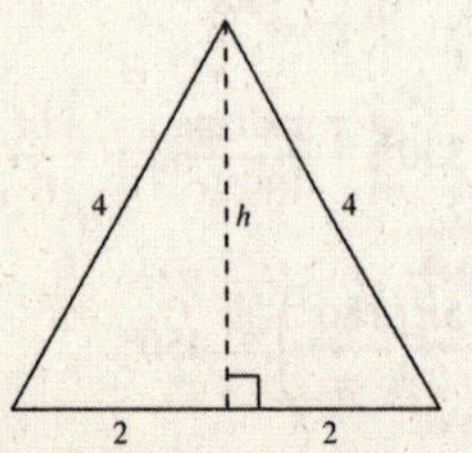

43. $h^2 + \left(\frac{5}{2}\right)^2 = 5^2$

$h^2 = \frac{75}{4}$

$h = \frac{5\sqrt{3}}{2}$

$A = \frac{1}{2}bh$

$= \frac{1}{2}(5)\left(\frac{5\sqrt{3}}{2}\right)$

$= \frac{25\sqrt{3}}{4}$ sq ft

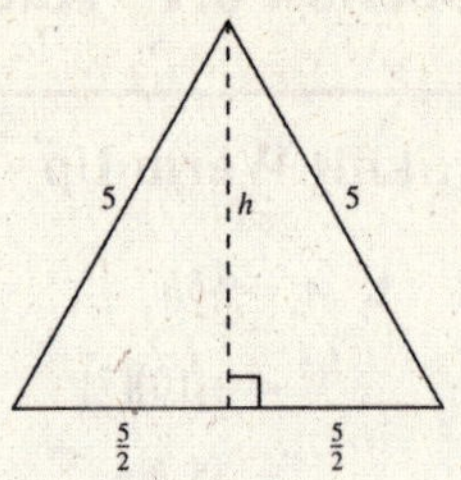

45. Using similar triangles, you have

$$\frac{h}{16 + 8} = \frac{6}{8}$$
$$8h = 144$$

which implies $h = 18$ feet.

47.

r	8 ft	15 in.	85 cm	24 in.	$\frac{12,963}{\pi}$ mi
s	12 ft	24 in.	$\frac{255\pi}{4}$ cm	96 in.	8642 mi
θ	$\frac{3}{2}$	$\frac{8}{5}$	$\frac{3\pi}{4}$	4	$\frac{2\pi}{3}$

49. $s = r\theta = \frac{7}{2}\left(\frac{5\pi}{6}\right) = \frac{35\pi}{12} \approx 9.16$ inches

51. (a) $\left(\frac{3142 \text{ rad}}{\text{min}}\right)\left(\frac{1 \text{ rev}}{2\pi \text{ rad}}\right) = \frac{1571}{\pi} \approx 500.06$ rev/min

(b) $d = rt$

$10{,}000 = \frac{1571}{\pi}t$

$\frac{10{,}000\pi}{1571} = t$

$t \approx 20$ min

53. $120^\circ\left(\frac{\pi \text{ radians}}{180^\circ}\right) = \frac{2\pi}{3}$ radians

$$A = \frac{1}{2}r^2\theta = \frac{1}{2}(70^2)\left(\frac{2\pi}{3}\right) = \frac{4900\pi}{3} \approx 5131.27 \text{ sq ft}$$

55. False. An obtuse angle is between 90° and 180°.

57. True. The angles would be 90°, 89°, and 1°.

59. Answers will vary.

Section 8.2 The Trigonometric Functions

Skills Warm Up

1. $315°\left(\frac{\pi \text{ radians}}{180°}\right) = \frac{7\pi}{4}$ radians

2. $-300°\left(\frac{\pi \text{ radians}}{180°}\right) = -\frac{5\pi}{3}$ radians

3. $-225°\left(\frac{\pi \text{ radians}}{180°}\right) = -\frac{5\pi}{4}$ radians

4. $390°\left(\frac{\pi \text{ radians}}{180°}\right) = \frac{13\pi}{6}$ radians

5. $x^2 - x = 0$
$x(x - 1) = 0$
$x = 0$
$x - 1 = 0 \Rightarrow x = 1$
$x = 0$ or $x = 1$

6. $2x^2 + x = 0$
$x(2x + 1) = 0$
$x = 0$
$2x + 1 = 0 \Rightarrow x = -\frac{1}{2}$
$x = 0$ or $x = -\frac{1}{2}$

7. $x^2 - 2x = 3$
$x^2 - 2x - 3 = 0$
$(x + 1)(x - 3) = 0$
$x + 1 = 0 \Rightarrow x = -1$
$x - 3 = 0 \Rightarrow x = 3$
$x = -1$ or $x = 3$

8. $x^2 - 5x = -6$
$x^2 - 5x + 6 = 0$
$(x - 3)(x - 2) = 0$
$x - 3 = 0 \Rightarrow x = 3$
$x - 2 = 0 \Rightarrow x = 2$
$x = 3$ or $x = 2$

9. $\frac{2\pi}{24}(t - 4) = \frac{\pi}{2}$
$t - 4 = 6$
$t = 10$

10. $\frac{2\pi}{12}(t - 2) = \frac{\pi}{4}$
$t - 2 = \frac{3}{2}$
$t = \frac{7}{2}$

11. $\frac{2\pi}{365}(t - 10) = \frac{\pi}{4}$
$t - 10 = \frac{365}{8}$
$t = \frac{445}{8}$

12. $\frac{2\pi}{12}(t - 4) = \frac{\pi}{2}$
$t - 4 = 3$
$t = 7$

1. Because $x = 4$ and $y = 3$, it follows that

$r = \sqrt{4^2 + 3^2} = \sqrt{25} = 5.$

So you have

$\sin\theta = \frac{3}{5}$ $\quad \csc\theta = \frac{5}{3}$

$\cos\theta = \frac{4}{5}$ $\quad \sec\theta = \frac{5}{4}$

$\tan\theta = \frac{3}{4}$ $\quad \cot\theta = \frac{4}{3}$.

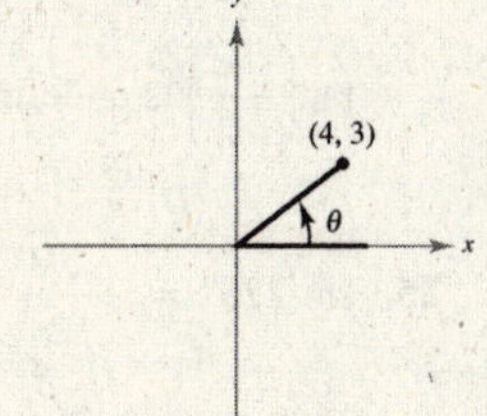

3. Because $x = -12$ and $y = -5$, it follows that

$r = \sqrt{(-12)^2 + (-5)^2} = 13.$

So you have

$\sin\theta = -\frac{5}{13}$ $\quad \csc\theta = -\frac{13}{5}$

$\cos\theta = -\frac{12}{13}$ $\quad \sec\theta = -\frac{13}{12}$

$\tan\theta = \frac{5}{12}$ $\quad \cot\theta = \frac{12}{5}$.

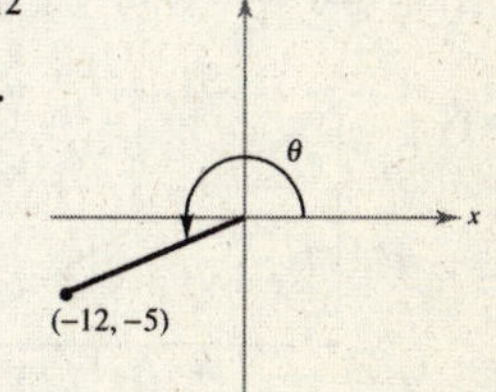

5. Because $x = -2$ and $y = 4$, it follows that

$r = \sqrt{(-2)^2 + 4^2} = \sqrt{20} = 2\sqrt{5}.$

So you have

$\sin\theta = \dfrac{2\sqrt{5}}{5}$ $\qquad$ $\csc\theta = \dfrac{\sqrt{5}}{2}$

$\cos\theta = -\dfrac{\sqrt{5}}{5}$ $\qquad$ $\sec\theta = -\sqrt{5}$

$\tan\theta = -2$ $\qquad$ $\cot\theta = -\dfrac{1}{2}.$

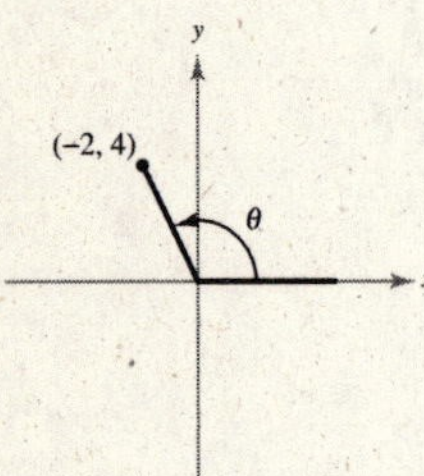

7. Because $y = 1$ and $r = 3$, the length of the third side of the triangle is

$x = \sqrt{2^2 - 1^2} = 2\sqrt{2}.$

So you have

$\sin\theta = \dfrac{1}{3}$ $\qquad$ $\csc\theta = 3$

$\cos\theta = \dfrac{2\sqrt{2}}{3}$ $\qquad$ $\sec\theta = \dfrac{3}{2\sqrt{2}} = \dfrac{3\sqrt{2}}{4}$

$\tan\theta = \dfrac{1}{2\sqrt{2}} = \dfrac{\sqrt{2}}{4}$ $\qquad$ $\cos\theta = 2\sqrt{2}.$

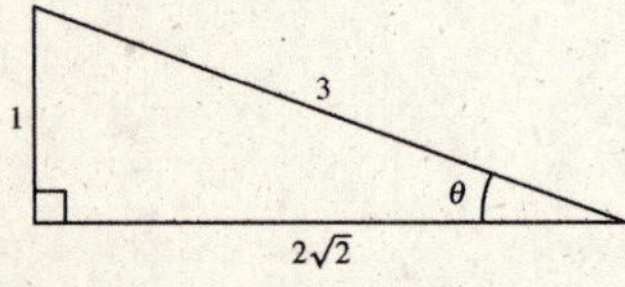

9. Because $x = 1$ and $r = 2$, the length of the opposite side is $y = \sqrt{2^2 - 1^2} = \sqrt{3}.$

So you have

$\sin\theta = \dfrac{\sqrt{3}}{2}$ $\qquad$ $\csc\theta = \dfrac{2}{\sqrt{3}} = \dfrac{2\sqrt{3}}{3}$

$\cos\theta = \dfrac{1}{2}$ $\qquad$ $\sec\theta = 2$

$\tan\theta = \sqrt{3}$ $\qquad$ $\cot\theta = \dfrac{1}{\sqrt{3}} = \dfrac{\sqrt{3}}{3}.$

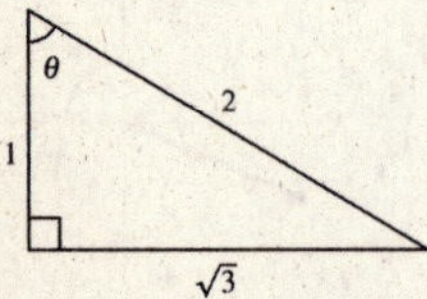

11. Because $x = 1$ and $y = 3$, the length of the hypotenuse is $r = \sqrt{1^2 + 3^2} = \sqrt{10}.$

So you have

$\sin\theta = \dfrac{3}{\sqrt{10}} = \dfrac{3\sqrt{10}}{10}$ $\qquad$ $\csc\theta = \dfrac{\sqrt{10}}{3}$

$\cos\theta = \dfrac{1}{\sqrt{10}} = \dfrac{\sqrt{10}}{10}$ $\qquad$ $\sec\theta = \sqrt{10}$

$\tan\theta = 3$ $\qquad$ $\cot\theta = \dfrac{1}{3}.$

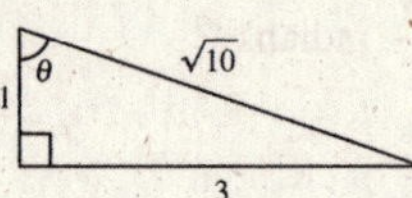

13. Because the sine is negative and the cosine is positive, θ must lie in Quadrant IV.

15. Because the sine is positive and the secant is positive, θ must lie in Quadrant I.

17. Because the cosecant is positive and the tangent is negative, θ must lie in Quadrant II.

19. $\sin 60° = \dfrac{\sqrt{3}}{2}$ $\qquad$ $\csc 60° = \dfrac{2\sqrt{3}}{3}$

$\cos 60° = \dfrac{1}{2}$ $\qquad$ $\sec 60° = 2$

$\tan 60° = \sqrt{3}$ $\qquad$ $\cot 60° = \dfrac{\sqrt{3}}{3}$

21. $\sin\left(\dfrac{\pi}{4}\right) = \dfrac{\sqrt{2}}{2}$ $\qquad$ $\csc\left(\dfrac{\pi}{4}\right) = \sqrt{2}$

$\cos\left(\dfrac{\pi}{4}\right) = \dfrac{\sqrt{2}}{2}$ $\qquad$ $\sec\left(\dfrac{\pi}{4}\right) = \sqrt{2}$

$\tan\left(\dfrac{\pi}{4}\right) = 1$ $\qquad$ $\cot\left(\dfrac{\pi}{4}\right) = 1$

23. $\sin\left(-\dfrac{\pi}{2}\right) = -1$ $\qquad$ $\csc\left(-\dfrac{\pi}{2}\right) = -1$

$\cos\left(-\dfrac{\pi}{2}\right) = 0$ $\qquad$ $\sec\left(-\dfrac{\pi}{2}\right)$ is undefined.

$\tan\left(-\dfrac{\pi}{2}\right)$ is undefined. $\qquad$ $\cot\left(-\dfrac{\pi}{2}\right) = 0$

25. $\sin 225° = -\dfrac{\sqrt{2}}{2}$ $\qquad$ $\csc 225° = -\sqrt{2}$

$\cos 225° = -\dfrac{\sqrt{2}}{2}$ $\qquad$ $\sec 225° = -\sqrt{2}$

$\tan 225° = 1$ $\qquad$ $\cot 225° = 1$

27. $\sin 300° = -\dfrac{\sqrt{3}}{2}$ $\qquad \csc 300° = -\dfrac{2\sqrt{3}}{3}$

$\cos 300° = \dfrac{1}{2}$ $\qquad \sec 300° = 2$

$\tan 300° = -\sqrt{3}$ $\qquad \cot 300° = -\dfrac{\sqrt{3}}{3}$

29. $\sin 750° = \dfrac{1}{2}$ $\qquad \csc 750° = 2$

$\cos 750° = \dfrac{\sqrt{3}}{2}$ $\qquad \sec 750° = \dfrac{2\sqrt{3}}{3}$

$\tan 750° = \dfrac{\sqrt{3}}{3}$ $\qquad \cot 750° = \sqrt{3}$

31. $\sin\left(\dfrac{10\pi}{3}\right) = -\dfrac{\sqrt{3}}{2}$ $\qquad \csc\left(\dfrac{10\pi}{3}\right) = -\dfrac{2\sqrt{3}}{3}$

$\cos\left(\dfrac{10\pi}{3}\right) = -\dfrac{1}{2}$ $\qquad \sec\left(\dfrac{10\pi}{3}\right) = -2$

$\tan\left(\dfrac{10\pi}{3}\right) = \sqrt{3}$ $\qquad \cot\left(\dfrac{10\pi}{3}\right) = \dfrac{\sqrt{3}}{3}$

33. $\sin 10° \approx 0.1736$

35. $\tan\left(\dfrac{\pi}{9}\right) \approx 0.3640$

37. $\cos(-110°) \approx -0.3420$

39. $\tan 240° \approx 1.7321$

41. $\sin(-0.65) \approx -0.6052$

43. Because $\tan 30° = \dfrac{1}{\sqrt{3}} = \dfrac{y}{100}$ it follows that

$$y = \frac{100}{\sqrt{3}} = \frac{100\sqrt{3}}{3}.$$

45. Because $\tan 60° = \dfrac{\sqrt{3}}{1} = \dfrac{25}{x}$, it follows that

$$x = \frac{25}{\sqrt{3}} = \frac{25\sqrt{3}}{3}.$$

47. Because $\sin 40° = \dfrac{10}{r}$, it follows that

$$r = \frac{10}{\sin 40°} \approx 15.5572.$$

49. $\theta = \dfrac{\pi}{6}$ or $\theta = \dfrac{5\pi}{6}$

51. $\theta = \dfrac{\pi}{3}$ or $\theta = \dfrac{4\pi}{3}$

53. $\theta = \dfrac{\pi}{3}$ or $\theta = \dfrac{2\pi}{3}$

55. $\theta = \dfrac{\pi}{3}$ or $\theta = \dfrac{5\pi}{3}$

57. $\theta = \dfrac{5\pi}{4}$ or $\theta = \dfrac{7\pi}{4}$

59. $\theta = \dfrac{\pi}{3}$ or $\theta = \dfrac{2\pi}{3}$

61. $2\sin^2\theta = 1$

$$\sin\theta = \pm\frac{\sqrt{2}}{2}$$

$$\theta = \frac{\pi}{4}, \frac{3\pi}{4}, \frac{5\pi}{4}, \frac{7\pi}{4}$$

63. $\tan^2\theta - \tan\theta = 0$

$\tan\theta(\tan\theta - 1) = 0$

$\tan\theta = 0$ $\qquad$ or $\tan\theta = 1$

$\theta = 0, \pi, 2\pi$ $\qquad \theta = \dfrac{\pi}{4}, \dfrac{5\pi}{4}$

65. $\sin 2\theta - \cos\theta = 0$

$2\sin\theta\cos\theta - \cos\theta = 0$

$\cos\theta(2\sin\theta - 1) = 0$

$\cos\theta = 0$ $\qquad$ or $2\sin\theta = 1$

$\theta = \dfrac{\pi}{2}, \dfrac{3\pi}{2}$ $\qquad \theta = \dfrac{\pi}{6}, \dfrac{5\pi}{6}$

67. $\sin\theta = \cos\theta$

Dividing both sides by $\cos\theta$ produces

$\tan\theta = 1$

$$\theta = \frac{\pi}{4}, \frac{5\pi}{4}.$$

69. $\cos^2\theta + \sin\theta = 1$

Using the identity $\cos^2\theta = 1 - \sin^2\theta$ produces

$1 - \sin^2\theta + \sin\theta = 1$

$\sin^2\theta - \sin\theta = 0$

$\sin\theta(\sin\theta - 1) = 0$

$\sin\theta = 0$ $\qquad$ or $\sin\theta = 1$

$\theta = 0, \pi, 2\pi$ $\qquad \theta = \dfrac{\pi}{2}.$

71. Let h be the height of the ladder against the house. Then

$$\sin 75° = \frac{h}{20} \text{ and}$$

$$h = 20 \sin 75°$$

$$\approx 19.32 \text{ feet.}$$

The ladder reaches about 19.32 feet up the house.

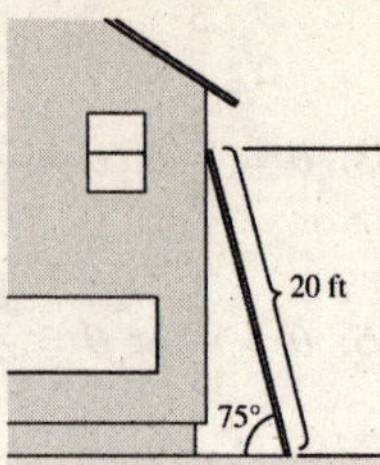

73.

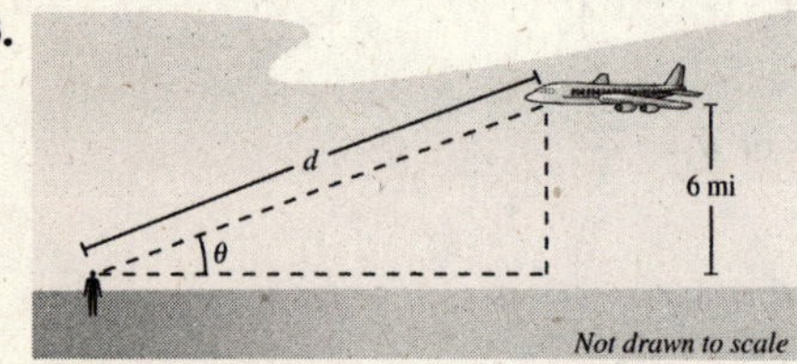

(a) $\sin 30° = \frac{6}{d} \Rightarrow d = \frac{6}{\sin 30°} = 12 \text{ mi}$

(b) $\sin 60° = \frac{6}{d} \Rightarrow d = \frac{6}{\sin 60°} = \frac{12}{\sqrt{3}} \approx 6.9 \text{ mi}$

(c) $\sin 90° = \frac{6}{d} \Rightarrow d = \frac{6}{\sin 90°} = 6 \text{ mi}$

75. (a)

$$\tan 82° = \frac{x}{45}$$

$$45 \tan 82° = x$$

$$320.19 \approx x$$

$$\text{Total height} \approx 123 + 320.19$$

$$= 443.19 \text{ m}$$

(b)

$$\cos 82° = \frac{45}{d}$$

$$d = \frac{45}{\cos 82°}$$

$$d \approx 323.34$$

The distance between you and your friend is about 323.34 meters.

77.

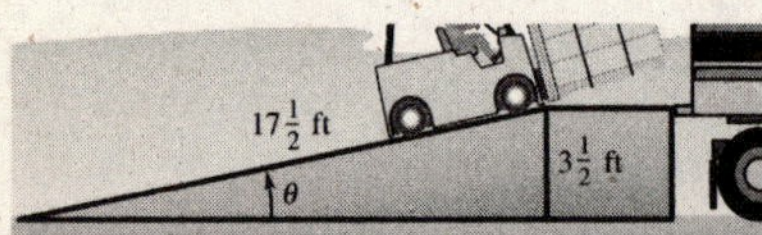

$$\sin \theta = \frac{3.5}{17.5}$$

$$\theta = \sin^{-1}\left(\frac{3.5}{17.5}\right) \approx 11.5°$$

79. Let h be the height of the mountain, and x be the distance from you and the mountain.

$$\tan 3.5° = \frac{h}{13 + x} \qquad \tan 9° = \frac{h}{x}$$

$$\tan 3.5°(13 + x) = h \qquad x \tan 9° = h$$

$$\tan 3.5°(13 + x) = x \tan 9°$$

$$13 \tan 3.5° + x \tan 3.5° = x \tan 9°$$

$$x \tan 3.5° - x \tan 9° = -13 \tan 3.5°$$

$$x(\tan 3.5° - \tan 9°) = -13 \tan 3.5°$$

$$x = \frac{-13 \tan 3.5°}{\tan 3.5° - \tan 9°}$$

$$\frac{-13 \tan 3.5°}{\tan 3.5° - \tan 9°}(\tan 9°) = h$$

$$1.3 \approx h$$

The mountain is about 1.3 miles high.

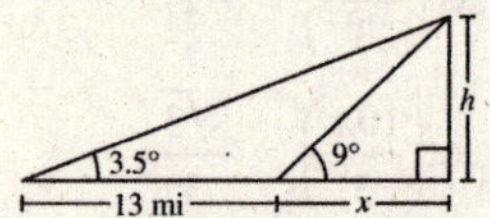

Not drawn to scale

81. (a) 10:00 P.M.:

$$T(0) = 98.6 + 4 \cos 0$$

$$= 98.6 + 4$$

$$= 102.6°\text{F}$$

(b) 4:00 A.M.:

$$T(6) = 98.6 + 4 \cos\left(\frac{6\pi}{36}\right)$$

$$= 98.6 + 4\left(\frac{\sqrt{3}}{2}\right)$$

$$\approx 102.1°\text{F}$$

(c) 10:00 A.M.:

$$T(12) = 98.6 + 4 \cos\left(\frac{12\pi}{36}\right)$$

$$= 98.6 + 4\left(\frac{1}{2}\right)$$

$$= 100.6°\text{F}$$

(d) The temperature returns to normal when

$$98.6 = 98.6 + 4 \cos\left(\frac{\pi t}{36}\right)$$

$$0 = \cos\left(\frac{\pi t}{36}\right)$$

$$\frac{\pi}{2} = \frac{\pi t}{36}$$

$$t = 18 \text{ or } 4 \text{ P.M.}$$

83.

x	0	2	4	6	8	10
$f(x)$	0	2.7021	2.7756	1.2244	1.2979	4

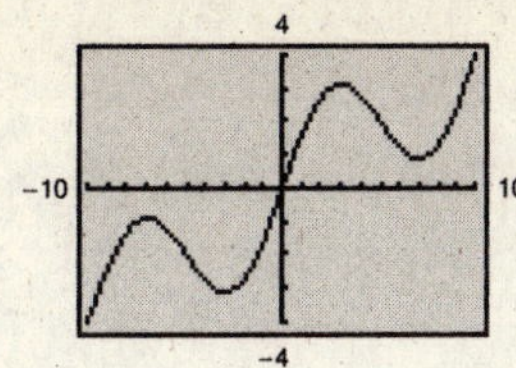

85. True

87. False

$$\sin^2 45° - \cos^2 45° = \left(\frac{\sqrt{2}}{2}\right)^2 - \left(\frac{\sqrt{2}}{2}\right)^2 = \frac{1}{2} - \frac{1}{2} = 0 \neq 1$$

Section 8.3 Graphs of Trigonometric Functions

Skills Warm Up

1. $\lim_{x\to 2}\left(x^2 + 4x + 2\right) = 2^2 + 4(2) + 2 = 14$

2. $\lim_{x\to 3}\left(x^3 - 2x^2 + 1\right) = 3^3 - 2(3)^2 + 1 = 10$

3. $\cos \frac{\pi}{2} = 0$

4. $\sin \pi = 0$

5. $\tan \frac{5\pi}{4} = 1$

6. $\cot \frac{2\pi}{3} = -\frac{\sqrt{3}}{3}$

7. $\sin \frac{11\pi}{6} = -\frac{1}{2}$

8. $\cos \frac{5\pi}{6} = -\frac{\sqrt{3}}{2}$

9. $\cos \frac{5\pi}{3} = \frac{1}{2}$

10. $\sin \frac{4\pi}{3} = -\frac{\sqrt{3}}{2}$

11. $\cos 15° \approx 0.9659$

12. $\sin 220° \approx -0.6428$

13. $\sin 275° \approx -0.9962$

14. $\cos 310° \approx 0.6428$

15. $\sin 103° \approx 0.9744$

16. $\cos 72° \approx 0.3090$

17. $\tan 327° \approx -0.6494$

18. $\tan 140° \approx -0.8391$

1. $y = 2 \sin 2x$

Period: $\frac{2\pi}{2} = \pi$

Amplitude: 2

3. $y = \frac{3}{2} \cos \frac{x}{2}$

Period: $\frac{2\pi}{1/2} = 4\pi$

Amplitude: $\frac{3}{2}$

5. $y = \frac{1}{2} \cos \pi x$

Period: $\frac{2\pi}{\pi} = 2$

Amplitude: $\frac{1}{2}$

7. $y = -\sin 3x$

Period: $\frac{2\pi}{3}$

Amplitude: 1

9. $y = -\dfrac{3}{2}\sin 6x$

Period: $\dfrac{2\pi}{6} = \dfrac{\pi}{3}$

Amplitude: $\dfrac{3}{2}$

11. $y = \dfrac{1}{2}\sin\dfrac{2x}{3}$

Period: $\dfrac{2\pi}{2/3} = 3\pi$

Amplitude: $\dfrac{1}{2}$

13. $y = 3\sin 4\pi x$

Period: $\dfrac{2\pi}{4\pi} = \dfrac{1}{2}$

Amplitude: 3

15. $y = 3\tan x$

Period: $\dfrac{2\pi}{1} = 2\pi$

17. $y = 3\sec 5x$

Period: $\dfrac{2\pi}{5}$

19. $y = \cot\dfrac{\pi x}{6}$

Period: $\dfrac{\pi}{\pi/6} = 6$

21. $y = \sec 2x$

The graph of this function has a period of π and matches graph (c).

23. $y = \cot\dfrac{\pi x}{2}$

The graph of this function has a period of 2 and matches graph (f).

25. $y = 2\csc\dfrac{x}{2}$

The graph of this function has a period of 4π and matches graph (b).

27. $y = \sin\dfrac{x}{2}$

Period: 4π

Amplitude: 1

x-intercepts:

$(0, 0), (2\pi, 0), (4\pi, 0)$

Maximum: $(\pi, 1)$

Minimum: $(3\pi, -1)$

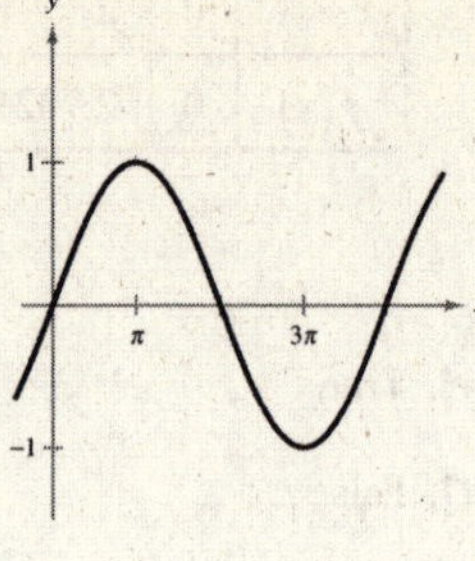

29. $y = 2\cos\dfrac{\pi x}{3}$

Period: 6

Amplitude: 2

x-intercepts:

$\left(-\frac{3}{2}, 0\right), \left(\frac{3}{2}, 0\right), \left(\frac{9}{2}, 0\right)$

Maxima: $(0, 2), (6, 2)$

Minimum: $(3, -2)$

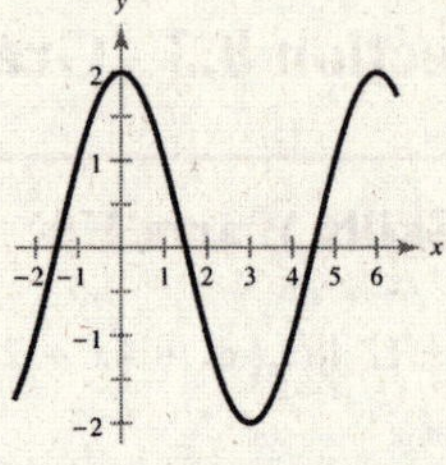

31. $y = -2\sin 6x$

Period: $\dfrac{\pi}{3}$

Amplitude: 2

x-intercepts:

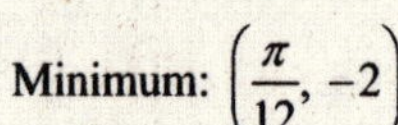

$(0, 0), \left(\dfrac{\pi}{6}, 0\right), \left(\dfrac{\pi}{3}, 0\right)$

Maximum: $\left(\dfrac{\pi}{4}, 2\right)$

Minimum: $\left(\dfrac{\pi}{12}, -2\right)$

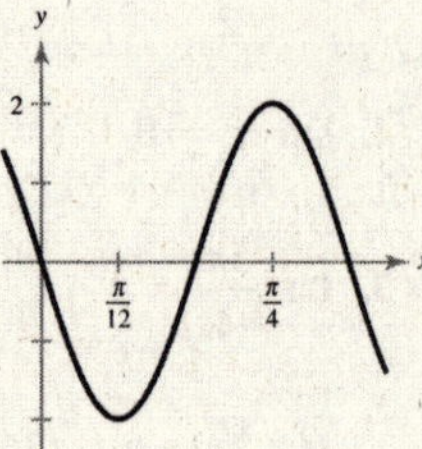

33. $y = \cos 2\pi x$

Period: 1

Amplitude: 1

x-intercepts: $\left(\frac{1}{4}, 0\right), \left(\frac{3}{4}, 0\right)$

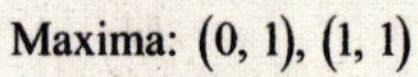

Maxima: $(0, 1), (1, 1)$

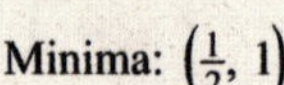

Minima: $\left(\frac{1}{2}, 1\right)$

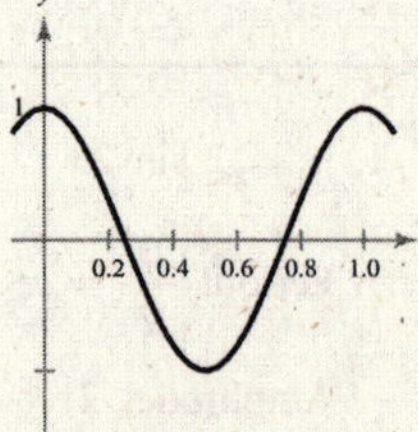

35. $y = 2\tan x$

Period: π

x-intercepts: $(0, 0), (\pi, 0)$

Asymptotes:

$x = -\dfrac{\pi}{2}, x = \dfrac{\pi}{2}, x = \dfrac{3\pi}{2}$

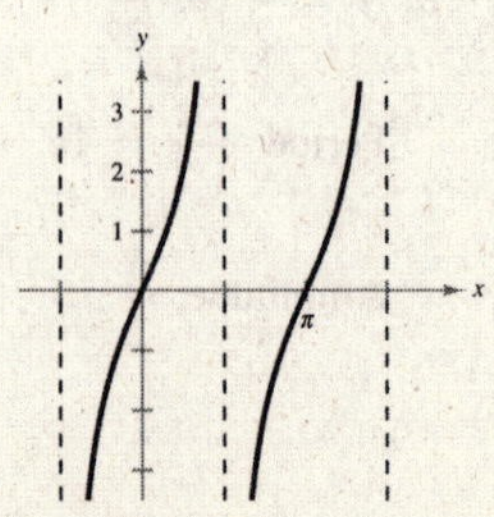

37. $y = \frac{1}{2} \tan \frac{\pi x}{2}$

Period: 2

x-intercepts: $(0, 0), (2, 0)$

Asymptotes:

$x = -1, x = 1, x = 3$

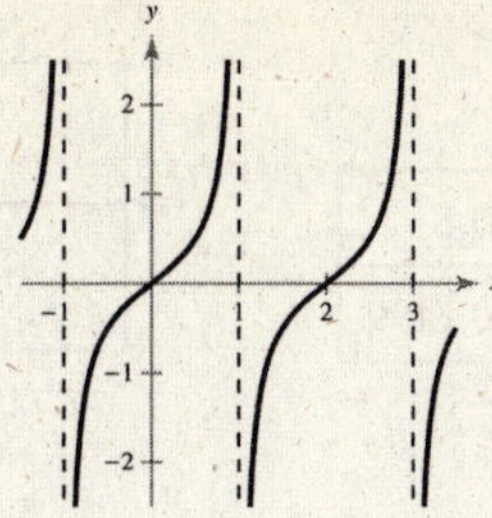

39. $y = 2 \sec 4x$

Period: $\frac{\pi}{2}$

Asymptotes: $x = \frac{\pi}{8}, x = \frac{3\pi}{8}, x = \frac{5\pi}{8}, x = \frac{7\pi}{8}$

Maximum: $\left(\frac{\pi}{4}, -2\right)$

Minimum: $\left(\frac{\pi}{2}, 2\right)$

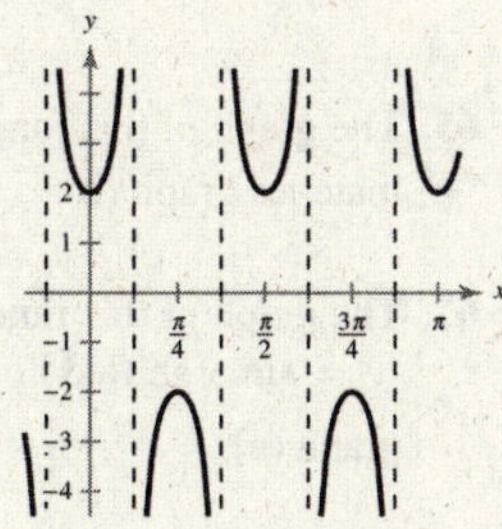

41. $y = -\sin \frac{2\pi x}{3}$

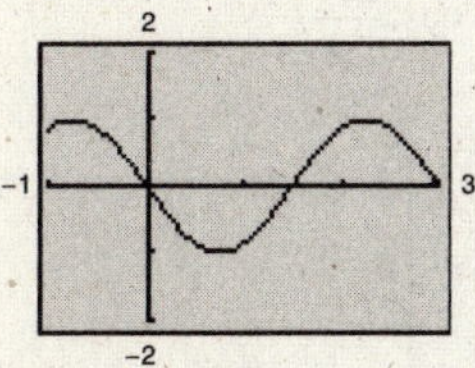

43. $y = \cot 2x$

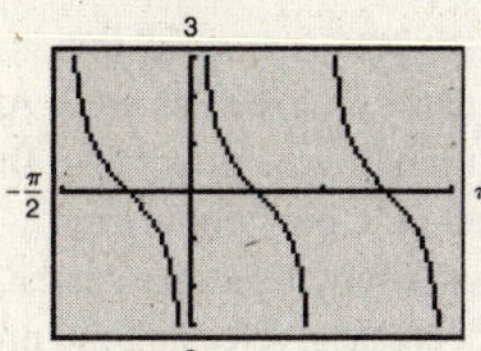

45. $y = \csc \frac{2x}{3}$

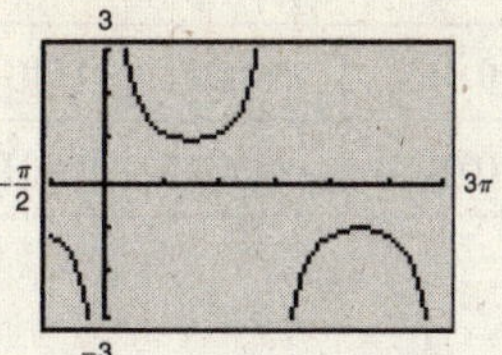

47. $y = 2 \sec 2x$

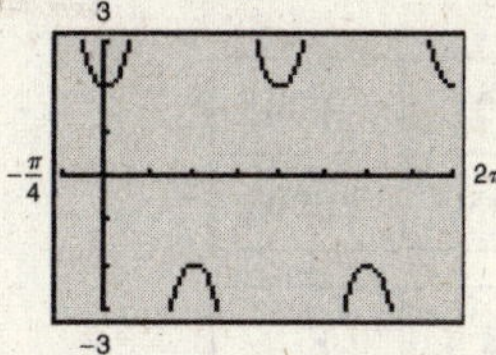

49. $y = \csc 2\pi x$

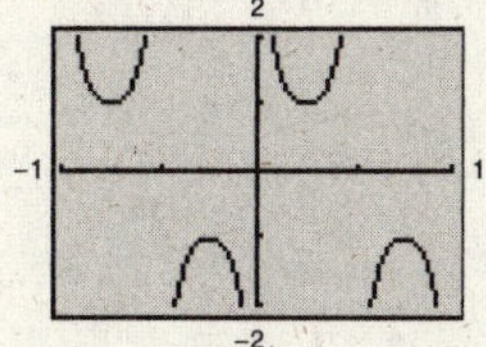

51. $f(x) = \frac{\sin 4x}{2x}$

x	−0.1	−0.01	−0.001
$f(x)$	1.9471	1.9995	2.0000

x	0.001	0.01	0.1
$f(x)$	2.0000	1.9995	1.9471

$$\lim_{x \to 0} \frac{\sin 4x}{2x} = 2$$

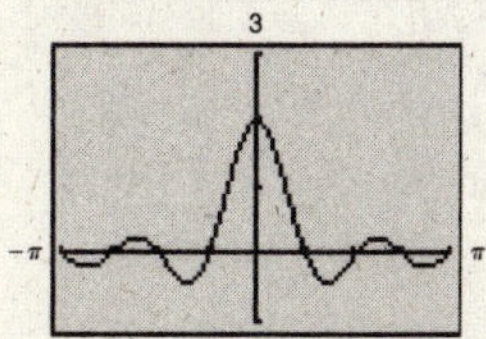

53. $y = \frac{\tan 3x}{\tan 4x}$

x	−0.1	−0.01	−0.001	0.001	0.01	0.1
$f(x)$	0.7317	0.7498	0.75	0.75	0.7498	0.7317

$$\lim_{x \to 0} \frac{\tan 3x}{\tan 4x} = 0.75$$

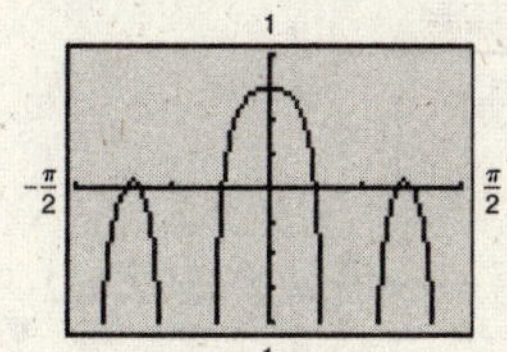

55. $f(x) = \dfrac{3(1 - \cos x)}{x}$

x	-0.1	-0.01	-0.001	0.001	0.01	0.1
$f(x)$	-0.1499	-0.0150	-0.0015	0.0015	0.0150	0.1499

$$\lim_{x\to 0} \frac{3(1 - \cos x)}{x} = 0$$

57. $f(x) = \dfrac{\tan 2x}{x}$

x	-0.1	-0.01	-0.001
$f(x)$	2.0271	2.0003	2.0000

x	0.001	0.01	0.1
$f(x)$	2.0000	2.0003	2.0271

$$\lim_{x\to 0} \frac{\tan 2x}{x} = 2$$

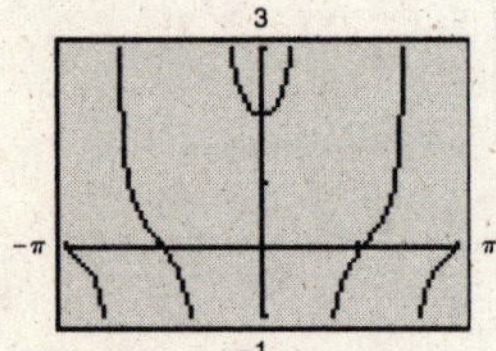

59. $f(x) = \dfrac{\sin^2 x}{x}$

x	-0.1	-0.01	-0.001
$f(x)$	-0.0997	-0.0100	-0.0010

x	0.001	0.01	0.1
$f(x)$	0.0010	0.0100	0.0997

$$\lim_{x\to 0} \frac{\sin^2 x}{x} = 0$$

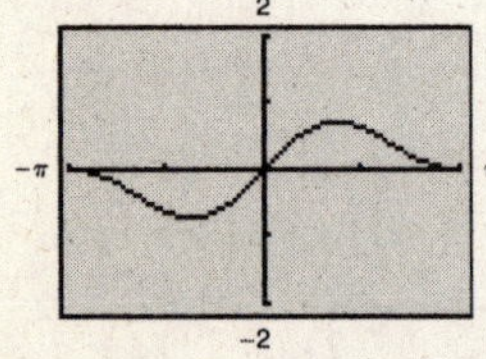

61. $f(x) = a \cos x + d$

Amplitude: $\frac{1}{2}[3 - (-1)] = 2 \Rightarrow a = 2$

Vertical shift one unit upward $\Rightarrow d = 1$

So, $f(x) = 2 \cos x + 1$.

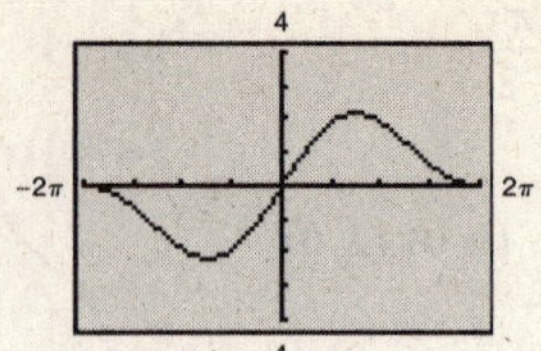

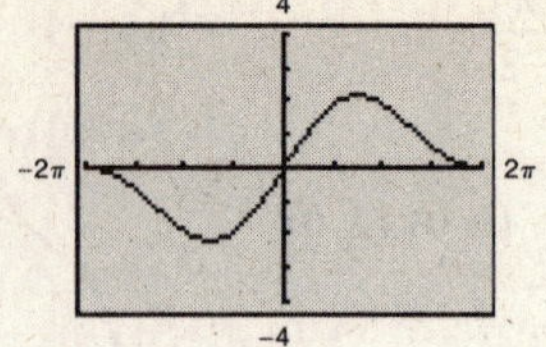

63. $f(x) = a \cos x + d$

Amplitude: $\frac{1}{2}[8 - 0] = 4 \Rightarrow a = 4$

Reflection in the x-axis: $-a = -4$

Vertical shift four units upward $\Rightarrow d = 4$

So, $f(x) = -4 \cos x + 4$.

65. The graph of this function has a period of 2π and matches graph (b).

67. The graph of this function is the graph of $y = \sin x$ shifted horizontally π units and matches graph (a).

69. (a) $P = 8000 + 2500 \sin \dfrac{2\pi t}{24}$

$p = 12{,}000 + 4000 \cos \dfrac{2\pi t}{24}$

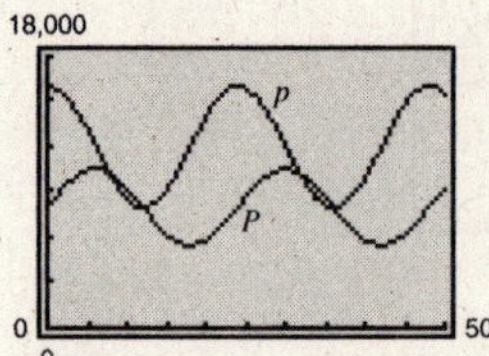

(b) As the population of the prey increases, the population of the predator increases as well. At some point, the predator eliminates the prey faster than the prey can reproduce, and the prey population decreases rapidly. As the prey becomes scarce, the predator population decreases, releasing the prey from predator pressure, and the cycle begins again.

71. $P = \sin \dfrac{2\pi t}{23}$, $E = \sin \dfrac{2\pi t}{28}$, and $I = \sin \dfrac{2\pi t}{33}$ where $t \ge 0$.

$$P(8930) = \sin \frac{2\pi(8930)}{23} \approx 0.9977$$

$$E(8930) = \sin \frac{2\pi(8930)}{28} \approx -0.4339$$

$$I(8930) = \sin \frac{2\pi(8930)}{33} \approx -0.6182$$

The physical cycle is about 0.9977, the emotional cycle is about -0.4339, and the intellectual cycle is about -0.6182.

73. $v = 0.9 \sin \dfrac{\pi t}{3}$

(a) Period $= \dfrac{2\pi}{\pi/3} = 6$ sec

(b) Cycles per minute $= \dfrac{60}{6} = 10$

(c)

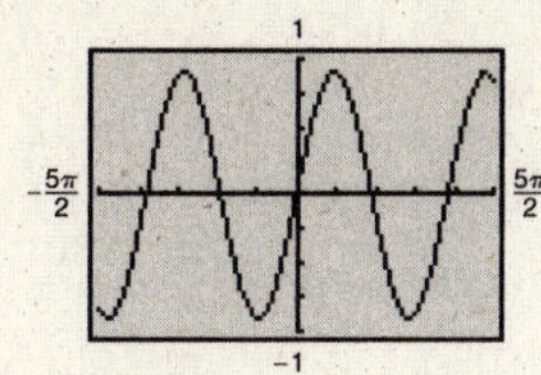

75. $y = 0.001 \sin 880\pi t$

(a) Period $= \dfrac{2\pi}{880\pi} = \dfrac{1}{440}$

(b) Frequency $= \dfrac{1}{\text{period}} = \dfrac{1}{1/440} = 440$

(c)

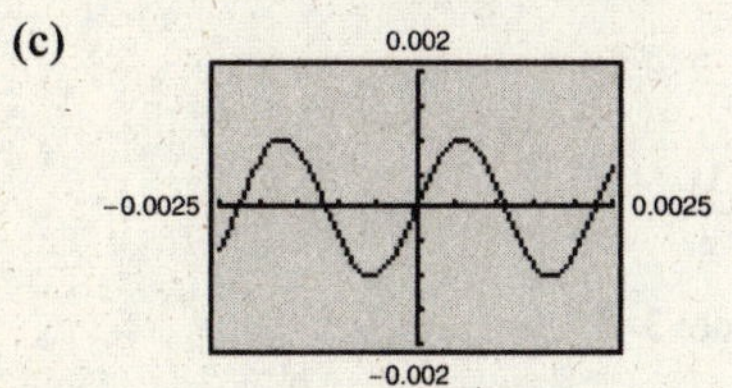

77. (a)

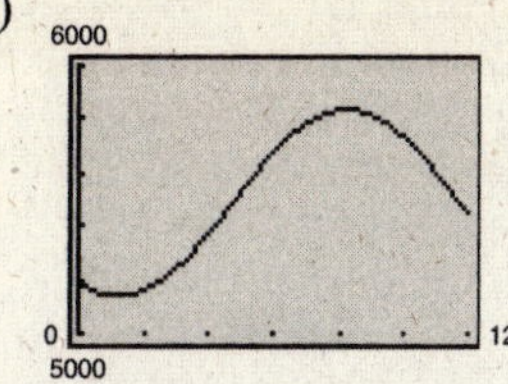

(b) Yes. The graph of W and $y = 5500$ intersect at $t \approx 4.81$ and $t \approx 11.64$. The number of construction workers exceeded 5.5 million from late April to mid November.

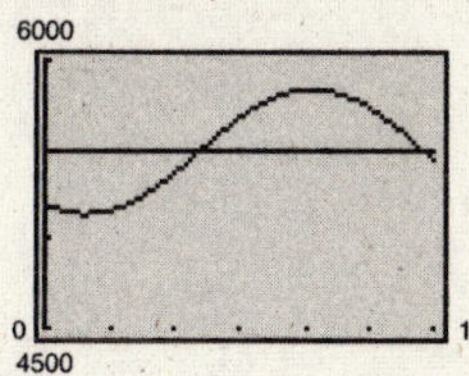

79. (a) A

(b) B

(c) B

(d) They are reciprocals of each other.

81. (a)

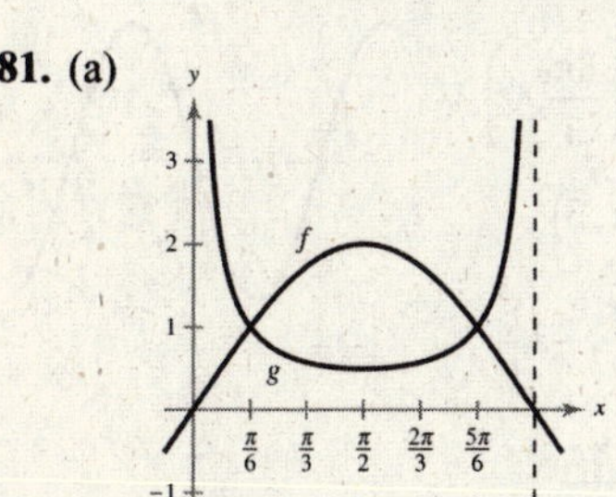

(b) $f > g$ over the interval $\dfrac{\pi}{x} < x < \dfrac{5\pi}{x}$.

(c) As x approaches π, $f(x) = 2 \sin x$ approaches 0. Because $g(x) = \dfrac{1}{f(x)}$, $g(x) = 0.5 \csc x$ approaches infinity as x approaches π.

83. False. The amplitude is $|-3| = 3$.

85. True

Chapter 8 Quiz Yourself

1. $15°\left(\dfrac{\pi \text{ radians}}{180°}\right) = \dfrac{\pi}{12}$ radian

2. $105°\left(\dfrac{\pi \text{ radians}}{180°}\right) = \dfrac{7\pi}{12}$ radians

3. $-80°\left(\dfrac{\pi \text{ radians}}{180°}\right) = -\dfrac{4\pi}{9}$ radians

4. $35°\left(\dfrac{\pi \text{ radians}}{180°}\right) = \dfrac{7\pi}{36}$ radian

5. $\frac{2\pi}{3}\left(\frac{180°}{\pi}\right) = 120°$

6. $\frac{4\pi}{15}\left(\frac{180°}{\pi}\right) = 48°$

7. $-\frac{4\pi}{3}\left(\frac{180°}{\pi}\right) = -240°$

8. $\frac{11\pi}{12}\left(\frac{180°}{\pi}\right) = 165°$

9. $\sin\left(-\frac{\pi}{4}\right) = -\frac{\sqrt{2}}{2}$

10. $\cos 240° = -\frac{1}{2}$

11. $\tan\frac{5\pi}{6} = -\frac{\sqrt{3}}{3}$

12. $\cot 45° = 1$

13. $\sec(-60°) = 2$

14. $\csc\frac{3\pi}{2} = -1$

15. $\tan\theta - 1 = 0$

$\tan\theta = 1$

$\theta = \frac{\pi}{4}, \frac{5\pi}{4}$

16. $\cos^2\theta - 2\cos\theta + 1 = 0$

$(\cos\theta - 1)(\cos\theta - 1) = 0$

$\cos\theta = 1$

$\theta = 0, 2\pi$

17. $\sin^2\theta = 3\cos^2\theta$

$\frac{\sin^2\theta}{\cos^2\theta} = 3$

Using the identity $\frac{\sin\theta}{\cos\theta} = \tan\theta$ produces the following.

$\tan^2\theta = 3$

$\tan\theta = \pm\sqrt{3}$

$\theta = \frac{\pi}{3}, \frac{2\pi}{3}, \frac{4\pi}{3}, \frac{5\pi}{3}$

18. The angle θ is $90° - 60° = 30°$.

Find a by using the Pythagorean Theorem.

$a = \sqrt{10^2 - 5^2} = \sqrt{75} = 5\sqrt{3}$

19. The angle θ is $90° - 50° = 40°$.

$\cos 50° = \frac{a}{16}$

$a = 16\cos 50°$

$a \approx 10.2846$

20. The angle θ is $90° - 25° = 65°$.

$\tan 25° = \frac{a}{12}$

$a = 12\tan 25°$

$a \approx 5.5957$

21. $\sin 35° = \frac{d}{500}$

$500\sin 35° = d$

$286.8 \text{ ft} \approx d$

22. $y = -3\sin\frac{3x}{4}$

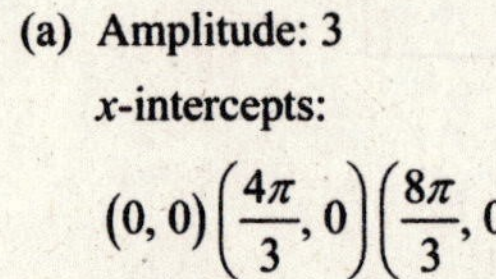

(a) Amplitude: 3

x-intercepts:

$(0, 0)\left(\frac{4\pi}{3}, 0\right)\left(\frac{8\pi}{3}, 0\right)$

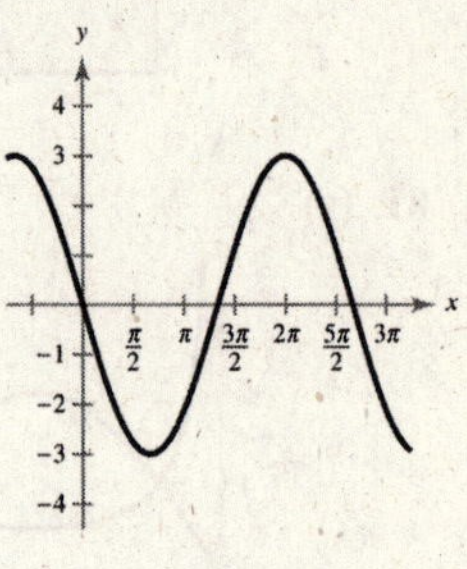

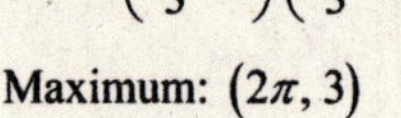

Maximum: $(2\pi, 3)$

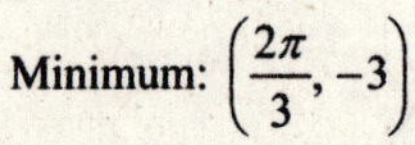

Minimum: $\left(\frac{2\pi}{3}, -3\right)$

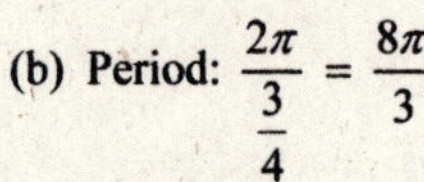

(b) Period: $\frac{2\pi}{\frac{3}{4}} = \frac{8\pi}{3}$

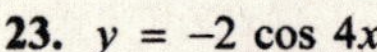

23. $y = -2\cos 4x$

(a) Amplitude: 2

x-intercepts:

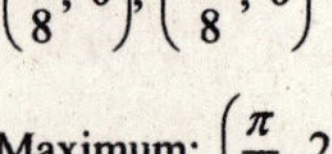

$\left(\frac{\pi}{8}, 0\right), \left(\frac{3\pi}{8}, 0\right)$

Maximum: $\left(\frac{\pi}{4}, 2\right)$

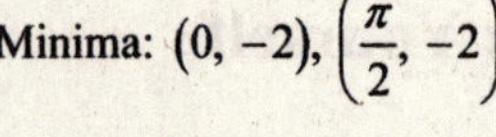

Minima: $(0, -2), \left(\frac{\pi}{2}, -2\right)$

(b) Period: $\frac{2\pi}{4} = \frac{\pi}{2}$

24. $y = \tan \dfrac{\pi x}{3}$

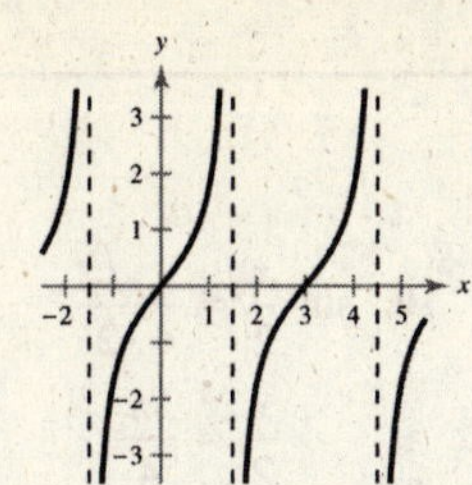

(a) x-intercepts:

$(0, 0), (3, 0)$

Asymptotes:

$x = -\dfrac{3}{2}, x = \dfrac{3}{2}, x = \dfrac{9}{2}$

(b) Period: $\dfrac{\pi}{\frac{\pi}{3}} = 3$

25. $S = 53.5 + 40.5 \cos \dfrac{\pi t}{6}$

(a)

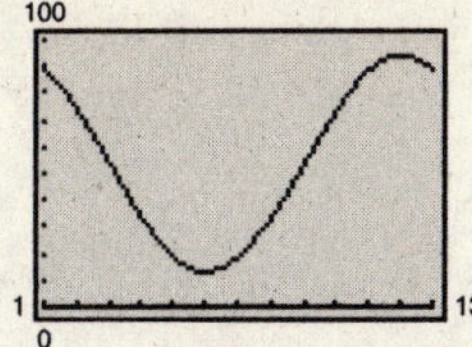

(b) Sales are maximum when $t = 12$ (December) and minimum when $t = 6$ (June).

Section 8.4 Derivatives of Trigonometric Functions

Skills Warm Up

1. $f(x) = 3x^3 - 2x^2 + 4x - 7$

$f'(x) = 9x^2 - 4x + 4$

2. $g(x) = (x^3 + 4)^4$

$g'(x) = 4(x^3 + 4)^3(3x^2)$

$= 12x^2(x^3 + 4)^3$

3. $f(x) = (x - 1)(x^2 + 2x + 3)$

$f'(x) = (x - 1)(2x + 2) + (x^2 + 2x + 3)$

$= 2x^2 - 2 + x^2 + 2x + 3$

$= 3x^2 + 2x + 1$

4. $g(x) = \dfrac{2x}{x^2 + 5}$

$$g'(x) = \frac{(x^2 + 5)(2) - 2x(2x)}{(x^2 + 5)^2} = \frac{2x^2 + 10 - 4x^2}{(x^2 + 5)^2} = \frac{-2x^2 + 10}{(x^2 + 5)^2} = \frac{2(5 - x^2)}{(x^2 + 5)^2}$$

5. $f(x) = x^2 + 4x + 1$

$f'(x) = 2x + 4 = 0$

Critical number: $x = -2$

Relative minimum: $(-2, -3)$

Interval	$-\infty < x < -2$	$-2 < x < \infty$
Sign of f'	$f' < 0$	$f' > 0$
Conclusion	Decreasing	Increasing

6. $f(x) = \frac{1}{3}x^3 - 4x + 2$

$f'(x) = x^2 - 4 = 0$

Critical numbers: $x = \pm 2$

Relative maximum: $\left(-2, \frac{22}{3}\right)$

Relative minimum: $\left(2, -\frac{10}{3}\right)$

Interval	$-\infty < x < -2$	$-2 < x < 2$	$2 < x < \infty$
Sign of f'	$f' > 0$	$f' < 0$	$f' > 0$
Conclusion	Increasing	Decreasing	Increasing

7. $\sin x = \dfrac{\sqrt{3}}{2}$

$x = \dfrac{\pi}{3}, \dfrac{2\pi}{3}$

8. $\cos x = -\dfrac{1}{2}$

$x = \dfrac{2\pi}{3}, \dfrac{4\pi}{3}$

Skills Warm Up —*continued*—

9. $\cos \frac{x}{2} = 0$

$\frac{x}{2} = \frac{\pi}{2}$

$x = \pi$

10. $\sin \frac{x}{2} = -\frac{\sqrt{2}}{2}$

$\frac{x}{2} = \frac{5\pi}{4}$

No solution in $0 \le x \le 2\pi$

1. $y = \sin\left(\frac{x}{3}\right)$

$y' = \cos\left(\frac{x}{3}\right)\left(\frac{1}{3}\right)$

$= \frac{1}{3}\cos\left(\frac{x}{3}\right)$

3. $y = \tan\left(4x^3\right)$

$y' = \sec^2\left(4x^3\right)\left(12x^2\right)$

$= 12x^2 \sec^2\left(4x^3\right)$

5. $f(t) = \tan(5t)$

$f'(t) = \sec^2(5t)(5)$

$= 5\sec^2(5t)$

7. $y = \sin^2 x = (\sin x)^2$

$y' = 2(\sin x)(\cos x)$

$= 2\sin x \cos x$

9. $y = \sec \pi x$

$y' = \pi \sec \pi x \tan \pi x$

11. $y = \cot(2x + 1)$

$y' = -\csc^2(2x + 1)(2)$

$= -2\csc^2(2x + 1)$

13. $y = \sqrt{\tan 2x} = (\tan(2x))^{1/2}$

$y' = \frac{1}{2}(\tan(2x))^{-1/2}\left(\sec^2(2x)(2)\right)$

$= \frac{\sec^2(2x)}{\sqrt{\tan(2x)}}$

15. $f(t) = t^2 \cos t$

$f'(t) = -t^2 \sin t + 2t \cos t$

17. $g(t) = \frac{\cos t}{t}$

$g'(t) = \frac{t(-\sin t) - (\cos t)(1)}{t^2}$

$= -\frac{t\sin t + \cos t}{t^2}$

19. $y = e^{x^2} \sec x$

$y' = e^{x^2}(\sec x \tan x) + 2xe^{x^2} \sec x$

$= e^{x^2} \sec x(\tan x + 2x)$

21. $y = \cos 3x + \sin^2 x$

$y' = -\sin 3x(3) + 2\sin x \cos x$

$= -3\sin 3x + 2\sin x \cos x$

23. $y = x \sin \frac{1}{x}$

$y' = x\cos\left(\frac{1}{x}\right)\left(-\frac{1}{x^2}\right) + \sin\frac{1}{x}$

$= \sin\frac{1}{x} - \frac{1}{x}\cos\frac{1}{x}$

25. $y = 2\tan^2 4x$

$y' = 4\tan(4x)\sec^2 4x(4) = 16\tan 4x \sec^2 4x$

Equivalently,

$y' = 16\tan 4x\left(1 + \tan^2 4x\right) = 16\tan 4x + 16\tan^3 4x.$

27. $y = e^{2x} \sin 2x$

$y' = e^{2x}(\cos 2x(2)) + 2e^{2x}\sin 2x$

$= 2e^{2x}(\cos 2x + \sin 2x)$

29. $y = (\cos x)^2$

$y' = 2(\cos x)(-\sin x)$

$= -2\cos x \sin x$

$= -\sin 2x$

31. $y = \cos^2 x - \sin^2 x$

$y' = -2 \cos x \sin x - 2 \sin x \cos x$

$= -4 \cos x \sin x = -2 \sin 2x$

33. $y = (\sin x)^2 - \cos 2x$

$y' = 2 \sin x \cos x + 2 \sin 2x$

$= \sin 2x + 2 \sin 2x$

$= 3 \sin 2x$

35. $y = \tan x - x$

$y' = \sec^2 x - 1 = \tan^2 x$

37. $y = \frac{1}{3}(\sin x)^3 - \frac{1}{5}(\sin x)^5$

$y' = (\sin x)^2 \cos x - (\sin x)^4 \cos x$

$= \sin^2 x \cos x(1 - \sin^2 x)$

$= \sin^2 x \cos x \cos^2 x$

$= \sin^2 x \cos^3 x$

39. $y = \ln(\sin^2 x)$

$y' = \frac{1}{\sin^2 x} 2 \sin x \cos x$

$= \frac{2 \cos x}{\sin x}$

$= 2 \cot x$

41. $y = \tan x, \quad \left(-\frac{\pi}{4}, -1\right)$

$y' = \sec^2 x, \quad y'\left(-\frac{\pi}{4}\right) = 2$

$y - (-1) = 2\left[x - \left(-\frac{\pi}{4}\right)\right]$

$y = 2x - 1 + \frac{\pi}{2}$

43. $y = \sin 4x, \quad (\pi, 0)$

$y' = 4 \cos 4x, \quad y'(\pi) = 4$

$y - 0 = 4(x - \pi)$

$y = 4x - 4\pi$

45. $y = \cot x, \quad \left(\frac{3\pi}{4}, -1\right)$

$y' = -\csc^2 x, \quad y'\left(\frac{3\pi}{4}\right) = -2$

$y - (-1) = -2\left(x - \frac{3\pi}{4}\right)$

$y = -2x - 1 + \frac{3\pi}{2}$

47. $y = \ln(\sin x + 2), \quad \left(\frac{3\pi}{2}, 0\right)$

$y' = \frac{1}{\sin x + 2}(\cos x), \quad y'\left(\frac{3\pi}{2}\right) = 0$

$y - 0 = 0\left(x - \frac{3\pi}{2}\right)$

$y = 0$

49. $\sin x + \cos 2y = 1$

$\cos x - 2 \sin 2y \frac{dy}{dx} = 0$

$\frac{dy}{dx} = \frac{\cos x}{2 \sin 2y}$

At $\left(\frac{\pi}{2}, \frac{\pi}{4}\right)$, you have $\frac{dy}{dx} = 0$.

51. $y = \sin \frac{5x}{4}$

$y' = \frac{5}{4} \cos \frac{5x}{4}$

The slope of the tangent line at $(0, 0)$ is $\frac{5}{4} = 1.25$. There is one complete cycle of the graph in the interval $[0, 2\pi]$.

53. $y = \sin 2x$

$y' = 2 \cos 2x$

The slope of the tangent line at $(0, 0)$ is 2. There are two complete cycles of the graph in the interval $[0, 2\pi]$.

55. $y = \sin x$

$y' = \cos x$

The slope of the tangent line at $(0, 0)$ is 1. There is one complete cycle of the graph in the interval $[0, 2\pi]$.

57. $y = \cos\left(\frac{\pi x}{2}\right)$

$y' = -\sin\left(\frac{\pi x}{2}\right)\left(\frac{\pi}{2}\right)$

$= -\frac{\pi}{2} \sin\left(\frac{\pi x}{2}\right)$

$-\frac{\pi}{2} \sin\left(\frac{\pi x}{2}\right) = 0$

$\sin\left(-\frac{\pi x}{2}\right) = 0$

Critical numbers: $x = 2, 4, 6$

Relative maximum: $(4, 1)$

Relative minima: $(2, -1), (6, -1)$

59. $y = \cos^2 x = (\cos x)^2$

$y' = 2(\cos x)(-\sin x) = -2\sin x\cos x$

$-2\sin x\cos x = 0$

$\sin x = 0$ or $\cos x = 0$

Critical numbers: $x = \frac{\pi}{2}, \pi, \frac{3\pi}{2}$

Relative maximum: $(\pi, 1)$

Relative minima: $\left(\frac{\pi}{2}, 0\right), \left(\frac{3\pi}{2}, 0\right)$

61. $y = 2\sin x + \sin 2x$

$y' = 2\cos x + 2\cos 2x$

$2\cos x + 2\cos 2x = 0$

$2\left[\cos x + 2\cos^2 x - 1\right] = 0$

$2\cos^2 x + \cos x - 1 = 0$

$(2\cos x - 1)(\cos x + 1) = 0$

$\cos x = \frac{1}{2}$ or $\cos x = -1$

Critical numbers: $x = \frac{\pi}{3}, \pi, \frac{5\pi}{3}$

Relative minimum: $\left(\frac{5\pi}{3}, -\frac{3\sqrt{3}}{2}\right)$

Relative maximum: $\left(\frac{\pi}{3}, \frac{3\sqrt{3}}{2}\right)$

63. $y = e^x\cos x$

$y' = -e^x\sin x + e^x\cos x$

$-e^x\sin x + e^x\cos x = 0$

$-e^x(\sin x - \cos x) = 0$

$\sin x = \cos x$

$\tan x = 1$

Critical numbers: $x = \frac{\pi}{4}, \frac{5\pi}{4}$

Relative minimum: $\left(\frac{5\pi}{4}, -\frac{\sqrt{2}}{2}e^{5\pi/4}\right)$

Relative maximum: $\left(\frac{\pi}{4}, \frac{\sqrt{2}}{2}e^{\pi/4}\right)$

65. $y = x - 2\sin x$

$y' = 1 - 2\cos x$

$1 - 2\cos x = 0$

$\cos x = \frac{1}{2}$

Critical numbers: $x = \frac{\pi}{3}, \frac{5\pi}{3}$

Relative minimum: $\left(\frac{\pi}{3}, \frac{\pi}{3} - \sqrt{3}\right)$

Relative maximum: $\left(\frac{5\pi}{3}, \frac{5\pi}{3} + \sqrt{3}\right)$

67. $y = x\sin x$

$y' = x(\cos x) + (\sin x)(1) = x\cos x + \sin x$

$y'' = x(-\sin x) + \cos x(1) + \cos x$

$= -x\sin x + 2\cos x$

69. $y = \cos(x^2)$

$y' = -\sin(x^2)(2x) = -2x\sin(x^2)$

$y'' = -2\left[x\cos(x^2)(2x) + \sin(x^2)(1)\right]$

$= -2\left(2x^2\cos(x^2) + \sin(x^2)\right)$

71. $h(t) = 0.20t + 0.03\sin 2\pi t$

$h'(t) = 0.20 + 0.06\pi\cos 2\pi t$

$0.20 + 0.06\pi\cos 2\pi t = 0$

$\cos 2\pi t = -\frac{10}{3\pi}$

$t = 0, \frac{1}{2}$

(a) $h'(t)$ is maximum when $t = 0$. The growth rate is a maximum at midnight.

(b) $h'(t)$ is minimum when $t = \frac{1}{2}$. The growth rate is a minimum at noon.

73. $W = 5488 + 347.6\sin(0.45t + 4.153)$

$W' = 347.6\cos(0.45t + 4.153)(0.45)$

$= 156.42\cos(0.45t + 4.153)$

$156.42\cos(0.45t + 4.153) = 0$

$0.45t + 4.153 = \frac{5\pi}{2}$

$t \approx 8.22 \Rightarrow$ August

During August ($t \approx 8.22$), the number of construction workers was approximately 5836 thousand or 5,836,000.

Note: The angle $0.45t + 4.153$ was set equal to $\frac{5\pi}{2}$ in order for t to lie within the interval $[1, 12]$.

75. $D = 12.12 + 1.87\cos\left(\dfrac{\pi(t + 5.83)}{6}\right)$

$$D' = 1.87\left[-\sin\left(\frac{\pi(t + 5.83)}{6}\right)\left(\frac{\pi}{6}\right)\right] = -\frac{1.87\pi}{6}\sin\left(\frac{\pi(t + 5.83)}{6}\right)$$

$$-\frac{1.87\pi}{6}\sin\left(\frac{\pi(t + 5.83)}{6}\right) = 0$$

$$\frac{\pi(t + 5.83)}{6} = 2\pi$$

$$t \approx 6.17 \Rightarrow \text{June}$$

The maximum number of daylight hours occurred in June $(t \approx 6.17)$ with 14 hours of daylight.

77. (a) $\dfrac{d}{dx}[\tan x] = \dfrac{d}{dx}\left[\dfrac{\sin x}{\cos x}\right]$

$$= \frac{(\cos x)(\cos x) - (\sin x)(-\sin x)}{(\cos x)^2}$$

$$= \frac{\cos^2 x + \sin^2 x}{\cos^2 x}$$

$$= \frac{1}{\cos^2 x}$$

$$= \sec^2 x$$

(b) $\dfrac{d}{dx}[\sec x] = \dfrac{d}{dx}\left[\dfrac{1}{\cos x}\right]$

$$= \frac{(\cos x)(0) - (1)(-\sin x)}{(\cos x)^2}$$

$$= \frac{\sin x}{\cos^2 x}$$

$$= \left(\frac{1}{\cos x}\right)\left(\frac{\sin x}{\cos x}\right)$$

$$= \sec x \tan x$$

(c) $\dfrac{d}{dx}[\cot x] = \dfrac{d}{dx}\left[\dfrac{\cos x}{\sin x}\right]$

$$= \frac{(\sin x)(-\sin x) - (\cos x)(\cos x)}{(\sin x)^2}$$

$$= \frac{-\sin^2 x - \cos^2 x}{\sin^2 x}$$

$$= \frac{-(\sin^2 x + \cos^2 x)}{\sin^2 x}$$

$$= -\frac{1}{\sin^2 x}$$

$$= -\csc^2 x$$

(d) $\dfrac{d}{dx}[\csc x] = \dfrac{d}{dx}\left[\dfrac{1}{\sin x}\right]$

$$= \frac{(\sin x)(0) - (1)(\cos x)}{(\sin x)^2}$$

$$= -\frac{\cos x}{\sin^2 x}$$

$$= -\left(\frac{1}{\sin x}\right)\left(\frac{\cos x}{\sin x}\right)$$

$$= -\csc x \cot x$$

79. (a)

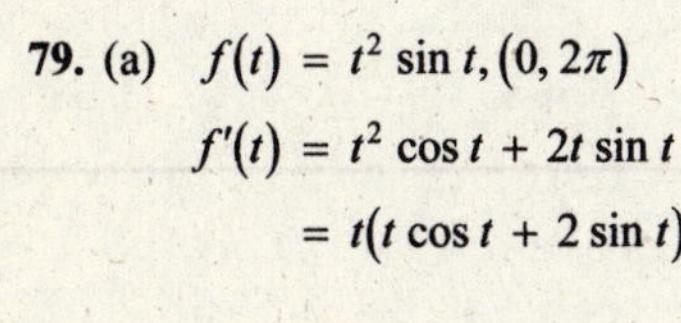

$f(t) = t^2 \sin t, (0, 2\pi)$

$f'(t) = t^2 \cos t + 2t \sin t$

$\quad = t(t \cos t + 2 \sin t)$

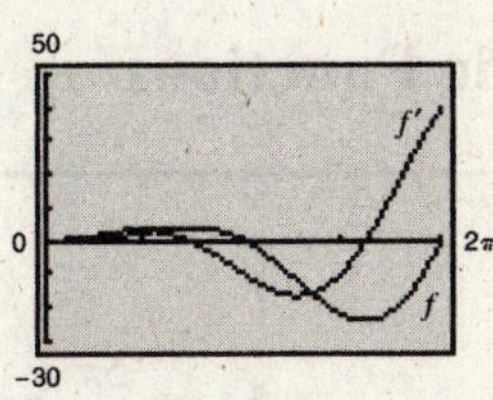

(b) $f'(t) = 0$ when $t = 0, t \approx 2.289$, and $t \approx 5.087$.

(c)

Interval	$(0, 2.289)$	$(2.289, 5.087)$	$(5.087, 2\pi)$
Sign of f'	$f' > 0$	$f' < 0$	$f' > 0$
Conclusion	f is increasing.	f is decreasing.	f is increasing.

(d) Relative maximum: $(2.289, 3.945)$

Relative minimum: $(5.087, -24.083)$

81. (a) $f(x) = \sin x - \frac{1}{3}\sin 3x + \frac{1}{5}\sin 5x, (0, \pi)$

$f'(x) = \cos x - \cos 3x + \cos 5x$

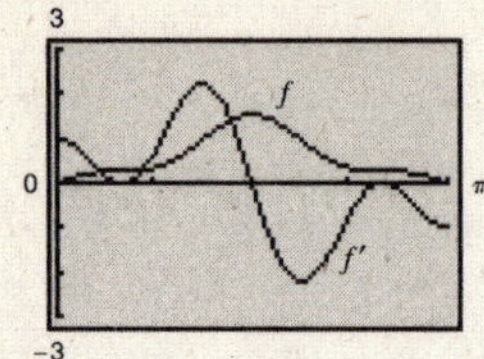

(b) $f'(x) = 0$ when $x \approx 0.524$, $x \approx 1.571$, and $x \approx 2.618$.

(c)

Interval	$(0, 0.524)$	$(0.524, 1.571)$	$(1.571, 2.618)$	$(2.618, \pi)$
Sign of f'	$f' > 0$	$f' > 0$	$f' < 0$	$f' < 0$
Conclusion	f is increasing.	f is increasing.	f is decreasing.	f is decreasing.

(d) Relative maximum: $\left(\frac{\pi}{2}, 1.533\right)$

83. (a) $f(x) = \sqrt{2x}\sin x, (0, 2\pi)$

$f'(x) = \sqrt{2x}\cos x + \frac{\sin x}{\sqrt{2x}}$

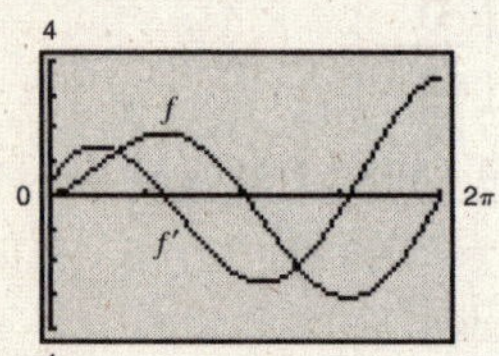

(b) $f'(x) = 0$ when $x \approx 1.837$ and $x \approx 4.816$.

(c)

Interval	$(0, 1.837)$	$(1.837, 4.816)$	$(4.816, 2\pi)$
Sign of f'	$f' > 0$	$f' < 0$	$f' > 0$
Conclusion	f is increasing.	f is decreasing.	f is increasing.

(d) Relative maximum: $(1.837, 1.849)$

Relative minimum: $(4.816, -3.087)$

85. False. If $y = (1 - \cos x)^{1/2}$, then

$y' = \frac{1}{2}(1 - \cos x)^{-1/2}(\sin x)$.

87. False. If $y = x\sin^2 x$, then

$y' = 2x\sin x\cos x + \sin^2 x = x\sin 2x + \sin^2 x$.

89. Answers will vary.

Section 8.5 Integrals of Trigonometric Functions

Skills Warm Up

1. $\cos\frac{5\pi}{4} = -\frac{\sqrt{2}}{2}$
2. $\sin\frac{7\pi}{6} = -\frac{1}{2}$
3. $\sin\left(-\frac{\pi}{3}\right) = -\frac{\sqrt{3}}{2}$
4. $\cos\left(-\frac{\pi}{6}\right) = \frac{\sqrt{3}}{2}$
5. $\tan\frac{5\pi}{6} = -\frac{\sqrt{3}}{3}$
6. $\cot\frac{5\pi}{3} = -\frac{\sqrt{3}}{3}$
7. $\sec\pi = -1$
8. $\cos\frac{\pi}{2} = 0$

Skills Warm Up —*continued*—

9. $\sin x \sec x = \sin x \cdot \frac{1}{\cos x} = \tan x$

10. $\csc x \cos x = \frac{1}{\sin x} \cdot \cos x = \cot x$

11. $\cos^2 x(\sec^2 x - 1) = \cos^2 x \tan^2 x$

$= \cos^2 x \cdot \frac{\sin^2 x}{\cos^2 x}$

$= \sin^2 x$

12. $\sin^2 x(\csc^2 x - 1) = \sin^2 x(\cot^2 x)$

$= \sin^2 x \cdot \frac{\cos^2 x}{\sin^2 x}$

$= \cos^2 x$

13. $\sec x \sin\left(\frac{\pi}{2} - x\right) = \sec x \cos x$

$= \frac{1}{\cos x} \cdot \cos x$

$= 1$

14. $\cot x \cos\left(\frac{\pi}{2} - x\right) = \cot x \sin x$

$= \frac{\cos x}{\sin x} \cdot \sin x$

$= \cos x$

15. $\cot x \sec x = \frac{\cos x}{\sin x} \cdot \frac{1}{\cos x}$

$= \frac{1}{\sin x}$

$= \csc x$

16. $\cot x(\sin^2 x) = \frac{\cos x}{\sin x} \cdot \sin^2 x = \cos x \sin x$

17. $\int_0^4 (x^2 + 3x - 4)\, dx = \left[\frac{1}{3}x^3 + \frac{3}{2}x^2 - 4x\right]_0^4$

$= \frac{64}{3} + 24 - 16 = \frac{88}{3}$

18. $\int_{-1}^1 (1 - x^2)\, dx = \left[x - \frac{1}{3}x^3\right]_{-1}^1$

$= \left(1 - \frac{1}{3}\right) - \left(-1 + \frac{1}{3}\right)$

$= \frac{4}{3}$

19. $\int_0^2 x(4 - x^2)\, dx = \int_0^2 (4x - x^3)\, dx$

$= \left[2x^2 - \frac{1}{4}x^4\right]_0^2$

$= 8 - 4$

$= 4$

20. $\int_0^1 x(9 - x^2)\, dx = \int_0^1 (9x - x^3)\, dx$

$= \left[\frac{9}{2}x^2 - \frac{1}{4}x^4\right]_0^1$

$= \frac{9}{2} - \frac{1}{4}$

$= \frac{17}{4}$

1. $\int 4 \sin x\, dx = -4 \cos x + C$

3. $\int \sin 2x\, dx = \frac{1}{2}\int 2 \sin 2x\, dx = -\frac{1}{2} \cos 2x + C$

5. $\int 4x^3 \cos(x^4)\, dx = \sin(x^4) + C$

7. $\int \sec^2\left(\frac{x}{5}\right) dx = 5 \int \frac{1}{5} \sec^2\left(\frac{x}{5}\right) dx = 5 \tan \frac{x}{5} + C$

9. $\int \sec(2x) \tan(2x)\, dx = \frac{1}{2} \int 2 \sec(2x) \tan(2x)\, dx$

$= \frac{1}{2} \sec(2x) + C$

11. $\int \tan^3 x \sec^2 x\, dx = \frac{1}{4} \tan^4 x + C$

13. $\int \frac{\sec^2 x}{\tan x}\, dx = \ln|\tan x| + C$

15. $\int \frac{\sec x \tan x}{\sec x - 1}\, dx = \ln|\sec x - 1| + C$

17. $\int \frac{\sin x}{1 + \cos x}\, dx = -\ln|1 + \cos x| + C$

19. $\int \cot \pi x\, dx = \frac{1}{\pi}\int \pi \cot \pi x\, dx$

$= \frac{1}{\pi} \ln|\sin \pi x| + C$

21. $\int \csc 2x\, dx = \frac{1}{2}\int 2 \csc 2x\, dx$

$= \frac{1}{2} \ln|\csc 2x - \cot 2x| + C$

23. $\int \tan\left(\frac{x}{4}\right)\sec^6\left(\frac{x}{4}\right)dx = \int \sec^5\left(\frac{x}{4}\right)\sec\left(\frac{x}{4}\right)\tan\left(\frac{x}{4}\right)dx$

$$= 4\int\left[\sec\left(\frac{x}{4}\right)\right]^5\left(\sec\left(\frac{x}{4}\right)\tan\left(\frac{x}{4}\right)\left(\frac{1}{4}\right)\right)dx$$

$$= \frac{4\left[\sec\left(\frac{x}{4}\right)\right]^6}{6} + C$$

$$= \frac{2}{3}\sec^6\left(\frac{x}{4}\right) + C$$

25. $\int \frac{\csc^2 x}{\cot^3 x}\,dx = -\int \cot^{-3} x\left(-\csc^2 x\right)dx$

$$= -\frac{\cot^{-2} x}{-2} + C$$

$$= \frac{1}{2}\tan^2 x + C$$

27. $\int e^x \sin e^x\,dx = -\cos e^x + C$

29. $\int e^{\sin \pi x}\cos \pi x\,dx = \frac{1}{\pi}\int e^{\sin \pi x}(\pi \cos \pi x)\,dx$

$$= \frac{1}{\pi}e^{\sin \pi x} + C$$

31. $\int(\sin x + \cos x)^2\,dx = \int\left(\sin^2 x + 2\sin x\cos x + \cos^2 x\right)dx$

$$= \int(1 + \sin 2x)\,dx$$

$$= \int dx + \tfrac{1}{2}\int 2\sin 2x\,dx$$

$$= x - \tfrac{1}{2}\cos 2x + C$$

$$= \sin^2 x + x + C \text{ or } -\cos^2 x + x + C$$

33. Let $u = x$ and $dv = \cos 2x\,dx$.

Then $du = dx$ and $v = \frac{1}{2}\sin 2x$.

$$\int x\cos 2x\,dx = \tfrac{1}{2}x\sin 2x - \int \tfrac{1}{2}\sin 2x\,dx$$

$$= \tfrac{1}{2}x\sin 2x - \tfrac{1}{4}\int \sin 2x(2)\,dx$$

$$= \tfrac{1}{2}x\sin 2x + \tfrac{1}{4}\cos 2x + C$$

35. Let $u = 6x$ and $dv = \sec^2 x\,dx$.

Then $du = 6\,dx$ and $v = \tan x$.

$$\int 6x\sec^2 x\,dx = 6x\tan x - \int 6\tan x\,dx$$

$$= 6x\tan x + 6\ln|\cos x| + C$$

37. Let $u = t$ and $dv = \csc 3t\cot 3t\,dt$.

Then $du = dt$ and $v = -\frac{1}{3}\csc 3t$.

$$\int t\csc 3t\cot 3t\,dt = -\tfrac{1}{3}t\csc 3t - \int -\tfrac{1}{3}\csc 3t\,dt$$

$$= -\tfrac{1}{3}t\csc 3t + \tfrac{1}{9}\ln|\csc 3t - \cot 3t| + C$$

39. $\int_0^{\pi/4}\cos\frac{4x}{3}\,dx = \frac{3}{4}\sin\frac{4x}{3}\Big]_0^{\pi/4}$

$$= \frac{3}{4}\left(\sin\frac{\pi}{3}\right)$$

$$= \frac{3\sqrt{3}}{8}$$

$$\approx 0.6495$$

41. $\int_{\pi/2}^{2\pi/3}\sec^2\frac{x}{2}\,dx = 2\tan\frac{x}{2}\Big]_{\pi/2}^{2\pi/3}$

$$= 2\left(\sqrt{3} - 1\right)$$

$$\approx 1.4641$$

43. $\displaystyle\int_{\pi/12}^{\pi/4} \csc 2x \cot 2x\, dx = -\frac{1}{2}\csc 2x\Big]_{\pi/12}^{\pi/4}$

$\displaystyle = -\frac{1}{2}\left[\csc\frac{\pi}{2} - \csc\frac{\pi}{6}\right]$

$\displaystyle = -\frac{1}{2}[1 - 2]$

$\displaystyle = \frac{1}{2}$

45. $\displaystyle\int_0^1 \tan(1 - x)\, dx = \ln\left|\cos(1 - x)\right|\Big]_0^1$

$= \ln(\cos 0) - \ln(\cos 1)$

≈ 0.6156

47. Area $\displaystyle = \int_0^{2\pi} \cos\frac{x}{4}\, dx$

$\displaystyle = 4\sin\frac{x}{4}\Big]_0^{2\pi}$

$= 4$ square units

49. Area $\displaystyle = \int_0^{\pi} (x + \sin x)\, dx$

$\displaystyle = \left[\frac{x^2}{2} - \cos x\right]_0^{\pi}$

$\displaystyle = \frac{\pi^2}{2} + 2$

≈ 6.9348 square units

51. Area $\displaystyle = \int_0^{\pi} (\sin x + \cos 2x)\, dx$

$\displaystyle = \left[-\cos x + \tfrac{1}{2}\sin 2x\right]_0^{\pi}$

$= 1 + 1$

$= 2$ square units

53.

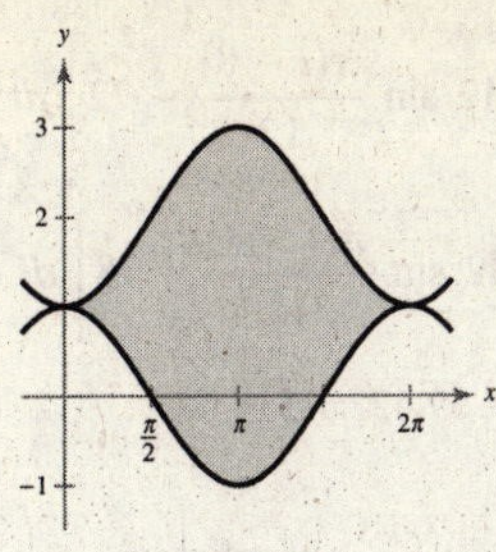

$\displaystyle A = \int_0^{2\pi} (2 - \cos x) - (\cos x)\, dx$

$\displaystyle = \int_0^{2\pi} (2 - 2\cos x)\, dx$

$\displaystyle = \left[2x - 2\sin x\right]_0^{2\pi}$

$= 4\pi$ square units

55.

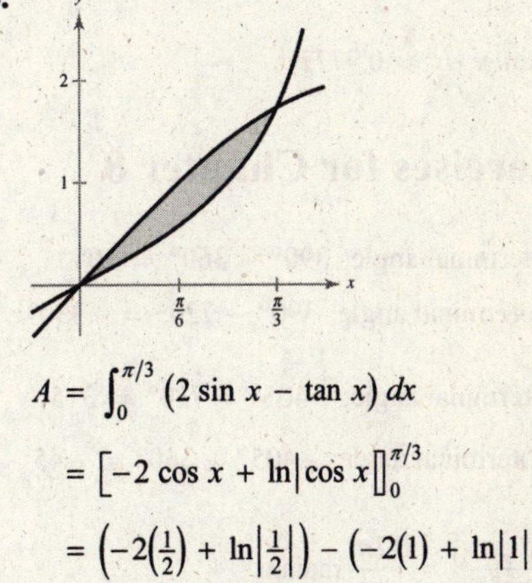

$\displaystyle A = \int_0^{\pi/3} (2\sin x - \tan x)\, dx$

$\displaystyle = \left[-2\cos x + \ln|\cos x|\right]_0^{\pi/3}$

$\displaystyle = \left(-2\left(\tfrac{1}{2}\right) + \ln\left|\tfrac{1}{2}\right|\right) - \left(-2(1) + \ln|1|\right)$

$\displaystyle = \ln\left(\tfrac{1}{2}\right) + 1$ square units

57. Total $\displaystyle = \int_0^{12} \left[2.47\sin(0.40t + 1.80) + 2.08\right] dt$

$\displaystyle = \left[-6.175\cos(0.40t + 1.80) + 2.08t\right]_0^{12}$

≈ 17.69 in.

59. $Q = 936 + 737.3\cos(0.31t + 0.928)$

(a) Average for first quarter $\displaystyle = \frac{1}{3}\int_0^3 \left(936 + 737.3\cos(0.31t + 0.928)\right) dt$

$\displaystyle = \frac{1}{3}\left[936t + \frac{737.3}{0.31}\sin(0.31t + 0.928)\right]_0^3 \approx 1061.8$ trillion Btu

(b) Average for fourth quarter $\displaystyle = \frac{1}{3}\int_9^{12} \left(936 + 737.3\cos(0.31t + 0.928)\right) dt$

$\displaystyle = \frac{1}{3}\left[936t + \frac{737.3}{0.31}\sin(0.31t + 0.928)\right]_9^{12} \approx 576.9$ trillion Btu

(c) Average for entire year $\displaystyle = \frac{1}{12}\int_0^{12} \left(936 + 737.3\cos(0.31t + 0.928)\right) dt$

$\displaystyle = \frac{1}{12}\left[936t + \frac{737.3}{0.31}\sin(0.31t + 0.928)\right]_0^{12} \approx 579.6$ trillion Btu

61. (a) $C = 0.3\int_8^{20}\left[72 + 12\sin\frac{\pi(t-8)}{12} - 72\right]dt \approx \$27.50$

(b) $C = 0.3\int_{10}^{18}\left[72 + 12\sin\frac{\pi(t-8)}{12} - 78\right]dt \approx \$9.42$

Savings $= \$27.50 - \$9.42 = \$18.08$

63. Total for the entire year $= \int_1^{365} 100{,}000\left[1 + \sin\left(\frac{2\pi(t-60)}{365}\right)\right]dt$

$$= \left[100{,}000\left(t - \left(\frac{365}{2\pi}\right)\cos\left(\frac{2\pi(t-60)}{365}\right)\right)\right]_1^{365}$$

$\approx 36{,}485{,}431.2$ lb

Average per day assuming 200 production days/yr $= \frac{\text{Total for year}}{200\text{ days}} = \frac{36{,}485{,}431.2}{200} \approx 182{,}427.2$ lb/day

65. $\int_0^{\pi/2}\sqrt{x}\sin x\,dx \approx 0.9777$

67. True. $\sin x$ is periodic with period 2π.

Review Exercises for Chapter 8

1. Positive coterminal angle: $390° - 360° = 30°$

Negative coterminal angle: $390° - 720° = -330°$

3. Positive coterminal angle: $-405° + 720° = 315°$

Negative coterminal angle: $-405° + 360° = -45°$

5. $340°\left(\frac{\pi\text{ radians}}{180°}\right) = \frac{17\pi}{9}$ radians

7. $-60°\left(\frac{\pi\text{ radians}}{180°}\right) = -\frac{\pi}{3}$ radians

9. $-480°\left(\frac{\pi\text{ radians}}{180°}\right) = -\frac{8\pi}{3}$ radians

11. $110°\left(\frac{\pi\text{ radians}}{180°}\right) = \frac{11\pi}{18}$ radians

13. $\frac{4\pi}{3}\left(\frac{180°}{\pi}\right) = 240°$

15. $-\frac{2\pi}{3}\left(\frac{180°}{\pi}\right) = -120°$

17. $b = \sqrt{8^2 - 4^2} = \sqrt{48} = 4\sqrt{3}$

$\theta = 90° - 30° = 60°$

19. $c = 5$, $\theta = 60°$ (equilateral triangle)

$$a = \sqrt{5^2 - \left(\frac{5}{2}\right)^2} = \frac{5\sqrt{3}}{2}$$

21. $h = \sqrt{16^2 - 4.4^2} = \sqrt{236.64} \approx 15.38$ feet

23. $r = \sqrt{(-3)^2 + 3^2} = \sqrt{18} = 3\sqrt{2}$

$\sin\theta = \frac{\sqrt{2}}{2}$ $\qquad \csc\theta = \sqrt{2}$

$\cos\theta = -\frac{\sqrt{2}}{2}$ $\qquad \sec\theta = -\sqrt{2}$

$\tan\theta = -1$ $\qquad \cot\theta = -1$

25. $\sin(-45°) = -\frac{\sqrt{2}}{2}$ $\qquad \csc(-45°) = -\sqrt{2}$

$\cos(-45°) = \frac{\sqrt{2}}{2}$ $\qquad \sec(-45°) = \sqrt{2}$

$\tan(-45°) = -1$ $\qquad \cot(-45°) = -1$

27. $\sin\frac{5\pi}{3} = -\frac{\sqrt{3}}{2}$ $\qquad \csc\frac{5\pi}{3} = -\frac{2\sqrt{3}}{3}$

$\cos\frac{5\pi}{3} = \frac{1}{2}$ $\qquad \sec\frac{5\pi}{3} = 2$

$\tan\frac{5\pi}{3} = -\sqrt{3}$ $\qquad \cot\frac{5\pi}{3} = -\frac{\sqrt{3}}{3}$

29. $\sin(-225°) = \frac{\sqrt{2}}{2}$ $\qquad \csc(-225°) = \sqrt{2}$

$\cos(-225°) = -\frac{\sqrt{2}}{2}$ $\qquad \sec(-225°) = -\sqrt{2}$

$\tan(-225°) = -1$ $\qquad \cot(-225°) = -1$

31. $\sin\left(-\frac{11\pi}{6}\right) = \frac{1}{2}$ $\qquad \csc\left(-\frac{11\pi}{6}\right) = 2$

$\cos\left(-\frac{11\pi}{6}\right) = \frac{\sqrt{3}}{2}$ $\qquad \sec\left(-\frac{11\pi}{6}\right) = \frac{2\sqrt{3}}{3}$

$\tan\left(-\frac{11\pi}{6}\right) = \frac{\sqrt{3}}{3}$ $\qquad \cot\left(-\frac{11\pi}{6}\right) = \sqrt{3}$

33. $\tan 33° \approx 0.6494$

35. $\sec\frac{12\pi}{5} = \frac{1}{\cos(12\pi/5)} = \sqrt{5} + 1 \approx 3.2361$

37. $\sin\left(-\frac{\pi}{9}\right) \approx -0.3420$

39. $\cos 105° \approx -0.2588$

41. $\cos 70° = \frac{50}{r} \Rightarrow r = \frac{50}{\cos 70°} \approx 146.19$

43. $\tan 20° = \frac{25}{x} \Rightarrow x = \frac{25}{\tan 20°} \approx 68.69$

45. $2\cos\theta + 1 = 0$

$$\cos\theta = -\frac{1}{2}$$

$$\theta = \frac{2\pi}{3}, \frac{4\pi}{3}$$

47. $2\sin^2\theta + 3\sin\theta + 1 = 0$

$(2\sin\theta - 1)(\sin\theta + 1) = 0$

$$\sin\theta = -\frac{1}{2} \text{ or } \sin\theta = -1$$

$$\theta = \frac{7\pi}{6}, \frac{11\pi}{6} \text{ or } \theta = \frac{3\pi}{2}$$

49. $\sec^2\theta - \sec\theta - 2 = 0$

$(\sec\theta - 2)(\sec\theta + 1) = 0$

$$\sec\theta = 2 \text{ or } \sec\theta = -1$$

$$\cos\theta = \frac{1}{2} \text{ or } \cos\theta = -1$$

$$\theta = \frac{\pi}{3}, \frac{5\pi}{3} \text{ or } \theta = \pi$$

51. $\tan 33° = \frac{h}{125}$

$h = 125 \tan 33° \approx 81.18$ feet

53. $y = -2\sin 4x$

Period: $\frac{2\pi}{4} = \frac{\pi}{2}$

Amplitude: $|-2| = 2$

55. $y = -2\cos\frac{3x}{2}$

Period: $\frac{2\pi}{\frac{3}{2}} = \frac{4\pi}{3}$

Amplitude: $|-2| = 2$

57. $y = 2\cos 6x$

Period: $\frac{2\pi}{6} = \frac{\pi}{3}$

Amplitude: 2

x-intercepts: $\left(\frac{\pi}{12}, 0\right), \left(\frac{\pi}{4}, 0\right)$

Relative minimum: $\left(\frac{\pi}{6}, -2\right)$

Relative maxima: $(0, 2), \left(\frac{\pi}{3}, 2\right)$

59. $y = \frac{1}{3}\tan x$

Period: π

Asymptotes:

$x = -\frac{\pi}{2}, x = \frac{\pi}{2}, x = \frac{3\pi}{2}$

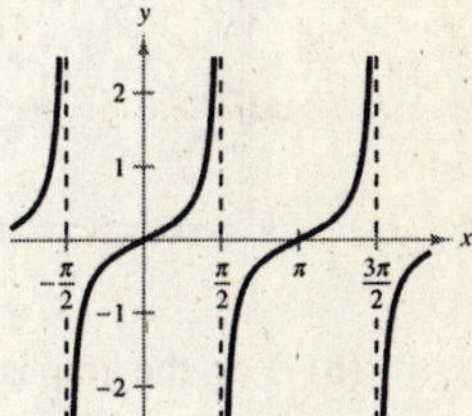

61. $y = 3\sin\frac{2x}{5}$

Period: $\frac{2\pi}{2/5} = 5\pi$

Amplitude: 3

x-intercepts: $(0, 0), \left(\frac{5\pi}{2}, 0\right), (5\pi, 0)$

Relative minimum: $\left(\frac{15\pi}{4}, -3\right)$

Relative maximum: $\left(\frac{5\pi}{4}, 3\right)$

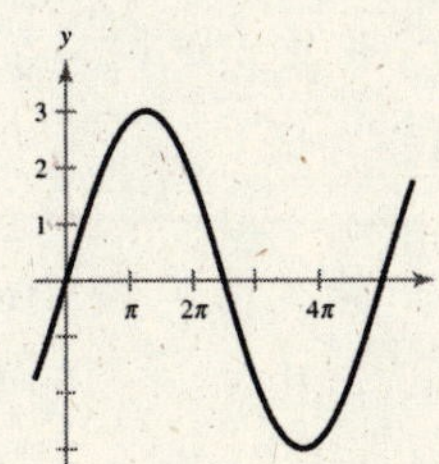

63. $y = \sec 2\pi x$

Period: $\dfrac{2\pi}{2\pi} = 1$

Asymptotes: $x = -\dfrac{1}{4}, x = \dfrac{1}{4}, x = \dfrac{3}{4}, x = \dfrac{5}{4}$

Relative minima: $(0, 1), (1, 1)$

Relative maximum: $\left(\dfrac{1}{2}, -1\right)$

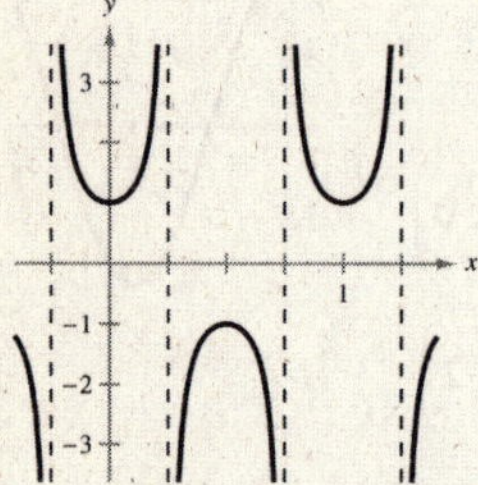

65. (a) 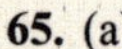

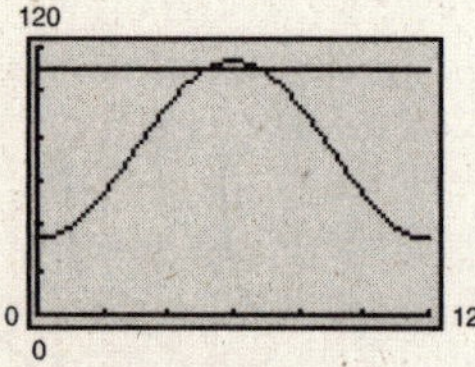

(b) Yes, the graphs of $S = 74 - 40\cos\left(\dfrac{\pi t}{6}\right)$ and $S = 110$ intersect at $t \approx 5.1$ and $t \approx 6.9$. So, during May and June, sales exceeded 110 units.

67. $y = \sin 5\pi x$

$y' = 5\pi \cos 5\pi x$

69. $y = \tan(3x^3)$

$y' = \sec^2(3x^3)(9x^2)$

$= 9x^2 \sec^2(3x^3)$

71. $y = \cot\left(\dfrac{3x^2}{5}\right)$

$y' = -\csc^2\left(\dfrac{3x^2}{5}\right)\left(\dfrac{6x}{5}\right)$

$= -\dfrac{6x}{5}\csc^2\left(\dfrac{3x^2}{5}\right)$

73. $y = (\cos(2\pi x))^{1/2}$

$y' = \dfrac{1}{2}(\cos(2\pi x))^{-1/2}(-\sin(2\pi x)(2\pi))$

$= -\dfrac{\pi \sin(2\pi x)}{\sqrt{\cos(2\pi x)}}$

75. $y = -x \tan x$

$y' = -x \sec^2 x - \tan x$

77. $y = \dfrac{\cos x}{x^2}$

$y' = \dfrac{x^2(-\sin x) - \cos x(2x)}{(x^2)^2} = -\dfrac{x \sin x + 2\cos x}{x^3}$

79. $y = \sin^2 x + x$

$y' = 2\sin x \cos x + 1$

$= \sin 2x + 1$

81. $y = \csc^4 x$

$y' = -4\csc^3 x \csc x \cot x$

$= -4\csc^4 x \cot x$

83. $y = e^x \cot x$

$y' = e^x(-\csc^2 x) + e^x \cot x$

$= e^x(\cot x - \csc^2 x)$

85. $y = \cos 2x, \quad \left(\dfrac{\pi}{4}, 0\right)$

$y' = -2\sin 2x, \quad y'\left(\dfrac{\pi}{4}\right) = -2$

$y - 0 = -2\left(x - \dfrac{\pi}{4}\right)$

$y = -2x + \dfrac{\pi}{2}$

87. $y = \csc x, \left(\dfrac{3\pi}{2}, -1\right)$

$y' = -\csc x \cot x, y'\left(\dfrac{3\pi}{2}\right) = 0$

$y - (-1) = 0\left(x - \dfrac{3\pi}{2}\right)$

$y = -1$

89. $y = \dfrac{1}{2}\sin^2 x, \quad \left(\dfrac{\pi}{2}, \dfrac{1}{2}\right)$

$y' = \sin x \cos x, \quad y'\left(\dfrac{\pi}{2}\right) = 0$

$y - \dfrac{1}{2} = 1(x - 0)$

$y = \dfrac{1}{2}$

91. $y = \sin \frac{\pi x}{4}$

$y' = \frac{\pi}{4} \cos \frac{\pi x}{4}$

$\frac{\pi}{4} \cos \frac{\pi x}{4} = 0$

$\cos \frac{\pi x}{4} = 0$

Critical numbers: $x = 2, 6$

Relative maximum: $(2, 1)$

Relative minimum: $(6, -1)$

93. $f(x) = \frac{x}{2} + \cos x$

$f'(x) = \frac{1}{2} - \sin x$

$\frac{1}{2} - \sin x = 0$

$-\sin x = -\frac{1}{2}$

$\sin x = \frac{1}{2}$

Critical numbers: $x = \frac{\pi}{6}, \frac{5\pi}{6}$

Relative minimum: $\left(\frac{5\pi}{6}, \frac{5\pi}{12} - \frac{\sqrt{3}}{2}\right)$

Relative maximum: $\left(\frac{\pi}{6}, \frac{\pi}{12} + \frac{\sqrt{3}}{2}\right)$

95. $f(x) = \sin^2 x + \sin x$

$f'(x) = 2 \sin x \cos x + \cos x$

$2 \sin x \cos x + \cos x = 0$

$\cos x(2 \sin x + 1) = 0$

$\cos x = 0$ or $2 \sin x + 1 = 0$

$2 \sin x = -1$

$\sin x = -\frac{1}{2}$

Critical numbers: $x = \frac{\pi}{2}, \frac{7\pi}{6}, \frac{3\pi}{2}, \frac{11\pi}{6}$

Relative minima: $\left(\frac{7\pi}{6}, -\frac{1}{4}\right), \left(\frac{11\pi}{6}, -\frac{1}{4}\right)$

Relative maxima: $\left(\frac{\pi}{2}, 2\right), \left(\frac{3\pi}{2}, 0\right)$

97. $S = 74 - 40 \cos\left(\frac{\pi t}{6}\right)$

$S' = \frac{240}{\pi} \sin\left(\frac{\pi t}{6}\right)$

$\frac{240}{\pi} \sin\left(\frac{\pi t}{6}\right) = 0$

$t = 6$, or June.

Maximum $(6, 114)$: The maximum number of jet skis sold was 114 in June.

99. $\int \sin(3x)\, dx = \frac{1}{3} \int \sin(3x)(3)\, dx$
$= -\frac{1}{3} \cos(3x) + C$

101. $\int \csc 5x \cot 5x\, dx = \frac{1}{5} \int 5 \csc 5x \cot 5x\, dx$
$= -\frac{1}{5} \csc 5x + C$

103. $\int x \csc^2(x^2)\, dx = \frac{1}{2} \int 2x \csc^2(x^2)\, dx$
$= -\frac{1}{2} \cot x^2 + C$

105. $\int \frac{\sec^2(2x)}{\tan(2x)}\, dx = \frac{1}{2} \int \frac{1}{\tan(2x)} (\sec^2(2x)(2))\, dx$
$= \frac{1}{2} \ln|\tan(2x)| + C$

107. $\int \tan(3x)\, dx = \frac{1}{3} \int \tan(3x)(3)\, dx$
$= -\frac{1}{3} \ln|\cos(3x)| + C$

109. $\int \sin^3 x \cos x\, dx = \frac{1}{4} \sin^4 x + C$

111. $\int_0^{\pi} (1 + \sin x)\, dx = \Big[x - \cos x\Big]_0^{\pi}$
$= \left[\pi - (-1)\right] - (-1)$
$= \pi + 2$

113. $\int_{-\pi/6}^{\pi/6} \sec^2 x\, dx = \tan x\Big]_{-\pi/6}^{\pi/6}$
$= \frac{1}{\sqrt{3}} - \left(-\frac{1}{\sqrt{3}}\right)$
$= \frac{2}{\sqrt{3}}$
$= \frac{2\sqrt{3}}{3}$

115. $\int_{-\pi/3}^{\pi/3} 4 \sec x \tan x\, dx = 4 \sec x\Big]_{-\pi/3}^{\pi/3}$
$= 4(2) - 4(2)$
$= 0$

117. $\int_{-\pi/2}^{\pi/2} (2x + \cos x)\, dx = \left[x^2 + \sin x\right]_{-\pi/2}^{\pi/2}$

$= \frac{\pi^2}{4} + 1 - \left(\frac{\pi^2}{4} - 1\right)$

$= 2$

119. $A = \int_0^{\pi/3} \sin 3x\, dx$

$= \frac{1}{3}\int_0^{\pi/3} 3 \sin 3x\, dx$

$= \left[-\frac{1}{3} \cos 3x\right]_0^{\pi/3}$

$= \frac{1}{3} + \frac{1}{3}$

$= \frac{2}{3}$ square unit

121. $A = \int_0^{\pi/2} (2 \sin x + \cos 3x)\, dx$

$= \left[-2 \cos x + \frac{1}{3} \sin 3x\right]_0^{\pi/2}$

$= \left(-\frac{1}{3}\right) - (-2) = \frac{5}{3}$ square units

123. Total $= \int_0^{12} \left(1.27 \sin(0.58t - 2.05) + 1.98\right) dt$

$= \left[-\frac{1.27}{0.58} \cos(0.58t - 2.05) + 1.98t\right]_0^{12}$

≈ 22.3 in.

125. Average $= \frac{1}{9} \int_0^9 \left(15.31 \sin(0.37t - 1.27) + 33.66\right) dt$

$= \frac{1}{9}\left[-\frac{15.31}{0.37} \cos(0.37t - 1.27) + 33.66t\right]_0^9$

$\approx$ \$37.2 billion

Chapter 8 Test Yourself

Function	θ (deg)	θ (rad)	Function value
1. sin	67.5°	$\frac{3\pi}{8}$	0.9239
2. cos	36°	$\frac{\pi}{5}$	0.8090
3. tan	15°	$\frac{\pi}{12}$	0.2679
4. cot	−30°	$-\frac{\pi}{6}$	$-\sqrt{3}$
5. sec	−40°	$-\frac{2\pi}{9}$	1.3054
6. csc	−225°	$-\frac{5\pi}{4}$	$\sqrt{2}$

7. $\cos 24° = \frac{25}{l}$

$l = \frac{25}{\cos 24°}$

$l \approx 27.4$ inches

8. $2 \sin \theta - \sqrt{2} = 0$

$2 \sin \theta = \sqrt{2}$

$\sin \theta = \frac{\sqrt{2}}{2}$

$\theta = \frac{\pi}{4}, \frac{3\pi}{4}$

9. $\cos^2 \theta - \sin^2 \theta = 0$

$\cos^2 \theta = \sin^2 \theta$

$1 = \frac{\sin^2 \theta}{\cos^2 \theta}$

$\pm 1 = \tan \theta$

$\theta = \frac{\pi}{4}, \frac{3\pi}{4}, \frac{5\pi}{4}, \frac{7\pi}{4}$

10. $\csc \theta = \sqrt{3} \sec \theta$

$\frac{\csc \theta}{\sec \theta} = \sqrt{3}$

$\frac{\cos \theta}{\sin \theta} = \sqrt{3}$

$\cot \theta = \sqrt{3}$

$\theta = \frac{\pi}{6}, \frac{7\pi}{6}$

11. $y = 3 \sin 2x$

Period: $\frac{2\pi}{2} = \pi$

Amplitude: 3

x-intercepts:

$(0, 0), \left(\frac{\pi}{2}, 0\right), (\pi, 0)$

Relative minimum: $\left(\frac{3\pi}{4}, -3\right)$

Relative maximum: $\left(\frac{\pi}{4}, 3\right)$

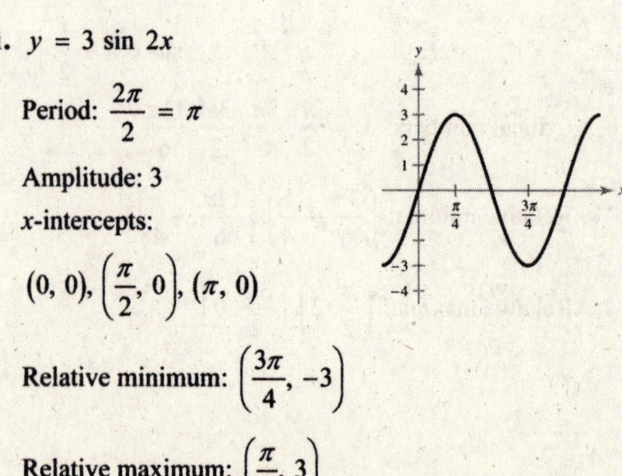

12. $y = 4\cos 3\pi x$

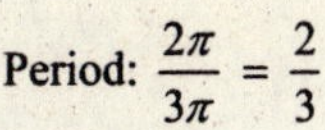

Period: $\dfrac{2\pi}{3\pi} = \dfrac{2}{3}$

Amplitude: 4

x-intercepts:

$\left(\dfrac{1}{6}, 0\right), \left(\dfrac{1}{2}, 0\right), \left(\dfrac{5}{6}, 0\right)$

Relative minimum: $\left(\dfrac{1}{3}, -4\right)$

Relative maxima: $(0, 4), \left(\dfrac{2}{3}, 4\right)$

13. $y = \cot \dfrac{\pi x}{5}$

Period: $\dfrac{\pi}{(\pi/5)} = 5$

x-intercepts: $\left(-\dfrac{5}{2}, 0\right), \left(\dfrac{5}{2}, 0\right)$

Asymptotes: $x = -5, x = 5$

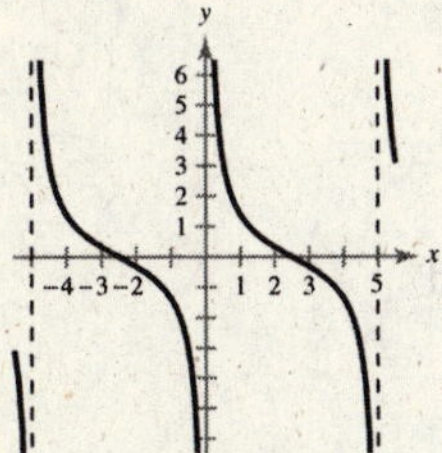

14. (a) $y = \cos x - \cos^2 x$

$y' = -\sin x + 2\cos x \sin x$

$y' = \sin 2x - \sin x$

(b) $-\sin x + 2\cos x \sin x = 0$

$-\sin x(1 - 2\cos x) = 0$

$\sin x = 0$ or $1 - 2\cos x = 0$

$-2\cos x = -1$

$\cos x = \dfrac{1}{2}$

Critical numbers: $x = \dfrac{\pi}{3}, \pi, \dfrac{5\pi}{3}$

Relative minimum: $(\pi, -2)$

Relative maxima: $\left(\dfrac{\pi}{3}, \dfrac{1}{4}\right), \left(\dfrac{5\pi}{3}, \dfrac{1}{4}\right)$

15. (a) $y = \sec\left(x - \dfrac{\pi}{4}\right)$

$y' = \sec\left(x - \dfrac{\pi}{4}\right)\tan\left(x - \dfrac{\pi}{4}\right)$

(b) $\sec\left(x - \dfrac{\pi}{4}\right)\tan\left(x - \dfrac{\pi}{4}\right) = 0$

$\sec\left(x - \dfrac{\pi}{4}\right) = 0$ or $\tan\left(x - \dfrac{\pi}{4}\right) = 0$

Critical numbers: $x = \dfrac{\pi}{4}, \dfrac{5\pi}{4}$

Relative minimum: $\left(\dfrac{\pi}{4}, 1\right)$

Relative maximum: $\left(\dfrac{5\pi}{4}, -1\right)$

16. (a) $y = \dfrac{1}{3 - \sin(x + \pi)}$

$y' = \dfrac{\cos(x + \pi)}{\left[3 - \sin(x + \pi)\right]^2}$

(b) $\cos(x + \pi) = 0$

Critical numbers: $x = \dfrac{\pi}{2}, \dfrac{3\pi}{2}$

Relative minimum: $\left(\dfrac{\pi}{2}, \dfrac{1}{4}\right)$

Relative maximum: $\left(\dfrac{3\pi}{2}, \dfrac{1}{2}\right)$

17. $\int \sin 5x\,dx = \frac{1}{5}\int 5\sin 5x\,dx = -\frac{1}{5}\cos 5x + C$

18. $\int \sec^2\left(\dfrac{x}{4}\right)dx = 4\int \sec^2\left(\dfrac{x}{4}\right)\left(\dfrac{1}{4}\right)dx = 4\tan\left(\dfrac{x}{4}\right) + C$

19. $\int x\csc x^2\,dx = -\frac{1}{2}\ln\left|\csc^2 x - \cot x^2\right| + C$

20. $\displaystyle\int_{1/4}^{1/2} \cos \pi x = \frac{1}{\pi}\int_{1/4}^{1/2} \pi\cos \pi x\,dx = \frac{1}{\pi}\sin \pi x\Big]_{1/4}^{1/2}$

$= \dfrac{1}{\pi}\left(1 - \dfrac{\sqrt{2}}{2}\right) = \dfrac{2 - \sqrt{2}}{2\pi} \approx 0.0932$

21. $\displaystyle\int_0^{\pi} \sec^2\frac{x}{3}\tan\frac{x}{3}\,dx = 3\int_0^{\pi}\frac{1}{3}\sec^2\frac{x}{3}\tan\frac{x}{3}\,dx$

$= \left[\dfrac{3}{2}\tan^2\dfrac{x}{3}\right]_0^{\pi}$

$= \dfrac{3}{2}(3 - 0)$

$= \dfrac{9}{2}$

22. $\int_{3\pi/4}^{5\pi/6} \csc(2x)\cot(2x)\,dx = \frac{1}{2}\int_{3\pi/4}^{5\pi/6} \csc(2x)\cot(2x)(2)\,dx$

$$= -\frac{1}{2}\Big[\csc(2x)\Big]_{3\pi/4}^{5\pi/6}$$

$$= -\frac{1}{2}\left(-\frac{2\sqrt{3}}{3} + 1\right)$$

$$= \frac{\sqrt{3}}{3} - \frac{1}{2}$$

23. $\int_{\pi/4}^{3\pi/8} \sin(4x)\cos(4x)\,dx = \frac{1}{4}\int\big(\sin(4x)\big)\big(\cos(4x)(4)\big)\,dx$

$$= \tfrac{1}{8}\Big[\sin^2(4x)\Big]_{\pi/4}^{3\pi/8}$$

$$= \tfrac{1}{8}(1 - 0)$$

$$= \tfrac{1}{8}$$

24. $S = 20.3 - 17.2\cos\dfrac{\pi t}{6},\ 0 \le t \le 12$

(a) Total sales $= \displaystyle\int_0^{12}\left(20.3 - 17.2\cos\frac{\pi t}{6}\right)dt = \left[20.3t - \frac{17.2}{(\pi/6)}\sin\frac{\pi t}{6}\right]_0^{12} = \$243.6$ thousand

The total sales during the year were $243,600.

(b) Average sales from April through October $= \displaystyle\frac{1}{10-3}\int_3^{10}\left(20.3 - 17.2\cos\frac{\pi t}{6}\right)dt$

$$= \frac{1}{7}\left[20.3t - \frac{17.2}{(\pi/6)}\sin\frac{\pi t}{6}\right]_3^{10} \approx \$29.06 \text{ thousand}$$

The average monthly sales from April through October are about $29,060.

CHAPTER 9
Probability and Calculus

CHAPTER 9
Probability and Calculus

Section 9.1 Discrete Probability

Skills Warm Up

1. $\frac{1}{9} + \frac{2}{3} + \frac{2}{9} = x$

$\frac{1}{9} + \frac{6}{9} + \frac{2}{9} = x$

$\frac{9}{9} = x$

$1 = x$

2. $\frac{1}{3} + \frac{5}{12} + \frac{1}{8} + \frac{1}{12} + \frac{x}{24} = 1$

$\frac{8}{24} + \frac{10}{24} + \frac{3}{24} + \frac{2}{24} + \frac{x}{24} = \frac{24}{24}$

$\frac{23 + x}{24} = \frac{24}{24}$

$23 + x = 24$

$x = 1$

3. $0\left(\frac{1}{16}\right) + 1\left(\frac{3}{16}\right) + 2\left(\frac{8}{16}\right) + 3\left(\frac{3}{16}\right) + 4\left(\frac{1}{16}\right) = \frac{32}{16} = 2$

4. $0\left(\frac{1}{12}\right) + 1\left(\frac{2}{12}\right) + 2\left(\frac{6}{12}\right) + 3\left(\frac{2}{12}\right) + 4\left(\frac{1}{12}\right) = \frac{24}{12} = 2$

5. $(0 - 1)^2\left(\frac{1}{4}\right) + (1 - 1)^2\left(\frac{1}{2}\right) + (2 - 1)^2\left(\frac{1}{4}\right) = \frac{1}{4} + \frac{1}{4} = \frac{1}{2}$

6. $(0 - 2)^2\left(\frac{1}{12}\right) + (1 - 2)^2\left(\frac{2}{12}\right) + (2 - 2)^2\left(\frac{6}{12}\right) + (3 - 2)^2\left(\frac{2}{12}\right) + (4 - 2)^2\left(\frac{1}{12}\right) = \frac{4}{12} + \frac{2}{12} + \frac{2}{12} + \frac{4}{12} = 1$

7. $\frac{3}{8} = 0.375 = 37.5\%$

8. $\frac{9}{11} \approx 0.8182 = 81.82\%$

9. $\frac{13}{24} \approx 0.5417 = 54.17\%$

10. $\frac{112}{256} = 0.4375 = 43.75\%$

1. (a) $S = \{HHH, HHT, HTH, THH, HTT, THT, TTH, TTT\}$

(b) $A = \{HHH, HHT, HTH, THH\}$

(c) $B = \{HTT, THT, TTH, TTT\}$

3. (a) $S = \{III, IIO, IIU, IOI, IUI, OII, UII, IOO, IOU, IUO, IUU, OIO, OIU, UIO, UIU, OOI,$
$OUI, UOI, UUI, OOO, OOU, OUO, UOO, OUU, UOU, UUO, UUU\}$

(b) $A = \{III, IIO, IIU, IOI, IUI, OII, UII\}$

(c) $B = \{III, IIO, IIU, IOI, IUI, OII, UII, IOU, IUO, IUU, OIU, UIO, UIU, OUI, UOI, UUI,$
$OUU, UOU, UUO, UUU\}$

5. (a)

Random variable	0	1	2	3	4
Frequency	1	4	6	4	1

(b)

Random variable, x	0	1	2	3	4
Probability, $P(x)$	$\frac{1}{16}$	$\frac{4}{16}$	$\frac{6}{16}$	$\frac{4}{16}$	$\frac{1}{16}$

7. (a) $P(\text{college}) = \frac{12}{72} + \frac{9}{72} = \frac{21}{72} = \frac{7}{24}$

(b) $P(\text{not college}) = 1 - \frac{7}{24} = \frac{17}{24}$

(c) $P(\text{girl, not college}) = \frac{32}{72} = \frac{4}{9}$

9. $P(3) = 1 - (0.20 + 0.35 + 0.15 + 0.05) = 0.25$

11. The table represents a probability distribution because each distinct value of x corresponds to a probability $P(x)$ such that $0 \le P(x) \le 1$, and the sum of all $P(x)$ is 1.

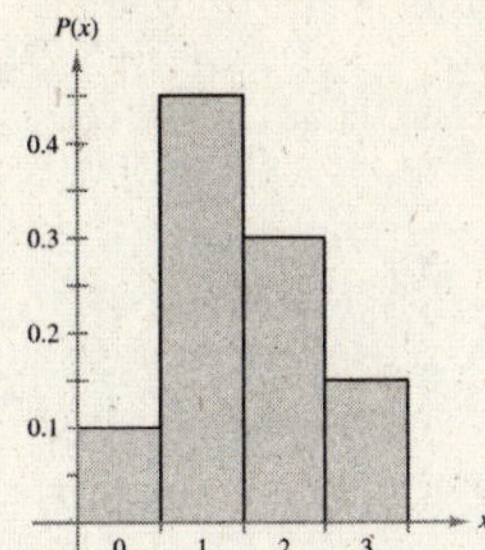

13. The table does not represent a probability distribution because $P(4)$ is not a probability satisfying the inequality $0 \le P(x) \le 1$, and because the sum of all values of $P(x)$ is not equal to 1.

15. (a) $P(1 \le x \le 3) = P(1) + P(2) + P(3)$

$$= \frac{3}{20} + \frac{6}{20} + \frac{6}{20}$$

$$= \frac{15}{20} = \frac{3}{4}$$

(b) $P(x \ge 2) = P(2) + P(3) + P(4)$

$$= \frac{6}{20} + \frac{6}{20} + \frac{4}{20}$$

$$= \frac{16}{20}$$

$$= \frac{4}{5}$$

17. (a) $P(x \le 3) = P(0) + P(1) + P(2) + P(3)$

$$= 0.041 + 0.189 + 0.247 + 0.326$$

$$= 0.803$$

(b) $P(x > 3) = P(4) + P(5)$

$$= 0.159 + 0.038$$

$$= 0.197$$

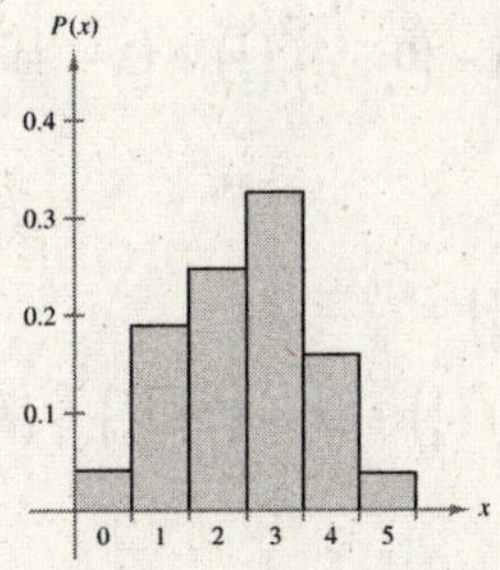

19. (a)

(b) $P(15 - 44) = P(15 - 24) + P(25 - 34) + P(35 - 44)$

$$= 0.198 + 0.262 + 0.255$$

$$= 0.715$$

(c) $P(a \ge 35) = P(35 - 44) + P(45 - 54) + P(55 - 64) + P(65 \text{ and over})$

$$= 0.255 + 0.195 + 0.069 + 0.017$$

$$= 0.536$$

(d) $P(a \le 24) = P(14 \text{ and under}) + P(15 - 24)$

$$= 0.004 + 0.198$$

$$= 0.202$$

21. (a) $S = \{gggg, gggb, ggbg, gbgg, bggg, ggbb, gbbg, gbgb, bgbg, bbgg, bggb, gbbb, bgbb, bbgb, bbbg, bbbb\}$

(b)

x	0	1	2	3	4
$P(x)$	$\frac{1}{16}$	$\frac{4}{16}$	$\frac{6}{16}$	$\frac{4}{16}$	$\frac{1}{16}$

(c)

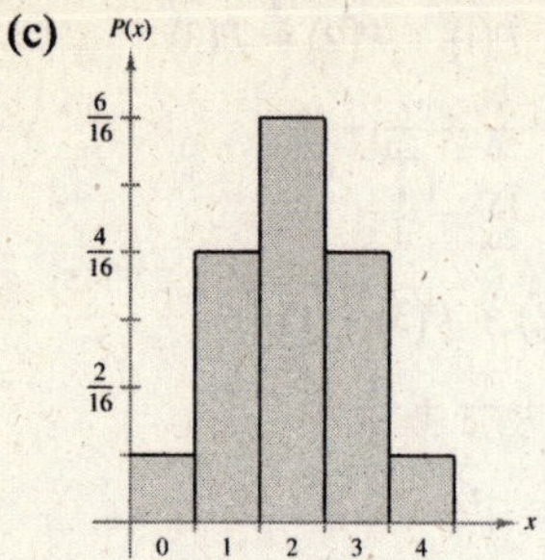

(d) Probability of at least one boy = 1 − probability of all girls

$$P = 1 - \frac{1}{16} = \frac{15}{16}$$

23. $E(x) = 1\left(\frac{1}{16}\right) + 2\left(\frac{3}{16}\right) + 3\left(\frac{8}{16}\right) + 4\left(\frac{3}{16}\right) + 5\left(\frac{1}{16}\right) = \frac{48}{16} = 3$

$V(x) = (1-3)^2\left(\frac{1}{16}\right) + (2-3)^2\left(\frac{3}{16}\right) + (3-3)^2\left(\frac{8}{16}\right) + (4-3)^2\left(\frac{3}{16}\right) + (5-3)^2\left(\frac{1}{16}\right) = 14\left(\frac{1}{16}\right) = \frac{7}{8} = 0.875$

$\sigma = \sqrt{V(x)} = 0.9354$

25. $E(x) = -3\left(\frac{1}{5}\right) + (-1)\left(\frac{1}{5}\right) + 0\left(\frac{1}{5}\right) + 3\left(\frac{1}{5}\right) + 5\left(\frac{1}{5}\right) = \frac{4}{5}$

$V(x) = \left(-3 - \frac{4}{5}\right)^2\left(\frac{1}{5}\right) + \left(-1 - \frac{4}{5}\right)^2\left(\frac{1}{5}\right) + \left(0 - \frac{4}{5}\right)^2\left(\frac{1}{5}\right) + \left(3 - \frac{4}{5}\right)^2\left(\frac{1}{5}\right) + \left(5 - \frac{4}{5}\right)^2\left(\frac{1}{5}\right) = 8.16$

$\sigma = \sqrt{V(x)} \approx 2.857$

27. (a) $E(x) = 1\left(\frac{1}{4}\right) + 2\left(\frac{1}{4}\right) + 3\left(\frac{1}{4}\right) + 4\left(\frac{1}{4}\right) = \frac{10}{4} = 2.5$

$V(x) = (1-2.5)^2\left(\frac{1}{4}\right) + (2-2.5)^2\left(\frac{1}{4}\right) + (3-2.5)^2\left(\frac{1}{4}\right) + (4-2.5)^2\left(\frac{1}{4}\right) = 1.25$

$\sigma = \sqrt{1.25} \approx 1.118$

(b) $E(x) = 2\left(\frac{1}{16}\right) + 3\left(\frac{2}{16}\right) + 4\left(\frac{3}{16}\right) + 5\left(\frac{4}{16}\right) + 6\left(\frac{3}{16}\right) + 7\left(\frac{2}{16}\right) + 8\left(\frac{1}{16}\right) = \frac{80}{16} = 5$

$V(x) = (2-5)^2\left(\frac{1}{16}\right) + (3-5)^2\left(\frac{2}{16}\right) + (4-5)^2\left(\frac{3}{16}\right) + (5-5)^2\left(\frac{4}{16}\right) + (6-5)^2\left(\frac{3}{16}\right) + (7-5)^2\left(\frac{2}{16}\right) + (8-5)^2\left(\frac{1}{16}\right)$

$= \frac{5}{2}$

$= 2.5$

$\sigma = \sqrt{2.5} \approx 1.581$

29. (a) $E(x) = 10(0.25) + 15(0.30) + 20(0.25) + 30(0.15) + 40(0.05) = 18.50$ thousand or 18,500

$V(x) = (10 - 18.50)^2(0.25) + (15 - 18.50)^2(0.30) + (20 - 18.50)^2(0.25) + (30 - 18.50)^2(0.15) + (40 - 18.50)^2(0.05)$

$= 65.25$

$\sigma = \sqrt{V(x)} \approx 8.078$

(b) Expected revenue: $R = \$4.95(18.50)(1000) = \$91,575$

31. $E(x) = 0(0.995) + 30,000(0.0036) + 60,000(0.0011) + 90,000(0.0003) = 201$

Each customer should be charged $201.

33. (a)

x	0	1	2	3	4
$P(x)$	$\frac{40}{160}$	$\frac{61}{160}$	$\frac{40}{160}$	$\frac{17}{160}$	$\frac{2}{160}$

(b)

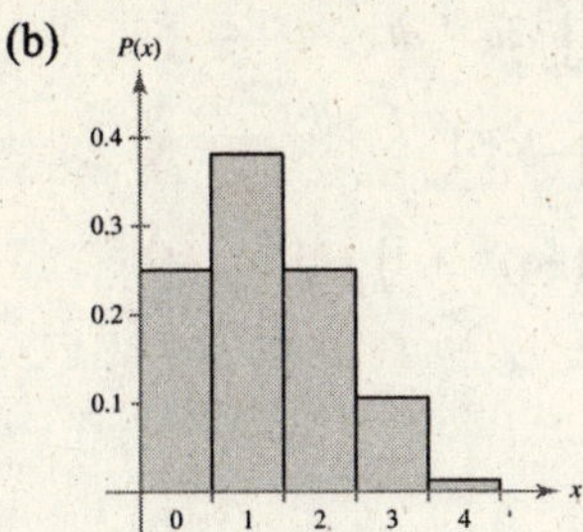

(c) $P(1 \le x \le 3) = P(1) + P(2) + P(3)$

$= \frac{61}{160} + \frac{40}{160} + \frac{17}{160}$

$= \frac{118}{160} = \frac{59}{80} = 0.7375$

(d) $E(x) = 0\left(\frac{40}{160}\right) + 1\left(\frac{61}{160}\right) + 2\left(\frac{40}{160}\right) + 3\left(\frac{17}{160}\right) + 4\left(\frac{2}{160}\right) = \frac{5}{4}$

$V(x) = \left(0 - \frac{5}{4}\right)^2\left(\frac{40}{160}\right) + \left(1 - \frac{5}{4}\right)^2\left(\frac{61}{160}\right) + \left(2 - \frac{5}{4}\right)^2\left(\frac{40}{160}\right) + \left(3 - \frac{5}{4}\right)^2\left(\frac{17}{160}\right) + \left(4 - \frac{5}{4}\right)^2\left(\frac{2}{160}\right) = \frac{39}{40} = 0.975$

$\sigma = \sqrt{\frac{39}{40}} = \sqrt{0.975} \approx 0.9874$

On average, you can expect the player to get 1.25 hits per game. The variance and standard deviation are measures of how spread out the data are.

35. $E(x) = 35\left(\frac{1}{38}\right) + (-1)\left(\frac{37}{38}\right) = -\frac{2}{38} \approx -\$0.05$ or $E(x) = 36\left(\frac{1}{38}\right) + 0\left(\frac{37}{38}\right) = \frac{36}{38}$

$\frac{36}{38} - 1 = -\$0.05$

37. City 1: Expected value $= 0.3(20) + 0.7(-4) = 3.2$ million

City 2: Expected value $= 0.2(50) + 0.8(-9) = 2.8$ million

The company should open the store in City 1.

39. Answers will vary.

Section 9.2 Continuous Random Variables

Skills Warm Up

1. The domain of the rational function $f(x) = \dfrac{1}{x}$ consists of all real numbers except $x = 0$. The value of $\dfrac{1}{x}$ is positive for all positive real numbers x. So, $f(x)$ is continuous and nonnegative on $[0, 1]$.

2. The polynomial function $f(x) = x^2 - 1$ is continuous at every real number. Because $f(0) = 0^2 - 1 = -1$, however, $f(x)$ is continuous but *not* nonnegative on $[0, 1]$.

3. The polynomial function $f(x) = 3 - x$ is continuous at every real number, but $f(5) = 3 - 5 = -2$. So, $f(x)$ is continuous but *not* nonnegative on $[1, 5]$.

4. The function $f(x) = e^{-x} = \dfrac{1}{e^x}$ is continuous at every real number and its value is positive on $[0, 1]$. So, $f(x)$ is continuous and nonnegative on $[0, 1]$.

5. $\int_0^4 \frac{1}{4}\,dx = \frac{1}{4}x\Big]_0^4 = 1$

Skills Warm Up —*continued*—

6. $\int_1^2 \frac{2-x}{2}\,dx = \int_1^2 \left(1 - \frac{x}{2}\right) dx$
$= \left[x - \frac{1}{4}x^2\right]_1^2$
$= 1 - \frac{3}{4}$
$= \frac{1}{4}$

7. $\int_0^\infty 3e^{-3t}\,dt = \lim_{b\to\infty} \int_0^b 3e^{-3t}\,dt$
$= \lim_{b\to\infty} \left[-e^{-3t}\right]_0^b$
$= \lim_{b\to\infty} \left(-e^{-3b} + 1\right) = 1$

1. 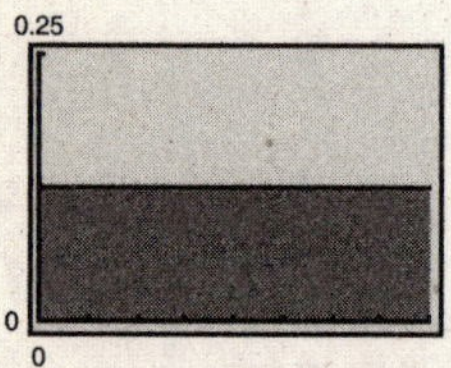

f is a probability density function because

$\int_0^8 \frac{1}{8}\,dx = \frac{1}{8}x\Big]_0^8 = 1$ and $f(x) = \frac{1}{8} \ge 0$ on $[0, 8]$.

3. 

f is a probability density function because

$\int_0^4 \frac{4-x}{8}\,dx = \frac{1}{8}\left(4x - \frac{x^2}{2}\right)\Big]_0^4 = 1$ and

$f(x) = \frac{4-x}{8} \ge 0$ on $[0, 4]$.

5. 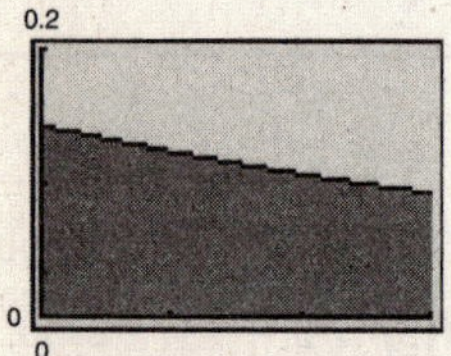

$f(x)$ is not a probability density function because

$\int_0^1 \frac{1}{7}e^{-x/7}\,dx = \left[-e^{-x/7}\right]_0^3 \approx 0.349 \ne 1$

7. 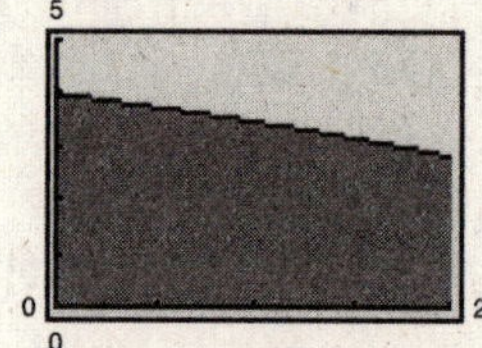

f is not a probability density function because

$\int_0^2 2\sqrt{4-x}\,dx = -\frac{4}{3}\left[(4-x)^{3/2}\right]_0^2 = -\frac{4}{3}\sqrt{8} + \frac{32}{3}$
$\approx 6.895 \ne 1.$

9. 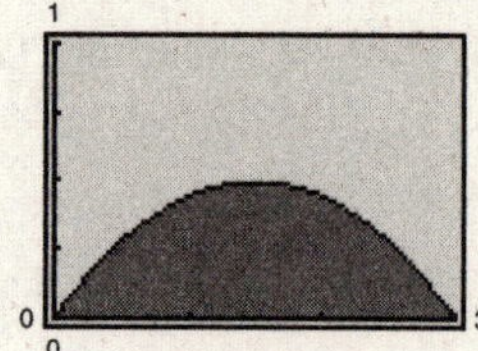

f is a probability density function because

$f(x) = \frac{2}{9}x(3-x) \ge 0$ on $[0, 3]$ and

$\int_0^3 \left(\frac{2}{3}x - \frac{2}{9}x^2\right) dx = \left[\frac{x^2}{3} - \frac{2x^3}{27}\right]_0^3 = 1.$

11. $\int_1^4 kx\,dx = \left[\frac{kx^2}{2}\right]_1^4 = \frac{15}{2}k = 1 \Rightarrow k = \frac{2}{15}$

13. $\int_{-2}^2 k(4 - x^2)\,dx = k\left(4x - \frac{x^3}{3}\right)\Big]_{-2}^2$
$= \frac{32k}{3} = 1 \Rightarrow k = \frac{3}{32}$

15. $\int_0^\infty ke^{-x/2}\,dx = \lim_{b\to\infty} \left[-2ke^{-x/2}\right]_0^b = 2k = 1 \Rightarrow k = \frac{1}{2}$

17. $\int_a^b \frac{1}{3}\,dx = \left.\frac{x}{3}\right]_a^b = \frac{b-a}{3}$

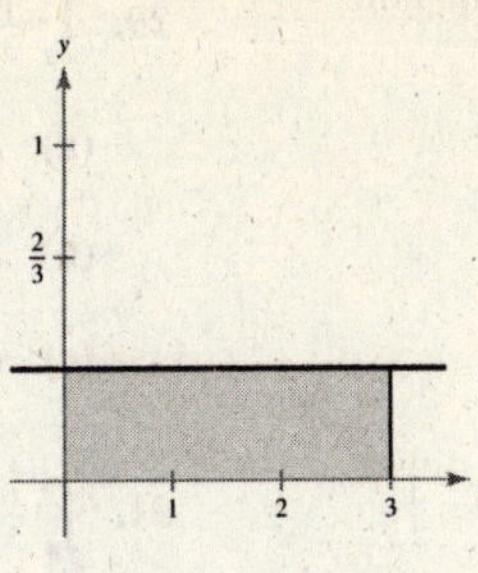

(a) $P(0 < x < 2) = \frac{2-0}{3} = \frac{2}{3}$

(b) $P(1 < x < 2) = \frac{2-1}{3} = \frac{1}{3}$

(c) $P(1 < x < 3) = \frac{3-1}{3} = \frac{2}{3}$

(d) $P(x = 2) = \frac{2-2}{3} = 0$

19. $\int_a^b \frac{x}{50}\,dx = \left[\frac{x^2}{100}\right]_a^b = \frac{b^2-a^2}{100}$

(a) $P(0 < x < 6) = \frac{36-0}{100} = \frac{9}{25}$

(b) $P(4 < x < 6) = \frac{36-16}{100} = \frac{1}{5}$

(c) $P(8 < x < 10) = \frac{100-64}{100} = \frac{9}{25}$

(d) $P(x \geq 2) = P(2 \leq x \leq 10) = \frac{100-4}{100} = \frac{24}{25}$

21. $\int_a^b \frac{3}{16}\sqrt{x}\,dx = \left.\left(\frac{3}{16}\right)\frac{2}{3}x^{3/2}\right]_a^b = \frac{1}{8}\left[b\sqrt{b} - a\sqrt{a}\right]$

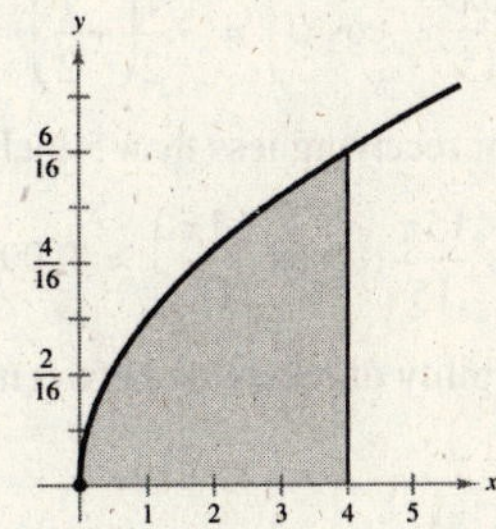

(a) $P(0 < x < 2) = \frac{\sqrt{2}}{4} \approx 0.354$

(b) $P(2 < x < 4) = 1 - \frac{\sqrt{2}}{4} \approx 0.646$

(c) $P(1 < x < 3) = \frac{1}{8}\left(3\sqrt{3} - 1\right) \approx 0.525$

(d) $P(x \leq 3) = \frac{3\sqrt{3}}{8} \approx 0.650$

23. $\int_a^b \frac{1}{3}e^{-t/3}\,dt = \left.e^{-t/3}\right]_a^b = e^{-a/3} - e^{-b/3}$

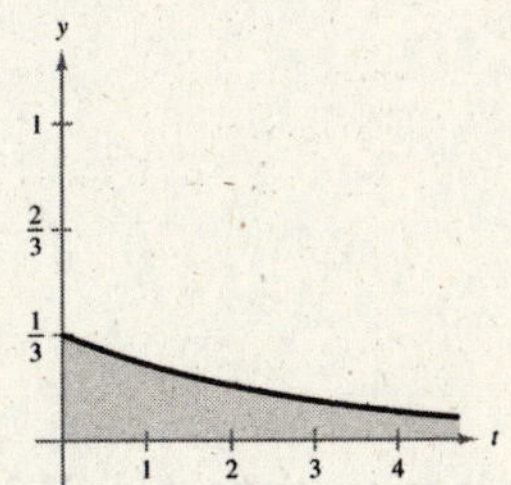

(a) $P(t < 2) = e^{-0/3} - e^{-2/3} \approx 0.4866$

(b) $P(t \geq 2) = e^{-2/3} - 0 \approx 0.5134$

(c) $P(1 < t < 4) = e^{-1/3} - e^{-4/3} \approx 0.4529$

(d) $P(t = 3) = 0$

25. $P(a < t < b) = \int_a^b \frac{1}{30}\,dt = \left.\frac{t}{30}\right]_a^b = \frac{b-a}{30}$

(a) $P(0 \leq t \leq 5) = \frac{5-0}{30} = \frac{1}{6}$

(b) $P(18 \leq t < 30) = \frac{30-18}{30} = \frac{2}{5}$

27. Using substitution, let $u = 9 - t$. Then $t = 9 - u$ and $dt = -du$.

$$\int \frac{5}{324}t\sqrt{9-t}\,dt = \frac{5}{324}\int (9-u)u^{1/2}(-du)$$
$$= \frac{5}{324}\int \left(u^{3/2} - 9u^{1/2}\right)du$$
$$= \frac{5}{324}\int \left(\frac{2}{5}u^{5/2} - 6u^{3/2}\right)du$$

$$\int_a^b \frac{5}{324}t\sqrt{9-t}\,dt = \frac{5}{324}\left[\frac{2}{5}(9-t)^{5/2} - 6(9-t)^{3/2}\right]_a^b$$

(a) $P(t < 3) = \int_0^3 f(t)\,dt = 1 - \frac{1}{3}\sqrt{6} \approx 0.1835$

(b) $P(4 < t < 8) = \int_4^8 f(t)\,dt = \frac{25}{81}\sqrt{5} - \frac{7}{81} \approx 0.6037$

29. $\int_a^b \frac{1}{3}e^{-t/3}\,dt = -e^{-t/3}\Big]_a^b = e^{-a/3} - e^{-b/3}$

(a) $P(0 < t < 2) = e^{-0/3} - e^{-2/3} = 1 - e^{-2/3} \approx 0.487$

(b) $P(2 < t < 4) = e^{-2/3} - e^{-4/3} \approx 0.250$

(c) $P(t > 2) = 1 - P(0 < t < 2) = e^{-2/3} \approx 0.513$

31. $\int_a^b \frac{1}{5}e^{-t/5}\,dt = -e^{-t/5}\Big]_a^b = e^{-a/5} - e^{-b/5}$

(a) $P(0 < t < 6) = 1 - e^{-6/5} \approx 0.699$

(b) $P(2 < t < 6) = e^{-2/5} - e^{-6/5} \approx 0.369$

(c) $P(t > 8) = e^{-8/5} - \lim_{b\to\infty} e^{-b/5} = e^{-8/5} \approx 0.202$

33. $P(a \le x \le b) = \int_a^b \frac{\pi}{30}\sin\frac{\pi x}{15}\,dx = \left[-\frac{1}{2}\cos\frac{\pi x}{15}\right]_a^b = -\frac{1}{2}\left(\cos\frac{\pi b}{15} - \cos\frac{\pi a}{15}\right)$

(a) $P(0 \le x \le 10) = -\frac{1}{2}\left(\cos\frac{10\pi}{15} - \cos 0\right) = -\frac{1}{2}\left(-\frac{3}{2}\right) = 0.75$

There is a 75% probability of receiving up to 10 inches of rain.

(b) $P(10 \le x \le 15) = -\frac{1}{2}\left(\cos\frac{15\pi}{15} - \cos\frac{10\pi}{15}\right) = -\frac{1}{2}\left(-\frac{1}{2}\right) = 0.25$

There is a 25% probability of receiving between 10 and 15 inches of rain.

(c) $P(0 \le x < 5) = -\frac{1}{2}\left(\cos\frac{5\pi}{15} - \cos 0\right) = -\frac{1}{2}\left(-\frac{1}{2}\right) = 0.25$

There is a 25% probability of receiving less than 5 inches of rain.

(d) $P(12 \le x \le 15) = -\frac{1}{2}\left(\cos\frac{15\pi}{15} - \cos\frac{12\pi}{15}\right) \approx 0.095$

There is about a 9.5% probability of receiving between 12 and 15 inches of rain.

35. $\int_a^b \frac{1}{36}xe^{-x/16}\,dx = \frac{-1}{6}e^{-x/6}(x+6)\Big]_a^b$

$$P(0 < x < b) = 0.90$$
$$-\frac{1}{6}e^{-x/6}(x+6)\Big]_0^b = 0.90$$
$$-\frac{1}{6}\left[e^{-b/6}(b+6) - 6\right] = 0.90$$
$$-\frac{1}{6}(b+6)e^{-b/6} = -0.10$$
$$(b+6)e^{-b/6} = 0.60$$

Solving for b with a graphing utility or using a computer, $b \approx 23.34$ tons.

Section 9.3 Expected Value and Variance

Skills Warm Up

1. $\int_0^m \frac{1}{16}\,dx = 0.5$

$\frac{1}{16}x\Big]_0^m = 0.5$

$\frac{1}{16}m = 0.5$

$m = 8$

2. $\int_0^m \frac{1}{3}e^{-t/3}\,dt = 0.5$

$-e^{-t/3}\Big]_0^m = 0.5$

$\left(-e^{-m/3} - e^0\right) = 0.5$

$-e^{-m/3} + 1 = 0.5$

$-e^{-m/3} = -0.5$

$e^{-m/3} = 0.5$

$-\frac{m}{3} = \ln 0.5$

$m = -3\ln 0.5 \approx 2.0794$

3. $\int_0^2 \frac{x^2}{2}\,dx = \frac{1}{6}x^3\Big]_0^2 = \frac{1}{6}\left(2^3 - 0^3\right) = \frac{8}{6} = \frac{4}{3}$

4. $\int_1^2 x(4 - 2x)\,dx = \int_1^2 \left(4x - 2x^2\right)dx$

$= \left[2x^2 - \frac{2}{3}x^3\right]_1^2 = \frac{8}{3} - \frac{4}{3} = \frac{4}{3}$

5. $P(a \le x \le b) = \int_a^b \frac{1}{8}\,dx = \frac{1}{8}x\Big]_a^b = \frac{b-a}{8}$

(a) $P(x \le 2) = P(0 \le x \le 2) = \frac{2-0}{8} = \frac{1}{4}$

(b) $P(3 < x < 7) = \frac{7-3}{8} = \frac{1}{2}$

6. $P(a \le x \le b) = \int_a^b \left(6x - 6x^2\right)dx = \left[3x^2 - 2x^3\right]_a^b = \left(3b^2 - 2b^3\right) - \left(3a^2 - 2a^3\right)$

(a) $P\left(x \le \frac{1}{2}\right) = P\left(0 \le x \le \frac{1}{2}\right) = \left[3\left(\frac{1}{2}\right)^2 - 2\left(\frac{1}{2}\right)^3\right] - \left[3(0)^2 - 2\left(0^3\right)\right] = \frac{1}{2}$

(b) $P\left(\frac{1}{4} \le x \le \frac{3}{4}\right) = \left[3\left(\frac{3}{4}\right)^2 - 2\left(\frac{3}{4}\right)^3\right] - \left[3\left(\frac{1}{4}\right)^2 - 2\left(\frac{1}{4}\right)^3\right] = \frac{27}{32} - \frac{5}{32} = \frac{11}{16}$

1. (a) $\mu = \int_a^b x f(x)\,dx = \int_0^3 x\left(\frac{1}{3}\right)dx = \frac{x^2}{6}\Big]_0^3 = \frac{3}{2}$

(b) $\sigma^2 = \int_a^b x^2 f(x)\,dx - \mu^2 = \int_0^3 x^2\left(\frac{1}{3}\right)dx - \left(\frac{3}{2}\right)^2 = \frac{x^3}{9}\Big]_0^3 - \frac{9}{4} = 3 - \frac{9}{4} = \frac{3}{4}$

(c) $\sigma = \sqrt{\frac{3}{4}} = \frac{\sqrt{3}}{2}$

(d)

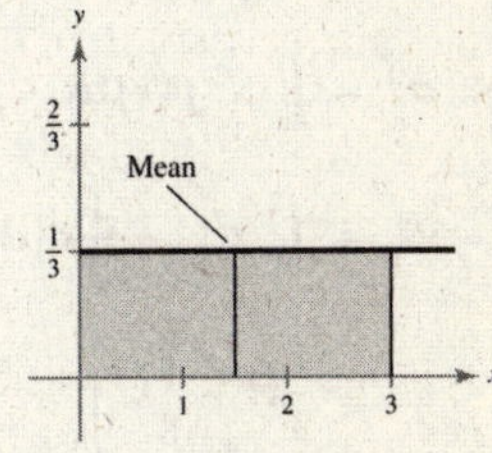

3. (a) $\mu = \int_a^b t f(t)\,dt = \int_0^6 t\left(\frac{t}{18}\right)dt = \frac{t^3}{54}\Big]_0^6 = 4$

(b) $\sigma^2 = \int_a^b t^2 f(t)\,dt - \mu^2 = \int_0^6 t^2\left(\frac{t}{18}\right)dt - 4^2 = \frac{t^4}{72}\Big]_0^6 - 4^2 = 18 - 16 = 2$

(c) $\sigma = \sqrt{2}$

(d)

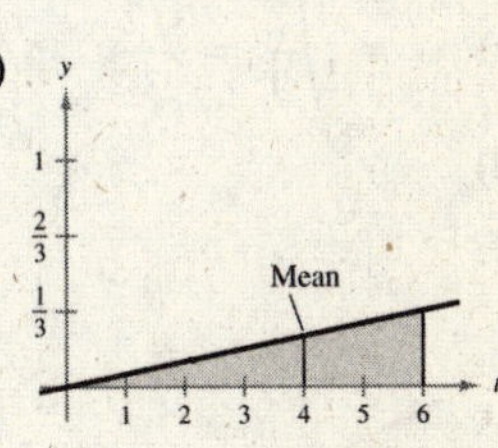

5. (a) $\mu = \int_a^b x f(x)\,dx$

$= \int_1^4 x\left(\frac{4}{3x^2}\right) dx$

$= \frac{4}{3}\int_1^4 \frac{1}{x}\,dx$

$= \frac{4}{3}\ln|x|\Big]_1^4$

$= \frac{4}{3}\ln 4 \approx 1.848$

(b) $\sigma^2 = \int_a^b x^2 f(x)\,dx - \mu^2$

$= \int_1^4 x^2\left(\frac{4}{3x^2}\right) dx - \left[\frac{4}{3}\ln 4\right]^2$

$= \frac{4}{3}x\Big]_1^4 - \frac{16}{9}(\ln 4)^2$

$= 4 - \frac{16}{9}(\ln 4)^2 \approx 0.583$

(c) $\sigma = \sqrt{0.583} \approx 0.76$

(d)

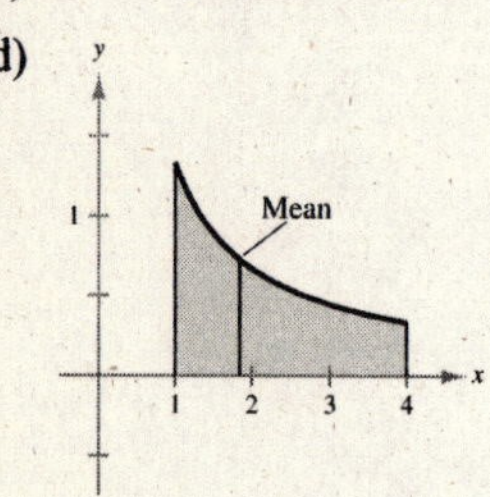

7. (a) $\mu = \int_a^b x f(x)\,dx$

$= \int_0^2 x\left(1 - \frac{1}{2}x\right) dx$

$= \frac{1}{2}x^2 - \frac{1}{6}x^3\Big]_0^2 = \frac{2}{3}$

(b) $\sigma^2 = \int_a^b x^2 f(x)\,dx - \mu^2$

$= \int_0^2 x^2\left(1 - \frac{1}{2}x\right) dx - \left(\frac{2}{3}\right)^2$

$= \frac{1}{3}x^3 - \frac{1}{8}x^4\Big]_0^2 - \frac{4}{9} = \frac{2}{9}$

(c) $\sigma = \sqrt{\frac{2}{9}} = \frac{\sqrt{2}}{3}$

(d)

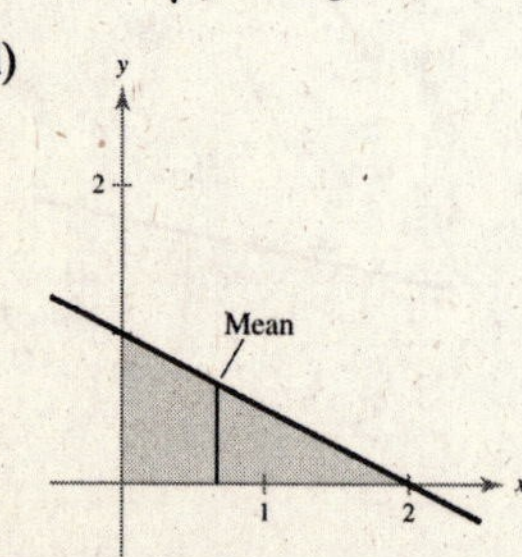

9. (a) $\mu = \int_a^b x f(x)\,dx = \int_0^1 x\left[6x(1 - x)\right] dx$

$= 2x^3 - \frac{3}{2}x^4\Big]_0^1 = \frac{1}{2}$

(b) $\sigma^2 = \int_a^b x^2 f(x)\,dx - \mu^2$

$= \int_0^1 x^2\left[6x(1 - x)\right] dx - \left(\frac{1}{2}\right)^2$

$= \frac{3}{2}x^4 - \frac{6}{5}x^5\Big]_0^1 - \frac{1}{4} = \frac{1}{20}$

(c) $\sigma = \sqrt{\frac{1}{20}} = \frac{1}{2\sqrt{5}} = \frac{\sqrt{5}}{10}$

(d)

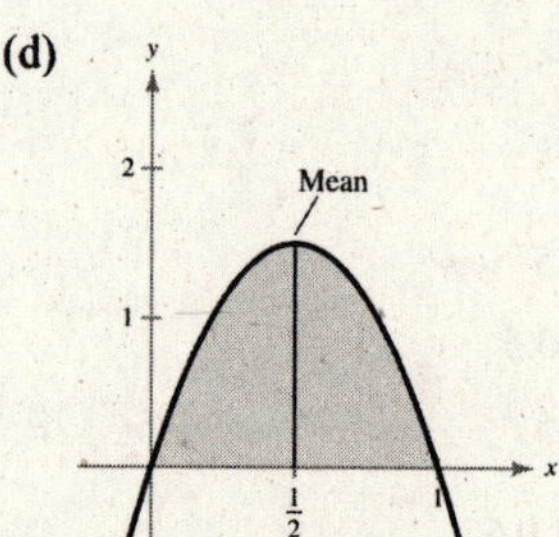

11. (a) $\mu = \int_a^b x f(x)\,dx$

$= \int_0^4 x\left(\frac{3}{16}\sqrt{4 - x}\right) dx$

$= \frac{3}{16}\int_0^4 x\sqrt{4 - x}\,dx$

$= -\frac{1}{40}(4 - x)^{3/2}(3x + 8)\Big]_0^4$

$= \frac{8}{5} = 1.6$

(b) $\sigma^2 = \int_a^b x^2 f(x)\,dx - \mu^2$

$= \int_0^4 x^2\left(\frac{3}{16}\sqrt{4 - x}\right) dx - \left(\frac{8}{5}\right)^2$

$= \frac{3}{16}\int_0^4 x^2\sqrt{4 - x}\,dx - \frac{64}{25}$

$= -\frac{1}{280}(4 - x)^{3/2}\left(15x^2 + 48x + 128\right)\Big]_0^4 - \frac{64}{25}$

$= \frac{128}{35} - \frac{64}{25}$

$= \frac{192}{175} \approx 1.097$

(c) $\sigma \approx 1.047$

(d)

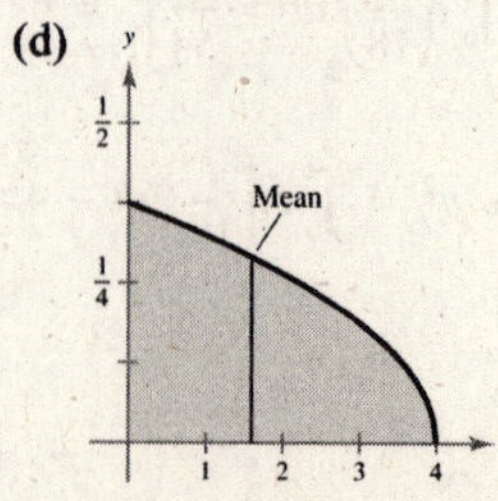

13. Mean $= \frac{1}{2}(0 + 11) = 5.5$

Median $= m$: $\int_0^m \frac{1}{11}\,dx = 0.5$

$$\frac{m}{11} = 0.5$$

$$m = 5.5$$

15. Mean $= \int_0^{1/2} x(4)(1 - 2x)\,dx = \frac{1}{6}$

Median $= m$: $\int_0^m 4(1 - 2x)\,dx = \frac{1}{2}$

$$4x - 4x^2\Big]_0^m = \frac{1}{2}$$

$$4m - 4m^2 = \frac{1}{2} \Rightarrow m \approx 0.1465$$

$\left(m \approx 0.8536 \text{ is not in the interval } \left[0, \frac{1}{2}\right]\right)$

17. Mean $= \int_0^\infty t\left(\frac{1}{9}e^{-t/9}\right)dt = -(t + 9)e^{-t/9}\Big]_0^\infty = 9$

Median $= \int_0^m \frac{1}{9}e^{-t/9}\,dt = -e^{-t/9}\Big]_0^m = 1 - e^{m/9} = \frac{1}{2}$

$$e^{-m/9} = \frac{1}{2} \Rightarrow -\frac{m}{9} = \ln\frac{1}{2}$$

$$m = -9\ln\frac{1}{2} \approx 6.238$$

19. $f(x) = \frac{1}{10}$, $[0, 10]$ is a uniform probability density function.

Mean: $\frac{a + b}{2} = \frac{0 + 10}{2} = 5$

Variance: $\frac{(b - a)^2}{12} = \frac{(10 - 0)^2}{12} = \frac{100}{12} = \frac{25}{3}$

Standard deviation: $\frac{b - a}{\sqrt{12}} = \frac{10 - 0}{\sqrt{12}} \approx 2.887$

21. $f(x) = \frac{1}{8}e^{-x/8}$, $[0, \infty)$ is an exponential probability density function with $a = \frac{1}{8}$.

Mean: $\frac{1}{a} = 8$

Variance: $\frac{1}{a^2} = 64$

Standard deviation: $\frac{1}{a} = 8$

23. $f(x) = \frac{1}{11\sqrt{2\pi}}e^{-(x-100)^2/242}$, $(-\infty, \infty)$ is a normal probability density function with $\mu = 100$ and $\sigma = 11$.

Mean: $\mu = 100$

Variance: $\sigma^2 = 121$

Standard deviation: $\sigma = 11$

25. Mean $= 0$

Standard deviation $= 1$

$P(0 \le x \le 0.85) \approx 0.3023$

27. Mean $= 6$

Standard deviation $= 6$

$P(x \ge 2.23) \approx 0.6896$

29. Mean $= 8$

Standard deviation $= 2$

$P(3 \le x \le 13) \approx 0.9876$

31. $\mu = 50$, $\sigma = 10$

(a) $P(x > 55) = \int_{55}^\infty \frac{1}{10\sqrt{2\pi}}e^{-(x-50)^2/2(10^2)}\,dx$

≈ 0.3085

(b) $P(x > 60) = \int_{60}^\infty \frac{1}{10\sqrt{2\pi}}e^{-(x-50)^2/2(10^2)}\,dx$

≈ 0.1587

(c) $P(x < 60) = \int_{-\infty}^{60} \frac{1}{10\sqrt{2\pi}}e^{-(x-50)^2/2(10^2)}\,dx$

≈ 0.8413

(d) $P(30 < x < 55) = \int_{30}^{55} \frac{1}{10\sqrt{2\pi}}e^{-(x-50)^2/2(10^2)}\,dx$

≈ 0.6687

33. (a) Since the arrival time t (in minutes) is uniformly distributed between 10:00 A.M. and 10:10 A.M., let $f(t) = \frac{1}{10}$ over $[0, 10]$, where $t = 0$ corresponds to 10:00 A.M.

(b) Mean $= \frac{10}{2} = 5$

The mean is 10:05 A.M.

Standard deviation $= \frac{10}{\sqrt{12}} \approx 2.887$ min

(c) $1 - \int_3^{10} \frac{1}{10}\,dx = 1 - \frac{7}{10} = \frac{3}{10} = 0.30$

35. (a) Because $\mu = 2$, $f(t) = \frac{1}{2}e^{-t/2}$, $0 \le t < \infty$.

(b) $P(0 < t < 1) = \int_0^1 \frac{1}{2}e^{-t/2}\, dt = -e^{-t/2}\Big]_0^1 = 1 - e^{-1/2} \approx 0.3935$

37. (a) Because $\mu = 5$, $f(t) = \frac{1}{5}e^{-t/5}$, $0 \le t < \infty$.

(b) $P(\mu - \sigma < t < \mu + \sigma) = P(0 < t < 10) = \int_0^{10} \frac{1}{5}e^{-t/5}\, dt = -e^{-t/5}\Big]_0^{10} = 1 - e^{-2} \approx 0.865 = 86.5\%$

39. (a) $\mu = \int_0^7 \frac{6}{343}x(7 - x)\, dx = \frac{6}{343}\int_0^7 \left(7x - x^2\right) dx = \frac{6}{343}\left[\frac{7}{2}x^2 - \frac{x^3}{3}\right]_0^7 = 1$

$$\sigma^2 = \int_0^7 \frac{6}{343}x^2(7 - x)\, dx - (1)^2 = \frac{6}{343}\int_0^7 \left(7x^2 - x^3\right) dx - 1 = \frac{6}{343}\left[\frac{7}{3}x^3 - \frac{x^4}{4}\right]_0^7 - 1 = \frac{7}{2} - 1 = \frac{5}{2}$$

$$\sigma = \sqrt{\frac{5}{2}} = \frac{\sqrt{10}}{2} \approx 1.581$$

(b) $\int_0^m \frac{6}{343}x(7 - x)\, dx = \frac{6}{343}\int_0^m \left(7x - x^2\right) dx = \frac{6}{343}\left[\frac{7}{2}x^2 - \frac{x^3}{3}\right]_0^m = \frac{6}{343}\left[\frac{7m^2}{2} - \frac{m^3}{3}\right] = \frac{1}{2}$

$$\frac{7m^2}{2} - \frac{m^3}{3} = \frac{343}{12} \Rightarrow 0 = (m - 3.5)\left(4m^2 - 28m - 98\right)$$

$$m = 3.5 \text{ or } m = \frac{7 \pm \sqrt{147}}{4}$$

In the interval $[0, 7]$, $m = 3.5$.

(c) $P(\mu - \sigma < x < \mu + \sigma) = P\left(1 - \frac{\sqrt{10}}{2} < x < 1 + \frac{\sqrt{10}}{2}\right) \approx P(-0.5811 < x < 2.581)$

$$= \int_{-0.5811}^{2.5811} \frac{6}{343}x(7 - x)\, dx = \frac{6}{343}\left[\frac{7}{2}x^2 - \frac{x^3}{3}\right]_{-0.5811}^{2.5811} \approx 0.286$$

41. $f(x) = \frac{2x + 1}{12}, \quad 0 \le x \le 3$

Expected value $= \int_0^3 x\left(\frac{2x + 1}{12}\right) dx = \frac{1}{12}\left[\frac{2x^3}{3} + \frac{x^2}{2}\right]_0^3 = \frac{15}{8}$

43. (a) $\mu = \int_0^\infty \frac{1}{9}x^2 e^{-x/3}\, dx$ (Use integration by parts.)

$$= \lim_{b \to \infty} -3\left[\frac{x^2}{9}e^{-x/3} - 2\left(-\frac{x}{3} - 1\right)e^{-x/3}\right]_0^b = 6$$

$\sigma^2 = \int_0^\infty \frac{1}{9}x^3 e^{-x/3}\, dx - (6)^2$ (Use integration by parts.)

$$= \lim_{b \to \infty}\left[-\frac{x^3}{3}e^{-x/3}\right]_0^b + 9\int_0^\infty \frac{1}{9}x^2 e^{-x/3}\, dx - 36$$

$= 0 + 9(6) - 36$

$= 18$ (Use part (a).)

$\sigma = \sqrt{18} = 3\sqrt{2} \approx 4.243$

(b) $P(x > 4) = 1 - P(x < 4) = 1 - \int_0^4 \frac{1}{9}xe^{-x/3}\, dx = 1 - \left[-\frac{1}{3}e^{-x/3}(x + 3)\right]_0^4 \approx 0.615$

45. (a) $\dfrac{x-\mu}{\sigma} = \dfrac{340-400}{24} = -2.5$

Your battery fell short of the expected life by 2.5 standard deviations.

(b) $f(x) = \dfrac{1}{24\sqrt{2\pi}}e^{-(x-400)^2/1152}$

$P(x > 340) = \displaystyle\int_{340}^{\infty} \frac{1}{24\sqrt{2\pi}}e^{-(x-400)^2/1152}\,dx \approx 0.9938$

Note: $1 - P(0 < x < 340) \approx 0.9938$

47. $u = 4.5,\ \sigma = 0.5$

(a) $P(4 < x < 5) = \displaystyle\int_4^5 \frac{1}{0.5\sqrt{2\pi}}e^{-(x-4.5)^2/2(0.5)^2}\,dx = \frac{2}{\sqrt{2\pi}}\int_4^5 e^{-2(x-4.5)^2}\,dx \approx 0.6827$

(b) $P(x < 3) = \displaystyle\int_0^3 \frac{1}{0.5\sqrt{2\pi}}e^{-(x-4.5)^2/2(0.5)^2}\,dx = \frac{2}{\sqrt{2\pi}}\int_0^3 e^{-2(x-4.5)^2}\,dx \approx 0.0013 = 0.13\%$

No, only about 0.13% of the batteries will last less than 3 years.

49. $f(x) = \dfrac{1}{16\sqrt{2\pi}}e^{-(x-266)^2/2(16)^2} = \dfrac{1}{40.106}e^{-(x-266)^2/512}$

(a)

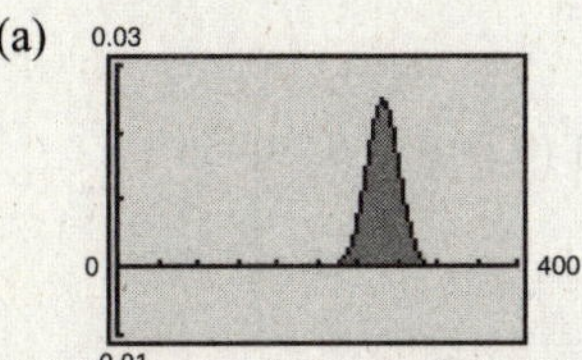

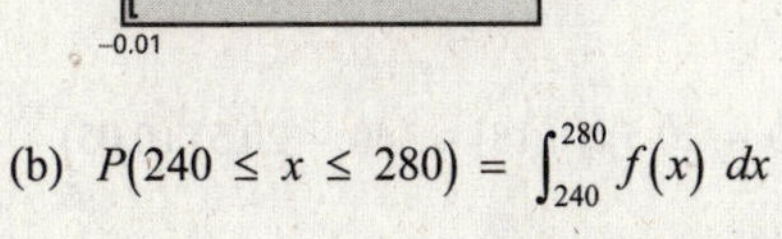

(b) $P(240 \le x \le 280) = \displaystyle\int_{240}^{280} f(x)\,dx$

≈ 0.757 or 76%

(c) $P(x > 280) = \displaystyle\int_{280}^{\infty} f(x)\,dx$

≈ 0.191 or 19.1%

51. (a)

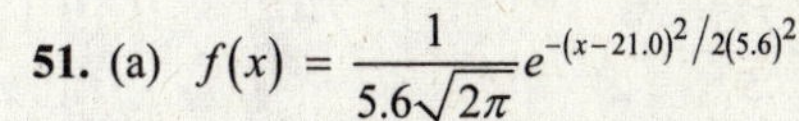

$f(x) = \dfrac{1}{5.6\sqrt{2\pi}}e^{-(x-21.0)^2/2(5.6)^2}$

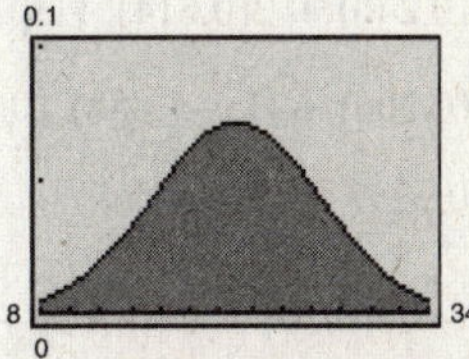

(b) $P(25 \le x \le 30) = \displaystyle\int_{25}^{30} \frac{1}{5.6\sqrt{2\pi}}e^{-(x-21.0)^2/2(5.6)^2}\,dx$

≈ 0.1835 or 18.35%

(c) $P(x < 18) = \displaystyle\int_0^{18} \frac{1}{5.6\sqrt{2\pi}}e^{-(x-21.0)^2/2(5.6)^2}\,dx$

≈ 0.2960 or 29.60%

Review Exercises for Chapter 9

1. (a) $S = \{1, 2, 3, 4, 5, 6, 7, 8, 9, 10, 11, 12\}$

(b) $A = \{9, 10, 11, 12\}$

(c) $B = \{1, 3, 5, 7, 9, 11\}$

3. (a) $S = \{000, 001, 002, 003, \ldots, 997, 998, 999\}$

The sample space consists of all distinct orders of three digits from 0 to 9.

(b) $A = \{100, 101, 102, 103, \ldots, 198, 199\}$

(c) $B = \{050, 100, 150, 200, 250, 300, 350, \ldots, 900, 950\}$

5.

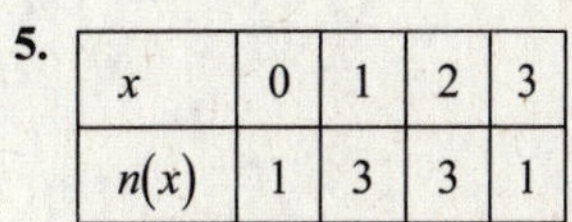

x	0	1	2	3
$n(x)$	1	3	3	1

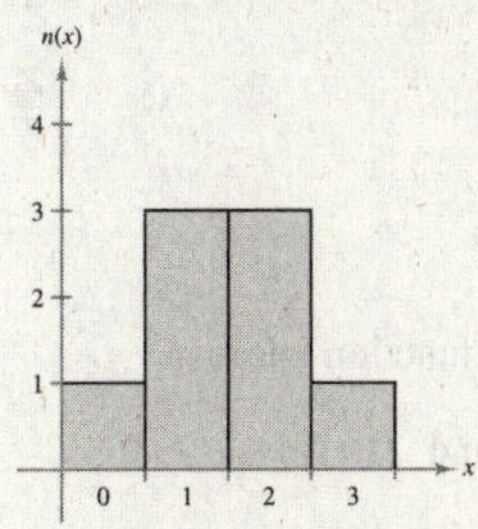

7.

x	0	1	2	3
$P(x)$	$\frac{1}{8}$	$\frac{3}{8}$	$\frac{3}{8}$	$\frac{1}{8}$

9. The table does not represent a probability distribution because the sum of the probabilities does not equal 1.

11. (a) $P(2 \le x \le 4) = P(2) + P(3) + P(4)$

$$= \frac{7}{18} + \frac{5}{18} + \frac{3}{18} = \frac{15}{18} = \frac{5}{6}$$

(b) $P(x \ge 3) = P(3) + P(4) + P(5)$

$$= \frac{5}{18} + \frac{3}{18} + \frac{2}{18} = \frac{10}{18} = \frac{5}{9}$$

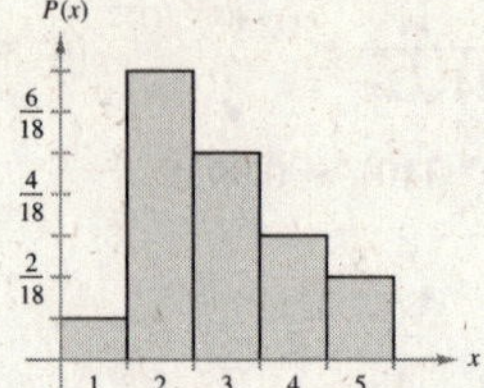

13. $E(x) = 0\left(\frac{1}{10}\right) + 1\left(\frac{3}{10}\right) + 2\left(\frac{2}{10}\right) + 3\left(\frac{3}{10}\right) + 4\left(\frac{1}{10}\right) = 2$

$V(x) = (0-2)^2\left(\frac{1}{10}\right) + (1-2)^2\left(\frac{3}{10}\right) + (2-2)^2\left(\frac{2}{10}\right) + (3-2)^2\left(\frac{3}{10}\right) + (4-2)^2\left(\frac{1}{10}\right) = \frac{7}{5}$

$\sigma = \sqrt{\frac{7}{5}} = \frac{\sqrt{35}}{5}$

15. $E(x) = 0(0.006) + 1(0.240) + 2(0.614) + 3(0.140) = 1.888$

$V(x) = (0 - 1.888)^2(0.006) + (1 - 1.888)^2(0.240) + (2 - 1.888)^2(0.614) + (3 - 1.888)^2(0.140) = 0.391456$

$\sigma = \sqrt{0.391456} \approx 0.6257$

17. (a) $E(x) = 10(0.10) + 15(0.20) + 20(0.50) + 30(0.15) + 40(0.05) = 20.5$

$V(x) = (10 - 20.5)^2(0.10) + (15 - 20.5)^2(0.2) + (20 - 20.5)^2(0.5) + (30 - 20.5)^2(0.15) + (40 - 20.5)^2(0.05)$

$= 49.75$

$\sigma = \sqrt{49.75} \approx 7.05$

(b) $R = 20.5(1000)(3.95) = \$80{,}975$

19. $E(x) = 0(0.10) + 1(0.28) + 2(0.39) + 3(0.17) + 4(0.04) + 5(0.02) = 1.83$

$V(x) = (0 - 1.83)^2(0.10) + (1 - 1.83)^2(0.28) + (2 - 1.83)^2(0.39) + (3 - 1.83)^2(0.17)$

$+ (4 - 1.83)^2(0.04) + (5 - 1.83)^2(0.02) \approx 1.1611$

$\sigma = \sqrt{V(x)} \approx 1.0775$

21.

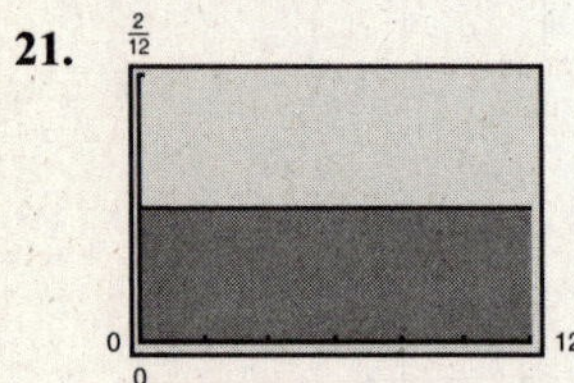

f is a probability density function because $\int_0^{12} \frac{1}{12}\,dx = \frac{1}{12}x\Big]_0^{12} = 1$ and $f(x) = \frac{1}{12} \ge 0$ on $[0, 12]$.

23.

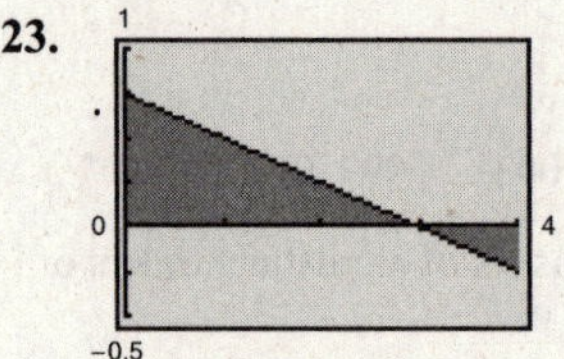

f is not a probability density function because $f(x) = \frac{1}{4}(3 - x)$ is negative on $(3, 4]$.

25.

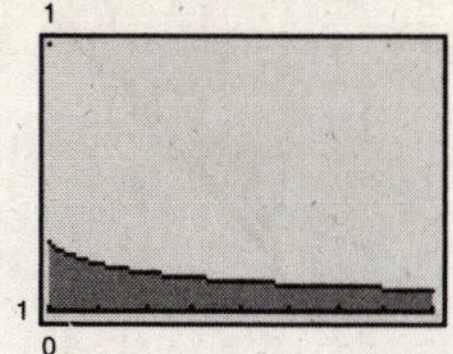

f is a probability density function because

$\int_1^9 \frac{1}{4\sqrt{x}}\,dx = \frac{1}{4}\left[2\sqrt{x}\right]_1^9 = 1$ and $f(x) = \frac{1}{4\sqrt{x}} \geq 0$

on $[1, 9]$.

27.

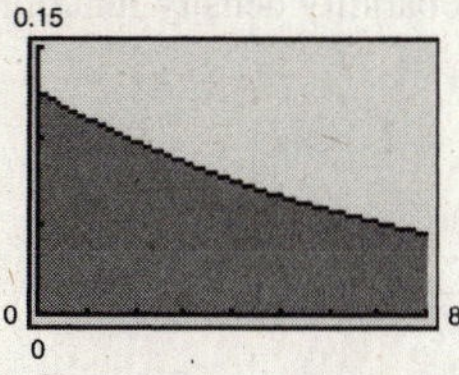

$f(x)$ is not a probability density function because

$\int_0^8 \frac{1}{8}e^{-x/8}\,dx = \left[-e^{-x/8}\right]_0^8 \approx 0.632 \neq 1.$

29.

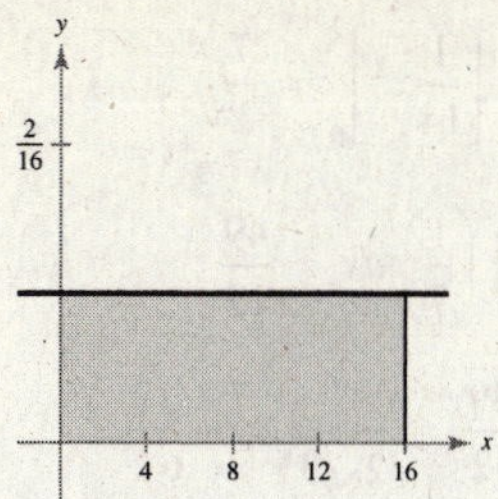

(a) $P(0 < x < 5) = \int_0^5 \frac{1}{16}\,dx = \frac{1}{16}x\Big]_0^5 = \frac{5}{16}$

(b) $P(12 < x < 13) = \int_{12}^{13} \frac{1}{16}\,dx = \frac{1}{16}x\Big]_{12}^{13} = \frac{1}{16}$

(c) $P(x \geq 5) = \int_5^{16} \frac{1}{16}\,dx = \frac{1}{16}x\Big]_5^{16} = \frac{11}{16}$

(d) $P(8 < x < 12) = \int_8^{12} \frac{1}{16}\,dx = \frac{1}{16}x\Big]_8^{12} = \frac{1}{4}$

31.

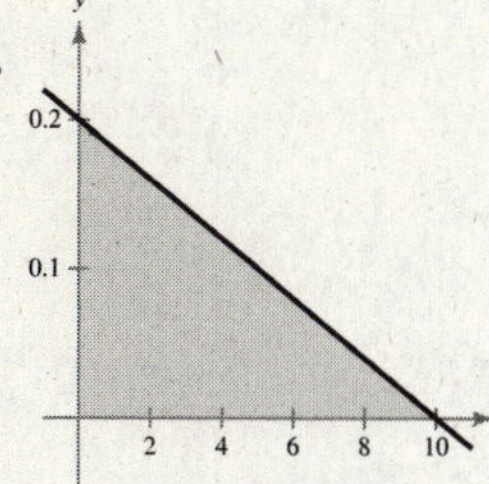

(a) $P(0 < x < 2) = \int_0^2 \frac{1}{50}(10 - x)\,dx = \left[\frac{1}{50}\left(10x - \frac{x^2}{2}\right)\right]_0^2 = \frac{9}{25}$

(b) $P(x \geq 7) = \int_7^{10} \frac{1}{50}(10 - x)\,dx = \left[\frac{1}{50}\left(10x - \frac{x^2}{2}\right)\right]_7^{10} = \frac{9}{100}$

(c) $P(x \leq 5) = \int_0^5 \frac{1}{50}(10 - x)\,dx = \left[\frac{1}{50}\left(10x - \frac{x^2}{2}\right)\right]_0^5 = \frac{3}{4}$

(d) $P(8 < x < 9) = \int_8^9 \frac{1}{50}(10 - x)\,dx = \left[\frac{1}{50}\left(10x - \frac{x^2}{2}\right)\right]_8^9 = \frac{3}{100}$

33.

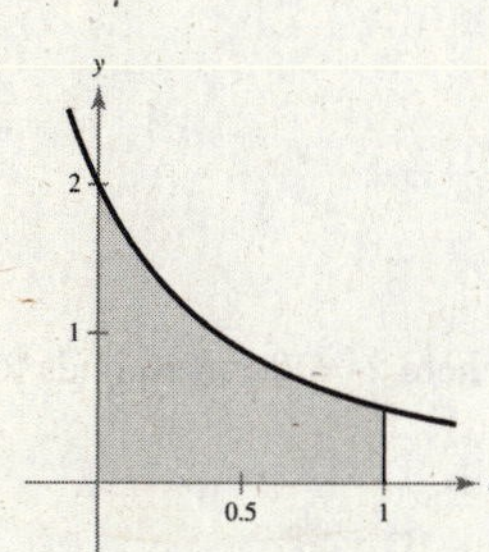

(a) $P\left(0 < x < \frac{1}{2}\right) = \int_0^{1/2} \frac{2}{(x + 1)^2}\,dx = \left[-\frac{2}{x + 1}\right]_0^{1/2} = \frac{2}{3}$

(b) $P\left(\frac{1}{4}x < \frac{3}{4}\right) = \int_{1/4}^{3/4} \frac{2}{(x + 1)^2}\,dx = \left[-\frac{2}{x + 1}\right]_{1/4}^{3/4} = \frac{16}{35}$

(c) $P\left(x \geq \frac{1}{2}\right) = \int_{1/2}^{1} \frac{2}{(x + 1)^2}\,dx = \left[-\frac{2}{x + 1}\right]_{1/2}^{1} = \frac{1}{3}$

(d) $P\left(\frac{1}{10} < x < \frac{3}{10}\right) = \int_{1/10}^{3/10} \frac{2}{(x + 1)^2}\,dx = \left[-\frac{2}{x + 1}\right]_{1/10}^{3/10} = \frac{40}{143}$

35. (a) $P(0 < t < 5) = \int_0^5 \frac{1}{12}e^{-t/12}\,dt = \left[-e^{-t/12}\right]_0^5 = 1 - e^{-5/12} \approx 0.341$

(b) $P(9 < t < 12) = \int_9^{12} \frac{1}{12}e^{-t/12}\,dt = \left[-e^{-t/12}\right]_9^{12} = e^{3/4} - e^{-1} \approx 0.104$

37. (a) $\mu = \int_0^7 x\left(\frac{1}{7}\right) dx = \left[\frac{1}{14}x^2\right]_0^7 = \frac{7}{2}$

(b) $V(x) = \int_0^7 \left(x - \frac{7}{2}\right)^2\left(\frac{1}{7}\right) dx = \frac{49}{12}$

(c) $\sigma = \sqrt{V|x|} = \sqrt{\frac{49}{12}} = \frac{7}{2\sqrt{3}} = \frac{7\sqrt{3}}{6}$

(d)

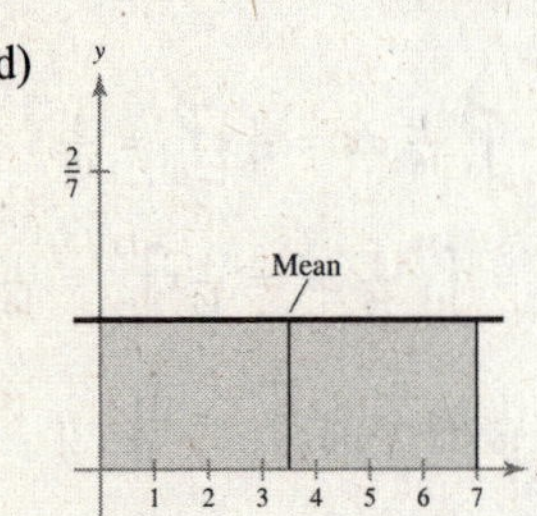

39. (a) $\mu = \int_1^3 x\left(\frac{3}{2x^2}\right) dx = \left[\frac{3}{2}\ln x\right]_1^3 = \frac{3}{2}\ln 3 \approx 1.648$

(b) $V(x) = \int_1^3 \left(x - \frac{3}{2}\ln 3\right)^2\left(\frac{3}{2x^2}\right) dx$

$= 3 - \left(\frac{3}{2}\ln 3\right)^2$

≈ 0.284

(c) $\sigma = \sqrt{V(x)} = \sqrt{3 - \left(\frac{3}{2}\ln 3\right)^2} \approx 0.533$

(d)

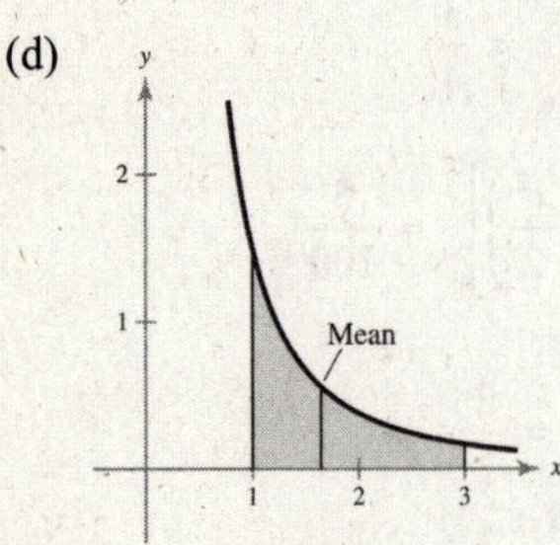

41. (a) $\mu = \int_0^3 x\left(\frac{2}{9}x(3 - x)\right) dx = \left[-\frac{x^4}{18} + \frac{2}{9}x^3\right]_0^3 = \frac{3}{2}$

(b) $V(x) = \int_0^3 \left(x - \frac{3}{2}\right)^2\left(\frac{2}{9}x(3 - x)\right) dx = \frac{9}{20}$

(c) $\sigma = \sqrt{V(x)} = \sqrt{\frac{9}{20}} = \frac{3}{2\sqrt{5}} = \frac{3\sqrt{5}}{10}$

(d)

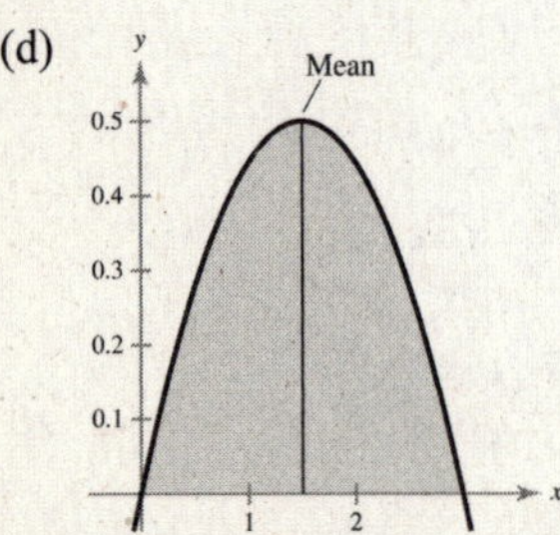

43. $\int_0^m \frac{1}{14}\, dx = \left[\frac{1}{14}x\right]_0^m = \frac{1}{14}m$

$\frac{1}{14}m = \frac{1}{2}$

$m = 7$

45. $\int_0^m 0.25e^{-x/4}\, dx = \frac{1}{2}$

$1 - e^{-m/4} = \frac{1}{2}$

$m \approx 2.7726$

47. $f(x) = \frac{1}{2}$ is a uniform probability density function.

$\mu = \frac{a + b}{2} = \frac{0 + 2}{2} = 1$

$V(x) = \frac{(b - a)^2}{12} = \frac{(2 - 0)^2}{12} = \frac{1}{3}$

$\sigma = \frac{b - a}{\sqrt{12}} = \frac{2 - 0}{\sqrt{12}} = \frac{2}{\sqrt{12}} = \frac{\sqrt{3}}{3}$

49. $f(x) = \frac{1}{6}e^{-x/6}$ is an exponential probability density function with $a = \frac{1}{6}$.

$\mu = \frac{1}{1/6} = 6$

$V(x) = \frac{1}{(1/6)^2} = 36$

$\sigma = \sqrt{36} = 6$

51. $f(x) = \frac{1}{3\sqrt{2\pi}}e^{-(x-16)^2/18}$ is a normal probability density function, with mean 16 and standard deviation 3.

$\mu = 16$

$V(x) = \sigma^2 = (3)^2 = 9$

$\sigma = 3$

53. (a) $f(t) = \frac{1}{20}$, $[0, 20]$, where $t = 0$ corresponds to 7:00 A.M.

(b) $\mu = \int_0^{20} t\left(\frac{1}{20}\right) dt = \left[\frac{t^2}{40}\right]_0^{20} = 10$ or 7:10 A.M.

$V(x) = \int_0^{20} (t - 10)^2\left(\frac{1}{20}\right) dt = \frac{100}{3}$

$\sigma = \sqrt{\frac{100}{3}} = \frac{10}{\sqrt{3}} = \frac{10\sqrt{3}}{3} \approx 5.8$ min

(c) $1 - \int_4^{20} \frac{1}{20}\, dt = 1 - \left[\frac{1}{20}t\right]_4^{20} = 1 - \frac{4}{5} = \frac{1}{5} = 0.2$

55. (a) $f(t) = \frac{1}{15}e^{-t/15}, 0 \le t < \infty$

(b) $P(0 < t < 10) = \int_0^{10} \frac{1}{15}e^{-t/15}\,dt = \left[-e^{-t/15}\right]_0^{10} = 1 - e^{-2/3} \approx 0.4866$

57. (a)

$$\mu = \int_0^8 x\left(\frac{3}{256}x(8 - x)\right)dx = \left[-\frac{3x^4}{1024} + \frac{x^3}{32}\right]_0^8$$

$$V(x) = \int_0^8 (x - 4)^2\left(\frac{3}{256}x(8 - x)\right)dx = \frac{16}{5}$$

$$\sigma = \sqrt{\frac{16}{5}} = \frac{4\sqrt{5}}{5}$$

(b) $\text{Median} = \int_0^m \left(\frac{3}{256}x(8 - x)\right)dx = \left[\frac{3x^2}{64} - \frac{x^3}{256}\right]_0^m = \frac{3m^2}{64} - \frac{m^3}{256}$

$$\frac{3m^2}{64} - \frac{m^3}{256} = \frac{1}{2}$$

$$m^3 - 12m^2 + 128 = 0$$

$$(m - 4)(m^2 - 8m - 32) = 0$$

$$m = 4 \text{ or } m = 4 \pm 4\sqrt{3}$$

In the interval $[0, 8]$, $m = 4$.

(c)

$$\begin{aligned} P(\mu - \sigma < x < \mu + \sigma) &= P\left(4 - \frac{4\sqrt{5}}{5} < x < 4 + \frac{4\sqrt{5}}{5}\right) \\ &\approx P(2.2111x < 5.7889) \\ &= \int_{2.2111}^{5.7889} \left(\frac{3}{256}x(8 - x)\right)dx \\ &= \left[\frac{3x^2}{64} - \frac{x^2}{256}\right]_{2.2111}^{5.7889} \\ &\approx 0.6261 \text{ or } 62.61\% \end{aligned}$$

59. $f(x) = \frac{1}{25\sqrt{2\pi}}e^{-(x-168)^2/2(25)^2} = \frac{1}{25\sqrt{2\pi}}e^{-(x-168)^2/1250}$

(a)

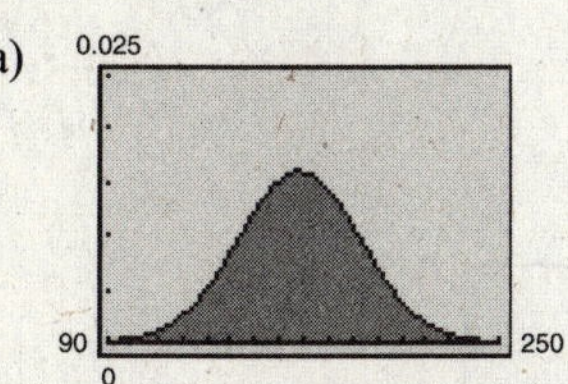

(b) $P(70 < x < 105) = \int_{70}^{105} f(x)\,dx \approx 0.0058 \text{ or } 0.58\%$

(c)

$$\begin{aligned} P(x > 120) &= 1 - P(0 < x < 120) \\ &= 1 - \int_0^{120} f(x)\,dx \\ &\approx 1 - 0.0274 \\ &= 0.9726 \text{ or } 97.26\% \end{aligned}$$

or

$$P(x > 120) = \int_{120}^{\infty} f(x)\,dx \approx 0.9726 \text{ or } 97.26\%$$

Chapter 9 Test Yourself

1. (a) $S = \{TTTT, TTTF, TTFT, TTFF, TFTT, TFTF, TFFT, TFFF, FTTT, FTTF, FTFT, FTFF, FFTT, FFTF, FFFT, FFFF\}$

(b)

Random variable, x	0	1	2	3	4
Frequency of x, $n(x)$	1	4	6	4	1

(c)

Random variable, x	0	1	2	3	4
Probability, $P(x)$	$\frac{1}{16}$	$\frac{4}{16}$	$\frac{6}{16}$	$\frac{4}{16}$	$\frac{1}{16}$

2. $P(\text{red and not a face card}) = \frac{20}{52} = \frac{5}{13}$

3. (a) $P(x < 3) = P(1) + P(2) = \frac{3}{16} + \frac{7}{16} = \frac{5}{8}$

(b) $P(x \geq 3) = P(3) + P(4) = \frac{1}{16} + \frac{5}{16} = \frac{3}{8}$

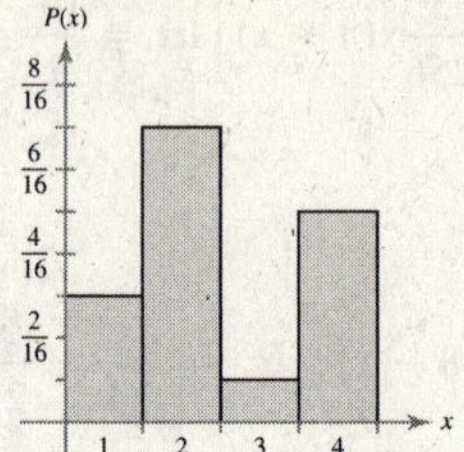

4. (a) $P(7 \leq x \leq 10) = P(7) + P(8) + P(9) + P(10) = 0.21 + 0.13 + 0.19 + 0.42 = 0.95$

(b) $P(x > 8) = P(9) + P(10) + P(11) = 0.19 + 0.42 + 0.05 = 0.66$

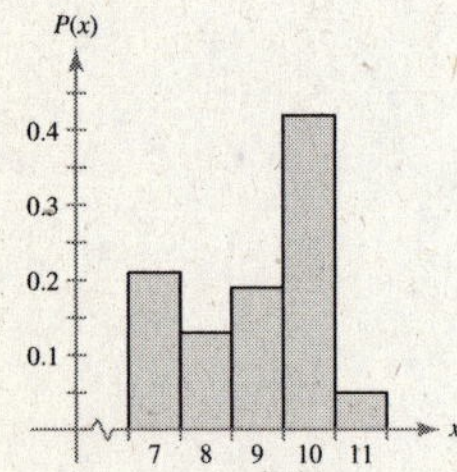

5. $E(x) = 0\left(\frac{2}{10}\right) + 1\left(\frac{1}{10}\right) + 2\left(\frac{4}{10}\right) + 3\left(\frac{3}{10}\right) = \frac{9}{5} = 1.8$

$V(x) = (0 - 1.8)^2\left(\frac{2}{10}\right) + (1 - 1.8)^2\left(\frac{1}{10}\right) + (2 - 1.8)^2\left(\frac{4}{10}\right) + (3 - 1.8)^2\left(\frac{3}{10}\right) = 1.16$

$\sigma = \sqrt{V(x)} \approx 1.077$

6. $E(x) = -2(0.141) + (-1)(0.305) + 0(0.257) + 1(0.063) + 2(0.234) = -0.056$

$V(x) = (-2 + 0.056)^2(0.141) + (-1 + 0.056)^2(0.305) + (0 + 0.056)^2(0.257) + (1 + 0.056)^2 0.063 + (2 + 0.056)^2(0.234)$
$= 1.864864$

$\sigma = \sqrt{V(x)} \approx 1.366$

7.

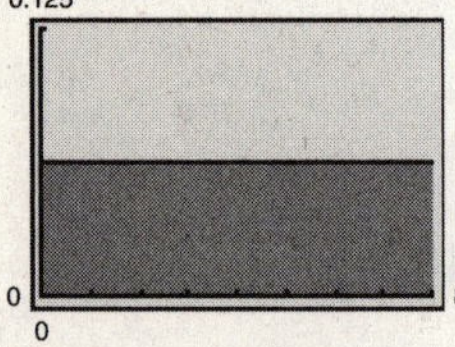

$f(x)$ is not a probability density function because

$$\int_0^8 \frac{1}{16}\,dx = \left[\frac{1}{16}x\right]_0^8 = \frac{1}{2} \neq 1.$$

8.

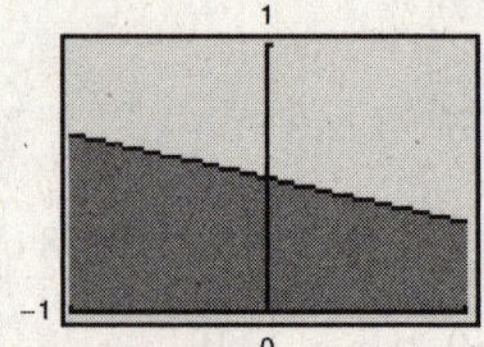

f is a probability density function because

$$\int_{-1}^1 \frac{3 - x}{6}\,dx = \left[\frac{1}{2}x - \frac{x^2}{12}\right]_{-1}^1 = 1.$$

9. 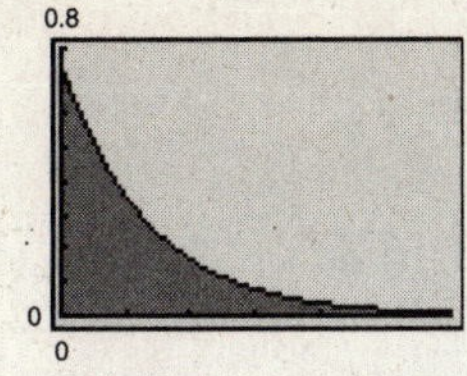

$f(x)$ is a probability density function.

$$\int_0^\infty \frac{3}{4}e^{-3x/4}\,dx = \left[-e^{-3x/4}\right]_8^\infty = 1$$

and $f(x) \geq 0$ over the interval $[0, \infty)$.

10. $\displaystyle\int_a^b \frac{2x}{9}\,dx = \left[\frac{1}{9}x^2\right]_a^b = \frac{1}{9}b^2 - \frac{1}{9}a^2$

(a) $P(0 \leq x \leq 1) = \frac{1}{9}(1)^2 - \frac{1}{9}(0)^2 = \frac{1}{9}$

(b) $P(2 \leq x \leq 3) = \frac{1}{9}(3)^2 - \frac{1}{9}(2)^2 = \frac{5}{9}$

11. $\displaystyle\int_a^b 4\left(x - x^3\right)dx = \int_a^b \left(4x - 4x^3\right)dx = \left[2x^2 - x^4\right]_a^b = \left(2b^2 - b^4\right) - \left(2a^2 - a^4\right)$

(a) $P(0 < x < 0.5) = \left[2(0.5)^2 - 0.5^4\right] - \left[2(0)^2 - 0^4\right] = 0.4375$

(b) $P(0.25 \leq x < 1) = \left[2(1)^2 - 1^4\right] - \left[2(0.25)^2 - 0.25^4\right] \approx 0.879$

12. $\displaystyle\int_a^b 2xe^{-x^2}\,dx = -e^{-x^2}\Big]_a^b = -\left(e^{-b^2} - e^{-a^2}\right)$

(a) $P(x < 1) = -\left(e^{-1^2} - e^{0^2}\right) = 1 - e^{-1} \approx 0.632$

(b) $\displaystyle P(x \geq 1) = \int_1^\infty 2xe^{-x^2}\,dx = \lim_{b\to\infty}\left[-e^{-x^2}\right]_1^b = \lim_{b\to\infty}\left(-e^{-b^2} + e^{-1}\right) = e^{-1} \approx 0.368$

13. $f(x) = \frac{1}{14}, [0, 14]$

$$\mu = \frac{a+b}{2} = \frac{0+14}{2} = 7$$

$$V(x) = \frac{(b-a)^2}{12} = \frac{(14-0)^2}{12} = \frac{49}{3}$$

$$\sigma = \frac{b-a}{\sqrt{12}} = \frac{14}{\sqrt{12}} \approx 4.041$$

14. $f(x) = 3x - \frac{3}{2}x^2, [0, 1]$

$$\mu = \int_0^1 x\left(3x - \tfrac{3}{2}x^2\right)dx = \int_0^1 \left(3x^2 - \tfrac{3}{2}x^3\right)dx$$

$$= \left[x^3 - \tfrac{3}{8}x^4\right]_0^1 = \tfrac{5}{8}$$

$$V(x) = \int_0^1 x^2\left(3x - \tfrac{3}{2}x^2\right)dx - \left(\tfrac{5}{8}\right)^2$$

$$= \int_0^1 \left(3x^3 - \tfrac{3}{2}x^4\right)dx - \tfrac{25}{64}$$

$$= \left[\tfrac{3}{4}x^4 - \tfrac{3}{10}x^5\right]_0^1 - \tfrac{25}{64}$$

$$= \tfrac{9}{20} - \tfrac{25}{64}$$

$$= \tfrac{19}{320}$$

$$\sigma = \sqrt{V(x)} \approx 0.244$$

15. $f(x) = e^{-x}, [0, \infty)$; f is an exponential probability density function with $a = 1$.

$$\mu = \frac{1}{a} = \frac{1}{1} = 1$$

$$V(x) = \frac{1}{a^2} = \frac{1}{1^2} = 1$$

$$\sigma = \sqrt{V(x)} = 1$$

16. $f(x) = \dfrac{1}{10\sqrt{2\pi}}e^{-(x-110)^2/2(10)^2}$; $\mu = 110$, $\sigma = 10$

$$P(\mu - \sigma < x < \mu + \sigma) = P(100 < x < 120)$$

$$= \int_{100}^{120} f(x)\,dx \approx 0.683$$

CHAPTER 10
Series and Taylor Polynomials

CHAPTER 10
Series and Taylor Polynomials

Section 10.1 Sequences

Skills Warm Up

1. $\lim_{x\to\infty} \frac{1}{x^3} = 0$

2. $\lim_{x\to\infty} \frac{2x^2}{x^2+1} = \frac{2}{1} = 2$

3. $\lim_{x\to\infty} \frac{x^3-1}{x^2+2} = \infty$

4. $\lim_{x\to\infty} \frac{1}{2^{x-1}} = 0$

5. $\frac{n^2-4}{n^2+2n} = \frac{(n+2)(n-2)}{n(n+2)} = \frac{n-2}{n},$

   $n \neq -2, n \neq 0$

6. $\frac{n^2+n-12}{n^2-16} = \frac{(n+4)(n-3)}{(n+4)(n-4)} = \frac{n-3}{n-4},$

   $n \neq -4, n \neq 4$

7. $\frac{3}{n} + \frac{1}{n^3} = \frac{3n^2+1}{n^3}, \ n \neq 0$

8. $\frac{1}{n-1} + \frac{1}{n+2} = \frac{n+2+n-1}{(n-1)(n+2)} = \frac{2n+1}{(n-1)(n+2)},$

   $n \geq 2$

1. $a_n = 2n - 1$

$a_1 = 2(1) - 1 = 1$

$a_2 = 2(2) - 1 = 3$

$a_3 = 2(3) - 1 = 5$

$a_4 = 2(4) - 1 = 7$

$a_5 = 2(5) - 1 = 9$

3. $a_n = 3^n$

$a_1 = 3^1 = 3$

$a_2 = 3^2 = 9$

$a_3 = 3^3 = 27$

$a_4 = 3^4 = 81$

$a_5 = 3^5 = 243$

5. $a_n = \frac{n}{n+1}$

$a_1 = \frac{1}{1+1} = \frac{1}{2}$

$a_2 = \frac{2}{2+1} = \frac{2}{3}$

$a_3 = \frac{3}{3+1} = \frac{3}{4}$

$a_4 = \frac{4}{4+1} = \frac{4}{5}$

$a_5 = \frac{5}{5+1} = \frac{5}{6}$

7. $a_n = \frac{3^n}{n!}$

$a_1 = \frac{3^1}{1!} = \frac{3}{1} = 3$

$a_2 = \frac{3^2}{2^1} = \frac{9}{2}$

$a_3 = \frac{3^3}{3!} = \frac{27}{6} = \frac{9}{2}$

$a_4 = \frac{3^4}{4!} = \frac{81}{24} = \frac{27}{8}$

$a_5 = \frac{3^5}{5!} = \frac{243}{120} = \frac{81}{40}$

9. $a_n = \frac{(-1)^n}{n^2}$

$a_1 = \frac{(-1)^1}{1^2} = -1$

$a_2 = \frac{(-1)^2}{2^2} = \frac{1}{4}$

$a_3 = \frac{(-1)^3}{3^2} = -\frac{1}{9}$

$a_4 = \frac{(-1)^4}{4^2} = \frac{1}{16}$

$a_5 = \frac{(-1)^5}{5^2} = -\frac{1}{25}$

11. This sequence converges because $\lim_{n\to\infty} \frac{5}{n} = 0$.

13. This sequence converges because $\lim_{n\to\infty} \left(3 - \frac{1}{2^n}\right) = 3$.

15. This sequence converges because $\lim_{n\to\infty} (0.5)^n = 0$.

17. This sequence converges because $\lim_{n\to\infty} \frac{n+1}{n} = 1$.

19. This sequence converges because

$$\lim_{n\to\infty} \frac{n^2 + 3n - 4}{2n^2 + n - 3} = \frac{1}{2}.$$

21. This sequence diverges because

$$\lim_{n\to\infty} \frac{n^2 - 25}{n + 5} = \lim_{n\to\infty} (n - 5) = \infty.$$

23. This sequence converges because $\lim_{n\to\infty} \frac{1 + (-1)^n}{n} = 0$.

25. This sequence diverges because

$$\lim_{n\to\infty} \frac{n!}{n} = \lim_{n\to\infty} (n-1)! = \infty.$$

27. This sequence diverges because

$$\lim_{n\to\infty} \frac{(n+1)!}{n!} = \lim_{n\to\infty} \frac{(n+1)n!}{n!} = \lim_{n\to\infty} (n+1) = \infty.$$

29. This sequence diverges because $\lim_{n\to\infty} (-1)^n \left(\frac{n}{n+1}\right)$ does not exist.

31. The sequence $a_n = (-1)^n + 2$ oscillates between 1 and 3, so a_n diverges. $\lim_{n\to\infty} a_n$ does not exist.

33. $a_n = 3n - 2$

35. $a_n = 5n - 6$

37. $a_n = \frac{n+1}{n+2}$

39. $a_n = \frac{(-1)^{n-1}}{2^{n-2}}$

41. $a_n = 1 + \frac{1}{n} = \frac{n+1}{n}$

43. $a_n = 2(-1)^n$

45. $a_n = \frac{4n}{n!}$

47. Because $a_n = 3n - 1$, the next two terms are $a_5 = 14$ and $a_6 = 17$.

49. Because $a_n = \frac{1}{3} + \frac{2n}{3}$, the next two terms are $a_5 = \frac{11}{3}$ and $a_6 = \frac{13}{3}$.

51. Because $a_n = 3\left(-\frac{1}{2}\right)^{n-1}$, the next two terms are $a_5 = \frac{3}{16}$ and $a_6 = -\frac{3}{32}$.

53. Because $a_n = 2(3^{n-1})$, the next two terms are 162 and 486.

55. Because $a_n = 20\left(\frac{1}{2}\right)^{n-1}$, the sequence is geometric.

57. Because $a_n = \frac{2}{3}n + 2$, the sequence is arithmetic.

59. One example is $a_n = \frac{3n+1}{4n}$.

61. $A_n = P\left[1 + \frac{r}{12}\right]^n = 9000\left[1 + \frac{0.06}{12}\right]^n = 9000(1.005)^n$

$A_1 = \$9045.00$ $\quad A_6 \approx \$9273.40$

$A_2 \approx \$9090.23$ $\quad A_7 \approx \$9319.76$

$A_3 \approx \$9135.68$ $\quad A_8 \approx \$9366.36$

$A_4 \approx \$9181.35$ $\quad A_9 \approx \$9413.20$

$A_5 \approx \$9227.26$ $\quad A_{10} \approx \$9460.26$

63. $A_n = 2000(11)\left[(1.1)^n - 1\right]$

(a) $A_1 = \$2200$

$A_2 = \$4620$

$A_3 = \$7282$

$A_4 = \$10{,}210.20$

$A_5 = \$13{,}431.22$

$A_6 \approx \$16{,}974.34$

(b) $A_{20} \approx \$126{,}005.00$

(c) $A_{40} \approx \$973{,}703.62$

65. (a) $a_n = 1.7n + 337$, $n = 0$ corresponds to 1980

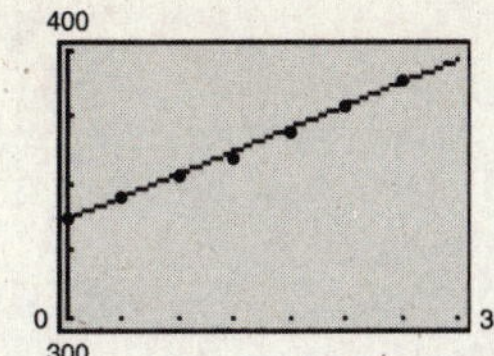

(b) $a_{40} = 1.7(40) + 337 = 405$ parts per million

67. (a) $a_n = 72.6n + 333$, $n = 0$ corresponds to 2000

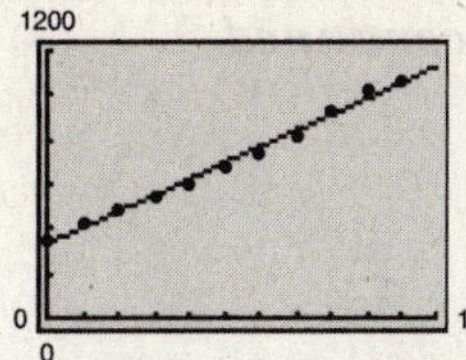

(b) $a_{16} = 72.6(16) + 333 \approx 1495$ stores

69. (a) $a_n = -0.01394n^3 + 0.2648n^2 - 0.573n + 9.52$, $n = 0$ corresponds to 2000

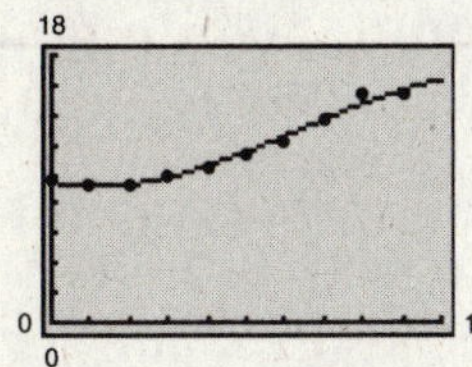

(b) $a_{15} = -0.01394(15)^3 + 0.2648(15)^2 - 0.573(15) + 9.52 \approx \$13.46$ billion

71. (a) $h_n = 12\left(\frac{2}{3}\right)^n$

(b) $h_1 = 12\left(\frac{2}{3}\right) = 8$ ft

$h_2 = 12\left(\frac{2}{3}\right)^2 = \frac{16}{3} \approx 5.3$ ft

$h_3 = 12\left(\frac{2}{3}\right)^3 = \frac{32}{9} \approx 3.6$ ft

$h_4 = 12\left(\frac{2}{3}\right)^4 = \frac{64}{27} \approx 2.4$ ft

$h_5 = 12\left(\frac{2}{3}\right)^5 = \frac{128}{81} \approx 1.6$ ft

$h_6 = 12\left(\frac{2}{3}\right)^6 = \frac{256}{243} \approx 1.1$ ft

(c) This sequence converges because

$\lim_{n\to\infty} 12\left(\frac{2}{3}\right)^n = 0.$

73. (a) $a_n = 32{,}800(1.05)^{n-1}$

$a_4 = 32{,}800(1.05)^{4-1} = \$37{,}970.10$

or

Year 1: \$32,800

Year 2: $(32{,}800)(1.05) = \$34{,}440$

Year 3: $(34{,}440)(1.05) = \$36{,}162$

Year 4: $(36{,}162)(1.05) = \$37{,}970.10$

(b) This sequence diverges because

$\lim_{n\to\infty} 32{,}800(1.05)^{n-1} = \infty.$

75. Answers will vary.

Section 10.2 Series and Convergence

Skills Warm Up

1. $\frac{1}{2} + \frac{1}{3} + \frac{1}{4} + \frac{1}{5} = \frac{77}{60}$

2. $1 + \frac{3}{4} + \frac{4}{6} + \frac{5}{8} = \frac{73}{24}$

3. $\dfrac{1 - \left(\frac{1}{2}\right)^5}{1 - \frac{1}{2}} = \dfrac{\frac{31}{32}}{\frac{1}{2}} = \dfrac{31}{16}$

4. $\dfrac{3\left[1 - \left(\frac{1}{3}\right)^4\right]}{1 - \frac{1}{3}} = \dfrac{\frac{80}{27}}{\frac{2}{3}} = \dfrac{40}{9}$

5. $\dfrac{2\left[1 - \left(\frac{1}{4}\right)^3\right]}{1 - \frac{1}{4}} = \dfrac{\frac{63}{32}}{\frac{3}{4}} = \dfrac{21}{8}$

Skills Warm Up —*continued*—

6. $\dfrac{\frac{1}{2}\left[1-\left(\frac{1}{2}\right)^5\right]}{1-\frac{1}{2}} = \dfrac{\frac{31}{64}}{\frac{1}{2}} = \dfrac{31}{32}$

7. $\lim_{n\to\infty} \dfrac{3n}{4n+1} = \dfrac{3}{4}$

8. $\lim_{n\to\infty} \dfrac{3n}{n^2+1} = 0$

9. $\lim_{n\to\infty} \dfrac{n!}{n!-3} = \dfrac{1}{1} = 1$

10. $\lim_{n\to\infty} \dfrac{2n!+1}{4n!-1} = \dfrac{2}{4} = \dfrac{1}{2}$

1. $2+4+6+8+10 = 2(1)+2(2)+2(3)+2(4)+2(5)$

$$= \sum_{n=1}^{5} 2n$$

3. $\dfrac{1}{1}+\dfrac{1}{3}+\dfrac{1}{5}+\dfrac{1}{7}+\dfrac{1}{9}+\dfrac{1}{11}+\dfrac{1}{13} = \dfrac{1}{2(1)-1}+\dfrac{1}{2(2)-1}+\dfrac{1}{2(3)-1}+\dfrac{1}{2(4)-1}+\dfrac{1}{2(5)-1}+\dfrac{1}{2(6)-1}+\dfrac{1}{2(7)-1}$

$$= \sum_{n=1}^{7} \frac{1}{2n-1}$$

5. $S_1 = 1$

$S_2 = 1+\frac{1}{4} = \frac{5}{4} = 1.25$

$S_3 = 1+\frac{1}{4}+\frac{1}{9} = \frac{49}{36} \approx 1.361$

$S_4 = 1+\frac{1}{4}+\frac{1}{9}+\frac{1}{16} = \frac{205}{144} \approx 1.424$

$S_5 = 1+\frac{1}{4}+\frac{1}{9}+\frac{1}{16}+\frac{1}{25} = \frac{5269}{3600} \approx 1.464$

7. $S_1 = 3$

$S_2 = 3-\frac{9}{2} = -\frac{3}{2} = -1.5$

$S_3 = 3-\frac{9}{2}+\frac{27}{4} = \frac{21}{4} = 5.25$

$S_4 = 3-\frac{9}{2}+\frac{27}{4}-\frac{81}{8} = -\frac{39}{8} = -4.875$

$S_5 = 3-\frac{9}{2}+\frac{27}{4}-\frac{81}{8}+\frac{243}{16} = \frac{165}{16} = 10.3125$

9. This series diverges by the Test for Convergence of a Geometric Series because $|r| = \left|\frac{5}{2}\right| > 1$.

11. This series diverges by the Test for Convergence of a Geometric Series because $|r| = |1.055| > 1$.

13. This series diverges by the *n*th-Term Test because

$$\lim_{n\to\infty} \frac{n}{n+1} = 1 \neq 0.$$

15. This series diverges by the *n*th-Term Test because

$$\lim_{n\to\infty} \frac{n^2}{n^2+1} = 1 \neq 0.$$

17. This series converges by the Test for Convergence of a Geometric Series because $|r| = \left|\frac{3}{4}\right| < 1$.

19. This series converges by the Test for Convergence of a Geometric Series because $|r| = |0.9| < 1$.

21. Because $a = 7$ and $r = \dfrac{1}{3}$, you have

$$S = \frac{7}{1-(1/3)} = \frac{21}{2}.$$

23. Because $a = 5$ and $r = \dfrac{3}{8}$, you have

$$S = \frac{5}{1-(3/8)} = 8.$$

25. $\displaystyle\sum_{n=0}^{\infty}\left(\frac{1}{2^n}-\frac{1}{3^n}\right) = \sum_{n=0}^{\infty}\left(\frac{1}{2}\right)^n - \sum_{n=0}^{\infty}\left(\frac{1}{3}\right)^n$

$$= \frac{1}{1-(1/2)} - \frac{1}{1-(1/3)}$$

$$= 2 - \frac{3}{2} = \frac{1}{2}$$

27. $\displaystyle\sum_{n=0}^{\infty}\left[(0.7)^n + (0.9)^n\right] = \sum_{n=0}^{\infty}(0.7)^n + \sum_{n=0}^{\infty}(0.9)^n$

$$= \frac{1}{1-0.7} + \frac{1}{1-0.9}$$

$$= \frac{10}{3} + 10 = \frac{40}{3}$$

29. Because $a = 3$ and $r = \frac{1}{8}$, you have

$$S_n = \frac{3\left(1 - (1/8)^{n+1}\right)}{1 - (1/8)} = \frac{3\left(1 - (1/8)^{n+1}\right)}{7/8} = \frac{3}{7}\left(8 - \left(\frac{1}{8}\right)^n\right).$$

$$S_4 = \frac{3}{7}\left(8 - \left(\frac{1}{8}\right)^4\right) \approx 3.428$$

$$S_6 = \frac{3}{7}\left(8 - \left(\frac{1}{8}\right)^6\right) \approx 3.429$$

$$S_{10} = \frac{3}{7}\left(8 - \left(\frac{1}{8}\right)^{10}\right) \approx 3.429$$

31. Because $a = 5$ and $r = -\frac{1}{2}$, you have

$$S_n = \frac{5\left(1 - (-1/2)^{n+1}\right)}{1 - (-1/2)} = \frac{5\left(1 - (-1/2)^{n+1}\right)}{3/2} = \frac{10}{3}\left(1 - \left(-\frac{1}{2}\right)^{n+1}\right).$$

$$S_5 = \frac{10}{3}\left(1 - \left(-\frac{1}{2}\right)^6\right) \approx 3.281$$

$$S_7 = \frac{10}{3}\left(1 - \left(-\frac{1}{2}\right)^8\right) \approx 3.320$$

$$S_{10} = \frac{10}{3}\left(1 - \left(-\frac{1}{2}\right)^{11}\right) \approx 3.335$$

33. This series diverges by the nth-Term Test because

$$\lim_{n\to\infty} \frac{n + 10}{10n + 1} = \frac{1}{10} \neq 0.$$

35. This series diverges by the nth-Term Test because

$$\lim_{n\to\infty} \frac{n! + 1}{n!} = 1 \neq 0.$$

37. This series diverges by the nth-Term Test because

$$\lim_{n\to\infty} \frac{3n - 1}{2n + 1} = \frac{3}{2} \neq 0.$$

39. This series diverges by the Test for Convergence of a Geometric Series because $|r| = 1.075 > 1$.

41. This series converges by the Test for Convergence of a Geometric Series because $\sum_{n=0}^{\infty} \frac{8}{5^n} = \sum_{n=0}^{\infty} 8\left(\frac{1}{5}\right)^n$ and $|r| = \left|\frac{1}{5}\right| < 1$.

43. $0.\overline{4} = \sum_{n=0}^{\infty} (0.4)(0.1)^n = \frac{0.4}{1 - 0.1} = \frac{0.4}{0.9} = \frac{4}{9}$

45. $0.\overline{81} = \sum_{n=0}^{\infty} 0.81(0.01)^n = \frac{0.81}{1 - 0.01} = \frac{0.81}{0.99} = \frac{9}{11}$

47. (a) $$\sum_{i=0}^{n-1} 8000(0.9)^i = \frac{8000\left[1 - (0.9)^{(n-1)+1}\right]}{1 - 0.9} = 80{,}000\left(1 - 0.9^n\right)$$

(b) $$\sum_{i=0}^{\infty} 8000(0.9)^i = \frac{8000}{1 - 0.9} = 80{,}000$$

49. $D_1 = 16$

$D_2 = 0.81(16) + 0.81(16) = 32(0.81)$

$D_3 = 32(0.81)^2$

$\vdots$

$$D = -16 + \sum_{n=0}^{\infty} 32(0.81)^n = -16 + \frac{32}{1 - 0.81} = -16 + \frac{32}{0.19} = \frac{2896}{19} \approx 152.42 \text{ feet}$$

51. (a) $$A = \sum_{n=1}^{60} 100\left(1 + \frac{0.09}{12}\right)^n = -100 + \sum_{n=0}^{60} \left(1 + \frac{0.09}{12}\right)^n$$

(b) $$A = -100 + \frac{100\left(1 - 1.0075^{61}\right)}{1 - 1.0075} \approx \$7598.98$$

53. $A = \sum_{n=1}^{N} P\left(1 + \frac{r}{12}\right)^n = -P + \sum_{n=0}^{N} P\left(1 + \frac{r}{12}\right)^n = -P + \frac{P\left[1 - \left(1 + \frac{r}{12}\right)^{N+1}\right]}{1 - \left(1 + \frac{r}{12}\right)}$

$= -P - \frac{12}{r}P\left[1 - \left(1 + \frac{r}{12}\right)^{N+1}\right] = P\left[-1 - \frac{12}{r} + \frac{12}{r}\left(1 + \frac{r}{12}\right)^{N+1}\right]$

$= P\left[-\left(1 + \frac{12}{r}\right) + \frac{12}{r}\left(1 + \frac{r}{12}\right)\left(1 + \frac{r}{12}\right)^N\right] = P\left[-\left(1 + \frac{12}{r}\right) + \left(\frac{12}{r} + 1\right)\left(1 + \frac{r}{12}\right)^N\right]$

$= P\left[-1 + \left(1 + \frac{r}{12}\right)^N\right]\left(1 + \frac{12}{r}\right) = P\left[\left(1 + \frac{r}{12}\right)^N - 1\right]\left(1 + \frac{12}{r}\right)$

55. $A = \sum_{n=0}^{\infty} 500(0.75)^n = \frac{500}{1 - 0.75} = \$2000$ million

57. $V_n = 225{,}000(1 - 0.3)^n = 225{,}000(0.7)^n$

$V_5 = 225{,}000(0.7)^5 = \$37{,}815.75$

59. $T_1 = 40{,}000$

$T_2 = 40{,}000(1 + 0.04)$

$T_3 = 40{,}000(1 + 0.04)^2$

$\vdots$

$T_n = 40{,}000(1 + 0.04)^{n-1}$

$\sum_{n=1}^{40} 40{,}000(1 + 0.04)^{n-1} = \frac{40{,}000(1 - 1.04^{40})}{(1 - 1.04)}$

$\approx \$3{,}801{,}020.63$

61. $\sum_{n=1}^{\infty} \left(\frac{1}{2}\right)^n = -1 + \sum_{n=0}^{\infty} \left(\frac{1}{2}\right)^n = -1 + \frac{1}{1 - \frac{1}{2}} = 1$

63. $P(n) = \frac{1}{2}\left(\frac{1}{2}\right)^n$

$P(2) = \frac{1}{2}\left(\frac{1}{2}\right)^2 = \frac{1}{8}$

$\sum_{n=0}^{\infty} \frac{1}{2}\left(\frac{1}{2}\right)^n = \frac{\frac{1}{2}}{1 - \frac{1}{2}} = \frac{\frac{1}{2}}{\frac{1}{2}} = 1$

65. At factory: 500

1 mile away: $(0.85)500$

2 miles away: $(0.85)^2 500$

12 miles away: $(0.85)^{12} 500 \approx 71.12$ ppm

67. $\sum_{n=1}^{\infty} n^2\left(\frac{1}{2}\right)^n = 6$

69. $\sum_{n=1}^{\infty} \frac{1}{(2n)!} \approx 0.5431$

71. False. For example, $\sum_{n=1}^{\infty} \frac{1}{n}$ diverges even though

$\lim_{n\to\infty} \frac{1}{n} = 0.$

Section 10.3 p-Series and the Ratio Test

Skills Warm Up

1. $\frac{n!}{(n+1)!} = \frac{1 \cdot 2 \cdot 3 \cdots (n-1) \cdot n}{1 \cdot 2 \cdot 3 \cdots (n-1) \cdot n \cdot (n+1)} = \frac{1}{n+1}$

2. $\frac{(n+1)!}{n!} = \frac{1 \cdot 2 \cdot 3 \cdots (n-1) \cdot n \cdot (n+1)}{1 \cdot 2 \cdot 3 \cdots (n-1) \cdot n} = n + 1$

3. $\frac{3^{n+1}}{n+1} \cdot \frac{n}{3^n} = \frac{3^n \cdot 3 \cdot n}{3^n(n+1)} = \frac{3n}{n+1}$

Skills Warm Up —continued—

4. $\dfrac{(n+1)^2}{(n+1)!}\cdot\dfrac{n!}{n^2} = \dfrac{1\cdot 2\cdot 3\cdots(n-1)\cdot n\cdot(n+1)\cdot(n+1)}{1\cdot 2\cdot 3\cdots(n-1)\cdot n\cdot(n+1)\cdot n^2} = \dfrac{n+1}{n^2}$

5. $\lim\limits_{n\to\infty}\dfrac{(n+1)^2}{n^2} = \lim\limits_{n\to\infty}\dfrac{n^2+2n+1}{n^2} = \lim\limits_{n\to\infty}\dfrac{1}{1} = 1$

6. $\lim\limits_{n\to\infty}\dfrac{5^{n+1}}{5^n} = \lim\limits_{n\to\infty}\dfrac{5^n\cdot 5}{5^n} = \lim\limits_{n\to\infty} 5 = 5$

7. $\lim\limits_{n\to\infty}\left(\dfrac{5}{n+1}\div\dfrac{5}{n}\right) = \lim\limits_{n\to\infty}\left(\dfrac{\frac{5}{n+1}}{\frac{5}{n}}\right) = \lim\limits_{n\to\infty}\dfrac{5n}{5(n+1)} = \lim\limits_{n\to\infty}\dfrac{5}{5} = 1$

8. $\lim\limits_{n\to\infty}\left(\dfrac{(n+1)^3}{3^{n+1}}\div\dfrac{n^3}{3^n}\right) = \lim\limits_{n\to\infty}\left[\dfrac{\frac{(n+1)^3}{3^{n+1}}}{\frac{n^3}{3^n}}\right] = \lim\limits_{n\to\infty}\dfrac{3^n(n+1)^3}{3^{n+1}(n^3)} = \lim\limits_{n\to\infty}\dfrac{3^n(n+1)^3}{3^n\cdot 3n^3} = \lim\limits_{n\to\infty}\dfrac{(n+1)^3}{3n^3} = \dfrac{1}{3}$

9. The series $\sum\limits_{n=1}^{\infty}\dfrac{1}{4^n}$ is geometric because $a = 1$ and $r = \dfrac{1}{4}$.

10. The series $\sum\limits_{n=1}^{\infty}\dfrac{1}{n^4}$ is not geometric because it is *not* of the form $\sum\limits_{n=1}^{\infty} ar^n$.

1. The series $\sum\limits_{n=1}^{\infty}\dfrac{1}{n^2}$ is a *p*-series with $p = 2$.

3. The series $\sum\limits_{n=1}^{\infty}\dfrac{1}{5^n} = \sum\limits_{n=1}^{\infty}1\left(\dfrac{1}{5}\right)^n$ is *not* a *p*-series. This series is geometric with $r = \dfrac{1}{5}$.

5. The series $\sum\limits_{n=1}^{\infty}\dfrac{1}{n^n} = 1 + \dfrac{1}{2^2} + \dfrac{1}{3^3} + \dfrac{1}{4^4} + \cdots$ is *not* a *p*-series. The exponent changes with each term.

7. The series $1 + \dfrac{1}{\sqrt[3]{2}} + \dfrac{1}{\sqrt[3]{3}} + \dfrac{1}{\sqrt[3]{4}} + \cdots = \sum\limits_{n=1}^{\infty}\dfrac{1}{\sqrt[3]{n}}$ is a *p*-series with $p = \dfrac{1}{3}$.

9. This series converges because $p = 3 > 1$.

11. This series diverges because $p = \frac{1}{5} < 1$.

13. This series converges because $p = 1.03 > 1$.

15. $1 + \dfrac{1}{\sqrt{2}} + \dfrac{1}{\sqrt{3}} + \dfrac{1}{\sqrt{4}} + \cdots = \sum\limits_{n=1}^{\infty}\dfrac{1}{\sqrt{n}} = \sum\limits_{n=1}^{\infty}\dfrac{1}{n^{1/2}}$

So, this series diverges because $p = \frac{1}{2} < 1$.

17. $1 + \dfrac{1}{\sqrt[3]{4}} + \dfrac{1}{\sqrt[3]{9}} + \dfrac{1}{\sqrt[3]{16}} + \cdots = \sum\limits_{n=1}^{\infty}\dfrac{1}{n^{2/3}}$

So, this series diverges because $p = \dfrac{2}{3} < 1$.

19. Because $a_n = \dfrac{6^n}{n!}$, you have

$$\lim_{n\to\infty}\left|\frac{a_{n+1}}{a_n}\right| = \lim_{n\to\infty}\left|\frac{6^{n+1}}{(n+1)!}\cdot\frac{n!}{6^n}\right| = \lim_{n\to\infty}\frac{6}{n+1} = 0 < 1$$

and the series converges.

21. Because $a_n = n(3/4)^n$, you have

$$\lim_{n\to\infty}\left|\frac{a_{n+1}}{a_n}\right| = \lim_{n\to\infty}\left|\frac{(n+1)3^{n+1}}{4^{n+1}}\cdot\frac{4^n}{n3^n}\right| = \lim_{n\to\infty}\frac{3(n+1)}{4n} = \frac{3}{4} < 1$$

and the series converges.

23. Because $a_n = n/4^n$, you have

$$\lim_{n\to\infty}\left|\frac{a_{n+1}}{a_n}\right| = \lim_{n\to\infty}\left|\frac{n+1}{4^{n+1}}\cdot\frac{4^n}{n}\right| = \lim_{n\to\infty}\frac{n+1}{4n} = \frac{1}{4} < 1$$

and the series converges.

25. Because $a_n = 2^n/n^5$, you have

$$\lim_{n\to\infty}\left|\frac{a_{n+1}}{a_n}\right| = \lim_{n\to\infty}\left|\frac{2^{n+1}}{(n+1)^5}\cdot\frac{n^5}{2^n}\right|$$

$$= \lim_{n\to\infty}\frac{2n^5}{(n+1)^5} = 2 > 1$$

and the series diverges.

27. Because $a_n = \dfrac{(-1)^n 2^n}{n!}$, you have

$$\lim_{n\to\infty}\left|\frac{a_{n+1}}{a_n}\right| = \lim_{n\to\infty}\left|\frac{(-1)^{n+1}2^{n+1}}{(n+1)!}\cdot\frac{n!}{(-1)^n 2^n}\right|$$

$$= \lim_{n\to\infty}\frac{2}{n+1} = 0 < 1$$

and the series converges.

29. Because $a_n = \dfrac{4^n}{3^n+1}$, you have

$$\lim_{n\to\infty}\left|\frac{a_{n+1}}{a_n}\right| = \lim_{n\to\infty}\left|\frac{4^{n+1}}{3^{n+1}+1}\cdot\frac{3^n+1}{4^n}\right|$$

$$= \lim_{n\to\infty}\frac{4(3^n)+4}{3(3^n)+1} = \frac{4}{3} > 1$$

and the series diverges.

31. Because $a_n = \dfrac{n5^n}{n!}$, you have

$$\lim_{n\to\infty}\left|\frac{a_{n+1}}{a_n}\right| = \lim_{n\to\infty}\left|\frac{(n+1)5^{n+1}}{(n+1)!}\cdot\frac{n!}{n5^n}\right|$$

$$= \lim_{n\to\infty}\frac{5}{n} = 0 < 1$$

and the series converges.

33. $$\lim_{n\to\infty}\left|\frac{a_{n+1}}{a_n}\right| = \lim_{n\to\infty}\left|\frac{1}{(n+1)^{3/2}}\cdot\frac{n^{3/2}}{1}\right|$$

$$= \lim_{n\to\infty}\left(\frac{n}{n+1}\right)^{3/2} = 1$$

So, the Ratio Test is inconclusive. (The series is a convergent *p*-series.)

35. $$\lim_{n\to\infty}\left|\frac{a_{n+1}}{a_n}\right| = \lim_{n\to\infty}\left|\frac{1}{(n+1)^3}\cdot\frac{n^3}{1}\right| = \lim_{n\to\infty}\left(\frac{n}{n+1}\right)^3 = 1$$

So, the Ratio Test is inconclusive. (The series is a convergent *p*-series.)

37. $$\sum_{n=1}^{\infty}\frac{1}{n^3} \approx \frac{1}{1}+\frac{1}{8}+\frac{1}{27}+\frac{1}{64} = \frac{2035}{1728} \approx 1.1777$$

The error is less than or equal to

$$\frac{1}{(p-1)N^{p-1}} = \frac{1}{(3-1)4^{3-1}} = \frac{1}{32}.$$

39. $$\sum_{n=1}^{\infty}\frac{1}{n^{7/2}} \approx 1+\frac{1}{2^{7/2}}+\frac{1}{3^{7/2}}+\frac{1}{4^{7/2}}+\frac{1}{5^{7/2}}+\frac{1}{6^{7/2}}+\frac{1}{7^{7/2}}+\frac{1}{8^{7/2}}+\frac{1}{9^{7/2}}+\frac{1}{10^{7/2}} \approx 1.1256$$

The error is less than or equal to

$$\frac{1}{(p-1)N^{p-1}} = \frac{1}{(7/2-1)10^{(7/2-1)}} = \frac{1}{(5/2)10^{5/2}} = \frac{2}{500\sqrt{10}} = \frac{1}{250\sqrt{10}} = \frac{\sqrt{10}}{2500} \approx 0.0013.$$

41. $$\sum_{n=1}^{\infty}\frac{2}{\sqrt[4]{n^3}} = \sum_{n=1}^{\infty}\frac{2}{n^{3/4}} = 2+\frac{2}{2^{3/4}}+\cdots$$

Diverges $\left(p\text{-series with } p = \frac{3}{4} < 1\right)$

Matches (a).

43. $$\sum_{n=1}^{\infty}\frac{2}{\sqrt{n^5}} = \sum_{n=1}^{\infty}\frac{2}{n^{5/2}} = 2+\frac{2}{2^{5/2}}+\cdots$$

Converges $\left(p\text{-series with } p = \frac{5}{2} > 1\right)$

Matches (e).

45. $$\sum_{n=1}^{\infty}\frac{2}{n\sqrt{n}} = \sum_{n=1}^{\infty}\frac{2}{n^{3/2}} = 2+\frac{2}{2^{3/2}}+\cdots$$

Converges $\left(p\text{-series with } p = \frac{3}{2} > 1\right)$

Matches (f).

47. This series diverges by the *n*th-Term Test because

$$\lim_{n\to\infty}\frac{2n}{n+1} = 2 \neq 0.$$

49. This series converges by the p-series test because $p = \frac{5}{4} > 1$.

$$\sum_{n=1}^{\infty} \frac{4}{n\sqrt[4]{n}} = \sum_{n=1}^{\infty} \frac{4}{n^{5/4}} \approx 18.38$$

51. This series converges by the Geometric Series Test because $|r| = \frac{3}{4} < 1$.

$$\sum_{n=1}^{\infty} \left(-\frac{3}{4}\right)^n = -1 + \sum_{n=0}^{\infty} \left(-\frac{3}{4}\right)^n = -1 + \frac{1}{1 - (-3/4)} = -1 + \frac{4}{7} = -\frac{3}{7} \approx -0.43$$

53. This series converges by the Geometric Series Test because $|r| = \left|-\frac{2}{3}\right| = \frac{2}{3} < 1$.

$$\sum_{n=0}^{\infty} \frac{(-1)^n 2^n}{3^n} = \frac{1}{1 - \left(-\frac{2}{3}\right)} = \frac{3}{5}$$

55. Both series are convergent p-series, so their difference is convergent.

$$\sum_{n=1}^{\infty} \left(\frac{1}{n^2} - \frac{1}{n^3}\right) \approx 0.4429$$

57. This series diverges by the Geometric Series Test because $|r| = \frac{5}{4} > 1$.

59. This series diverges by the Ratio Test because

$a_n = \dfrac{n!}{3^{n-1}}$ and

$$\lim_{n\to\infty} \left|\frac{a_{n+1}}{a_n}\right| = \lim_{n\to\infty} \left|\frac{(n+1)!}{3^n} \cdot \frac{3^{n-1}}{n!}\right| = \lim_{n\to\infty} \frac{n+1}{3} = \infty.$$

61. This series diverges by the nth-Term Test because

$$\lim_{n\to\infty} \frac{n}{\sqrt{n^2+1}} = 1 \neq 0.$$

63. $\displaystyle\sum_{n=1}^{\infty} \frac{2^n}{5^{n-1}} = 5\sum_{n=1}^{\infty} \left(\frac{2}{5}\right)^n$ converges by the Geometric Series Test because $r = \frac{2}{5} < 1$.

$$5\sum_{n=1}^{\infty} \left(\frac{2}{5}\right)^n = \frac{5(2/5)}{1 - (2/5)} = \frac{10}{3}.$$

65. No, although the terms approach zero, the series diverges because the partial sums approach infinity.

Chapter 10 Quiz Yourself

1. $a_n = \left(\frac{-1}{4}\right)^n$

$a_1 = \left(\frac{-1}{4}\right)^1 = -\frac{1}{4}$

$a_2 = \left(\frac{-1}{4}\right)^2 = \frac{1}{16}$

$a_3 = \left(\frac{-1}{4}\right)^3 = -\frac{1}{64}$

$a_4 = \left(\frac{-1}{4}\right)^4 = \frac{1}{256}$

$a_5 = \left(\frac{-1}{4}\right)^5 = -\frac{1}{1024}$

2. $a_n = \dfrac{n+1}{n+3}$

$a_1 = \frac{1+1}{1+3} = \frac{2}{4} = \frac{1}{2}$

$a_2 = \frac{2+1}{2+3} = \frac{3}{5}$

$a_3 = \frac{3+1}{3+3} = \frac{4}{6} = \frac{2}{3}$

$a_4 = \frac{4+1}{4+3} = \frac{5}{7}$

$a_5 = \frac{5+1}{5+3} = \frac{6}{8} = \frac{3}{4}$

3. $a_n = 5(-1)^n$

$a_1 = 5(-1)^1 = -5$

$a_2 = 5(-1)^2 = 5$

$a_3 = 5(-1)^3 = -5$

$a_4 = 5(-1)^4 = 5$

$a_5 = 5(-1)^5 = -5$

4. $a_n = \dfrac{n-2}{n!}$

$a_1 = \dfrac{1-2}{1!} = -1$

$a_2 = \dfrac{2-2}{2!} = 0$

$a_3 = \dfrac{3-2}{3!} = \dfrac{1}{6}$

$a_4 = \dfrac{4-2}{4!} = \dfrac{2}{24} = \dfrac{1}{12}$

$a_5 = \dfrac{5-2}{5!} = \dfrac{3}{120} = \dfrac{1}{40}$

5. This sequence converges because $\lim_{n\to\infty} \dfrac{3}{\sqrt{n}} = 0.$

6. This sequence converges because $\lim_{n\to\infty} \dfrac{n}{2n+3} = \dfrac{1}{2}.$

7. This sequence converges because $\lim_{n\to\infty} \dfrac{2}{(n+1)!} = 0.$

8. This sequence diverges because $\lim_{n\to\infty} \dfrac{(-1)^n}{2}$ does not exist.

9. Sample answer: $a_n = \dfrac{n-1}{n^2}$

10. Sample answer: $a_n = (-1)^n 3^{1/n}$

11. Sample answer: $a_n = 2^{(-1)^n}$

12. $S_1 = 0$

$S_2 = 0 + \frac{1}{2} = \frac{1}{2} = 0.5$

$S_3 = 0 + \frac{1}{2} + \frac{1}{3} = \frac{5}{6} \approx 0.8333$

$S_4 = 0 + \frac{1}{2} + \frac{1}{3} + \frac{1}{8} = \frac{23}{24} \approx 0.9583$

$S_5 = 0 + \frac{1}{2} + \frac{1}{3} + \frac{1}{8} + \frac{1}{30} = \frac{119}{120} \approx 0.99167$

13. $S_1 = 1$

$S_2 = 1 - 1 = 0$

$S_3 = 1 - 1 + \frac{3}{4} = \frac{3}{4} = 0.75$

$S_4 = 1 - 1 + \frac{3}{4} - \frac{1}{2} = \frac{1}{4} = 0.25$

$S_5 = 1 - 1 + \frac{3}{4} - \frac{1}{2} + \frac{5}{16} = \frac{9}{16} = 0.5625$

14. $\displaystyle\sum_{n=0}^{\infty} 4\left(\frac{2}{3}\right)^n = \frac{4}{1-(2/3)} = 12$

15. $\displaystyle\sum_{n=0}^{\infty}\left(\frac{1}{3^n} - \frac{1}{3^{n+1}}\right) = \sum_{n=0}^{\infty}\left(\frac{1}{3}\right)^n - \sum_{n=0}^{\infty}\frac{1}{3\cdot 3^n}$

$\displaystyle = \sum_{n=0}^{\infty}\left(\frac{1}{3}\right)^n - \frac{1}{3}\sum_{n=0}^{\infty}\left(\frac{1}{3}\right)^n$

$\displaystyle = \frac{1}{1-\frac{1}{3}} - \frac{1}{3}\left(\frac{1}{1-\frac{1}{3}}\right) = 1$

16. $\displaystyle 5 + 0.5 + 0.05 + 0.005 + \cdots = \sum_{n=0}^{\infty} 5(0.1)^n$

$\displaystyle = \frac{5}{1-0.1} = \frac{50}{9}$

17. This series diverges by the nth-Term Test because $\displaystyle\lim_{n\to\infty}\frac{2n^2-1}{n^2+1} = 2 \neq 0.$

18. This series converges by the Geometric Series Test because $|r| = \frac{1}{2} < 1.$

19. This series diverges by the Geometric Series Test because $|r| = \frac{5}{3} > 1.$

20. This series converges by the p-series test because $p = \frac{5}{3} > 1.$

21. This series converges by the Ratio Test because

$\displaystyle a_n = \frac{n}{(n-1)!}$ and $\displaystyle\lim_{n\to\infty}\left|\frac{a_{n+1}}{a_n}\right| = \lim_{n\to\infty}\left|\frac{n+1}{n!}\cdot\frac{(n-1)!}{n}\right|$

$\displaystyle = \lim_{n\to\infty}\frac{n+1}{n^2} = 0 < 1.$

22. This series diverges by the Ratio Test because

$\displaystyle a_n = \left(\frac{2}{3}\right)^n n!$ and

$\displaystyle\lim_{n\to\infty}\left|\frac{a_{n+1}}{a_n}\right| = \lim_{n\to\infty}\left|\frac{2^n\cdot 2\cdot(n+1)!}{3^n\cdot 3}\cdot\frac{3^n}{2^n\cdot n!}\right|$

$\displaystyle = \lim_{n\to\infty}\frac{2(n+1)}{3} = \infty.$

23. (a) $A = \sum_{n=1}^{36} 200\left(1 + \frac{0.06}{12}\right)^n = \sum_{n=1}^{36} 200(1.005)^n$

(b) $\sum_{n=1}^{36} 200(1.005)^n = -200 + \sum_{n=0}^{36} 200(1.005)^n = -200 + \frac{200(1 - 1.005^{37})}{1 - 1.005} \approx \$7906.56$

Section 10.4 Power Series and Taylor's Theorem

Skills Warm Up

1. $f(x) = x^2,\ g(x) = x - 1$

$f(g(x)) = (x - 1)^2$

$g(f(x)) = x^2 - 1$

2. $f(x) = 3x,\ g(x) = 2x + 1$

$f(g(x)) = 3(2x + 1) = 6x + 3$

$g(f(x)) = 2(3x) + 1 = 6x + 1$

3. $f(x) = \sqrt{x + 4},\ g(x) = x^2$

$f(g(x)) = \sqrt{x^2 + 4}$

$g(f(x)) = \left(\sqrt{x + 4}\right)^2 = x + 4,\ x \ge -4$

4. $f(x) = e^x,\ g(x) = x^2$

$f(g(x)) = e^{x^2}$

$g(f(x)) = (e^x)^2 = e^{2x}$

5. $f(x) = 5e^x$

$f'(x) = 5e^x$

$f''(x) = 5e^x$

$f'''(x) = 5e^x$

$f^{(4)}(x) = 5e^x$

6. $f(x) = \ln x$

$f'(x) = \frac{1}{x}$

$f''(x) = -\frac{1}{x^2}$

$f'''(x) = \frac{2}{x^3}$

$f^{(4)}(x) = -\frac{6}{x^4}$

7. $f(x) = 3e^{2x}$

$f'(x) = 6e^{2x}$

$f''(x) = 12e^{2x}$

$f'''(x) = 24e^{2x}$

$f^{(4)}(x) = 48e^{2x}$

8. $f(x) = \ln 2x$

$f'(x) = \frac{2}{x}$

$f''(x) = -\frac{2}{x^2}$

$f'''(x) = \frac{4}{x^3}$

$f^{(4)}(x) = -\frac{12}{x^4}$

9. $\frac{3^n}{n!} \div \frac{3^{n+1}}{(n+1)!} = \frac{3^n}{n!} \cdot \frac{(n+1)!}{3^{n+1}} = \frac{3^n \cdot 1 \cdot 2 \cdot 3 \cdots (n-1) \cdot n \cdot (n+1)}{3^n \cdot 3 \cdot 1 \cdot 2 \cdot 3 \cdots (n-1) \cdot n} = \frac{n+1}{3}$

10. $\frac{n!}{(n+2)!} \div \frac{(n+1)!}{(n+3)!} = \frac{n!}{(n+2)!} \cdot \frac{(n+3)!}{(n+1)!}$

$= \frac{[1 \cdot 2 \cdot 3 \cdots (n-1) \cdot n] \cdot [1 \cdot 2 \cdot 3 \cdots (n-1) \cdot n \cdot (n+1) \cdot (n+2) \cdot (n+3)]}{[1 \cdot 2 \cdot 3 \cdots (n-1) \cdot n \cdot (n+1) \cdot (n+2)] \cdot [1 \cdot 2 \cdot 3 \cdots (n-1) \cdot n \cdot (n+1)]} = \frac{n+3}{n+1}$

1. The series $\sum_{n=0}^{\infty}\left(\frac{x}{4}\right)^n$ is centered at 0.

$$\sum_{n=0}^{\infty}\left(\frac{x}{4}\right)^n = 1 + \frac{x}{4} + \left(\frac{x}{4}\right)^2 + \left(\frac{x}{4}\right)^3 + \left(\frac{x}{4}\right)^4 + \cdots$$

3. The series $\sum_{n=0}^{\infty}\frac{(-1)^{n+1}(x+1)^n}{n!}$ is centered at -1.

$$\sum_{n=0}^{\infty}\frac{(-1)^{n+1}(x+1)^n}{n!} = -1 + (x+1) - \frac{(x+1)^2}{2} + \frac{(x+1)^3}{6} - \frac{(x+1)^4}{24} + \cdots$$

5. $\lim_{n\to\infty}\left|\frac{a_{n+1}x^{n+1}}{a_n x^n}\right| = \lim_{n\to\infty}\left|\frac{(x/2)^{n+1}}{(x/2)^n}\right| = \lim_{n\to\infty}\left|\frac{x}{2}\right| < 1 \Rightarrow |x| < 2$

Radius $= 2$

7. $\lim_{n\to\infty}\left|\frac{a_{n+1}x^{n+1}}{a_n x^n}\right| = \lim_{n\to\infty}\left|\frac{(-1)^{n+1}x^{n+1}/(n+1)}{(-1)^n x^n/n}\right| = \lim_{n\to\infty}\left|\frac{xn}{n+1}\right| < 1 \Rightarrow |x| < 1$

Radius $= 1$

9. $\lim_{n\to\infty}\left|\frac{a_{n+1}x^{5n+5}}{a_n x^{5n}}\right| = \lim_{n\to\infty}\left|\frac{\frac{x^{5n+5}}{(n+1)!}}{\frac{x^{5n}}{n!}}\right| = \lim_{n\to\infty}\left|\frac{x^5}{n+1}\right| = 0 \Rightarrow -\infty < x < \infty$

Radius $= \infty$

11. $\lim_{n\to\infty}\left|\frac{a_{n+1}x^{n+1}}{a_n x^n}\right| = \lim_{n\to\infty}\left|\frac{(n+1)!x^{n+1}/2^{n+1}}{n!x^n/2^n}\right| = \lim_{n\to\infty}\left|\frac{(n+1)x}{2}\right| = \infty$

This series converges only at $x = 0$.

Radius $= 0$

13. $\lim_{n\to\infty}\left|\frac{a_{n+1}x^{n+1}}{a_n x^n}\right| = \lim_{n\to\infty}\left|\frac{(-1)^{n+2}x^{n+1}/4^{n+1}}{(-1)^{n+1}x^n/4^n}\right| = \lim_{n\to\infty}\left|\frac{x}{4}\right| < 1 \Rightarrow |x| < 4$

Radius $= 4$

15. $\lim_{n\to\infty}\left|\frac{a_{n+1}(x-5)^{n+1}}{a_n(x-5)^n}\right| = \lim_{n\to\infty}\left|\frac{(-1)^{n+2}(x-5)^{n+1}/(n+1)5^{n+1}}{(-1)^{n+1}(x-5)^n/n5^n}\right|$

$= \lim_{n\to\infty}\left|\frac{x-5}{5}\cdot\frac{n+1}{n}\right| \Rightarrow |x-5| < 5 \Rightarrow \left|\frac{x-5}{5}\right| < 1 \Rightarrow |x-5| < 5$

Radius $= 5$

17. $\lim_{n\to\infty}\left|\frac{a_{n+1}(x-1)^{n+1}}{a_n(x-1)^n}\right| = \lim_{n\to\infty}\left|\frac{(-1)^{n+2}(x-1)^{n+2}/(n+2)}{(-1)^{n+1}(x-1)^{n+1}/(n+1)}\right| = \lim_{n\to\infty}\left|(x-1)\frac{n+1}{n+2}\right| = \lim_{n\to\infty}|x-1| < 1 \Rightarrow |x-1| < 1$

Radius $= 1$

19. $\lim_{n\to\infty}\left|\frac{a_{n+1}(x-3)^{n+1}}{a_n(x-3)^n}\right| = \lim_{n\to\infty}\left|\frac{(x-3)^n/3^n}{(x-3)^{n-1}/3^{n-1}}\right| = \lim_{n\to\infty}\left|\frac{x-3}{3}\right| < 1 \Rightarrow |x-3| < 3$

Radius $= 3$

21. $\lim_{n\to\infty}\left|\dfrac{a_{n+1}x^{n+1}}{a_n x^n}\right| = \lim_{n\to\infty}\left|\dfrac{(n+1)(-2x)^n/(n+2)}{n(-2x)^{n-1}/(n+1)}\right| = \lim_{n\to\infty}\left|\dfrac{-2x(n+1)^2}{n(n+2)}\right| < 1 \Rightarrow |-2x| < 1 \Rightarrow -\dfrac{1}{2} < x < \dfrac{1}{2}$

Radius $= \dfrac{1}{2}$

23. $\lim_{n\to\infty}\left|\dfrac{a_{n+1}x^{n+1}}{a_n x^n}\right| = \lim_{n\to\infty}\left|\dfrac{x^{2n+3}/(2n+3)!}{x^{2n+1}/(2n+1)!}\right| = \lim_{n\to\infty}\left|\dfrac{x^2}{(2n+3)(2n+2)}\right| = 0 \Rightarrow -\infty < x < \infty$

Radius $= \infty$

25.
$$\begin{aligned} f(x) &= e^x & f(1) &= e\\ f'(x) &= e^x & f'(1) &= e\\ f''(x) &= e^x & f''(1) &= e\\ &\vdots & &\vdots\\ f^{(n)}(x) &= e^x & f^{(n)}(1) &= e \end{aligned}$$

The power series for f is

$$\begin{aligned} e^x &= f(1) + f'(1)x + \frac{f''(1)x^2}{2!} + \cdots\\ &= e + \frac{e(x-1)}{1!} + \frac{e(x-1)^2}{2!} + \frac{e(x-1)^3}{3!} + \cdots + \frac{e(x-1)^n}{n!} + \cdots\\ &= e\sum_{n=0}^{\infty}\frac{(x-1)^n}{n!}. \end{aligned}$$

$$\lim_{n\to\infty}\left|\frac{(x-1)^{n+1}/(n+1)!}{(x-1)^n/n!}\right| = \lim_{n\to\infty}\left|\frac{x-1}{n+1}\right| = 0 \Rightarrow R = \infty$$

27.
$$\begin{aligned} f(x) &= e^{3x} & f(0) &= 1\\ f'(x) &= 3e^{3x} & f'(0) &= 3\\ f''(x) &= 9e^{3x} & f''(0) &= 9\\ f'''(x) &= 27e^{3x} & f'''(0) &= 27\\ & & &\vdots\\ & & f^{(n)}(0) &= 3^n \end{aligned}$$

The power series for f is

$$\begin{aligned} e^{3x} &= f(0) + f'(0)x + \frac{f''(0)x^2}{2!} + \cdots\\ &= 1 + 3x + \frac{9x^2}{2!} + \frac{27x^3}{3!} + \cdots + \frac{(3x)^n}{n!} + \cdots\\ &= \sum_{n=0}^{\infty}\frac{(3x)^n}{n!}. \end{aligned}$$

$$\lim_{n\to\infty}\left|\frac{(3x)^{n+1}/(n+1)!}{(3x)^n/n!}\right| = \lim_{n\to\infty}\left|\frac{3x}{n+1}\right| = 0 \Rightarrow R = \infty$$

29.
$$\begin{aligned} f(x) &= \frac{1}{x+1} & f(0) &= 1\\ f'(x) &= -\frac{1}{(x+1)^2} & f'(0) &= -1\\ f''(x) &= \frac{2}{(x+1)^3} & f''(0) &= 2\\ f'''(x) &= -\frac{6}{(x+1)^4} & f'''(0) &= -6\\ & & &\vdots\\ & & f^{(n)}(0) &= (-1)^n n! \end{aligned}$$

The power series for f is

$$\begin{aligned} \frac{1}{x+1} &= f(0) + f'(0)x + \frac{f''(0)x^2}{2!} + \cdots\\ &= 1 - x + \frac{2x^2}{2!} - \frac{6x^3}{3!} + \cdots + \frac{(-1)^n n! x^n}{n!} + \cdots\\ &= 1 - x + x^2 - x^3 + \cdots\\ &= \sum_{n=0}^{\infty}(-1)^n x^n. \end{aligned}$$

$$\lim_{n\to\infty}\left|\frac{(-1)^{n+1}x^{n+1}}{(-1)^n x^n}\right| = \lim_{n\to\infty}|x| < 1 \Rightarrow R = 1$$

31. $f(x) = \sqrt{x}$ $\qquad f(1) = 1$

$f'(x) = \dfrac{1}{2\sqrt{x}}$ $\qquad f'(1) = \dfrac{1}{2}$

$f''(x) = -\dfrac{1}{4x\sqrt{x}}$ $\qquad f''(1) = -\dfrac{1}{4}$

$f'''(x) = \dfrac{3}{8x^2\sqrt{x}}$ $\qquad f'''(1) = \dfrac{3}{8}$

$f^{(4)}(x) = -\dfrac{15}{16x^3\sqrt{x}}$ $\qquad f^{(4)}(1) = -\dfrac{15}{16}$

The general pattern (for $n \geq 2$) is

$$f^{(n)}(1) = (-1)^{n-1}\frac{1 \cdot 3 \cdot 5 \cdots (2n-3)}{2^n}.$$

The power series for $f(x) = \sqrt{x}$ is

$$f(1) + f'(1)(x-1) + \frac{f''(1)(x-1)^2}{2!} + \cdots = 1 + \frac{1}{2}(x-1) - \frac{1}{8}(x-1)^2 + \frac{1}{16}(x-1)^3 + \cdots$$

$$= 1 + \frac{1}{2}(x-1) + \sum_{n=2}^{\infty}\frac{(-1)^{n+1}1 \cdot 3 \cdot 5 \cdot \cdots \cdot (2n-3)}{2^n n!}(x-1)^n.$$

$$\lim_{n\to\infty}\left|\frac{1 \cdot 3 \cdot 5 \cdot \cdots \cdot (2n-1)(x-1)^{n+1}/2^{n+1}(n+1)!}{1 \cdot 3 \cdot 5 \cdot \cdots \cdot (2n-3)(x-1)^n/2^n n!}\right| = \lim_{n\to\infty}\left|\frac{(2n-1)(x-1)}{2(n+1)}\right| \Rightarrow |x-1| < 1 \Rightarrow R = 1$$

33. $\dfrac{1}{(1+x)^2} = (1+x)^{-2}$

$$= 1 + (-2)x + \frac{(-2)(-3)x^2}{2!} + \frac{(-2)(-3)(-4)x^3}{3!} + \cdots$$

$$= 1 - 2x + 3x^2 - 4x^3 + \cdots$$

$$= \sum_{n=0}^{\infty}(-1)^n(n+1)x^n$$

$$\lim_{n\to\infty}\left|\frac{(-1)^{n+1}(n+2)x^{n+1}}{(-1)^n(n+1)x^n}\right| = \lim_{n\to\infty}\left|\frac{x(n+2)}{n+1}\right| \Rightarrow |x| < 1 \Rightarrow R = 1$$

35. $\dfrac{1}{(1+x)^3} = (1+x)^{-3}$

$$= 1 + (-3)x + \frac{(-3)(-4)x^2}{2!} + \frac{(-3)(-4)(-5)x^3}{3!} + \cdots$$

$$= 1 - 3x + \frac{12x^2}{2!} - \frac{60x^3}{3!} + \cdots$$

$$= \sum_{n=0}^{\infty}\frac{(-1)^n(n+1)!x^n}{2 \cdot n!} = \sum_{n=0}^{\infty}\frac{(-1)^n(n+1)(n+2)x^n}{2}$$

$$\lim_{n\to\infty}\left|\frac{(-1)^{n+1}(n+2)(n+3)x^{n+1}/2}{(-1)^n(n+1)(n+2)x^n/2}\right| = \lim_{n\to\infty}\left|\frac{x(n+3)}{n+1}\right| \Rightarrow |x| < 1 \Rightarrow R = 1$$

37. (a) $f(x) = \sum_{n=0}^{\infty}\left(\frac{x}{2}\right)^n = \sum_{n=0}^{\infty}\frac{x^n}{2^n}$

$$\lim_{n\to\infty}\left|\frac{(x/2)^{n+1}}{(x/2)^n}\right| \Rightarrow \left|\frac{x}{2}\right| < 1 \Rightarrow |x| < 2$$

$R = 2$

(b) $f'(x) = \sum_{n=1}^{\infty}\frac{nx^{n-1}}{2^n}$

$$\lim_{n\to\infty}\left|\frac{(n+1)x^n/2^{n+1}}{nx^{n-1}/2^n}\right| = \lim_{n\to\infty}\left|\frac{n+1}{2n}x\right| \Rightarrow \left|\frac{x}{2}\right| < 1 \Rightarrow |x| < 2$$

$R = 2$

(c) $f''(x) = \sum_{n=2}^{\infty}\frac{n(n-1)x^{n-2}}{2^n}$

$$\lim_{n\to\infty}\left|\frac{(n+1)nx^{n-1}/2^{n+1}}{n(n-1)x^{n-2}/2^n}\right| = \lim_{n\to\infty}\left|\frac{n+1}{n-1}\cdot\frac{x}{2}\right| \Rightarrow \left|\frac{x}{2}\right| < 1 \Rightarrow |x| < 2$$

$R = 2$

(d) $\int f(x)\,dx = \sum_{n=0}^{\infty}\frac{x^{n+1}}{(n+1)2^n} + C$

$$\lim_{n\to\infty}\left|\frac{x^{n+2}/(n+2)2^{n+1}}{x^{n+1}/(n+1)2^n}\right| = \lim_{n\to\infty}\left|\frac{n+1}{n+2}\cdot\frac{x}{2}\right| \Rightarrow \left|\frac{x}{2}\right| < 1 \Rightarrow |x| < 2$$

$R = 2$

39. (a) $f(x) = \sum_{n=0}^{\infty}\frac{(x+1)^{n+1}}{n+1}$

$$\lim_{n\to\infty}\left|\frac{(x+1)^{n+2}/(n+2)}{(x+1)^{n+1}/(n+1)}\right| = \lim_{n\to\infty}\left|\frac{n+1}{n+2}(x+1)\right| \Rightarrow |x+1| < 1$$

$R = 1$

(b) $f'(x) = \sum_{n=0}^{\infty}(x+1)^n$

$$\lim_{n\to\infty}\left|\frac{(x+1)^{n+1}}{(x+1)^n}\right| \Rightarrow |x+1| < 1$$

$R = 1$

(c) $f''(x) = \sum_{n=1}^{\infty}n(x+1)^{n-1}$

$$\lim_{n\to\infty}\left|\frac{(n+1)(x+1)^n}{n(x+1)^{n-1}}\right| = \lim_{n\to\infty}\left|\frac{n+1}{n}(x+1)\right| \Rightarrow |x+1| < 1$$

$R = 1$

(d) $\int f(x)\,dx = \sum_{n=0}^{\infty}\frac{(x+1)^{n+2}}{(n+2)(n+1)} + C$

$$\lim_{n\to\infty}\left|\frac{(x+1)^{n+3}/(n+3)(n+2)}{(x+1)^{n+2}/(n+2)(n+1)}\right| = \lim_{n\to\infty}\left|\frac{(x+1)(n+1)}{(n+3)}\right| \Rightarrow |x+1| < 1$$

$R = 1$

41. Because the power series for e^x is

$$e^x = \sum_{n=0}^{\infty} \frac{x^n}{n!},$$

it follows that the power series for e^{x^3} is

$$e^{x^3} = \sum_{n=0}^{\infty} \frac{\left(x^3\right)^n}{n!} = \sum_{n=0}^{\infty} \frac{x^{3n}}{n!}.$$

43. $$3x^2 e^{x^3} = \frac{d}{dx}\left[e^{x^3}\right] = \sum_{n=1}^{\infty} \frac{3nx^{3n-1}}{n!} = 3\sum_{n=1}^{\infty} \frac{x^{3n-1}}{(n-1)!} = 3\sum_{n=0}^{\infty} \frac{x^{3n+2}}{n!}$$

45. Because the power series for $1/(1+x)$ is

$$f(x) = \frac{1}{1+x} = \sum_{n=0}^{\infty} (-1)^n x^n,$$

it follows that the power series for $1/\left(1+x^4\right)$ is

$$f\left(x^4\right) = \frac{1}{1+x^4} = \sum_{n=0}^{\infty} (-1)^n x^{4n}.$$

47. Because the power series for $\dfrac{1}{1+x} = \sum_{n=0}^{\infty} (-1)^n x^n$, it follows that the power series for $\dfrac{1}{1+x^2}$ is

$$f\left(x^2\right) = \frac{1}{1+x^2} = \sum_{n=0}^{\infty} (-1)^n \left(x^2\right)^n = \sum_{n=0}^{\infty} (-1)^n x^{2n}.$$

Then the power series for

$$\frac{3x}{1+x^2} \text{ is } 3x\, f\left(x^2\right) = 3x\sum_{n=0}^{\infty} (-1)^n x^{2n} = \sum_{n=0}^{\infty} (-1)^n (3x) x^{2n} = \sum_{n=0}^{\infty} (-1)^n 3x^{2n+1}.$$

49. Because the power series for $\dfrac{1}{x}$ is $\sum_{n=0}^{\infty} (-1)^n (x-1)^n$, it follows that $\ln x = \int \dfrac{1}{x}\, dx = \sum_{n=0}^{\infty} \dfrac{(-1)^n (x-1)^{n+1}}{n+1}$.

51. Because the power series for $-\dfrac{1}{x}$ is $\sum_{n=0}^{\infty} (-1)^{n+1} (x-1)^n$, it follows that $\dfrac{d}{dx}\left[-\dfrac{1}{x}\right] = \dfrac{1}{x^2} = \sum_{n=1}^{\infty} (-1)^{n+1} n(x-1)^{n-1}$.

53. (a) Let $f(x) = \sum_{n=0}^{\infty} \left(\dfrac{x}{4}\right)$, then

$$f\left(\frac{5}{2}\right) = \sum_{n=0}^{\infty} \left(\frac{5/2}{4}\right)^n = \sum_{n=0}^{\infty} \left(\frac{5}{8}\right)^n.$$

So, $S = \dfrac{a}{1-r} = \dfrac{1}{1-(5/8)} = \dfrac{8}{3}$.

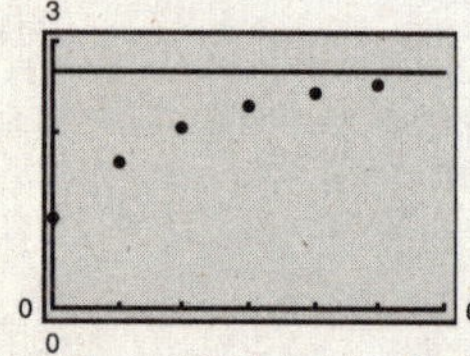

(b) Let $f(x) = \sum_{n=0}^{\infty} \left(\dfrac{x}{4}\right)^n$, then

$$f\left(-\frac{5}{2}\right) = \sum_{n=0}^{\infty} \left(\frac{-5/2}{4}\right)^n = \sum_{n=0}^{\infty} \left(-\frac{5}{8}\right)^n.$$

So, $S = \dfrac{a}{1-r} = \dfrac{1}{1-(-5/8)} = \dfrac{8}{13}$.

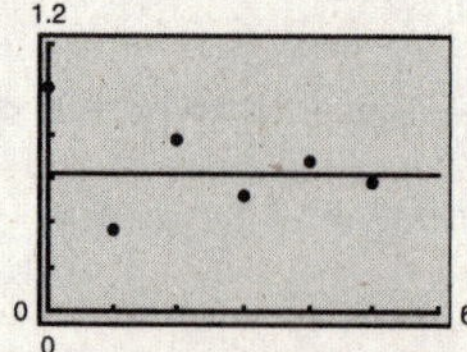

(c) In part (a), the rate of convergence is slower but all of the partial sums approach the line from below. In part (b), the partial sums approach the sum more quickly, but they oscillate above and below the line.

Section 10.5 Taylor Polynomials

Skills Warm Up

1. $f(x) = e^{3x} = 1 + (3x) + \frac{(3x)^2}{2!} + \frac{(3x)^3}{3!} + \frac{(3x)^4}{4!} + \cdots = \sum_{n=0}^{\infty} \frac{3^n x^n}{n!}$

2. $f(x) = e^{-3x} = 1 + (-3x) + \frac{(-3x)^2}{2!} + \frac{(-3x)^3}{3!} + \frac{(-3x)^4}{4!} + \cdots = \sum_{n=0}^{\infty} (-1)^n \left(\frac{3^n x^n}{n!}\right)$

3. $f(x) = \frac{4}{x} = 4\left[1 - (x-1) + (x-2)^2 - (x-1)^3 + \cdots\right] = 4\sum_{n=0}^{\infty} (-1)^n (x-1)^n$

4. $f(x) = \ln 5x = \ln 5 + \ln x = \ln 5 + (x-1) - \frac{(x-1)^2}{2} + \frac{(x-1)^3}{3} - \frac{(x-1)^4}{4} + \cdots = \ln 5 + \sum_{n=1}^{\infty} \frac{(-1)^{n-1}(x-1)^n}{n}$

5. $f(x) = (1+x)^{1/4} = 1 + \frac{x}{4} - \frac{3x^2}{16 \cdot 2!} + \frac{3 \cdot 7x^3}{64 \cdot 3!} - \frac{3 \cdot 7 \cdot 11x^4}{256 \cdot 4!} + \cdots$

6. $f(x) = \sqrt{1+x} = (1+x)^{1/2} - 1 + \frac{x}{2} + \frac{x^2}{4 \cdot 2!} + \frac{1 \cdot 3x^3}{8 \cdot 3!} + \frac{1 \cdot 3 \cdot 5x^4}{16 \cdot 4!} + \cdots$

7. $\int_0^1 \left(1 - x + x^2 - x^3 + x^4\right) dx = \left[x - \frac{1}{2}x^2 + \frac{1}{3}x^3 - \frac{1}{4}x^4 + \frac{1}{5}x^5\right]_0^1 = \frac{47}{60}$

8. $\int_0^{1/2} \left(1 + \frac{x}{3} - \frac{x^2}{9} + \frac{5x^3}{27}\right) dx = \left[x + \frac{x^2}{6} - \frac{x^3}{27} + \frac{5x^4}{108}\right]_0^{1/2} = \frac{311}{576}$

9. $\int_1^2 \left[(x-1) - \frac{(x-1)^2}{2} + \frac{(x-1)^3}{3}\right] dx = \left[\frac{x^2}{2} - x - \frac{x^3}{6} + \frac{x^2}{2} - \frac{x}{2} + \frac{x^4}{12} - \frac{x^3}{3} + \frac{x^2}{2} - \frac{x}{3}\right]_1^2$

$$= \left[\frac{x^4}{12} - \frac{x^3}{2} + \frac{3x^2}{2} - \frac{11x}{6}\right]_1^2 = -\frac{1}{3} - \left(-\frac{3}{4}\right) = \frac{5}{12}$$

10. $\int_1^{3/2} \left[1 - (x-1) + (x-1)^2 - (x-1)^3\right] dx = \left[x - \frac{x^2}{2} + x + \frac{x^3}{3} - x^2 + x - \frac{x^4}{4} + x^3 - \frac{3x^2}{2} + x\right]_1^{3/2}$

$$= \left[-\frac{x^4}{4} + \frac{4x^3}{3} - 3x^2 + 4x\right]_1^{3/2} = \frac{159}{64} - \frac{25}{12} = \frac{72}{192}$$

1.

x	0	$\frac{1}{4}$	$\frac{1}{2}$	$\frac{3}{4}$	1
$f(x) = e^{x/2}$	1	1.1331	1.2840	1.4550	1.6487
$1 + \frac{x}{2}$	1	1.125	1.25	1.375	1.5
$1 + \frac{x}{2} + \frac{x^2}{8}$	1	1.1328	1.2813	1.4453	1.625
$1 + \frac{x}{2} + \frac{x^2}{8} + \frac{x^3}{48}$	1	1.1331	1.2839	1.4541	1.6458
$1 + \frac{x}{2} + \frac{x^2}{8} + \frac{x^3}{48} + \frac{x^4}{384}$	1	1.1331	1.2840	1.4549	1.6484

3. $e^x = \sum_{n=0}^{\infty} \frac{x^n}{n!}$

(a) $S_1(x) = 1 + x$

(b) $S_2(x) = 1 + x + \frac{x^2}{2}$

(c) $S_3(x) = 1 + x + \frac{x^2}{2} + \frac{x^3}{6}$

(d) $S_4(x) = 1 + x + \frac{x^2}{2} + \frac{x^3}{6} + \frac{x^4}{24}$

5. $e^{2x} = \sum_{n=0}^{\infty} \frac{2^n x^n}{n!}$

(a) $S_1(x) = 1 + 2x$

(b) $S_2(x) = 1 + 2x + 2x^2$

(c) $S_3(x) = 1 + 2x + 2x^2 + \frac{4}{3}x^3$

(d) $S_4(x) = 1 + 2x + 2x^2 + \frac{4}{3}x^3 + \frac{2}{3}x^4$

7. $\ln(x + 1) = \sum_{n=1}^{\infty} \frac{(-1)^{n-1} x^n}{n}$

(a) $S_1(x) = x$

(b) $S_2(x) = x - \frac{x^2}{2}$

(c) $S_3(x) = x - \frac{x^2}{2} + \frac{x^3}{3}$

(d) $S_4(x) = x - \frac{x^2}{2} + \frac{x^3}{3} - \frac{x^4}{4}$

9. $\sqrt{x + 1} = 1 + \frac{x}{2 \cdot 1!} - \frac{x^2}{4 \cdot 2!} + \frac{3x^3}{8 \cdot 3!} - \frac{3 \cdot 5x^4}{16 \cdot 4!}$

(a) $S_1(x) = 1 + \frac{x}{2}$

(b) $S_2(x) = 1 + \frac{x}{2} - \frac{x^2}{8}$

(c) $S_3(x) = 1 + \frac{x}{2} - \frac{x^2}{8} + \frac{x^3}{16}$

(d) $S_4(x) = 1 + \frac{x}{2} - \frac{x^2}{8} + \frac{x^3}{16} - \frac{5x^4}{128}$

11. $\frac{1}{(x + 1)^2} = \sum_{n=0}^{\infty} (-1)^n (n + 1)x^n$

(a) $S_1(x) = 1 - 2x$

(b) $S_2(x) = 1 - 2x + 3x^2$

(c) $S_3(x) = 1 - 2x + 3x^2 - 4x^3$

(d) $S_4(x) = 1 - 2x + 3x^2 - 4x^3 + 5x^4$

13. $f(x) = \frac{x}{x + 1} = 1 - \frac{1}{x + 1} = 1 - \sum_{n=1}^{\infty} (-1)^n x^n$

(a) $S_1(x) = x$

(b) $S_2(x) = x - x^2$

(c) $S_3(x) = x - x^2 + x^3$

(d) $S_4(x) = x - x^2 + x^3 - x^4$

15. $\frac{1}{1 + x^2} = \sum_{n=0}^{\infty} (-1)^n x^{2n}$

(a) $S_2(x) = 1 - x^2$

(b) $S_4(x) = 1 - x^2 + x^4$

(c) $S_6(x) = 1 - x^2 + x^4 - x^6$

(d) $S_8(x) = 1 - x^2 + x^4 - x^6 + x^8$

17. $f(x) = \frac{1}{\sqrt[3]{x + 1}}$

$$S_4(x) = 1 - \frac{x}{3} + \frac{2x^2}{9} - \frac{14x^3}{81} + \frac{35x^4}{243}$$

19. $y = -\frac{1}{2}x^2 + 1$ is a parabola through (0, 1); matches (d).

21. $y = e^{-1/2}[(x + 1) + 1]$ is a line; matches (a).

23. $f(x) = e^{-x}$, $c = 0$

$$S_6(x) = 1 - x + \frac{x^2}{2} - \frac{x^3}{6} + \frac{x^4}{24} - \frac{x^5}{120} + \frac{x^6}{720}$$

$$f\left(\frac{1}{2}\right) \approx 1 - \frac{1}{2} + \frac{1}{8} - \frac{1}{48} + \frac{1}{384} - \frac{1}{3840} + \frac{1}{46{,}080}$$

$$\approx 0.607$$

$$|R_6| \le \frac{(1/2)^7}{7!}(e^0) \approx 0.00000155$$

25. $f(x) = \ln x,\ c = 2$

$$S_6(x) = \ln 2 + \frac{1}{2}(x-2) - \frac{1}{8}(x-2)^2 + \frac{1}{24}(x-2)^3 - \frac{1}{64}(x-2)^4 + \frac{1}{160}(x-2)^5 - \frac{1}{384}(x-2)^6$$

$$f\left(\frac{3}{2}\right) = \ln 2 + \frac{1}{2}\left(-\frac{1}{2}\right) - \frac{1}{8}\left(\frac{1}{4}\right) + \frac{1}{24}\left(-\frac{1}{8}\right) - \frac{1}{64}\left(\frac{1}{16}\right) + \frac{1}{160}\left(-\frac{1}{32}\right) - \frac{1}{384}\left(\frac{1}{64}\right) \approx 0.4055$$

$$|R_6| \le \frac{(1/2)^7}{7!}\left(\frac{720}{(3/2)^7}\right) \approx 0.000065$$

27. $|R_5| \le \frac{f^{(6)}(z)}{6!}x^6 = \frac{e^{-z}}{6!}x^6$

In the interval $[0, 1]$, it follows that

$$R_5 \le \frac{1}{6!} \approx 0.00139.$$

29. The $(n+1)$ derivative of $f(x) = e^x$ is e^x. The maximum value of $\left|f^{n+1}(x)\right|$ on the interval $[0, 2]$ is $e^2 < 8$.

So, the nth remainder is bounded by

$$|R_n| \le \left|\frac{8}{(n+1)!}(x-1)^{n+1}\right|,\ 0 \le x \le 2$$

$$|R_n| \le \frac{8}{(n+1)!}(1).$$

When $n = 7$, $\frac{8}{(7+1)!} = 1.98 \times 10^{-4} \approx 0.000198 < 0.001$, so $n = 7$ will approximate e^x with an error less than 0.001 in $[0, 2]$.

31. $S_6(x) = 1 - x^3 + \frac{x^6}{2}$

$$\int_0^1 e^{-x^3}\,dx \approx \int_0^1\left(1 - x^3 + \frac{x^6}{2}\right)dx = \left[x - \frac{x^4}{4} + \frac{x^7}{14}\right]_0^1 = \frac{23}{28} \approx 0.82143$$

33. $S_6(x) = 1 - \frac{1}{2}x^2 + \frac{3}{8}x^4 - \frac{5}{16}x^6$

$$\int_0^{1/2} \frac{1}{\sqrt{1+x^2}}\,dx \approx \int_0^{1/2}\left(1 - \frac{1}{2}x^2 + \frac{3}{8}x^4 - \frac{5}{16}x^6\right)dx = \left[x - \frac{x^3}{6} + \frac{3x^5}{40} - \frac{5x^7}{112}\right]_0^{1/2} \approx 0.481$$

35. (a) For $f(x) = e^x$,

$$P_4(x) = 1 + x + \frac{x^2}{2} + \frac{x^3}{6} + \frac{x^4}{24}.$$

For $g(x) = xe^x$,

$$Q_5(x) = x + x^2 + \frac{x^3}{2} + \frac{x^4}{6} + \frac{x^5}{24}.$$

Therefore, $Q_5(x) = xP_4(x)$.

(b) For $h(x) = x^2e^x$,

$P_6(x) = x^2 + x^3 + \frac{x^4}{2} + \frac{x^5}{6} + \frac{x^6}{24}$ is of degree 6.

(c) For $n(x) = e^x/x$,

$P_3(x) = \frac{1}{x} + 1 + \frac{x}{2} + \frac{x^2}{6} + \frac{x^3}{24}$ is of degree 3.

Section 10.6 Newton's Method

Skills Warm Up

1. $f(x) = x^2 - 2x - 1,\ f(2.4) = -0.04$

   $f'(x) = 2x - 2,\ f'(2.4) = 2.8$

2. $f(x) = x^3 - 2x^2 + 1,\ f(-0.6) = 0.064$

   $f'(x) = 3x^2 - 4x,\ f'(-0.6) = 3.48$

3. $f(x) = e^{2x} - 2,\ f(0.35) = e^{0.7} - 2 \approx 0.014$

   $f'(x) = 2e^{2x},\ f'(0.35) = 2e^{0.7} \approx 4.028$

4. $f(x) = e^{x^2} - 7x + 3,\ f(1.4) = e^{1.96} - 6.8 \approx 0.299$

   $f'(x) = 2xe^{x^2} - 7,\ f'(1.4) = 2.8e^{1.96} - 7 \approx 12.878$

5. $|x - 5| \le 0.1$

   $-0.1 \le x - 5 \le 0.1$

   $4.9 \le x \le 5.1$

6. $|4 - 5x| \le 0.01$

   $-0.01 \le 4 - 5x \le 0.01$

   $-4.01 \le -5x \le -3.99$

   $0.802 \ge x \ge 0.798$

7. $\left|2 - \dfrac{x}{3}\right| \le 0.01$

   $-0.01 \le 2 - \dfrac{x}{3} \le 0.01$

   $-2.01 \le -\dfrac{x}{3} \le -1.99$

   $6.03 \ge x \ge -5.97$

8. $|2x + 7| \le 0.01$

   $-0.01 \le 2x + 7 \le 0.01$

   $-7.01 \le 2x \le -6.99$

   $-3.505 \le x \le -3.495$

9. $y = x^2 - x - 2,\ y = 2x - 1$

   $x^2 = x - 2 = 2x - 1$

   $x^2 - 3x - 1 = 0$

   $$x = \frac{3 \pm \sqrt{9 - 4(1)(-1)}}{2} = \frac{3 \pm \sqrt{13}}{2}$$

   Points of intersection:

   $$\left(\frac{3 + \sqrt{13}}{2}, 2 + \sqrt{13}\right), \left(\frac{3 - \sqrt{13}}{2}, 2 - \sqrt{13}\right)$$

10. $y = x^2,\ y = x + 1$

    $x^2 = x + 1$

    $x^2 - x - 1 = 0$

    $$x = \frac{1 \pm \sqrt{1 - 4(1)(-1)}}{2} = \frac{1 \pm \sqrt{5}}{2}$$

    Points of intersection:

    $$\left(\frac{1 + \sqrt{5}}{2}, \frac{3 + \sqrt{5}}{2}\right), \left(\frac{1 - \sqrt{5}}{2}, \frac{3 - \sqrt{5}}{2}\right)$$

1. $f(x) = x^2 - 5, f'(x) = 2x$

   $$x_2 = x_1 - \frac{f(x_1)}{f'(x_1)} = 2 - \frac{(2)^2 - 5}{2(2)} = 2.25$$

   $$x_3 = x_2 - \frac{f(x_2)}{f'(x_2)} = 2.25 - \frac{(2.25)^2 - 5}{2(2.25)} \approx 2.2361$$

3. $f(x) = x^3 + x - 1,\ f'(x) = 3x^2 + 1$

n	x_n	$f(x_n)$	$f'(x_n)$	$\dfrac{f(x_n)}{f'(x_n)}$	$x_n - \dfrac{f(x_n)}{f'(x_n)}$
1	0.5	–0.375	1.75	–0.2143	0.7143
2	0.7143	0.0787	2.5306	0.0311	0.6832
3	0.6832	0.0021	2.4002	0.0009	0.6823

Approximation: $x \approx 0.682$

5. $f(x) = 5\sqrt{x-1} - 2x$

$$f'(x) = \frac{5}{2\sqrt{x-1}} - 2$$

n	x_n	$f(x_n)$	$f'(x_n)$	$\frac{f(x_n)}{f'(x_n)}$	$x_n - \frac{f(x_n)}{f'(x_n)}$
1	1.2	−0.1639	3.5902	−0.0457	1.2457
2	1.2457	−0.0130	3.0436	−0.0043	1.2500
3	1.2500	0	3	0	1.25

Approximation: $x \approx 1.25$

7. $f(x) = \ln x + x$, $f'(x) = \frac{1}{x} + 1$

n	x_n	$f(x_n)$	$f'(x_n)$	$\frac{f(x_n)}{f'(x_n)}$	$x_n - \frac{f(x_n)}{f'(x_n)}$
1	0.6	0.0892	2.6667	0.0334	0.5666
2	0.5666	−0.0015	2.7649	−0.0005	0.5671
3	0.5671	−0.0001	2.7634	−0.00004	0.5675

Approximation: $x \approx 0.568$

9. $f(x) = e^{x/2} + x^2 - 3$, $f'(x) = \frac{1}{2}e^{x/2} + 2x$

n	x_n	$f(x_n)$	$f'(x_n)$	$\frac{f(x_n)}{f'(x_n)}$	$x_n - \frac{f(x_n)}{f'(x_n)}$
1	−1.5	−0.2776	−2.7638	0.1005	−1.6005
2	−1.6005	0.0108	−2.9764	−0.0036	−1.5969
3	−1.5969	0.0001	−2.9688	−0.00004	−1.5969

Approximation: $x \approx -1.5969$

n	x_n	$f(x_n)$	$f'(x_n)$	$\frac{f(x_n)}{f'(x_n)}$	$x_n - \frac{f(x_n)}{f'(x_n)}$
1	1.0	−0.3513	2.824	−0.1244	1.1244
2	1.1244	0.0188	3.1261	0.0060	1.1184
3	1.1184	0.00009	3.1114	0.00003	1.1184

Approximation: $x \approx 1.1184$

11. $f(x) = x^4 + x^3 - 1,\ f'(x) = 4x^3 + 3x^2$

n	x_n	$f(x_n)$	$f'(x_n)$	$\dfrac{f(x_n)}{f'(x_n)}$	$x_n - \dfrac{f(x_n)}{f'(x_n)}$
1	1	1	7	$\frac{1}{7}$	$\frac{6}{7}$
2	$\frac{6}{7}$	0.1695	4.7230	0.0359	0.8213
3	0.8213	0.0089	4.2396	0.0021	0.8192
4	0.8192	0.0001	4.2123	0.00002	0.8192

Approximation: $x \approx 0.819$

n	x_n	$f(x_n)$	$f'(x_n)$	$\dfrac{f(x_n)}{f'(x_n)}$	$x_n - \dfrac{f(x_n)}{f'(x_n)}$
1	−1	−1	−1	1	−2
2	−2	7	−20	−0.35	−1.65
3	−1.65	1.9199	−9.801	−0.1959	−1.4541
4	−1.4541	0.3962	−5.9550	−0.0665	−1.3876
5	−1.3876	0.0356	−4.9106	−0.0072	−1.3804
6	−1.3804	0.00006	−4.8049	−0.00008	−1.3803

Approximation: $x \approx -1.380$

13. Let $3 - x = \dfrac{1}{(x^2 + 1)}$ and define

$h(x) = \dfrac{1}{x^2 + 1} + x - 3$. Then $h'(x) = -\dfrac{2x}{(x^2 + 1)^2} + 1$.

n	x_n	$h(x_n)$	$h'(x_n)$	$\dfrac{h(x_n)}{h'(x_n)}$	$x_n - \dfrac{h(x_n)}{h'(x_n)}$
1	3	0.1	0.94	0.1064	2.8936
2	2.8936	0.0003	0.9341	0.0003	2.8933

Approximation: $x \approx 2.893$

15. Let $x = e^{-x}$ and define $h(x) = x - e^{-x}$. Then $h'(x) = 1 + e^{-x}$.

n	x_n	$h(x_n)$	$h'(x_n)$	$\dfrac{h(x_n)}{h'(x_n)}$	$x_n - \dfrac{h(x_n)}{h'(x_n)}$
1	0.5	−0.1065	1.6065	−0.0663	0.5663
2	0.5663	−0.0013	1.5676	−0.0008	0.5671

Approximation: $x \approx 0.567$

17. From the graph there is one real zero near $x = 12$. Let $x_1 = 12$.

$$f(x) = \frac{1}{4}x^3 - 3x^2 + \frac{3}{4}x - 2$$

$$f'(x) = \frac{3}{4}x^2 - 6x + \frac{3}{4}$$

$$x_{n+1} = x_n - \frac{(1/4)x_n^3 - 3x_n^2 + (3/4)x_n - 2}{(3/4)x_n^2 - 6x_n + (3/4)}$$

$$= x_n - \frac{x_n^3 - 12x_n^2 + 3x_n - 8}{3x_n^2 - 24x_n + 3}$$

$$= \frac{2x_n^3 - 12x_n^2 + 8}{3x_n^2 - 24x_n + 3}$$

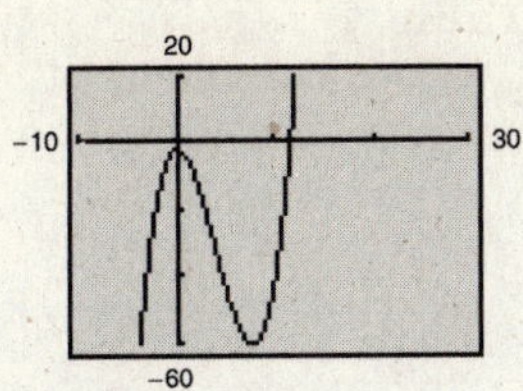

n	1	2	3	4
x_n	12	11.8095	11.8033	11.8033

Zero: $x \approx 11.803$

19. From the graph, there are two real zeros near $x = -1$ and $x = 2$.

$$f(x) = \frac{1}{2}x^4 - 3x - 3$$

$$f'(x) = 2x^3 - 3$$

$$x_{n+1} = x_n - \frac{(1/2)x_n^4 - 3x_n - 3}{2x_n^3 - 3} = \frac{3x_n^4 + 6}{4x_n^3 - 6}$$

n	1	2	3	4
x_n	−1	−0.9	−0.8937	−0.8937

n	1	2	3	4
x_n	2	2.0769	2.0720	2.0720

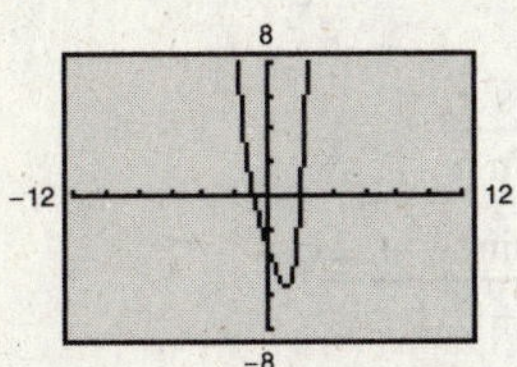

Zeros: $x \approx -0.894, 2.072$

21. From the graph, there are three real zeros near $x = 0.8$, $x = 1.1$, and $x = 1.8$.

$$f(x) = x^3 - 3.9x^2 + 4.79x - 1.881$$

$$f'(x) = 3x^2 - 7.8x + 4.79$$

$$x_{n+1} = x_n - \frac{x_n^3 - 3.9x_n^2 + 4.79x_n - 1.881}{3x_n^2 - 7.8x_n + 4.79}$$

$$= \frac{2000x_n^3 - 3900x_n^2 + 1881}{3000x_n^2 - 7800x_n + 4790}$$

n	1	2	3	4	5	6
x_n	0.8	0.8702	0.8959	0.8999	0.9	0.9

n	1	2
x_n	1.1	1.1

n	1	2	3	4	5
x_n	1.8	1.9340	1.9023	1.9000	1.9

0.10
0.6
2.2
−0.15

Zeros: $x = 0.9, 1.1, 1.9$

23. From the graph, there are two real zeros near $x = 1.1$ and $x = 8$.

$$f(x) = 3\sqrt{x-1} - x$$

$$f'(x) = \frac{3}{2\sqrt{x-1}} - 1$$

$$x_{n+1} = x_n - \frac{6(x_n - 1) - 2x_n\sqrt{x_n - 1}}{3 - 2\sqrt{x_n - 1}}$$

$$= \frac{3x_n - 6}{2\sqrt{x_n - 1} - 3}$$

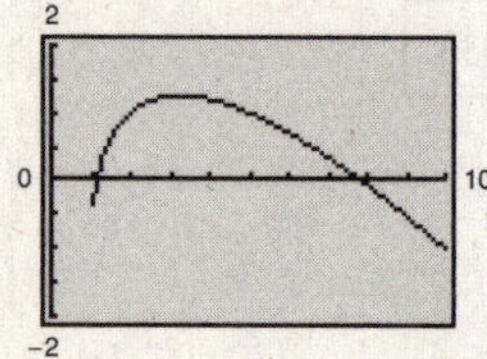

Zeros: $x \approx 1.146, 7.854$

n	1	2	3	4
x_n	1.1	1.1404	1.1458	1.1459

n	1	2	3	4
x_n	8	7.8551	7.8541	7.8541

25. From the graph, there are two real zeros near $x = -2$ and $x = 0.5$.

$$f(x) = 2x + 4 - e^{3x}$$

$$f'(x) = 2 - 3e^{3x}$$

$$x_{n+1} = x_n - \frac{2x_n + 4 - e^{3x_n}}{2 - 3e^{3x_n}}$$

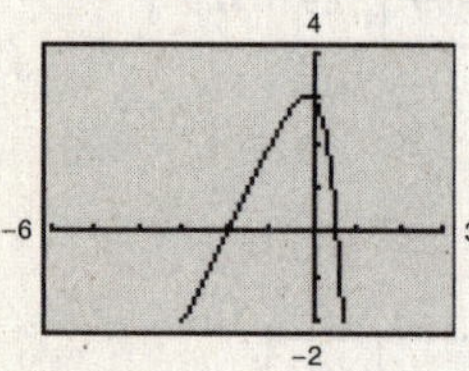

Zeros: $x \approx -1.999, 0.542$

n	1	2
x_n	-2	-1.9988

n	1	2	3	4
x_n	0.5	0.5453	0.5421	0.5420

27. From the graph, there is one real zero near $x = 2$.

$$f(x) = x - 3 + \ln x$$

$$f'(x) = 1 + \frac{1}{x}$$

$$x_{n+1} = x_n - \frac{x_n(x_n - 3 + \ln x_n)}{x_n + 1} = \frac{4x_n - x_n \ln x_n}{x + 1}$$

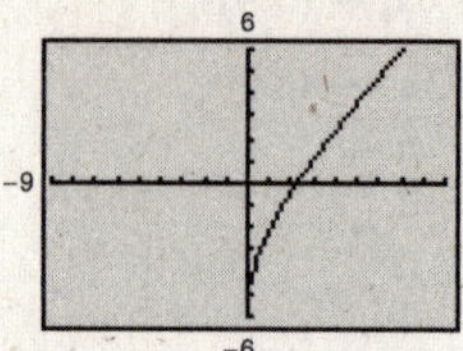

Zero: $x \approx 2.208$

n	1	2	3	4
x_n	2	2.2046	2.2079	2.2079

29. From the graph, there is one real zero near $x = 1$.

$$f(x) = x^3 - \cos x$$

$$f'(x) = 3x^2 + \sin x$$

$$x_{n+1} = x_n - \frac{x_n^3 - \cos x_n}{3x_n^2 + \sin x_n}$$

$$= \frac{\cos x_n + x_n \sin x_n + 2x_n^3}{\sin x_n + 3x_n^2}$$

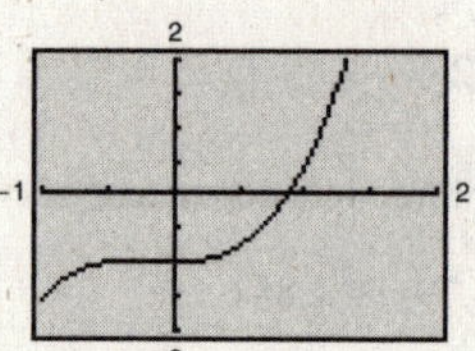

Zero: $x \approx 0.866$

n	1	2	3	4
x_n	1	0.8803	0.8657	0.8655

31. $y = 2x^3 - 6x^2 + 6x - 1$

$y' = 6x^2 + 12x + 6$

Newton's Method fails because $f'(x_1) = 0$.

n	x_n	$f(x_n)$	$f'(x_n)$	$\frac{f(x_n)}{f'(x_n)}$	$x_n - \frac{f(x_n)}{f'(x_n)}$
1	1	1	0	undefined	undefined

33. $y = -x^3 + 3x^2 - x + 1$

$y' = -3x^2 + 6x - 1$

Newton's Method fails because the function fails to converge.

n	x_n	$f(x_n)$	$f'(x_n)$	$\frac{f(x_n)}{f'(x_n)}$	$x_n - \frac{f(x_n)}{f'(x_n)}$
1	1	2	2	1	0
2	0	1	−1	−1	1
3	1	2	2	1	0
4	0	1	−1	−1	1

35. Let $f(x) = x^2 - a$. Then $f'(x) = 2x$.

$$x_{n+1} = x_n - \frac{x_n^2 - a}{2x_n} = \frac{x_n^2 + a}{2x_n}$$

37. Let $f(x) = x^2 - 7$. Then $f'(x) = 2x$.

$$x_{n+1} = \frac{x_n^2 + 7}{2x_n}$$

n	1	2	3	4	5
x_n	2	2.75	2.6477	2.6458	2.6458

Approximation: $\sqrt{7} \approx 2.646$

39. Let $f(x) = x^4 - 6$. Then $f'(x) = 4x^3$.

$$x_{n+1} = \frac{3x_n^4 + 6}{4x_n^3}$$

n	1	2	3	4	5
x_n	2	1.6875	1.5778	1.5652	1.5651

Approximation: $\sqrt[4]{6} \approx 1.565$

41. Let $f(x) = (1/x) - a$. Then $f'(x) = -1/x^2$.

$$x_{n+1} = x_n - \frac{(1/x_n) - a}{-1/(x_n^2)} = x_n(2 - ax_n)$$

43. $d = \sqrt{(x-1)^2 + (y-0)^2}$

$= \sqrt{(x-1)^2 + (4-x^2)^2}$

$= \sqrt{x^2 - 2x + 1 + 16 - 8x^2 + x^4}$

Find critical numbers of

$f(x) = x^4 - 7x^2 - 2x + 16.$

$f'(x) = 4x^3 - 14x - 2$

$f''(x) = 12x^2 - 14$

$$x_{n+1} = x_n - \frac{4x_n^3 - 14x_n - 2}{12x_n^2 - 14} = \frac{4x_n^3 + 1}{6x_n^2 - 7}$$

n	1	2	3	4
x_n	2	1.9412	1.9385	1.9385

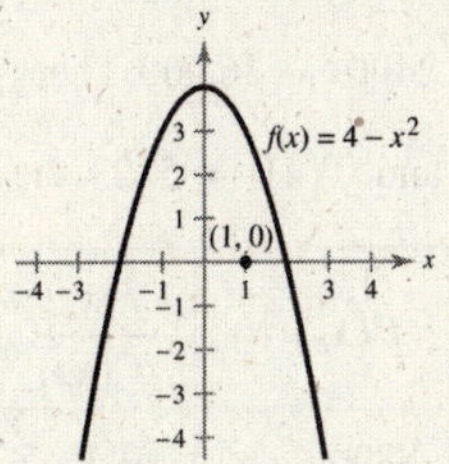

The point closest to $(1, 0)$ is approximately $(1.939, 0.240)$.

45. The time is $T = \dfrac{\sqrt{x^2 + 4}}{3} + \dfrac{\sqrt{x^2 - 6x + 10}}{4}$.

To minimize the time, set dT/dx equal to zero and solve for x. This produces the equation

$7x^4 - 42x^3 + 43x^2 + 216x - 324 = 0.$

Let $f(x) = 7x^4 - 42x^3 + 43x^2 + 216x - 324$. Then $f'(x) = 28x^3 - 126x^2 + 86x + 216$.

Because $f(1) = -100$ and $f(2) = 56$, the solution is in the interval $(1, 2)$.

n	x_n	$f(x_n)$	$f'(x_n)$	$\frac{f(x_n)}{f'(x_n)}$	$x_n - \frac{f(x_n)}{f'(x_n)}$
1	1.7	19.5887	135.624	0.1444	1.5556
2	1.5556	−1.0480	150.2780	−0.0070	1.5626
3	1.5626	0.0014	149.591	0.000009	1.5626

Approximation: $x \approx 1.563$ miles

You should row towards a point approximately 1.563 miles down the coast from the boat's nearest point to the coast.

47. $C = 0.0001x + 0.02x^2 + 0.4x + 800$

Average cost: $\overline{C} = \dfrac{C}{x} = 0.0001x^2 + 0.02x + 0.4 + \dfrac{800}{x}$

To minimize $\overline{C}$, set $d\overline{C}/dx$ equal to zero and solve for x.

This produces the equation $\overline{C}' = 0.0002x + 0.02 - \dfrac{800}{x^2} = \dfrac{0.0002x^3 + 0.02x^2 - 800}{x^2} = 0.$

Let $f(x) = 0.0002x^3 + 0.02x^2 - 800$. Then $f'(x) = 0.0006x^2 + 0.04x$.

Because $f(130) = -22.6$ and $f(132) = 8.4736$, the solution is in the interval $(130, 132)$.

n	x_n	$f(x_n)$	$f'(x_n)$	$\dfrac{f(x_n)}{f'(x_n)}$	$x_n - \dfrac{f(x_n)}{f'(x_n)}$
1	131	−7.1618	15.5366	−0.4610	131.4610
2	131.4610	0.0215	15.6276	0.0014	131.4596
3	131.4596	−0.0003	15.6274	−0.00002	131.4596

Approximation: $x \approx 131$ units

To minimize the average cost per unit, make the production level 131 units.

49. $C = 100\left(\dfrac{200}{x^2} + \dfrac{x}{x + 30}\right)$

To minimize C, set dC/dt equal to zero and solve for t. This produces the equation

$$C' = 100\left[-\frac{400}{x^3} + \frac{30}{(x + 30)^2}\right] = 1000\left[\frac{3x^3 - 40x^2 - 2400x - 36{,}000}{x^3(x + 30)^2}\right] = 0.$$

Let $f(x) = 3x^3 - 40x^2 - 2400x - 36{,}000$. Then $f'(x) = 9x^2 - 80x - 2400$.

Because $f(39) = -12{,}483$ and $f(41) = 5123$, the solution is in the interval $(39, 41)$.

n	x_n	$f(x_n)$	$f'(x_n)$	$\dfrac{f(x_n)}{f'(x_n)}$	$x_n - \dfrac{f(x_n)}{f'(x_n)}$
1	40	−4000	8800	−0.4545	40.4545
2	40.4545	65.9841	9092.7391	0.0073	40.4472

Approximation: $x \approx 40.45$ units

To minimize cost, the order size should be 4045 units.

51. False. Let $f(x) = \dfrac{x^2 - 1}{x - 1}$.

Review Exercises for Chapter 10

1. $a_n = 3n + 4$

$a_1 = 3(1) + 4 = 7$

$a_2 = 3(2) + 4 = 10$

$a_3 = 3(3) + 4 = 13$

$a_4 = 3(4) + 4 = 16$

$a_5 = 3(5) + 4 = 19$

3. $a_n = \left(\frac{3}{2}\right)^n$

$a_1 = \left(\frac{3}{2}\right)^1 = \frac{3}{2}$

$a_2 = \left(\frac{3}{2}\right)^2 = \frac{9}{4}$

$a_3 = \left(\frac{3}{2}\right)^3 = \frac{27}{8}$

$a_4 = \left(\frac{3}{2}\right)^4 = \frac{81}{16}$

$a_5 = \left(\frac{3}{2}\right)^5 = \frac{243}{32}$

5. $a_n = \dfrac{4^n}{n!}$

$a_1 = \dfrac{4^1}{1!} = 4$

$a_2 = \dfrac{4^2}{2!} = \dfrac{16}{2} = 8$

$a_3 = \dfrac{4^3}{3!} = \dfrac{64}{6} = \dfrac{32}{3}$

$a_4 = \dfrac{4^4}{4!} = \dfrac{256}{24} = \dfrac{32}{3}$

$a_5 = \dfrac{4^5}{5!} = \dfrac{1024}{120} = \dfrac{128}{15}$

7. This sequence converges because $\lim_{n\to\infty} \dfrac{n+1}{n^2} = 0$.

9. This sequence diverges because $\lim_{n\to\infty} \dfrac{n^3}{n^2+1} = \infty$.

11. This sequence converges because $\lim_{n\to\infty} \left(5 + \dfrac{1}{3^n}\right) = 5$.

13. This sequence converges because $\lim_{n\to\infty} \dfrac{1}{n^{4/3}} = 0$.

15. This sequence diverges because

$$\lim_{n\to\infty} \frac{n!}{(n-3)!} = \lim_{n\to\infty} (n)(n-1)(n-2) = \infty.$$

17. $a_n = 2n + 5, n = 1, 2, 3, \ldots$

19. $a_n = \dfrac{1}{n!}, n = 1, 2, 3, \ldots$ OR

$a_n = \dfrac{1}{(n+1)!}, n = 0, 1, 2, \ldots$

21. $a_n = (-1)^{n-1}\left(\dfrac{2^{n-1}}{3^n}\right), n = 1, 2, 3, \ldots$ OR

$a_n = (-1)^n\left(\dfrac{2^n}{3^{n+1}}\right), n = 0, 1, 2, \ldots$

23. $A_n = 1000(1 + 0.0025)^n$

$A_1 = \$1002.50$

$A_2 = \$1005.01$

$A_3 = \$1007.52$

$A_4 = \$1010.04$

$A_5 = \$1012.56$

$A_6 = \$1015.09$

$A_7 = \$1017.63$

$A_8 = \$1020.18$

$A_9 = \$1022.73$

$A_{10} = \$1025.28$

25. (a) $h_n = 16\left(\frac{3}{4}\right)^n$

(b) $h_1 = 12$ ft

$h_2 = 9$ ft

$h_3 = \frac{27}{4} = 6.75$ ft

$h_4 = \frac{81}{16} \approx 5.06$ ft

$h_5 = \frac{243}{64} \approx 3.80$ ft

(c) The sequence converges; $\lim_{n\to\infty} 16\left(\frac{3}{4}\right)^n = 0$.

27. $S_1 = \frac{3}{2} = 1.5$

$S_2 = \frac{3}{2} + \frac{9}{4} = \frac{15}{4} = 3.75$

$S_3 = \frac{3}{2} + \frac{9}{4} + \frac{27}{8} = \frac{57}{8} = 7.125$

$S_4 = \frac{3}{2} + \frac{9}{4} + \frac{27}{8} + \frac{81}{16} = \frac{195}{16} \approx 12.188$

$S_5 = \frac{3}{2} + \frac{9}{4} + \frac{27}{8} + \frac{81}{16} + \frac{243}{32} = \frac{633}{32} \approx 19.781$

29. $S_1 = \dfrac{1}{2} = 0.5$

$S_2 = \dfrac{1}{2} - \dfrac{1}{24} = \dfrac{11}{24} \approx 0.4583$

$S_3 = \dfrac{1}{2} - \dfrac{1}{24} + \dfrac{1}{720} = \dfrac{331}{720} \approx 0.4597$

$S_4 = \dfrac{1}{2} - \dfrac{1}{24} + \dfrac{1}{720} - \dfrac{1}{40{,}320} = \dfrac{18{,}535}{40{,}320} \approx 0.4597$

$S_5 = \dfrac{1}{2} - \dfrac{1}{24} + \dfrac{1}{720} - \dfrac{1}{40{,}320} + \dfrac{1}{3{,}628{,}800} = \dfrac{1{,}668{,}151}{3{,}628{,}800} \approx 0.4597$

31. This series diverges by the nth-Term Test because

$$\lim_{n\to\infty} \frac{n^2+1}{n(n+1)} = 1 \neq 0.$$

33. This geometric series converges because $|r| = 0.25 < 1$.

35. $\lim_{n\to\infty} \frac{2n}{n+5} = 2 \neq 0$, and the series diverges.

37. $\lim_{n\to\infty} \frac{n^2}{n^2+1} = 1 \neq 0$, and the series diverges.

39. Because $a = 2$ and $r = \frac{2}{3}$, you have

$$S_n = \frac{2\left[1-(2/3)^{n+1}\right]}{1-2/3} = \frac{2\left[1-(2/3)^{n+1}\right]}{1/3}.$$

$$S_3 = 6\left[1-(2/3)^4\right] = \frac{130}{27} \approx 4.8148$$

$$S_5 = 6\left[1-(2/3)^6\right] = \frac{1330}{243} \approx 5.4733$$

$$S_{10} = 6\left[1-(2/3)^{11}\right] = \frac{350,198}{59,049} \approx 5.9306$$

41. $\lim_{n\to\infty} \left(\frac{5}{4}\right)^n = \infty \neq 0$, and the series diverges.

43. $\lim_{n\to\infty} \frac{1}{4}(4)^n \neq 0$, and the series diverges.

45. $\sum_{n=0}^{\infty}\left[(0.5)^n + (0.2)^n\right] = \frac{1}{1-0.5} + \frac{1}{1-0.2} = 2 + \frac{5}{4} = \frac{13}{4}$

47. (a) After 1 year: 9000

$$\begin{aligned}
\text{2 years: } & 9000 + 9000(0.85) = 9000(1+0.85)\\
\text{3 years: } & 9000 + 0.85\left[9000(1+0.85)\right]\\
&= 9000\left(1+0.85+0.85^2\right)\\
&\vdots\\
n\text{ years: } & 9000\left[1+0.85+0.85^2+\cdots+0.85^{n-1}\right]\\
&= 9000\left(\frac{1-(0.85)^n}{1-0.85}\right)\\
&= 60,000\left(1-0.85^n\right)
\end{aligned}$$

So $60,000\left(1-0.85^n\right)$ units will be in use after n years.

(b) 60,000 units is the stabilization level.

49. 1st person: $500(0.75) = 375$

2nd person: $500(0.75)^2 = 281.25$

3rd person: $500(0.75)^3 = 210.94$

$\vdots$

nth person: $500(0.75)^n$

$$\sum_{n=1}^{\infty} \frac{500}{1-0.75} = \$2000$$

51. 1st person: $600(0.725) = 435$

2nd person: $600(0.725)^2 \approx 315.38$

3rd person: $600(0.725)^3 \approx 228.65$

$\vdots$

nth person: $600(0.725)^n$

$$\sum_{n=1}^{\infty} \frac{600}{1-0.725} \approx \$2181.82$$

53. (a) $V_n = 120,000(0.7)^n$

(b) $V_5 = 120,000(0.7)^5 = \$20,168.40$

55. $\sum_{n=1}^{\infty} \frac{1}{n^4}$ converges by the p-series test because $p = 4 > 1$.

57. $\sum_{n=1}^{\infty} \frac{1}{n\sqrt[4]{n}} = \sum_{n=1}^{\infty} \frac{1}{n^{5/4}}$ converges by the p-series test because $p = \frac{5}{4} > 1$.

59. This series converges by the Ratio Test because

$$\lim_{n\to\infty} \left|\frac{(n+1)4^{n+1}}{(n+1)!} \cdot \frac{n!}{n4^n}\right| = \lim_{n\to\infty} \frac{4}{n} = 0 < 1.$$

61. This series diverges by the Ratio Test because

$$\lim_{n\to\infty}\left|\frac{(-1)^{n+1}3^{n+1}}{n+1}\cdot\frac{n}{(-1)^n 3^n}\right| = \lim_{n\to\infty}\left|\frac{3n}{n+1}\right| = 3 > 1.$$

63. This series converges by the Ratio Test because

$$\lim_{n\to\infty}\left|\frac{(n+1)2^{n+1}}{(n+1)!}\cdot\frac{n!}{n2^n}\right| = \lim_{n\to\infty}\frac{2}{n} = 0 < 1.$$

65. $\sum_{n=1}^{\infty}\frac{3}{\sqrt{n}}$ matches (b). The series diverges by the p-series test because $p = \frac{1}{2} < 1$.

67. $\sum_{n=1}^{\infty}\frac{3}{n}$ matches (a). The series diverges by the p-series test because $p = 1$ (the harmonic series).

69. The series $\sum_{n=1}^{\infty}\frac{(x-2)^n}{2^n}$ is centered at $c = 2$.

$$\sum_{n=1}^{\infty}\frac{(x-2)^n}{2^n} = \frac{x-2}{2} + \frac{(x-2)^2}{4} + \frac{(x-2)^3}{8} + \frac{(x-2)^4}{16} + \frac{(x-2)^5}{32} + \cdots$$

71. The series $\sum_{n=1}^{\infty}\frac{(-1)^n x^n}{3^n}$ is centered at $c = 0$.

$$\sum_{n=1}^{\infty}\frac{(-1)^n x^n}{3^n} = -\frac{x}{3} + \frac{x^2}{9} - \frac{x^3}{27} + \frac{x^4}{81} - \frac{x^5}{243} + \cdots$$

73. $\lim_{n\to\infty}\left|\frac{(x/10)^{n+1}}{(x/10)^n}\right| = \lim_{n\to\infty}\left|\frac{x}{10}\right| \Rightarrow \left|\frac{x}{10}\right| < 1 \Rightarrow |x| < 10$

$R = 10$

75. $\lim_{n\to\infty}\left|\frac{4^{n+1}x^{n+1}}{(n+1)!}\cdot\frac{n!}{4^n x^n}\right| = \lim_{n\to\infty}\left|\frac{4x}{n+1}\right| = |4x|(0) < 1$, for all x.

$R = \infty$

77. $\lim_{n\to\infty}\left|\frac{3^{n+1}(x-2)^{n+1}/(n+1)}{3^n(x-2)^n/n}\right| = \lim_{n\to\infty}\left|\frac{3n(x-2)}{n+1}\right| \Rightarrow 3|x-2| < 1 |x-2| < \frac{1}{3}$

$R = \frac{1}{3}$

79. $\lim_{n\to\infty}\left|\frac{(x-2)^{n+1}/2^{n+1}}{(x-2)^n/2^n}\right| = \lim_{n\to\infty}\left|\frac{(x-2)}{2}\right| < 1 \Rightarrow |x-2| < 2$

$R = 2$

81.

$$\begin{aligned}
f(x) &= e^{-0.5x} & f(0) &= 1\\
f'(x) &= -\frac{1}{2}e^{-0.5x} & f'(0) &= -\frac{1}{2}\\
f''(x) &= \frac{1}{4}e^{-0.5x} & f''(0) &= \frac{1}{4}\\
f'''(x) &= -\frac{1}{8}e^{-0.5x} & f'''(0) &= -\frac{1}{8}\\
&& &\vdots\\
&& f^{(n)}(0) &= \left(-\frac{1}{2}\right)^n
\end{aligned}$$

The power series for f is $e^{-0.5x} = 1 - \frac{1}{2}x + \frac{1}{4}\cdot\frac{x^2}{2!} - \frac{1}{8}\cdot\frac{x^3}{3!} + \cdots = \sum_{n=0}^{\infty}\left(-\frac{1}{2}\right)^n\frac{x^n}{n!}$.

$$\lim_{n\to\infty}\left|\frac{(-1)^{n+1}x^{n+1}}{2^{n+1}(n+1)!}\cdot\frac{2^n n!}{(-1)^n x^n}\right| = \lim_{n\to\infty}\left|\frac{x}{2(n+1)}\right| = \left|\frac{x}{2}\right|(0) < 1, \text{ for all } x.$$

$R = \infty$

83. $f(x) = x^{-1/2}$ $\quad f(1) = 1$

$f'(x) = -\frac{1}{2}x^{-3/2}$ $\quad f'(1) = -\frac{1}{2}$

$f''(x) = \frac{3}{4}x^{-5/2}$ $\quad f''(1) = \frac{3}{4} = \frac{1 \cdot 3}{2^2}$

$f'''(x) = -\frac{15}{8}x^{-7/2}$ $\quad f'''(1) = -\frac{15}{8} = -\frac{1 \cdot 3 \cdot 5}{2^3}$

$\vdots$

$$f^{(n)}(1) = \frac{(-1)^{n+1} 1 \cdot 3 \cdot 5 \cdots (2n-1)}{2^n}, \ (n \ge 1)$$

The power series for f is $\frac{1}{\sqrt{x}} = 1 - \frac{1}{2}(x-1) + \frac{3}{2^2 2!}(x-1)^2 - \frac{3 \cdot 5}{2^3 3!}(x-1)^3 + \cdots = 1 + \sum_{n=1}^{\infty} \frac{(-1)^n 1 \cdot 3 \cdot 5 \cdots (2n-1)(x-1)^n}{2^n n!}$.

$$\lim_{n\to\infty}\left|\frac{(-1)^{n+1} 1 \cdot 3 \cdot 5 \cdots (2n+1)(x-1)^{n+1}}{2^{n+1}(n+1)!} \cdot \frac{2^n \cdot n!}{(-1)^n 1 \cdot 3 \cdot 5 \cdots (2n-1)(x-1)^n}\right| = \lim_{n\to\infty}\left|\frac{(2n+1)(x-1)}{2(n+1)}\right| = |x-1| < 1 \Rightarrow R = 1$$

85. $f(x) = (1+x)^{1/4}$ $\quad f(0) = 1$

$f'(x) = \frac{1}{4}(1+x)^{-3/4}$ $\quad f'(0) = \frac{1}{4}$

$f''(x) = -\frac{3}{16}(1+x)^{-7/4}$ $\quad f''(0) = -\frac{3}{16} = -\frac{3}{4^2}$

$f'''(x) = \frac{21}{64}(1+x)^{-11/4}$ $\quad f'''(0) = \frac{21}{64} = \frac{3 \cdot 7}{4^3}$

$\vdots$

$$f^{(n)}(0) = \frac{(-1)^n 3 \cdot 7 \cdot 11 \cdots (4n-5)}{4^n}, \ (n \ge 2)$$

The power series for f is

$$\sqrt[4]{1+x} = 1 + \frac{1}{4}x - \frac{3}{4^2 2!}x^2 + \frac{3 \cdot 7}{4^3 3!}x^3 - \frac{3 \cdot 7 \cdot 11}{4^4 4!}x^4 + \cdots = 1 + \frac{x}{4} - \sum_{n=2}^{\infty} \frac{(-1)^n 3 \cdot 7 \cdot 11 \cdots (4n-5)x^n}{4^n n!}.$$

$$\lim_{n\to\infty}\left|\frac{(-1)^{n+1} 3 \cdot 7 \cdot 11 \cdots (4n-1)x^{n+1}}{4^{n+1}(n+1)!} \cdot \frac{4^n \cdot n!}{(-1)^n 3 \cdot 7 \cdot 11 \cdots (4n-5)x^n}\right| = \lim_{n\to\infty}\left|\frac{(4n-1)x}{4(n+1)}\right| = |x| < 1 \Rightarrow R = 1$$

87. $\ln(x+2) = \ln\left[2\left(\frac{x}{2}+1\right)\right] = \ln 2 + \ln\left(\frac{x}{2}+1\right) = \ln 2 + \frac{1}{2}x - \frac{1}{8}x^2 + \frac{1}{24}x^3 - \frac{1}{64}x^4 + \cdots = \ln 2 + \sum_{n=1}^{\infty}(-1)^{n+1}\frac{x^n}{2^n n}$

89. $(1+x^2)^2 = 1 + 2x^2 + \frac{2(1)x^4}{2!} = 1 + 2x^2 + x^4$

91. $f(x) = x^2 e^x$

$= x^2 \sum_{n=0}^{\infty} \frac{x^n}{n!}$

$= x^2\left[1 + x + \frac{x^2}{2!} + \cdots\right]$

$= \sum_{n=0}^{\infty} \frac{x^{n+2}}{n!}$

93. $f(x) = \frac{x^2}{x+1}$

$= x^2 \sum_{n=0}^{\infty} (-1)^n x^n$

$= x^2\left[1 - x + x^2 - x^3 + \cdots\right]$

$= \sum_{n=0}^{\infty} (-1)^n x^{n+2}$

95. $f(x) = e^{-4x}$ $\quad f(0) = 1$ $\quad a_0 = \dfrac{1}{0!} = 1$

$f'(x) = -4e^{-4x}$ $\quad f'(0) = -4$ $\quad a_1 = \dfrac{-4}{1!} = -4$

$f''(x) = 16e^{-4x}$ $\quad f''(0) = 16$ $\quad a_2 = \dfrac{16}{2!} = 8$

$f'''(x) = -64e^{-4x}$ $\quad f'''(0) = -64$ $\quad a_3 = \dfrac{-64}{3!} = -\dfrac{32}{3}$

$f^{(4)}(x) = 256e^{-4x}$ $\quad f^{(4)}(0) = 256$ $\quad a_4 = \dfrac{256}{4!} = \dfrac{32}{3}$

(a) $S_1(x) = 1 - 4x$

(b) $S_2(x) = 1 - 4x + 8x^2$

(c) $S_3(x) = 1 - 4x + 8x^2 - \dfrac{32}{3}x^3$

(d) $S_4(x) = 1 - 4x + 8x^2 - \dfrac{32}{3}x^3 + \dfrac{32}{3}x^4$

97. $f(x) = \ln(x + 2)$ $\quad f(0) = \ln 2$ $\quad a_0 = \dfrac{\ln 2}{0!} = \ln 2$

$f'(x) = \dfrac{1}{x + 2}$ $\quad f'(0) = \dfrac{1}{2}$ $\quad a_1 = \dfrac{\frac{1}{2}}{1!} = \dfrac{1}{2}$

$f''(x) = -\dfrac{1}{(x + 2)^2}$ $\quad f''(0) = -\dfrac{1}{4}$ $\quad a_2 = \dfrac{-\frac{1}{4}}{2!} = -\dfrac{1}{8}$

$f'''(x) = \dfrac{2}{(x + 2)^3}$ $\quad f'''(0) = \dfrac{1}{4}$ $\quad a_3 = \dfrac{\frac{1}{4}}{3!} = \dfrac{1}{24}$

$f^{(4)}(x) = -\dfrac{6}{(x + 2)^4}$ $\quad f^{(4)}(0) = -\dfrac{3}{8}$ $\quad a_4 = \dfrac{-\frac{3}{8}}{4!} = -\dfrac{1}{64}$

(a) $S_1(x) = \ln 2 + \dfrac{x}{2}$

(b) $S_2(x) = \ln 2 + \dfrac{x}{2} - \dfrac{x^2}{8}$

(c) $S_3(x) = \ln 2 + \dfrac{x}{2} - \dfrac{x^2}{8} + \dfrac{x^3}{24}$

(d) $S_4(x) = \ln 2 + \dfrac{x}{2} - \dfrac{x^2}{8} + \dfrac{x^3}{24} - \dfrac{x^4}{64}$

99. $f(x) = \dfrac{1}{(x + 3)^2}$ $\quad f(0) = \dfrac{1}{9}$ $\quad a_0 = \dfrac{\frac{1}{9}}{0!} = \dfrac{1}{9}$

$f'(x) = -\dfrac{2}{(x + 3)^3}$ $\quad f'(0) = -\dfrac{2}{27}$ $\quad a_1 = \dfrac{-\frac{2}{27}}{1!} = -\dfrac{2}{27}$

$f''(x) = \dfrac{6}{(x + 3)^4}$ $\quad f''(0) = \dfrac{2}{27}$ $\quad a_2 = \dfrac{\frac{2}{27}}{2!} = \dfrac{1}{27}$

$f'''(x) = -\dfrac{24}{(x + 3)^5}$ $\quad f'''(0) = -\dfrac{8}{81}$ $\quad a_3 = \dfrac{-\frac{8}{81}}{3!} = -\dfrac{4}{243}$

$f^{(4)}(x) = \dfrac{120}{(x + 3)^6}$ $\quad f^{(4)}(0) = \dfrac{40}{243}$ $\quad a_4 = \dfrac{\frac{40}{243}}{4!} = \dfrac{5}{729}$

(a) $S_1(x) = \dfrac{1}{9} - \dfrac{2}{27}x$

(b) $S_2(x) = \dfrac{1}{9} - \dfrac{2}{27}x + \dfrac{1}{27}x^2$

(c) $S_3(x) = \dfrac{1}{9} - \dfrac{2}{27}x + \dfrac{1}{27}x^2 - \dfrac{4}{243}x^3$

(d) $S_4(x) = \dfrac{1}{9} - \dfrac{2}{27}x + \dfrac{1}{27}x^2 - \dfrac{4}{243}x^3 + \dfrac{5}{729}x^4$

101. $f(x) = e^{x/3},\ c = 0$

$$f(x) \approx 1 + \frac{x}{3} + \frac{x^2}{18} + \frac{x^3}{162} + \frac{x^4}{1944} + \frac{x^5}{29{,}160} + \frac{x^6}{524{,}880}$$

$$f(1.25) \approx 1 + \frac{1.25}{3} + \frac{(1.25)^2}{18} + \frac{(1.25)^3}{162} + \frac{(1.25)^4}{1944} + \frac{(1.25)^5}{29{,}160} + \frac{(1.25)^6}{524{,}880} \approx 1.5169$$

$$\text{Error} = R_n = \frac{f^{(n+1)}(z)}{(n+1)!}(x-c)^{n+1}$$

$$f(x) = e^{x/3} \Rightarrow f^{(7)}(x) = \frac{e^{x/3}}{2187}$$

$$R_6 = \frac{f^{(7)}(1.25)(1.25)^7}{7!} \approx 0.00000066$$

103. $f(x) = \ln(1 + x),\ c = 1$

$$f(x) \approx \ln 2 + \frac{x-1}{2} - \frac{(x-1)^2}{8} + \frac{(x-1)^3}{24} - \frac{(x-1)^4}{64} + \frac{(x-1)^5}{160} - \frac{(x-1)^6}{384}$$

$$f(1.5) \approx \ln 2 + \frac{1.5-1}{2} - \frac{(1.5-1)^2}{8} + \frac{(1.5-1)^3}{24} - \frac{(1.5-1)^4}{64} + \frac{(1.5-1)^5}{160} - \frac{(1.5-1)^6}{384} \approx 0.9163$$

$$\text{Error} = R_n = \frac{f^{(n+1)}(z)}{(n+1)!}(x-c)^{n+1}$$

$$f(x) = \ln(1+x) \Rightarrow f^{(7)}(x) = \frac{720}{(x+1)^7}$$

$$R_6 = \frac{f^{(7)}(1)(0.5)^7}{7!} \approx 0.0000087$$

105. $\sqrt{1 + x^2} = 1 + \dfrac{x^3}{2} - \dfrac{x^6}{8} + \cdots$

$$\int_0^{0.3} \sqrt{1 + x^3}\,dx \approx \left[x + \frac{x^4}{8} - \frac{x^7}{56}\right]_0^{0.3} \approx 0.301$$

107. $\ln(x^2 + 1) = x^2 - \dfrac{x^4}{2} + \dfrac{x^6}{3} - \cdots$

$$\int_0^{0.75} \ln(x^2+1)\,dx \approx \int_0^{0.75}\left(x^2 - \frac{x^4}{2} + \frac{x^6}{3}\right)dx = \left[\frac{x^3}{3} - \frac{x^5}{10} + \frac{x^7}{21}\right]_0^{0.75} \approx 0.12325$$

109. $f(x) = x^3 + 3x - 1$

$f'(x) = 3x^2 + 3$

n	x_n	$f(x_n)$	$f'(x_n)$	$\dfrac{f(x_n)}{f'(x_n)}$	$x_n - \dfrac{f(x_n)}{f'(x_n)}$
1	0.25	–0.2344	3.1875	–0.0735	0.3235
2	0.3235	0.0044	3.3140	0.0013	0.3222
3	0.3222	0.00005	3.3114	0.00001	0.3222

Approximation: $x \approx 0.322$

111. $f(x) = \sqrt{x + 4} + x$

$f'(x) = \dfrac{1}{2\sqrt{x + 4}} + 1$

n	x_n	$f(x_n)$	$f'(x_n)$	$\dfrac{f(x_n)}{f'(x_n)}$	$x_n - \dfrac{f(x_n)}{f'(x_n)}$
1	−1.75	−0.2500	1.3333	−0.1875	−1.5625
2	−1.5625	−0.0013	1.3203	−0.0009	−1.5616
3	−1.5616	−0.00006	1.3202	−0.00005	−1.5616

Approximation: $x \approx -1.562$

113. $f(x) = x^4 + x - 3$

$f'(x) = 4x^3 + 1$

n	x_n	$f(x_n)$	$f'(x_n)$	$\dfrac{f(x_n)}{f'(x_n)}$	$x_n - \dfrac{f(x_n)}{f'(x_n)}$
1	−1.5	0.5625	−12.5	−0.045	−1.4550
2	−1.4550	0.0268	−11.3211	−0.0024	−1.4526
3	−1.4526	−0.0003	−11.2602	0.00003	−1.4526

Approximation: $x \approx -1.453$

n	x_n	$f(x_n)$	$f'(x_n)$	$\dfrac{f(x_n)}{f'(x_n)}$	$x_n - \dfrac{f(x_n)}{f'(x_n)}$
1	1.1	−0.4359	6.324	−0.0689	1.1689
2	1.1689	0.0358	7.3884	0.0048	1.1641
3	1.1641	0.0005	7.3100	0.00006	1.1640

Approximation: $x \approx 1.164$

115. Let $x^5 = (x + 3)$ and define $h(x) = x^5 - x - 3$.

Then $h'(x) = 5x^4 - 1$.

$$x_{n+1} = x_n - \frac{f(x_n)}{f'(x_n)} = x_n - \frac{x_n^{\,5} - x_n - 3}{5x_n^{\,4} - 1}$$

n	x_n	$h(x_n)$	$h'(x_n)$	$\dfrac{h(x_n)}{h'(x_n)}$	$x_n - \dfrac{h(x_n)}{h'(x_n)}$
1	1.25	−1.1982	11.2070	−0.1069	1.3569
2	1.3569	0.2429	15.9497	0.0152	1.3417
3	1.3417	0.0062	15.2029	0.0004	1.3413

Approximation: $x \approx 1.341$

117. Let $x^3 = e^{-x}$ and define $h(x) = x^3 - e^{-x}$. Then $h'(x) = 3x^2 + e^{-x}$.

n	x_n	$h(x_n)$	$h'(x_n)$	$\dfrac{h(x_n)}{h'(x_n)}$	$x_n - \dfrac{h(x_n)}{h'(x_n)}$
1	0.9	0.3224	2.8366	0.1137	0.7863
2	0.7863	0.0306	2.3103	0.0133	0.7730
3	0.7730	0.0003	2.2542	0.0001	0.7729

Approximation: $x \approx 0.773$

119. From the graph there is one real zero near $x = -3$.

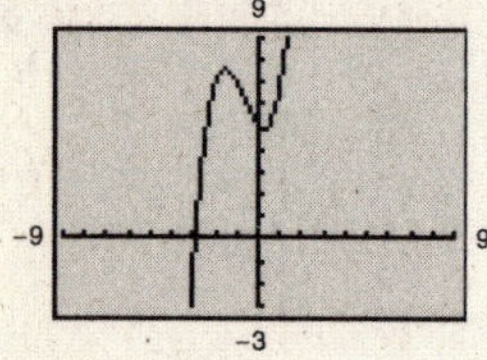

n	1	2	3	4
x_n	−3.0	−2.9286	−2.9259	−2.9259

Zero: $x \approx -2.926$

$$f(x) = x^3 + 2x^2 - x + 5$$

$$f'(x) = 3x^2 + 4x - 1$$

$$x_{n+1} = x_n - \frac{f(x_n)}{f'(x_n)} = \frac{x_n - x_n^3 + 2x_n^2 - x_n + 5}{3x_n^2 + 4x_n - 1}$$

Chapter 10 Test Yourself

1. $a_n = 4_n - 2$

$a_1 = 4(1) - 2 = 2$

$a_2 = 4(2) - 2 = 6$

$a_3 = 4(3) - 2 = 10$

$a_4 = 4(4) - 2 = 14$

$a_5 = 4(5) - 2 = 18$

2. $a_n = 2^n$

$a_1 = 2^1 = 2$

$a_2 = 2^2 = 4$

$a_3 = 2^3 = 8$

$a_4 = 2^4 = 16$

$a_5 = 2^5 = 32$

3. $a_n = \dfrac{n + 3}{n}$

$a_1 = \dfrac{1 + 3}{1} = 4$

$a_2 = \dfrac{2 + 3}{2} = \dfrac{5}{2}$

$a_3 = \dfrac{3 + 3}{3} = 2$

$a_4 = \dfrac{4 + 3}{4} = \dfrac{7}{4}$

$a_5 = \dfrac{5 + 3}{5} = \dfrac{8}{5}$

4. $a_n = \dfrac{(-1)^n}{n!}$

$a_1 = \dfrac{(-1)^1}{1!} = -1$

$a_2 = \dfrac{(-1)^2}{2!} = \dfrac{1}{2}$

$a_3 = \dfrac{(-1)^3}{3!} = -\dfrac{1}{6}$

$a_4 = \dfrac{(-1)^4}{4!} = \dfrac{1}{24}$

$a_5 = \dfrac{(-1)^5}{5!} = -\dfrac{1}{120}$

5. The sequence converges to 0, because $\lim_{n\to\infty} \left(\frac{3}{5}\right)^n = 0.$

6. The sequence converges to $\frac{1}{3}$ because

$$\lim_{n\to\infty} \frac{n^2}{3n^2 + 4} = \frac{1}{3}.$$

7. The sequence diverges because the

$\lim_{n\to\infty} \frac{(-1)^{n+1}}{6}$ oscillates between $\frac{1}{6}$ and $\frac{-1}{6}$.

8. The sequence converges to 0, because

$$\lim_{n\to\infty} \frac{4n}{(n-1)!} = 0.$$

9. $a_n = \frac{(-1)^{n+1}n}{n^2+1}$

10. This series converges by the Ratio Test because

$a_n = \frac{4^n}{n!}$ and

$$\lim_{n\to\infty}\left|\frac{a_n + 1}{a_n}\right| = \lim_{n\to\infty}\left|\frac{4^{n+1}}{(n+1)!}\cdot\frac{n!}{4^n}\right| = \lim_{n\to\infty}\left|\frac{4}{n}\right| = 0 < 1.$$

11. This series diverges by the nth-Term Test because

$$\lim_{n\to\infty} \frac{n+1}{n-3} = 1 \neq 0.$$

12. This series converges by the Test for Convergence of a Geometric Series because $\sum_{n=0}^{\infty} \frac{2}{5^n} = \sum_{n=0}^{\infty} 2\left(\frac{1}{5}\right)^n$ and

$r = \frac{1}{5} < 1.$

13. This series diverges by the p-series test because

$\sum_{n=1}^{\infty} \frac{\sqrt[3]{n}}{\sqrt{n}} = \sum_{n=1}^{\infty} \frac{n^{1/3}}{n^{1/2}} = \sum_{n=1}^{\infty} \frac{1}{n^{1/6}}$ and $p = \frac{1}{6} < 1.$

14. This series diverges by the Geometric Series Test because $\sum_{n=0}^{\infty} 2\left(\frac{5}{3}\right)^n$ and $|r| = \frac{5}{3} > 1.$

15. This series converges by the p-series test because

$\sum_{n=1}^{\infty} \frac{4}{\sqrt[4]{n^5}} = \sum_{n=1}^{\infty} \frac{4}{n^{5/4}}$ and $p = \frac{5}{4} > 1.$

16. This series converges by the Geometric Series Test because $\sum_{n=0}^{\infty} 5\left(-\frac{1}{6}\right)^n$ and $|r| = \frac{1}{6} < 1.$

17. This series diverges by the Ratio Test because $a_n = \frac{(n+1)!}{5^n}$ and

$$\lim_{n\to\infty}\left|\frac{a_{n+1}}{a_n}\right| = \lim_{n\to\infty}\left|\frac{(n+2)!}{5^{n+1}}\cdot\frac{5^n}{(n+1)!}\right| = \lim_{n\to\infty}\left|\frac{n+2}{5}\right| = \infty > 1.$$

18. $$\sum_{n=0}^{\infty}\left(\frac{2}{5^n} - \frac{1}{7^{n+1}}\right) = \sum_{n=0}^{\infty} 2\left(\frac{1}{5}\right)^n - \sum_{n=0}^{\infty} \frac{1}{7}\left(\frac{1}{7}\right)^n = \frac{2}{1-\frac{1}{5}} - \frac{\frac{1}{7}}{1-\frac{1}{7}} = \frac{5}{2} - \frac{1}{6} = \frac{7}{3}$$

19. $\sum_{n=0}^{\infty} (-1)^{n+1}\left(\frac{x}{3}\right)^n$

(a) Centered at $c = 0$

(b) $S_0 = -1$

$S_1 = -1 + \frac{x}{3}$

$S_2 = -1 + \frac{x}{3} - \frac{x^2}{9}$

$S_3 = -1 + \frac{x}{3} - \frac{x^2}{9} + \frac{x^3}{27}$

$S_4 = -1 + \frac{x}{3} - \frac{x^2}{9} + \frac{x^3}{27} - \frac{x^4}{81}$

(c) $\lim_{n\to\infty}\left|\frac{a_{n+1}x^{n+1}}{a_n x^n}\right| = \lim_{n\to\infty}\left|\frac{(-1)^{n+2}(x/3)^{n+1}}{(-1)^{n+1}(x/3)^n}\right| = \lim_{n\to\infty}\left|\frac{x}{3}\right| < 1 \Rightarrow |x| < 3$

Radius $= 3$

20. $\displaystyle\sum_{n=0}^{\infty} \frac{x^n}{(n+1)!}$

(a) Centered at $c = 0$

(b) $S_0 = 1$

$$S_1 = 1 + \frac{x}{2}$$

$$S_2 = 1 + \frac{x}{2} + \frac{x^2}{6}$$

$$S_3 = 1 + \frac{x}{2} + \frac{x^2}{6} + \frac{x^3}{24}$$

$$S_4 = 1 + \frac{x}{2} + \frac{x^2}{6} + \frac{x^3}{24} + \frac{x^4}{120}$$

(c) $\displaystyle\lim_{n\to\infty}\left|\frac{a_{n+1}x^{n+1}}{a_n x^n}\right| = \lim_{n\to\infty}\left|\frac{x^{n+1}/(n+2)!}{x^n/(n+1)!}\right| = \lim_{n\to\infty}\left|\frac{x}{n+2}\right| = 0 \Rightarrow -\infty < x < \infty$

Radius $= \infty$

21. $\displaystyle\sum_{n=0}^{\infty} \frac{(-1)^n (x-3)^n}{(n+4)^2}$

(a) Centered at $c = 3$

(b) $S_0 = \dfrac{1}{16}$

$$S_1 = \frac{1}{16} - \frac{(x-3)}{25}$$

$$S_2 = \frac{1}{16} - \frac{(x-3)}{25} + \frac{(x-3)^2}{36}$$

$$S_3 = \frac{1}{16} - \frac{(x-3)}{25} + \frac{(x-3)^2}{36} - \frac{(x-4)^3}{49}$$

$$S_4 = \frac{1}{16} - \frac{(x-3)}{25} + \frac{(x-3)^2}{36} - \frac{(x-4)^3}{49} + \frac{(x-4)^4}{64}$$

(c) $\displaystyle\lim_{n\to\infty}\left|\frac{a_{n+1}x^{n+1}}{a_n x^n}\right| = \lim_{n\to\infty}\left|\frac{(-1)^{n+1}(x-3)^{n+1}/(n+5)^2}{(-1)^n(x-3)^n/(n+4)^2}\right| = \lim_{n\to\infty}\left|\frac{(n+4)^2(x-3)}{(n+5)^2}\right| = \lim_{n\to\infty}|x-3| < 1 \Rightarrow |x-3| < 1$

Radius $= 1$

22.

$$\begin{aligned}
f(x) &= e^{3x+1} & f(0) &= e & a_0 &= e/0! \\
f'(x) &= 3e^{3x+1} & f'(0) &= 3e & a_1 &= 3e/1! \\
f''(x) &= 9e^{3x+1} & f''(0) &= 9e & a_2 &= 9e/2! \\
f'''(x) &= 27e^{3x+1} & f'''(0) &= 27e & a_3 &= 27e/3! \\
&\vdots & &\vdots & &\vdots \\
f^{(n)}(x) &= 3^n e^{3x+1} & f^{(n)}(0) &= 3^n e & a_n &= \frac{3^n e}{n!}
\end{aligned}$$

The power series for f is

$$e^{3x+1} = \sum_{n=0}^{\infty} \frac{3^n e x^n}{n!} = e\sum_{n=0}^{\infty} \frac{3^n x^n}{n!}.$$

23. $\displaystyle(1+x)^{2/3} = 1 + \frac{2}{3}x - \frac{2x^2}{9\cdot 2!} + \frac{8x^3}{27\cdot 3!} - \frac{56x^4}{81\cdot 4!} + \cdots = 1 + \frac{2x}{3} - \frac{x^2}{9} + \frac{4x^3}{81} - \frac{7x^4}{243} + \cdots$

24. $f(x) = e^{x/2}$ $\quad f(0) = 1 \quad a_0 = \frac{1}{0!} = 1$

$f'(x) = \frac{1}{2}e^{x/2} \quad f'(0) = \frac{1}{2} \quad a_1 = \frac{\frac{1}{2}}{1!} = \frac{1}{2}$

$f''(x) = \frac{1}{4}e^{x/2} \quad f''(0) = \frac{1}{4} \quad a_2 = \frac{\frac{1}{4}}{2!} = \frac{1}{8}$

$f'''(x) = \frac{1}{8}e^{x/2} \quad f'''(0) = \frac{1}{8} \quad a_3 = \frac{\frac{1}{8}}{3!} = \frac{1}{48}$

$f^{(4)}(x) = \frac{1}{16}e^{x/2} \quad f^{(4)}(0) = \frac{1}{16} \quad a_4 = \frac{\frac{1}{16}}{4!} = \frac{1}{384}$

(a) $S_1(x) = 1 + \frac{x}{2}$

(b) $S_2(x) = 1 + \frac{x}{2} + \frac{x^2}{8}$

(c) $S_3(x) = 1 + \frac{x}{2} + \frac{x^2}{8} + \frac{x^3}{48}$

(d) $S_4(x) = 1 + \frac{x}{2} + \frac{x^2}{8} + \frac{x^3}{48} + \frac{x^4}{384}$

25. $f(x) = \frac{2}{x+2} \quad f(0) = 1 \quad a_0 = \frac{1}{0!} = 1$

$f'(x) = -\frac{2}{(x+2)^2} \quad f'(0) = -\frac{1}{2} \quad a_1 = \frac{-\frac{1}{2}}{1!} = -\frac{1}{2}$

$f''(x) = \frac{4}{(x+2)^3} \quad f''(0) = \frac{1}{2} \quad a_2 = \frac{\frac{1}{2}}{2!} = \frac{1}{4}$

$f'''(x) = -\frac{12}{(x+2)^4} \quad f'''(0) = -\frac{3}{4} \quad a_3 = \frac{-\frac{3}{4}}{3!} = -\frac{1}{8}$

$f^{(4)}(x) = \frac{48}{(x+2)^5} \quad f^{(4)}(0) = \frac{3}{2} \quad a_4 = \frac{\frac{3}{2}}{4!} = \frac{1}{16}$

(a) $S_1(x) = 1 - \frac{x}{2}$

(b) $S_2(x) = 1 - \frac{x}{2} + \frac{x^2}{4}$

(c) $S_3(x) = 1 - \frac{x}{2} + \frac{x^2}{4} - \frac{x^3}{8}$

(d) $S_4(x) = 1 - \frac{x}{2} + \frac{x^2}{4} - \frac{x^3}{8} + \frac{x^4}{16}$

26. $f(x) = \frac{1}{(x+4)^2} \quad f(0) = \frac{1}{16} \quad a_0 = \frac{\frac{1}{16}}{0!} = \frac{1}{16}$

$f'(x) = -\frac{2}{(x+4)^3} \quad f'(0) = -\frac{1}{32} \quad a_1 = \frac{-\frac{1}{32}}{1!} = -\frac{1}{32}$

$f''(x) = \frac{6}{(x+4)^4} \quad f''(0) = \frac{3}{128} \quad a_2 = \frac{\frac{3}{128}}{2!} = \frac{3}{256}$

$f'''(x) = -\frac{24}{(x+4)^5} \quad f'''(0) = -\frac{3}{128} \quad a_3 = \frac{-\frac{3}{128}}{3!} = -\frac{1}{256}$

$f^{(4)}(x) = \frac{120}{(x+4)^6} \quad f^{(4)}(0) = \frac{15}{512} \quad a_4 = \frac{\frac{15}{512}}{4!} = \frac{5}{4096}$

(a) $S_1(x) = \frac{1}{16} - \frac{x}{32}$

(b) $S_2(x) = \frac{1}{16} - \frac{x}{32} + \frac{3x^2}{256}$

(c) $S_3(x) = \frac{1}{16} - \frac{x}{32} + \frac{3x^2}{256} - \frac{x^3}{256}$

(d) $S_4(x) = \frac{1}{16} - \frac{x}{32} + \frac{3x^2}{256} - \frac{x^3}{256} + \frac{5x^4}{4096}$

27. (a) $a_n = 2.7n + 283$, $n = 1$ corresponds to 2001

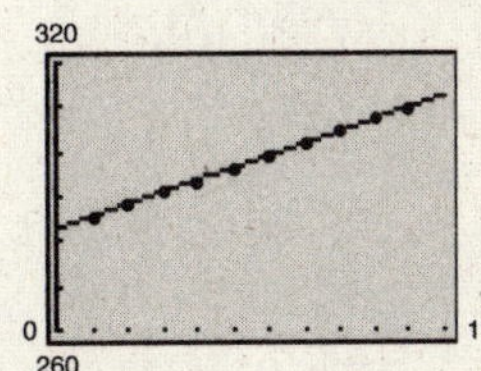

(b) 2020: $a_{20} = 2.7(20) + 283 = 337$ million

28. From the graph, there is one real zero near $x = 1$.

$f(x) = x^3 + x - 3$

$f'(x) = 3x^2 + 1$

$x_{n+1} = x_n - \frac{x_n^3 + x_n - 3}{3x_n^2 + 1} = \frac{2x_n^3 + 3}{3x_n^2 + 1}$

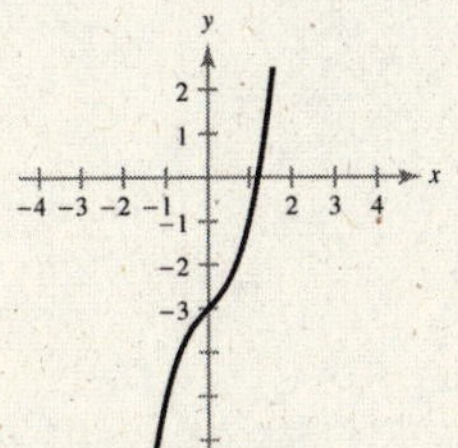

Zero: $x \approx 1.213$

Graphing utility: $x \approx 1.213$

n	1	2	3	4	5
x_n	1	1.25	1.2143	1.2134	1.2134

CHAPTER 11
Differential Equations

CHAPTER 11
Differential Equations

Section 11.1 Solutions of Differential Equations

Skills Warm Up

1. $y = 3x^2 + 2x + 1$
$y' = 6x + 2$
$y'' = 6$

2. $y = -2x^3 - 8x + 4$
$y' = -6x^2 - 8$
$y'' = -12x$

3. $y = -3e^{2x}$
$y' = -6e^{2x}$
$y'' = -12e^{2x}$

4. $y = -3e^{x^2}$
$y' = -6xe^{x^2}$
$y'' = -6x\left(2xe^{x^2}\right) + e^{x^2}(-6) = -12x^2e^{x^2} - 6e^{x^2}$

5. $0.5 = 9 - 9e^{-k}$
$-8.5 = -9e^{-k}$
$\frac{17}{18} = e^{-k}$
$\ln \frac{17}{18} = -k$
$-\ln \frac{17}{18} = k$
$0.0572 \approx k$

6. $14.75 = 25 - 25e^{-2k}$
$-10.25 = -25e^{-2k}$
$0.41 = e^{-2k}$
$\ln 0.41 = -2k$
$-\frac{1}{2} \ln 0.41 = k$
$0.4458 \approx k$

1. By differentiation, you have

$$\frac{dy}{dx} = 4Ce^{4x} = 4y.$$

3. $y = 2x^3$

$y' = 6x^2$ and $y' - \frac{3}{x}y = 6x^2 - \frac{3}{x}\left(2x^3\right) = 0$

5. Because $y' = 2Cx - 3$, you have $xy' - 3x - 2y = x(2Cx - 3) - 3x - 2\left(Cx^2 - 3x\right) = 0$.

7. Because $y' = \ln x + 1 + C$, you have $x(y' - 1) - (y - 4) = x(\ln x + 1 + C - 1) - (x \ln x + Cx + 4 - 4) = 0$.

9. $y = x^2$
$y' = 2x$
$y'' = 2$ and $x^2y'' - 2y = x^2(2) - 2\left(x^2\right) = 0$

11. Because $y' = C_1 \cos x + C_2 \sin x$ and $y'' = -C_1 \sin x + C_2 \cos x$, you have
$y'' + y = -C_1 \sin x + C_2 \cos x + C_1 \sin x - C_2 \cos x = 0$.

13.
$y = e^{-2x}$
$y' = -2e^{-2x}$
$y'' = 4e^{-2x}$
$y''' = -8e^{-2x}$
$y^{(4)} = 16e^{-2x}$
$y^{(4)} - 16y = 16e^{-2x} - 16\left(e^{-2x}\right) = 0$

So, it is a solution of the differential equation.

15.
$y = 4x^{-1}$
$y' = -4x^{-2}$
$y'' = 8x^{-3}$
$y''' = -24x^{-4}$
$y^{(4)} = 96x^{-5}$
$y^{(4)} - 16y = 96x^{-5} - 16\left(4x^{-1}\right) \neq 0$

So, it is *not* a solution of the differential equation.

17. $y = \frac{2}{9}xe^{-2x}$

$y' = -\frac{4}{9}xe^{-2x} + \frac{2}{9}e^{-2x}$

$y'' = \frac{8}{9}xe^{-2x} - \frac{8}{9}e^{-2x}$

$y''' = -\frac{16}{9}xe^{-2x} + \frac{24}{9}e^{-2x}$

$y''' - 3y' + 2y = \left(-\frac{16}{9}xe^{-2x} + \frac{24}{9}e^{-2x}\right) - 3\left(-\frac{4}{9}xe^{-2x} + \frac{2}{9}e^{-2x}\right) + 2\left(\frac{2}{9}xe^{-2x}\right) = 2e^{-2x}$

So, it is *not* a solution of the differential equation.

19. $y = xe^x$

$y' = xe^x + e^x$

$y'' = xe^x + 2e^x$

$y''' = xe^x + 3e^x$

$y''' - 3y' + 2y = \left(xe^x + 3e^x\right) - 3\left(xe^x + e^x\right) + 2\left(xe^x\right)$

$= 0$

So, it is a solution of the differential equation.

21. Because $y' = -2Ce^{-2x} = -2y$, it follows that $y' + 2y = 0$. To find the particular solution, use the fact that $y = 3$ when $x = 0$. That is, $3 = Ce^0 = C$. So, $C = 3$ and the particular solution is $y = 3e^{-2x}$.

23. Because $y' = C_2(1/x)$ and $y'' = -C_2(1/x^2)$, it follows that $xy'' + y' = 0$. To find the particular solution, use the fact that $y = 5$ and $y' = 1/2$ when $x = 1$. That is,

$\frac{1}{2} = C_2\frac{1}{1} \quad \Rightarrow C_2 = \frac{1}{2}$

$5 = C_1 + \frac{1}{2}(0) \Rightarrow C_1 = 5.$

So, the particular solution is $y = 5 + \frac{1}{2}\ln|x|$.

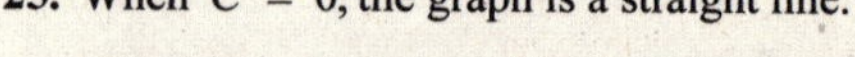

25. When $C = 0$, the graph is a straight line.

When $C = 1$, the graph is a parabola opening upward with a vertex at $(-2, 0)$.

When $C = -1$, the graph is a parabola opening downward with a vertex at $(-2, 0)$.

When $C = 2$, the graph is a parabola opening upward with a vertex at $(-2, 0)$.

When $C = -2$, the graph is a parabola opening downward with a vertex at $(-2, 0)$.

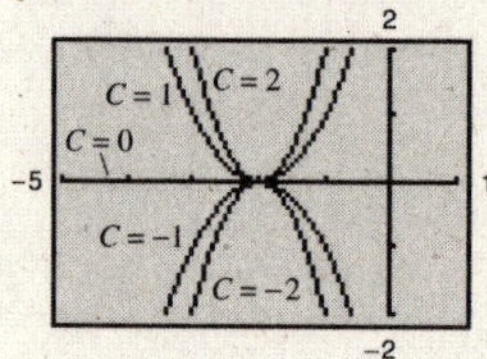

27. $y = \int 3x^2\,dx = x^3 + C$

29. $y = \int \frac{1}{1 + x}\,dx = \ln|1 + x| + C$

31. $y = \int x\sqrt{x^2 + 6}\,dx = \frac{1}{2}\int 2x\left(x^2 + 6\right)^{1/2}\,dx = \frac{1}{3}\left(x^2 + 6\right)^{3/2} + C$

33. $y = \int \cos 4x\,dx = \frac{1}{4}\sin 4x + C$

35. Because $y = 4$ when $x = 4$, you have $4^2 = C4^3$, which implies that $C = \frac{1}{4}$, and the particular solution is $y^2 = \frac{1}{4}x^3$.

37. Because $y = 3$ when $x = 0$, you have $3 = Ce^0$, which implies that $C = 3$, and the particular solution is $y = 3e^x$.

39. (a) Because $N = 100$ when $t = 0$, it follows that $C = 650$. So, the population function is $N = 750 - 650e^{-kt}$. Also, because $N = 160$ when $t = 2$, it follows that

$160 = 750 - 650e^{-2k}$

$e^{-2k} = \frac{59}{65}$

$k = -\frac{1}{2}\ln\frac{59}{65} \approx -0.0484.$

So, the population function is $N = 750 - 650e^{-0.0484t}$.

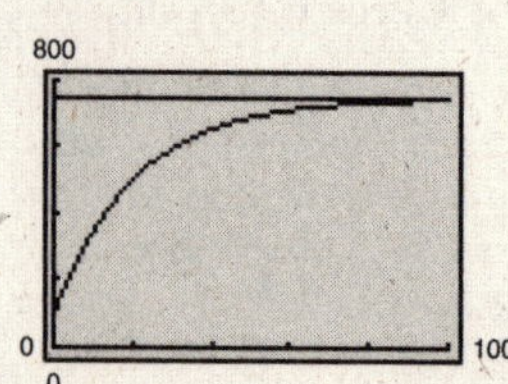

(b) See accompanying graph.

(c) When $t = 4$, $N \approx 214$.

41. If $s = 25 - \frac{13}{\ln 3}\ln\frac{h}{2} = 25 - \frac{13}{\ln 3}(\ln h - \ln 2)$, then differentiate with respect to h to obtain

$$\frac{ds}{dh} = -\frac{13}{\ln 3}\left(\frac{1}{h}\right)$$

$$\frac{ds}{dh} = \frac{(-13/\ln 3)}{h}, \text{ which implies } k = -\frac{13}{\ln 3}.$$

43.

$$y = a + Ce^{k(1-b)t}$$

$$\frac{dy}{dt} = Ck(1-b)e^{k(1-b)t}$$

$$a + b(y-a) + \frac{1}{k}\frac{dy}{dt} = a + b\left[\left(a + Ce^{k(1-b)t}\right) - a\right] + \frac{1}{k}\left[Ck(1-b)e^{k(1-b)t}\right]$$

$$= a + bCe^{k(1-b)t} + C(1-b)e^{k(1-b)t} = a + Ce^{k(1-b)t}\left[b + (1-b)\right] = a + Ce^{k(1-b)t} = y$$

45. True

Section 11.2 Separation of Variables

Skills Warm Up

1. $\int x^{3/2}\,dx = \frac{2}{5}x^{5/2} + C$

$$\frac{d}{dx}\left(\frac{2}{5}x^{5/2} + C\right) = x^{3/2}$$

2. $\int\left(t^3 - t^{1/3}\right)dt = \frac{1}{4}t^4 - \frac{3}{4}t^{4/3} + C$

$$\frac{d}{dt}\left(\frac{1}{4}t^4 - \frac{3}{4}t^{4/3} + C\right) = t^3 - t^{1/3}$$

3. $\int\frac{2}{x-5}\,dx = 2\ln|x-5| + C$

$$\frac{d}{dx}\left(2\ln|x-5| + C\right) = \frac{2}{x-5}$$

4. $\int\frac{y}{2y^2+1}\,dy = \frac{1}{4}\int\frac{4y}{2y^2+1}\,dy$

$$= \frac{1}{4}\ln\left|2y^2+1\right| + C$$

$$= \frac{1}{4}\ln\left(2y^2+1\right) + C$$

$$\frac{d}{dy}\left(\frac{1}{4}\ln\left(2y^2+1\right) + C\right) = \frac{1}{4}\frac{4y}{2y^2+1} = \frac{y}{2y^2+1}$$

5. $\int e^{2y}\,dy = \frac{1}{2}\int 2e^{2y}\,dy = \frac{1}{2}e^{2y} + C$

$$\frac{d}{dy}\left(\frac{1}{2}e^{2y} + C\right) = \frac{1}{2}(2)e^{2y} = e^{2y}$$

6. $\int xe^{1-x^2}\,dx = -\frac{1}{2}\int -2xe^{1-x^2}\,dx = -\frac{1}{2}e^{1-x^2} + C$

$$\frac{d}{dx}\left(-\frac{1}{2}e^{1-x^2} + C\right) = -\frac{1}{2}(-2x)e^{1-x^2} = xe^{1-x^2}$$

7. $(3)^2 - 6(3) = 1 + C$

$$-9 = 1 + C$$

$$-10 = C$$

8. $(-1)^2 + (-2)^2 = C$

$$1 + 4 = C$$

$$5 = C$$

9.

$$10 = 2e^{2k}$$

$$5 = e^{2k}$$

$$\ln 5 = 2k$$

$$\tfrac{1}{2}\ln 5 = k$$

$$0.8047 \approx k$$

10. $(6)^2 - 3(6) = e^{-k}$

$$18 = e^{-k}$$

$$\ln 18 = -k$$

$$-\ln 18 = k$$

$$-2.8904 \approx k$$

1. Yes, $\dfrac{dy}{dx} = \dfrac{x}{y+3}$

$$(y+3)\,dy = x\,dx$$

3. Yes, $\dfrac{dy}{dx} = \dfrac{1}{x} + 1$

$$dy = \left(\frac{1}{x} + 1\right)dx$$

5. No. the variables cannot be separated.

7. $\dfrac{dy}{dx} = 2x$

$$\int dy = \int 2x\,dx$$

$$y = x^2 + C$$

9. $\dfrac{dr}{ds} = 0.05r$

$$\int \frac{dr}{r} = \int 0.05\,ds$$

$$\ln|r| = 0.05s + C_1$$

$$r = e^{0.05s + C_1} = e^{C_1}e^{0.05s} = Ce^{0.05s}$$

11. $\dfrac{dy}{dx} = \dfrac{x-1}{y^3}$

$$\int y^3\,dy = \int (x-1)\,dx$$

$$\frac{1}{4}y^4 = \frac{1}{2}x^2 - x + C_1$$

$$y^4 = 2x^2 - 4x + C$$

13. $3y^2\dfrac{dy}{dx} = 1$

$$\int 3y^2\,dy = \int dx$$

$$y^3 = x + C$$

$$y = \sqrt[3]{x + C}$$

15. $x^2 + 4y\dfrac{dy}{dx} = 0$

$$4y\frac{dy}{dx} = -x^2$$

$$\int 4y\,dy = \int \left(-x^2\right)dx$$

$$2y^2 = -\frac{1}{3}x^3 + C_1$$

$$y^2 = -\frac{1}{6}x^3 + C$$

17. $y' - xy = 0$

$$\frac{dy}{dx} = xy$$

$$\int \frac{1}{y}\,dy = \int x\,dx$$

$$\ln|y| = \frac{1}{2}x^2 + C_1$$

$$y = e^{(x^2/2)+C_1} = e^{C_1}e^{x^2/2} = Ce^{x^2/2}$$

19. $e^y\dfrac{dy}{dt} = 3t^2 + 1$

$$\int e^y\,dy = \int \left(3t^2 + 1\right)dt$$

$$e^y = t^3 + t + C$$

$$y = \ln\left|t^3 + t + C\right|$$

21. $\dfrac{dy}{dx} = \sqrt{1-y}$

$$\int (1-y)^{-1/2}\,dy = \int dx$$

$$-2(1-y)^{1/2} = x + C_1$$

$$\sqrt{1-y} = \frac{-x}{2} + C$$

$$1 - y = \left(C - \frac{x}{2}\right)^2$$

$$y = 1 - \left(C - \frac{x}{2}\right)^2$$

23. $(2+x)y' = 2y$

$$(2+x)\frac{dy}{dx} = 2y$$

$$\int \frac{1}{2y}\,dy = \int \frac{1}{2+x}\,dx$$

$$\frac{1}{2}\ln|y| = \ln\left|C_1(2+x)\right|$$

$$\sqrt{y} = C_1(2+x)$$

$$y = C(2+x)^2$$

25. $y\dfrac{dy}{dx} = \sin x$

$$\int y\,dy = \int \sin x\,dx$$

$$\tfrac{1}{2}y^2 = -\cos x + C_1$$

$$y^2 = -2\cos x + C$$

27. (a) $\dfrac{dy}{dx} = x$

$$\int dy = \int x\,dx$$

$$y = \frac{1}{2}x^2 + C$$

(b)

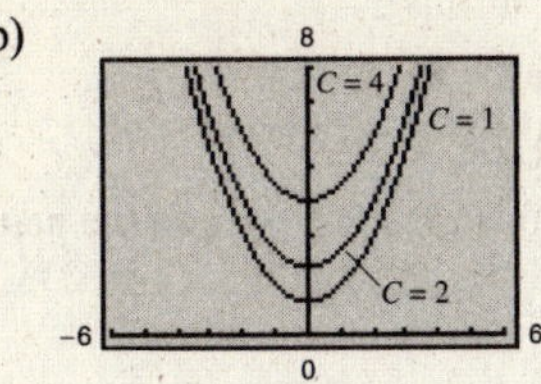

29. (a) $\dfrac{dy}{dx} = y + 3$

$$\int\left(\frac{1}{y+3}\right)dy = \int dx$$

$$\ln|y + 3| = x + C_1$$

$$y + 3 = e^{x+C_1}$$

$$y = e^{x+C_1} - 3$$

$$y = e^{C_1}e^x - 3$$

$$y = Ce^x - 3$$

(b)

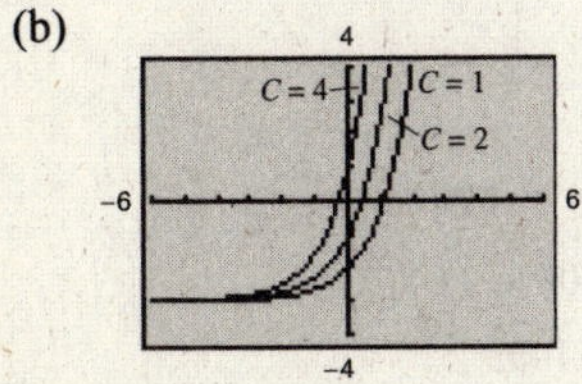

31. $yy' - e^x = 0$

$$y\frac{dy}{dx} = e^x$$

$$\int y\,dy = \int e^x\,dx$$

$$\frac{y^2}{2} = e^x + C$$

When $x = 0$, $y = 4$. So, $C = 7$ and the particular solution is $y^2 = 2e^x + 14$.

33. $x(y + 4) + y' = 0$

$$\frac{dy}{dx} = -x(y + 4)$$

$$\int\frac{1}{y+4}\,dy = \int -x\,dx$$

$$\ln|y + 4| = -\frac{x^2}{2} + C_1$$

$$|y + 4| = e^{-x^2/2+C_1}$$

$$y = -4 + Ce^{-x^2/2}$$

When $x = 0$, $y = -5$. So, $C = -1$ and the particular solution is $y = -4 - e^{-x^2/2}$.

35. $\sqrt{x^2 - 16}\,y' = 5x$

$$\frac{dy}{dx} = \frac{5x}{\sqrt{x^2 - 16}}$$

$$\int dy = \int\frac{5x}{\sqrt{x^2 - 16}}\,dx$$

$$y = 5\left(x^2 - 16\right)^{1/2} + C$$

When $x = 5$, $y = -2$.

So, $-2 = 5(9)^{1/2} + C \Rightarrow C = -17$ and the particular solution is $y = 5\sqrt{x^2 - 16} - 17$.

37. $\dfrac{dy}{dx} = y\cos x$

$$\int\frac{1}{y}\,dy = \int\cos x\,dx$$

$$\ln|y| = \sin x + C_1$$

$$y = e^{\sin x + C_1}$$

$$y = Ce^{\sin x}$$

When $x = 0$, $y = 1$. So, $C = 1$ and the particular solution is $y = e^{\sin x}$.

39. $\dfrac{dy}{dx} = -\dfrac{9x}{16y}$

$$\int 16y\,dy = \int -9x\,dx$$

$$8y^2 = -\frac{9}{2}x^2 + C_1$$

$$y^2 = -\frac{9}{16}x^2 + C$$

When $x = -1$, $y = 1$.

So, $(1)^2 = -\dfrac{9}{16}(-1)^2 + C \Rightarrow C = \dfrac{25}{16}$ and the particular equation is $y^2 = -\dfrac{9}{16}x^2 + \dfrac{25}{16}$

or $\dfrac{9}{16}x^2 + y^2 = \dfrac{25}{16} \Rightarrow 9x^2 + 16y^2 = 25.$

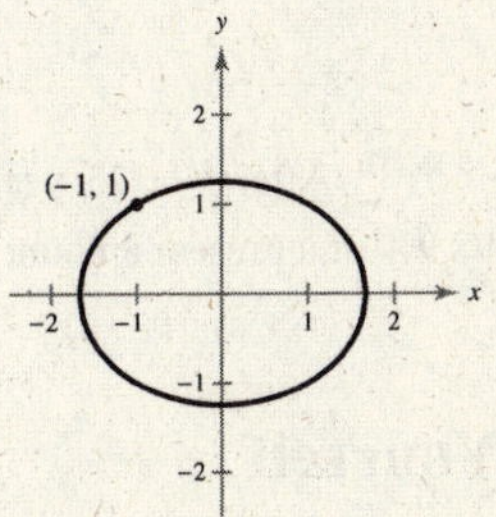

41. $$\frac{dv}{dt} = 3.456 - 0.1v$$

$$\int \frac{dv}{3.456 - 0.1v} = \int dt$$

$$-10 \ln|3.456 - 0.1v| = t + C_1$$

$$(3.456 - 0.1v)^{-10} = C_2e^t$$

$$3.456 - 0.1v = Ce^{-0.1t}$$

$$v = -10Ce^{-0.1t} + 34.56$$

When $t = 0$, $v = 0$. So, $C = 3.456$ and the solution is $v = 34.56(1 - e^{-0.1t})$.

43. $$\frac{dV}{dt} = kV^{2/3}$$

$$\int V^{-2/3}\,dV = \int k\,dt$$

$$3V^{1/3} = kt + C_1$$

$$V^{1/3} = \frac{1}{3}kt + C$$

$$V = \left(\frac{1}{3}kt + C\right)^3$$

45. Let y represent the mass of the radium and let t represent the time (in years).

$$\frac{dy}{dt} = ky$$

$$\int \frac{1}{y}\,dy = \int k\,dt$$

$$\ln y = kt + C_1$$

$$y = e^{kt + C_1}$$

$$y = Ce^{kt}$$

The half-life is 1599 years.

$$\tfrac{1}{2}C = Ce^{k(1599)}$$

$$\tfrac{1}{2} = e^{1599k}$$

$$\ln \tfrac{1}{2} = 1599k$$

$$\frac{1}{1599} \ln \frac{1}{2} = k$$

$$-0.000433 \approx k$$

The equation is $y = Ce^{-0.000433t}$. After $t = 25$ years, $y \approx 0.989C$. So, about 98.9% of a present amount will remain.

47. $$\frac{dw}{dt} = k(1200 - w)$$

$$\int \frac{1}{1200 - w}\,dw = \int k\,dt$$

$$-\ln|1200 - w| = kt + C_1$$

$$\ln|1200 - w| = -kt + C_2$$

$$1200 - w = e^{-kt + C_2}$$

$$w = 1200 - Ce^{-kt}$$

When $t = 0$, $w = 60$. So, $C = 1140$ and the particular solution is $w = 1200 - 1140e^{-kt}$.

(a) 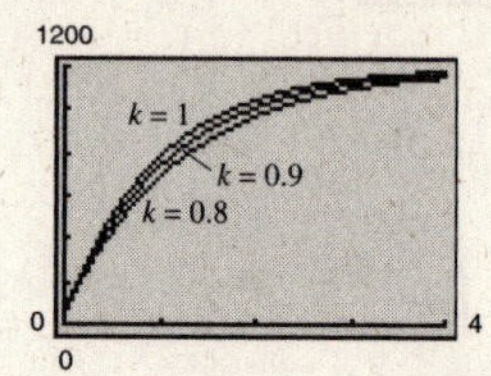

(b) For $k = 0.8$ and $w = 800$:

$$800 = 1200 - 1140e^{-0.8t}$$

$$-400 = 1140e^{-0.8t}$$

$$\tfrac{20}{57} = e^{-0.8t}$$

$$\ln \tfrac{20}{57} = -0.8t$$

$$-1.25 \ln \tfrac{20}{57} = t$$

$$1.31 \text{ years} \approx t$$

For $k = 1$ and $w = 800$:

$$800 = 1200 - 1140e^{-t}$$

$$-400 = -1140e^{-t}$$

$$\tfrac{20}{57} = e^{-t}$$

$$\ln \tfrac{20}{57} = -t$$

$$-\ln \tfrac{20}{57} = t$$

$$1.05 \text{ years} \approx t$$

For $k = 0.9$ and $w = 800$:

$$800 = 1200 - 1140e^{-0.9t}$$

$$-400 = -1140e^{-0.9t}$$

$$\tfrac{20}{57} = e^{-0.9t}$$

$$\ln \tfrac{20}{57} = -0.9t$$

$$-\tfrac{10}{9}\ln \tfrac{20}{57} = t$$

$$1.16 \text{ years} \approx t$$

(c) The weight of the calf approaches 1200 pounds as t increases for each of the models.

Chapter 11 Quiz Yourself

1. Because $y' = -\left(\frac{1}{2}\right)Ce^{-x/2}$, you have $2y' + y = 2\left(-\frac{1}{2}Ce^{-x/2}\right) + Ce^{-x/2} = 0$.

2. Because $y' = -C_1 \sin x + C_2 \cos x$ and $y'' = -C_1 \cos x - C_2 \sin x$, you have $y'' + y = -C_1 \cos x - C_2 \sin x + C_1 \cos x + C_2 \sin x = 0$.

3. Because $y' = -x^{-2} = -\dfrac{1}{x^2}$ and $y'' = 2x^{-3} = \dfrac{2}{x^3}$,

you have $xy'' + 2y' = x\left(\dfrac{2}{x^3}\right) + 2\left(-\dfrac{1}{x^2}\right) = \dfrac{2}{x^2} - \dfrac{2}{x^2} = 0$.

4. Because $y' = \dfrac{3}{5}x^2 - 1 + \dfrac{C}{2\sqrt{x}}$,

you have $2x\left(\dfrac{3}{5}x^2 - 1 + \dfrac{C}{2\sqrt{x}}\right) - \left(\dfrac{x^3}{5} - x + C\sqrt{x}\right) = \dfrac{6x^3}{5} - 2x + C\sqrt{x} - \dfrac{x^3}{5} + x - C\sqrt{x} = x^3 - x$.

5. Because $y' = 3C_1 \cos 3x - 3C_2 \sin 3x$ and $y'' = -9C_1 \sin 3x - 9C_2 \cos 3x$, it follows that $y'' + 9y = -9C_1 \sin 3x - 9C_2 \cos 3x + 9(C_1 \sin 3x + C_2 \cos 3x) = 0$. To find the particular solution, use the fact that $y = 2$ and $y' = 1$ when $x = \dfrac{\pi}{6}$. That is,

$$C_1 \sin \frac{3\pi}{6} + C_2 \cos \frac{3\pi}{6} = 2 \quad \Rightarrow C_1 = 2$$

$$3C_1 \cos \frac{3\pi}{6} - 3C_2 \sin \frac{3\pi}{6} = 1 \Rightarrow C_2 = -\frac{1}{3}.$$

So, the particular solution is $y = 2 \sin 3x - \frac{1}{3} \cos 3x$.

6. Because $y' = C_1 + 3C_2x^2$ and $y'' = 6C_2x$, it follows that $x^2y'' - 3xy' + 3y = 6C_2x^3 - 3x\left(C_1 + 3C_2x^2\right) + 3\left(C_1x + C_2x^3\right) = 0$. To find the particular solution, use the fact that $y = 0$ and $y' = 4$ when $x = 2$.

That is, $2C_1 + 8C_2 = 0$

$C_1 + 12C_2 = 4$.

Solving this system, you obtain $C_1 = -2$ and $C_2 = \frac{1}{2}$. So, the particular solution is $y = -2x + \frac{1}{2}x^3$.

7.
$$\frac{dy}{dx} = -4x + 4$$
$$\int dy = \int(-4x + 4)\,dx$$
$$y = -2x^2 + 4x + C$$

8.
$$y' = (x + 2)(y - 1)$$
$$\frac{1}{y-1}\frac{dy}{dx} = x + 2$$
$$\int \frac{dy}{y-1} = \int (x + 2)\,dx$$
$$\ln|y - 1| = \frac{(x + 2)^2}{2} + C_1$$
$$y - 1 = e^{(x+2)^2/2 + C_1}$$
$$y = 1 + Ce^{(x+2)^2/2}$$

9.
$$y\frac{dy}{dx} = \frac{1}{2x + 1}$$
$$\int y\,dy = \int\left(\frac{1}{2x + 1}\right)dx$$
$$\frac{1}{2}y^2 = \frac{1}{2}\ln|2x + 1| + C_1$$
$$y^2 = \ln|2x + 1| + C$$

10.
$$\frac{dy}{dx} = \frac{x}{3y^2 + 1}$$
$$\int\left(3y^2 + 1\right)dy = \int x\,dx$$
$$y^3 + y = \frac{1}{2}x^2 + C$$

11. (a)
$$\frac{dy}{dx} = \frac{x^2 + 1}{2y}$$
$$\int 2y\,dy = \int\left(x^2 + 1\right)dx$$
$$y^2 = \frac{1}{3}x^3 + x + C$$
$$y = \pm\sqrt{\frac{1}{3}x^3 + x + C}$$

(b)

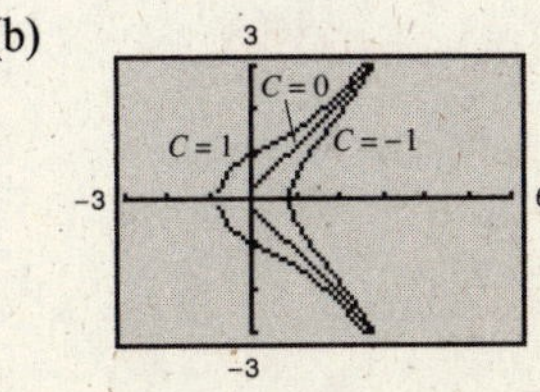

12. (a)
$$\frac{dy}{dx} = \frac{y}{x-3}$$
$$\int \frac{1}{y}\,dy = \int \frac{1}{x-3}\,dx$$
$$\ln|y| = \ln|x-3| + \ln|C_1|$$
$$\ln|y| = \ln|C_1(x-3)|$$
$$y = C(x-3)$$

(b)

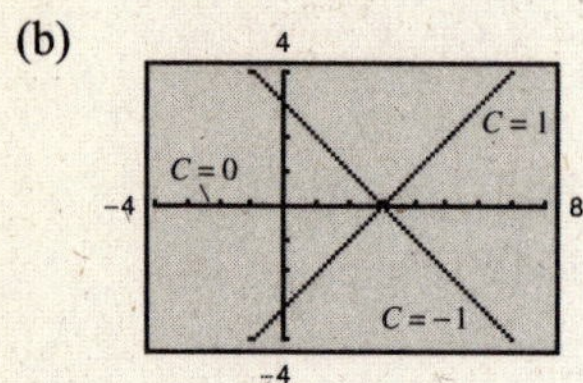

13.
$$y' + 2y - 1 = 0$$
$$\frac{dy}{dx} = 1 - 2y$$
$$\int \frac{1}{1-2y}\,dy = \int dx$$
$$-\frac{1}{2}\ln|1-2y| = x + C_1$$
$$\ln|1-2y| = -2x + C_2$$
$$1 - 2y = e^{-2x+C_2}$$
$$-2y = e^{-2x+C_2} - 1$$
$$y = -\tfrac{1}{2}\left(Ce^{-2x} - 1\right)$$

When $x = 0$, $y = 1$. So, $C = -1$ and the particular solution is $y = \frac{1}{2}\left(e^{-2x} + 1\right)$.

14.
$$\frac{dy}{dx} = y\sin \pi x$$
$$\int \frac{1}{y}\,dy = \int \sin \pi x\,dx$$
$$\ln y = -\frac{1}{\pi}\cos \pi x + C_1$$
$$y = e^{-1/\pi \cos \pi x + C_1}$$
$$y = Ce^{-(\cos \pi x)/\pi}$$

When $x = \frac{1}{2}$, $y = -3$. So, $C = -3$ and the particular solution is $y = -3e^{-(\cos \pi x)/\pi}$.

15.
$$y' = 3x^2 y$$
$$\frac{dy}{dx} = 3x^2 y$$
$$\int \frac{1}{y}\,dy = \int 3x^2\,dx$$
$$\ln|y| = x^3 + C_1$$
$$y = e^{x^3 + C_1}$$
$$y = e^{C_1}e^{x^3}$$
$$y = Ce^{x^3}$$

When $x = 0$, $y = 2$. So, $2 = Ce^{(0)^3} \Rightarrow C = 2$ and the particular equation is $y = 2e^{x^3}$.

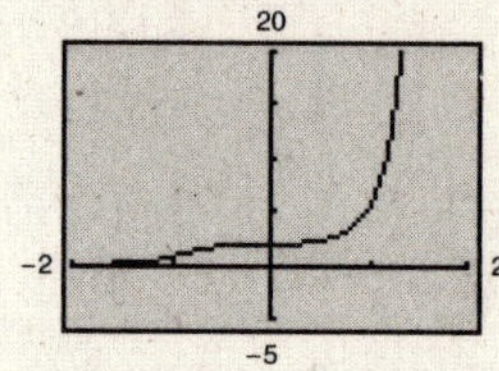

16.
$$\frac{dv}{dt} = k(20 - v)$$
$$\int \frac{1}{20 - v}\,dv = \int k\,dt$$
$$-\ln|20 - v| = kt + C_1$$
$$\ln|20 - v| = -kt + C_2$$
$$20 - v = e^{-kt + C_2}$$
$$20 - v = Ce^{-kt}$$
$$v = 20 - Ce^{-kt}$$

Because the boat is starting at rest, you have $t = 0$ and $v = 0$. So, $C = 20$.

When $t = 0.5$ and $v = 10$, you have
$$10 = 20 - 20e^{-0.5k}$$
$$e^{-0.5k} = \tfrac{1}{2}$$
$$-0.5k = \ln \tfrac{1}{2}$$
$$k \approx 1.386.$$

So, $v = 20 - 20e^{-1.386t}$.

Section 11.3 First-Order Linear Differential Equations

Skills Warm Up

1. $e^{-x}(e^{2x} + e^x) = e^{-x}e^{2x} + e^{-x}e^x = e^x + 1$

2. $\frac{1}{e^{-x}}(e^{-x} + e^{2x}) = \frac{e^{-x}}{e^{-x}} + \frac{e^{2x}}{e^{-x}} = 1 + e^{3x}$

3. $e^{-\ln x^3} = \frac{1}{e^{\ln x^3}} = \frac{1}{x^3}$

4. $e^{2\ln x + x} = e^{2\ln x}e^x = e^{\ln x^2}e^x = x^2e^x$

5. $\int 4e^{2x}\,dx = 4\left(\frac{1}{2}\right)\int e^{2x}(2)\,dx = 2e^{2x} + C$

6. $\int xe^{3x^2}\,dx = \frac{1}{6}\int e^{3x^2}(6x)\,dx = \frac{1}{6}e^{3x^2} + C$

7. $\int \frac{1}{2x+5}\,dx = \frac{1}{2}\int \frac{2}{2x+5}\,dx = \frac{1}{2}\ln|2x+5| + C$

8. $\int \frac{x+1}{x^2+2x+3}\,dx = \frac{1}{2}\int \frac{2x+2}{x^2+2x+3}\,dx$
$= \frac{1}{2}\ln|x^2+2x+3| + C$

9. $\int (4x-3)^2\,dx = \frac{1}{4}\int 4(4x-3)^2\,dx$
$= \frac{1}{4}\left(\frac{1}{3}\right)(4x-3)^3 + C$
$= \frac{1}{12}(4x-3)^3 + C$

10. $\int x(1-x^2)^2\,dx = -\frac{1}{2}\int (1-x^2)^2(-2x)\,dx$
$= -\frac{1}{6}(1-x^2)^3 + C = \frac{1}{6}(x^2-1)^3 + C$

1. $x^3 - 2x^2y' + 3y = 0$
$-2x^2y' + 3y = -x^3$
$y' - \frac{3}{2x^2}y = \frac{x}{2}$

3. $xy' + y = xe^x$
$y' + \frac{1}{x}y = e^x$

5. $y + 1 = (x-1)y'$
$(1-x)y' + y = -1$
$y' + \frac{1}{1-x}y = \frac{1}{x-1}$

7. For this linear differential equation, you have $P(x) = 3$ and $Q(x) = 6$. So, the integrating factor is $u(x) = e^{\int 3\,dx} = e^{3x}$ and the general solution is $y = \frac{1}{u(x)}\int Q(x)u(x)\,dx = e^{-3x}\int 6e^{3x}\,dx = e^{-3x}(2e^{3x} + C) = 2 + Ce^{-3x}$.

9. For this linear differential equation, you have $P(x) = -1$ and $Q(x) = e^{4x}$. So, the integrating factor is $u(x) = e^{\int(-1)\,dx} = e^{-x}$ and the general solution is $y = \frac{1}{u(x)}\int Q(x)u(x)\,dx = e^x\int e^{4x}e^{-x}\,dx = e^x\int e^{3x}\,dx = e^x\left(\frac{1}{3}e^{3x} + C\right) = \frac{1}{3}e^{4x} + Ce^x$.

11. $\frac{dy}{dx} = \frac{x^2+3}{x}$
$\int dy = \int\left(x + \frac{3}{x}\right)dx$
$y = \frac{1}{2}x^2 + 3\ln|x| + C$

13. For this linear differential equation, you have $P(x) = 2x$ and $Q(x) = 10x$. So, the integrating factor is $u(x) = e^{\int 2x\,dx} = e^{x^2}$ and the general solution is $y = \frac{1}{u(x)}\int Q(x)\,u(x)\,dx = e^{-x^2}\int 10xe^{x^2}\,dx = e^{-x^2}(5e^{x^2} + C) = 5 + Ce^{-x^2}$.

15. For this linear differential equation, $y' + y\left(\frac{1}{x-1}\right) = x + 1$, you have $P(x) = \frac{1}{x-1}$ and $Q(x) = x + 1$.

So, the integrating factor is $u(x) = e^{\int 1/(x-1)\,dx} = e^{\ln(x-1)} = x - 1$ and the general solution is

$$y = \frac{1}{u(x)}\int Q(x)u(x)\,dx = \frac{1}{x-1}\int (x+1)(x-1)\,dx = \frac{1}{x-1}\left(\frac{x^3}{3} - x + C_1\right) = \frac{x^3 - 3x + C}{3(x-1)}.$$

17. For this linear differential equation, $y' + \frac{2}{x^3}y = \frac{1}{x^3}e^{1/x^2}$, you have $P(x) = \frac{2}{x^3}$ and $Q(x) = \frac{1}{x^3}e^{1/x^2}$.

So, the integrating factor is $u(x) = e^{\int(2/x^3)\,dx} = e^{-1/x^2}$ and the general solution is

$$y = \frac{1}{u(x)}\int Q(x)u(x)\,dx = e^{1/x^2}\int \frac{1}{x^3}e^{1/x^2}e^{-1/x^2}\,dx = e^{1/x^2}\int \frac{1}{x^3}\,dx = e^{1/x^2}\left(-\frac{1}{2x^2} + C\right).$$

19. Separation of Variables:

$$\frac{dy}{dx} = 4 - y$$
$$\int \frac{dy}{4-y} = \int dx$$
$$-\ln|4 - y| = x + C_1$$
$$4 - y = e^{-(x+C_1)}$$
$$4 - y = C_2e^{-x}$$
$$y = 4 - C_2e^{x} = 4 + Ce^{-x}$$

First-Order Linear:

$P(x) = 1,\ Q(x) = 4$

$u(x) = e^{\int 1\,dx} = e^x$

$$y = \frac{1}{e^x}\int 4e^x\,dx = \frac{1}{e^x}\left[4e^x + C\right] = 4 + Ce^{-x}$$

21. Separation of Variables:

$$\frac{dy}{dx} = 2x(1 + y)$$
$$\int \frac{dy}{1+y} = \int 2x\,dx$$
$$\ln|1 + y| = x^2 + C_1$$
$$1 + y = e^{x^2 + C_1}$$
$$y = Ce^{x^2} - 1$$

First-Order Linear:

$P(x) = -2x,\ Q(x) = 2x$

$u(x) = e^{\int -2x\,dx} = e^{-x^2}$

$$y = \frac{1}{e^{-x^2}}\int 2xe^{-x^2}\,dx$$
$$= e^{x^2}\left[-e^{-x^2} + C\right] = Ce^{x^2} - 1$$

23. $y' - 2x = 0$ matches (c) because $y = x^2 + C \Rightarrow y' = 2x \Rightarrow y' - 2x = 0$.

25. $y' - 2xy = 0$ matches (a) because $y = Ce^{x^2} \Rightarrow y' = Ce^{x^2}(2x) = 2xy \Rightarrow y' - 2xy = 0$.

27. Because $P(x) = 1$ and $Q(x) = 6e^x$, the integrating factor is $u(x) = e^{\int dx} = e^x$ and the general solution is

$y = \frac{1}{e^x}\int 6e^x(e^x)\,dx = \frac{1}{e^x}\left[3e^{2x} + C\right] = 3e^x + Ce^{-x}$. Because $y = 3$ when $x = 0$, it follows that $C = 0$ and the particular solution is $y = 3e^x$.

29. Because $P(x) = \frac{1}{x}$ and $Q(x) = 0$, the integrating factor is $u(x) = e^{\int 1/x\,dx} = e^{\ln x} = x$ and the general solution is

$y = \frac{1}{x}\int 0\,dx = \frac{C}{x}$. Because $y = 2$ when $x = 2$, it follows that $C = 4$ and the particular solution is $y = \frac{4}{x}$ or $xy = 4$.

31. Because $P(x) = 3x^2$ and $Q(x) = 3x^2$, the integrating factor is $u(x) = e^{\int 3x^2\,dx} = e^{x^3}$ and the general solution is

$y = e^{-x^3}\int 3x^2e^{x^3}\,dx = e^{-x^3}\left(e^{x^3} + C\right) = 1 + Ce^{-x^3}$. Because $y = 6$ when $x = 0$, it follows that $C = 5$ and the particular solution is $y = 1 + 5e^{-x^3}$.

33. $xy' + 2y = 3x^2 - 5x$

$$y' + \frac{2}{x}y = 3x - 5$$

Because $P(x) = \frac{2}{x}$ and $Q(x) = 3x - 5$, the integrating factor is $u(x) = e^{\int 2/x\,dx} = e^{2\ln|x|} = e^{\ln x^2} = x^2$.

The general solution is $y = \frac{1}{x^2}\int(3x - 5)x^2\,dx = \frac{1}{x^2}\int(3x^3 - 5x^2)\,dx = \frac{1}{x^2}\left(\frac{3}{4}x^4 - \frac{5}{3}x^3 + C\right) = \frac{3}{4}x^2 - \frac{5}{3}x + \frac{C}{x^2}$

Because $y = 3$ when $x = -1$, you have $3 = \frac{3}{4}(-1)^2 - \frac{5}{3}(-1) + \frac{C}{(-1)^2}$, and it follows that $C = \frac{7}{12}$. The particular solution is $y = \frac{3}{4}x^2 - \frac{5}{3}x + \frac{7}{12x^2}$.

35. $\frac{dS}{dt} = 0.2(100 - S) + 0.2t$

$$S' + 0.2S = 20 + 0.2t$$

Because $P(t) = 0.2$ and $Q(t) = 20 + 0.2t$, the integrating factor is $u(t) = e^{\int 0.2\,dt} = e^{t/5}$ and the general solution is

$S = e^{-t/5}\int e^{t/5}\left(20 + \frac{t}{5}\right)dt$. Using integration by parts, the integral is

$S = e^{-t/5}\left(100e^{t/5} + te^{t/5} - 5e^{t/5} + C\right) = 100 + t - 5 + Ce^{-t/5} = 95 + t + Ce^{-t/5}$. Because $S = 0$ when $t = 0$, it follows that $C = -95$ and the particular solution is $S = t + 95\left(1 - e^{-t/5}\right)$. During the first 10 years, the sales are as follows.

t	0	1	2	3	4	5	6	7	8	9	10
S	0	18.22	33.32	45.86	56.31	65.05	72.39	78.57	83.82	88.30	92.14

37. $\frac{dv}{dt} = kv - 9.8$

$$v' - kv = -9.8$$

Because $P(t) = -k$ and $Q(t) = -9.8$, the integrating factor is $u(t) = e^{\int -k\,dt} = e^{-kt}$ and the general solution is

$v = e^{kt}\int -9.8e^{-kt}\,dt = e^{kt}\left(\frac{9.8}{k}e^{-kt} + C_1\right) = \frac{9.8}{k} + C_1e^{kt}$ or $v = \frac{1}{k}\left(9.8 + Ce^{kt}\right)$.

39. (a) $\frac{dN}{dt} = k(40 - N)$

$$N' = 40k - kN$$

$$N' + kN = 40k$$

Because $P(t) = k$ and $Q(t) = 40k$, the integrating factor is $u(t) = e^{\int k\,dt} = e^{kt}$ and the general solution is

$N = e^{-kt}\int 40ke^{kt}\,dt = e^{-kt}\left(40e^{kt} + C\right) = 40 + Ce^{-kt}$.

(b) For $t = 1$, $N = 10$:

$10 = 40 + Ce^{-kt} \Rightarrow -30 = Ce^{-k}$

For $t = 20$, $N = 19$:

$19 = 40 + Ce^{-20k} \Rightarrow -21 = Ce^{-20k}$

$$\frac{30}{21} = \frac{e^{-k}}{e^{-20k}}$$

$$\frac{10}{7} = e^{19k}$$

$$\ln\left(\frac{10}{7}\right) = 19k \Rightarrow k = \frac{1}{19}\ln\left(\frac{10}{7}\right) \approx 0.0188$$

$$-30 = Ce^{-k} \Rightarrow C = -30e^{k} \approx -30.57$$

$$N = 40 - 30.57e^{-0.0188t}$$

41. Answers will vary.

Section 11.4 Applications of Differential Equations

Skills Warm Up

1. $\dfrac{dy}{dx} = 3x$

$\int dy = \int 3x\,dx$

$y = \dfrac{3}{2}x^2 + C$

2. $2y\dfrac{dy}{dx} = 3$

$\int y\,dy = \int \dfrac{3}{2}\,dx$

$\dfrac{1}{2}y^2 = \dfrac{3}{2}x + C_1$

$y^2 = 3x + C$

3. $\dfrac{dy}{dx} = 2xy$

$\int \dfrac{1}{y}\,dy = \int 2x\,dx$

$\ln|y| = x^2 + C_1$

$y = e^{x^2+C_1}$

$y = Ce^{x^2}$

4. $\dfrac{dy}{dx} = \dfrac{x-4}{4y^3}$

$\int 4y^3\,dy = \int (x-4)\,dx$

$y^4 = \dfrac{1}{2}(x-4)^2 + C$

5. For this linear differential equation, you have $P(x) = 2$ and $Q(x) = 4$. So, the integrating factor is $u(x) = e^{\int 2\,dx} = e^{2x}$ and the general solution is

$$y = \frac{1}{u(x)}\int Q(x)u(x)\,dx = e^{-2x}\int 4e^{2x}\,dx$$
$$= e^{-2x}\left(2e^{2x} + C\right) = 2 + Ce^{-2x}.$$

6. For this linear differential equation, you have $P(x) = 2$ and $Q(x) = e^{-2x}$. So, the integrating factor is $u(x) = e^{\int 2\,dx} = e^{2x}$ and the general solution is

$$y = \frac{1}{u(x)}\int Q(x)u(x)\,dx = e^{-2x}\int e^{-2x}e^{2x}\,dx$$
$$= e^{-2x}\int dx = e^{-2x}(x + C).$$

7. For this linear differential equation, you have $P(x) = x$ and $Q(x) = x$. So, the integrating factor is $u(x) = e^{\int x\,dx} = e^{x^2/2}$ and the general solution is

$$y = \frac{1}{u(x)}\int Q(x)u(x)\,dx = e^{-x^2/2}\int xe^{x^2/2}\,dx$$
$$= e^{-x^2/2}\left(e^{x^2/2} + C\right) = 1 + Ce^{-x^2/2}.$$

8. For this linear differential equation, you have $P(x) = \dfrac{2}{x}$ and $Q(x) = x$. So, the integrating factor is $u(x) = e^{\int 2/x\,dx} = e^{2\ln x} = e^{\ln x^2} = x^2$ and the general solution is

$$y = \frac{1}{u(x)}\int Q(x)u(x)\,dx = \frac{1}{x^2}\int x^3\,dx$$
$$= \frac{1}{x^2}\left(\frac{1}{4}x^4 + C\right) = \frac{1}{4}x^2 + \frac{C}{x^2}.$$

9. $\dfrac{dy}{dx} = kx^2$

10. $\dfrac{dx}{dt} = k(x - t)$

1. The general solution is $y = \dfrac{-1}{kt + C}$. Because $y = 45$ when $t = 0$, it follows that $45 = \dfrac{-1}{C}$ and $C = \dfrac{-1}{45}$.

So, $y = -\dfrac{1}{kt - (1/45)} = \dfrac{45}{1 - 45kt}$.

Because $y = 4$ when $t = 2$, you have $4 = \dfrac{45}{1 - 45k(2)} \Rightarrow k = -\dfrac{45}{360}$.

So, $y = \dfrac{45}{1 + (41/8)t} = \dfrac{360}{8 + 41t}$.

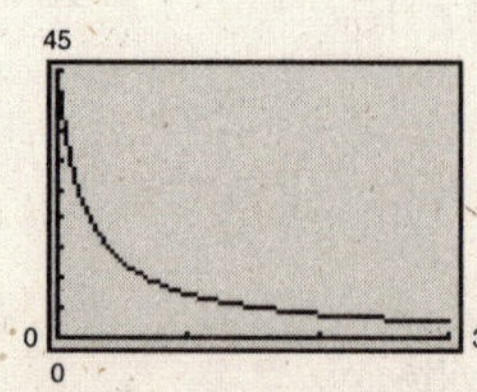

3. The general solution is $y = 1 - Ce^{-kt}$. Because $y = 0$ when $t = 0$, it follows that $C = 1$. Because $y = 0.75$ when $t = 1$, you have

$$\begin{aligned} 0.75 &= 1 - e^{-k} \\ -0.25 &= -e^{-k} \\ -\ln 0.25 &= k \\ k &\approx 1.386. \end{aligned}$$

So, $y = 1 - e^{-1.386t}$.

5. The general solution is $y = 200e^{-Ce^{-kt}}$.

Because $y = 100$ when $t = 0$, you have

$$\begin{aligned} 100 &= 200e^{-Ce^{0}} \\ \frac{1}{2} &= e^{-C} \\ \ln\left(\frac{1}{2}\right) &= -C \\ -\ln\left(\frac{1}{2}\right) &= C \\ C &\approx 0.6931. \end{aligned}$$

Because $y = 150$ when $t = 2$, you have

$$\begin{aligned} 150 &= 200e^{-0.6931e^{-2k}} \\ \frac{3}{4} &= e^{-0.6931e^{-2k}} \\ \ln\left(\frac{3}{4}\right) &= -0.6931e^{-2k} \\ -\frac{\ln(3/4)}{0.6931} &= e^{-2k} \\ \ln\left(-\frac{\ln(3/4)}{0.6931}\right) &= -2k \\ -\frac{1}{2}\ln\left(-\frac{\ln(3/4)}{0.6931}\right) &= k \\ k &\approx 0.4397. \end{aligned}$$

So, $y = 200e^{-0.6931e^{-0.4397t}}$.

7. The general solution is $\dfrac{y(2 - y)}{(1 - y)^2} = Ce^{2kt}$.

Because $y = 0.1$ when $t = 0$, you have

$$\begin{aligned} \frac{0.1(2 - 0.1)}{(1 - 0.1)^2} &= Ce^{0} \\ \frac{19}{81} &= C. \end{aligned}$$

Because $y = 0.4$ when $t = 4$, you have

$$\begin{aligned} \frac{0.4(2 - 0.4)}{(1 - 0.4)^2} &= \frac{19}{81}e^{8k} \\ \frac{16}{9} &= \frac{19}{81}e^{8k} \\ \frac{144}{19} &= e^{8k} \\ \ln\left(\frac{144}{19}\right) &= 8k \\ \frac{1}{8}\ln\left(\frac{144}{19}\right) &= k \\ k &\approx 0.25317. \end{aligned}$$

So, $\dfrac{y(2 - y)}{(1 - y)^2} = \dfrac{19}{81}e^{0.50634t}$ is the particular solution.

9. The general solution is $y = Ce^{kx}$. Because $y = 1$ when $x = 0$, it follows that $C = 1$. So, $y = e^{kx}$. Because $y = 2$ when $x = 3$, it follows that $2 = e^{3k}$, which implies that $k = \dfrac{\ln 2}{3} \approx 0.2310$. So, the particular solution is $y \approx e^{0.2310x}$.

11. The general solution is $y = Ce^{kx}$. Because $y = 4$ when $x = 0$, it follows that $C = 4$. So, $y = 4e^{kx}$. Because $y = 1$ when $x = 4$, it follows that $\frac{1}{4} = e^{4k}$, which implies that $k = \frac{1}{4}\ln\frac{1}{4} \approx -0.3466$. So, the particular solution is $y \approx 4e^{-0.3466x}$.

13. The general solution is $y = Ce^{kx}$. Because $y = 2$ when $x = 2$ and $y = 4$ when $x = 3$, it follows that $2 = Ce^{2k}$ and $4 = Ce^{3k}$. By equating C-values from these two equations, you have

$$\begin{aligned} 2e^{-2k} &= 4e^{-3k} \\ \tfrac{1}{2} &= e^{-k} \Rightarrow k = \ln 2 \approx 0.6931. \end{aligned}$$

This implies that

$$C = 2e^{-2\ln 2} = 2e^{\ln(1/4)} = 2\left(\tfrac{1}{4}\right) = \tfrac{1}{2}.$$

So, the particular solution is $y = \frac{1}{2}e^{x\ln 2} \approx \frac{1}{2}e^{0.6931x}$.

15. $\dfrac{dy}{dt} = ky$

$$\int \frac{1}{y}\,dy = \int k\,dt$$
$$\ln|y| = kt + C_1$$
$$y = e^{kt+C_1}$$
$$y = Ce^{kt}$$

Because the initial amount is 20 grams, you know that $C = 20$. Because $y = 16$ when $t = 1$, it follows that

$$16 = 20e^k$$
$$\frac{4}{5} = e^k$$
$$\ln\left(\frac{4}{5}\right) = k$$
$$-0.2231 \approx k.$$

So, the particular solution is $y = 20e^{-0.2231t}$.

17. $\dfrac{dP}{dt} = kP$

$$\int \frac{dP}{P} = \int k\,dt$$
$$\ln|P| = kt + C_1$$
$$P = e^{kt+C_1}$$
$$P = Ce^{kt} \qquad (t = 0 \text{ corresponds to } 2000)$$

Because $P = 200{,}000$ when $t = 0$, you have

$$200{,}000 = Ce^0$$
$$200{,}000 = C.$$

Because the constant of proportionality is $k = 0.015$, you have

$$P = 200{,}000e^{0.015t}.$$

The population in 2020 $(t = 20)$ is

$$P = 200{,}000e^{0.015(20)} \approx 269{,}972 \text{ people.}$$

19. The general solution is $y = Ce^{kt}$. Because $y = 0.60C$ when $t = 1$, you have $0.60C = Ce^k \Rightarrow k = \ln 0.60 \approx -0.5108$. So, $y = Ce^{-0.5108t}$. When $y = 0.20C$, you have

$$0.20C = Ce^{-0.5108t}$$
$$\ln 0.20 = -0.5108t$$
$$t \approx 3.15 \text{ hours.}$$

21. $\dfrac{dS}{dt} = k(L - S)$

$$\int \frac{dS}{L - S} = \int k\,dt$$
$$-\ln|L - S| = kt + C_1$$
$$\ln|L - S| = -kt + C_2$$
$$L - S = e^{-kt+C_2}$$
$$L - S = Ce^{-kt}$$

So, $S = L - Ce^{-kt}$ is a general solution.

Because $S = 0$ when $t = 0$, you have

$$0 = L - Ce^0$$
$$C = L.$$

So, $S = L - Le^{-kt} \Rightarrow S = L(1 - e^{-kt})$.

23. From Example 3, the general solution is $y = 60e^{-Ce^{-kt}}$.

Because $y = 8$ when $t = 0$,

$$8 = 60e^{-C} \Rightarrow C = \ln\frac{15}{2} \approx 2.0149.$$

Because $y = 15$ when $t = 3$,

$$15 = 60e^{-2.0149e^{-3k}}$$
$$\frac{1}{4} = e^{-2.0149e^{-3k}}$$
$$\ln\frac{1}{4} = -2.0149e^{-3k}$$
$$k = -\frac{1}{3}\ln\left(\frac{\ln 1/4}{-2.0149}\right) \approx 0.1246.$$

So, $y = 60e^{-2.0149e^{-0.1246t}}$.

When $t = 10$, $y \approx 34$ beavers.

25. (a)
$$\frac{dN}{dt} = kN(500 - N)$$
$$\int \frac{dN}{N(500 - N)} = \int k\,dt$$
$$\frac{1}{500}\int\left[\frac{1}{N} + \frac{1}{500 - N}\right]dN = \int k\,dt$$
$$\ln|N| - \ln|500 - N| = 500(kt + C_1)$$
$$\frac{N}{500 - N} = e^{500kt + C_2} = Ce^{500kt}$$
$$N = \frac{500Ce^{500kt}}{1 + Ce^{500kt}}$$

When $t = 0$, $N = 100$.

So, $100 = \dfrac{500C}{1 + C} \Rightarrow C = 0.25.$

So, $N = \dfrac{125e^{500kt}}{1 + 0.25e^{500kt}}.$

When $t = 4$, $N = 200$.

So,
$$200 = \frac{125e^{2000k}}{1 + 0.25e^{2000k}} \Rightarrow k = \frac{\ln(8/3)}{2000} \approx 0.00049.$$

So, $N = \dfrac{125e^{0.2452t}}{1 + 0.25e^{0.2452t}} = \dfrac{500}{1 + 4e^{-0.2452t}}.$

(b) When $t = 1$:
$$N = \frac{500}{1 + 4e^{-0.2452}} \approx 121 \text{ deer}$$

(c) When $N = 350$:
$$350 = \frac{500}{1 + 4e^{-0.2452t}}$$
$$1 + 4e^{-0.2452t} = \frac{10}{7}$$
$$4e^{-0.2452t} = \frac{3}{7}$$
$$e^{-0.2452t} = \frac{3}{28}$$
$$-0.2452t = \ln\left(\frac{3}{28}\right)$$
$$t = \frac{\ln\left(\frac{3}{28}\right)}{-0.2452}$$
$$t \approx 9.1 \text{ years}$$

27. (a)
$$\frac{dQ}{dt} = -\frac{Q}{20}$$
$$\int \frac{dQ}{Q} = \int -\frac{1}{20}\,dt$$
$$\ln|Q| = -\frac{1}{20}t + C_1$$
$$Q = e^{-(1/20)t + C_1} = Ce^{-(1/20)t}$$

Because $Q = 25$ when $t = 0$, you have $25 = C$.

So, the particular solution is $Q = 25e^{-(1/20)t}$.

(b) When $Q = 15$, you have $15 = 25e^{-(1/20)t}$.
$$\frac{3}{5} = e^{-(1/20)t}$$
$$\ln\left(\frac{3}{5}\right) = -\frac{1}{20}t$$
$$-20\ln\left(\frac{3}{5}\right) = t$$
$$t \approx 10.217 \text{ minutes}$$

29.
$$\frac{dP}{dt} = kP + N$$
$$\int \frac{1}{kP + N}\,dP = \int dt$$
$$\frac{1}{k}\ln|kP + N| = t + C_1$$
$$kP + N = C_2e^{kt}$$
$$P = Ce^{kt} - \frac{N}{k}$$

31.
$$\frac{dA}{dt} = rA + P$$
$$\int \frac{1}{rA + P}\,dA = \int dt$$
$$\frac{1}{r}\ln|rA + P| = t + C_1$$
$$rA + P = Ce^{rt}$$
$$A = \frac{1}{r}\left(Ce^{rt} - P\right)$$

Because $A = 0$ when $t = 0$, it follows that $C = P$.

So, you have $A = \dfrac{P}{r}\left(e^{rt} - 1\right)$.

33.
$$A = \frac{P}{r}\left(e^{rt} - 1\right)$$
$$\frac{Ar}{e^{rt} - 1} = P$$

Because $A = 120{,}000{,}000$ when $t = 8$ and $r = 0.08$, you have $P = \dfrac{(120{,}000{,}000)(0.08)}{e^{(0.08)(8)} - 1} \approx \$10{,}708{,}538.49.$

35. (a) $\dfrac{dC}{dt} = \left(-\dfrac{R}{V}\right)C$

$\displaystyle\int \frac{dC}{C} = \int -\frac{R}{V}\, dt$

$\ln|C| = -\dfrac{R}{V}t + K_1$

$C = Ke^{-Rt/V}$

Because $C = C_0$ when $t = 0$, it follows that $K = C_0$ and the function is $C = C_0e^{-Rt/V}$.

(b) Finally, as $t \to \infty$, you have

$\lim\limits_{t\to\infty} C = \lim\limits_{t\to\infty} C_0e^{-Rt/V} = 0.$

37. (a) $\dfrac{dC}{dt} = \dfrac{Q}{V} - \left(\dfrac{R}{V}\right)C$

$\displaystyle\int \frac{1}{Q - RC}\, dC = \int \frac{1}{V}\, dt$

$-\dfrac{1}{R}\ln|Q - RC| = \dfrac{t}{V} + K_1$

$Q - RC = e^{-R[(t/V)+K_1]}$

$C = \dfrac{1}{R}\left(Q - e^{-R[(t/V)+K_1]}\right)$

$= \dfrac{1}{R}\left(Q - Ke^{-Rt/V}\right)$

Because $C = 0$ when $t = 0$, it follows that $K = Q$ and you have $C = \dfrac{Q}{R}\left(1 - e^{-Rt/V}\right)$.

(b) As $t \to \infty$, the limit of C is Q/R.

Review Exercises for Chapter 11

1. Because $y = Ce^{x/2}$ and $y' = \dfrac{Ce^{x/2}}{2}$, you have

$2\left(\dfrac{C}{2}e^{x/2}\right) = Ce^{x/2} = y.$

3. Because $y = 3e^{x^2}$ and $y' = 6xe^{x^2}$, you have

$6xe^{x^2} - 2x\left(3e^{x^2}\right) = 6xe^{x^2} - 6xe^{x^2} = 0.$

5. Because $y = x^2$ and $y' = 2x$, you have $\dfrac{1}{2}y' - \dfrac{1}{x}y = \dfrac{1}{2}(2x) - \dfrac{1}{x}\left(x^2\right) = 0.$

7. Because $y = \dfrac{x^3}{5} - x + C\sqrt{x}$ and $y' = \dfrac{3}{5}x^2 - 1 + \dfrac{C}{2\sqrt{x}}$, you have

$$2x\left(\frac{3}{5}x^2 - 1 + \frac{C}{2\sqrt{x}}\right) - \left(\frac{x^3}{5} - x + C\sqrt{x}\right) = \frac{6x^3}{5} - 2x + C\sqrt{x} - \frac{x^3}{5} + x - C\sqrt{x} = x^3 - x.$$

9. Because $y' = C_1e^x - C_2e^{-x}$ and $y'' = C_1e^x + C_2e^{-x}$, you have $y'' - y = \left(C_1e^x + C_2e^{-x}\right) - \left(C_1e^x + C_2e^{-x}\right) = 0.$

11. $y = x^2$

$y' = 2x$

$xy' - 2y = x(2x) - 2\left(x^2\right) = 0 \neq x^3e^x$

So, the function is *not* a solution of the differential equation.

13. $y = x^2e^x$

$y' = x^2e^x + 2xe^x$

$xy' - 2y = x\left(x^2e^x + 2xe^x\right) - 2x^2e^x = x^3e^x$

So, the function is a solution of the differential equation.

15. Because $y' = -5Ce^{-5x} = -5y$, it follows that $y' + 5y = 0$. To find the particular solution, use the fact that $y = 1$ when $x = 0$. That is, $1 = Ce^0 \Rightarrow C = 1$. So, the particular solution is $y = e^{-5x}$.

17. When $C = 1$, the graph is the exponential curve $y = e^{3/x}$.

When $C = 2$, the graph is the exponential curve $y = 2e^{3/x}$.

When $C = 4$, the graph is the exponential curve $y = 4e^{3/x}$.

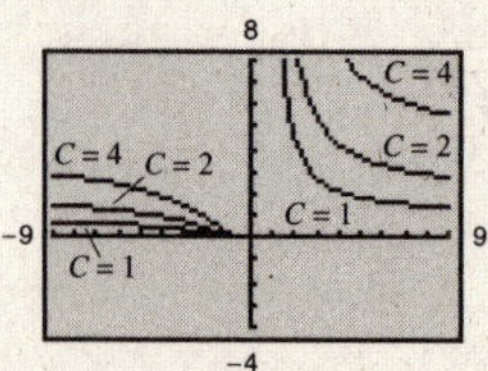

19. Because $y = 1$ when $x = 1$, you have $1 = C\left(1^2 + 1\right)$, which implies that $C = \frac{1}{2}$, and the particular solution is $y = \frac{1}{2}\left(x^2 + 1\right)$.

21. $y = \int (2x^2 + 5)\, dx = \frac{2}{3}x^3 + 5x + C$

23. $y = \int (3 \cos x)\, dx = 3 \sin x + C$

25. $y = \int \left(1 + \frac{3}{x}\right) dx = x + 3 \ln|x| + C$

27. Yes, $\frac{dy}{dx} = \frac{y}{x + 3}$

$$\frac{1}{y}\, dy = \frac{1}{x + 3}\, dx.$$

29. No, the variables cannot be separated.

31.
$$\frac{dy}{dx} = 4x$$
$$\int dy = \int 4x\, dx$$
$$y = 2x^2 + C$$

33.
$$4y^3 \frac{dy}{dx} = 5$$
$$\int 4y^3\, dy = \int 5\, dx$$
$$y^4 = 5x + C$$

35. $y' + 2xy^2 = 0$
$$\frac{dy}{dx} = -2xy^2$$
$$\int \frac{1}{y^2}\, dy = \int -2x\, dx$$
$$-\frac{1}{y} = -x^2 + C_1$$
$$\frac{1}{y} = x^2 + C$$
$$y = \frac{1}{x^2 + C}$$

37.
$$\frac{dy}{dx} = (x + 1)(y + 1)$$
$$\int \frac{1}{y + 1}\, dy = \int (x + 1)\, dx$$
$$\ln|y + 1| = \frac{1}{2}x^2 + x + C_1$$
$$y + 1 = e^{(1/2)x^2 + x + C_1}$$
$$y = Ce^{(x^2/2) + x} - 1$$

39.
$$\frac{dy}{dx} = -\frac{y + 2}{2x^3}$$
$$\int \frac{1}{y + 2}\, dy = \int -\frac{1}{2x^3}\, dx$$
$$\ln|y + 2| = \frac{1}{4x^2} + C_1$$
$$y + 2 = e^{1/(4x^2) + C_1}$$
$$y = Ce^{1/(4x^2)} - 2$$

41.
$$\frac{dy}{dx} = \frac{\cos x}{y}$$
$$\int y\, dy = \int \cos x\, dx$$
$$\frac{1}{2}y^2 = \sin x + C_1$$
$$y^2 = 2 \sin x + C$$

43. (a)
$$\frac{dy}{dx} = 3x^2$$
$$\int dy = \int 3x^2\, dx$$
$$y = x^3 + C$$

(b)

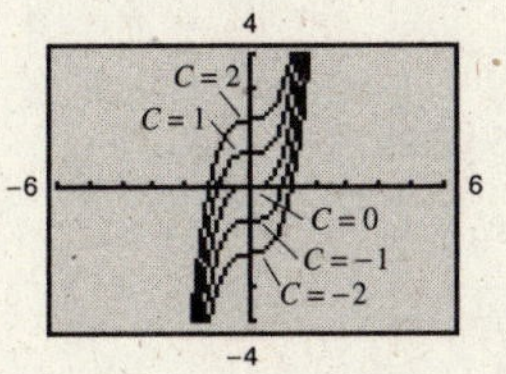

4
C = 2
C = 1
−6
6
C = 0
C = −1
C = −2
−4

45. $yy' + e^x = 0$
$$\int y\, dy = \int -e^x\, dx$$
$$\frac{1}{2}y^2 = -e^x + C_1$$
$$y^2 = -2e^x + C$$

When $x = 0$, $y = 2$. So, $C = 6$ and the particular solution is $y^2 = -2e^x + 6$.

47.
$$\frac{dy}{dx} = 2xy \cos x^2$$
$$\int \frac{1}{y}\, dy = \int 2x \cos x^2\, dx$$
$$\ln|y| = \sin x^2 + C_1$$
$$y = e^{\sin x^2 + C_1} = Ce^{\sin x^2}$$

When $x = 0$, $y = 1$. So, $C = 1$ and the particular solution is $y = e^{\sin x^2}$.

49.
$$\frac{dy}{dx} = \frac{6x}{5y}$$
$$\int y\, dy = \int \frac{6}{5}x\, dx$$
$$\frac{1}{2}y^2 = \frac{3}{5}x^2 + C_1$$
$$y^2 = \frac{6}{5}x^2 + C_2$$
$$6x^2 - 5y^2 = C$$

When $x = 1, y = 1$.

So, $6(1)^2 - 5(1)^2 = C \Rightarrow C = 1$.

The particular equation is $6x^2 - 5y^2 = 1$.

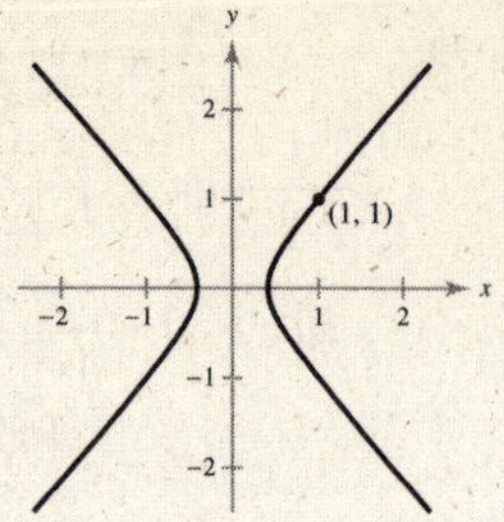

51. Let y represent the mass of the carbon and let t represent the time (in years).
$$\frac{dy}{dt} = ky$$
$$\int \frac{1}{y}\, dy = \int k\, dt$$
$$\ln y = kt + C_1$$
$$y = e^{kt + C_1}$$
$$y = Ce^{kt}$$

The half-life is 5715 years.
$$\frac{1}{2}C = Ce^{k(5715)}$$
$$\frac{1}{2} = e^{5715k}$$
$$\ln\frac{1}{2} = 5715k$$
$$\frac{1}{5715}\ln\frac{1}{2} = k$$
$$-0.000121 \approx k$$

The equation is $y = Ce^{-0.000121t}$. After $t = 1000$ years, $y \approx 0.886C$. So, about 88.6% of a present amount will remain.

53.
$$x^4 + 4x^2y' - 4y = 0$$
$$4x^2y' - 4y = -x^4$$
$$y' - \frac{1}{x^2}y = -\frac{1}{4}x^2$$

55.
$$x = 2x^3(y' - y)$$
$$2x^3y' - 2x^3y = x$$
$$y' - y = \frac{1}{2x^2}$$

57. For this linear differential equation, you have $P(x) = 4$ and $Q(x) = 8$. So, the integrating factor is $u(x) = e^{\int 4\,dx} = e^{4x}$ and the general solution is
$$y = \frac{1}{u(x)}\int Q(x)u(x)\, dx$$
$$= e^{-4x}\int 8e^{4x}\, dx$$
$$= e^{-4x}\left(2e^{4x} + C\right)$$
$$= 2 + Ce^{-4x}.$$

59. For this linear differential equation, you have $P(x) = -\frac{1}{x}$ and $Q(x) = 2x - 3$. So, the integrating factor is $u(x) = e^{\int -1/x\,dx} = e^{-\ln x} = x^{-1} = \frac{1}{x}$ and the general solution is
$$y = \frac{1}{u(x)}\int Q(x)u(x)\, dx$$
$$= x\int (2x - 3)\left(\frac{1}{x}\right) dx$$
$$= x\int\left(2 - \frac{3}{x}\right) dx$$
$$= x(2x - 3\ln x + C)$$
$$= 2x^2 - 3x\ln|x| + Cx.$$

61. For this linear differential equation, you have $P(x) = 6$ and $Q(x) = e^{2x}$. So, the integrating factor is $u(x) = e^{\int 6\,dx} = e^{6x}$ and the general solution is
$$y = \frac{1}{u(x)}\int Q(x)u(x)\, dx = e^{-6x}\int e^{2x}e^{6x}\, dx = e^{-6x}\int e^{8x}\, dx = e^{-6x}\left(\frac{1}{8}e^{8x} + C\right) = \frac{1}{8}e^{2x} + Ce^{-6x}.$$

63. For this linear differential equation, you have $P(x) = \frac{1}{x}$ and $Q(x) = 3x + 4$. So, the integrating factor is $u(x) = e^{\int 1/x\,dx} = e^{\ln x} = x$ and the general solution is
$$y = \frac{1}{u(x)}\int Q(x)u(x)\, dx = \frac{1}{x}\int (3x + 4)x\, dx = \frac{1}{x}\int\left(3x^2 + 4x\right) dx = \frac{1}{x}\left(x^3 + 2x^2 + C\right) = x^2 + 2x + \frac{C}{x}.$$

65. For this linear differential equation, you have $P(x) = 1$ and $Q(x) = 6$. So, the integrating factor is $u(x) = e^{\int dx} = e^x$ and the general solution is

$$y = \frac{1}{u(x)}\int Q(x)\,u(x)\,dx$$
$$= e^{-x}\int 6e^x\,dx$$
$$= e^{-x}(6e^x + C)$$
$$= 6 + Ce^{-x}.$$

When $x = 0$, $y = 3$. So, $3 = 6 + Ce^0 \Rightarrow C = -3$.

The particular solution is $y = 6 - 3e^{-x}$.

67. $$\frac{ds}{dh} = \frac{k}{h}$$
$$\int ds = \int \frac{k}{h}\,dh$$
$$s = k\ln|h| + C$$

When $h = 2, s = 25$. So, $25 = k\ln|2| + C$.

When $h = 6, s = 12$. So, $12 = k\ln|6| + C$.

Because $C = 25 - k\ln 2$ and $C = 12 - k\ln 6$,

$$25 - k\ln 2 = 12 - k\ln 6$$
$$13 = k\ln 2 - k\ln 6$$
$$13 = k(\ln 2 - \ln 6)$$
$$13 = k\ln\left(\frac{1}{3}\right)$$
$$k = \frac{13}{\ln\left(\frac{1}{3}\right)} = -\frac{13}{\ln 3}.$$

So, $25 = -\dfrac{13}{\ln 3}\ln 2 + C \Rightarrow 25 + \dfrac{13\ln 2}{\ln 3} = C.$

The particular solution is

$$s = -\frac{13}{\ln 3}\ln h + 25 + \frac{13\ln 2}{\ln 3} \text{ or}$$
$$s = 25 - \frac{13}{\ln 3}(\ln h - \ln 2) \text{ or}$$
$$s = 25 - \frac{13}{\ln 3}\ln\left(\frac{h}{2}\right),\ 2 < h < 15.$$

69. $$\frac{dy}{dx} = -k\frac{y}{x}$$
$$\int \frac{1}{y}\,dy = \int -\frac{k}{x}\,dx$$
$$\ln|y| = -k\ln|x| + C_1$$
$$y = e^{-k\ln|x| + C_1}$$
$$= Cx^{-k}$$

71. $$\frac{dy}{dt} = k\sqrt[3]{y}$$
$$\int \frac{1}{y^{1/3}}\,dy = \int k\,dt$$
$$\frac{3}{2}y^{2/3} = kt + C_1$$
$$y^{2/3} = \frac{2}{3}kt + C$$

Because $y = 27$ when $t = 0$, you know that $C = 9$.
Because $y = 8$ when $t = 1$, it follows that

$$\left(8^{2/3}\right) = \frac{2}{3}k + 9$$
$$4 = \frac{2}{3}k + 9$$
$$-5 = \frac{2}{3}k$$
$$-7.5 = k.$$

So, the particular solution is $y^{2/3} = -5t + 9$ or $y = (-5t + 9)^{3/2}$.

The following is used for Exercise 73.

$$\frac{dT}{dt} = k(T - T_0)$$
$$\int \frac{dT}{T - T_0} = \int k\,dt$$
$$\ln|T - T_0| = kt + C_1$$
$$T - T_0 = e^{kt + C_1}$$
$$T - T_0 = Ce^{kt}$$
$$T = T_0 + Ce^{kt}$$

73. $T = T_0 + Ce^{kt}$

When $t = 0$, $T = 1500$. So,

$1500 = 90 + Ce^0 \Rightarrow C = 1410.$

When $t = 1$, $T = 1120$.

So, $1120 = 90 + 1410e^k$

$$1030 = 1410e^k$$
$$\tfrac{103}{141} = e^k$$
$$k = \ln\left(\tfrac{103}{141}\right) \approx -0.3140.$$

So, $T = 90 + 1410e^{-0.3140t}$.

When $t = 5$,

$T = 90 + 1410e^{(-0.3140)(5)} \approx 383.3°$ F.

75. $y = 64e^{-Ce^{-kt}}$

When $t = 0$, $y = 12$.

$$12 = 64e^{-Ce^0}$$
$$\frac{3}{16} = e^{-C}$$
$$-C = \ln\left(\frac{3}{16}\right)$$
$$C = -\ln\left(\frac{3}{16}\right) \approx 1.674$$

When $t = 3$, $y = 28$.

$$28 = 64e^{-1.674e^{-3k}}$$
$$\frac{7}{16} = e^{-1.674e^{-3k}}$$
$$\ln\left(\frac{7}{16}\right) = -1.674e^{-3k}$$
$$\frac{\ln\left(\frac{7}{16}\right)}{-1.674} = e^{-3k}$$
$$\ln\left(\frac{\ln\left(\frac{7}{16}\right)}{-1.674}\right) = -3k$$
$$k = -\frac{1}{3}\ln\left(\frac{\ln\left(\frac{7}{16}\right)}{-1.674}\right) \approx 0.2352$$

So, $y = 64e^{-1.674e^{-0.2352t}}$ is the particular solution.

After 8 years $(t = 8)$,

$$y = 64e^{-1.674e^{(-0.2352)(8)}} \approx 49.6 \text{ or } 50 \text{ pelicans.}$$

77. (a) $\dfrac{dy}{dt} = -2\left(\dfrac{y}{20}\right) + 1$

$$\frac{dy}{dt} + \frac{1}{10}y = 1$$
$$y' + \frac{1}{10}y = 1$$
$$P(t) = \frac{1}{10},\ Q(t) = 1$$
$$u(t) = e^{\int 1/10\,dt} = e^{(1/10)t}$$
$$y = e^{-t/10}\int e^{t/10}\,dt$$
$$= e^{-t/10}\left(10e^{t/10} + C\right)$$
$$= 10 + Ce^{-t/10}$$

Because $y = 2$ when $t = 0$, where y is the amount of alcohol in gallons,

$$2 = 10 + Ce^0$$
$$-8 = C.$$

So, $y = 10 - 8e^{-t/10}$ is the particular solution.

(b) Find y when $t = 10$.

$$y = 10 - 8e^{-10/10} = 10 - 8e^{-1} \approx 7.1 \text{ gallons}$$

Chapter 11 Test Yourself

1. Because $y' = -2e^{-2x}$, you have

$$3y' + 2y = 3\left(-2e^{-2x}\right) + 2\left(e^{-2x}\right) = -4e^{-2x}.$$

2. Because $y' = -\dfrac{1}{(x+1)^2}$ and $y'' = \dfrac{2}{(x+1)^3}$, you have

$$y'' - \frac{2y}{(x+1)^2} = \frac{2}{(x+1)^3} - \frac{2\left(\frac{1}{x+1}\right)}{(x+1)^2} = 0.$$

3.
$$yy' = x$$
$$\int y\,dy = \int x\,dx$$
$$\frac{1}{2}y^2 = \frac{1}{2}x^2 + C_1$$
$$y^2 = x^2 + C$$

4.
$$\frac{dy}{dx} = 8x(y+2)$$
$$\int \frac{dy}{y+2} = \int 8x\,dx$$
$$\ln|y+2| = 4x^2 + C_1$$
$$y + 2 = e^{4x^2 + C_1}$$
$$y + 2 = Ce^{4x^2}$$
$$y = Ce^{4x^2} - 2$$

5.
$$\frac{dy}{dx} = \frac{\cos \pi x}{3y^2}$$
$$\int 3y^2\,dy = \int \cos \pi x\,dx$$
$$y^3 = \frac{1}{\pi}\sin \pi x + C$$

6. For this linear differential equation, you have $P(x) = 5$ and $Q(x) = 2$.

So, the integrating factor is $u(x) = e^{\int 5\,dx} = e^{5x}$ and the general solution is

$$\begin{aligned} y &= \frac{1}{u(x)}\int Q(x)u(x)\,dx \\ &= e^{-5x}\int 2e^{5x}\,dx \\ &= e^{-5x}\left(\frac{2}{5}e^{5x} + C\right) \\ &= \frac{2}{5} + Ce^{-5x}. \end{aligned}$$

7. For this linear differential equation, you have $P(x) = -2$ and $Q(x) = e^{2x}$. So, the integrating factor is $u(x) = e^{\int -2\,dx} = e^{-2x}$ and the general solution is

$$\begin{aligned} y &= \frac{1}{u(x)}\int Q(x)u(x)\,dx \\ &= e^{2x}\int e^{2x}\left(e^{-2x}\right)dx \\ &= e^{2x}\int dx \\ &= e^{2x}(x + C) \\ &= xe^{2x} + Ce^{2x}. \end{aligned}$$

8. For this linear differential equation, you have $P(x) = \frac{1}{x}$ and $Q(x) = x$. So, the integrating factor is $u(x) = e^{\int 1/x\,dx} = e^{\ln x} = x$ and the general solution is

$$\begin{aligned} y &= \frac{1}{u(x)}\int Q(x)u(x)\,dx \\ &= \frac{1}{x}\int x^2\,dx \\ &= \frac{1}{x}\left(\frac{1}{3}x^3 + C\right) \\ &= \frac{1}{3}x^2 + \frac{C}{x}. \end{aligned}$$

9. $y' + x^2y - x^2 = 0$

$$\begin{aligned} \frac{dy}{dx} &= x^2(1 - y) \\ \int \frac{1}{1 - y}\,dy &= \int x^2\,dx \\ -\ln|1 - y| &= \frac{1}{3}x^3 + C_1 \\ \ln|1 - y| &= -\frac{1}{3}x^3 + C_2 \\ 1 - y &= e^{(-x^3/3) + C_2} \\ y &= 1 - Ce^{-x^3/3} \end{aligned}$$

When $x = 0$, $y = 1$. So, $C = 1$ and the particular solution is $y = 1 - e^{-x^3/3}$.

10. $y'e^{-x^2} = 2xy$

$$\begin{aligned} \frac{dy}{dx} &= 2xe^{x^2}y \\ \int \frac{1}{y}\,dy &= \int 2xe^{x^2}\,dx \\ \ln|y| &= e^{x^2} + C_1 \\ y &= e^{e^{x^2} + C_1} = Ce^{e^{x^2}} \end{aligned}$$

When $x = 0$, $y = e$. So, $C = 1$ and the particular solution is $y = e^{e^{x^2}}$.

11. $x\frac{dy}{dx} = \frac{1}{7}\ln x$

$$\begin{aligned} \int 7\,dy &= \int \frac{1}{x}\ln x\,dx \\ 7y &= \frac{1}{2}(\ln x)^2 + C_1 \\ y &= \frac{1}{14}(\ln x)^2 + C \end{aligned}$$

When $x = 1$, $y = -2$. So, $C = -2$ and the particular solution is $y = \frac{1}{14}(\ln x)^2 - 2$.

12. $\dfrac{dw}{dt} = k(200 - w)$

(a) $\displaystyle\int \frac{dw}{200 - w} = \int k\,dt$

$$-\ln|200 - w| = kt + C_1$$
$$-\ln|200 - w| = -kt + C_2$$
$$200 - w = e^{-kt + C_2}$$
$$200 - w = Ce^{-kt}$$
$$w = 200 - Ce^{-kt}$$

Because $w = 10$ when $t = 0$, you have

$10 = 200 - Ce^{\circ} \Rightarrow C = 190.$

So, $w = 200 - 190e^{-kt}$.

(b)

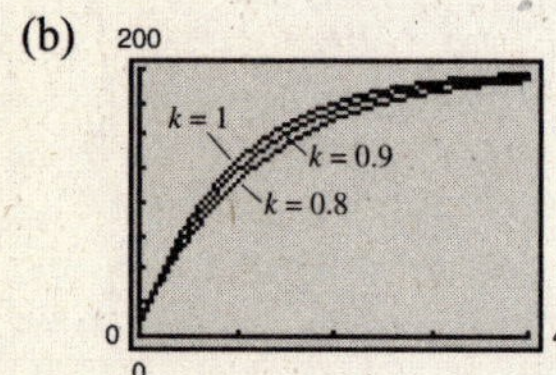

(c) For $k = 0.8$, $w = 200 - 190e^{-0.8t}$.

Let $w = 150$:

$$150 = 200 - 190e^{-0.8t}$$
$$-50 = -190e^{-0.8t}$$
$$\frac{5}{19} = e^{-0.8t}$$
$$\ln\left(\frac{5}{19}\right) = -0.8t$$
$$t = -\frac{1}{0.8}\ln\left(\frac{5}{19}\right) \approx 1.7 \text{ years}$$

For $k = 0.9$, $w = 200 - 190e^{-0.9t}$.

Let $w = 150$:

$$150 = 200 - 190e^{-0.9t}$$
$$-50 = -190e^{-0.9t}$$
$$\frac{5}{19} = e^{-0.9t}$$
$$\ln\left(\frac{5}{19}\right) = -0.9t$$
$$t = -\frac{1}{0.9}\ln\left(\frac{5}{19}\right) \approx 1.5 \text{ years}$$

For $k = 1$, $w = 200 - 190e^{-t}$.

Let $w = 150$:

$$150 = 200 - 190e^{-t}$$
$$-50 = -190e^{-t}$$
$$\frac{5}{19} = e^{-t}$$
$$\ln\left(\frac{5}{19}\right) = -t$$
$$t = -\ln\left(\frac{5}{19}\right) \approx 1.3 \text{ years}$$

(d) The maximum weight can be found for all three models as the limit of w as $t \to \infty$.

$$\lim_{t\to\infty} w = \lim_{t\to\infty}\left(200 - 190e^{-kt}\right) = 200 - 0 \Rightarrow 200 \text{ pounds}$$

13. $\dfrac{dy}{dt} = \dfrac{1 - y}{4}$

(a)

$$\int \frac{dy}{1 - y} = \int \frac{1}{4}\,dt$$
$$-\ln|1 - y| = \frac{1}{4}t + C_1$$
$$\ln|1 - y| = -\frac{1}{4}t + C_2$$
$$1 - y = e^{(-1/4)t + C_2}$$
$$1 - y = Ce^{-t/4}$$
$$y = 1 - Ce^{-t/4}$$

Because $y = 0$ when $t = 0$, you have

$0 = 1 - Ce^{\circ} \Rightarrow C = 1.$

So, $y = 1 - e^{-t/4}$.

(b) Let $y = 0.5$ and solve for t.

$$0.5 = 1 - e^{-t/4}$$
$$-0.5 = -e^{-t/4}$$
$$0.5 = e^{-t/4}$$
$$\ln(0.5) = -\frac{t}{4}$$
$$t = -4\ln(0.5) \approx 2.77 \Rightarrow 3 \text{ years}$$

(c) Let $t = 4$ and solve for y.

$$y = 1 - e^{-4/4} = 1 - e^{-1} \approx 0.632 \text{ or } 63.2\%$$

APPENDICES

APPENDIX A
A Precalculus Review

Section A.1 The Real Number Line and Order

1. Because $0.25 = \frac{1}{4}$, it is rational.

3. Because π is irrational, $\dfrac{3\pi}{2}$ is irrational.

5. Because $4.3\overline{451}$ has a repeating decimal expansion, it is rational.

7. Because $\sqrt[3]{64} = 4$, it is rational.

9. Because 60 is not the cube of a rational number, $\sqrt[3]{60}$ is irrational.

11. $5x - 12 > 0$

$5x > 12$

$x > \frac{12}{5}$

(a) Yes, if $x = 3$, then $x = 3 = \frac{15}{5}$ is greater than $\frac{12}{5}$.

(b) No, if $x = -3$, then $x = -3 = -\frac{15}{5}$ is not greater than $\frac{12}{5}$.

(c) Yes, if $x = \frac{5}{2}$, then $x = \frac{5}{2} = \frac{25}{10}$ is greater than $\frac{12}{5} = \frac{24}{10}$.

13. $0 < \dfrac{x-2}{4} < 2$

$0 < x - 2 < 8$

$2 < x < 10$

(a) Yes, if $x = 4$, then $2 < x < 10$.

(b) No, if $x = 10$, then x is not less than 10.

(c) No, if $x = 0$, then x is not greater than 2.

15. $x - 5 \geq 7$

$x - 5 + 5 \geq 7 + 5$

$x \geq 12$

17. $4x + 1 < 2x$

$4x + 1 - 2x - 1 < 2x - 2x - 1$

$2x < -1$

$\frac{1}{2}(2x) < \frac{1}{2}(-1)$

$x < -\frac{1}{2}$

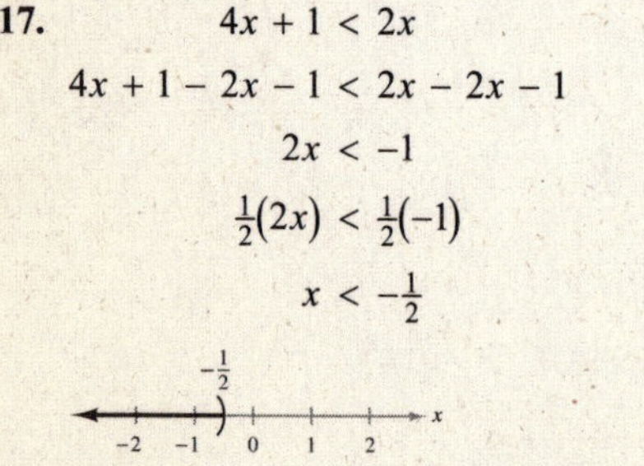

19. $4 - 2x < 3x - 1$

$4 - 2x - 4 - 3x < 3x - 1 - 4 - 3x$

$-5x < -5$

$-\frac{1}{5}(-5x) > -\frac{1}{5}(-5)$

$x > 1$

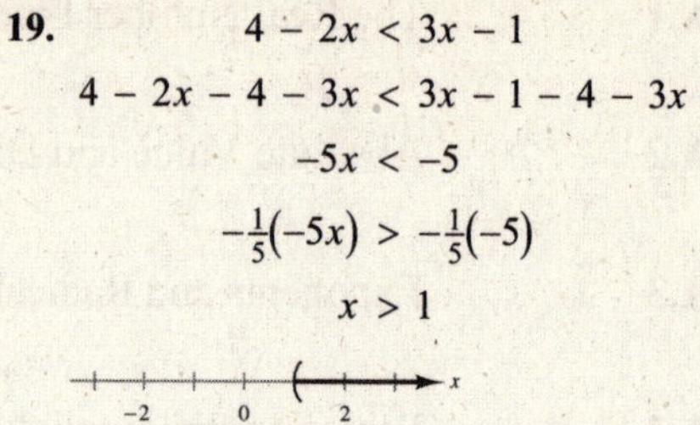

21. $-4 < 2x - 3 < 4$

$-4 + 3 < 2x - 3 + 3 < 4 + 3$

$-1 < 2x < 7$

$\dfrac{-1}{2} < \dfrac{2x}{2} < \dfrac{7}{2}$

$-\dfrac{1}{2} < x < \dfrac{7}{2}$

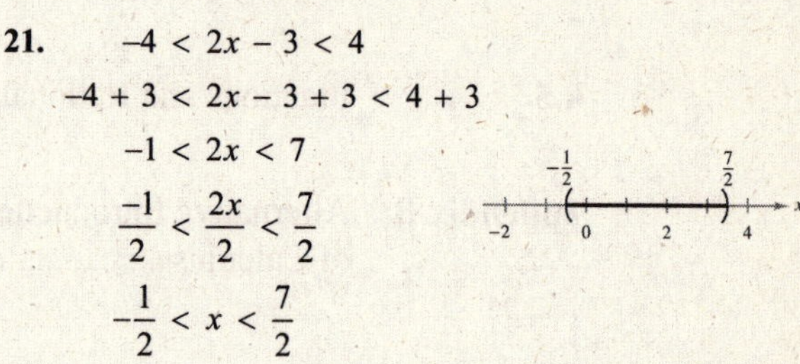

23. $\frac{3}{4} > x + 1 > \frac{1}{4}$

$\frac{3}{4} - 1 > x + 1 - 1 > \frac{1}{4} - 1$

$-\frac{1}{4} > x > -\frac{3}{4}$

$-\frac{3}{4} < x < -\frac{1}{4}$

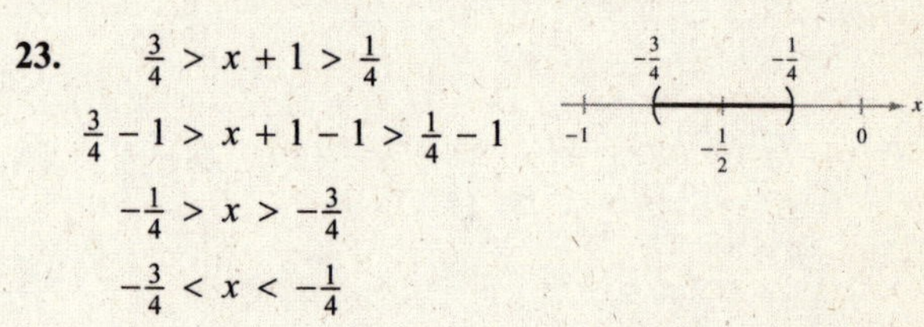

25. $\dfrac{x}{2} + \dfrac{x}{3} > 5$

$6\left(\dfrac{x}{2}\right) + 6\left(\dfrac{x}{3}\right) > 6(5)$

$3x + 2x > 30$

$5x > 30$

$\dfrac{1}{5}(5x) > \dfrac{1}{5}(30)$

$x > 6$

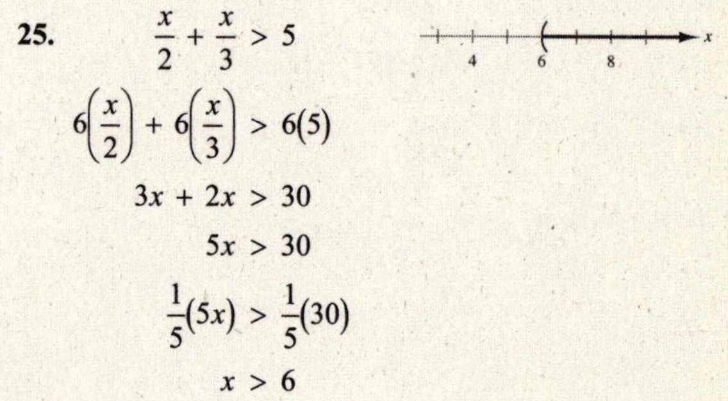

27. $2x^2 - x < 6$

$2x^2 - x - 6 < 0$

$(2x + 3)(x - 2) < 0$

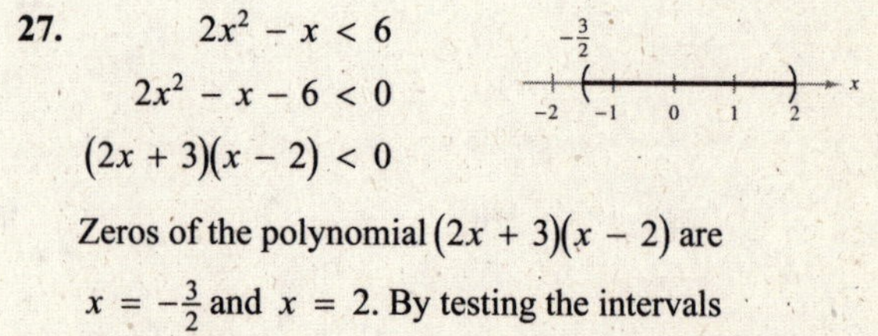

Zeros of the polynomial $(2x + 3)(x - 2)$ are $x = -\frac{3}{2}$ and $x = 2$. By testing the intervals $\left(-\infty, -\frac{3}{2}\right)$, $\left(-\frac{3}{2}, 2\right)$, and $(2, \infty)$, the solution set is $-\frac{3}{2} < x < 2$.

29. Let E represent the earnings per share, in dollars. Then $4.1 \leq E \leq 4.25$.

31. Let p represent the percent of Americans who conduct banking transactions online. Then $p \le 40$.

33. $A = 20$ and $r = 220 - A = 200$. Let T be the target heart rate. Then

$$(0.60)(200) \le T \le (0.90)(200)$$
$$120 \le T \le 180$$

So, the target heart rate for a 20-year-old is between 120 beats per minute and 180 beats per minute.

35. $R = 115.95x$ and $C = 95x + 750$ and because $R > C$, you can write

$$115.95x > 95x + 750$$
$$20.95x > 750$$
$$x > \frac{750}{20.95} = 35.7995...$$
$$x \ge 36.$$

So, this product will return a profit if $x \ge 36$ units.

37. (a) False. Because $a < b$, $-2a > -2b$.

(b) True. Because $a < b$, $a + 2 < b + 2$.

(c) True. Because $a < b$, $6a < 6b$.

(d) False, if $ab > 0$, then $\frac{1}{a} > \frac{1}{b}$.

Section A.2 Absolute Value and Distance on the Real Number Line

1. (a) $d = |126 - 75| = 51$

(b) $d = 75 - 126 = -51$

(c) $d = 126 - 75 = 51$

3. (a) $d = |9.34 - (-5.65)| = 14.99$

(b) $d = -5.65 - 9.34 = -14.99$

(c) $d = 9.34 - (-5.65) = 14.99$

5. (a) $d = \left|\frac{16}{5} - \frac{112}{75}\right| = \frac{128}{75}$

(b) $d = \frac{112}{75} - \frac{16}{5} = -\frac{128}{75}$

(c) $d = \frac{16}{5} - \frac{112}{75} = \frac{128}{75}$

7. $|x| \le 2$

9. $|x| > 2$

11. $|x - 5| \le 3$

13. $|x - 2| > 2$

15. $|x - 5| < 3$

17. $|y - a| \le 2$

19. $-4 < x < 4$

−6 −4 −2 0 2 4 6 x

21. $\frac{x}{2} < -3$ or $\frac{x}{2} > 3$

$2\left(\frac{x}{2}\right) < 2(-3)$ $\quad$ $2\left(\frac{x}{2}\right) > 2(3)$

$x < -6$ $\quad$ $x > 6$

−6 0 6 x

23.

$$-2 < x - 5 < 2$$
$$-2 + 5 < x - 5 + 5 < 2 + 5$$
$$3 < x < 7$$

0 1 2 3 4 5 6 7 8 x

25. $\frac{x-3}{2} \le -5$ or $\frac{x-3}{2} \ge 5$

$\frac{x-3}{2}(2) \le -5(2)$ $\quad$ $\frac{x-3}{2}(2) \ge 5(2)$

$x - 3 \le -10$ $\quad$ $x - 3 \ge 10$

$x - 3 + 3 \le -10 + 3$ $\quad$ $x - 3 + 3 \ge 10 + 3$

$x \le -7$ $\quad$ $x \ge 13$

−7 13 −10 0 10 x

27. $10 - x < -4$ or $10 - x > 4$

$10 - x - 10 < -4 - 10$ $\quad$ $10 - x - 10 > 4 - 10$

$-x < -14$ $\quad$ $-x > -6$

$x > 14$ $\quad$ $x < 6$

6 10 14 x

29.

$$-1 < 9 - 2x < 1$$
$$-1 - 9 < 9 - 2x - 9 < 1 - 9$$
$$-10 < -2x < -8$$
$$-\tfrac{1}{2}(-10) > -\tfrac{1}{2}(-2)x > -\tfrac{1}{2}(-8)$$
$$5 > x > 4$$
$$4 < x < 5$$

2 4 6 x

31.

$$-b \le x - a \le b$$
$$-b + a \le x - a + a \le b + a$$
$$a - b \le x \le a + b$$

$a - b$ $\quad$ a $\quad$ $a + b$ x

33. $$-2b < \frac{3x - a}{4} < 2b$$

$$4(-2b) < 4\left(\frac{3x - a}{4}\right) < 4(2b)$$

$$-8b < 3x - a < 8b$$

$$-8b + a < 3x - a + a < 8b + a$$

$$\frac{1}{3}(a - 8b) < x < \frac{1}{3}(a + 8b)$$

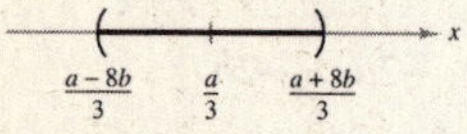

35. $\text{Midpoint} = \dfrac{8 + 24}{2} = 16$

37. $\text{Midpoint} = \dfrac{-6.85 + 9.35}{2} = 1.25$

39. $\text{Midpoint} = \dfrac{-\frac{1}{2} + \frac{3}{4}}{2} = \dfrac{\frac{1}{4}}{2} = \dfrac{1}{8}$

41. $|p - 33.15| \le 2$

43. $|x - 20| \le 0.75$

$$-0.75 \le x - 20 \le 0.75$$

$$19.25 \le x \le 20.75$$

The lowest and highest acceptable weights for a 20-ounce cereal box are 19.25 ounces and 20.75 ounces.

45. (a) $|E - 4750| \le 500$

$$0.05(4750) = 237.50$$

$$|E - 4750| \le 237.50$$

(b) \$5116.37 is not within 5% of the specified budgeted amount; at variance.

47. (a) $|E - 20{,}000| \le 500$

$$0.05(20{,}000) = 1000$$

$$|E - 20{,}000| \le 1000$$

(b) \$22,718.35 is not within \$500 of the specified budgeted amount; at variance.

49. Let r be the percent of defective units.

$$0.0005 - 0.0001 \le r \le 0.0005 + 0.0001$$

$$0.0004 \le r \le 0.0006$$

Number of defective units $= x = 150{,}000r$

$$0.0004(150{,}000) \le 150{,}000r \le 0.0006(150{,}000)$$

$$60 \le x \le 90$$

Cost of refunds $= C = 195.99x$

$$60(195.99) \le 195.99x \le 90(195.99)$$

$$11{,}759.40 \le C \le 17{,}639.10$$

The total cost of refunds should be between \$11,759.40 and \$17,639.10.

Section A.3 Exponents and Radicals

1. $-2(3)^3 = -2(27) = -54$

3. $4(2)^{-3} = 4\left(\frac{1}{8}\right) = \frac{1}{2}$

5. $\dfrac{1 + 3^{-1}}{3^{-1}} = \dfrac{1 + 1/3}{1/3} = \dfrac{4/3}{1/3} = 4$

7. $3(-2)^2 - 4(-2)^3 = 3(4) - 4(-8) = 12 + 32 = 44$

9. $6(10)^0 - [6(10)]^0 = 6(1) - (60)^0 = 6 - 1 = 5$

11. $\sqrt[3]{27^2} = \sqrt[3]{729} = 9$

13. $4^{-1/2} = \dfrac{1}{\sqrt{4}} = \dfrac{1}{2}$

15. $(-32)^{-2/5} = \dfrac{1}{(-32)^{2/5}} = \dfrac{1}{\left(\sqrt[5]{-32}\right)^2} = \dfrac{1}{(-2)^2} = \dfrac{1}{4}$

17. $500(1.01)^{60} \approx 908.3483$

19. $\sqrt[3]{-54} \approx -3.7798$

21. $$6y^{-2}(2y^4)^{-3} = 6y^{-2}(2^{-3}y^{-12})$$

$$= 6\left(\frac{1}{2^3}\right)y^{-2-12}$$

$$= 6\left(\frac{1}{8}\right)y^{-14}$$

$$= \frac{3}{4y^{14}}$$

23. $10(x^2)^2 = 10x^4$

25. $\dfrac{7x^2}{x^{-3}} = 7x^{2+3} = 7x^5$

27. $\dfrac{10(x + y)^3}{4(x + y)^{-2}} = \dfrac{5}{2}(x + y)^{3+2} = \dfrac{5}{2}(x + y)^5$

29. $\dfrac{3x\sqrt{x}}{x^{1/2}} = \dfrac{3x(x^{1/2})}{x^{1/2}} = 3x^{1+(1/2)-(1/2)} = 3x$

31. $\sqrt{8} = \sqrt{4 \cdot 2} = \sqrt{4}\sqrt{2} = 2\sqrt{2}$

33. $\sqrt[3]{54x^5} = \sqrt[3]{(27x^3)(2x^2)} = \sqrt[3]{27x^3}\sqrt[3]{2x^2} = 3x\sqrt[3]{2x^2}$

35. $\sqrt[3]{144x^9y^{-4}z^5} = \sqrt[3]{(8x^9y^{-3}z^3)(18y^{-1}z^2)}$

$= 2x^3y^{-1}z\sqrt[3]{18y^{-1}z^2}$

$= \dfrac{2x^3z}{y}\sqrt[3]{\dfrac{18z^2}{y}}$

37. $3x^3 - 12x = 3x(x^2 - 4) = 3x(x + 2)(x - 2)$

39. $2x^{5/2} + x^{-1/2} = x^{-1/2}(2x^3 + 1) = \dfrac{2x^3 + 1}{x^{1/2}}$

41. $3x(x + 1)^{3/2} - 6(x + 1)^{1/2} = 3(x + 1)^{1/2}(x(x + 1) - 2)$

$= 3(x + 1)^{1/2}(x^2 + x - 2)$

$= 3(x + 1)^{1/2}(x - 1)(x + 2)$

43. $\dfrac{5x^6 + x^3}{3x^2} = \dfrac{1}{3}\left(\dfrac{5x^6}{x^2} + \dfrac{x^3}{x^2}\right)$

$= \dfrac{1}{3}(5x^4 + x)$

$= \dfrac{1}{3}x(5x^3 + 1)$

45. $\sqrt{x - 4}$ is defined when $x \ge 4$.
Therefore, the domain is $[4, \infty)$.

47. $\sqrt{x^2 + 3}$ is defined for all real numbers.
Therefore, the domain is $(-\infty, \infty)$.

49. $\dfrac{1}{\sqrt[3]{x - 4}}$ is defined for all real numbers except $x = 4$.
Therefore, the domain is $(-\infty, 4) \cup (4, \infty)$.

51. $\dfrac{\sqrt{x + 2}}{1 - x}$

The numerator is defined when $x \ge -2$.
The denominator is defined when $x \ne 1$.
Therefore, the domain is $[-2, 1) \cup (1, \infty)$.

53. $A = 10{,}000\left(1 + \dfrac{0.065}{12}\right)^{120} \approx \$19{,}121.84$

55. $A = 5000\left(1 + \dfrac{0.055}{4}\right)^{60} \approx \$11{,}345.46$

57. $T = 2\pi\sqrt{\dfrac{L}{32}} = 2\pi\sqrt{\dfrac{4}{32}} = 2\pi\sqrt{\dfrac{1}{8}} = 2\pi\dfrac{1}{2\sqrt{2}}$

$= \dfrac{\pi}{\sqrt{2}} = \dfrac{\pi\sqrt{2}}{2} \approx 2.22$ seconds

Section A.4 Factoring Polynomials

1. $a = 6, b = -7$, and $c = 1$

$x = \dfrac{7 \pm \sqrt{49 - 24}}{12} = \dfrac{7 \pm 5}{12}$

So, $x = \dfrac{7 + 5}{12} = 1$ or $x = \dfrac{7 - 5}{12} = \dfrac{1}{6}$.

3. $a = 4, b = -12$, and $c = 9$

$x = \dfrac{12 \pm \sqrt{144 - 144}}{8} = \dfrac{12}{8} = \dfrac{3}{2}$

5. $a = 1, b = 4$, and $c = 1$

$y = \dfrac{-4 \pm \sqrt{16 - 4}}{2} = \dfrac{-4 \pm 2\sqrt{3}}{2} = -2 \pm \sqrt{3}$

7. $a = 2, b = 3$, and $c = -4$

$x = \dfrac{-3 \pm \sqrt{9 + 32}}{4} = \dfrac{-3 \pm \sqrt{41}}{4}$

9. $x^2 - 8x + 16 = (x - 4)^2$

11. $4x^2 + 4x + 1 = (2x + 1)^2$

13. $3x^2 - 4x + 1 = (3x - 1)(x - 1)$

15. $3x^2 - 5x + 2 = (3x - 2)(x - 1)$

17. $x^2 - 4xy + 4y^2 = (x - 2y)^2$

19. $81 - y^4 = (9 + y^2)(9 - y^2)$

$= (9 + y^2)(3 + y)(3 - y)$

21. $x^3 - 8 = x^3 - 2^3$

$= (x - 2)(x^2 + 2x + 4)$

23. $y^3 + 64 = y^3 + 4^3$

$= (y + 4)(y^2 - 4y + 16)$

25. $x^3 - y^3 = (x - y)(x^2 + xy + y^2)$

27. $x^3 - 4x^2 - x + 4 = x^2(x - 4) - (x - 4)$
$= (x - 4)(x^2 - 1)$
$= (x - 4)(x + 1)(x - 1)$

29. $2x^3 - 3x^2 + 4x - 6 = x^2(2x - 3) + 2(2x - 3)$
$= (2x - 3)(x^2 + 2)$

31. $2x^3 - 4x^2 - x + 2 = 2x^2(x - 2) - (x - 2)$
$= (x - 2)(2x^2 - 1)$

33. $x^4 - 15x^2 - 16 = (x^2 - 16)(x^2 + 1)$
$= (x - 4)(x + 4)(x^2 + 1)$

35. $x^2 - 5x = 0$
$x(x - 5) = 0$
$x = 0, 5$

37. $x^2 - 9 = 0$
$(x + 3)(x - 3) = 0$
$x = -3, 3$

39. $x^2 - 3 = 0$
$(x + \sqrt{3})(x - \sqrt{3}) = 0$
$x = \pm\sqrt{3}$

41. $(x - 3)^2 - 9 = 0$
$x^2 - 6x + 9 - 9 = 0$
$x(x - 6) = 0$
$x = 0, 6$

43. $x^2 + x - 2 = 0$
$(x + 2)(x - 1) = 0$
$x = -2, 1$

45. $x^2 - 5x + 6 = 0$
$(x - 2)(x - 3) = 0$
$x = 2, 3$

47. $3x^2 + 5x + 2 = 0$
$(3x + 2)(x + 1) = 0$
$x = -\frac{2}{3}, -1$

49. $x^3 + 64 = 0$
$x^3 = -64$
$x = \sqrt[3]{-64} = -4$

51. $x^4 - 16 = 0$
$x^4 = 16$
$x = \pm\sqrt[4]{16} = \pm 2$

53. $x^3 - x^2 - 4x + 4 = 0$
$x^2(x - 1) - 4(x - 1) = 0$
$(x - 1)(x^2 - 4) = 0$
$(x - 1)(x - 2)(x + 2) = 0$
$x = 1, \pm 2$

55. Because $\sqrt{x^2 - 4} = \sqrt{(x + 2)(x - 2)}$, the roots are $x = \pm 2$. By testing points inside and outside the interval $[-2, 2]$, we find that the expression is defined when $x \le -2$ or $x \ge 2$. So, the domain is $(-\infty, -2] \cup [2, \infty)$.

57. Because $\sqrt{x^2 - 7x + 12} = \sqrt{(x - 3)(x - 4)}$, the roots are $x = 3$ and $x = 4$. By testing points inside and outside the interval $[3, 4]$, we find that the expression is defined when $x \le 3$ or $x \ge 4$. So, the domain is $(-\infty, 3] \cup [4, \infty)$.

59. Because $\sqrt{5x^2 + 6x + 1} = \sqrt{(5x + 1)(x + 1)}$, the roots are $x = -\frac{1}{5}$ and $x = -1$. By testing the intervals $(-\infty, -1)$, $\left(-1, -\frac{1}{5}\right)$, and $\left(-\frac{1}{5}, \infty\right)$, we find that the expression is defined when $x \le -1$ or $x \ge -\frac{1}{5}$. So, the domain is $(-\infty, -1] \cup \left[-\frac{1}{5}, \infty\right)$.

61.
$$\begin{array}{r|rrrr} -1 & 1 & -3 & -6 & -2 \\ & & -1 & 4 & 2 \\ \hline & 1 & -4 & -2 & 0 \end{array}$$

So, the factorization is
$x^3 - 3x^2 - 6x - 2 = (x + 1)(x^2 - 4x - 2)$.

63.
$$\begin{array}{r|rrrr} -1 & 2 & -1 & -2 & 1 \\ & & -2 & 3 & -1 \\ \hline & 2 & -3 & 1 & 0 \end{array}$$

So, the factorization is
$2x^3 - x^2 - 2x + 1 = (x + 1)(2x^2 - 3x + 1)$.

65. Possible rational zeros: $\pm 8, \pm 4, \pm 2, \pm 1$

Use synthetic division for $x = -1$.

$$\begin{array}{r|rrrr} -1 & 1 & -1 & -10 & -8 \\ & & -1 & 2 & 8 \\ \hline & 1 & -2 & -8 & 0 \end{array}$$

So,

$$\begin{aligned} x^3 - x^2 - 10x - 8 &= 0 \\ (x + 1)(x^2 - 2x - 8) &= 0 \\ (x + 1)(x + 2)(x - 4) &= 0 \\ x &= -1, -2, 4. \end{aligned}$$

67. Possible rational roots: $\pm 1, \pm 2, \pm 3, \pm 6$

Use synthetic division for $x = 1$.

$$\begin{array}{r|rrrr} 1 & 1 & -6 & 11 & -6 \\ & & 1 & -5 & 6 \\ \hline & 1 & -5 & 6 & 0 \end{array}$$

So,

$$\begin{aligned} x^3 - 6x^2 + 11x - 6 &= 0 \\ (x - 1)(x^2 - 5x + 6) &= 0 \\ (x - 1)(x - 2)(x - 3) &= 0 \\ x &= 1, 2, 3. \end{aligned}$$

69. Possible rational zeros:

$\pm 6, \pm 3, \pm 2, \pm 1, \pm\frac{3}{2}, \pm\frac{1}{2}, \pm\frac{2}{3}, \pm\frac{1}{3}, \pm\frac{1}{6}$

Use synthetic division for $x = 3$.

$$\begin{array}{r|rrrr} 3 & 6 & -11 & -19 & -6 \\ & & 18 & 21 & 6 \\ \hline & 6 & 7 & 2 & 0 \end{array}$$

So,

$$\begin{aligned} 6x^3 - 11x^2 - 19x - 6 &= 0 \\ (x - 3)(6x^2 + 7x + 2) &= 0 \\ (x - 3)(3x + 2)(2x + 1) &= 0 \\ x &= 3, -\tfrac{2}{3}, -\tfrac{1}{2}. \end{aligned}$$

71. Possible rational roots: $\pm 1, \pm 2, \pm 4$

Use synthetic division for $x = 4$.

$$\begin{array}{r|rrrr} 4 & 1 & -3 & -3 & -4 \\ & & 4 & 4 & 4 \\ \hline & 1 & 1 & 1 & 0 \end{array}$$

So,

$$\begin{aligned} x^3 - 3x^2 - 3x - 4 &= 0 \\ (x - 4)(x^2 + x + 1) &= 0. \end{aligned}$$

Because $x^2 + x + 1$ has no real solutions, $x = 4$ is the only real solution.

73. Possible rational zeros: $\pm\frac{1}{4}, \pm\frac{1}{2}, \pm 1, \pm 2$

Use synthetic division for $x = -1$.

$$\begin{array}{r|rrrr} -1 & 4 & 11 & 5 & -2 \\ & & -4 & -7 & 2 \\ \hline & 4 & 7 & -2 & 0 \end{array}$$

So,

$$\begin{aligned} 4x^3 + 11x^2 + 5x - 2 &= 0 \\ (x + 1)(4x^2 + 7x - 2) &= 0 \\ (x + 1)(4x - 1)(x + 2) &= 0 \\ x &= -1, \tfrac{1}{4}, -2. \end{aligned}$$

75. The equation is a second-degree equation, so it has two solutions.

$$\begin{aligned} 0.0003x^2 - 1200 &= 0 \\ 0.0003x^2 &= 1200 \\ x^2 &= 4{,}000{,}000 \\ x &= \pm 2000 \end{aligned}$$

There cannot be a negative number of units, so $x = -2000$ should be rejected. A production level of 2000 units will minimize the average cost.

Section A.5 Fractions and Rationalization

1. $$\begin{aligned} \frac{x^2 - 7x + 12}{x^2 + 3x - 18} &= \frac{(x - 4)(x - 3)}{(x + 6)(x - 3)} \\ &= \frac{x - 4}{x + 6}, \ x \neq 3 \end{aligned}$$

3. $$\begin{aligned} \frac{x^2 + 3x - 10}{2x^2 - x - 6} &= \frac{(x + 5)(x - 2)}{(2x + 3)(x - 2)} \\ &= \frac{x + 5}{2x + 3}, \ x \neq 2 \end{aligned}$$

5. $$\frac{x}{x - 2} + \frac{3}{x - 2} = \frac{x + 3}{x - 2}$$

7. $$\begin{aligned} x - \frac{3}{x} &= \frac{x^2}{x} - \frac{3}{x} \\ &= \frac{x^2 - 3}{x} \end{aligned}$$

9. $$\frac{2}{x-3}+\frac{5x}{3x+4}=\frac{2(3x+4)}{(x-3)(3x+4)}+\frac{5x(x-3)}{(x-3)(3x+4)}=\frac{6x+8+5x^2-15x}{(x-3)(3x+4)}=\frac{5x^2-9x+8}{(x-3)(3x+4)}$$

11. $$\frac{2}{x^2-4}-\frac{1}{x-2}=\frac{2}{(x-2)(x+2)}-\frac{1}{x-2}\cdot\frac{(x+2)}{(x+2)}=\frac{2-(x+2)}{(x-2)(x+2)}=-\frac{x}{x^2-4}$$

13. $$\frac{x}{x^2+x-2}-\frac{1}{x+2}=\frac{x}{(x+2)(x-1)}-\frac{1}{x+2}=\frac{x}{(x+2)(x-1)}-\frac{1(x-1)}{(x+2)(x-1)}=\frac{x-(x-1)}{(x+2)(x-1)}=\frac{1}{(x+2)(x-1)}$$

15. $$\frac{2}{x^2+1}-\frac{1}{x}+\frac{1}{x(x^2+1)}=\frac{2x-(x^2+1)+1}{x(x^2+1)}=\frac{2x-x^2-1+1}{x(x^2+1)}=\frac{-x^2+2x}{x(x^2+1)}=\frac{-x(x-2)}{x(x^2+1)}=\frac{-(x-2)}{x^2+1}$$

17. $$\frac{-x}{(x+1)^{3/2}}+\frac{2}{(x+1)^{1/2}}=\frac{-x}{(x+1)^{3/2}}+\frac{2(x+1)}{(x+1)^{3/2}}=\frac{x+2}{(x+1)^{3/2}}$$

19. $$\frac{2-t}{2\sqrt{1+t}}-\sqrt{1+t}=\frac{2-t}{2\sqrt{1+t}}-\frac{\sqrt{1+t}}{1}\cdot\frac{2\sqrt{1+t}}{2\sqrt{1+t}}=\frac{(2-t)-2(1+t)}{2\sqrt{1+t}}=\frac{-3t}{2\sqrt{1+t}}$$

21. $$\left(2x\sqrt{x^2+1}-\frac{x^3}{\sqrt{x^2+1}}\right)\div(x^2+1)=\left(\frac{2x(x^2+1)}{\sqrt{x^2+1}}-\frac{x^3}{\sqrt{x^2+1}}\right)\frac{1}{x^2+1}=\frac{2x^3+2x-x^3}{\sqrt{x^2+1}}\cdot\frac{1}{x^2+1}=\frac{x^3+2x}{\sqrt{x^2+1}(x^2+1)}=\frac{x(x^2+2)}{(x^2+1)^{3/2}}$$

23. $$\frac{(x^2+2)^{1/2}-x^2(x^2+2)^{-1/2}}{x^2}=\frac{(x^2+2)^{-1/2}\left[(x^2+2)-x^2\right]}{x^2}=\frac{2}{x^2\sqrt{x^2+2}}$$

25. $\dfrac{\dfrac{\sqrt{x+1}}{\sqrt{x}} - \dfrac{\sqrt{x}}{\sqrt{x+1}}}{2(x+1)} = \left(\dfrac{(x+1)}{\sqrt{x}\sqrt{x+1}} - \dfrac{x}{\sqrt{x}\sqrt{x+1}}\right)\dfrac{1}{2(x+1)} = \dfrac{1}{2\sqrt{x}(x+1)^{3/2}}$

27. $\dfrac{-x^2}{(2x+3)^{3/2}} + \dfrac{2x}{(2x+3)^{1/2}} = \dfrac{-x^2}{(2x+3)^{3/2}} + \dfrac{2x(2x+3)}{(2x+3)^{3/2}} = \dfrac{3x^2+6x}{(2x+3)^{3/2}} = \dfrac{3x(x+2)}{(2x+3)^{3/2}}$

29. $\dfrac{2}{\sqrt{10}} = \dfrac{2}{\sqrt{10}} \cdot \dfrac{\sqrt{10}}{\sqrt{10}} = \dfrac{2\sqrt{10}}{10} = \dfrac{\sqrt{10}}{5}$

31. $\dfrac{4x}{\sqrt{x-1}} = \dfrac{4x}{\sqrt{x-1}} \cdot \dfrac{\sqrt{x-1}}{\sqrt{x-1}} = \dfrac{4x\sqrt{x-1}}{x-1}$

33.
$$\begin{aligned}\frac{49(x-3)}{\sqrt{x^2-9}} &= \frac{49(x-3)}{\sqrt{x^2-9}} \cdot \frac{\sqrt{x^2-9}}{\sqrt{x^2-9}}\\ &= \frac{49(x-3)\sqrt{x^2-9}}{(x+3)(x-3)}\\ &= \frac{49\sqrt{x^2-9}}{x+3},\ x \neq 3\end{aligned}$$

35.
$$\begin{aligned}\frac{5}{\sqrt{14}-2} &= \frac{5}{\sqrt{14}-2} \cdot \frac{\sqrt{14}+2}{\sqrt{14}+2}\\ &= \frac{5\left(\sqrt{14}+2\right)}{14-4}\\ &= \frac{\sqrt{14}+2}{2}\end{aligned}$$

37.
$$\begin{aligned}\frac{1}{\sqrt{6}+\sqrt{5}} &= \frac{1}{\sqrt{6}+\sqrt{5}} \cdot \frac{\sqrt{6}-\sqrt{5}}{\sqrt{6}-\sqrt{5}}\\ &= \frac{\sqrt{6}-\sqrt{5}}{6-5}\\ &= \sqrt{6}-\sqrt{5}\end{aligned}$$

39.
$$\begin{aligned}\frac{2}{\sqrt{x}+\sqrt{x-2}} &= \frac{2}{\sqrt{x}+\sqrt{x-2}} \cdot \frac{\sqrt{x}-\sqrt{x-2}}{\sqrt{x}-\sqrt{x-2}}\\ &= \frac{2\left(\sqrt{x}-\sqrt{x-2}\right)}{x-(x-2)}\\ &= \sqrt{x}-\sqrt{x-2}\end{aligned}$$

41.
$$\begin{aligned}\frac{\sqrt{x+2}-\sqrt{2}}{x} &= \frac{\sqrt{x+2}-\sqrt{2}}{x} \cdot \frac{\sqrt{x+2}+\sqrt{2}}{\sqrt{x+2}+\sqrt{2}}\\ &= \frac{x+2-2}{x\left(\sqrt{x+2}+\sqrt{2}\right)}\\ &= \frac{1}{\sqrt{x+2}+\sqrt{2}},\ x \neq 0\end{aligned}$$

43. $P = 10{,}000,\ r = 0.075\ N = 60$

$$M = 10{,}000\left[\frac{0.075/12}{1-\left(\dfrac{1}{(0.075/12)+1}\right)^{60}}\right] \approx \$200.38$$

APPENDIX B
Alternative Introduction to the Fundamental Theorem of Calculus

1. The subintervals are $[0, 1]$, $[1, 2]$, $[2, 3]$, $[3, 4]$, and $[4, 5]$.

The midpoints of each subinterval are $\frac{1}{2}, \frac{3}{2}, \frac{5}{2}, \frac{7}{2}$, and $\frac{9}{2}$.

Because the width of each rectangle is 1, the sum of the areas of the five rectangles is

$$S = (1)f\left(\frac{1}{2}\right) + (1)f\left(\frac{3}{2}\right) + (1)f\left(\frac{5}{2}\right) + (1)f\left(\frac{7}{2}\right) + (1)f\left(\frac{9}{2}\right) = \frac{35}{2} = 17.5 \text{ square units.}$$

3. Left Riemann sum: 0.518
Right Riemann sum: 0.768

5. Left Riemann sum: 0.746
Right Riemann sum: 0.646

7. Left Riemann sum: 0.859
Right Riemann sum: 0.659

9. 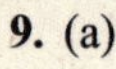 (a)

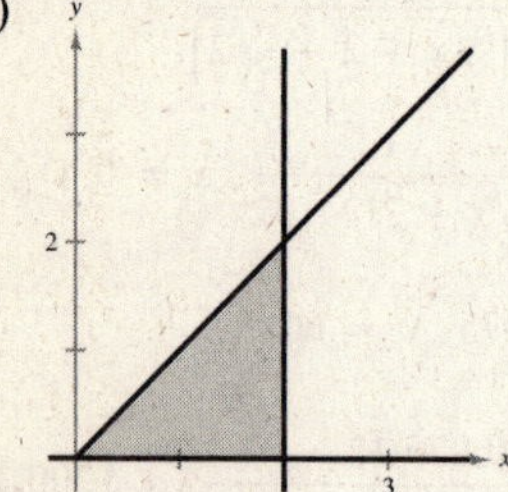

(b) $\Delta x = \dfrac{2 - 0}{n} = \dfrac{2}{n}$

Endpoints: $0 < 0 + \dfrac{2}{n} < 0 + \dfrac{4}{n} < \cdots < 0 + \dfrac{2n}{n} < 2$

$$0 < 1\left(\frac{2}{n}\right) < 2\left(\frac{2}{n}\right) < \cdots < (n-1)\left(\frac{2}{n}\right) < n\left(\frac{2}{n}\right)$$

(c) Because $y = x$ is increasing, $f(m_i) = f(x_{i-1})$ on $[x_{i-1}, x_i]$.

$$s(n) = \sum_{i=1}^{n} f(x_{i-1})\,\Delta x = \sum_{i=1}^{n} f\left[(i-1)\left(\frac{2}{n}\right)\right]\left(\frac{2}{n}\right) = \sum_{i=1}^{n}\left[(i-1)\left(\frac{2}{n}\right)\right]\left(\frac{2}{n}\right)$$

(d) $f(M_i) = f(x_i)$ on $[x_{i-1}, x_i]$

$$S(n) = \sum_{i=1}^{n} f(x_i)\,\Delta x = \sum_{i=1}^{n} f\left[i\left(\frac{2}{n}\right)\right]\left(\frac{2}{n}\right) = \sum_{i=1}^{n}\left[i\left(\frac{2}{n}\right)\right]\left(\frac{2}{n}\right)$$

(e)

n	5	10	50	100
Left sum, S_L	1.6	1.8	1.96	1.98
Right sum, S_R	2.4	2.2	2.04	2.02

(f) $\displaystyle\lim_{n\to\infty} \sum_{i=1}^{n}\left[(i-1)\left(\frac{2}{n}\right)\right]\left(\frac{2}{n}\right) = \lim_{n\to\infty} \frac{2n+2}{n} = 2$

$$\lim_{n\to\infty} \sum_{i=1}^{n}\left[i\left(\frac{2}{n}\right)\right]\left(\frac{2}{n}\right) = \lim_{n\to\infty}\left[2 + \frac{2}{n}\right] = 2$$

11. $\int_0^5 3\,dx$

13. $\int_{-4}^{4} \left(4 - |x|\right) dx = \int_{-4}^{0} (4 + x)\,dx + \int_0^4 (4 - x)\,dx$

15. $\int_{-2}^{2} \left(4 - x^2\right) dx$

17. $\int_0^2 \sqrt{x + 1}\,dx$

19. $A = 12$

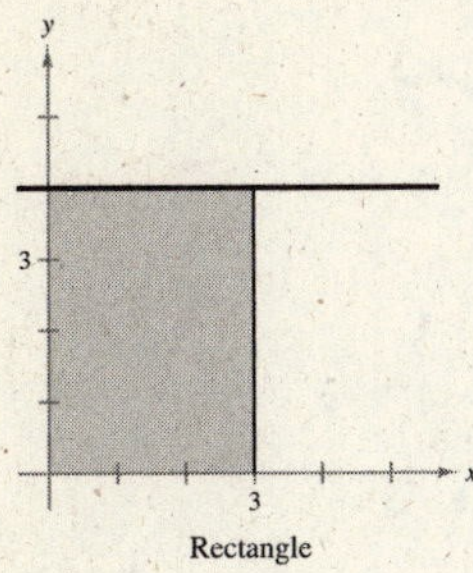
Rectangle

21. $A = 8$

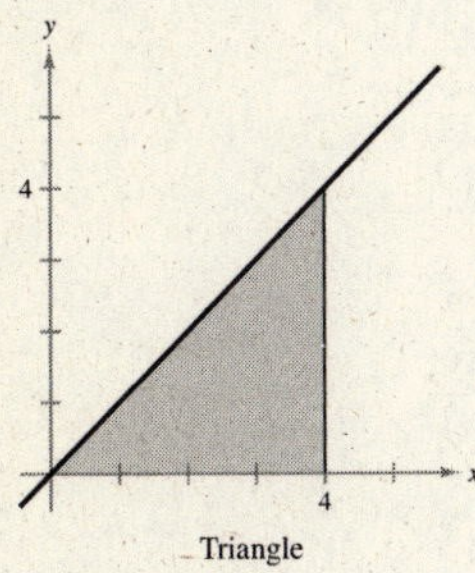
Triangle

23. $A = 14$

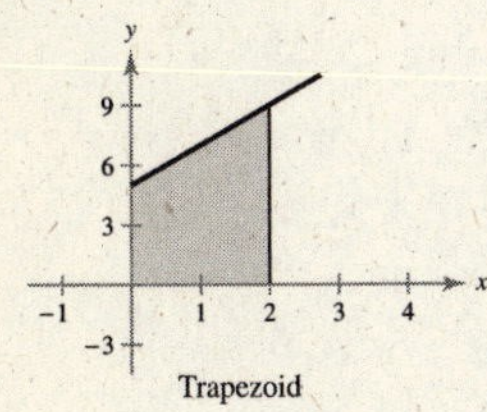
Trapezoid

25. $A = 1$

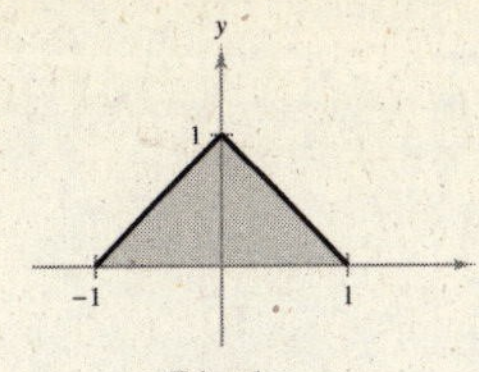
Triangle

27. $A = \dfrac{9\pi}{2}$

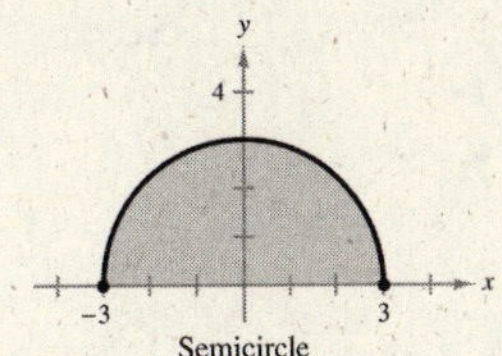
Semicircle

29. $S = 1 + 2 + 3 + \cdots + (n - 1) + n$

$S = n + (n - 1) + \cdots + 3 + 2 + 1$

Adding,

$2S = (n + 1) + (n + 1) + \cdots + (n + 1) \quad (n \text{ times})$

$2S = n(n + 1)$

$S = \dfrac{n(n + 1)}{2}$

31. $\sum_{i=1}^{n} f(x_i)\,\Delta x > \int_1^5 f(x)\,dx$